Developments in Earth Surface Processes 1

PALEOKARST

A Systematic and Regional Review

Developments in Earth Surface Processes

Developments in Earth Surface Processes 1

PALEOKARST

a systematic and regional review

Edited by

PAVEL BOSÁK
Geoindustria, Prague, Czechoslovakia

DEREK C. FORD
Department of Geography, McMaster University,
Hamilton, Canada

JERZY GŁAZEK
Institute of Geology, Warsaw University,
Warsaw, Poland

IVAN HORÁČEK
Institute of Geology and Geotechnics, Czechoslovak
Academy of Sciences, Laboratory of Quaternary
Geology,
Prague, Czechoslovakia

ELSEVIER
Amsterdam - Oxford - New York - Tokyo 1989

Scientific Editor
Prof. Dr. Václav Král, DrSc.

Reviewers
Doc. Dr. Vladimír Panoš, Csc.
Dr. Vojen Ložek, DrSc.

Published in co-edition with Academia, Publishing House
of the Czechoslovak Academy of Sciences, Prague

Distribution of this is being handled by the following publishers

for the U.S.A. and Canada
Elsevier Science Publishing Company, Inc.
655 Avenue of the Americas
New York, New York 10010

for the East European Countries, China, Northern Korea, Cuba,
Vietnam and Mongolia
Academia, Publishing House of the Czechoslovak Academy of Sciences, Prague, Czechoslovakia

for all remaining areas
Elsevier Science Publishers
Sara Burgerhartstraat 25
P.O. Box 21, 1000 AE Amsterdam, The Netherlands

ISBN 0–444–98874–2 (ELSEVIER)
ISBN 80–200–0032–1 (ACADEMIA)

Library of Congress Cataloging-in-Publication Data

Paleokarst: a systematic and regional review / edited by Pavel Bosák
... [et al.].
p. cm. -- (Developments in earth surface processes; 1)
Bibliography: p.
Includes indexes.
ISBN 0-444-98874-2 (U.S.)
1. Karst. I. Bosák, Pavel. II. Series.
GB600.P35 1989
551.4'47--dc19 89–1170
CIP

Printed in Czechoslovakia

This book is dedicated to the memory of the late Professor Joseph Newell Jennings (1916–1984)

CONTENTS

CONTRIBUTORS:

Bárdossy, György
HUNGALU, Hungarian Aluminium Corporation, P.O.Box 30, H-1387 Budapest, 62, Hungary
Bleahu, Marcian
51, str. Maria Rosetti, 70234 Bucuresti, Romania
Böcker, Tivadár
HUNGALU, Hungarian Aluminium Corporation, P.O.Box 30, H-1387 Budapest 62, Hungary
Boni, Maria
Istituto di Geologia e Geofisica, Dipartimento di Scienze della Terra, Università degli Studi di Napoli, Largo San Marcellino, 10, I-80138 Napoli, Italy
Bosák, Pavel
Geoindustria, Komunardů 6, 170 04 Praha 7, Czechoslovakia
Bourrouilh-Le Jan, Françoise G.
Laboratoire de Géologie-Sédimentologie comparée et appliquée, Université de Bordeaux I, Avenue des Facultés, 33405 Talence, France – Associate researcher of Muséum National d'Histoire Naturelle de Paris – Opunoku Centre EPHE, Moorea, Tahiti, French Polynesia
D'Argenio, Bruno
Istituto di Geologia e Geofisica, Dipartimento di Scienze della Terra, Università degli Studi di Napoli, Largo San Marcellino, 10, I-80138 Napoli, Italy
Dżułyński, Stanisław
Geological Institute, Polish Academy of Sciences, Oleandry 2a, 30–063, Poland
Eraso, Adolfo
Escuela technica superior de ingenieros de minas, Departamento de ingenieria geologica, Catedra de hidrogeologia. Ríos Rosas, 21, 28 003 Madrid, Spain
Ford, Derek C.
Department of Geography, McMaster University, 1280 Main Street West, Hamilton, Ontario L8S 4K1, Canada
Ford, Trevor D.
Department of Geology, University of Leicester, University Road, Leicester LE1 7RH, Great Britain
Fuchs, Yves
Laboratoire de Minéralogie Experimentale et Appliquée, Université Pierre et Marie Curie – Paris VI, 4, Place Jussieu 25, 75252 Paris Cedex 05, France
Gavrilović, Dušan
Department of Geography and Space Planning, University of Beograd, Studentski trg 3, Beograd, Yugoslavia
Głazek, Jerzy
Institute of Geology, University of Warsaw, al. Żwirki i Wigury 93, 02-089 Warszawa, Poland

Horáček, Ivan
Laboratory of Quaternary Geology, Institute of Geology and Geotechnics, Czechoslovak Academy of Sciences, V Holešovičkách 41, 182 09 Praha 8, Czechoslovakia
Kordos, László
Hungarian Geological Survey, Népstadion ut 14,H–1143 Budapest, Hungary
Lauritzen, Stein-Erik
Department of Geology, Section B, University of Bergen, Allegt 41, N–5000 Bergen, Norway
Müller, Pál
Hungarian Geological Survey, Népstadion ut 14,H–1143 Budapest, Hungary
Palmer, Arthur N.
Department of Earth Sciences, State University College, Oneonta, New York 13820, U.S.A.
Palmer, Margaret L.
Department of Earth Sciences, State University College, Oneonta, New York 13820, U.S.A.
Panoš, Vladimír
Geographical Institute, Czechoslovak Academy of Sciences, Mendlovo náměstí 1, 662 82 Brno, Czechoslovakia
Quinif, Yves
Faculté Polytechnique de Mons, Rue de Houdain, B-7000 Mons, Belgium
Sass-Gustkiewicz, Maria
Institute of Geology and Mineral Deposits, Academy of Mining and Metallurgy of Stanisław Staszic, Al. Mickiewicza 30, 30-059 Kraków, Poland
Stanev, Ivan Stefanov
Research Institute of Mineral Deposits, 23, Khristo Kabaciev Bul., 1505 Sofia, Bulgaria
Touahri, B.
E.N.Re.M, Wilaya d'Alger, Algeria
Trashliev, Stoyan Stefanov
Research Institute of Mineral Deposits, 23, Khristo Kabaciev Bul., 1505 Sofia, Bulgaria
Tsykin, Rostislav Alekseevich
Institute of Non-Ferrous Metals, Vuzovsky pereulok 3, 660025 Krasnoyarsk, U.S.S.R.
Vizy, Béla
HUNGALU, Hungarian Aluminium Corporation, P.O.Box 30, H-1387 Budapest 62, Hungary
Wilk, Zbigniew
Institute of Hydrogeology and Engineering Geology, Academy of Mining and Metallurgy of Stanisław Staszic, Al. Mickiewicza 30, 30-059 Kraków, Poland
Zhang Shouyue
Karst Research Group, Institute of Geology, Academia Sinica, P.O.Box 634, Beijing, China
Zötl, Josef G.
Institut für Geothermie und Hydrogeologie, Joanneum Forschungszentrum Graz, Elisabethstrasse 16, A-8010 Graz, Austria

PREFACE

The growing importance of karst-related research regularly leads to new developments in paleokarst studies. Although *paleokarstology* has not existed as a clearly defined scientific discipline, numerous publications have appeared dealing with paleokarst, its paleontological contents, mineral deposits or the problems of hydrogeology and engineering geology, etc. that it creates. Paleokarst studies tend to differ from studies of recent and modern karst landforms. Studies of present karst landforms are mostly carried out by physical geographers and geomorphologists and are little known to most geologists, whereas paleokarst research has tended to be performed by specialists in the geological sciences. This book, with the exception of two contributions, is written by professional geologists or paleontologists. Our efforts are directed at showing the importance of paleokarst as a conserver of records of the geological past, although morphological and geomorphological topics are dealt with, too.

The growing interest in paleokarst studies has resulted not only from the economic importance of the phenomena, but also from their scientific importance. As is shown in Part 5 of this book, paleokarst records provide many missing links in the chronostratigraphy of particular areas, and regional geological structures or units. The contribution of the study of paleokarst to lithostratigraphy, sedimentology, biostratigraphy, plant and animal phylogeny and paleoecology etc., helps both in paleogeographic and paleotectonic reconstructions of the evolution of the Earth's surface and in specific applied studies.

The growing number of paleokarst sites studied in some detail, of recognised paleokarst-related mineral deposits and of problems connected with their exploitation, led us to attempt to collect the available data on paleokarst and to summarise it together with some general review and discussion. The effort would have been impossible without the active participation of all of the contributors. We thank them for their written cooperation and care in preparation of their contributions. The realization of our plans was facilitated by the cooperation of the Academia and Elsevier Science Publishers. The results, which were neither so simple nor easy to achieve as originally supposed and which are somewhat different from the initial broader scheme, are presented here in Paleokarst.

Acknowledgements: We thank all contributors for their willing cooperation. The Publishing House Academia together with the Elsevier Company took the risk of publishing this book devoted entirely to paleokarst. The Czech Speleological Society sponsored the editorial work and correspondence expenses and we gratefully acknowledge their support.

Discussion about problems of Paleokarst have been held with many colleagues over the last few years. We cannot mention all here, but our book has benefited particularly from discussion on concepts and definitions of general terms with Bernard Gèze (Paris, France), Hubert Trimmel (Vienna, Austria), and Vladimír Panoš (Brno, Czechoslovakia). We are especially grateful to György Bárdossy (Budapest, Hungary) for his help during the preparation of Part 3, "Mineral deposits connected with karst". We are grateful to Josef Rubín, Editor-in-Chief of the Earth Sciences Department of the Academia Publishing House for his help, encouragement and patience during the preparation of the book. Václav Král (Charles University, Prague, Czechoslovakia) was the Scientific Editor of the book, and Vojen Ložek (Czechoslovak Academy of Sciences, Prague, Czechoslovakia) and Vladimír Panoš (Czechoslovak Academy of Sciences, Brno, Czechoslovakia) were reviewers of the texts; we are grateful for their many constructive notes and helpful ideas. The manuscript was kindly retyped by the technical staff of McMaster University, Hamilton, Canada. Last but not least are our thanks to our families for their patience and encouragement when the work encountered the snags that must crop up in lengthy, multi-disciplinary and international efforts such as this.

Pavel Bosák

President of the Commission
on Speleochronology and Paleokarst
of the International
Speleological Union

FOREWORD

In gathering together information concerning most regions where paleokarst is known, this book is the first serious attempt to summarize that subject.

Two fundamental aspects are developed. The first one deals with mineral fillings of paleokarst; Pb and Zn deposits, Fe deposits, phosphates, bauxites etc. These are often of economic importance. The second one discusses the difficult problems created during civil engineering works and opencast mining in paleokarst. The specialist will find here, not only accounts of the distribution of paleokarst in space and time, but of its scientific magnitude and its practical consequences.

The importance of the book, however, is not only contained within its contents. It will lead to new directions in future research work on karst. The reason is that according to concepts which are in development, karst definition may be the result of the organization of subterranean drainage in soluble rocks. Karstification means the development of this organization. The process is irreversible. In any particular karst area, karst operates as an open thermodynamic system driven by the physical potential for water circulation (represented by the hydraulic gradient) and by the chemical potential for rock dissolution. The resistance in the system is caused by geological properties of the rock, lithological as well as structural.

During the development of a karst system, its entropy is growing due to the activation of the karstification potential and to the diminishing of system resistance by dissolution of rock. Ultimately, it can be said that the system reaches a stationary state. Paleokarst represents such a stationary state when the dissipation function of the system, which represents the velocity of entropy production reaches a minimum.

The dissipation function has two components: internal dissipation determined by the production of entropy inside the system and external dissipation determined by the production of entropy outside of it. In an open thermodynamic system, a stationary state is reached when the effect of external dissipation reaches a minimum. This is the case for paleokarst.

This way of describing karst as a dissipative structure, gives us the opportunity to apply to its study, all of Prigonine's postulates and theory about generalized thermodynamics in which Lyapounov's functions predict the future evolution of the system. The problem in the case of karst is centered upon the difficulties of quantifying the potential of karstification.

We see, however, that man's intervention during engineering works in paleokarst areas often supplies the stimulus necessary to increase rapidly the external dissipation function. This phenomenon is evidenced by previously fossil karst being re-activated.

The study of the following specific cases of paleokarst gives an opportunity to understand better its operation in a quantitative way. That can lead to predictions of karst performance and more effective intervention in this interesting natural phenomenon.

Adolfo Eraso

President of the International Speleological Union (1981–1986)

PART I

INTRODUCTION

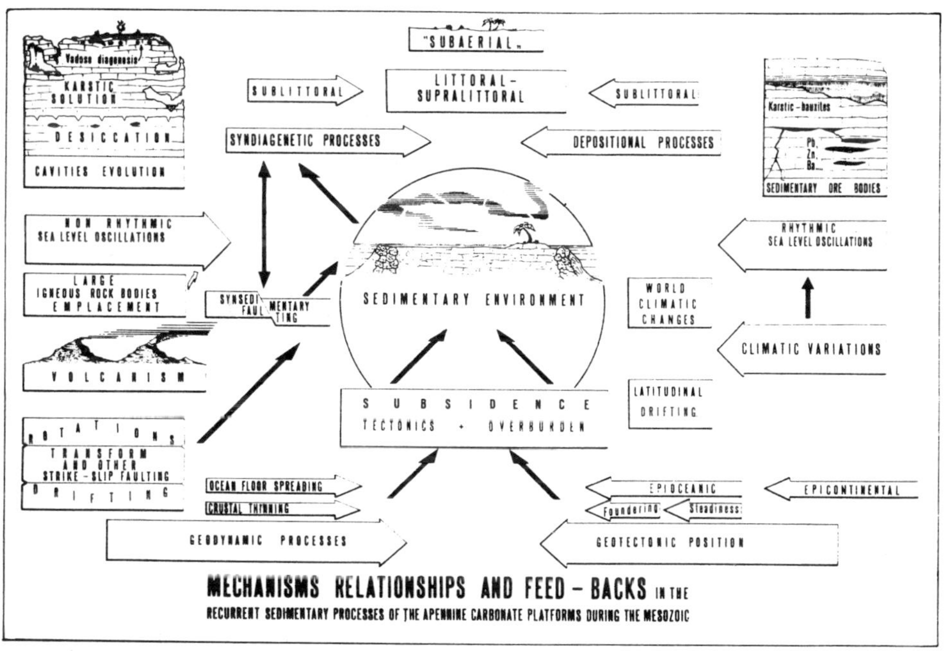

MECHANISMS RELATIONSHIPS AND FEED-BACKS IN THE RECURRENT SEDIMENTARY PROCESSES OF THE APENNINE CARBONATE PLATFORMS DURING THE MESOZOIC

INTRODUCTION

Karst is commonly treated as a distinctive geomorphic process and/or landform region displaying a specific underground drainage and surface morphology (JENNINGS, 1985). In most academic textbooks, karst is thought of only in terms of present geomorphic processes or the modern morphology that is forming. Few authors are concerned with the karst processes active in the geological past or the features that they created. However, many deposits of importance in economic geology are associated with or result from past karst activity. Among these are some of the world's richest deposits of bauxite, Pb-Zn ores and phosphates. In karst areas there are great practical problems with water supply and civil engineering works. Many of these problems are not caused by modern or young (Quaternary) karstification at all, but karst development in the more distant geological past. There is also much misunderstanding and terminological confusion; for instance the sharp differences between karst- and fissure- or porous-aquifers is not even noted in many hydrogeological textbooks, where karst is treated as a superficial process confined to the top of the saturation zone and/or limited by the present base level of erosion. The terms *paleokarst* and *fossil karst* are widely used, sometimes being treated as synonyms, and sometimes used to indicate opposites.

The Scope and Aim of this Book

Published studies specifically focussed on paleokarst are relatively few. However, a great deal of information about paleokarst phenomena has accumulated as peripheral observation in studies of other aspects of geology and geography. This information is dispersed in hundreds of journals and conference proceedings, is published in many languages, and generally is scarcely accessible to many scholars.

P. Bosák, D. C. Ford, J. Głazek and I. Horáček (Editors):
Paleokarst. A Systematic and Regional Review. – Elsevier and Academia. Amsterdam and Praha. 1989.

The study of paleokarst is of ever-increasing importance in the Earth Sciences and in the utilization of the natural environment. Its relationship to traditional approaches to karst study is not always clearly understood. As a result, there has been both uncertainty and some distrust regarding the use of the term and the concept. It has become desirable to review the terminology and to illustrate it with examples of paleokarst phenomena. This book therefore attempts to fill an important gap and, in so doing, to reconcile the often conflicting views of specialists and to give new impetus to further studies in this fascinating domain.

At the 7th International Congress of Speleology, the need for co-ordinated paleokarst studies was suggested by J. Głazek, and the scope of the UIS Commission on Speleochronology was extended to become the Commission on Speleochronology and Paleokarst. At this Congress, the idea of compiling information on paleokarst in all member countries was first discussed. This was then elaborated in Commission meetings in 1980 (Sofia, Bulgaria) and 1981 (Bowling Green, Kentucky, U.S.A.). Publication problems were solved in 1984 when ACADEMIA Publishing House (Prague, Czechoslovakia) accepted a proposal presented by the Czech Speleological Society on behalf of the UIS Commission and itself. Editors were selected, the coedition with ELSEVIER Science Publisher (Amsterdam, The Netherlands) was arranged and authors from all UIS member countries were formally invited. This book is the result of an idea born in the UIS Commission on Speleochronology and Paleokarst and the initiative of the Central Scientific Commission of the Czech Speleological Society.

It is not feasible to review the paleokarst of the whole world, but the paleokarst of countries represented in this volume, ranging around the globe from China to the U.S.A., from Canada to the South Pacific islands — appears to encompass all types of paleokarst development. It is a set of case studies in a wide range of geological and geographical settings.

The editors are not charged with the responsibility of unifying all local terminological differences in karst and related subjects such as stratigraphy. Such differences in usage are unavoidable in the regional part of the book through the different development of science in different countries. Also, we cannot present all papers with a similar style and content because of the differing styles and emphases in investigations in the nations represented here. However, we believe that the book presents accurate information on the present status of studies, and may serve as a collection of models for further development of investigations in other countries.

Some desirable abbreviations and language corrections have been undertaken by the editors. Papers by authors with a native language other than English were corrected by Professor Derek Ford, President of the UIS. There have been quite a lot of changes to improve the text and styles. However, at all times he has tried to retain the author's original meaning precisely; there has been no intention to change the sense of what they wrote. Sometimes, correspondence and consultation of the texts in the native languages of authors were needed to clear up some editorial problems. A few controversial statements have been commented upon by means of editorial footnotes, but the chapter authors are responsible for the scope and contents of their particular texts.

The book has two principal aims: to present regional reviews, and to demonstrate some of the important economic or scientific problems associated with paleokarst. Part 2; Regional reviews (the longest part) contains submitted contributions devoted to separate countries and arranged in geographical order: Eurasia from west to east and including North Africa, America from north to south. This arrangement keeps neighbouring countries close to one another, and seems more natural than any other. The later parts present case studies and reviews of mineral deposits in karst (Part 3), hydrogeological aspects of paleokarst and engineering hazards in paleokarst areas (Part 4) and some concluding remarks by editors. We regret that we were not able to find authors in some countries where paleokarst is known or is expected to occur. Some invited specialists accepted our invitation but were not able to prepare their contributions before the final acceptable dead-line e. g. France, FRG. Some gaps in the scope of the book we try to fill by remarks included in the general considerations. All papers included here were obtained by editors at the end of 1985 or beginning of 1986, as the book was completed at March 1986.

We hope that the book will prove to be a useful review of the present state of knowledge in the domain of paleokarst and that it will support new studies all over the world, both theoretical and for practical purposes.

The Editors

PALEOKARST AS A PROBLEM

Jerzy Głazek, Pavel Bosák and Derek C. Ford

Traditionally, karst studies have been concerned with phenomena believed to be of recent to subrecent origin. However, surprisingly many modern karst features are products of a long and complicated, polycyclic development and thus, in the broad sense, they fall into the paleokarst domain. This is the case in the Classical (Dinaric) Karst, once treated as a global model for typical recent karst (see GAVRILOVIĆ, 201–216). If, in older work, only striking ancient karst phenomena were definitely distinguished as paleokarst or fossil karst (e.g. REICHENBACH, 1834; SAWICKI, 1909; VAN DEN BROECK, MARTEL and RAHIR, 1910; SAMSONOWICZ, 1934; GÈZE, 1938; BRETZ, 1940), so nowadays it is necessary everywhere to show that particular karst forms can have developed in the very limited time span close to recent epoch. Still very little is known about many geomorphological characteristics of paleokarst. Most field work has been concerned with delineating the extent of mineral deposits in the karst rather than in understanding the origin of the host karst features, although knowledge of the latter would certainly aid in predicting much ore distribution. Furthermore, great efforts have been made in the study of karst groundwaters but there, also, little attention has been given to evidently paleokarst conduits in which even at depths greater than 2 km there is quite vigorous modern flow (see ZÖTL, 506–509).

Complicated paleokarst conditions have caused many catastrophic collapses and ground subsidence throughout the world (LEGGET, 1973; CALEMBERT et al., 1975). Reactivation of paleokarst systems by natural or artificial changes of karst water levels are the prime causes of these events. Excessive water pumping for water supply, mine dewatering etc. lowers the buoyant support of karst cover. On the other hand, dam construction is responsible for the rise of the hydraulic head, resulting in some huge water losses and even in substantial damage to the structures themselves.

For these reasons paleokarst is a complicated scientific problem with major economic importance.

P. Bosák, D. C. Ford, J. Głazek and I. Horáček (Editors):
Paleokarst. A Systematic and Regional Review. – Elsevier and Academia. Amsterdam and Praha. 1989.

TERMINOLOGY

Pavel Bosák, Derek C. Ford and Jerzy Głazek

There are numerous terms to describe the many different karst phenomena that developed in the geological past. They are often used in a disputed manner. Many terms originally describing one particular phenomenon have later come to be used in a different sense. The terminological jungle is clearly seen when comparing glossaries of karst and speleological terms (e.g. Anelli, 1959; Trimmel, 1965; Fénelon, 1968; Monroe, 1970; Fink, 1973; Panoš, 1978) and also in many textbooks. Jennings (1971, p. 3) describes the result as "a patchwork quilt" in which it is often difficult to discriminate among different terms though it is desirable to retain them as names. A similar simplification was proposed by Gèze (1973). Since then other authors have introduced new terms to clarify particular situations; this endeavour has often led to further misunderstandings (e.g. Quinlan, 1972; Panoš, 1978). *Paleokarst* and *fossil karst* belong to the list of early terms (cf. De Martonne, 1910). They have been used freely since then in broader or narrower senses, together with new terms that have similar specifications. Moreover, different terms are applied in different countries and/or languages to describe the same feature.

Review

Paleokarst and fossil karst. Both terms are defined similarly — "as features formed by agents active in the karst environment in the past" (Laznicka, 1985), "under morphogenetic conditions differing from the present" (Gams, 1973; Panoš, 1978). They are treated as synonyms (De Martonne, 1910; Jennings, 1971). However, either the first or the second of these terms may be used preferentially or exclusively in particular languages, e.g. in many Slavonic languages *paleokarst* is hardly acceptable and etymologic equivalents of *fossil karst* or *old karst* are preferred (*drevnii* or *iskopaemyi karst* in Russian; *fosilní kras* in

P. Bosák, D. C. Ford, J. Głazek and I. Horáček (Editors):
Paleokarst. A Systematic and Regional Review. – Elsevier and Academia. Amsterdam and Praha. 1989.

Czech or Slovenian; *kras kopalny* in Polish). In German a similar term is also used — *Urkarst*. In other languages *paleokarst* is preferred or both terms are acceptable. In these latter cases, there have been many attempts to differentiate between these two general terms. Generally, these differentiations were related to: (1) any present karst activity; (2) the geological age of major karstification, and (3) the relationship to the present surface.

Some authors use *paleokarst* to describe surface karst that is inactive today (e.g. GAMS, 1973; PANOŠ, 1978; BOSÁK, 1981b; and in this volume ZHANG SHOUYUE, and GAVRILOVIĆ). There is similar inconsistency in the age relationships; paleokarst is treated as older and strictly determined in its age (e.g. GAVRILOVIĆ, p. 201, and ZHANG SHOUYUE, p. 303), whereas fossil karst may be young — Late Cenozoic (e.g. TRIMMEL, 1965; GOSPODARIĆ, 1981). In relation to the modern land surface, GAMS (1973) suggests fossil karst for forms found only at depth in geological sections, while others call such features — paleokarst (MONROE, 1970; BAUD and MASSON, 1975; T. D. FORD, p. 51).

GÈZE (1981) has suggested that paleokarst be used for strictly scientific purposes, and fossil karst more freely. Some authors have divided paleokarst or fossil karst into different categories. SWEETING (1972) divided *fossil karst* into *relict* and *exhumed* sub-types. QUINLAN (1972) divided *paleokarst* into *relict* and *interstratal* karst. PALMER and PALMER (p. 337) following QUINLAN (1978) suggest two categories of *paleokarst* — *buried* and *intrastratal*.

After the manuscript was finished, the JENNINGS'S (1985) textbook appeared. Here, on page 7, JENNINGS noted:"the term paleokarst has been variously applied by different authorities, and now is perhaps best used in a broad way to comprise any parts of a karst landscape that are of considerable age and are not adjusted to the present controlling factors. It is by no means always easy to distinguish between active, relict, subjacent, buried and exhumed components in karst. This may be due to a paucity of stratigraphical sections but often is because of compound histories. Thus, both buried dolines and filled caves are likely places for a later resumption of solution and for fresh inputs of sediments. What was simply a buried feature may become also subjacent in origin. There is also the possibility of confusion with structural landforms in the sense of landforms reflecting geological structure".

Buried karst is that covered by later consolidated rocks or sediments (MONROE, 1970; JENNINGS, 1971; QUINLAN, 1972; GVOZDETSKY, 1972; PANOŠ, 1972; in this volume D. C. FORD, T. D. FORD, ZHANG SHOUYUE, PALMER and PALMER). Post-karst mantles lie unconformably or disconformably on the karst relief (JENNINGS, 1971; ZHANG SHOUYUE, p. 303). Buried karst phenomena were created at or near a former land surface by the action of meteoric waters (PALMER and PALMER, p. 337) and developed before the cover strata were laid down (JENNINGS, 1971; PANOŠ, 1978). Their development has been stopped

(PANOŠ, 1978) so that no reflection of further development is detectable in the superficial relief (GVOZDETSKY, 1972). They constitute no part of the contemporary landscape (QUINLAN, 1972). Caves were completely filled during the burial (JENNINGS, 1971). Some authors note that the cover of the buried karst should be impermeable (PANOŠ, 1978). For QUINLAN (1972), buried karst is one category of paleokarst. For JENNINGS (1971) and GAMS (1973), buried karst, fossil karst and paleokarst are synonyms. For MONROE (1970), covered karst and paleokarst are synonyms of buried karst.

Covered karst. Definitions of the term, *covered karst* in the literature are based on several distinct criteria (1) continuity of the cover, (2) nature of the cover and (3) time relationships of the cover to the karstification; i.e. this term does not have an uniform meaning. Soluble rocks may be covered by a mantle partly or entirely (JENNINGS, 1971), continuously (PANOŠ, 1978) or covered without detailed specification (GVOZDETSKY, 1972; in this volume T. D. FORD, ZHANG SHOUYUE, TSYKIN). The cover consists of soil (MONROE, 1970), soil or other superficial unconsolidated deposits (T. D. FORD, p. 51), any loose sediments (ZHANG SHOUYUE, p. 303), thick permeable autochthonous or allochthonous weathering sedimentary and soil mantles (PANOŠ, 1978) or any insoluble mantle without genetic connection with the karst rocks, such as clastic marine deposits, moraine, glaciofluvial or terrace sediments (GVOZDETSKY, 1972). Some karst forms are believed to develop under such covers (MONROE, 1970; GVOZDETSKY, 1972; PANOŠ, 1978; TSYKIN, 1980; T. D. FORD, p. 51). TSYKIN (1980) reported simultaneous karstification and accumulation of the cover. The projection of subsurface karst forms into any cover strata (for instance, by subsidence dolines) is possible due to cover permeability and ongoing karstification of limestones (PANOŠ, 1978), through the washing (suffosion) of clayey-sandy material from the cover into widened joints in the bedrock (GVOZDETSKY, 1972), or perhaps even to crustal flexure during deglaciation (D. C. FORD, p. 313–336). For MONROE (1970), buried karst is a synonym of *covered karst.*

Intrastratal versus interstratal karst. *Interstratal karst* (QUINLAN, 1972; T. D. FORD, p. 51) develops along a bedding plane or unconformity. PALMER and PALMER (p. 337) report that most solutional processes at depth appear not be to limited to such boundaries between strata and therefore they prefer the term, *intrastratal karst.* These types of karst are younger than their immediate cover (QUINLAN, 1972) and so are formed by deep-seated processes in the subsurface. PALMER and PALMER (p. 337) distinguish two types of intrastratal karst: (1) *subjacent karst* (cf. QUINLAN, 1978) associated with meteoric waters and (2) *hypogene karst,* formed by solutional aggressiveness produced by deep chemical reactions (i.e. thermal water), not including those where mixing causes rejuvenation of aggressiveness in waters derived originally from surface sources. For QUINLAN (1972), interstratal karst is a category of paleokarst. For

PALMER and PALMER (p. 337) intrastratal karst is "a category of paleokarst only if it is out of adjustment with the present geological setting". They note, that this "criterion is often difficult to apply, for intrastratal processes tend to operate over a long time span, and many features that qualify as intrastratal karst were formed by a process still operating at a diminished rate today".

For ZÖTL (p. 509), underground karst waterways (i.e. intrastratal or subjacent karst) are paleokarst features only if "developed under other geological and/or climatic conditions. In many cases modern water flows through such passages". Recent hydrogeological influence can be "insignificant or have led to a total change of the paleokarst".

Subjacent karst. There are two meanings of the term. The first designates subsurface karst developed beneath a cover of non-karstic rocks, (see also MARTIN, 1965; JENNINGS, 1971; SWEETING, 1972). Such forms develop at depth only because of the relative insolubility of overlying strata (QUINLAN, 1978; PALMER and PALMER, p. 337). For some authors (e.g. ZHANG SHOUYUE, p. 303) subjacent karst is a synonym for intrastratal karst. A special kind of the subjacent karst is FINK's (1976) *kryptokarst* developed in carbonate rocks with non-carbonate hanging walls and footwalls.

The second definition was adopted especially by GAMS (1973) and PANOŠ (1978). According to them, subjacent karst is a semiburied karst type and comprises surface and underground karst forms beneath autochthonous or allochthonous impermeable or partly permeable mantles that cover the majority of the karst relief and protect it against further development. Such mantles, because they are comparatively thin, may be locally disturbed. This permits further evolution of the underlying or footwall karst. *Semi-buried karst* is a combination of fossil karst, partly exhumed, and of recent karst (PANOŠ, 1978, p. 95).

Relict karst. „Without having been buried, some karst forms are relict from previous morphogenetic conditions that are no longer operative" (JENNINGS, 1971; see also SWEETING, 1972). They now appear as unmodified to highly modified relicts of original older forms (PANOŠ, 1978). Relict karst is treated as a category of paleokarst (GAVRILOVIĆ, p. 201) or of fossil karst (SWEETING, 1972).

Exhumed karst describes features exposed by the erosion of former covering strata (MONROE, 1970; JENNINGS, 1971) which fossilized the karst for some time (PANOŠ, 1978). According to GAMS (1973), the cover should be impermeable; according to SWEETING (1972) it should not be limestone. The karst forms are usually newly exposed in changed climatic conditions (PANOŠ, 1978) so that the macroforms are of fossil origin while the microforms can be modern (GAMS, 1973). SWEETING (1972) classes exhumed karst as a category of fossil karst.

Other terms adopted to describe or to specify some special paleokarst/fossil karst features were summarized by PANOŠ (1978). Example include *healed karst, choked karst, mineralised karst, reactivated* or *rejuvenated karst.* The last term has a broader sense than does exhumed karst. LLOPIS LLADO (1953) used „*karst holofossile completo*" (completely fossilized), „*karst holofossile incompleto*" and „*karst merofossile*" (superficial forms partly filled). D'ARGENIO (in GUIDA, 1978) introduced „*paleocarsismo*" and „*neocarsismo*" (paleokarst and neokarst) to summarize the geological evolution of carbonate platforms in Southern Italy. Paleokarst is associated with morphogenesis attributed to paleotectonic processes, whilst neokarst evolution took place during subsequent uplift of the mountain chains. Paleo- and neokarst here are defined to represent two periods in the general evolution of sedimentary carbonate bodies and they do not include the effects of any different processes.

Hydrothermal karst. Many karst features encountered in outcrops, mines and boreholes were evidently produced by the action of hot solutions. They are termed hydrothermal karst (see DŻUŁYŃSKI and SASS-GUSTKIEWICZ, p. 377–397). Such features include all forms of solutional cavities as well as precipitated mineral formations. These phenomena are sometimes termed thermal karst or thermomineral karst. Sub-types are recognized. Deep subjacent karst is created where the water is heated during deep circulation in a confined karst aquifer (e.g. Hungarian Plain — ALFÖLDI and LORBERER, 1976 JAKUCS, 1977b; MÜLLER and SÁRVÁRY, 1977). Such dissolution may be responsible for the redistribution and deposition of ores in carbonate traps (DŻUŁYŃSKI and SASS-GUSTKIEWICZ; T. D. FORD; PALMER and PALMER; FUCHS *all in this volume*). Some hydrothermal karst has been created by very deep-seated soluis termed *hypogene karst* (PALMER and PALMER, p. 337; TSYKIN, p. 266–267). One special category recognised in hydrothermal karst are solutional forms created by convection in mixtures of hot ascending and cold infiltrating waters in caves (RUDNICKI, 1979).

Time of karstification. Former subaerial paleokarst may be assigned to a span of geologic time roughly defined as being older than the burial but younger than the karstified rocks. More accurate dating may be suggested indirectly by paleogeographical criteria. Moreover, in such time spans, several or many shorter karstification episodes may be recognised by applying various criteria (climatic change, rise and fall of sea level, etc.). Such lesser episodes are frequently termed *phases of karstification* (HERAK and STRINGFIELD, 1972). Some authors believe that this is not precise enough. GŁAZEK (1973) distinguished *karstification phases* and *karstification stages.* He used „karstification phase" for long periods of continental weathering and erosion, normally contained between two thick sedimentary, usually marine, groups in the stratigraphic sequence. In such phases, many different karst features were formed, including the big ones. BOSÁK (1981b) suggested that a karstification phase must be distinguishable over a large area of uniform geological (geotectonic) evolution. The upper and lower limits of the phase should be quite clear and the products of karstification should be present in substantial quantity and some variety. Within a given karstification phase, GŁAZEK (1973) recognised different *karstification stages.* These may be expressed in the development of larger forms and/or in different generations of small karst forms. A karstification stage registers some significant change of climate and relief during one karstification phase.

This review shows that there is inconsistency in the terms adopted to describe the phenomena of paleokarst. The same terms have been used in different manners and different terms applied to the same feature. This situation has resulted from the work of many individuals and in some cases it arises from the differing traditions of separate karstological schools. The introduction of some of the recent terms has perhaps confused an already unclear terminology. For example, paleokarst (paleocarsismo) and neokarst (neocarsismo) as defined by D'ARGENIO (in GUIDA, 1978) describe local conditions in southern Italy and are fixed to the lower local boundary of „neotectonic" movements. but many neokarst forms can be fully fossilized (see, for example, the chapters by GŁAZEK; BÁRDOSSY and KÓRDOS; BOSÁK, HORÁČEK and PANOŠ) and consequently belong to the category of paleokarst. Ancient karst forms can be still active as well, (cf. D. C. FORD, p. 313–336; ZÖTL, p. 483–509). These terms may be useful in local analysis but they should not be adopted for inter-regional correlation.

Although some of these definitions result in static treatment of paleokarst, perhaps it should be treated as an open thermodynamic system (BAKALOWICZ, BALVOUX and MANGIN, 1974; ERASO, 1975; MANGIN, 1976; QUINIF, 1983) that sometimes reaches a stationary state (GŁAZEK and SZYNKIEWICZ, 1979). In any particular karst area, in any geological period, karst functions as an open thermodynamic system expending the physical potential of water circulation and the chemical potential of rock dissolution. Resistance within the system is created by geological structure, morphology and the lithological properties of the rocks. During the development of a karst system its entropy is growing by the decrease of potentials and reduction of resistance by rock removal. According to Prigonine's principle (NICOLIS and PRIGONINE, 1977) a system reaches its stationary state when the dissipation (entropy production) attains a minimum. Dissipation may be divided into: (1) internal dissipation, represented by entropy remaining inside the system and (2) external dissipation, of entropy leaving the system. In the stationary state in any thermodynamic system, external dissipation reaches a minimum. In a karst system at the stationary state, any change of external dissipation caused by tectonic movement, climatic change and, especially, by engineering, may give a new stimulus to the external dissipation process. In other words, the system is rapidly brought to a state far from the possible minimum rate of external dissipation. When dissipation abruptly increases like this we call it "rejuvenation", or in engineering works "an unforeseen catastrophe". Rejuvenation of karst water circulation is mainly dissipated not by the development of new karst channels in solid rocks but by the removal of unconsolidated infill from existing karst channels, which frequently is a much faster process. Such changes are numerous in the history of any ancient karst system.

Definitions Proposed Here

Some standardised definitions of paleokarst terms were considered by P. Bosák, D. C. Ford, B. Gèze, J. Głazek, V. Panoš and H. Trimmel at Maletín (Czechoslovakia) in April, 1984. For the future we recommend the following usage: **Paleokarst** refers to karst developed largely or entirely during past geological periods. In English language publications we prefer the use of paleokarst to that of „fossil karst". It is less ambiguous. Paleokarst is divided into (1) buried karst, (2) some intrastratal karsts. **Buried karst** describes karst phenomena formed at the surface of the Earth and then covered by later rocks. This karst is clearly older than the cover strata and there is a disconformity or an unconformity. This is the best recognized, most widely agreed category of paleokarst. **Intrastratal karst** is formed within rocks already buried by younger strata; the

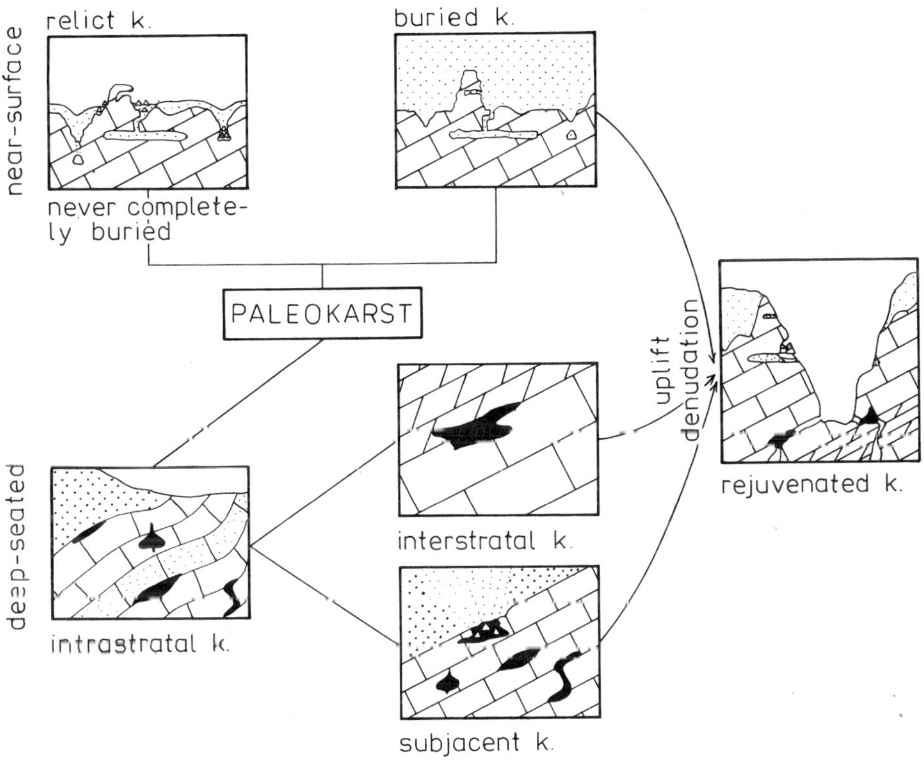

1-1. Types of paleokarst.

karstification is *younger* than the cover. Subjacent karst and interstratal karst are special categories of intrastratal karst. **Subjacent karst** refers to karst that develops beneath less soluble or insoluble strata; it is the most common type of intrastratal karst. **Interstratal karst** is developed along bedding planes or unconformities within soluble rocks. All such karst features have not been part of any former karst landscapes. Flow up-dip is common in intrastratal karst where there is artesian confinement; the waters often carry considerable dissolved loads (brines) and/or thermal energy (hydrothermal karst).

We recognize that there is abundant modern (active) intrastratal karstification in progress. In a simple definition, paleo-intrastratal karst is now entirely inactive or inert; it may be represented by Mississippi Valley-type massive sulphide fillings, for example. In practice, intermediate examples can be expected such as largely inert intrastratal karst that is still active in zones of highest hydraulic gradient, or ancient karst that has been rejuvenated by glacial action (see D. C. FORD, p. 328–336).

There are some examples of karst landforms that were created at the Earth's surface under one set of morphogenetic conditions and which survive at the surface under a present, different set of conditions. Normally these will display modern superimposed small forms (e.g. rill karren) or will be degraded from their original form. These are defined as **relict karst.** Cave systems that are drained and abandoned by the groundwater streams that created them, as a consequence of the lowering of springs, are subterranean relict karst. **Exhumed karst** is that exposed by the erosion of cover strata that buried it. Karstification can start immediately after sedimentation; this case was defined as **syngenetic karst,** formed in soluble sediments as they are being lithified (JENNINGS, 1971).

Karst period is proposed here to define long-lasting times of continental weathering and groundwater circulation which were terminated by an ensuing marine transgression. Karst periods will be recognized by upper unconformities or disconformities. In such periods many different karst forms were probably created. They may be divisible into several generations. Therefore the karst period often can be divided into two or more **karst phases.** Individual phases in a region are caused by geodynamic or major climatic change, e.g. uplift or downwarping, a phase of permafreezing, etc. Rather big changes that occurred within a brief time span may be called **karst events,** e.g. catastrophic collapse of a surface, the infill of a cave passage by the deposit of just one flood of a few hours' duration, etc.

PART II
REGIONAL REVIEW

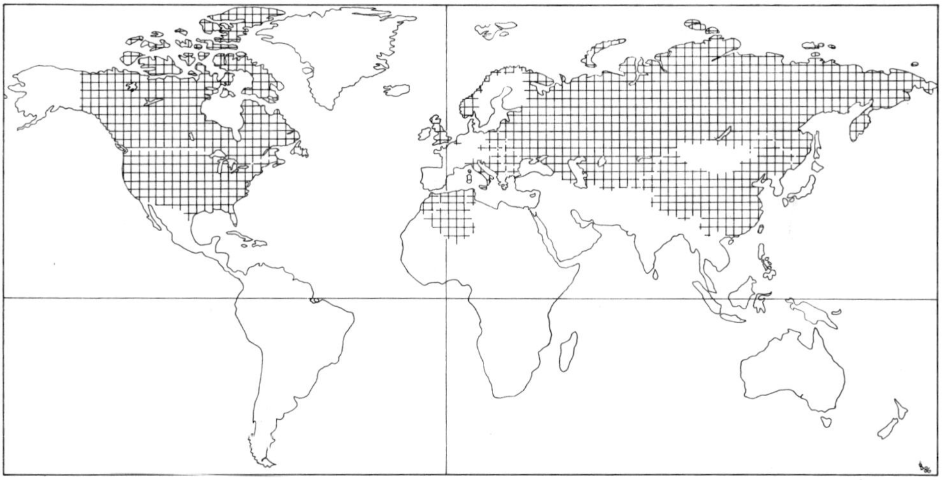

PALEOKARST OF BELGIUM

Yves Quinif

Introduction

Belgian paleokarsts developed either as a consequence of the Variscan Orogeny, where examples include early (Fammenian) paleokarsts in Visé and the mid-orogenic (Lower Namurian) paleokarsts, or they developed during later continental evolution as in the case of the natural shafts of Houiller. The majority of the paleokarst phenomena are developed in Carboniferous (Dinantian) limestones (EK and POTY, 1982).

Ancient karst phenomena have been encountered often in outcrops, in boreholes or in mining works but only a few have been interpreted in a karst paleogeographical context. Pb-Zn ore bodies belong to this latter category. The greatest tonnages are obtained from ore bodies that were reconcentrations within karst cavities or depressions, but this karst relationship was very late in being adopted as a prospecting tool.

Geological Framework

Pre-Permian Evolution

The Caledonian massifs in the Ardenne (Rocroi, Stavelot, Serpont, Givonne) and in Brabant consist of non-karstic rocks of Cambrian and Silurian age. The Devonian (Lower Gedinian in the South and Couvinian in the North) began with a basal conglomerate upon an angular unconformity. The cover rocks on these Caledonian massifs then include all stages until the Westphalian. Karst limestones occur in the Middle Devonian (Couvinian-Givetian-Frasnian) and in the Carboniferous (Dinantian; Fig. 2–2).

P. Bosák, D. C. Ford, J. Głazek and I. Horáček (Editors):
Paleokarst. A Systematic and Regional Review. – Elsevier and Academia. Amsterdam and Praha. 1989.

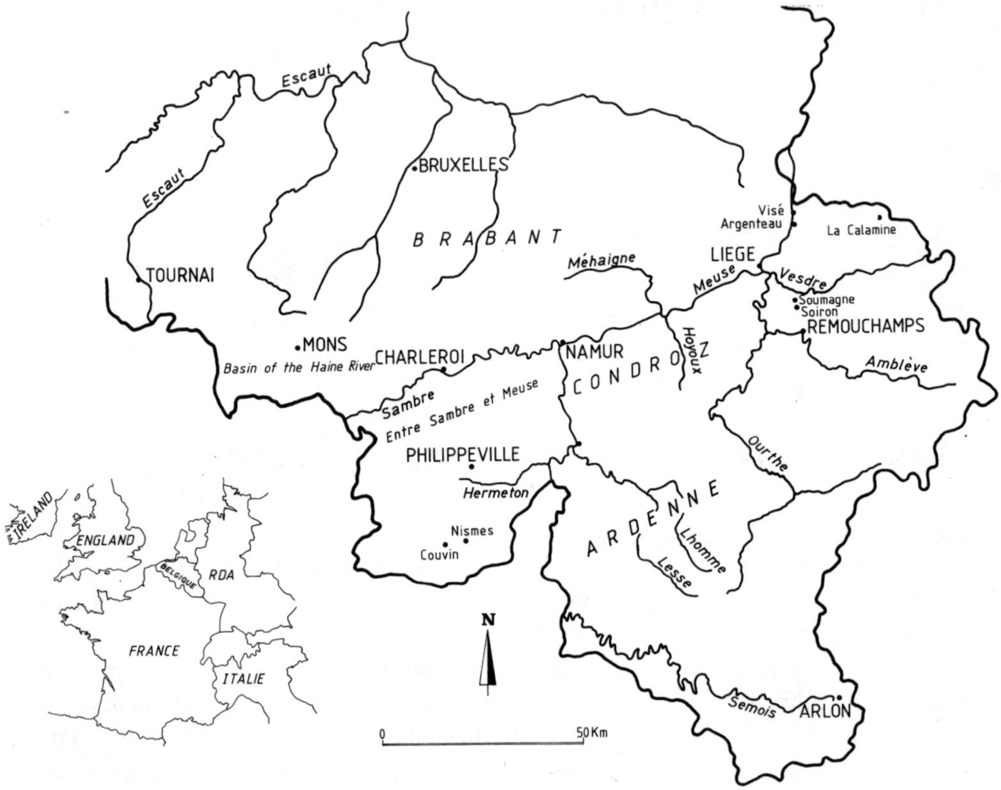

2–1. Localization map of principle sites.

In the Upper Silurian, the principal phase of the Variscan Orogeny was marked by superimposed folding of the Caledonian platforms of Ardenne and of Brabant, overthrusting of the Dinant block over the Namur synclinorium and folding of the cover strata in the Dinant synclinorium. The main tectonic units thus created were: (1) the Cambro-Silurian platform (Ardenne and Brabant) which has no karst, (2) the fold system of the Ardenne, lacking karst strata in the Lower Devonian parts but with limestones of Middle Devonian and Carboniferous in the Dinant synclinorium, (3) the Namur synclinorium (an autochthon) which also contains karst limestones, mostly Dinantian. The northern part of this unit rests against the Brabant platform (MICHOT 1980; PRODRÔME, 1954; ROBASZYNSKI and DUPUIS, 1983; Fig. 2–2).

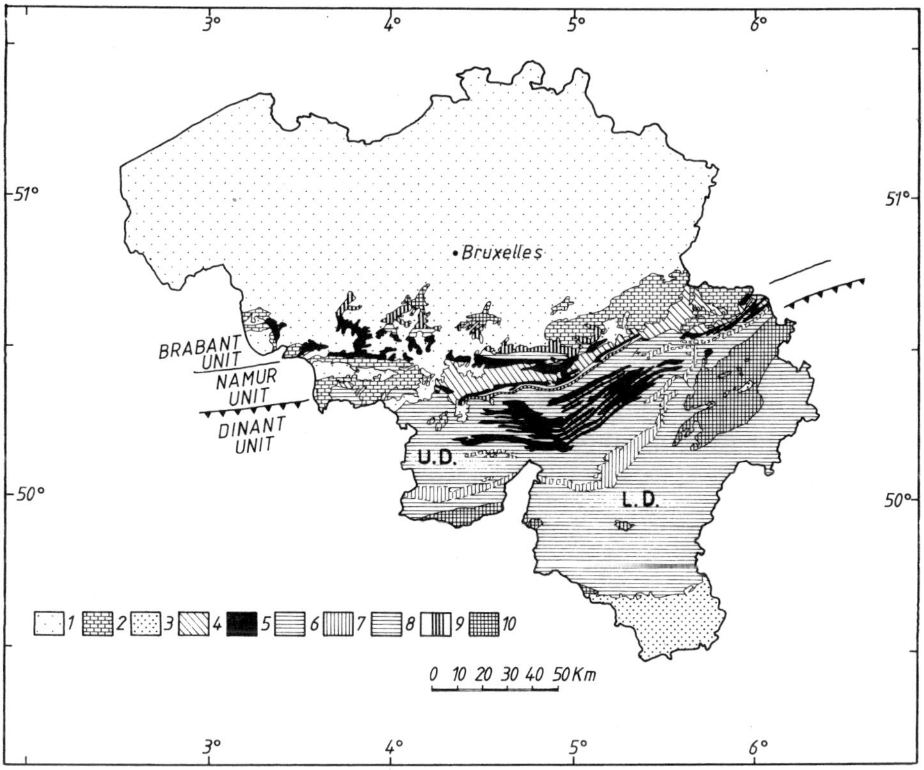

2–2. Geological framework of Belgium. The Dinant unit of the Ardenne Massif overlaps the Namur unit lying on the Caledonian platform of the Brabant unit. 1. Tertiary cover; 2. Cretaceous (essentially chalks); 3. Jurassic and Triassic; 4. Namurian; 5. Dinantian limestones; 6. Upper Devonian (U.D.); 7. Middle Devonian (with Frasnian, Givetian and Couvinian limestones); 8. Lower Devonian (L. D.); 9. Silurian; 10. Cambrian.

Post-Variscan Evolution

Erosion of the Variscan relief ended with the formation of a peneplain with Triassic and Jurassic sediments accumulating on the southern margin of the Ardenne. In the subsided Haine River Basin around Mons and further east around Liège, Cretaceous sediments were deposited; they consist primarily of chalks, without karstification. The evidence of a pre-Maestrichtian erosion surface is observable to the East, in the high Ardennes.

During the Tertiary, there was clastic sedimentation in the northern part of Belgium, resulting in very complex sequences of Paleogene deposits. In the Ardennes, we find a succession of marine transgressions and regressions and the formation of an erosion surface under tropical climatic conditions during the Eocene and Oligocene. Remnants of the sands and clays are trapped in some karst depressions. The first river channel trends of the Ardennes developed on a cover of soft rocks dating probably from Oligocene times (ALEXANDRE, 1976; LAURANT, 1976). Following them and until the Pliocene, a series of other planation surfaces developed under a savannah climate which became more and more temperate.

Terminology

The fundamental definition of *paleokarst* used here is that it refers to paleomorphological features created by karst processes of the past and now fossilized within a geological formation. The paleokarsts of the Visé region are a type example. This concept is extended to include *deep karsts* that have always been active and evolving beneath a thick cover of unkarstifiable strata that is disrupted by the deep karst action today and has been in the past.

The Principal Paleokarst Phenomena

Pre-Fammenian Paleokarsts

EK (1976) described the only pre-Fammenian karst feature. It is an example of intra-Frasnian karren visible on a wall of Remouchamps Cave. It measures ten to twenty centimetres horizontally and vertically and is limited at its bottom and top by flat surfaces which distinguish it from stylolites.

Fammenian Paleokarsts of the Visé Region

Geological Structure (BLESS, BOUCKAERT and PAPROTH, 1980)

The Visé region is on the SE border of the Brabant Massif (Fig. 2–1). The Cambro-Silurian platform is covered by conglomerates and biostromal carbonates of Midddle Devonian, Givetian shales and sandstones and Frasnian biostromal limestones (Lustin Formation). The limestones are directly overlain by Visean carbonates. The Fammenian is missing. Upper Visean (*V3c*) deposits cover the whole area.

The Frasnian/Carboniferous contact is very rugged. Two extreme situations occur: (1) silicified carbonates of the upper *V3c* cover undisturbed Frasnian limestones that are perforated by karst cavities, (2) Upper Visean limestones cover a cyclopean breccia of Frasnian limestone elements, which are sometimes covered by a layer of Upper Frasnian shales.

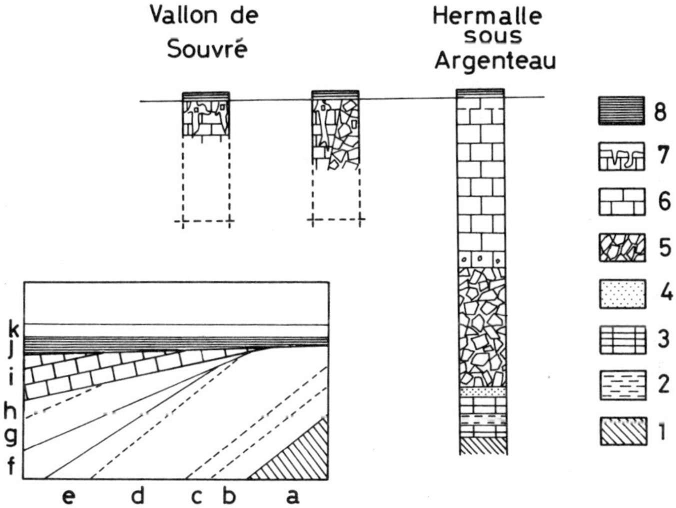

2–3. Stratigraphy of the Visé site (after Poty, 1982). Three lithostratigraphic columns show three types of extreme stratigraphic contact between Visean and Frasnian. In the Vallon de Souvré lydites (Upper Visean, V3c) there are fossilised karst cavities developed in Frasnian limestones and filled with Tournaisian deposits. The Hermalle-sous-Argenteau borehole shows Upper Visean limestones on a cyclopean breccia built of Frasnian elements. In the centre, the third column represents a quarry outcrop with a visible lateral contact of two stratigraphic units. 1. (a) Lower Paleozoic; 2. Givetian shales; 3. Givetian limestones; 4. (e) Frasnian sandstones; 5. Frasnian solution collapse breccia; 6. (i) Upper Visean limestones; 7. karst cavities in Frasnian limestones with Tournaisian infill; 8. (j) Upper Visean lydites; b – Givetian; d – Lustin Formation (limestones and breccia); f – Tournaisian; g – Lower Visean; h – Middle Visean; k – Namurian.

Paleogeography of the Sites

Tectonic movements at the beginning of the Fammenian led to the formation of a brachyanticline in the Visé region. Tension jointing was well developed in limestones of the Lustin Formation which, where emerged, were karstified in two distinct styles: (a) in the vadose zone, corrosion created a mogote relief with large collapses and formation of cavities, (b) in the phreatic zone, solution-collapse caused the formation of cyclopean breccia.

Several transgressions fossilized this relief during the Dinantian. A third transgression (Upper Tournaisian) invaded the Souvré Valley. Karst mogotes were bevelled and buried by the Upper Visean (*V3c*) strata.

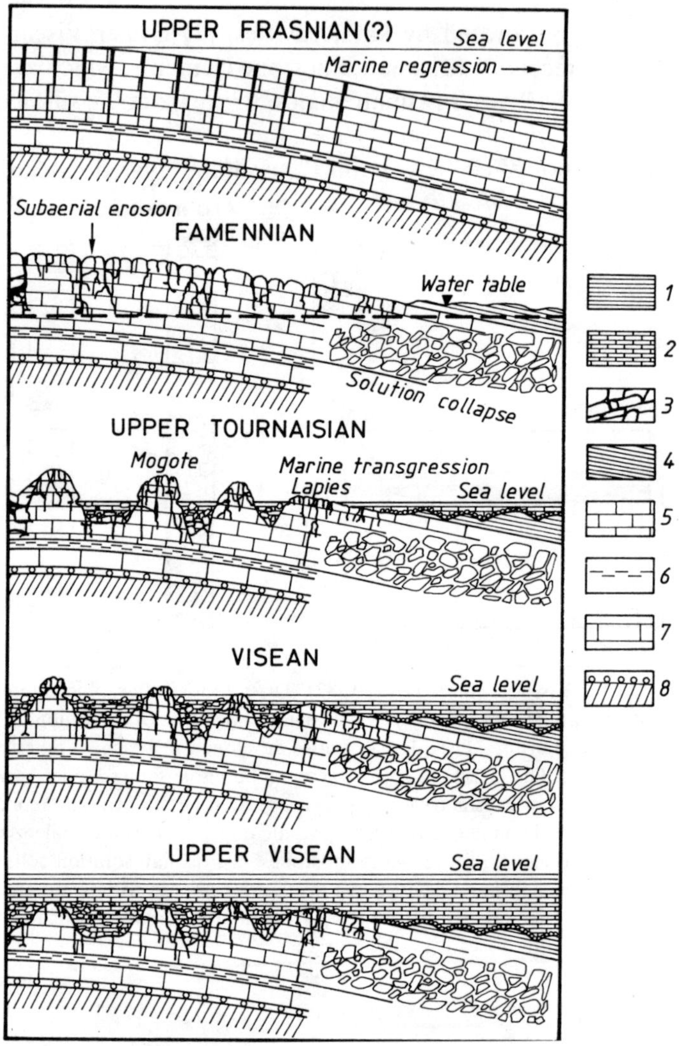

2–4. Visean paleokarst; schematic evolution from the Upper Frasnian to Upper Visean (after POTY, 1982). 1. Upper Visean lydites (V3c); 2. Dinantian limestones; 3. cyclopean breccia; 4. Upper Frasnian; 5. Lower Frasnian limestones; 6. Lower Frasnian sandstones; 7. Givetian; 8. Lower Paleozoic landforms with basal conglomerate.

Karst Cavities of the Souvré Valley (POTY, 1980, 1982)

Frasnian limestones of the Lustin Formation are perforated by cavities ranging from fissures a few cm wide to pipes of several decimetres, directed along bedding planes and extension joints. Three types of deposits fill the voids: (1) wall coverings of millimetric to centimetric fluoroapatite crust, (2) marine deposits ordered in fining-upward sequences with calcareous microbreccia at the base and clayey micrites at the top. These deposits belong to the Upper Tournaisian and Lower Visean. Vertical fissures descending from the surface are filled with coarser sediments than horizontal caves, the latter being poorly connected; (3) voids which were not filled by (1) and (2) contain layered calcite.

The Cyclopean Breccia

Several cyclopean breccias were recognized in boreholes (e.g. 82 m in the Hermalle-sous-Argenteau hole) and in outcrops. These breccias are composed of angular elements of Frasnian limestones without stratification or other sedimetary structure. They display all the characteristics of solution-collapse breccia.

Dinantian and Lower Namurian Paleokarsts

Geological Framework

Meanwhile, deposition in the Dinant block was continuous from the base to the top of the Dinantian. In the Namur autochthon and on the margins of the Brabant Massif, there were several episodes of emergence where karstification took place. After a Late Fammenian regression, a series of transgressions and regressions affected these areas. Lower Visean transgression followed, and during Lower Namurian "Sudeten" fold phase emergence occurred again.

Intra-Visean Paleokarsts

DELEPINE (1911), DAMIEAN (1956), PIRLET (1970) described structures which disturbed the deposition of Upper Visean limestones due to the presence of underlying karst in Lower Tournaisian grey dolostone. At Chockier many decimetric to metric cavities are filled with Upper Visean limestones and radial fibrous calcite. The paleokarst is located in the top 50 m of dolostone. At Marchempré, a karst chamber 40 m wide is developed in Middle Visean limestone and filled with a breccia of local blocks (CALEMBERT and VAN LECKWIJCK, 1943).

Lower Namurian Paleokarsts

At Soumagne and Soiron, boreholes revealed the presence of cavities in Visean limestones that were filled with a black breccia of marine shales from the basal Namurian (*E1*) and covered by continental sandstones of the Lower Namurian (*E2*) before passing upwards to Middle Namurian marine shales (GRAULICH, 1962).

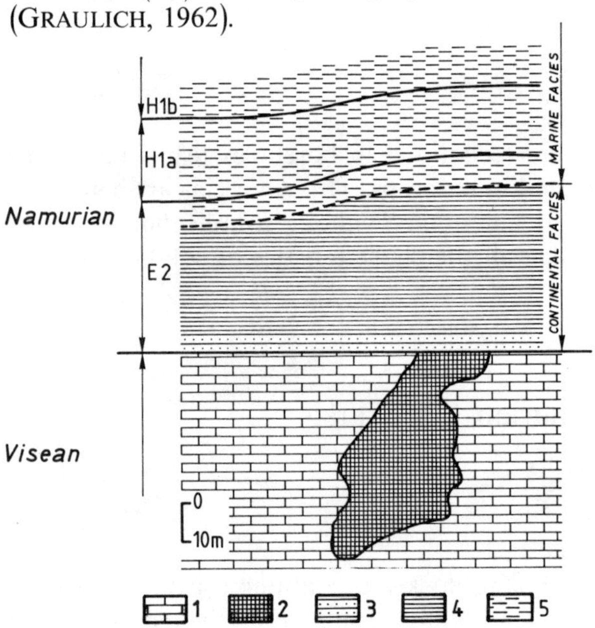

2–5. Lower Namurian paleokarst (after GRAULICH, 1962). A karst pocket filled with basal Namurian marine deposits is fossilized by continental Lower Namurian. The geological section is reconstructed from the Soumagne and Soiron boreholes. 1. limestones; 2. black marine shales; 3. sandstones; 4. continental shales and sandstones; 5. marine shales.

In Lovegnée Tunnel further east, the Visean/Namurian contact is always marked by the presence of karst pockets with clay filling older than the Lower Namurian (*E2*) in an inverse sequence, which proves that karstification occurred before the inversion (transgression) and was thus in the Namurian.

The Visé region also shows karst pockets in Visean limestones, filled with black shales of Lower Namurian age fossilized before the Middle Namurian.

In conclusion, marine pyritiferous black shales of basal to Lower Namurian age were deposited on Visean limestones in the Namur autochthon. Emergence led to karstification during the Lower Namurian, with the erosion of all shales except those trapped in karst pockets. This karstified structure was then covered by the Middle Namurian.

Deep Karst in Hainaut and Mesozic Paleokarsts

Summary

The exploitation of coal in Hainaut has revealed for more than a century that there are vertical, cylindrical structures where coal seams have dropped and in which are more recent geological strata. The coal-miners call these structures *circular faults* or *natural pits*. Thanks to recent work based primarily on deep borehole data, it is known that the *natural pits* are a manifestation of karst dating from the Lower Cretaceous.

Geological Structure

The natural pits occur between Charleroi and the French frontier around Mons. This basin is a part of the northern border of the Paris Basin where the post-Paleozoic cover is of Albian to Quaternary age and fills a subsided section of the Namur block.

After the Variscan Orogeny, a continental phase began, probably during the Stephanian. Sand and clay channel deposits of Neocomian age represent the Wealden facies which plays an important role in the paleokarst. Following them, transgression that lasted from the Albian to the Turonian saw different clastic sequences laid down before the onset of Senonian chalk deposition (ROBASZYNSKI, 1979).

The Natural Pits

These cylindrical structures are irregular in detail but their outlines are rounded. Their horizontal dimensions are varied but always at least several tens of metres (36 × 57 m at a depth of 660 m in pit No. 58 or 75 × 100 m at –260 m in pit No. 60; DELMER and VAN WICHELEN, 1980). Vertically the pits extend for several hundred of metres.

The pit fillings are of two types: (1) marl and clay breccias with lignite and pyrite, derived from post-Paleozoic formations. Several studies have recognised numerous examples of the Wealden facies. MARLIÈRE (1932) described at the Bernissart natural pit a polygenetic breccia composed of small clasts of Albian to Lower Cenomanian shale and large clasts of coal, all being cemented with calcite; (2) Westphalian deposits as breccias are the second type of filling.

The geometry of the pits is not well known; their tops have not been excavated and the bases are below the mining works. Blind terminations have been proven

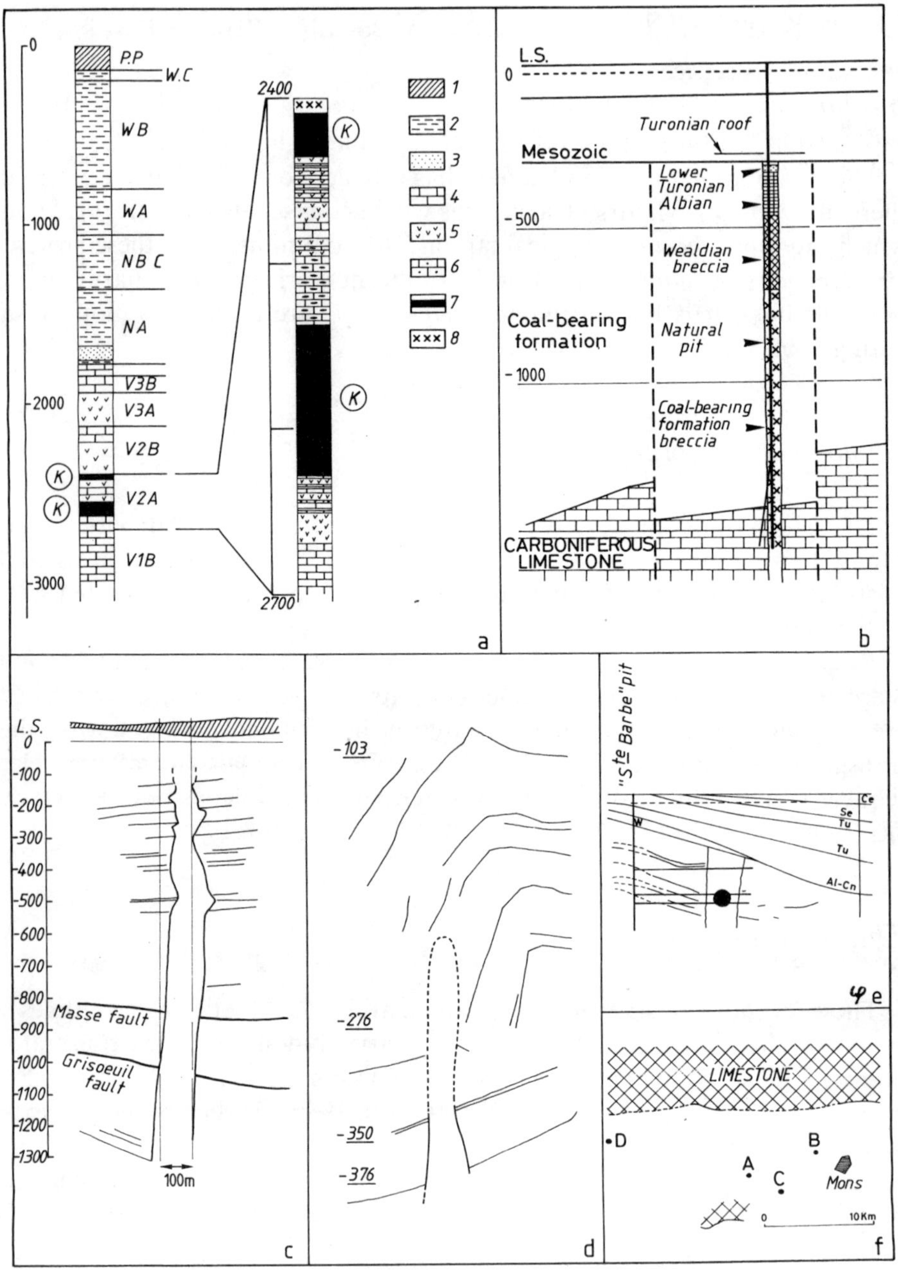
P.P
W.C
WB
WA
NB C
NA
V3B
V3A
V2B
V2A
V1B
K
0
1000
2000
3000
2400
2700
1
2
3
4
5
6
7
8
a
L.S.
0
Turonian roof
Mesozoic
Lower Turonian
Albian
-500
Wealdian breccia
Coal-bearing formation
Natural pit
-1000
Coal-bearing formation breccia
CARBONIFEROUS LIMESTONE
b
L.S.
0
-100
-200
-300
-400
-500
-600
-700
-800
-900
-1000
-1100
-1200
-1300
Masse fault
Grisoeuil fault
100m
c
-103
-276
-350
-376
d
"Ste Barbe" pit
Ce
Se
Tu
Tu
Al-Cn
W
e
LIMESTONE
D
A
B
C
Mons
0
10 Km
f

in the pits found in the Courcelles Nord Mine near Charleroi, proving the operation of deep genetic processes.

A borehole at Ghlin (DELMER et al., 1982) penetrated a natural pit to the base, giving a stratigraphic column of its filling. The cavity extends through 104 m of Cenozoic and 628 m of Cretaceous strata. The lower part of the Cretaceous comprised 45 m of disturbed Lower Turonian, 140 m of glauconitic and silicified Albian sandstone, 218 m of breccia and debris of Wealden facies and 800 m of brecciated coal measures before penetrating Carboniferous limestones at a depth of 1 585 m. Angled offset drilling at −1 298 m passed out of the pit and into coal-bearing strata in situ.

Let us end by recalling the most famous of these pits, Bernissart, in which a deposit of Iguanodons was found. Two horizontal galleries at depths of 322 and 366 m entered a pit 80 m wide with a filling of Wealden clay where skeletons of *Iguanodon* were distributed in bone beds at four levels over a total thickness of 34 m (QUINET, 1969).

Paleokarst Explanation of the Natural Pits

The St. Ghislain borehole supplied the information needed to understand the genesis of the pits (DELMER, 1977). Between 2 408 and 2 728 m the drill revealed a large karstified zone, with voids of a metre or more, in an alternation of Visean limestones and anhydrite (GROESSENS, CONIL and HENNEBERT, 1979), thus proving the existence of deep karst under the Mons Coal Basin.

The natural pits are the result of deep karst dissolution with upward stoping the Neocomian (Wealden) in an environment of rivers and swamps. The Ghlin borehole indicates that the process was still active in the Albian. It is difficult to determine how long the collapse processes remained active. There is no characteristic depression morphology in the modern surface. Probably this paleokarst must be attributed to the Neocomian with some extensions into the Albian. It was fossilized by the Turonian transgression.

2–6. Paleokarst of the natural pits at Hainaut. a. St. Ghislain borehole with detail of the karstified zone between 2 400 and 2 700 m (after DELMER, 1977; GROESSENS, CONIL and HENNEBERT, 1979). 1. PP=post-Paleozoic formations; 2. coal-bearing formations (shales, sandstones, coal, etc); 3. sandstones; 4. limestones; 5. anhydrite; 6. limestones with cherts; 7. karstified zones (K); 8. dolomites; WA, WB, WC: Westphalian; NA, NB, NC: Namurian; V1, V2, V3,: Visean. b. Ghlin borehole with natural pit (after DELMER et al., 1982). c. Natural pit of the Flenu Coal Mine. Lines represent the exploitation level. d. Blind natural pit of the Courcelles Coal Mine. e. *Iguanodon* accumulation (black dot) in the natural pit of Bernissart (c. to e. after DELMER and VAN WICHELEN, 1980). W. Wealden; Al–Cn. Albian-Cenomanian; Tu. Turonian; Se. Senonian; Ce. Cenozoic. f. Localities of sites. A. Saint-Ghislain borehole; B. Ghlin borehole; C. Flenu Coal Mine; D. Bernissart.

Tertiary Paleokarsts

Depressions of Condroz and Entre-Sambre-et-Meuse (Calembert, 1954; Soyer, 1979; Girolimetto, 1982; Thorez and Bourguignon, 1984)

(1) Morphology. The 1 : 160 000 geological map of the region of Condroz and Entre-Sambre-et-Meuse between Charleroi and Liège shows many Tertiary deposits with irregular forms, most of which lie on Dinantian limestones. The exploitation of sands, clays and iron ore at some of these sites has shown their pocket-like geometry. Some examples attain a depth of one hundred metres and are several hundred metres in diameter. Some depressions contain rounded limestone pinnacles.

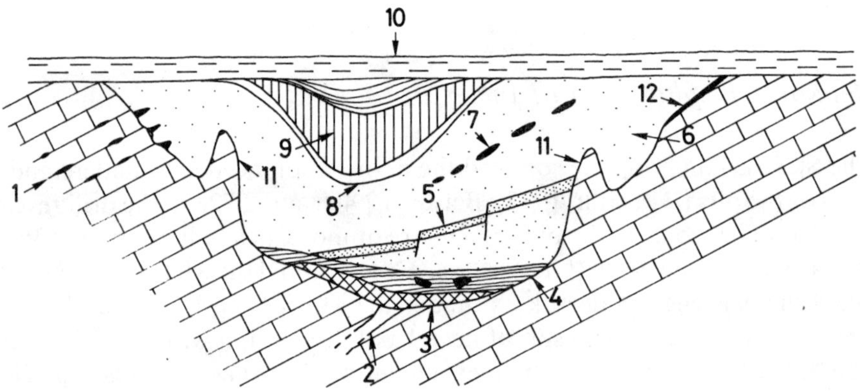

2–7. Synthetic section through a Tertiary paleokarst depression in the Condroz (after Calembert, 1954; Thorez and Bourguignon, 1984). 1. Visean limestones with cherts; 2. evacuation channel; 3. iron accumulation; 4. residual clay (kaolinite predominates) with silicified nodules; 5. stratified sands with collapse faults; 6. Tertiary sands; 7. silicification; 8. lacustrine sediments; 9. lignitic clays; 10. Quaternary deposits; 11. residual limestone peak; 12. silicified edge of depression.

(2) Fillings. Generally the series begins with marine sands of unknown thickness upon which lie a rhythmic suite of 0.1 to 1.0 m beds of plastic clays, sands, clayey sands, paludal and lacustrine sediments, sometimes with great masses of lignite clays rich in plant remains. After deposition the fillings were modified by compaction, faulting and landsliding. In addition, geochemical processes changed the nature of the sediments: these included depigmentation and rubefaction, formation of iron and manganese oxide nodules, silicification and accumulation of iron ore at the contact with limestone. Clays are kaolinised to a greater or lesser extent. Part of the kaolinite comes from leaching of overlying soils.

(3) Paleogeographical context. During the Tertiary, transgressions and regressions alternated with continental periods in the Ardennes. Oligocene deposits constitute the majority of basin fill in the Condroz. Depressions in Entre-Sambre-et-Meuse also contain Eocene sands. Palynological investigations show that in the basal Oligocene, paludal forms were developed under a subtropical continental climate. Thick kaolin beds developed on silicified rocks, attesting to the existence of a hot and wet climate favourable to karstification. The great transgression of the Lower Oligocene followed. In the Condroz, this transgression is attributed to the Chatian.

The "Abannets" of the Couvin-Nismes Area

Around Nismes and Couvin, the tops of limestone hills contain depressions of complicated form, several dozen metres deep and one or two hundred metres in diameter. In the past, they were filled with sediments similar to those just described above. They were excavated for their iron ore and sands. They were interpreted by VAN DEN BROECK, MARTEL and RAHIR (1910) to be pre-Tertiary ponors, but recent studies suggest they represent Tertiary pedologic karren forms developing under the cover. FOURNEAU (1972) attributed the same origin to them as to the *Grand Lapies* of Landelies.

Fossil Polje of Nismes

Some fragments of planation surfaces on limestones in the Couvin-Nismes area are interpreted by SOUCHEZ (1963) as remnants of a Tertiary polje which was destroyed by the incision of the modern hydrographic network. These surfaces are mantled by clays with quartz grains and plant remains, perhaps deposited by temporary floods in hot climate conditions.

Conclusions

Continental phases with a hot and wet climate during the Tertiary led to the development of karst no Devonian and Carboniferous limestones. Continental and marine deposits filled depressions (dolines, uvalas and poljes) associated with denudation surfaces. The relief was gentle, a slightly undulating plateau. Karst forms, particularly in basins, must have been associated with a phreatic network. Perhaps a part of the caves actually penetrable today were generated during the Tertiary in the phreatic zone, but no correlative deposits have been found to prove this assertion.

Practical Consequences

Geothermal Energy in Hainaut

The deep karst discovered by boreholes in Hainaut supports an enormous aquifer of hot water. For example the water temperature is 74 °C in the St. Ghislain borehole. Pumping tests have been made and an urban regional heating network is in the process of establishment.

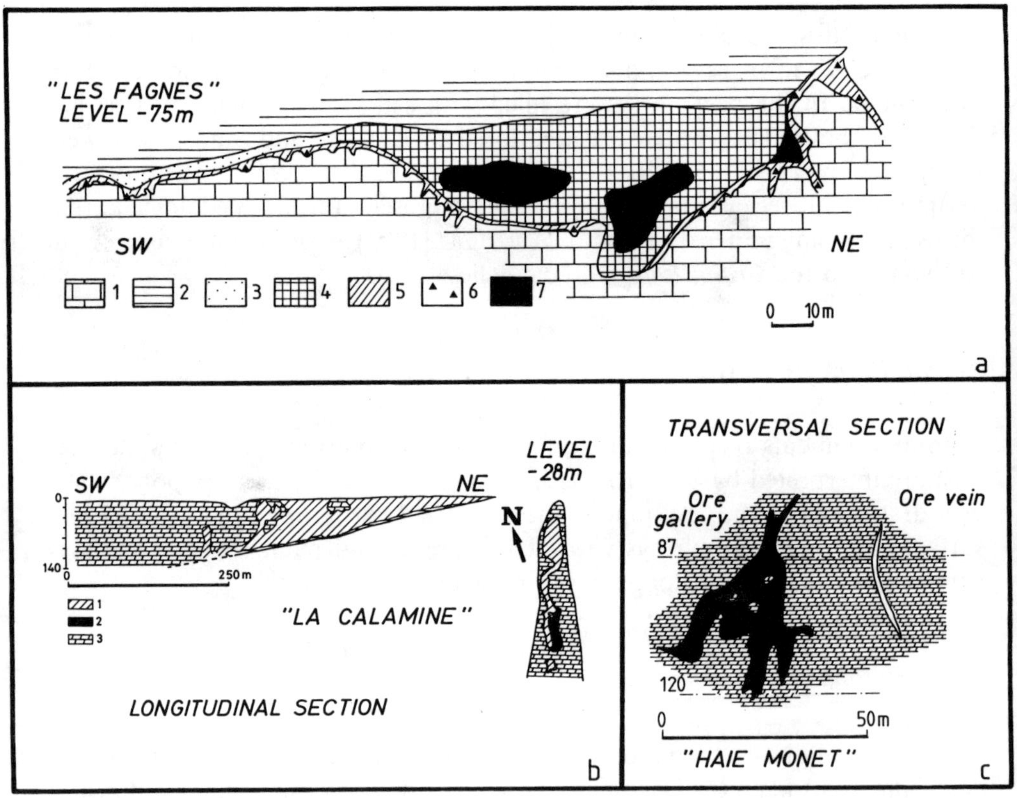

2–8. Examples of mineralized paleokarsts (after BALCON, 1981; DEJONGHE and JANS, 1983). a. Lower Namurian paleokarst with depressions in Visean limestones. 1. Visean limestones; 2. Namurian shales; 3. ochre clay; 4. iron and zinc sulphides; 5. smithsonite; 6. galena; 7. black clays sometimes with sulphides. b. Zinc orebody La Calamine. 1. calamine (mixture of zinc oxides, carbonates and silicates); 2. clays with zinc; 3. Tournaisian dolostones. c. Mineralized paleokarst gallery Haie-Monet.

Lead-Zinc Ore Deposits

Dinantian limestones of the northeastern and eastern parts of the Namur unit are rich in lead-zinc orebodies. Original abundances of elements in the Dinantian are poorly known (BALCON, 1981). A first phase of concentration occurred with deposition in veins, principally in the dolostone series. Afterwards, the Dinantian and particularly, the Lower Namurian, karst phases led to the formation of the principal orebodies, called *amas* by the old miners. Here we summarise three representative examples.

Les Fagnes is a pocket where the walls are covered by a crust of smithsonite and iron ore with galena. The pocket filling is of black clay with iron and zinc sulfides and containing root remains. The pocket developed in Visean limestones and was covered by Lower Namurian shales.

Haie-Monet is a network of galleries controlled by bedding and a joint system, filled with breccia cemented by calcite and smithsonite above the water table and with stratified blende, galena and marcasite below it. Although there are no data to allow us to date this mineralized paleokarst precisely, it seems likely that it is connected with the Lower Visean emergence.

La Calamine is an orebody of calamine (a mixture of smithsonite, willemite and hemimorphite) with clays containing zinc (DEJONGHE and JANS, 1983). The orebody, contained within a syncline of Tournaisian dolostone, is 450 m long, 65 m wide and known to a depth of 110 m, where it remains oxidized. The ore deposit can be interpreted as a paleokarst associated with the Paleozoic/Mesozoic unconformity (the Maestrichtian surface which is developed in the eastern part of Belgium).

Uranium Accumulations (GRAULICH, 1962; HEREBOSCH, DE WITTE and PREAT, 1979)

Pockets in the Namurian paleokarst contain some small concentrations of uranium which derive from the leaching of Lower Namurian black shales. Oxidation of pyrite in these shales during the karstification yielded sulphuric acid which dissolved the uranium. The sulphate solutions drained down into the karst cavities where they were precipitated on carbonate walls together with the phosphatic crusts. Uranium concentrations never reach economic levels.

The Iron of the Abannets

The *Abannets,* karst depressions in the Nismes area, were exploited in the past for their iron. The orebodies originated from glauconite in the Tertiary sands in depression fillings. The iron was dissolved, migrated down and precipitated against the limestone walls.

Conclusions

Belgian paleokarsts are developed principally in Dinantian limestones and to a lesser extent in Devonian limestones. A first series of paleokarsts developed with the emergence of the Namur block during the Fammenian, Dinantian and Lower Namurian. They are the consequence of the early Variscan movements.

The natural pits in Hainaut are the result of deep interstratal karstification under a thick cover during the Lower Cretaceous. Finally, the Tertiary paleokarsts display corrosion under an overburden of sands and clays with the formation of depressions. The climate was hot and wet.

For pure science, paleokarst supplies much information concerning the past climates and paleogeography. In practical terms, paleokarst may host orebodies of metals such as lead, zinc and iron.

PALEOKARST OF BRITAIN

Trevor D. Ford

Introduction

If paleokarst is defined as the record of development of karstic landforms, either surface or underground, in any erosion cycle prior to the present, then it is widespread in the Carboniferous Limestone outcrop areas of Britain. Paleokarstic features are particularly well developed in the Peak District of Derbyshire, in Wales and the Mendip Hills of Somerset (Fig. 2–9). Published studies of paleokarst in these areas are few and studies have generally been made as a side-line to other aspects of stratigraphic or sedimentary history. A number of other areas have more limited paleokarsts in rocks ranging in age from Cambrian to Cretaceous. Each area is considered separately.

Karst surfaces may be classified as (1) *bare* or *uncovered karst;* that which develops on a fully exposed limestone surface; (2) *covered karst;* that which develops under a cover of soil or other superficial deposits; (3) *interstratal karst;* that which develops by solution along a bedding plane or unconformity after the overlying beds have been deposited, and (4) *buried karst;* any of above buried by younger strata after development. Most of the paleokarsts discussed herein fall within category (2) in their genesis though they are now seen as within (4) owing to their being within the stratigraphic record. Some instances of interstratal karst are also considered herein.

The Peak District of Derbyshire

Erosion surfaces within the Carboniferous Limestone and paleokarstic features developed subsequently in the Peak District have been described previously by the present author (T. D. FORD, 1964, 1969, 1972, 1977, 1984); by WALSH et al. (1972), WALKDEN (1974, 1977, 1982) and OAKMAN (1984) (Fig. 2–10). The outcrop is some 40 km long from north to south and 15 km wide. The strati-

P. Bosák, D. C. Ford, J. Głazek and I. Horáček (Editors):
Paleokarst. A Systematic and Regional Review. – Elsevier and Academia. Amsterdam and Praha. 1989.

graphic sequence of beds exposed totals some 500 m and the base is nowhere visible. The exposed sequence presents a wide variety of lithofacies broadly representing a paleogeography of marginal reef complexes around the fringes of the outcrop and lagoonal calcarenites in the centre. Dips are generally low and

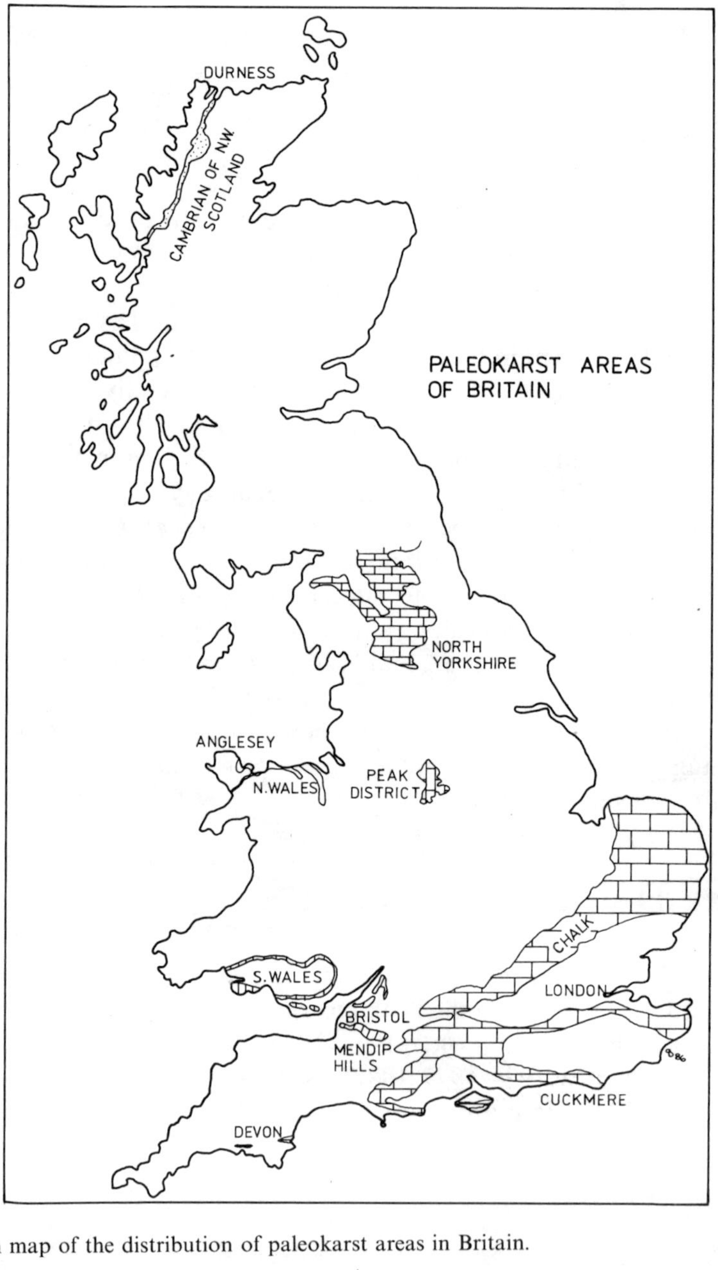

2–9. Sketch map of the distribution of paleokarst areas in Britain.

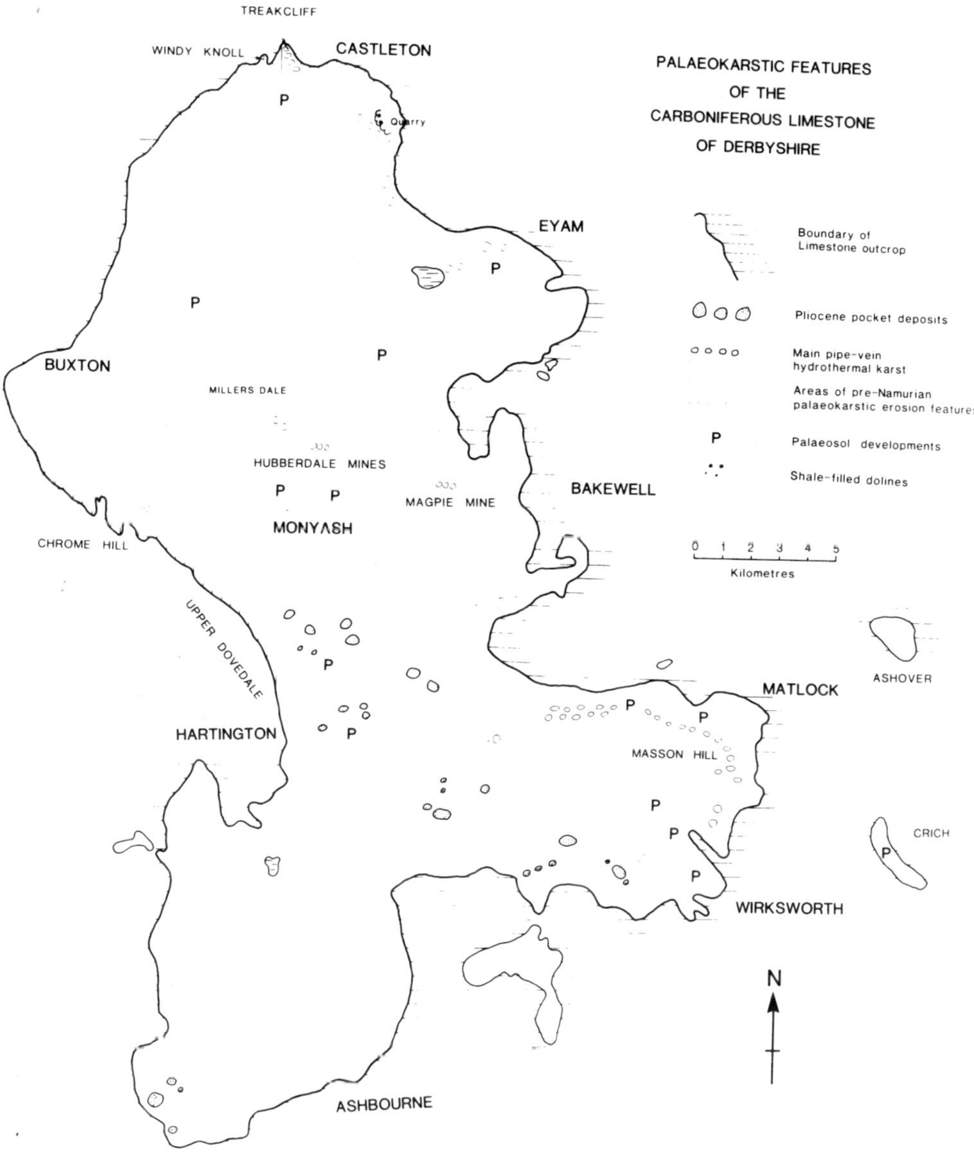

2–10. Sketch map showing the distribution of paleokarst features in the Carboniferous Limestone of the Peak District, Derbyshire. Encircled areas mark limestone outcrops (also in Figs. 2–14, 2–17 and 2–10).

do not often exceed 15 °, but gentle folding on east-west axes in the east, and north-south axes in the west yields a complex outcrop pattern. Intermittent volcanicity is manifested as a series of basaltic lava flows and tuffs interbedded with the limestone (WALTERS and INESON, 1981; INESON and WALTERS, 1983).

Thin tuff horizons, usually less than 30 cm thick, are frequently present in the limestones between the thicker lavas (WALKDEN, 1974).

Limestone sedimentation in the British Carboniferous was demonstrated by RAMSBOTTOM (1973) to have proceeded as a series of cyclic transgressions and

2–11. Paleokarst surface with 1 m deep potholes revealed by removal of 30 cm tuff layer. Bennett's Quarry, Grangemill, Derbyshire (Photo by T. D. Ford).

regressions reflected in cyclic changes in lithological characters and fossil content. Each cycle in theory can, at the time of maximum regression of the sea, result in exposure of the lime sediment surface, with the development of karren, dolines and possible cenotes. Paleosols have been shown to exist by the presence of sedimentary textures, rhizoliths, hard grounds and early diagenetic fabrics such as an indurated caliche-like topmost few cms of each limestone bed. Many paleokarst surfaces are associated with thin tuff horizons and these too provide evidence of paleosols sometimes with evidence of a plant cover. Soil acids seem to have percolated into the underlying limestone locally with potholes up to 30 cm in diameter and 1 m deep (Fig. 2–11). In a few localities there are pits up to 10 m wide and deep, full of collapsed lava and limestone boulders and these may represent ancient cenotes.

Elsewhere in Derbyshire ADAMS (1980) noted paleosol development and diagenetic textures in limestones near Monyash as evidence of minor emergence of late Brigantian reefs. Calcarenite cyclothems in the Crich inlier are similarly capped with paleosols (BRIDGES, 1982).

Whilst no major present day karstic landforms can be interpreted as being inherited from the above processes, these stratigraphic interruptions to a continuous limestone sequence have undoubtedly provided pathways for percolation water in later erosion cycles and cave development has ensued at some levels. Initiation of some major phreatic tubes in the Castleton area may well have utilised such ancient paleosols. For example, the Peak Cavern controlling bedding plane is on a clay some 5 cm thick in places with a trace of coal.

Intermittent tectonic uplift of the limestone massif during sedimentation also resulted in disconformities with more strongly developed *potholed surfaces*. WALKDEN (1977) described scoured erosion hollows in Millers Dale associated with local unconformities. Whilst they are paleokarstic surfaces, they have not been shown to be of any wide extent and have not apparently been related to later cave development.

Post-Dinantian Uplift and Erosion

At the end of Dinantian times, carbonate sedimentation in the Peak District was brought to an abrupt halt by uplift with considerable erosion of parts of the limestone massif before being covered by the advancing deltaic sediments of Namurian age. In both the Castleton and Upper Dovedale areas perhaps as

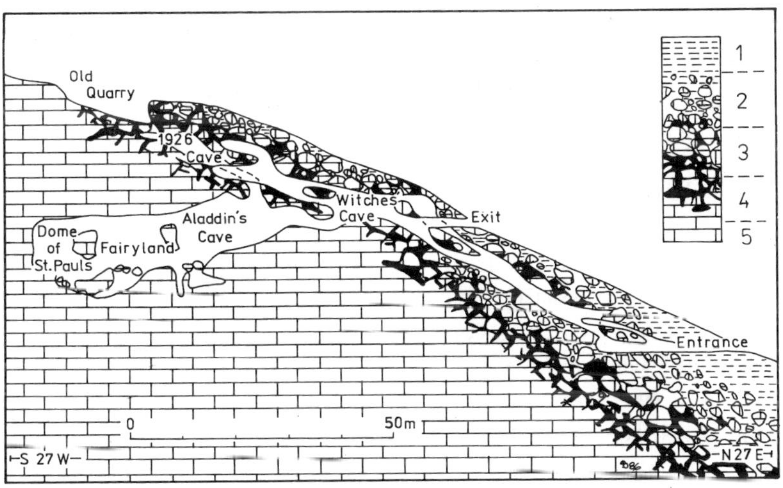

2–12. Diagrammatic section of Treak Cliff, Castleton, Derbyshire, showing the mid Carboniferous boulder bed with voids filled with Blue John fluorspar. Ancient phreatic caves extending into the eroded reef limestone were similarly lined with Blue John (modified after SIMPSON and BROADHURST, 1969). 1. Namurian shale; 2. limestone boulders with shale or calcite matrix; 3. limestone boulders mostly with Blue John fluorspar matrix; 4. fissures and phreatic caves in fore-reef limestone with Blue John infills; 5. Dinantian fore-reef limestone.

much as 100 m of the upper beds of limestone were eroded from the marginal reef complex though further south the contemporary lagoonal sediments were preserved in tectonic downwarps. The eroded surface of the reef limestone complex around Castleton has fissures full of limestone boulders in a matrix variously of finely comminuted limestone or of shale. They can be regarded as deep grikes filled in during the Namurian transgression. Lower down the slopes, massive boulder beds accumulated as the debris from erosion of reef crests slid down into the deeper waters of the basin (SIMPSON and BROADHURST, 1969) (Fig. 2–12). The reef limestones beneath the erosion surface were at the same time subject to karstic drainage with the development of small phreatic tube cave systems.

Two kilometres east of Castleton, in the Blue Circle Cement works quarry, quarrying has breached a number of cavities in the limestone surface. These penetrate downwards through the topmost beds by as much as 10 m. They are generally vertical sided and are filled with brecciated Namurian shale, some weathered to a brown colour and some still fresh. The mode of transport of the shale is unclear and the occurrences may signify either the penecontemporary fill of dolines in the limestone surface or interstratal solution causing solution collapse of the shale cover.

Both around Castleton and elsewhere in Derbyshire rare patches of an insoluble residue of chert gravel suggest either Late Dinantian or pre-Namurian solutional lowering of the limestone surface or of interstratal solution at a later stage. Both paleokarstic and interstratal karstic solution collapse structures which have been mineralized with fluorspar and galena have been described from the Castleton and Bradwell areas by BUTCHER and HEDGES (1987).

Post-Carboniferous Uplift and Erosion

Following the first main phase of upfolding of the South Pennine anticline in Late Carboniferous times, the region was subjected to erosion during Permian and Triassic times, in which a large part of the Upper Carboniferous delta-swamp sediments were removed. No actual landforms can be attributed to the Permian Period, but Triassic beds do rest directly on the limestone in a small area of the south-western part of the limestone massif. The unconformity is probably a paleokarstic surface but it has not been described in detail.

At Breedon-on-the-Hill on the Derbyshire/Leicestershire border two inliers of dolomitized Carboniferous Limestone project through the cover of Upper Triassic Mercia Mudstone. The inliers were effectively islands in a playa lake and are variously flanked by coarse boulder breccias, screes of dolomite clasts and by mudstone. A few fissures are filled with mudstone, sometimes with local enrichments of hydrothermal minerals, and a few have yielded sparse reptile bones.

Hydrothermal Paleokarst

The galena-sphalerite-fluorite-baryte-calcite deposits of Derbyshire are thought to have originated from hydrothermal solutions penetrating (a) into fault fractures and joints resulting from the Late Carboniferous earth movements, (b) into favourable lithologies of limestone to yield replacement ore deposits and (c) into pre-existing cavities. The last two categories are commonly confused in Derbyshire under the local terms of *flats* and *pipes*. Broadly, a flat is a bedding-controlled ore-body, either by replacement or by filling a bedding cavity, whilst a pipe is any other form of cavity lined or filled by the mineral suite, often with associated metasomatic replacement. A review of processes of mineralisation has been published recently by INESON and FORD (1982). What is more important from the paleokarst point of view is the origin of the cavities and the processes of their formation. Temperatures of mineralisation as shown by fluid inclusion studies were around 80 ° to 100 °C and these, together with structural evidence, suggest that the limestone massif was still buried by Upper Carboniferous strata at the time of mineralisation to a depth at least 2 km and possibly 3 km. The chemistry of the mineralising solutions indicates that they were extremely dilute, and the paleokarstic implication is of large bodies of hot

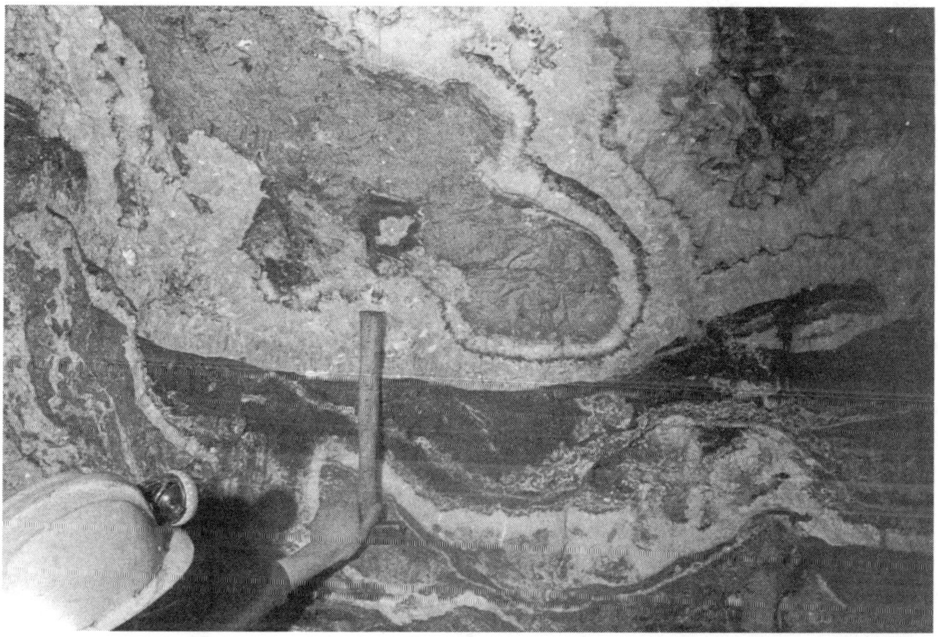

2–13. Section through hydrothermal paleokarst cavity showing: a. calcite lining at the bottom of the ancient cavity; b. dark catenary precipitate of sulphide minerals on floor of cavity; c. final infill of calcite with thin layers of sphalerite partly encrusting a limestone roof pendant in centre. Blende Vein, Magpie Mine, Sheldon, Derbyshire (Photo by T. D. Ford).

water moving through the limestone mass, probably escaping eventually at the surface as hot springs. The widespread deposition of calcite, often in freely grown crystals of large size, suggests that the hydrothermal waters were not aggressive during mineralisation, but that does not exclude an early, pre-mineralisation, aggressive phase being responsible for pipe-vein caverns.

Apart from deposition in faults and joints the hydrothermal minerals occur in (a) the voids between the boulders of the pre-Namurian boulder bed, now partly filled with the Blue John variety of fluorite deposits in Treak Cliff at Castleton (T. D. FORD, 1969; Fig. 2–12); (b) calcite-lined solution caverns at the boundary of porous dolomitised limestones resting on relatively impermeable unaltered limestone as in Masson Hill and the Golconda Mine (T.D. FORD and KING, 1965, 1966); (c) solution cavities developed along the contacts with tuff horizons and (d) cavities developed from primary porosity in buried reef limestones as in the Blende Vein cavern in Magpie Mine (WORLEY, 1976; Fig. 2–13). Circulation of groundwater in these has, at various stages, resulted in their re-utilisation as phreatic drainage pathways, with resultant collapse of the mineral linings into alluvial breccias in the bottoms of pipes, often mixed with inwashed sediments at a later date.

Hydrothermally mineralized caves of type (c) are common in Masson Hill overlooking the gorge of the River Derwent at Matlock. Subsequent re-utilisation of these old caves by phreatic water movement in Pleistocene (and possibly Pliocene) times was followed by the inwashing of fluvio-glacial fills (NOEL, SHAW and FORD, 1984).

The Sub-Pliocene Pockets and Their Deposits

From post-Triassic to mid-Cenozoic times, the Carboniferous Limestone massif of the Peak District was almost certainly buried under a cover of Jurassic and Cretaceous sediments of which no trace survives. By mid-Cenozoic times the Pennine fold was uplifted again and karstic erosion was rejuvenated. By the end of Miocene times an escarpment of the Lower Triassic Sherwood Group conglomerates was retreating southwards from the southern margin of the limestone outcrop. Fans of pebble gravel, sand and clays spread northwards at the foot of the scarp into a braided, low gradient river system draining the South Pennines leaving a sheet of clastic sediments, the Brassington Formation (BOULTER et al., 1971; WALSH et al., 1972), resting on the limestone. Relics of this sheet are now preserved only in a series of solution collapse structures and were formerly known as *Pocket Deposits* (YORKE, 1954–1961). The processes of dismemberment of the formerly continuous sheet have been described by T. D. FORD (1967, 1969, 1972, 1977) and WALSH et al. (1972, 1980). Quarrying operations for the refractory materials have revealed paleokarstic phenomena at the margins of the pits. Patches of angular chert gravel up to 5 m thick resting on the dolomitised limestone surfaces and dragged down by collapse indicate the former presence of an insoluble chert residue over parts of the limestone surface. Resting on the chert gravel in a few cases are patches of Namurian shale, again about 5 m thick, and showing a weathering profile from unweathered blue-black shale below to reddish purple above, characteristic of warm temperate conditions. Whilst these relics are small and scattered, they indicate the former presence of a combined covered and interstratal karst landscape in a warm climate — a paleokarst having both a veneer of a former cover and of the insoluble residue present in the region before the deposition of the Brassington Formation sands and clays. The fossil plants in the clays are of Mio-Pliocene boundary age (BOULTER, 1971), so that this modified covered paleokarst is probably of Miocene or even earlier age.

The margins of the pockets show joints etched out and filled with sand due to post-depositional collapse and the sagging of the Brassington Formation sands into interstratal karst cavities. The sub-surface solution necessary for the development of these in turn requires a hydrological drainage system resurging nearby at lower altitude (T. D. FORD, 1972). Relics of the possible intermediate cave drainage have been discovered in caves penetrated by the Golconda lead and baryte mine (T. D. FORD and KING, 1965, 1966).

Thus a Mio-Pliocene covered paleokarst is demonstrated by the presence of the Brassington Formation, by the residual chert gravels and shale patches, and by the caves of the Golconda Mine.

The same general area as the pits with the Brassington Formation sands and clays is also characterised by the presence of dolomite tors (T. D. FORD, 1963, 1969). These crags are thought to have been developed by periglacial circum-denudation of decalcified, partly dolomitised limestone during the later Pleistocene, but some may be survivors from a comparable glacial/periglacial/interglacial cycle in earlier Pleistocene times.

The Mendip Hills of Somerset and the Bristol Area

The Carboniferous Limestone of the Mendip Hills and Bristol area (Fig. 2–14) also shows transgressive-regressive cycles (RAMSBOTTOM, 1973). Evidence of emergence is stronger in the Avon Gorge and Chipping Sodbury areas to the west and north of Bristol but exposure to sub-aerial processes is shown by the presence of stromatolitic horizons, mud-cracked bedding planes and very occasional thin coals in both the Bristol and Mendip areas. No detailed study of these has yet been published.

The limestone of the Mendip Hills lies in a series of asymmetric east-west-trending periclines with considerable faulting. The folds have Devonian sand

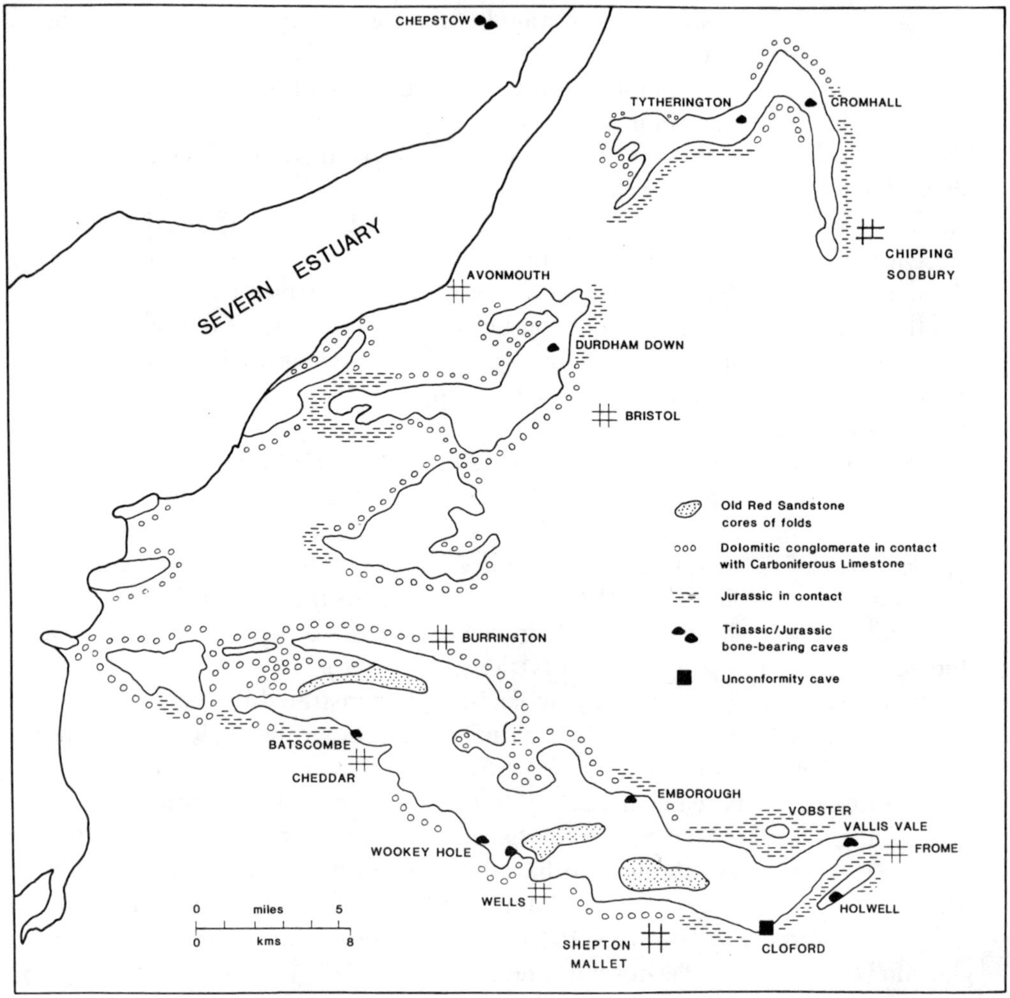

2–14. Sketch map showing the distribution of paleokarst features in the Mendip Hills and Bristol area.

stone and Silurian volcanic cores. Folding took place at the end of Carboniferous times and there was no Permian sedimentation. By Triassic times the Upper Carboniferous cover had been eroded off and Upper Triassic sediments were deposited unconformably on the deeply eroded folds. Though much of the Triassic cover has been eroded off, relics occur in three forms: conglomerate and breccia sheets; overlapping Triassic mudstones and fissures filled with bone-bearing sediments. In the extreme east of the Mendip Hills the Trias was overlapped by marine Jurassic limestones, which rest directly on a planed-off

Carboniferous surface. There is little evidence of marine karstic features, though oysters cemented themselves and mollusc borings penetrate a few centimetres downwards. Similar denuded folds of Carboniferous Limestone flanked by Triassic and Jurassic cover sediments occur around Bristol and Chipping Sodbury to the north of the Mendip Hills (Fig. 2–14). An outline of the Triassic paleokarst was given by ROBINSON (1957) as a background to her studies of the vertebrate faunas in the fissures. These were also described by HALSTEAD and NICOLL (1971). ALABASTER (1982) added comments relevant to his studies of Mendip mineralisation.

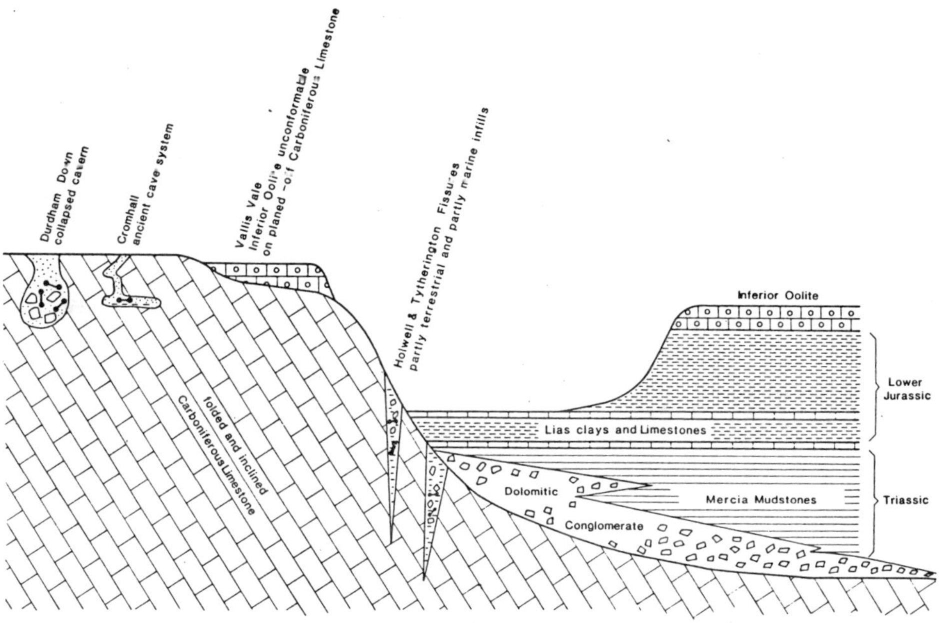

2–15. Diagrammatic sketch to show the relationships of ancient caves to overlapping covers of Triassic and Jurassic sediments (partly after HALSTEAD and NICOLLS, 1971; and ALABASTER, 1982).

The Dolomitic Conglomerate is essentially a series of Late Triassic lithified scree deposits variously flanking the Carboniferous Limestone hills or lying in ancient wadis. Sections of the latter are to be seen in the Avon Gorge at Bristol, in Burrington Combe on the north flank of the Mendips, and around Wookey Hole on the south flank. The clasts are largely Carboniferous Limestone set in a patchily dolomitised sandy matrix. Away from the limestone hills the breccia fans pass into alluvial material, gravelly at first and progressively finer-grained further from the source. The implication of such deposits is that hills of folded limestones constituted a paleokarst surface of arid to semi-arid character with little soil cover. Marine transgression seems to have occurred episodically in Rhaetian, Liassic and Middle Jurassic times. Limestones of Liassic age showing a littoral facies aspect are banked against the Carboniferous Limestone near Shepton Mallet in east Mendip. These overlap the earlier Dolomitic Conglome-

2–16. Karren (clints and grikes) developed in steeply dipping Carboniferous Limestone beneath a thin cover of Middle Jurassic Inferior Oolite (grass-covered in background). Vobster Quarry, Somerset (Photo by T. D. Ford).

rate but in turn are overlapped by Middle Jurassic Inferior Oolite which lies on a marine planation surface across folded Carboniferous Limestone near Frome, well seen in Vallis Vale (Fig. 2–15) and at Vobster (Fig 2–16). The Triassic climatic regime is likely to have been one of infrequent but heavy rainfall, not one which would normally lead to the development of cave systems. Progressive filling of adjacent lowland basins would, however, cause a rise in the water table and slow phreatic solution could then ensue.

Ancient caves or fissures with bone-bearing sediments have been intersected in a number of quarries (Fig. 2–15). HALSTEAD and NICOLL (1971) distinguished three types: (a) ancient cave systems, e.g. Cromhall; (b) collapsed cavern, e.g. Durdham Down; and (c) neptunian dykes, e.g. Holwell. Other collapsed caverns have been noted at Emborough and Batscombe. On the basis of the fossil contents ROBINSON (1957, 1971) distinguished two types of fissure: (a) those containing a Late Triassic terrestrial reptilian fauna only and (b) those containing Rhaetian to Lower Liassic vertebrates, both reptilian and mammalian, with some evidence of marine input in a few cases. The fissures themselves are of two types — tectonically opened fractures and solutionally widened joints. The faunas have been taken to suggest that they lived on upland areas well above sea-level but the discovery of glauconite grains and marine palynomorphs in the silts at Tytherington (MARSHALL and WHITESIDE, 1980; WHITESIDE and ROBINSON, 1983) suggest that the limestones may have formed low-lying islands and some fissures may have been effectively dolines, cenotes or even "Blue Holes" like those of the Bahamas.

The present upland surface of the Mendip Hills is thus a rejuvenated Triassic island surface, which has been subjected to slow phreatic solution, with marine planation in the Middle Jurassic, and some Pleistocene trimming.

There is little evidence of lateral caves being developed off the fissures in Triassic times, though a few quarry sections suggest the formation of caves followed by roof collapse, as at Slickstones Quarry at Cromhall, north of Bristol. Fossil reptilian remains occur both in the silty sediments of the cave floors and in between the collapsed boulders. Reptile remains are locally dispersed in re-deposited derived Carboniferous Limestone microbreccias (FRASER and WALKDEN, 1983).

The well-known Wookey Hole cave is partly developed in the Dolomitic Conglomerate but extends without any obvious change in morphology into the Carboniferous Limestone behind. The high permeability of the Dolomitic Conglomerate may have been a factor in speleogenesis but here it is not in itself evidence of a subterranean Triassic paleokarst.

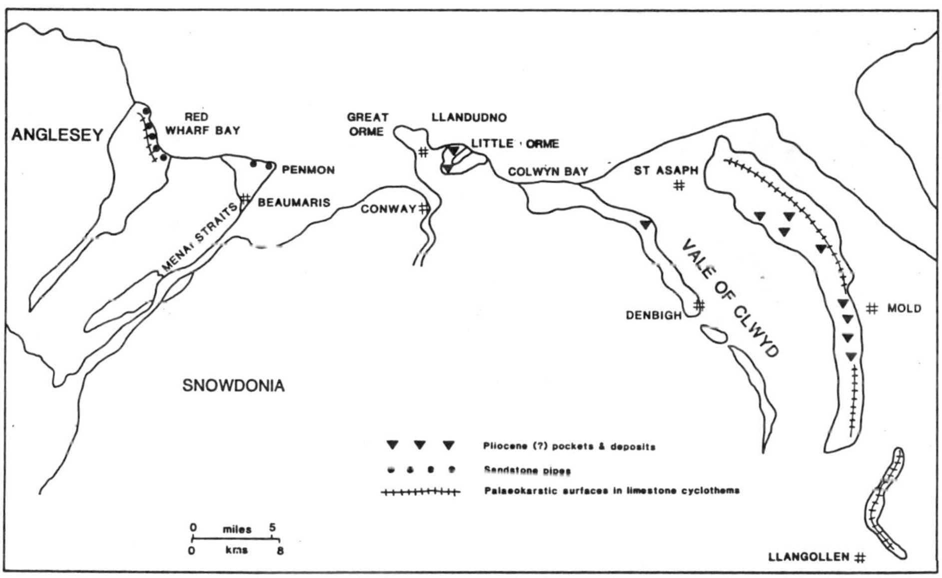

2–17. Sketch map showing the distribution of paleokarst features in North Wales and Anglesey.

A minor example of interstratal karst is the cave in Cloford Quarry developed along the unconformity between the Middle Jurassic Inferior Oolite and the underlying Carboniferous Limestone (DREW and INGLE SMITH, 1972). The age of initiation is unknown but it may well be no older than Quaternary.

Anglesey

The coastal sections of the Carboniferous Limestone of Red Wharf Bay on the east coast of Anglesey have long been famed for their sandstone-filled potholes in the limestone (Figs. 2–17 and 2–18). Noted by GREENLY (1901) they have been described by BAUGHAN and WALSH (1980) and by WALKDEN and

2–18. Sandstone-filled paleokarst pipe in the Carboniferous Limestone of Trwyn Dwlban point, Anglesey (Photo by T. D. Ford).

DAVIES (1983). The coast near Trwyn Dwlban provides sections of several hundred sandstone pipes, which WALKDEN and DAVIES have shown to occur at nine different stratigraphic horizons, each terminating a regressive carbonate cyclothem. The pipes are commonly 5 or 6 m deep and 1 or 2 m wide and contain sandstone showing bedding to various degrees. Most of the sandstone bodies are truncated on the foreshore by modern marine erosion but in the cliffs various relationships can be seen. Many are truncated and the next limestone bed lies directly on top, but a few show a thin sandstone bed passing from one pipe to another a few metres away. Some evidence of channelling of cyclothems into each other is also present.

WALKDEN and DAVIES (1983) have concluded that the pipes are paleokarstic phenomena due to overbank solution when regression took place at the end of a cycle. Some such overbank solution was beneath a sheet of fluvio-deltaic sand, presumably with a soil providing acidic meteoric waters. Removal of large parts of the sand sheet took place during meander channel migration and was completed in some cycles by the succeeding marine transgression.

The pipes and related bedding surfaces may be classified as covered paleokarst.

North Wales

As in other areas, the Carboniferous Limestone sequence is marked by a number of paleokarstic or paleosol surfaces at the termination of cyclothems (SOMERVILLE, 1979a, b) but these do not seem to have any related speleogenetic significance (Fig. 2–17).

Mining records in northeast Wales suggest the presence of hydrothermal karst there but no details have been described.

The outcrops of Carboniferous Limestone around Llandudno and in Flintshire are marked by some 20 solution subsidence outliers containing sands and clays with occasional lignite. Described by WALSH and BROWN (1971), these bear close comparision with the pockets and Brassington Formation of Derbyshire and seem to have formed in an analogous way during the Pliocene.

South Wales

Carboniferous strata are folded into a large synclinal basin trending east-west so that the limestone outcrops in a narrow band on North and South Crops on either side of the coalfield. The South Crop is steeply-dipping and has been greatly denuded and locally transgressed by Triassic strata. To the south of the coalfield *islands* of folded Carboniferous Limestone project through the unconformable cover of Triassic and Jurassic sediments (Fig. 2–19).

There is a general thinning of the Carboniferous Limestone towards the north-east in the eastern part of the North Crop. This is partly due to overstep by the lowest beds of the Millstone Grit across the various units of the limestone sequence, and partly due to progressive thinning of the units themselves. These thin limestone cyclothems are often bounded by surfaces characterised by paleosols and paleokarsts. V. P. WRIGHT et al. (1981) and V. P. WRIGHT (1982, 1984a, b) have argued that these are largely evidence of intra-Dinantian covered karst. Paleosols are represented by textures including rhizoliths, glaebules and vadose cements overlying phreatic cements. Rubbly calcareous clays on such surfaces appear to represent collapsed kavornossen karren (V. P. WRIGHT, 1982).

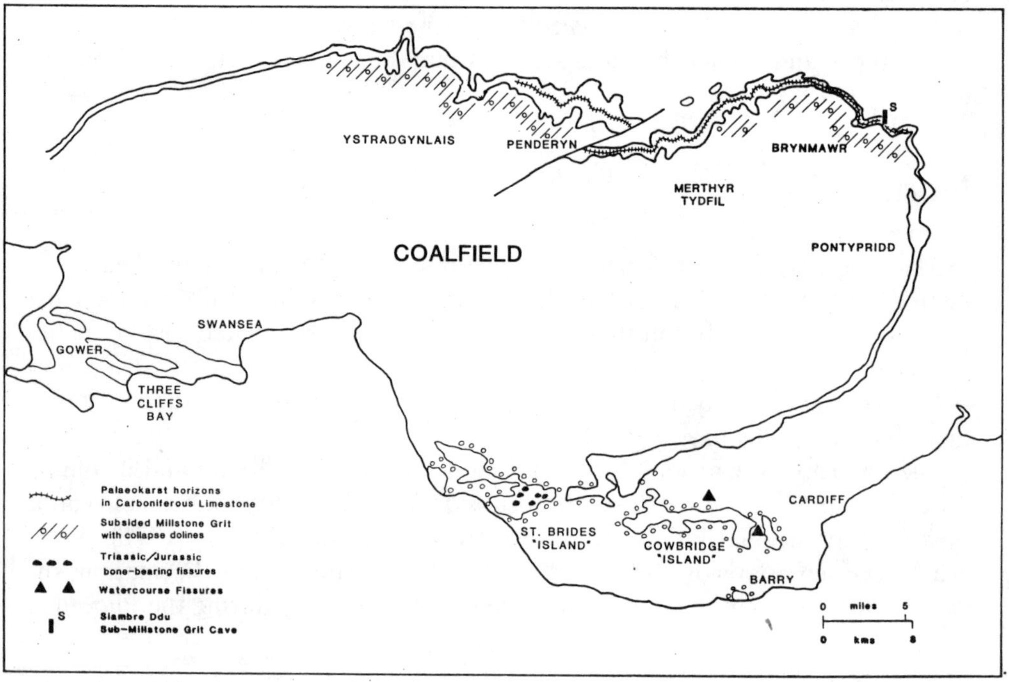

2–19. Sketch map showing the distribution of paleokarst features in South Wales.

Needle-fibre cements suggest a relationship to microbiological soil activity possibly by fungi (V. P. WRIGHT, 1984b). A paleokarstic surface at the top of the Gilwern Oolite shows extensive solution piping. Mammilated and pot-holed surfaces have also been noted in the Gower Peninsula.

The folded Carboniferous Limestone of South Glamorgan was deeply eroded in Permo-Triassic times before being buried by Late Triassic and Early Jurassic sediments. The former are marginal to the Triassic playa lake Mercia Mudstones and include lithified fossil scree slopes, talus fans and coarse alluvial deposits grading outwards into sands, silts and muds (TUCKER, 1977). Like the Mendip Hills of Somerset the *islands* were relict upland areas in a semi-arid to arid climate. Apart from its significance as a source of supply for the screes and alluvium, the paleogeomorphology has not been analysed but notes on paleokarstic surfaces have been published by THOMAS (1952) and fissures containing bone-bearing sediments have been noted (e.g. ROBINSON, 1957, 1971). As in the Mendip Hills these contain a mixture of small reptile bones, with a few early mammals and fish remains in a fine clastic matrix and pitfall trapping mechanisms have been invoked. Fragments of coniferous wood and spores were also found (LEWARNE and PALLOT, 1957; HARRIS, 1957). These suggest a Rhaetic to Lowest Lias age and HARRIS suggested that their charcoal-like appearance indicated their possibly having been burnt in a scrub fire and then washed into the fissure.

To the west of Glamorgan, comparable folds in the Carboniferous Limestone of the Gower Peninsula and South Pembrokeshire are characterised by small areas of coarse *gash-breccia* regarded by some as relics of the Triassic transgression but there is no direct evidence of either age or origin of the breccias. THOMAS (1971) regarded them as being of tectonic origin related to Late Carboniferous folding, occasionally with solutional collapse feature superimposed.

On the North Crop the Carboniferous Limestone, generally some 200–300 m thick, is overlain unconformably by the basal Millstone Grit quartzitic sandstones. The junction is an unconformity with the base of the sandstones transgressing almost completely across the limestone outcrop in the extreme east. The contact is poorly exposed but where it is visible, minor solution features occur. OWEN and JONES (1966) drew attention to small scale cave-like features beneath the unconformity filled with quartzitic sand and pebbles and THOMAS (1973, p. 74) suggested that they might represent a karst landscape now buried below the Millstone Grit. However, the evidence is largely masked by subsequent collapse having led to extensive interstratal karst. Described by THOMAS (1963, 1973, 1974) the features include an extensive collapse doline field, shallow subsidence depressions and foundered masses of the quartzitic sandstone. These features have resulted from the solution of limestone under a cover of sandstones up to 200 m thick and THOMAS has proposed that the only possible mechanism for the chaotic jumble of quartzite blocks is collapsed caverns of vast size. None of these has yet been demonstrated to exist, though the Siambre Ddu cave, near Brynmawr, with its sandstone roof, may be an imminent collapse of this type (THOMAS, 1974, p. 150; Fig. 2–19).

The collapse dolines number some thousands spread over about 60 km length of limestone outcrop. Individual dolines may be as small as 25m diameter but many are more than 100 m across. Foundered masses of Millstone Grit may be as much as 3 km in length. The depth of collapse is uncertain but THOMAS thought that some foundered masses may have gone through the whole thickness of the Carboniferous Limestone, i.e. more than 200 m. The chronology of solution and collapse has not been worked out but THOMAS suggested that many of the features were so freshly developed on a surface glaciated in the Devensian that they were in effect post-glacial and hence not paleokarst. However, he also suggested that some isolated foundered masses of sandstone on the limestone dip-slopes well-removed from parent Millstone Grit outcrops could be regarded as *fossil interstratal karst*. Once isolated, they could be planed off by glacial erosion and may date back to early in the Pleistocene or perhaps even earlier.

Pockets in the limestone surface filled with deeply weathered material derived from the Millstone Grit have been described in the Vaynor and Cwar yr Ystrad quarries of the North Crop by BATTIAU-QUENEY (1980, 1984a, b). She has argued that the limestones between the pockets resemble buried tower karst and that the fill in the pockets between the towers was derived from tropical soils of ferralitic and ferruginous types such as would be generated in a relatively hot and wet climate. Whilst no dating is available BATTIAU-QUENEY has suggested an Early Tertiary age for the tower karst and a later (Early Neogene) age for the burial and fill stage.

Northern England

Paleokarst in the classic caving area of the North Pennines is less obvious, being restricted to paleosols at the end of transgression-regression cyclothems and to solution-pitting beneath shale bands (WALTHAM, 1970, 1971, 1974). These pit features have only been seen underground where they have been intersected by caves. WALTHAM has observed cross-sections of depressions in the base of shale bands as much as one metre deep and 3 m wide. Since they can only be seen in two dimensions it cannot be proved whether they are cross-sections of circular pits like those in Anglesey or whether they are sections through channel-fills. A few shale-filled fissures suggest the former presence of grikes in exposed erosion surfaces. A few of the shale bands carry thin coal seams indicating the former presence of terrestrial soil and vegetation cover on exposed limestone platforms.

Northwest Scotland

The variously dolomitised Cambro-Ordovician Durness Limestones of the extreme northwest of Scotland have recently been shown to contain a stratigraphic break encompassing most of Middle and Upper Cambrian time (T. J.

PALMER, MCKERROW and COWIE, 1980; S. C. WRIGHT, 1983). The Sangomore Formation rests on a paleokarst surface of the Sailmhor and Eilean Dubh Formations on the south shore of Balnakeil Bay. Breccia-filled fissures pass down from the paleokarst surface into the lower formations by as much as 150 m. The breccias have apparently yielded a few uppermost Cambrian marine fossils whose significance has not yet been understood. The breccias have been attributed to the former presence of evaporites and collapse due to solution thereof. No detailed description is yet available.

South Devon

The folded and faulted Devonian limestones of Berry Head and the adjacent Shoalstone Beach near Brixham contain numerous sandstone dykes (RICHTER, 1966). Although these give the impression of being true paleokarstic features, detailed studies have shown that they are tectonic fissures opened, sometimes several times, by the stresses of folding. Some fissures close upwards, whilst others cut through earlier fissure-fills. The fills are of almost structureless red sandstone, often with radial calcite growing inwards from the walls. The calcite probably indicates a stage of phreatic growth in open fissures and the sand was virtually sucked in from the unconformable cover of Permian fluvial sediments. A single instance of a true cave filled with red sandstones has been noted by RICHTER but otherwise no paleokarstic features have been recognized.

The Chalk "Potholes"

Throughout southern and eastern England the Upper Cretaceous Chalk is covered unconformably by fluvial or marine Paleogene sands. Considerable thicknesses of chalk were eroded before deposition of the cover on a gently inclined surface. In almost every exposure of the contact there are sand-filled "pipes" extending down into the chalk to a depth of more than 10 m (KIRKALDY, 1950). They can be classified as inter-stratal karst and are generally thought to have originated by sub-surface solution of the soluble chalk below the unconformity. Particularly fine examples have been described at South Mimms to the north of London (THOREZ et al., 1971). Others occur in the cliffs of Studland Bay in east Dorset and at Cuckmere Haven in Sussex (CASTLEDEN, 1982; Fig. 2–20). At the last locality, pipes about 1.5 to 2 m, in diameter can be seen to penetrate the whole depth of the chalk cliff, some 20 m, into the foreshore platform. The cover is a series of sand and gravels, which can be seen to sag into the pipes. The chalk margins of the pipes, particularly those seen on the foreshore, are indurated by deposition of a cryptocrystalline cement through a thickness of about 15 cm. Cylinders of indurated chalk from which the fill has been removed stand as much as a metre high on the wave-cut platform.

2–20. Interstratal karst solution pipe in Chalk filled with Late Cenozoic sands and gravels, Cuckmere Haven, Sussex (Photo by T. D. Ford).

Jurassic Hardgrounds

There are numerous horizons in the sequence of Jurassic limestones and clays throughout Britain where hardgrounds occur. These are surfaces within limestone/clay successions characterised by encrustations of oysters and other bivalves and bearing vertical burrows by crustaceans, "worms" and other organisms (T. J. PALMER and FÜRSICH, 1974; SELLWOOD, 1978). The underlying limestones show evidence of early lithification but otherwise they indicate only a period of non-sedimentation with little evidence of emergence and the formation of sub-aerial paleokarsts. .

PALEOKARST OF NORWAY

Stein-Erik Lauritzen

Introduction

No systematic morphogenetic study of the paleokarsts in Norway has yet been done. However, unconformities of a karstic origin have been recorded *en passant* in stratigraphic studies. Most of the limestone caves known in the country are relict, evidently of pre-Holocene age, and a few of them can be distinguished as probably belonging to pre-Pleistocene genetic regimes.

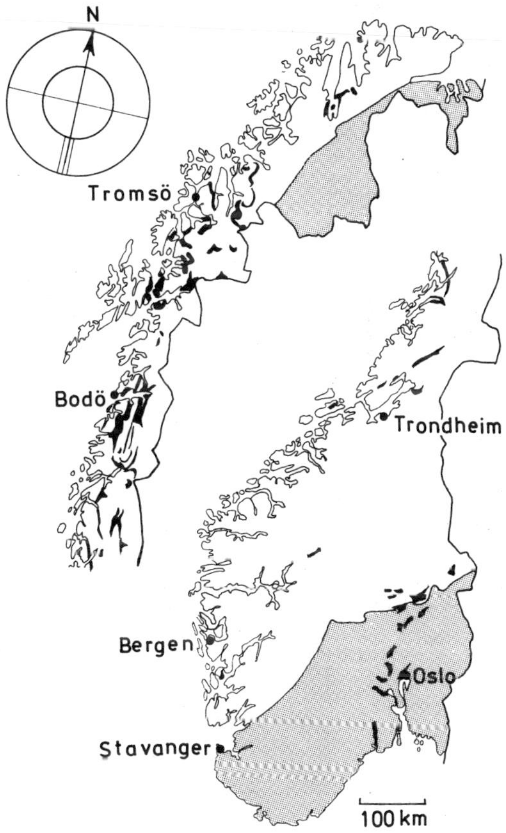

2–21. Distribution of carbonate outcrops in Norway. Black: carbonate rocks. Unshaded areas: the Caledonian orogenic belt.

P. Bosák, D. C. Ford, J. Głazek and I. Horáček (Editors):
Paleokarst. A Systematic and Regional Review. – Elsevier and Academia. Amsterdam and Praha. 1989.

Geological and Regional Setting of the Carbonate Rocks

The geographic and stratigraphic distribution of limestones and phases of karstification are shown in Figs. 2–21; 2–22. Carbonate deposits exist throughout most of the sedimentary column recorded in the country. Limestones are

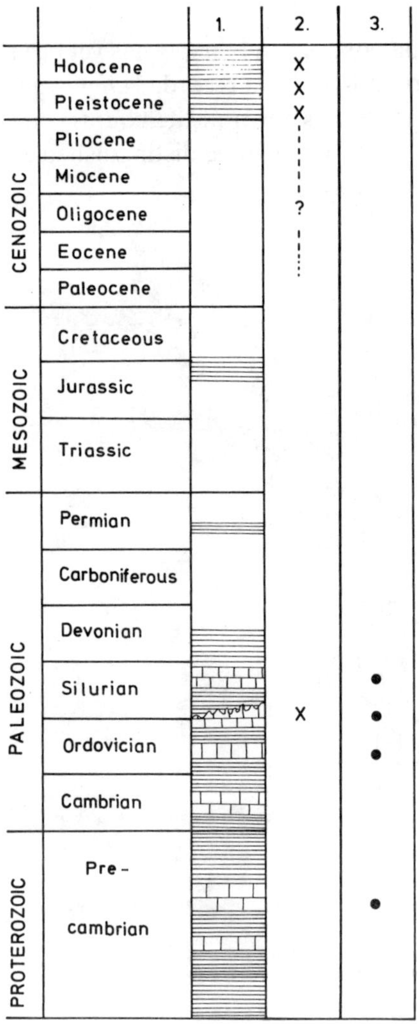

2–22. Stratigraphic distribution of sedimentary rocks and karstification in Norway. 1. occurrence of sedimentary deposits on the continental land surface. The relative thickness of carbonate/non-carbonate is not to scale; 2. phases of karstification through time; 3. distribution of lithologies that show affinity to Pleistocene and Holocene karstification.

represented in the Precambrian, and in the Cambro-Silurian system. Carbonates within the Caledonian orogenic belt show several phases of plastic deformation and profuse recrystallization into marbles. However, in a few locations, these rocks are less altered, and provide fossils which indicate a Cambro-Silurian age (STRAND, 1960; OFTEDAHL, 1980). Because of the strong deformation, possible original karstic unconformities are lost within most of the Caledonides.

Pre-Tertiary Buried and Subjacent Karst

Karstic unconformities are documented from the Ordovician/Silurian boundary, and from the Lower Silurian. SKJESETH (1963) and HENNINGSMOEN (1960) refer to a karstic contact between Middle Ordovician limestone and Lower Silurian quartzite. This is called the Mjøsa Limestone/Helgøya Quartzite unconformity. The section shows a paleokarstic relief with grikes and small caves within a relief of 3 m (Fig. 2–23).

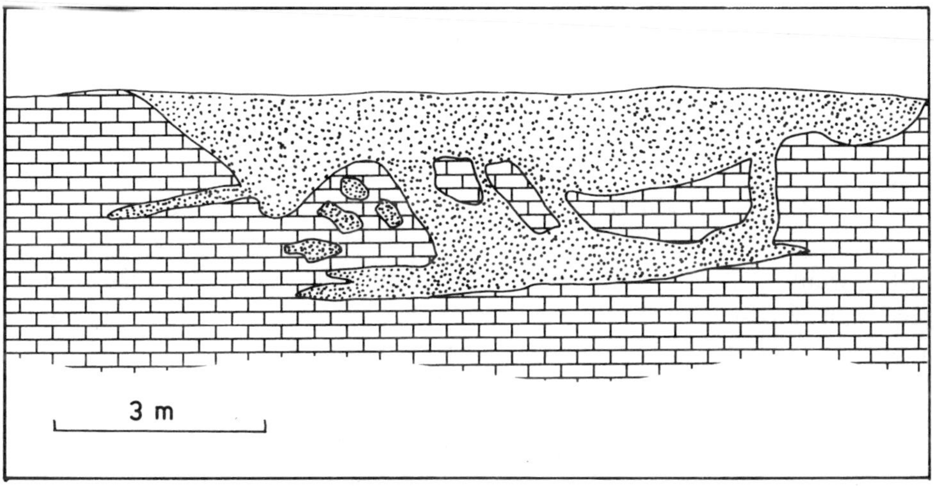

2–23. Buried karst at the Ordovician/Silurian boundary. The Mjøsa Limestone is deeply karstified with infilling Silurian quartzite (after HENNINGSMOEN, 1960).

OPALINSKI (1977) described this irregular contact in more detail. It is characterised as a series of pipes, cracks, bowl-like depressions, "floating" limestone blocks and deeper cave conduits, all filled with the Lower Silurian sandstone. The vertical pipes resemble grikes or, possibly, a jagged littoral karst. They exhibit serrated walls and narrow towards the base. The grike forms are tens of centimetres wide, whilst the kamenitza-type forms attain a width of 1–1.5 m and a depth of 70–80 cm. Sandstone-filled caves were detected as deep as 6 m below the top of the limestone. The forms and their stratigraphy were interpreted as a phase of karstification prior to stage *6c* (Middle Llandovery; OPALINSKI and HARLAND, 1981).

HANKEN (1974) also refers to karren-like channels and scallop-like forms in the Upper Ordovician of Ringerike, South Norway. The channels were 20–30 cm wide and several metres long; they were filled with shale and siltstone.

From the Lower Silurian dolostones of Southern Troms, North Norway, BJØRLYKKE and OLAUSSEN (1981) report a cave-filling breccia with a bimodal size distribution and a tendency to coarsen upward. This breccia was probably formed during the Silurian by leaching of anhydrite and gypsum resulting in collapse and brecciation of the overlying dolostone. It is therefore possible that this paleokarst is of the subjacent type. Lead-zinc mineralization is associated with the dolostone breccia. Lead isotope composition in one sample of galena suggests an age of 430 Ma, which is near the supposed age of the host-rock (MOORBATH and VOKES, 1963).

It should be borne in mind that true terrestrial facies for the known paleokarsts have not been proven for any of these deposits. Rather, BJØRLYKKE and OLAUSSEN (1981) suggest an intertidal environment with periodic supratidal conditions. In my opinion, neither the features described by SKJESETH (1963) nor OPALINSKI (1977) can yet be distinguished from karstic forms characteristic of a littoral environment.

Tertiary Landforms

The youngest continental pre-Quaternary deposits known in Norway are the few small Jurassic-Cretaceous outcrops of Andøya, North Norway (OFTEDAHL 1980). On the other hand, the existence of thick Mesozoic-Cenozoic deposits on the continental shelf demonstrate that the continental land surface has been subjected to denudation since Early Tertiary times. The remnants of this denudation is named *Norway's paleic* or *pre-glacial surface* (GJESSING, 1967).

On the larger scale, the paleic surface is characterized by extensive peneplanation, wide valleys or basins, and a smooth, undulating hilly topography. Glacial valleys and fjords are superimposed onto these landforms and truncate them. The paleic landforms were ascribed to a warm, probably semi-arid climate (GJESSING, 1967; and references therein).

On the smaller scale, the surface bedrock in Norway is glacially polished with few sites of preserved pre-glacial sediments. However, relic weathering profiles are sometimes encountered, exhibiting a mineralogic assemblage (kaolinite, gibbsite), which may be ascribed to a pre-glacial, warm, humid climatic regime (REUSCH, 1903; GJEMS, 1963).

ROALDSET et al. (1982) describe gibbsite, smectite and kaolinite from a profile of *in situ* weathered granitic and gabbroic rocks in West Norway. The mineralogy supports the assumption of lateritic/bauxitic weathering in a warm and humid climate. Such conditions have not prevailed since Late Tertiary times.

Because of the lack of dateable deposits, no decisive proof of pre-glacial karst has yet been presented. However, 60 million years or less of deep weathering should almost certainly have produced a deep karstic relief in concert with the other paleic landforms. It is unlikely that these macroforms have not been able

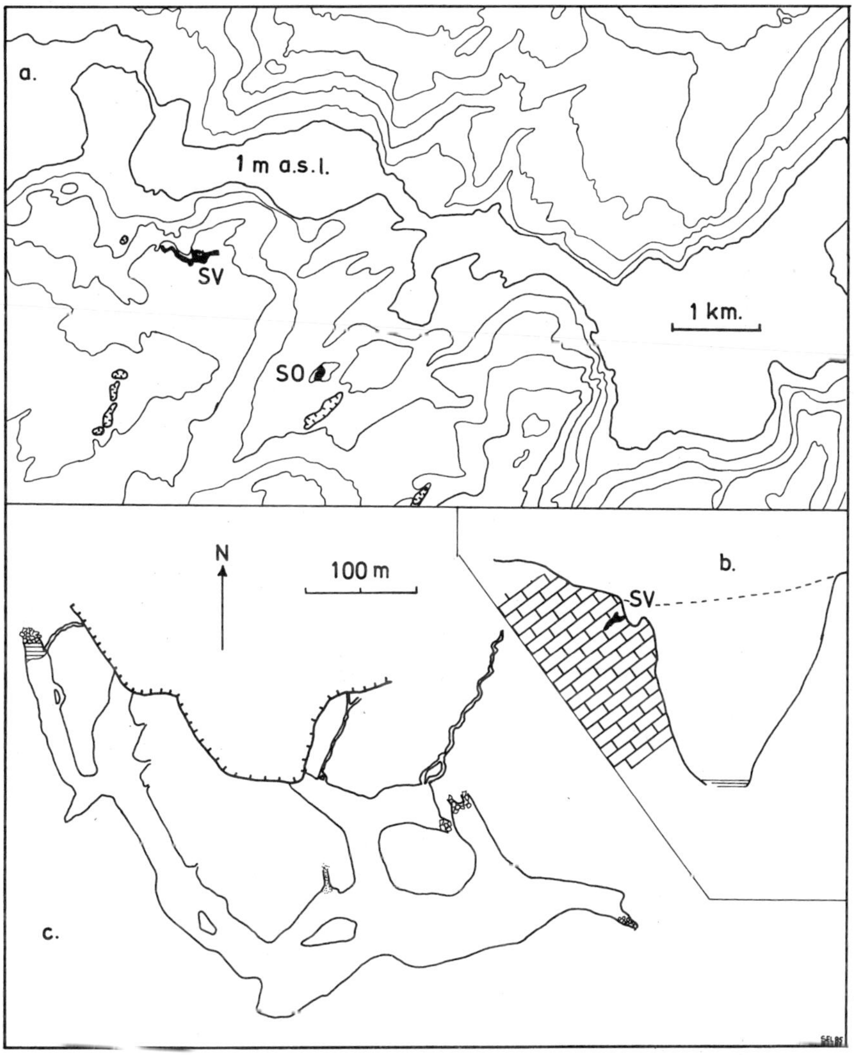

2–24. Probable paleokarst cave fragments along the Fauske Fjord-lake, northern Norway. a. topographic position of the caves. SV: Svarthamarhola; SO: Solvikhulen. Contour intervals 100 m; b. schematic cross-section of the fjord and Paleic surface, showing the position of the conduits. Dotted line: inferred floor of the Paleic valley. Vertical scale 5 times horizontal; c. survey of Svarthamarhola. The large conduits are extensively modified by collapse, but the original cross-sections exceed 2 000 m^2 in area. Survey plan modified from HEAP (1970).

to survive the Pleistocene glaciations, if only in a fragmentary form. By analogy with cave morphology in the semi-arid or warm humid areas of today, it is likely that such paleic karst existed (LAURITZEN, 1983).

Most relict cave conduits in Norway are comparatively small, with cross-sectional areas rarely exceeding 10 m^2. Uranium Series dating of speleothem deposits demostrate that such cave fragments are severral glacial cycles old (LAURITZEN and GASCOYNE, 1980; LAURITZEN, 1984). Based on kinetic and hydraulic grounds, it is likely that such small caves may well have developed within the glacial/interglacial sequences of the Pleistocene.

On the other hand, a few exceptionally large cave fragments are situated in hanging positions only a little below the paleic surrface (Fig. 2–24). Their cross-sectional area exceeds 2 000 m^2; the gigantic size and the topographic position of these conduits distinguish them from all other caves in the country and suggest a great age. The association of the paleic plain surface and the large conduits is strikingly similar to, for instance, the situation found in the Nullarbor Plain and its caves in Southern Australia (JENNINGS, 1967). A subsequent glacial incision into this plain would leave cave fragments in a very similar position to the Norwegian ones.

Concluding Remarks

The buried types of paleokarst known in Norway belong to the Ordovician/Silurian boundary. It is not definitely proven whether these unconformities were truly terrestrial or only littoral/supratidal types. In North Norway paleokarst breccias are associated with Pb-Zn mineralization. Other mineralization within the Caledonian orogenic belt may also be associated with paleokarstic facies, although the alternative model that some sulphide deposits are of primary sedimentary origin has been demonstrated (JUVE, 1967).

Tertiary paleokarst is not yet proven by direct stratigraphic or chronological evidence, but some macroforms are very suggestive of a pre-glacial age. Systematic karst research is only at its beginning in Norway, and most of the carbonate outcrops have not yet been studied in detail from this point of view.

PALEOKARST OF POLAND

Jerzy Głazek

Introduction

Numerous *fossil karst* phenomena have been reported in Poland. However, this information has generally been obtained during other investigations in geology, geomorphology, paleontology or archeology. Only a few paleokarst localities have been thoroughly studied in their own right.

Data on the fossil karst of Poland have been reviewed many times (GILEWSKA, 1964, 1971; GRADZIŃSKI and WÓJCIK, 1966; GŁAZEK, DĄBROWSKI and GRADZIŃSKI, 1972; GŁAZEK, 1973, 1975). Some of the reviews discuss the difficulties in interpreting *fossil karst* that are essentially caused by the possibility of karstification taking place beneath less soluble or insoluble strata (GRADZIŃSKI and WÓJCIK, 1966; GŁAZEK, 1973). Nevertheless, paleokarst studies have considerably supplemented our knowledge on the paleogeographic development of different areas, especially of platform regions where the current paleogeographic syntheses are constructed by interpreting marine or large continental formations that occupy less than 50 % of the geological time span. Within the time stratigraphic gaps, deposits are often preserved as a fill of karst forms. These have recorded very important data for paleogeographic speculation, but they need special care during investigations (QUINLAN, 1972; GŁAZEK, 1973). In earlier reviews (GILEWSKA, 1964; GRADZIŃSKI and WÓJCIK, 1966) paleokarst phenomena were treated differently in each present-day geomorphic unit of Poland. The present author introduced paleogeographic treatment of the entire country, in which paleokarst forms are a very important source of information that enables reconstructions of past scenery (GŁAZEK, DĄBROWSKI and GRADZIŃSKI, 1972; GŁAZEK, 1973; GŁAZEK and SZYNKIEWICZ, 1987).

Among the papers concerning paleokarst features in Poland, the first outstanding example is the description of a vertical shaft in Węże that was filled with Pliocene sediments containing a rich collection of vertebrate fossils (SAMSONOWICZ, 1934). Before that, in the 19th Century, vertebrate remains and artifacts excavated in caves were thoroughly studied (ROMER, 1883; OSSOWSKI,

P. Bosák, D. C. Ford, J. Głazek and I. Horáček (Editors):
Paleokarst. A Systematic and Regional Review. – Elsevier and Academia. Amsterdam and Praha. 1989.

1882). Since 1950 there have been descriptions of (i) deposits containing reptile bones in a Triassic cave at Stare Gliny (LIS and WÓJCIK, 1960; TARLO, 1959) and in Czatkowice (PASZKOWSKI and WIECZOREK, 1982), (ii) the polyphase development of underground karst forms in the Cracow Upland (GRADZIŃSKI, 1962), (iii) orefilled hydrothermal karst forms in Triassic carbonates of the Cracow-Silesian ore district (BOGACZ, DŻUŁYŃSKI and HARAŃCZYK, 1970; DŻUŁYŃSKI and SASS-GUSTKIEWICZ, 1982, p. 377–397), (iv) vertebrate and beetle-bearing Miocene karst deposits at Przeworno (GŁAZEK, OBERC and SULIMSKI, 1971; GALEWSKI and GŁAZEK, 1973; GŁAZEK, GALEWSKI and WYSOCZAŃSKI-MINKOWICZ, 1977; GŁAZEK et al., 1975). Very useful data for the reconstruction of the Late Pleistocene landscapes have been obtained from interdisciplinary studies of archeological excavations in caves (MADEYSKA, 1981, 1982).

Biostratigraphic dating of karst deposits was introduced by SAMSONOWICZ (1934) and later broadly developed (KOWALSKI, 1962, 1964; SULIMSKI, 1964; GŁAZEK, OBERC and SULIMSKI, 1971; BOSÁK et al., 1982; GŁAZEK and SZYNKIEWICZ, 1987). It was supplemented by semiquantitative bone-dating using complex fluorine-chlorine-apatite and collagen methods (WYSOCZAŃSKI-MINKOWICZ, 1969). The first isotope dating of charcoal from Late Pleistocene cave deposits was done by the ^{14}C method for archeological purposes in 1960 (Gro-2118, *fide* MADEYSKA-NIKLEWSKA, 1969). Later, it was extended for speleothem dating by the $^{230}Th/^{234}U$ method (GŁAZEK and HARMON, 1981, in press). Currently speleothem and bone datings using ^{14}C, thermoluminescence (TL) and electron spin resonance (ESR) methods are being introduced (HERCMAN et al., 1987).

Interest in paleokarst has been much increased in Poland because of mining, foundation engineering and dam construction problems (BAŻYŃSKI, 1960; KRASOŃ and WÓJCIK, 1965; WILK, MOTYKA and NIEWDANA, 1973; GŁAZEK and SZYNKIEWICZ, 1979; MOTYKA and WILK, 1984; WILK, p. 513–531).

Remarks on Terminology

In Polish scientific literature only one term — *kras kopalny* is used. It is translated as *fossil karst* but may be treated as a synonym for *paleokarst* in the English language literature. Currently, all forms evidently older than the last distinguishable glacial deposits in an area are treated as paleokarst in Poland. This criterion is diachronous because in southern and central Poland the youngest glacial deposits are of Middle Pleistocene age, while in the northern part of the country and in the highest mountains on the southern border, the latest glaciation is Late Upper Pleistocene (Vistulian, Würm).

The German term *subrosion* has been adopted in Polish literature for the underground solution of evaporites (POBORSKI, 1957); it is a type of subjacent karst. The term *hydrothermal karst* was introduced for "underground karst features produced by the action of hydrothermal solutions" by BOGACZ, DŻUŁYŃSKI and HARAŃCZYK (1970) in their discussion of the origin of the Cracow-Silesian zinc-lead deposits.

The term *phase of karstification* was used in a broad sense (GŁAZEK, DĄBROWSKI and GRADZIŃSKI, 1972; GŁAZEK, 1973, 1975) that is equivalent to the term *period of karstification* adopted in this book. Similarly, *stage of karstification* was used where *phase of karstification* is being used here.

The Geologic Frame of Karst Development

In Poland all the main tectonic realms of Europe are brought into contact (Fig. 2–25). The younger overlap the older (PEIVE et al., 1981, 1982; KSIĄŻKIEWICZ, OBERC and POŻARYSKI, 1978). In the north-eastern part of the country is the SW slope of the ancient East European Craton (Platform; abbreviated EP); the southern part belong to the Alpine Orogen of the Carpathians and the south-west part to the young epi-Variscan Central European Platform (CEP). This general division is widely accepted but the boundaries between these principal units are not universally agreed. The published opinions were strongly influenced by the now rejected "geosynclinal" theory. In the present writer's view, between the well established realms mentioned above there exists a stable continental margin of the European Craton that has been slightly affected by every younger tectonic event known in the development of the present European continent.

The undisputed EP is north-east of a deep fracture zone called Teisseyre-Tornquist Line (TTL) and consists of Proterozoic (Gothian?) crystalline basement covered with 0.5 to 5.0 km of flatlying, slightly faulted and tilted sediments of Late Precambrian to Cenozoic age. Scattered among the dominant siliciclastics are some interbedded carbonates and sulfates as well as numerous breaks in deposition. Only a few karst landforms are known at outcrop, in the Late Cretaceous chalk and limestones (MARUSZCZAK, 1966). Some subsurface karst features have been found in boreholes and mines in Devonian, Lower Carboniferous and Upper Jurassic limestones (ŻELICHOWSKI, JUSKOWIAKOWA and MIŁACZEWSKI, 1974; SKOMPSKI, 1985).

The CEP, which forms the north-eastern slope of the Bohemian Massif extends to a deep fracture zone termed the Dolsk Line (DL; GUTERCH, 1977; SPECZIK, 1985) and to the western boundary of the Upper Silesian Coal Basin (Fig. 2–25). The pre–Namurian metamorphic basement of the Bohemian Massif (VAN BREEMEN et al., 1982) which crops out in the Sudetic Mountains and fore-Sudetic Block (FSB), contains many graben with siliciclastics of Carboniferous, Permian, Triassic and Cretaceous age. From the Odra Deep Fracture (OF), this basement deepens to the north-east under epicontinental Late Carboniferous to Cretaceous sedimentary cover.

Within the complicated structure of the Late Precambrian and Paleozoic metamorphic sequences of the Sudetes and FSB, steep and fractured marble lenses occur that are strongly karstified. Flat post-Variscan cover on the north-eastern slope of this area contains the Zechstein evaporite-carbonate sequence, Muschelkalk carbonate-sulfate rocks and some limestones of Upper Jurassic and Upper Cretaceous age. These rocks are karstified to a considerable depth (KRASOŃ and WÓJCIK, 1965), the karstification having taken place during several periods of subaerial weathering.

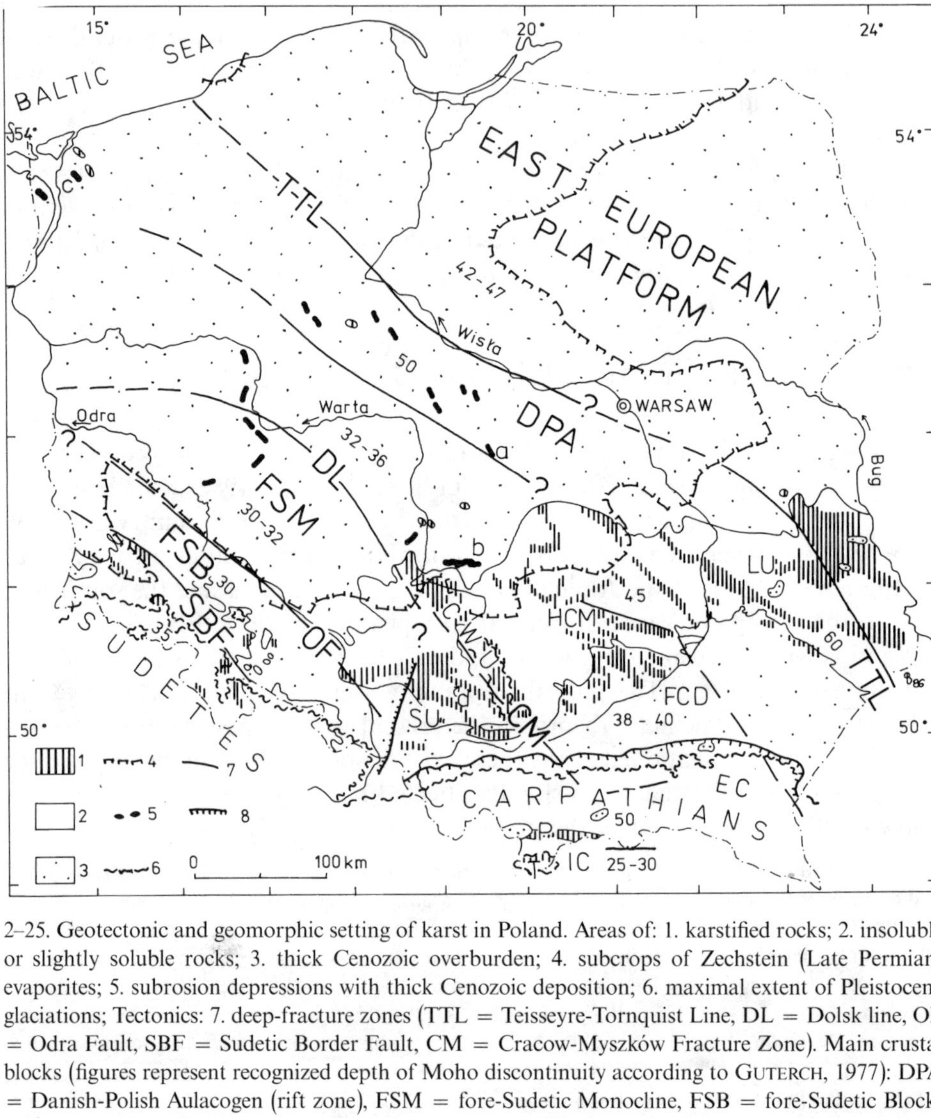

2–25. Geotectonic and geomorphic setting of karst in Poland. Areas of: 1. karstified rocks; 2. insoluble or slightly soluble rocks; 3. thick Cenozoic overburden; 4. subcrops of Zechstein (Late Permian) evaporites; 5. subrosion depressions with thick Cenozoic deposition; 6. maximal extent of Pleistocene glaciations; Tectonics: 7. deep-fracture zones (TTL = Teisseyre-Tornquist Line, DL = Dolsk line, OF = Odra Fault, SBF = Sudetic Border Fault, CM = Cracow-Myszków Fracture Zone). Main crustal blocks (figures represent recognized depth of Moho discontinuity according to GUTERCH, 1977): DPA = Danish-Polish Aulacogen (rift zone), FSM = fore-Sudetic Monocline, FSB = fore-Sudetic Block; 8. frontal overthrusts of Variscan and Alpine Orogens. Geomorphic units: Uplands of Meta-Carpathian Arch: SU = Silesian Upland, CWU = Cracow-Wieluń Upland, HCM = Holy Cross Mountains, LU = Lublin Upland; FCD = fore-Carpathian Deep; Carpathian Orogen: EC = External Carpathians, IC = Internal Carpathians, P = Pieniny Klippen Belt, T = Tatra Mountains.

Between the TTL and DL zones the basement is poorly known. It is believed to be of Late Precambrian age because slightly metamorphosed shales are found in numerous boreholes in the foreland of the Carpathians, beneath the Upper Silesian Coal Basin and Carpathian frontal overthrust (ŚLĄCZKA, 1976). The shales are covered by flat-lying fossiliferous Lower Cambrian rocks (ORŁOWSKI, 1975) and pass to the south-west (beneath the West Carpathians and in Moravia) into the Brno crystalline massif of the same age (VAN BREEMEN et al., 1982). This whole basement (called

Bruno-Vistulicum by DUDEK, 1980) may be treated as the block-faulted (mobilized) margin of the ancient European Craton. Within its boundaries, along the south-western side of the TTL there is a trough named the Danish-Polish (–Dobrogean) Aulacogen (DPA; GUTERCH, 1977). The trough existed from the Early Permian until Late Cretaceous times (KUTEK and GŁAZEK, 1972; GŁAZEK, TRAMMER and ZAWIDZKA, 1973; POŻARYSKI and BROCHWICZ–LEWIŃSKI, 1978). Probably it originated much earlier and may be considered an *aborted* Cadomian rift on which were superimposed the similar younger rifts of the Alpine cycle. Slight folding and faulting occurred there during every younger tectonic event that affected the development of the European continent. Finally, the aulacogen was inverted into the Mid-Polish Anticlinorium at the end of Cretaceous and beginning of Paleogene times (KUTEK and GŁAZEK, 1972).

The platform was invaded by the sea several times. From the Lower Cambrian onwards these transgressions left epicontinental siliciclastics or carbonate deposits and some evaporites. A thick epicontinental carbonate sequence with some sulphates and dolostones in its lower part was deposited from Early Middle Devonian to Early Carboniferous (GŁAZEK et al., 1982; SZULCZEWSKI, 1971; NARKIEWICZ, 1979, 1985; BEŁKA, 1987; SKOMPSKI,1985). Syndepositional faults were active (SZULCZEWSKI, 1973, 1978) and the area was differentiated into uplifted and subsided blocks, then gradually uplifted together with the Variscan Orogen in Late Carboniferous-Early Permian times. At that period karstification was common. It started as a syngenetic process on the rising blocks (SKOMPSKI, 1985; BEŁKA, 1987; Fig. 2–26). Although this karstification period was long and general throughout the whole country, evidence of it is poor due initially to the clastic overburden of Late Carboniferous regressive and deltaic deposits and then to the prevailing arid climate in the Permian (MROCZKOWSKI, 1982) and finally to later erosion.

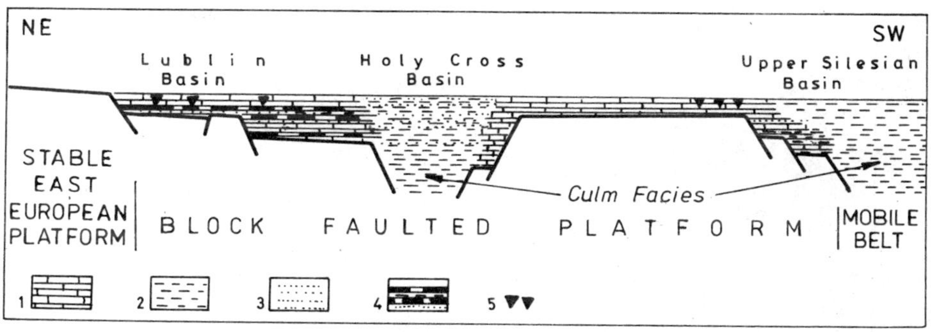

2–26. Schematic section showing the pattern of Late Visean marine deposits in Poland, not to scale (modified after SKOMPSKI, 1985). 1. carbonates; 2. claystones; 3. sandstones; 4. coal measures; 5. traces of syngenetic karst forms

The next transgression followed Variscan movements in the Late Permian times. There was catastrophic flooding of the Variscan foredeep (D. B. SMITH, 1970; MROCZKOWSKI, 1982) by the Zechstein Sea which left a thick carbonate-evaporite sequence passing gradually into the continental siliciclastics of the Lower Triassic. They were gradually submerged and covered by the limestone-dolostone-sulphate deposits of the Röt and Muschelkalk seas (GŁAZEK, TRAMMER and ZAWIDZKA, 1973). After gradual retreat in the late Middle Triassic, continental conditions prevailed until the widespread Late Jurassic transgression began. The Upper Jurassic limestones are the second outstanding karst formation, after that of the Devonian. In the Tithonian, the sea retreated again and the Early Cretaceous was a period of continental erosion. Karstification from that time is poorly evidenced, probably due to the clayey-marly character of the uppermost Jurassic and Berriasian-Hauterivian rocks which, in addition, are preserved only in a limited belt along the DPA axis (KUTEK and GŁAZEK, 1972; GŁAZEK, 1973).

The Albian-Cenomanian transgression covered the whole platform area and left mainly the low permeability, microporous chalk and marl deposits of the Upper Cretaceous. They are only feebly karstified (MARUSZCZAK, 1966).

In general, thicknesses and facies patterns on the platform of Permian-Mesozoic times suggest that interference from two structural directions affected the sedimentation and vertical movements. First there is an axis of subsidence along the DPA, with rapid thinning of sediments to the north-east and moderate thinning to the south-west of it. The deepest facies and more complete sequences occur along the subsidence axis, while there are shallower facies and longer breaks on the limbs of the trough (KUTEK and GŁAZEK, 1972). On the north-eastern limb and from the north along the axis of the DPA, the massive terrigenous influx from elevated parts of the EP is remarkable, while on the

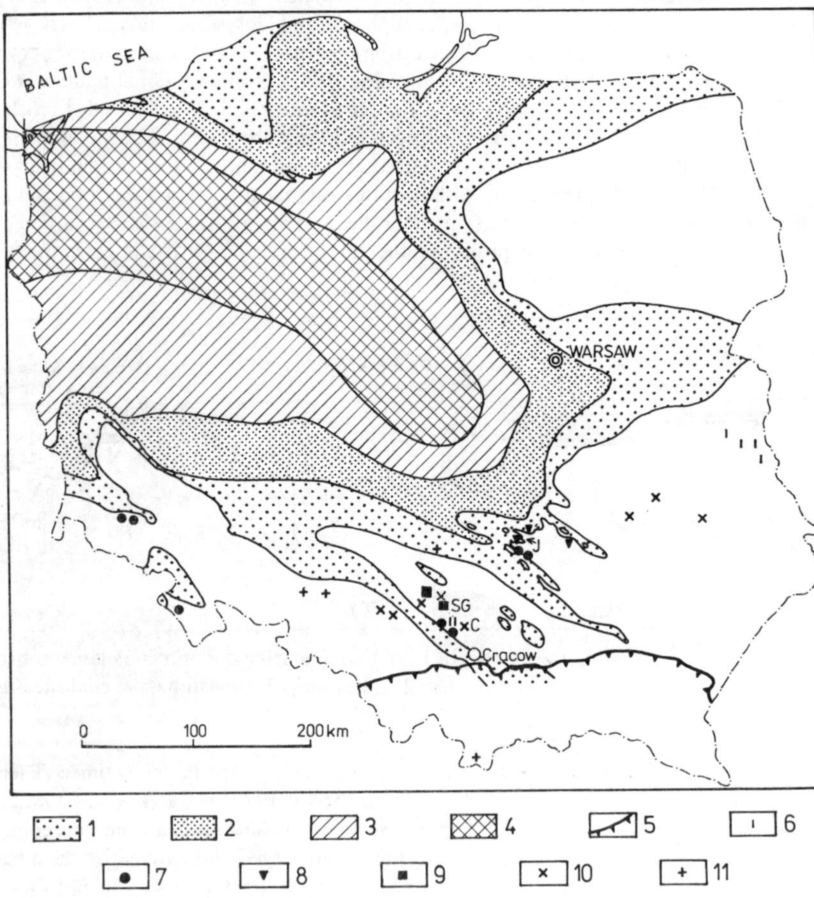

2–27. Distribution of Late Paleozoic and Mesozoic karst in Poland. 1. Extent of Permian deposits: Rotliegendes and thin Zechstein (200 m), showing NW-SE rift pattern controlling the Early Permian continental sedimentation; 2–4. thickness of Zechstein evaporite sequence showing typical thickness pattern of Late Permian and Mesozoic deposits (slightly altered by later erosion): 2. 200–500 m, 3. 500-1 000 m, 4. over 1 000 m; 5. northern margin of the Carpathian nappes; 6. syngenetic Visean karst; 7–11. subaerial karst buried with deposits of: 7. Permian, 8. Bunter, 9. Middle Triassic, 10. Late Triassic-Middle Jurassic, 11. Albian-Cenomanian.

south-eastern limb rather open marine carbonate deposition dominated. Clastic material carried from elevations of the Bohemian Massif is visible in the Upper Cretaceous, absent in the Upper Jurassic and only feebly present in the Middle Triassic.

The Permian-Mesozoic cover is stratigraphically more complete along the subsidence axis than further southwest and north-east. Some deposits preserved near the axis wedge out completely to the southwest and northeast of it. The Mid-Polish Anticlinorium began to rise in Late Maestrichtian times, together with broad regression on its southwestern limb. Regressive deposits of Late Maestrichtian and Paleocene age are known in the northern and eastern parts of the country, where remnant basins persisted until the Montian. Brown coal deposits of Late Paleocene to Early Eocene age (CIUK, 1974) are known in subrosion depressions on the Mid-Polish Anticlinorium and on its southwestern limb (cf. Fig. 2–25).

The second structural pattern is associated with the evolution of the Tethyan realm and is thus marked by east-west axes. The Mesozoic subsidence axis of the Central European Basin lies a little north of Warsaw, but the uplift axis migrated from south to north during the growth of the Tethyan stable continental margin. The earlier Permo-Triassic site of this axis (known as the Vindelician Arch) was covered by Carpathian overthrusts and partly consumed in subduction beneath the Carpathian Orogen. During subsequent transgressions, the arch was last to be inundated and so was covered with relatively thin and shallow deposits. The latest position of this arch (after the regression of the Sarmatian Sea) is called the Meta-Carpathian Arch and is visible in the present morphology as a Mid-Polish Upland Belt stretching from the Bohemian Massif through the Sudetes, Silesian-Cracow Upland, Holy Cross Mountains and Lublin Upland to the Volhynia and Podolia. Numerous paleokarst features of different ages are visible in this belt of uplands (Figs. 2–25, 2–27 and 2–36).

On the southern slope of the Vindelician Arch in the area of the Inner Carpathians (IC) (south of the Pieniny Klippen Belt, CP) sedimentation began on the crystalline Variscan basement. Upper Permian or Lower Triassic clastics pass upward into shallow marine carbonates deposited on a stable continental margin after the late Early Triassic. These carbonates are the oldest karst formation in the Tatra Mountains. Late Triassic to Middle Jurassic extension of the lithosphere led to strong facies differentiation and local breaks in the region. As a result, from the Middle Jurassic onwards, deep sea facies (radiolarites) were deposited in the troughs (the Pieniny and Subtatric sequences), while on the elevated blocks of the High Tatric sequence thick pelagic limestones were laid down. Deposition of this sequence continued until Early Cretaceous times, when it was replaced by a shallow Urgonian facies. These limestones are the second in stratigraphic order but are the most important karst formation in the Tatra Mountains. On the top of the Urgonian limestones traces of syngenetic karstification were noted by PASSENDORFER (1930). Immediately after its deposition, in the Albian times, the Urgonian carbonate platform was drowned and covered with pelagic limestones, marls and distal turbidites which terminate in the Turonian. Later, the area was folded and elevated. Subsequent erosion lasted until Middle Eocene times when, following brief deposition of Eocene conglomerates and carbonates, turbidites of Oligocene age covered the whole Tatric area. Lenses of red continental conglomerates at the base of the Eocene sequence are known but no apparent paleokarst forms have been described. However, some karstification is highly likely and frequently suggested by many authors. From the Late Jurassic to the Early Miocene, the Inner Carpathians were bordered on the north by a zone of siliciclastic turbidite sedimentation from the Outer Carpathians (EC).

Epicontinental seas connected with the North Sea Basin and the Tethyan Basin of the Middle Eocene invaded most of the platform areas in Poland (POŻARYSKA and ODRZYWOLSKA-BIEŃKOWA, 1977). They covered the older erosional landscape with its continental deposits retained in subrosion depressions with marly, glauconitic sands and silts. Regression at the Eocene-Oligocene boundary reduced this sea to a remnant in the northeastern corner of Poland (ODRZYWOLSKA-BIEŃKOWA, POŻARYSKA and MARTINI, 1978), while the rest of the country was covered either by tidal flat and delta deposits or was eroded. Numerous karst forms developed in central and southern Poland.

Some of them are well dated by vertebrate remains and the Lower Badenian overburden (Fig. 2–36). Thick brown coal deposits accumulated within subrosion depressions in western and central Poland prior to the Badenian transgression. The paleogeographic pattern of the Middle Miocene was governed by the subsidence of the Carpathian Foredeep. The sea invaded most of the platform part of Poland and deposited thick clays and silts in the south. Shallow marine deposits in the Carpathian Foredeep passed to the north in to alluvial plain deposits derived from Scandinavian rivers. These covered the Oligocene-Lower Miocene karst landscapes (Głazek and Szynkiewicz, 1987). Between the alluvial plain and the open marine facies, biocalcarenites and algal-vermetid reefs were deposited on the southern slope of the Mid-Polish Uplands while sands were laid down around river mouths (Radwański, 1969, 1973; Karnkowski, 1985; Pisera, 1985). A local salinity crisis in the Middle Badenian during a slow subsidence of the continental margin resulted in gypsum sedimentation in the northern part of the marine basin and halite precipitation near the Carpathian margin.

Lower Badenian marine sediments as well as slightly older and youger continental deposits were also deposited on the EC and then slightly folded and thrusted by forces associated with the Carpathian nappes (Bałuk, 1970). The Carpathian Orogenic Belt was strongly folded and thrust on to its foredeep during the Middle Miocene times. These movements probably began beneath sea level during Early Miocene times. The Orogenic Belt was uplifted and its erosion began (Burchart, 1972).

Later, during Late Badenian and Sarmatian times, alluvial plain deposits gradually overlapped the marine deposits of the Carpathian Foredeep. Such a situation is seen in Lower Silesia, while on the more elevated uplands only outliers of Middle Miocene deposits are preserved on hill-tops, in buried valleys or trapped in karst depressions (Fig. 2–36). Strong erosion and exhumation of the pre-Middle Miocene relief took place after deposition of the brackish Sarmatian strata. This erosion was accelerated by the Messinian crisis, when the erosional base level in the Black Sea area was lowered to –1 600 m (Hsü and Giovanoli, 1979) and then by Pliocene uplift of whole area (Głazek and Szynkiewicz, 1987). A new river pattern developed on the flat surface of the Middle Miocene deposits (Dżułyński et al., 1966). This pattern was partly affected by the pre-Middle Miocene depressions after the Miocene strata had been eroded. Because of this, numerous Late Neogene river valleys were abandoned and became filled with the so-called *preglacial deposits* (clastics) of Pliocene and Early Quaternary age. Just before the continental glaciers reached Poland, and as a result of further uplift of the Carpathian Orogen with its foreland, a new river pattern consequent to the newly created Baltic depression was formed and drainage to the Black Sea ceased (Dżułyński et al., 1968).

The Polish Lowland has been covered by continental ice sheets several times since the beginning of the Brunhes Epoch. The maximum extent of erratic boulders and till reaches the Carpathian slopes (Fig. 2–25). This maximum glaciation is called the Sanian Glaciation and correlates with the Elsterian to the West. Not less than three younger and less extensive glaciations (Odranian, Wartanian and Vistulian) are preserved at the surface. In addition, from subsurface evidence, a further one or two glaciations older than the Sanian and one or two younger (between the Sanian and the Odranian) are suggested but much disputed (Różycki, 1979; Lindner, 1984; Mojski, 1985). Well-dated cave deposits of glacial and interglacial origin contribute considerably to this discussion (Głazek, Lindner, Wysoczański-Minkowicz, 1976; Głazek et al., 1976, 1977; Głazek and Harmon, 1981).

The highest mountains on the southern border were also glaciated. There were at least three glaciations in the Tatras (Fig. 2–25). The latest one was well dated in the Magura Cave using the ^{14}C method at between 23 and 11 ka (HERCMAN et al., 1987). In the Karkonosze Group of the Sudetic Mountains traces of one glaciation are visible, probably the youngest one.

Periods of Karstification

The intricate geologic development of the Polish territory has resulted in several periods of karstification and, within these general periods, many particular phases or events of karstification can be distinguished.

First Period of Karstification; Early Carboniferous to Middle Triassic

The oldest period of karstification followed the decline of carbonate sedimentation in the Late Visean and lasted until the inundation of the post-Variscan landscape by the Muschelkalk sea. So far, the oldest phase identified within that period is Late Visean. It is recognized in the subsurface of the Lublin Upland and in the vicinity of Cracow, on elevated blocks of tectonically extended carbonate platforms (Figs. 2–26 and 2–27). Syngenetic karst activity is demonstrated by minor corrosion, vadose cements, and paleosoils with burrows and rhizoliths (SKOMPSKI, 1985; BEŁKA, 1987).

The next phase of karstification is characterized by red breccias, conglomerates, sandstones and siltstones filling sinkholes, vertical shafts and horizontal caves. These are common and described from: (i) the Holy Cross Mountains in Devonian carbonates (cf. GUIDE, 1978); (ii) the vicinity of Cracow in Visean limestones (SIEDLECKI and WIESER, 1948; PASZKOWSKI and WIECZOREK, 1982); (iii) the Sudetes in (?) Paleozoic marbles (GIERWIELANIEC and SZTUK, 1977; Z. BARANOWSKI and S. LORENC pers. comm., 1985).

In Czatkowice Quarry (C in Fig. 2–27) layered rhyodacite tuffites filling a large cave are intercalated with calcite and silica flowstone layers (PASZKOWSKI and WIECZOREK, 1982). All these forms are found near the edges of similar Rotliegendes deposits filling synsedimentary troughs (Fig. 2–27). The tuffites may be correlated with violent volcanic activity along the faults of neighbouring graben that filled with fanglomerates and rhyodacite tuffs during the Lower Permian. The deposits are found in corrosional caves and shafts of similar or slightly older age. Their age seems to be younger than the Westfalian coal-bearing sediments but older than the Zechstein transgression.

Within the Zechstein evaporite sequence in the subsurface of the Polish Lowland, borehole and seismic data plus a few observations in mines, reveal anomalous thinning or wedging out of the evaporite layers. These phenomena were interpreted as products of major subaerial karstification between particular salt cycles (KRASOŃ and WÓJCIK, 1965; POBORSKI, 1975). However, there are great difficulties in discriminating between the effects of early (syngenetic) and considerably younger (subjacent, in this case subrosional) karstification. Moreover, great changes of thickness in salt layers were caused by subaquatic dissolution (resulting from brine dilution in the lagoonal environment as a consequence of freshwater influx) and by salt migration. Nevertheless, detailed studies of borehole profiles have produced evidences of synsedimentary dissolution, soil formation and vadose (intertidal ?) diagenesis in the Zechstein rocks (PERYT, 1984). That means that at least some phases of syngenetic karstification occurred during the Zechstein sedimentation. These are contemporaneous with one karstification phase outside of the Zechstein basin.

2–28. Solution runnel on the Givetian limestones filled with Bunter siltstones at Jaworznia Quarry (J in Fig. 2–27), scale is 1 cm. (Photo by K. Boruta).

A subsequent karstification phase is recognized beneath Bunter deposits on the top of Devonian carbonates in several localities in the Holy Cross Mountains. At Jaworznia Quarry (J in Fig. 2–27) there are sinkholes, meandering solution runnels (Fig. 2–28) and solution pans filled with red siltstones (GŁAZEK and ROMANEK, in GUIDE, 1978). This phase is clearly separated from the older one only in the NW part of the HCM area where along the subsidence axis of the DPA, Bunter deposits were laid down with lithologic continuity onto the Zechstein.There, during the Bunter sedimentation, some Devonian carbonates became exposed to karstification along the uplifted edges of synsedimentary tilted blocks (KUTEK and GŁAZEK, 1972; GŁAZEK and ROMANEK, in GUIDE, 1978).

The last phase of karstification occurred along the Cracow-Myszków fracture zone. It is best represented at Stare Gliny Quarry (SG in Fig. 2–27 and Fig. 2–29) where a cave with bone breccia in a fossil (?) mogote of Devonian dolostones has been described (LIS and WÓJCIK, 1960). The fauna includes fish teeth and scales (*Gyrolepis, Saurichtys, Coelacantiformes*), as well as reptilian bones and skulls from the small archosaurs *Macrocnemus* and (dominant in this assemblage) *Nothosaurus* (TARLO, 1959). The bone-bearing layer is covered with a breccia of angular blocks of Devonian dolostones in a ground mass of Middle Muschelkalk (Late Anisian) *Diplopora* dolostones. This littoral cave was quickly destroyed and covered with shallow water deposits during structural remodelling of the Muschelkalk sea in the Illyrian (GŁAZEK, 1973; GŁAZEK, TRAMMER and ZAWIDZKA, 1973).

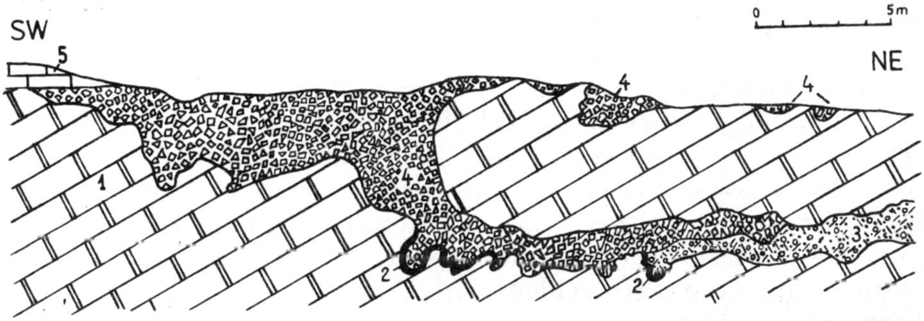

2–29. Section of the western wall of the Stare Gliny Quarry (SG in fig. 2–27) showing a paleokarst cave of Triassic age (after LIS and WÓJCIK, 1960). Middle Devonian: 1. dolostones; Middle Triassic: 2. residual clays, 3. cave deposits with reptilian bones, 4. breccia of Devonian dolostones within *Diplopora* dolostone (Middle Muschelkalk), 5. *Diplopora* dolostones.

The paleogeographic conditions were variable during this first karstification period. A wet and hot climate, resembling the present tropics, may be suggested for the flat coastal plains of the Lower Carboniferous. Probably similar conditions were dominant on the emerged edges of the tilted blocks of Frasnian carbonates in the HCM area. The two following phases developed under a semi-arid or arid climate, similar to recent deserts, with torrential mudflows and ephemeral streams in diversified, hilly landscapes. The last phase is represented by isolated dolostone islands that emerged from the shallow marine Middle Muschelkalk basin with its hypersaline characteristics. These hills were interpreted as inundated mogotes overlooking a corrosional plain on the Devonian carbonate rocks. However, they can be interpreted as emerged edges of tilted blocks thrown up during the Middle Triassic extension of the continental margin. Hot and rather dry conditions may be ascribed to this phase.

Second Period of Karstification; Late Triassic to Middle Jurassic

This period is represented at the surface in the Silesian and Cracow Upland and in the subsurface in the Lublin Upland. On the northeastern slopes of the Silesian Upland, in Triassic carbonates and beneath Quaternary deposits, there are narrow depressions up to 30 m deep and 2 km long (GÓRZYŃSKI, 1963;

GILEWSKA, 1965). They are filled with variegated clays, silts and sands. Among the deposits are (i) thin intercalations (10 cm) of black carbonaceous clays yielding lowermost Jurassic spores; (ii) halloysite beds up to 50 cm thick, and (iii) layers of boehmite-alunite clays up to 50 cm thick (GÓRZYŃSKI, 1963). These deposits are interpreted as Lower Jurassic fillings of karst features. However, they are hardly distinguishable from Tertiary subjacent karst forms with subsided Lower Jurassic rocks (GILEWSKA, 1965).

In Czatkowice Quarry (C in Fig. 2–27) there are karst cavities filled with grey sandstones and siltstones intercalated with thick, coarse crystalline flowstones. They contain reptilian bones and skulls that are not fully studied yet, but are of uppermost Triassic-Lower Jurassic appearance (PASZKOWSKI and WIECZOREK, 1982). The trends of these cavities are parallel to a neighbouring fault of pre-Callovian age. Similar grey sandy deposits with reptilian bones were found later in a debris heap at Stare Gliny Quarry (K. ZAWIDZKA, pers. comm., 1985). According to F. WESTPHAL (pers. comm., 1985) these bones represent younger reptiles than those described earlier by TARLO (1959). It is very probable that they were excavated from a small karst depression noted in the top of *Diplopora* dolostones beneath Callovian deposits (LIS and WÓJCIK, 1960).

The karstified top of the Devonian carbonates was noted in some deep boreholes in the Lublin Upland. Variegated clays with chaotic blocks of Devonian limestones are covered by flat-lying Bathonian (?) transgressive deposits (GŁAZEK and MATYJA, unpubl. 1974; ŻELICHOWSKI, JUSKOWIAKOWA and MIŁACZEWSKI, 1974). In this area a pattern of river valleys cut in the Paleozoic strata and filled with Lower Jurassic clastics was found (SZYDEŁ and SZYDEŁ, 1981). Thus, a buried karst of Lower to early Middle Jurassic age is very probable in this region.

The second period of karstification commenced after regression of the Middle Triassic sea and ended with the Middle Jurassic transgression. This karstification was caused by the old Cimmerian movements. It was well developed on both limbs of the DPA, where regressive Keuper deposits were eroded prior to the Middle Jurassic (Bathonian-Callovian) transgression. Rhaetian and/or Lower Jurassic continental to near-shore deposits are absent or considerably less well developed than along the axis of the DPA.

At that time, in a hot and humid climate, a landscape of gentle hills developed on the flanks of the DPA. Probably this period should be divided into two or more distinct phases of karstification but stratigraphic evidence is insufficient as yet.

It is believed that the subsurface processes that formed the zinc-lead deposits of the Upper Silesian-Cracow ore district belong to this period, especially the earlier stages distinguished by SASS-GUSTKIEWICZ (1985): "dolomitization of Muschelkalk limestones" and the "non-integrated circulation of ore-forming fluids" (DŻUŁYŃSKI and SASS-GUSTKIEWICZ, p. 377–397). Probably the next phase of "phreatic integrated solution transfer" also occurred before erosion of the clayey Keuper overburden. After erosion of the Keuper clays, travertines (Woźniki Limestones) were deposited in places on the surface among the Rhaetian variegated clays. These travertines are believed to have been deposited at springs which discharged the ore-depositing fluids (BOGACZ, DŻUŁYŃSKI and HARAŃCZYK, 1970). Their age may therefore indicate the late phases of the ore deposition, namely "sudden drop of pressure and temperatures" and the beginning of "vadose ore precipitation" (DŻUŁYŃSKI and SASS-GUSTKIEWICZ, p. 377–397). Thus, the period of principal ore formation may be ascribed to the time span between the mid-Ladinian facies change from calcareous to marly and clayey regressive deposits (GŁAZEK, TRAMMER and ZAWIDZKA, 1973) and the mid-Rhaetian precipitation of the Woźniki Limestones. That is nearly 25 Ma according to ODIN's (1982) time-scale. The hydraulic head responsible for this deep groundwater circulation, its pressure and temperature, can be explained by infiltration from the northern slope of the Vindelician Arch into older Paleozoic carbonates and sandstones already exposed by the Late Carboniferous-Early Triassic erosion. This deep, confined water circulation supposedly dissolved heavy metal compounds dispersed in Paleozoic bituminous and coal-bearing sediments as well as in igneous rocks, and carried them up-dip along the fractures of the Cracow-Myszków zone to the confined Muschelkalk carbonate aquifer. This hypothesis is supported by the more diversified ore mineralization noted along some fracture lines in the Paleozoic rocks beneath ore-bearing dolostones of the Triassic (GÓRECKA, 1972).

With the erosion of the Keuper rocks, the metalliferous aquifer in the Muschelkalk became partly unconfined. This probably caused the drop of pressure and temperature and, as a consequence, the rapid sulphide precipitation. At that time, "vadose" conditions were probably established for the first time in the ore-bearing strata and some dripstones may have been deposited. However, there was no major infiltration of oxidized water because the sulphides remained stable. Some further stages in the development of the ore deposits, with alternating phreatic and vadose conditions, resulted in insignificant ore redistribution (SASS-GUSTKIEWICZ, 1985).

LIPIARSKI (1971) described some intercalations of coal among internal sediments in the ore-bearing cavities. These coals contain arthropod and plant remains of Mesozoic appearance. They suggest infiltration of surface water into the ore deposits. Later, in Middle Jurassic times, the whole area was buried by transgressive sediments and the zinc-lead deposits were probably inert until the next period of karstification.

Finally, some authors have suggested that there was Late Triassic-Middle Jurassic karstification in the High Tatric sequence. However, there are numerous neptunian dykes of Middle Jurassic age as well as Fe–Mn encrustations on the drowned Triassic carbonate platforms and no evident paleokarst forms.

Generally, changing climatic conditions may be suggested during this period of karstification, from warm and semi-arid in the Late Triassic to warm and humid in Early and Middle Jurassic times.

Third Period of Karstification; (?) Tithonian-Cenomanian

Late Cimmerian movements caused broad regression of the Upper Jurassic sea in the platform area of Poland and extensive subsidence of its southern margin where, from the Tithonian until the Lower Miocene, huge turbidite sequences were deposited at the EC. Regressive marly deposits (Kimmeridgian and Tithonian) were laid down on both limbs of the DPA. Oxfordian limestones and, on the southwestern limb of the DPA, Triassic and also older soluble rocks, were exposed to karstification. This period was terminated by the universal transgression of Albian-Cenomanian age.

Numerous surface karst landforms are believed to have developed in this period (GILEWSKA, 1964; GRADZIŃSKI and WÓJCIK, 1966). However, most of them are subsided pocket outliers of Mid-Cretaceous transgressive clastics that were created together with removal of Upper Cretaceous deposits during the Tertiary (GRADZIŃSKI, 1962; GŁAZEK, 1973). Probably there were several specific causes for the insignificant extent of karst development during this period: (1) the low relief landscape until the Aptian/Lower Albian times was unfavourable for deep groundwater circulation; (2) the persistence of clayey, impermeable, regressive Upper Jurassic sediments, only slightly weathered at the top beneath the mid-Cretaceous transgression; (3) strong secondary silicification of exposed Upper Jurassic limestones forming gentle hills on the pre-Albian surface in the Cracow-Wieluń Upland (MARCINOWSKI, 1970; GŁAZEK, 1973), and (4) significant uplift and erosion of older rocks occurring only in late Early Cretaceous times, as determined from fission-track cooling-ages (105–95 Ma) of crystalline formations in the Sudetes (JARMOŁOWICZ-SZULC, 1984).

Unequivocal pre-Albian or pre-Cenomanian karst forms are known only in (1) Julianka and Mokrzesz (Cracow-Wieluń Upland); (2) Święta Anna Mount (Silesian Upland), and (3) in the Tatra Mountains (Fig. 2–27).

In Julianka Quarry a hill of completely silicified Upper Oxfordian limestones contains remnants of a cave that developed prior to silicification and submergence beneath mid-Cretaceous clastics. Nearby, in Mokrzesz, MARCINOWSKI (1970) described a pocket filled with silicified limestone debris and covered with Albian-Cenomanian clastics on a similar hill of Oxfordian limestones.

Further to the West, in an old quarry on Święta Anna Mount, the author found a sinkhole filled with grey silts and a layer of limestone boulders bored by lithophages, at the boundary with transgressive clastics. The sinkhole is developed in Middle Triassic limestones.

Traces of pre-Albian syngenetic karstification are known in the High Tatric sequence (PASSENDORFER, 1930; KRAJEWSKI, 1984). At the top of the Urgonian limestones, vadose diagenetic features and minute hollows are known. They are filled with yellow silt that is phosphatized and encrusted with Mn-Fe hydroxides at the top and covered with the condensed pelagic deposits of the Albian.

Paleokarst porosity developed in Oxfordian limestones on the slopes of the DPA. This caused catastrophic floods into coal mines being developed in the Lublin Upland in late Seventies. The floods occurred when the shafts reached the Cenomanian-Oxfordian boundary.

There was minor ore mineralization in Oxfordian limestones along some fracture zones in the Silesia-Cracow zinc-lead ore district (BEDNAREK, GÓRECKA and ZAPAŚNIK, 1985). It may be suggested that these ores were derived from underlying Triassic deposits and spread into the limestones beneath Upper Oxfordian marls. The hydraulic head responsible for such deep phreatic circulation was probably caused by uplift of the Sudetic area during Albian times, when Middle Triassic carbonates became exposed.

Thus, during this third period, only one phase (in the late Lower Cretaceous) is well established. It lies in areas relatively far from the DPA subsidence axis. A subtropical, savannah-like climate may be suggested for this phase.

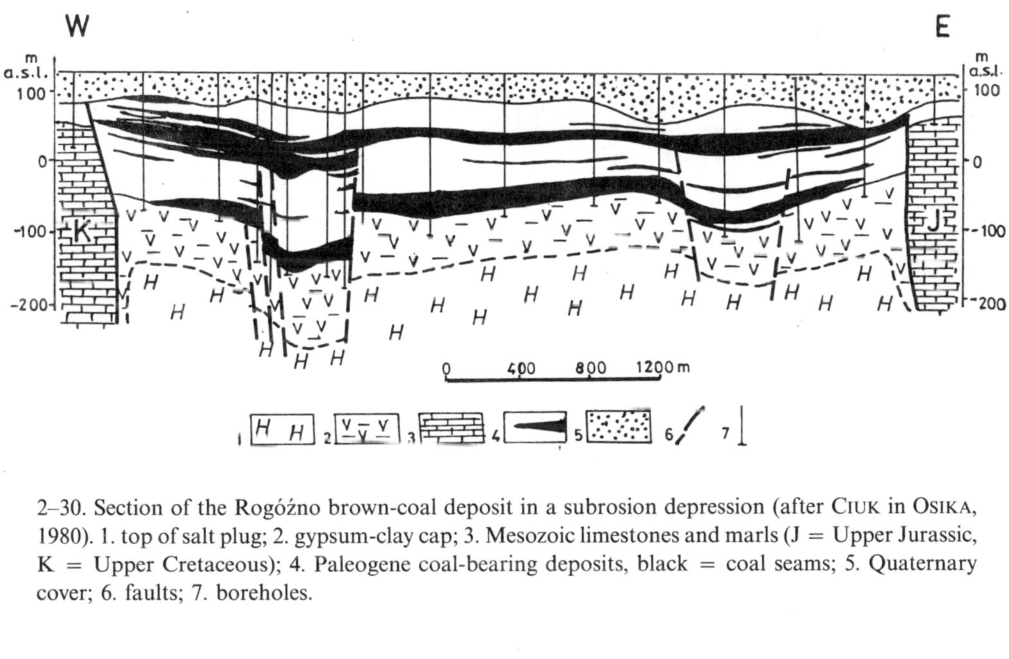

2–30. Section of the Rogóźno brown-coal deposit in a subrosion depression (after CIUK in OSIKA, 1980). 1. top of salt plug; 2. gypsum-clay cap; 3. Mesozoic limestones and marls (J = Upper Jurassic, K = Upper Cretaceous); 4. Paleogene coal-bearing deposits, black = coal seams; 5. Quaternary cover; 6. faults; 7. boreholes.

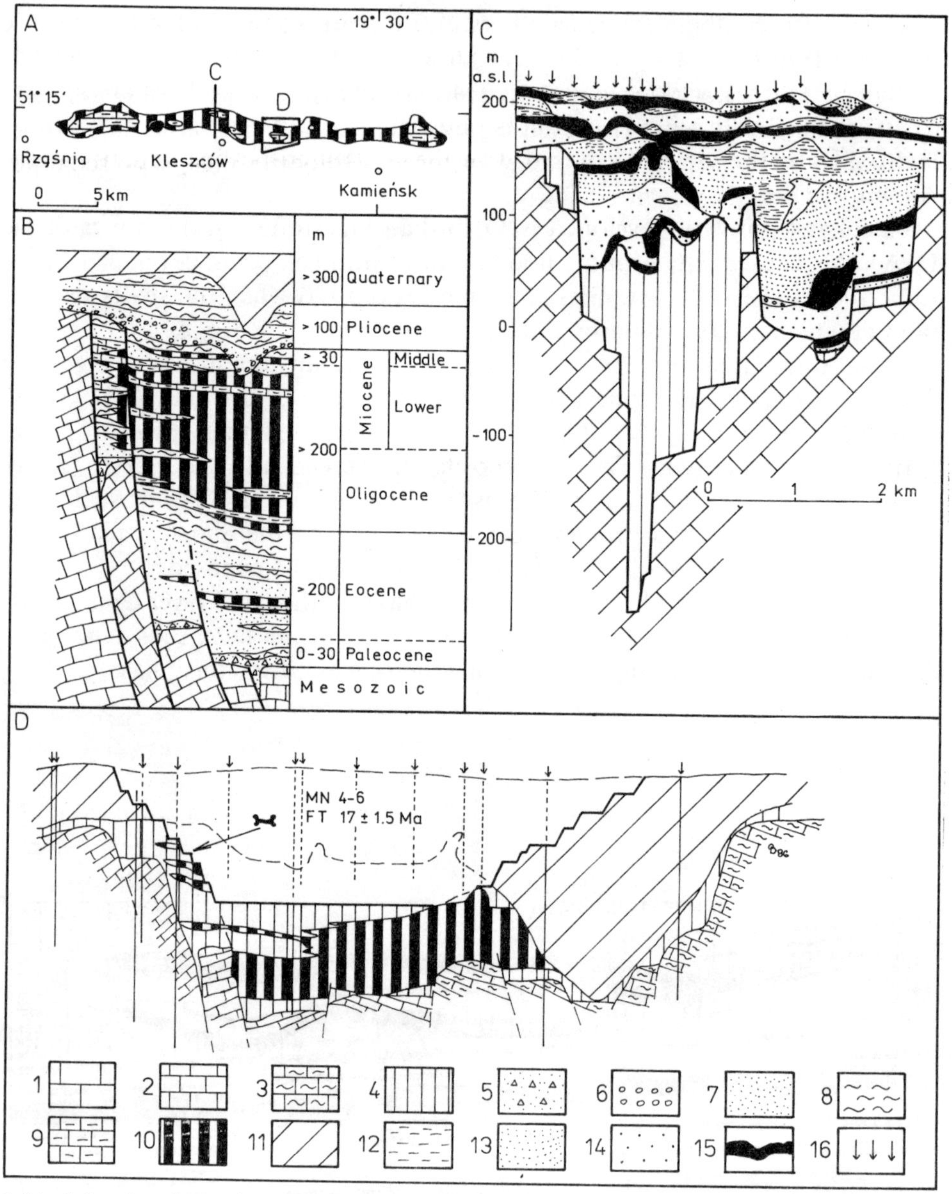

2–31. Subrosion depression with brown-coal deposits south of Bełchatów, Central Poland. A. Outline-map of deposit (after CIUK and PIWOCKI, 1980) showing distribution of peat and "lacustrine chalk" sedimentation filling the subsidence maxima. Salt dome (black spot) and open-cast mine (contour), as well as section lines (C–D) are marked. B. Stratigraphic scheme of Tertiary deposits (modified after CIUK and PIWOCKI, 1980). C. Section of Quaternary deposits, vertical scale 8 times horizontal (simplified after BARANIECKA and SARNACKA, 1971) showing two periods of maximum subsidence, namely Tertiary (Oligocene-Lower Miocene) and Middle Pleistocene (between two older and two younger glaciations). D. Section through the open-cast mine (after KUSZNERUK, 1984) showing the dated localities, not exaggerated. Mesozoic rocks: 1. undivided, 2. Upper Jurassic

Fourth Period of Karstification; Late Cretaceous – Holocene

The final karstification began with the Upper Cretaceous emergence and has persisted to modern times. However, here it will be treated only until the Middle Pleistocene glaciation, i.e. the Late Pleistocene-Holocene phases or separate Holocene phases if these are discernible are excluded. This long period encompasses many phases that are more or less clearly separated by deposits from marine incursions or from glaciations in the Pleistocene.

The oldest phase is represented by the development of deep subrosion depressions (more than 300 m) in northwestern Poland, along the crest of the Mid-Polish Anticlinorium or to the southwest of its axis (Fig. 2–25). These depressions are filled with Tertiary brown coal formations and are located directly above salt domes (Fig. 2–30; Fig. 2–25 site a) or along fracture zones (Fig. 2–31; Fig. 2–25 site b). In the latter case they are frequently separated from the top of the surviving evaporites by more than 2 km of Mesozoic deposits that display anomalous dips and fracturing.

Some of the depressions, especially those in the north-west, are covered by marine Middle Eocene-Oligocene deposits (Fig. 2–25 site c). They contain Late Paleocene–Early Eocene brown coal formations (CIUK, 1975) and appear to have been inert since at least the Middle Eocene. These depressions mark the first phase of deep intrastratal karst (subrosion). This phase developed as a result of the Laramide elevation of southwestern Poland (FSM, FSB and Sudetes in Fig. 2–25) when Permian clastics and carbonates were exposed by erosion on a sub-Late Eocene surface. Supposedly, groundwater travelled long distances via a combination of older primary (in the clastics) and younger, secondary vadose and diagenetic carbonate porosity (PERYT, 1984) plus Early Cenozoic intrastratal porosity.

No other karst features of this phase are proven. However, many regoliths are ascribed to this time and the Middle Eocene transgression buried an uneven erosional surface. Traces of it were probably later exhumed and incorporated into the younger landscapes, even the Recent one.

A second phase of karstification commenced with the Early Oligocene regression and was caused by considerable uplift of the continental margin facing the Paleogene Tethys Ocean. It was terminated by Middle Miocene (Badenian) transgression from the Carpathian Foredeep. Deep valley patterns were carved at that time and are preserved beneath the Badenian deposits.

The sub-Badenian morphology in the Carpathian Foredeep exceeds 600 m in relief and is recognized as buried valleys extending across the Middle Polish

---

limestones, 3. Upper Cretaceous marls; Tertiary: 4. undivided, 5. weathering rubble waste, 6. gravel, 7. sand, 8. silt and clay, 9. "lacustrine chalk", 10. brown coal; Quaternary: 11. undivided, 12. clay and silt, 13. fine sand, 14. coarse sand, 15. boulder clays and pavements; 16. boreholes.

Uplands from Lower Silesia to the Lublin Upland, as well as in central Poland (STAHL, 1932; RADWAŃSKI, 1969; CZARNIK, 1972; BŁASZAK et al., 1976; PANEK and SZUWARZYŃSKI, 1976; BUŁA and JURA, 1985).

Many surface karst forms developed during this phase (Fig. 2–36) and became fossilized with karst deposits or marine Badenian deposits (GILEWSKA, 1964, 1965; GRADZIŃSKI and WÓJCIK, 1966). Some of them are filled with subsided and decalcified Eocene glauconitic sands (KARASZEWSKI, 1966); many subrosion brown coal-bearing depressions also contain Eocene glauconitic sands at their base (PIWOCKI, 1975). An excellent example of a buried valley and associated group of sinkholes, all filled with local waste and marine Lower Badenian deposits, is described by PANEK and SZUWARZYŃSKI (1976) at the Pb-Zn mine near Cracow (Fig. 2–36 a). This is the evidence that the hydrothermal ore-forming activities were completed before this karstification phase.

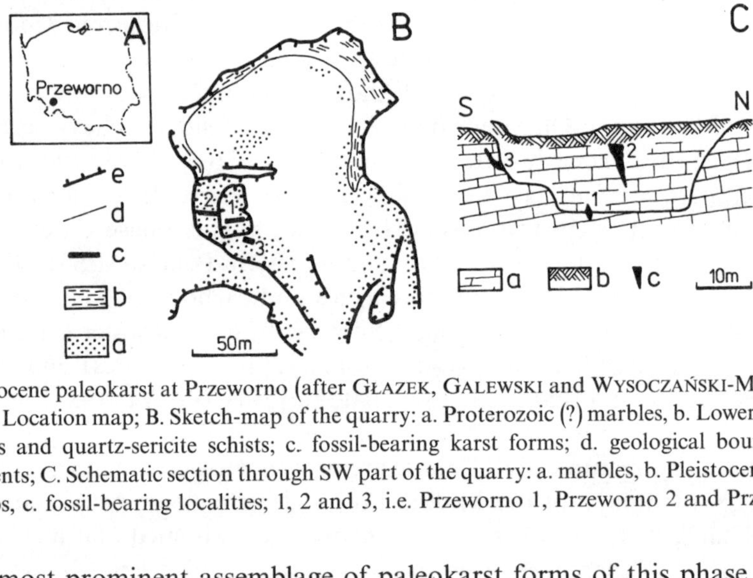

2–32. Miocene paleokarst at Przeworno (after GŁAZEK, GALEWSKI and WYSOCZAŃSKI-MINKOWICZ, 1977). A. Location map; B. Sketch-map of the quarry: a. Proterozoic (?) marbles, b. Lower Devonian quartzites and quartz-sericite schists; c. fossil-bearing karst forms; d. geological boundaries; e. escarpments; C. Schematic section through SW part of the quarry: a. marbles, b. Pleistocene deposits and heaps, c. fossil-bearing localities; 1, 2 and 3, i.e. Przeworno 1, Przeworno 2 and Przeworno 3.

The most prominent assemblage of paleokarst forms of this phase was discovered at Przeworno (GŁAZEK, OBERC and SULIMSKI, 1971; GALEWSKI and GŁAZEK, 1973; Figs. 2–32 and 2–36) in an isolated marble hill surrounded by Middle Miocene alluvial plain deposits. Three sites are known here:

(1) The oldest (?), Przeworno 1, is a horizontal karst channel filled with grey kaolinitic-illitic clays containing dispersed vertebrate bones and teeth transported by water. The presence of *Conohyus simorrensis* (LARTET) points to the Badenian (KUBIAK, 1982). According to MEIN's (1976) Neogene mammalian zonation (abbr. MN) it belongs to zones MN 5-6 (GŁAZEK and SZYNKIEWICZ, 1987).

(2) Przeworno 2 is a vertical fissure filled with illitic clay containing marble blocks and vertebrate remains slightly cemented with amorphous silica. Occur-

2–33. Middle Miocene siliceous flowstone in the cave floor of site Przeworno 3, coin diameter 3 cm (Photo by J. Głazek).

2–34. Middle Miocene diving beetle *Acilius cf. sulcatus* (L.) in the siliceous flowstone from the site Przeworno 3, x4 (Photo by B. Drozd).

rence of *Pliopithecus antiquus* (BLAINVILLE) and *Conohyus simorrensis* (LARTET) points to zone MN 6 (GŁAZEK and SZYNKIEWICZ, 1987).

(3) Przeworno 3 is a steep cave filled with siliceous flowstones (Fig. 2–33) containing diving beetles (Fig. 2–34), fishes, dragon-fly larvae and green algae.

These exceptional deposits were covered with laminated clays containing Middle Miocene pollen (GALEWSKI and GŁAZEK, 1973, 1978; GŁAZEK, GALEWSKI and WYSOCZAŃSKI-MINKOWICZ, 1977; SADOWSKA, 1977). The cave was probably a sinkhole draining the marshy surface of a Middle Miocene alluvial plain on which alkaline volcanic waters rich in silica became mixed with acid peat waters that carried the faunal remains; note the silicified mould inside a beetle carapace and the missing extremities (Fig. 2–34)

The Lower Badenian transgression onto the southern slope of the Holy Cross Mountains covered a diversified morphology and created Dalmatian type of shore line there for a short time (RADWAŃSKI, 1969). A small littoral cave is preserved and was exhumed at Lubania (GŁAZEK and RADWAŃSKI, 1970; Fig. 2–35).

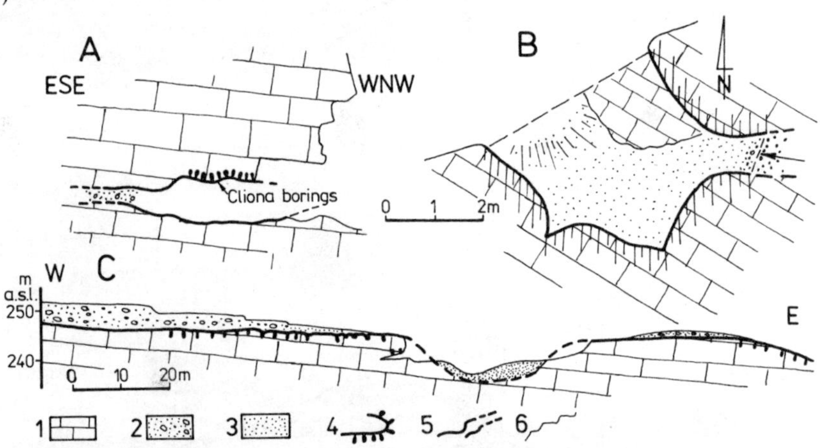

2–35. Littoral cave of Early Badenian age at Lubania on the southern slope of Holy Cross Mountains (cf. Fig. 2–36.; after GŁAZEK and RADWAŃSKI, 1970). A. Section; B. Plan (arrow marks section line); C. Geological section 1. Oxfordian limestones, 2. Badenian lithothamnian limestones, 3. Quaternary alluvial deposits, 4. Badenian borings on the abraded Oxfordian limestones; surfaces of limestones: 5. original, 6. secondary (due to subsequent erosion or excavation).

Numerous subrosion depressions filled with brown coal deposits of Oligocene and Lower Miocene age are associated with this pre-Badenian morphology (Fig. 2–25). The depressions were created or reactivated shortly after uplift as a consequence of groundwater flow down-dip via older paleokarst porosity with newly developed intrastratal porosity.

This situation is well illustrated and was recently dated in Bełchatów Coal Mine (Fig. 2–31). At the top of the coal measures were two layers of tephra intercalated with "lacustrine chalk" and a thin uppermost coal seam. The tephras were dated by the fission-track method by J. BURCHART (pers. comm., 1985) and by micromammals extracted from the "chalk" studied by V. FAHLBUSCH (pers. comm., 1985; BURCHART et al., in prep.). The age is approx. 17± 1.5 Ma (MN zones 4-6?). Together with numerous occurrences of Badenian foraminifera in clays overlying the coal-bearing deposits or erosional features (ŁUCZKOWSKA and DYJOR, 1971), that fixes the age of the upper boundary of this karstification phase (Fig. 2–36).

Borehole and speleological explorations in the Cracow-Wieluń Upland show

2–36. Neogene paleokarst localities in relation to contemporaneous deposits. 1. pre-Neogene rocks, 2. Oligocene-Lower Miocene fluviatile brown coal-bearing deposits of subsided continental margin, 3. alluvial-coastal plain deposits of Middle Miocene, 4. marine Middle Miocene of Carpathian Foredeep, 5. frontal overthrust of Carpathian nappes; fossiliferous karst localities: 6. Badenian, 7. Pliocene, 8. Lower Pleistocene, 9. remnants of caves buried by Badenian marine deposits, 10. sinkholes buried by Miocene deposits; a. Zn – Pb mine at Chrzanów, b. littoral cave at Lubania.

that karst water circulation was well organized in this phase up to the base of the Upper Jurassic limestones (i.e. over 300 m below surface and over 100 m b.s.l.); sinkholes filled with variegated siliciclastics exceed 100 m in depth and vertical caves display 50 m of relief. All these cavities, surface depressions and nearly all morphology of the Middle Polish Uplands were then filled with fine-grained clastics and covered with marine to alluvial plain deposits in Middle Miocene times (DŻUŁYŃSKI et al., 1966; GŁAZEK and SZYNKIEWICZ, 1987).

There is no site among the one hundred already known that contains vertebrate remains belonging to zones, MN 8–13, i.e. Upper Sarmatian-Pontian (GŁAZEK and SZYNKIEWICZ, 1987).

The third karst phase followed erosion of the Middle Miocene overburden as a result of the Messinian crisis and Pliocene uplift. Numerous exhumed caves reveal complicated sequences of bone-bearing deposits from Lower Pliocene to Early Pleistocene age (MN 14-17 and Q 1-2). Lengthy investigations of material excavated by J. Samsonowicz demonstrate the cyclic climatic changes that occurred in the Pliocene and Early Pleistocene (Fig. 2–37) before continental glaciations reached the Mid-Polish Uplands (SAMSONOWICZ, 1934; SULIMSKI, 1964; GŁAZEK, SULIMSKI and WYSOCZAŃSKI-MINKOWICZ, 1975).

Numerous karst features in the uplands are filled with red deposits and contain similar vertebrate faunas of Ruscinian (MN 14-15), Villányian (MN 16-17) or Biharian (Q 1-2) age, but in lesser quantities and in one or two layers (cf. Fig. 2–36).

Sulphur deposits in the northern slope of the Carpathian Foredeep also belong to this phase of karstification. A Middle Badenian gypsum was altered into cavernous marls in which the secondary porosity is partly filled with sulphur. This alteration is restricted to zones where the gypsum is below present river levels and covered by Middle Miocene clays 10–180 m thick. The sulphur was concentrated and preserved in a sequence of three distinct steps: (1) reduction of the original gypsum to H_2S by bitumens carried by groundwaters; (2) oxidation of H_2S to S by aqueous oxygen supplied in limited quantity or for a very short time, and (3) invasion by phreatic stagnant waters, which excluded the further oxidation of sulphur to gypsum (OSMÓLSKI, 1976). These steps can be interpreted as follows: (1) initial reduction to the Late Miocene, when the overthrust of the Carpathians onto Middle Miocene deposits expelled connate waters containing bitumens northwards into the shallow zone of the foredeep where the gypsum is interbedded with clastics and limestones; (2) slight oxidation to the Messinian entrenchment of rivers along the northern part of the foredeep, which reversed the direction of groundwater flow in its shallow parts via pre-existing paleokarst porosity, and (3) restoration of phreatic conditions to the Pliocene, when the joint effects of refilling the Black Sea Basin and the uplift of the Carpathian Foredeep halted the flow of fore-Carpathian rivers to the Black Sea.

The ensuing Pleistocene glacial epoch is characterized by short duration changes in the conditions of karst development. Each glaciation abruptly fossilized older karst forms or excavated them, depending upon the contemporaneous and local dynamics of water circulation. There is insufficient evidence

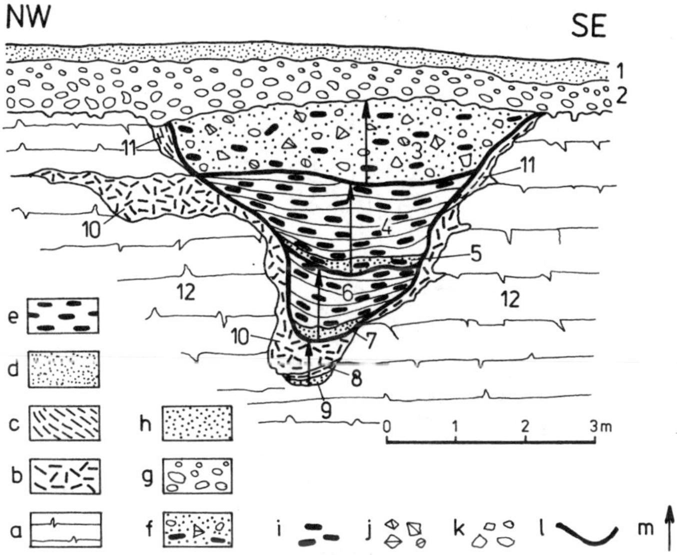

2–37. Cross-section of the Węże 1 site (modified after SAMSONOWICZ, 1934, and GŁAZEK, SULIMSKI and WYSOCZAŃSKI-MINKOWICZ, 1975). 1–12. Samsonowicz's bed numbering; a. Oxfordian limestones (12), b. stratified flowstone calcite (10), c. pink flowstone (8, 11), d. red calcareous sandstone (5, 7, 9), e. grey bone-bearing flowstone (4, 6), f. pink breccia with blocks of Oxfordian limestones and grey bone breccia (3), g. periglacial cover of rock debris (2), h. sandy soils (1), i. bones, j. blocks of grey bone-bearing flowstone, k. blocks of Oxfordian limestones, 1. erosional boundaries of sedimentary cycles in cave deposits, m. subsequent cycles of cave deposits.

for reasonably accurate correlation between karst and glacial events or to distinguish which karst phases represent particular interglacials. However, there are enough data to show that some spectacular changes in karst processes were caused by glaciations, as well as evidence in the karst that contributes significantly to our knowledge of the glacial events that occurred in Poland.

The karst relief and karst groundwater circulation were mostly fossilized during the glaciations. However, there were some prominent effects of meltwater attack and/or interglacial karstification.

The oldest glaciation recognized in cave deposits at Kozi Grzbiet (Figs. 2–36 and 2–38) belongs to the Upper Biharian in the European mammalian stratigraphy and to the lowermost Brunhes paleomagnetic epoch (GŁAZEK, LINDNER and WYSOCZAŃSKI-MINKOWICZ, 1976; GŁAZEK et al., 1977; BOSÁK et al., 1982). The first ice sheet reaching that particular area substantially changed the

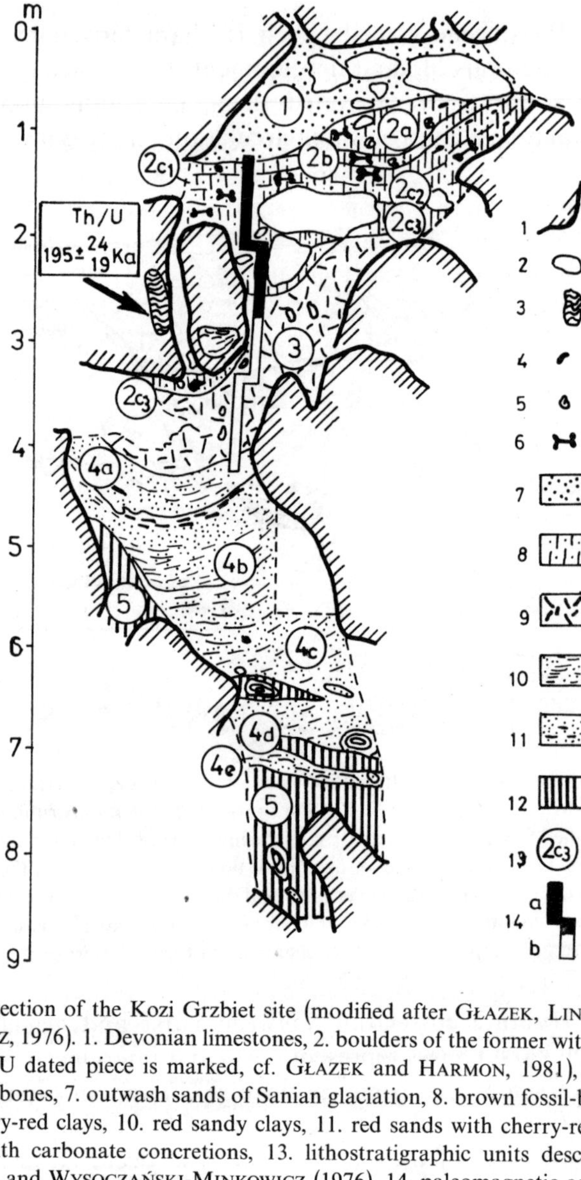

2–38. Geological section of the Kozi Grzbiet site (modified after GŁAZEK, LINDNER and WYSOCZAŃSKI-MINKOWICZ, 1976). 1. Devonian limestones, 2. boulders of the former within cave sediments, 3. flowstones (Th/U dated piece is marked, cf. GŁAZEK and HARMON, 1981), 4. older flowstone debris, 5. snails, 6. bones, 7. outwash sands of Sanian glaciation, 8. brown fossil-bearing cave loams, 9. brecciated cherry-red clays, 10. red sandy clays, 11. red sands with cherry-red clay streaks, 12. brownish clays with carbonate concretions, 13. lithostratigraphic units described in detail by GŁAZEK, LINDNER and WYSOCZAŃSKI-MINKOWICZ (1976), 14. paleomagnetic column (cf. GŁAZEK et al., 1977): a. normal magnetization (Brunhes epoch), b. reversed (Matuyama?). Note: beneath the material fossilizing the cave, which consists of fluvioglacial sands accompanied by frost-shattered limestone blocks (Unit 1), there are interglacial fauna-bearing layers (Unit 2) with marked climatic fluctuations (the warmer phases are 2b and $2c_2$) which contain Scandinavian material. Frost-shattering is visible as limestone blocks in the base of Unit 2 (layer $2c_3$) and brecciation of red clays (Unit 3). Units 3–5 represent Pliocene (?) filling of a vertical cave of Messinian or pre- Badenian age.

composition of local clastic deposits as a result of the admixture of Scandinavian material derived from crystalline rocks. It serves as a prominent bench-mark in the differentiation of Pleistocene and Pliocene karst fillings. The other bench-mark is the common red colour caused by hematite in the Pliocene fillings, while a brown colour caused by goethite characterizes the Pleistocene fills.

After two continental glaciations marked by boulder clays, strong subrosion activity caused parallel subsidence along the northern side of the Bełchatów brown coal deposit (Fig. 2–31 C). The subsidence was active before the last three glaciations (two of which covered this area but the last, the Vistulian, did not extend so far). This structure, when discovered, was first interpreted as an "interglacial valley", then as evidence of an "endogene tectonic phase" during the Middle Pleistocene interglacial! Similar "grabens" (deeper than 100 m) are found on salt domes in central Poland. They display an effect of glacial deep water circulation systems that follow paleokarst paths.

Creative action by glacial meltwater is represented by the development of Szachownica Cave in the northern part of the Cracow-Wieluń Upland (GŁAZEK, RUDNICKI and SZYNKIEWICZ, 1977; GŁAZEK et al., 1979). This cave system has a total passage length of more than one kilometre. It developed in a fracture net during the melt of the Wartanian glacier on the top of a limestone

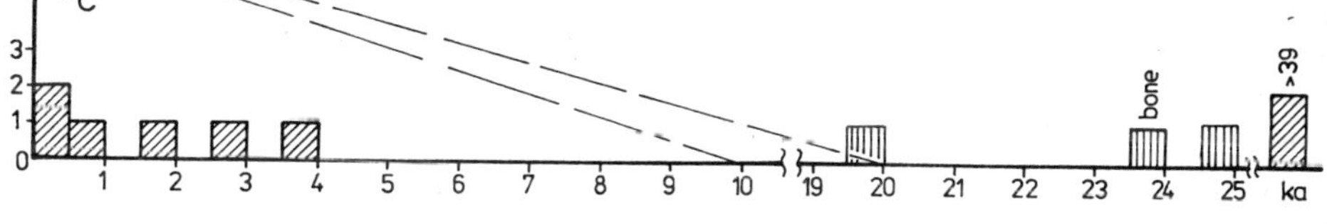

2–39. Isotope dating of speleothems and bones from caves of Poland. Uranium series dating of speleothems from all karst areas (after GŁAZEK and HARMON, 1981, in print) show discontinuous Pleistocene speleothem precipitation. Radiocarbon dating of flowstones and bones from the Magura Cave, Tatra Mountains (after HERCMAN et al., 1987) shows the last glacial non-deposition between 11 and 23 Ka.

hill that was nearly completely buried beneath outwash sands. The cave was fossil during the Upper Pleistocene and Holocene.

Available results of speleothem dating (Fig. 2–39) from caves scattered throughout the Polish karst areas (GŁAZEK and HARMON, 1981, in press; T. C.

ATKINSON, pers. comm., 1985) show discontinuous speleothem deposition and some temporal constraints on the development of the karst landscape, as well as suggesting timing of glacial events.

Spacious caves of the Cracow Upland that were thoroughly studied for archeological purposes (MADEYSKA, 1969, 1981, 1982) have revealed major episodes of excavation in the Pleistocene. These caves of supposedly Middle Miocene age contain only remnants of red allochthonous sediments of the Pliocene in vadose entrenchments, and huge Upper Pleistocene deposits. The latter were dated by the ^{14}C technique to 38 160 $\pm$ 1 250 y B.P. (GrN 2181) and contain broken stalagmites dated by the $^{230}Th/^{234}U$ technique to 110–130 ka (GŁAZEK and HARMON, 1981, in press).

The Pleistocene glaciations in the High Tatra Mountains caused numerous changes in the hydrologic environments of caves but only a few of these changes are dated and correlated with glacial events. However, available speleothem and bone dating (GŁAZEK and HARMON, 1981; GŁAZEK, 1984; HERCMAN et al., 1987; RÓŻAŃSKI, pers. comm., 1985) sedimentological data (WÓJCIK, 1966), and speleological reconstructions (RUDNICKI, 1967; GŁAZEK, RUDNICKI and SZYNKIEWICZ, 1977) reveal that: (1) the last (and most severe) glaciation in these mountains took place between 11 and 23 ka (cf. Fig. 2–39); (2) since speleothems older than this glaciation occur in active caves near valley bottoms and speleothems older than 300 ka occur approx. 40 m above the valley bottoms, these horizontal caves are considerably older than was expected and the glacial entrenchment of the valleys is considerably less than was supposed; (3) stagnant glacial water filled many high-lying, older caves (up to 200 m above the valley bottom); (4) many vertical caves were developed through older horizontal caves as a result of the drainage of stagnant and proglacial waters.

Last but not least, because the Tatra crystalline rocks have fission-track cooling-ages of approx. 15 Ma it seems that the general erosional dissection of the mountains was of Messinian and Pliocene age, while the glacial Pleistocene was only responsible for minor landscape modification of the already dissected mountains!

Paleogeographic conditions were highly variable during this last, long karst period. Wet and mildly moderate, passing into subtropical climates may be suggested for the first phase (Paleocene to Lower Eocene). The next karst phase (Oligocene to Lower Miocene) began with a strong deterioration of the climate (BUCHART, 1978; CAVELIER et al., 1981). Probably it was a wet, moderate climate gradually passing in the Early Miocene into mild subtropical conditions. Both Tertiary climatic optima (Middle Eocene and Middle Miocene) coincide with marine transgressions and so were not important for karstification in Poland.

The third karstification phase began in Late Miocene times with considerable drying and fall of temperature. The climate was continental, even semi-arid, during the Messinian as is shown by the presence of typical inhabitants of the African desert margins (*Gerbillinae, Hystrix*) in the lowermost Pliocene faunas (MN 14, e.g. Węże 1 and Podlesice sites, KOWALSKI, 1974; SULIMSKI, 1960). It was the Messinian dessication and the rise of the Alpine-Carpathian barrier that together created markedly continental climates throughout the whole of Europe, with the introduction of African and Asiatic semi-desert and steppe micromammal fauna. They dominate in the bone-bearing Pliocene karst sites of Poland (GŁAZEK and SZYNKIEWICZ, 1987).

The last karstification phase is characterized by strongly alternating climates, moderate in interglacials and cold in glacials.

Practical Consequences

Many important mineral deposits in Poland are associated with the operation of paleokarst processes. Hydrothermal karst of Mesozoic age is responsible for the most important zinc-lead deposits in the Cracow-Silesia ore district. These are discussed by DŻUŁYŃSKI and SASS-GUSTKIEWICZ, p. 377–397.

Important sulphur deposits in the Carpathian Foredeep were developed by paleokarst water circulation during the Neogene. Most of the brown coal deposits in Poland were also associated with paleokarst intrastratal dissolution of Permian evaporites. These deposits filled subrosion subsidences that resulted from deep regional groundwater circulation conveyed principally within the paleokarst porosity of the Zechstein carbonates.

Some small oil and gas deposits have been found in karstified Oxfordian limestones. These limestones form buried "hills" beneath Badenian deposits in the Carpathian Foredeep. Other minor deposits of oil and gas are found in the Zechstein carbonates in the Polish Lowland. The oil accumulated in secondary porosity that developed as a result of syngenetic and post-diagenetic paleokarst.

Many small deposits of lead, copper and iron ores were exploited over the centuries in karstified Devonian carbonate host-rocks in the Holy Cross Mountains (RUBINOWSKI, 1971). Numerous deposits of clays, moulding sands and iron ores were or are still exploited in karst depressions of Tertiary age in different regions of Poland (BOSÁK et al., 1979).

Some catastrophic subterranean floods have been caused by long distance groundwater systems flowing through paleokarst cavities in: (1) Lower Silesian copper mines (KRASOŃ and WÓJCIK, 1965); (2) the coal mines of the Lublin Upland, and (3) the zinc-lead mines in Cracow-Silesia ore district (see WILK, p. 513–531) and many open-cast mines and quarries.

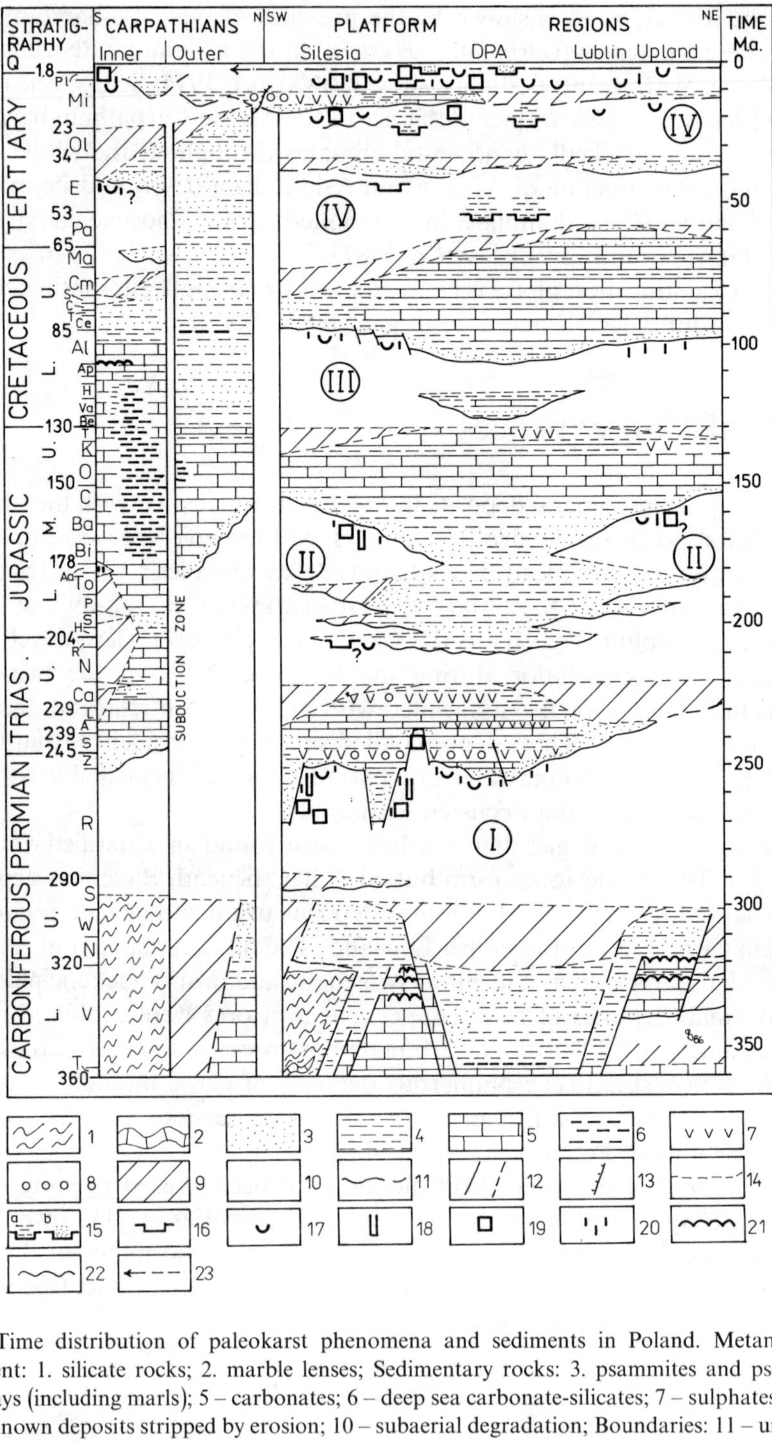

2–40. Time distribution of paleokarst phenomena and sediments in Poland. Metamorphosed basement: 1. silicate rocks; 2. marble lenses; Sedimentary rocks: 3. psammites and psephites; 4. silts-clays (including marls); 5 – carbonates; 6 – deep sea carbonate-silicates; 7 – sulphates; 8 – salts; 9 – unknown deposits stripped by erosion; 10 – subaerial degradation; Boundaries: 11 – unconform-

Conclusion

Paleokarst is a rich and complex phenomenon in Poland. Its study has a long tradition and is of growing importance. The most striking paleokarst in this country developed in post-Variscan (until Middle Triassic) and in Tertiary times. The distribution of paleokarst phenomena in the geological past and in geographic space is shown in Figs. 2–25, 2–27, 2–36 and 2–40.

---

able cover; 12 – synsedimentary faults; 13 – synsedimentary overthrusts; 14 – supposed limits of sedimentation; 15 – subrosion depressions filled with contemporaneous deposits (a. brown coal – bearing, b. drift deposits); 16 – poljes; 17 – sinkholes; 18 – shafts; 19 – caves; 20 – minor solution forms; 21 – syngenetic karst; 22 – karst corrosion surfaces; 23 – maximal extent of Pleistocene glaciers (Scandinavian ice-sheet in platform regions); I, II, III and IV are periods of karstification; time scale according to ODIN (1982).

PALEOKARST OF CZECHOSLOVAKIA

Pavel Bosák, Ivan Horáček and Vladimír Panoš

Introduction

The modern topography of the karst regions of Czechoslovakia consists of landforms that have been intricately and unequally developed during many phases of changing climate and of tectonic activity in the geological past. The problem of the dating of these forms has been studied intensively. The complete bibliography of Czechoslovakian paleokarst contains more than 300 entries dealing mostly with the paleokarst of the Bohemian Massif. Paleokarst research in the West Carpathians is only just beginning. The very complex form of the karst and paleokarst, with relict, fossil, recent, active or rejuvenated features and in periglacial conditions transformed karst, led PANOŠ (1964a) to define it as the *Central European Type* of polycyclic karst.

The investigation of paleokarst and related fills started with a study by ANDRÉ (1821) on clays, sands and iron ores in deposits known as the Rudice Formation in the Moravian Karst. REICHENBACH (1834) was the first to recognize these deposits as being karst-related (see also BOSÁK, p. 439–441). SAWICKI (1908) was first to interpret phases of development in the Slovak Karst.

There are no general reviews of the paleokarst except for the eastern margins of the Bohemian Massif (PANOŠ, 1962, 1964a, 1965) and the very short synthesis of BOSÁK and HORÁČEK (1981).

Terminology

In Czechoslovakia the term *fossil karst*, has been adopted to describe ancient karst forms. It has been understood in two ways; fossil karst *sensu stricto* and *sensu lato*. Fossil karst *s.s.* was defined by PANOŠ (1978) as karst forms which originated in the past, completely fossilized without any traces of further development. Fossil karst *s.l.* designates karst forms of possible polycyclic origin

P. Bosák, D. C. Ford, J. Głazek and I. Horáček (Editors):
Paleokarst. A Systematic and Regional Review. – Elsevier and Academia. Amsterdam and Praha. 1989.

with more phases of development, fossilization and rejuvenation, including also relict karst landforms. This broader definition of fossil karst will be used in our discussion of the Czechoslovak paleokarst.

The Bohemian Massif

Czechoslovakia consists of two important Central European geological and geotectonic units; the Bohemian Massif in the West and the West Carpathians in the East. Their boundary at the surface crosses Moravia. At depth, the structure of the East European Platform, the Bruno-vistulicum (DUDEK, 1980), extends farther southeast below the Carpathians, as a consequence of Alpine nappe overthrusting over the epi-Variscan platform in the foreland (GRECULA and ROTH, 1978) (Fig. 2–41).

The geological development of the Bohemian Massif was complex and of long duration. It culminated in the Variscan Orogeny with the emplacement of numerous plutons and extensive regional metamorphism. The Massif consists of three structural stages: (1) Precambrian (Cadomian) basement; (2) the Variscan (Paleozoic) unit, and (3) a post-Variscan platform cover (SUK et al., 1984).

Post-Variscan platform evolution (i.e. since Upper Permian or Lower Triassic times) is of dominant importance for the origin and preservation of paleokarst (Fig. 2–50). Pre- or syn-orogenic karstification phases appear to be less important. The stages of evolution of the platform cover reflect the individual orogenic phases in the adjacent Alpine-Carpathian domain and are as follows: (1) Upper Jurassic, corresponding to the platform development of the West Carpathians; (2) Upper Cretaceous, corresponding with the principal Alpine folding; (3) Paleogene, characterized by intense planation; (4) Lower to Middle Miocene, associated with the folding of the Outer (Flysch) Carpathians, and (5) Upper Miocene to Quaternary, the stage of platform cover in the Carpathians (MALKOVSKÝ, 1979). These stages of platform cover development correlate closely with individual karstification periods or phases.

The Bohemian Massif experienced a long period of emerged continental conditions after the great Permo-Carboniferous deposition. The period lasted until the Upper Cretaceous (Albian-Cenomanian). This was an anorogenic period characterized by fault tectonics. The Upper Jurassic transgression and sedimentation was only a short episode in the Massif, except at its southeastern margin. There, up to 2 000 m of carbonates and marls were deposited (ELIÁŠ and ELIÁŠOVÁ, 1984). The post-Cretaceous, (i.e. post-Santonian) period has been characterized by more intensive tectonics since the Upper Eocene. There are several phases of volcanic activity (35–17 Ma, 9–6 Ma, 2.8–0.86 Ma; KOPECKÝ in SUK et al., 1984) accompanied by tectonic movements, which broke a levelled surface of Mesozoic-Lower Paleogene age. Beginning in the early

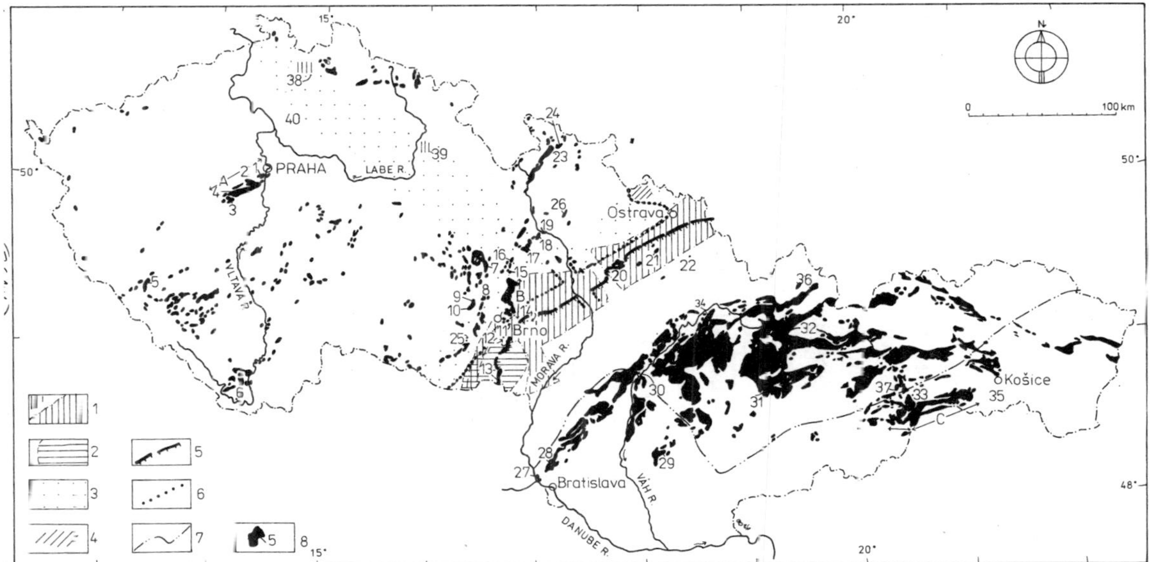

2–41. Karst areas in Czechoslovakia. 1 – subcrops of Devonian to Lower Carboniferous carbonates; 2 – subcrops of Upper Jurassic carbonates; 3 – hypothetical extent of the Upper Jurassic sea strait (after ELIÁŠ in SUK et al., 1984); 4 – Miocene gypsum; 5 – boundary of the Outer Carpathians (i.e. of flysch nappes); 6 – western boundary of Miocene Carpathian Foredeep; 7 – limits of North European Platform below the Carpathians (after GRECULA and ROTH, 1978); 8 – karst areas, numbers refer to sites discussed in the text; A = Bohemian Karst; B = Moravian Karst; C = Slovak Karst.

Savian phase, the Bohemian Massif was strongly uplifted as a whole and thereafter individual blocks were affected by differential movements until Quaternary times. There was just one marine transgression, during the Badenian, which invaded the eastern margins of the Massif only.

Karst Rocks

Karst areas, measured on outcrops of carbonate rocks, in the Bohemian Massif total nearly 300 km^2 (Fig. 2–41). The oldest limestone and dolostone are of Upper Proterozoic age and belong to the Varied Group of the Moldanubicum. At the present time, however, there is discussion concerning the precise age of the Group due to indications of Lower Paleozoic microfossils (cf. SUK et al., 1984). These carbonates occur mostly in South Bohemia and South Moravia. The most extensive karst rocks are Lower Paleozoic limestones and subordinate dolostones. They are Upper Silurian-Middle Devonian (Přídolian to Givetian) in Bohemia and Middle Devonian-Lower Carboniferous ((?) Eifelian to Tournaisian) in Moravia. Some are metamorphosed; those in North Moravia, North Bohemia and partly also in Central Bohemia. Upper Tertiary carbonates and evaporites (Karpatian and Badenian) occupy some small areas in Moravia (Fig. 2–41).

Large subcrops of Middle Devonian-Lower Carboniferous (ZUKALOVÁ and CHLUPÁČ, 1982) and Upper Jurassic (Callovian-Upper Tithonian; ELIÁŠ, 1981) limestones and dolostones are covered by up to 5 km of Tertiary flysch nappes and foredeep siliciclastics in the eastern margin of the Massif (Fig. 2–41). Exploration drilling for oil here proves several thousand square kilometres of subcrop with 1 500 m thick Devonian and 2 000 m thick Jurassic carbonates. Some Upper Devonian limestones have also been drilled beneath the Bohemian Cretaceous Basin at Hradec Králové (site 39, Fig. 2–41) in East Bohemia (CHLUPÁČ and ZIKMUNDOVÁ, 1976) and Česká Lípa (site 38) in North Bohemia. Finally, a part of the Lower Paleozoic in the Barrandian is covered by Upper Cretaceous rocks.

Paleokarst of the Bohemian Massif

The origin, development, fossilization and rejuvenation of the karst features have been affected by lithological and structural conditions, and by (1) the prevalence of a continental regime since the Permian; (2) marine transgressions in Late Jurassic, Late Cretaceous and Middle Miocene times; (3) increasing tectonic activity since the Late Paleogene, which caused also; (4) disappearance of the Paratethys Sea; (5) increasing relief and ruggedness resulting in a considerable increase of river erosion rates, so causing exhumation of old surfaces and rejuvenation of older karst features, and (6) a continual fall of the base level of erosion following stabilisation of the river systems in the Pliocene (Bosák and Horáček, 1981).

Middle Devonian Paleokarst

The oldest known paleokarst forms were detected in boreholes in the **Moravian Karst** (Josefov HV 104) and in the **Carpathian Foredeep** (Švábenice 1, site 14, fig. 2–41; Dvořák, 1978). Cyclic Lower Givetian deposition permitted the brief emergence of limestones. There are traces of freshwater vadose diagenesis and some surficial corrosion forms resembling coastal karren. They were buried by the siliciclastics of the next marine cycle.

Near Koněprusy village (site 3) in the **Bohemian Karst**, there are numerous vertical cracks up to 60 m deep in a Lower Devonian (Pragian) reef. They are filled

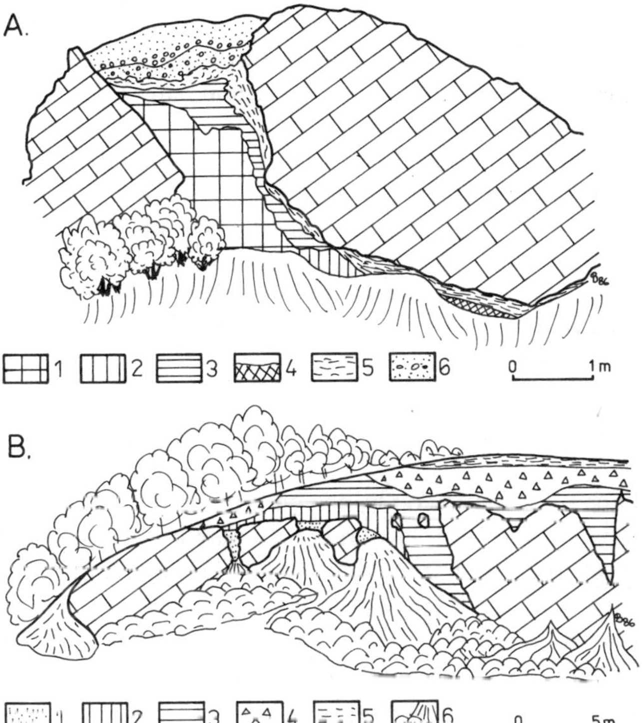

2–42. Polycyclic karst forms near Čebín in the Tišnov Karst (after Bosák, 1980a, 1983). A. The Čebínka Hill. 1. Lower Carboniferous greywackes; 2. Lower Carboniferous sandstones; 3. Permian red sandstones; 4. Permian red clays; 5. Badenian calcareous clays; 6. Quaternary deposits. B. The Dálky Hill. 1. Permian red clastics; 2. ? Paleogene weathering products; 3. Badenian sands and calcareous clays; 4. Quaternary deposits; 5. soils; 6. heaps.

with Lower to Middle Devonian (Dalejan to Givetian) limestones. The fissures have been interpreted as the results of gravity slumping of the reef without any effects of karst corrosion (CHLUPÁČ, 1967). A large system of neptunian dykes and extensive flat stromatactis in Lower Devonian (Dalejan) limestones are found in the same area. They have not been studied yet in detail, but dissolution by freshwater lenses or in salt/freshwater mixing zones cannot be excluded.

Upper Devonian to Lower Carboniferous Paleokarst

Regional regression with the emergence of individual tectonic blocks started with the early phases of the Variscan Orogeny within the Devonian depositional area in Moravia. The regression appears in the North (Ostrava region) and in the West (Tišnov region) at the Frasnian/Fammenian boundary, in the Moravian Karst during the Fammenian and in the Carpathian Foredeep in the Early Tournaisian. The tropical or subtropical climate was favourable for karstification. Irregular karst depressions and dolines developed and were filled with shales and greywackes of Lower Visean (Tišnov area, site 8; BOSÁK, 1983; Fig. 2–42A) or Upper to Late Visean (Ostrava region, site 22; ADÁMEK, DVOŘÁK and KALVODA, 1980). In the Moravian Karst (area B) they are filled with Upper Tournaisian organodetrital limestones and limestone breccias with phosphatic nodules dated by conodonts (VALKOVIČOVÁ, 1979; DVOŘÁK and FRIÁKOVÁ, 1981). This brief karstification followed the deposition of thick reef limestones and took place before its deep submergence beneath flysch deposits. This history has been described by SCHLAGER (1981) as a characteristic of drowned carbonate platforms that is very widely distributed in geological history.

Late Carboniferous to Permian Paleokarst

Since the changing climatic conditions with marked tropical humid and subtropical arid phases were not favourable for karstification, only minor karst features have been preserved from this period. Corroded and slightly widened fissures plus traces of irregular karst surfaces are noted around Čebínka Hill in the **Tišnov area** (site 8; BOSÁK, 1980a, 1983). The karst surface is covered by continental red beds of the Boskovice Trough (Stephanian C to Autunian; Fig. 2–42B).

Hydrothermal paleokarst is associated with the invasion of hot solutions at the end of the Variscan Orogeny. A small cave filled with limonite was described at Květnice Hill in the Tišnov area (site 9; SLEZÁK, 1954). Hydrothermal activity is responsible also for thick baryte-calcite-quartz-ankerite vein fills in Květnice and Dřínová Hill and for the origin of metasomatic sphalerite-galena mineralization in limestones near Heroltice (site 10). Some baryte veins fill substantially widened fissures in the limestones. A role for karst solution in the emplacement of local lead-zinc ores has not been demonstrated.

In **Bohemia**, this karstification phase has not been distinguished. A former, greater extent of Upper Silurian-Middle (and in places Upper) Devonian carbonates is proven by the mass occurrence of carbonate pebbles in the Permo--Carboniferous continental sediments of the Central Bohemians Coal Basins (namely, in the Mšeno Basin, site 40). The Paleozoic carbonates were destroyed by intense, deep erosion; probably there was also karstification within this period of erosion, as indicated by pebbles of speleothems.

Permian to Upper Jurassic Paleokarst

The upper limit of this period of subaerial planation is represented by an uppermost Middle Jurassic (Callovian) transgression in the eastern margins of the Bohemian Massif. Jurassic deposits also filled a supposed, relatively narrow corridor extending from Moravia northwest into Germany, the Moravian Strait which connected the Tethyan and Boreal realms. In emerged areas (Fig. 2–41) karstification lasted until the Upper Cretaceous transgression began. It is this phase, in which several macroforms originated, that can be reliably recognized as the oldest inherited karst forms still affecting the relief of the Bohemian Massif.

The **Moravian Karst** represents one of these basic macroforms. Before the Callovian marine transgression, a large even-bottomed longitudinal depression, with the characteristics of a complex semi-polje, had developed due to differential erosion of the Devonian-Lower Carboniferous sedimentary mantle which overlies the eastern flank of the domed granodioritic body between Brno and Boskovice (PANOŠ, 1962–63). The depression was excavated in Lower Carboniferous shales and greywackes and the underlying Devonian limestones were exposed in its levelled bottom. It became the southern part of the Callovian sea corridor (channel) mentioned above. Upper Jurassic, mostly shallow marine, carbonates (BOSÁK, 1978b; HANZLÍKOVÁ and BOSÁK, 1978) buried this depression with a continuous cover whose thickness is estimated to be 200 m (KETTNER, 1960).

At the **Hranice Karst** (site 20), very old forms developed in a large tectonic block of Devonian to Lower Carboniferous limestones. Another macroform of Mesozoic or even Upper Paleozoic age is represented here by a buried karst plateau which extends over a large neotectonic block of anticlinally folded Paleozoic at the southeastern margin of the Moravian Gate near Hranice (KODYM, 1960). The fossil levelled karst surface, covered with original kaolinitic weathering products, is furrowed by deep canyon-like karst valleys, marginal and internal poljes and large dolines surrounded by groups of karst towers or isolated inselbergs (Fig. 2–43A). The deep dissection of the karst plateau during Early Cretaceous or Early Paleogene times cannot be excluded. The typical humid tropical paleokarst landscape was completely fossilized by Karpatian and Badenian transgressions. It has attained the form of a mature holokarst.

Pre-Autunian karst phenomena have not yet been detected in the **Bohemian Karst** (area A) although several breaks in deposition of Paleozoic sediments are known. There are no paleontological data that indicate Permian, Triassic or Jurassic karstifications. TURNOVEC and PECHOVÁ (1980) described a large paleo-depression near Zadní Kopanina village (site 2), southwest of Prague,

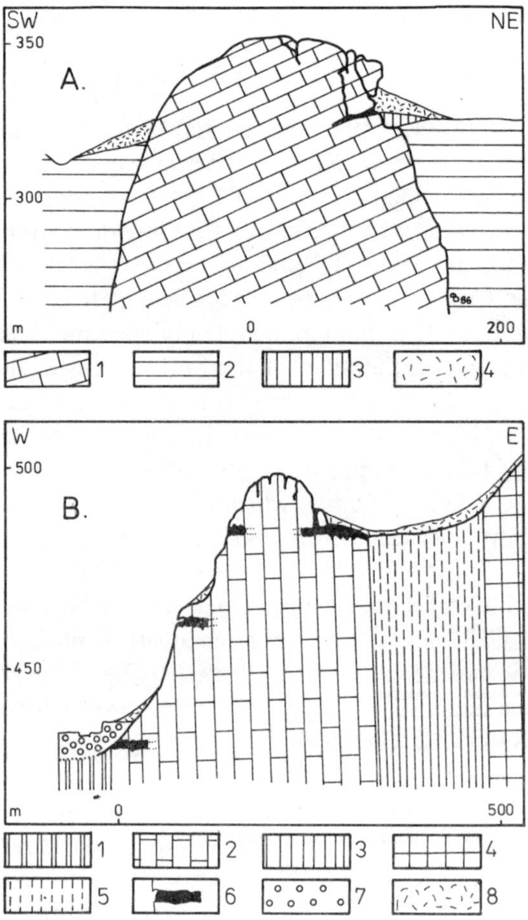

2–43. Cross-sections of fossil karst towers. A. Velká Kobylanka in the Hranice Karst (modified after TYRÁČEK, 1962). 1. Devonian limestones; 2. Miocene (Badenian) sediments; 3. Late Pleistocene loess; 4. debris. B. Vratíkov (after PANOŠ, 1964a). 1. Devonian shales; 2. Devonian limestones; 3. Devonian to Lower Carboniferous shales with radiolarites; 4. Lower Carboniferous greywackes; 5. laterite weathering products of shales (Mesozoic weathering); 6. kaolinitic red clays (Paleogene?) 7. Pleistocene terrace sediments; 8. Pleistocene to Holocene loess and debris.

partly filled with fossil colluvial deposits (scree). Their age was interpreted as Carboniferous and Permo-Triassic. It is evident that the development of karst was long-lasting here but the majority of surviving forms are associated with the Lower Cretaceous karstification.

Lower Cretaceous Paleokarst

A majority of the pre-Cenomanian paleokarst features originated during the Late Jurassic to Early Cretaceous period. The acceleration of karstification was caused by young Cimmerian movements which rejuvenated the levelled Upper Paleozoic-Mesozoic surface. Also, the climatic conditions changed substantially to the sub-tropical or tropical, warm and humid type.

In the **Bohemian Karst** (area A), Cenomanian continental and marine and Turonian marine deposits fill and bury depressions in the morphologically diverse surface of the Paleozoic fill of the Barrandian Basin. Individual paleo-depressions have diameters of tens to hundreds of metres and depths of up to 30 m. They are known southeast of Prague (e.g. site 2; KODYM, 1923; TURNOVEC, 1980; ZELENKA, 1980). Boreholes at Dívčí hrady Hill in Prague (site 1) discovered a shaft-like depression in limestones more than 120 m deep and some tens of metres wide. It was filled with Cenomanian continental clays and sands (ZELENKA, 1984). The pre-Cenomanian relief in places has the character of a tropical cockpit karst with deep and morphologically complex depressions separated by sharply peaked hills and ridges. The infill, except for the above mentioned Upper Cretaceous deposits which are dated paleontologically, is

2–44. Paleokarst depression filled with kaolinitic sand in the lower part and Upper Cretaceous sediments in the upper part. Na Parapleti Quarry, Bohemian Karst (Photo by V. Lysenko).

comprised of variegated tropical weathering products. They have the character of kaolins, kaolinitic clays and sands that are similar to the Rudice Formation in the Moravian Karst. They are described at several places in the Bohemian Karst (Fig. 2–44; site 4; HOMOLA, 1950; RÖHLICH and CHLUPÁČ, 1949–50).

HOMOLA (1950) and LOŽEK (1963) did not exclude the possibility of cave formation in this period but there is no record of caves having been preserved. However, very deep vertical fissures filled with Cenomanian clays are known at numerous places in the Bohemian Karst (LOŽEK, 1963). The karstification was interrupted by the Cenomanian transgression. The extent of pre-Cenomanian karstification was possibly greater than visible in the recent times, the features either being buried by Cretaceous deposits or destroyed by post-Mesozoic planation.

Lower Cretaceous paleokarst in the **Moravian Karst** (area B) has been the subject of nearly 60 scientific studies. Intense tropical weathering and karstification started with the Upper Jurassic (Kimmeridgian) regression. Jurassic carbonates that infilled the pre-Callovian complex semi-polje were destroyed; only small relicts are preserved in Blansko Graben, which was already active at this time. A cupola-shaped karst surface with local conical hills and cockpits as well as sand pipes or geological organs, developed (PANOŠ, 1962–63; DVOŘÁK, 1953; BURKHARDT, 1974; BOSÁK, 1978a, 1980b, 1981a). Sediments of the Rudice Formation infilled these forms (BOSÁK et al., 1979; BOSÁK, 1980b). Old mine workings indicate the existence of horizontal caves also filled with the sands and clays, and with iron ores. This supports the conjecture that the Lower Cretaceous paleokarst at the base of the pre-Callovian depression drained northwest

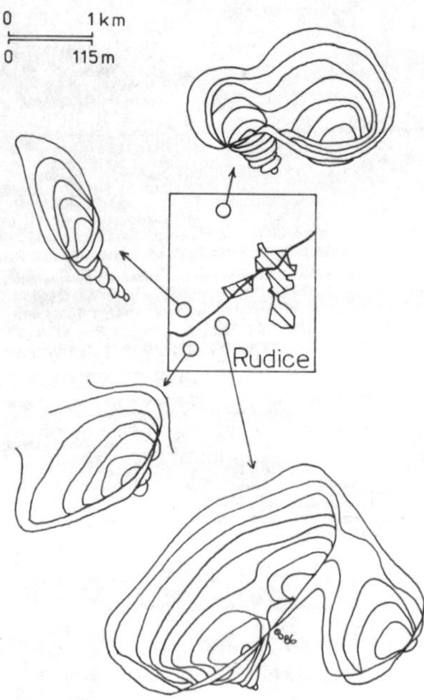

2–45. Supposed shape of some Lower Cretaceous paleokarst depressions filled with the Rudice Formation near Rudice, Moravian Karst (after BOSÁK, 1980b).

into the Blansko Graben and south towards Brno where relics of Aptian-Albian marine deposits with *Hedbergella* were found (Kuřim, near site 8; KRYSTEK, 1972), thus proving the existence of an ancient valley system there.

The deepening of the karst depressions was synchronous with deposition of the Rudice Formation. Prior to its deposition, there were only some shallow sinkholes filled with fossil scree and soil (Fig. 2–45). There was some subsidence after Late Cretaceous times, also, as proven by Cenomanian clays that subsided into some depressions (PELÍŠEK, 1976; BOSÁK, 1980b). The extent of the Rudice Formation was much larger than it is today. It covered most of the pre-Callovian depression, at least in the present central and northern part of the Moravian Karst.

A morphologically diversified relief, probably of Upper Paleozoic to Mesozoic age, developed in the zone of the supposed Callovian sea corridor, and is preserved on Devonian limestones between **Němčice** (site 15) and **Vratíkov** (site 16), some kilometres north of the Moravian Karst. These lateritic weathering layers cover a Lower Cretaceous or older conical karst. The karst cones are up to 50 m high and there are laterite layers several tens of metres thick (PANOŠ, 1962, 1964a; Fig. 2–44B). The ancient karst surface is dissected by deep pipes or geological organs with red kaolinitic clays and iron ores at the base. The iron was mined in the past. A part of these ancient karst landforms is covered by Cenomanian deposits of the Valchov Graben, which are redeposited sediments from the Rudice Formation, indicating that the Formation extended to the North of the Moravian Karst.

In the vicinity of Čebín (site 8) in the **Tišnov Karst** there are relics of an exhumed Lower Cretaceous or older karst plateau with cockpit-like depressions and rounded small hills. Relics of caves are rare. Depression fillings are sandy-clayey deposits similar to the Rudice Formation, plus some filled by variegated weathering products. Some kilometres to the south, near **Moravský Krumlov** (site 25) there are small pockets in Devonian limestones filled with kaolinitic sediments and limonite iron ore at the base. They are interpreted as equivalents of the Rudice Formation (BURKHARDT, 1962).

Some kilometres north of Čebín, near **Kunštát** and **Olešnice** (site 7), vertical corroded fissures, fossil karren and other small cavities are filled with limonite-hematite iron ores, manganese oxides and hydroxides, covered by red clayey soil and kaolinitic clays. PELÍŠEK (1975) suggests that they are pre-Cenomanian in age.

In the **Carpathian Foredeep**, autochthonous Devonian and Upper Jurassic carbonates cover a large area of the crystalline basement (the Bruno-vistulicum). They are buried under Tertiary sediments and flysch nappes. In the frontal parts of the nappes, there are in places some thrust slices of Upper Jurassic carbonate rocks that have been displaced several tens of kilometres to the west. The autochthonous cover as well as the slices of carbonates show interesting karst forms.

Lower Cretaceous paleokarst occurs in **Turold Hill** (site 13) in the Pavlovské vrchy Hills near Mikulov. The karst surface displays shallow depressions and limonite-manganese crusts and is developed in highly dolomitized and pulverized carbonates (BOSÁK et al., 1983, 1984). This surface is disconformably covered by Upper Turonian-Coniacian siliciclastic deposits. Similar weathering horizons at the top of Jurassic carbonates were also noted at several other places below the Tertiary cover (Z. ROTH, pers. comm., 1979).

Lower Cretaceous karst developed on Upper Jurassic to Lower Cretaceous limestones at **Kotouč Hill** (site 21) near Štramberk. Two phases of Lower Cretaceous karstification are distinguished here (HOUŠA, 1976, 1978): (1) Upper Valanginian with minor karst forms, such as coastal karren; (2) Uppermost Hauterivian to Aptian with diversified coastal karst landforms similar to the recent coastal zone of Cuba (PANOŠ, 1964b; HOUŠA, 1976). A complex system of channel-like caves which become narrower inside the rock massif formed in emerged areas. Smaller cavities are filled with fossil speleothems or bedded calcareous internal sediment. Wave abrasion during the Albian transgression partly destroyed the karst forms and surviving depressions were filled with detrital conglomerates as a consequence of surf action (PROSOVÁ, 1952; PANOŠ, 1964b; HOUŠA, 1976) [N.B.: there is a long-lasting controversy between HOUŠA, whose model is given above, and ELIÁŠ (e.g. 1983). ELIÁŠ proposes that the Kotouč Jurassic to Lower Cretaceous is a large breccia of limestone blocks accumulated at the mouth of a submarine canyon. The Lower Cretaceous rocks, for that reason, represent a matrix between the blocks and not the fillings of a Lower Cretaceous coastal karst].

Autochthonous Devonian and Jurassic carbonates under the Tertiary cover are highly karstified in places. They show conical karst topography. Exploration drilling for oil proved the existence of large buried karst towers in the Devonian near **Měnín** (site 12) and on Jurassic carbonates in the southeast suburbs of **Brno** (site 11). Here the summits of two extensive karst inselbergs (buttes) with nearly vertical walls more than 50 m high, project from the Badenian cover (BEZVODOVÁ and ZEMAN, 1983). Similar karstic elevations of the buried surface over the autochthonous Mesozoic strata in the area of the Carpathian and Alpine Foredeep are described by SCHOTT and STOPPEL (1976) in Austrian territory. The origin of this relief from karstification is here interpreted to be from the Early Cretaceous karstification phase, although its continuous development into the Paleogene cannot be excluded.

Paleogene to Lower Miocene Paleokarst

A large part of the Bohemian Massif was covered by Upper Cretaceous (Cenomanian to Santonian), mostly clastic deposits. However, the deep tropical weathering, indicated by intensive kaolinization and lateritization (STÖRR, KUŽVART and NEUŽIL, 1978) at least partly destroyed the Upper Cretaceous cover.

A long period of planation between the Paleocene and the Oligocene created a new Paleogene levelled surface in exhumed areas (DEMEK et al., 1965; SUK et all., 1984). Later, Upper Eocene to Oligocene and Upper Oligocene to Lower Miocene tectonic block movements broke and dislocated the levelled surface. This created the conditions necessary for a new phase of accelerated karstification; it increased relief enabling effective subsurface water circulation. The development of karst features during this phase was very complex and it remains unclear from many perspectives.

In the **Bohemian Karst**, new paleontological and geological data indicate that the main phase of evolution of the known caves occurred during the Oligo-Miocene period. Pollen and relicts of volcanic ash discovered in silicified speleothems in the Koněprusy Caves (site 3; HALBICHOVÁ and JANČAŘÍK, 1982–83) are dated to Oligocene-Early Miocene. This indicates that the largest cave system of the region, totalling 2 km in length, already existed at this time. Volcanic ash was transported probably by air from the Doupov stratovolcano, 80 km southwest of Koněprusy. The volcanic activity is dated as 35–17 Ma (KOPECKÝ, 1978) with the main eruptive phase being during Oligocene times.

LYSENKO (1982) suggests that a uniform cave level developed throughout the whole Bohemian Karst at this time. The oldest speleothems, Koněprusy Rosettes, result from deposition in a subterranean stagnant lake. The speleothems were syngenetically or early postgenetically silicified by uranium-bearing opal. It is difficult to date this process precisely. However, it must have been connected with intensive silica release by superficial weathering. The pollen mentioned above indicates Early Miocene speleothem formation and silicification.

The beginning of the karstification phase in the **Moravian Karst** is similarly characterized by deep weathering and erosion of the Upper Cretaceous cover (PANOŠ, 1962–63). There are two rather controversial models of cave development in the area between Brno and Boskovice (the Moravian, Němčice and

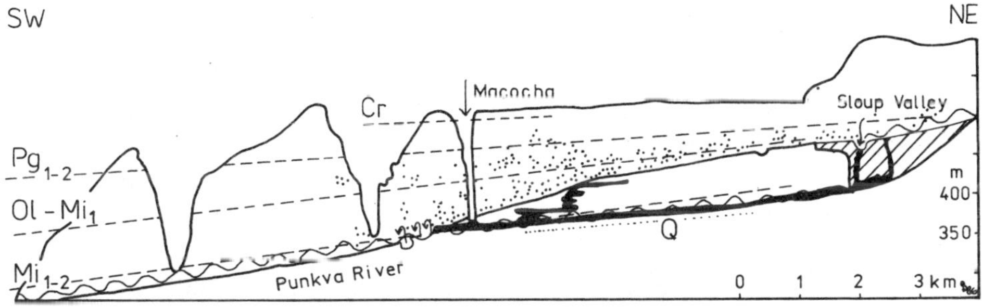

2–46. Schematic section of the northern part of the Moravian Karst along the Punkva River and the outline of concepts of the Cenozoic cave development (modified after BOSÁK and HORÁČEK, 1981). dots = cave openings; wavy line = active river bed of the Punkva River; dashed line = base of erosion in the respective period (Cr = Lower Cretaceous, Pg_{1-2} = Lower to Middle Paleogene; $Ol\text{-}Mi_1$ = Oligocene to Lower Miocene, Mi_{1-2} = Lower to Middle Miocene, Q = Quaternary).

Vratíkov Karsts). The first model, proposed by PANOŠ (e.g. 1962–63, 1970) describes the sequential development of cave levels and the origin of inserted cave systems inside older levels during the exhumation of the Badenian valley fill. HYPR (1982), and BOSÁK and HORÁČEK (1981) supposed rather more continuous development of the cave levels (Fig. 2–46). However, this model is schematic and does not explain completely the positive and negative changes in the base level of erosion in the developing Carpathian Foredeep.

2–47. Kůlna Cave, Moravian Karst (Photo by J. Přibyl).

The general character of the Moravian Karst — a fluviokarst — started to develop by the appearance of valley semi-poljes, large depressions near Sloup, Holštejn, Březina, Jedovnice, etc. They were drained by surficial streams even in the first stage of valley entrenchment. The nearly flat limestone bottom of the depressions represents the surface of the karst pediment originated by lateral corrosion. Shorter or longer relict caves (Kůlna, Fig. 2–47, and many others) were formed during lateral corrosion and erosion. Later lowering of the base level of erosion caused the dissection of the karst pediment by valleys. Deep canyon-like valleys are undoubtedly of pre-Karpatian age. They developed in four different stages, reflecting the tectonic regime of the Carpathians. Two main cave levels are situated 20 m apart. They originated during Late Oligocene to

Early Miocene times. The development of the valleys was characterised by stagnant periods, during which valley entrenchment halted. However, entrenchment did not stop along the whole length of a valley, since development continued from springs in its lower parts. Here deep canyon-like segments originated which are characterised by upstream enlargement in the second to fourth stage of valley development. Simultaneously, deeper and deeper entrenchments of blind valleys dissected the valley semi-poljes.

HYPR (1982) distinguished four phases of valley evolution associated with cave formation which started in Paleogene and halted in Karpathian. The upper level of horizontal caves is connected with the second stage of development and the lower level of the caves with the third stage. The Karpathian valley entrenchment caused "rejuvenation" of the main lower cave levels. A majority of the main underground cave networks belong to this lower level. Systems of the Amatérská Cave are developed with more than 30 km of passages, and the Býčí skála — Rudické propadání Cave system is over 12 km long. These long systems are correlated by PANOŠ (1963) with his lower cave level of Early Miocene age

Paleogene to Lower Miocene karstification probably rejuvenated some exhumed areas already karstified during Early Cretaceous times. This is the case of the Macocha Chasm where the upper entrance is believed to be a part of a Cretaceous or Paleogene karst depression. An extensive marginal semi-polje at the northern and eastern margin of the Moravian Karst (PANOŠ, 1963) can probably be correlated to the Lower Cretaceous deeply karstified zone as well.

Tišnov Karst is one of the most conspicuous examples of polycyclic karst (BOSÁK, 1983). Paleokarst forms of Paleogene to Miocene age are fossilized here either by Badenian marine or Karpatian freshwater deposits. In the vicinity of the villages of Čebín and Malhostovice (site 8), there is a buried karst pediplain with karst towers or inselbergs, partly exhumed from the Badenian cover. Their upper parts show traces of surf action with abrasion notches or caves (BOSÁK, 1980a). In Badenian abrasion breccias there is a relict of a syngenetic cave.

Large Paleogene to Lower Miocene paleokarst is covered by a complex freshwater sequence near Lažánky and Maršov (site 10). Geological organs and complex doline-like depressions are filled with several series of clays, kaolinitic clays, kaolins separated either by lignite seams in the upper part or by fossil loam at the base. The loam contains fragments of baryte, sometimes including sphalerite and galena crystals, and limonitic brecciated quartzite or, in places, limestone blocks. This horizon is rich in manganese minerals and limonitic concretions with vanadium minerals. The upper part of the fill contains molluscan shells, Characeans and fish remains of Karpatian age. The karst depressions are several hundreds of metres in diameter and more than 60 m deep in places. SKÁCEL (1954) noted also that some cave passages are filled with kaolinitic sediments. The limestone surface is dissected by deep karren.

In the **Hranice Karst**, a morphologically diversified relief with relict tropical karst was covered by Karpatian and Badenian deposits. Older positive landforms such as karst cones and towers were rounded by Miocene marine abrasion (TYRÁČEK, 1962) and display traces of burrowing coastal organisms. Cave systems are believed to have formed prior to the Miocene (PANOŠ, 1964a). They were filled with sediments, were later exhumed and in the Plio-Pleistocene were invaded by thermal waters. An example is Zbrašovské Caves (site 20; KUNSKÝ, 1957).

In the northern part of the **Drahany Upland** , numerous occurrences of tower karst and related relict karst forms have been reported by PANOŠ (1962, 1964a, 1965). Precise dating of the origin and evolution of this karst relief is not straight-forward. Some features are covered by Badenian marine deposits, some are buried by freshwater sediments of the Upper Miocene and/or Pliocene, some are completely exhumed. Fossil cone or tower karst is known from Červenka near Litovel (site 19), buried by ?freshwater Pannonian and Quaternary cover. Complex cone karst near Měrotín (site 18) is covered by Pleistocene loesses and terrace sediments. Geological organs near Mladeč, Javoříčko and Hvozdečko (sites 17 and 18) are filled with Badenian marine, ?Pannonian freshwater sediments and Quaternary cover debris. These karst forms originated on isolated inselbergs of Devonian carbonates cropping out from Lower Carboniferous rocks.

Javoříčské Caves (site 17) probably formed in the Paleogene to Early Miocene by subjacent karstification under a thin cover of Lower Carboniferous and Upper Cretaceous rocks. Caves in Třesín Hill result from the pre-Late Miocene subterranean flow of the Morava River (site 18). Most of these caves were fossilized at the end of the Pliocene (HORÁČEK, 1983).

The **North Moravian Karst** is similar. Some karst forms interpreted as Paleogene to Neogene occur in isolated carbonate outcrops. Karst cones near Supíkovice (site 24) are separated by depressions filled by *in situ* kaolinized sediments. They are buried by glacial deposits (CZUDEK and DEMEK, 1970). Features near Branná, Vápenná, etc. (site 23) filled with variegated kaolinitic weathering products are interpreted as Paleogene or Neogene (PANOŠ, 1964a). A majority of these landforms have been affected by periglacial mechanical weathering and destruction.

The presence of karst inselbergs southeast of Brno (site 11) in the **Carpathian Foredeep** was noted above. Evolution of this cone karst relief continued during Paleogene to Early Miocene times. Traces of pre–Karpatian karstification of Devonian and Upper Jurassic carbonates are numerous (ZEMAN, 1980; BEZVODOVÁ and ZEMAN, 1983). Deeply karstified carbonates were encountered by oil drilling in many places. Surficial or subterranean forms are filled with Karpatian or Badenian clastics, or their filling is composed of fine sediments without fossils. Such fill is interpreted as pre-Badenian. The detected depth of karstification reaches 80 m below unconformities.

Karstification was influenced also by the deep entrenchment of pre-Badenian valleys in the eastern margins of the Bohemian Massif. They started to develop in the Early Cretaceous and during the Paleogene attained depths of 200 to 1 000 m (ZEMAN, 1980). They represented a continuation of the surface drainage of adjacent areas. The deep valleys in the Moravian, Tišnov and Hranice Karsts, all partly or completely filled with Badenian deposits, are part of this system. The relatively sparse network of drilling cannot give enough data to evaluate properly the extent of karstified zones associated with these canyons.

Middle Miocene to Pleistocene Karstification

The youngest phase of karstification in the Bohemian Massif constitutes a complex sequence of creation, rejuvenation, reactivation, fossilization and /or destruction of karst forms. These effects were closely associated with the tectonic mobility of individual structural blocks and with vigorously changing climatic conditions, namely with periglacial climatic episodes since the Late Pliocene. Relatively short interglacial periods appear to have influenced the karst forms only by minor retouches rather than by any distinctive combination of karstification processes (BOSÁK and HORÁČEK, 1981). The Quaternary was also characterized by accumulation and erosion phases which were climate-controlled, for example the origin of secondary higher level in river caves (the trop-plein). It appears that Late Pliocene/Early Pleistocene fossilization processes were very significant in the paleokarst fossilization that was occurred in the Massif.

Examples of young, Miocene to Quaternary, karstification are numerous. Karst landforms of differing sizes are found even in the smallest occurrence of karst rocks. Most of these features are filled with sediments containing Pleistocene to Holocene faunal remains. There are many kryptokarst subterranean features, especially in metamorphic carbonates in South and West Bohemia (Bližná, site 6, and Loreta near Klatovy, site 5). The abundance of young karst forms does not allow us to discuss them in detail here.

In the **Bohemian Karst**, fossilization of the Koněprusy Caves (site 3) was completed in the late Middle Miocene to Early Pliocene. HORÁČEK (1982) discovered a fluviatile fill of sands, gravels and clays rich in vertebrate fossils of Upper Sarmatian age (Suchomasty 3, Fig. 2–49). The sediments infilled a horizontal cave and vertical chimney. The fauna consists of aquatic and terrestrial forms living in tropical to sub-tropical climates. An Early to Middle Pliocene bat assemblage in Nová propast Chasm, Koněprusy Caves, indicates that the cave was completely developed prior to the Pliocene, without any further important processes of destruction, deposition, rejuvenation or karstification occurring in later periods (HORÁČEK, 1980b).

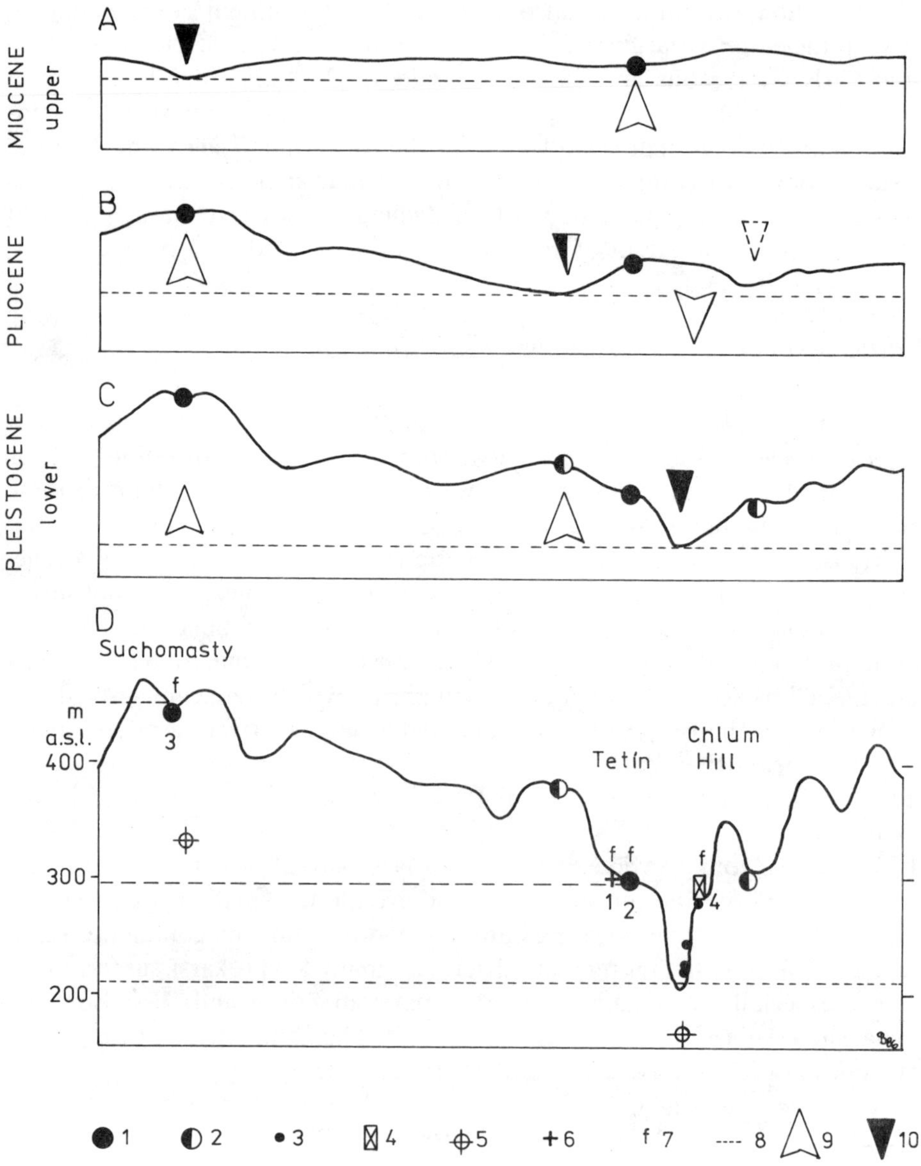

2–48. Schematic representation of relief evolution in the central part of the Bohemian Karst along the line Suchomasty-Doutnáč Hill and position of some discussed sites (A-C). The recent stage is presented in the lower part of the figure (D; after HORÁČEK, 1980a, completed). 1. paleontologically dated fluviatile Miocene deposits; 2. Neogene terrace deposits without faunal evidence; 3. Quaternary terrace deposits; 4. slope sequence with fauna overlying terrace deposits; 5. the deepest karst cavities proved by boreholes; 6. Upper Cretaceous deposits; 7. faunal evidence; 8. erosional base; 9. tectonic movements; 10. the position of the river bed, a place of terrace accumulation. Localities: 1. Kruhový Quarry, 2. Tetín, 3. Suchomasty, 4. Chlum Hill.

Miocene tectonic activity, mostly the post-Sarmatian movements, caused dissection of the relief in the Bohemian Karst. The consequence of this was the displacement of Oligo-Miocene caves to different elevations (LYSENKO, 1982). Paleontological data indicate that movement in the Koněprusy area has been active since the Sarmatian, whilst the Srbsko-Chlum Hill-Tetín block moved only between the Pliocene and Early Quaternary (HORÁČEK, 1980a; Fig. 2–48). Movement of this latter block created a graben-like structure along the Berounka River (LYSENKO, 1982). Displacements were such that the Early Pleistocene terrace +90 m above the Berounka River is at the same elevation as occurrences of Lower Miocene (Burdigalian = Ottnangian) mollusc-bearing terrestric deposits and Upper Turonian-Coniacian marls close by. This disproportionality was discovered by PETRBOK (1950), who described the "chronological paradox of river terraces". He knew only the Burdigalian locality; the Cretaceous sediments were not discovered until 1985. The "correct" position of Neogene sediments is expected to be at approximately +180 m above the river and the Upper Cretaceous should be even higher. All the data show that the dissection of Oligo-Miocene caves occurred during Miocene tectonic phases, when the lower levels of some caves also were created, e.g. in Koněprusy Caves. The development of caves in later periods is generally linked with the development of river systems and terraces (Berounka River).

2–49. Suchomasty 3 site with Upper Sarmatian fluviatile fill (Photo by I. Horáček).

The next karst genetic phase is associated with the rejuvenation of neotectonic activity that culminated in the Early Quaternary, during the Biharian. In the Koněprusy area, this phase saw roof destruction in some caves, with the re-opening of fossilized connecting shafts, etc., and the excavation of their pre-Quaternary sedimentary fill. This process took place also at Chlum Hill, but somewhat earlier. Mass deposition in open karst forms (dolines, etc.) took place during glacial cycles G to H at Koněprusy and H to J at Chlum Hill. During this period the Berounka River rapidly entrenched itself from a position +80 m above its present level (HORÁČEK, 1980a). The Berounka canyon, in fact, was deeper than it is now. It was partly filled with sands and gravels during younger Pleistocene phases.

In the **Moravian karst**, pre-Badenian relief was fossilized by Badenian clayey sediments and other fine clastics. The Badenian transgression covered extensive areas of the eastern slopes of the Bohemian Massif, far to the West. The last known relicts of the Badenian deposits occur near Česká Třebová. The exhumation of the Badenian marine fill from karst valleys and surface karst plateaus took place successively, probably in several phases. Corresponding to phases of the exhumation, inserted cave levels originated between the main, Oligo-Miocene cave levels. This process was influenced by local conditions of individual karst valley and cave systems, therefore the inserted horizontal net of caves could not originate in the uniform way. Subhorizontal underground drainage of marginal semi-poljes exhumed some filled caves depending on their relation to the contemporary position of karst springs. The local situation was complicated by the independent and different tectonic movements of at least six individual blocks of the Moravian Karst.

The process of exhumation of old and also partly inserted cave levels occurred for the first time after the Badenian and then during the Pleistocene, when the continental coarse-grained clastics of probably Pliocene to Pleistocene age originated. The process of exhumation of these young fills was similar to the previous exhumation and cave reactivation. Pliocene-Pleistocene fill of blind valley and marginal depressions is preserved until now, because deep vertical drainage developed connecting the surface of the sediments with exhumed horizontal cave systems.

The processes involved in the origin of the cave levels in the Moravian Karst and adjacent areas therefore were very complex. The gradual exhumation of older caves in the middle and lower course of the deep valley took place in at least two cycles. Secondary cave levels therefore occur at different relative and absolute altitudes (PANOŠ, 1963, 1964a).

To cite some other examples, there was important fossilization in the **Javoříčské Caves** (site 17) at the end of the Tertiary. The fossilization, the deposition of complex speleothem sequences interbedded with fossiliferous clastics, started at the end of the MN 16 (3.5 Ma B.P.) and finished not later than zone MN 17, the Early Villanyian. These fills are dissected by cracks. Infilling and fossilization in the region of the **Mladečské Caves** (site 18) and at Sovinec (site 26) took place during the mid-Early Pleistocene times since faunal finds correspond to zone Q 1 (HORÁČEK, 1983).

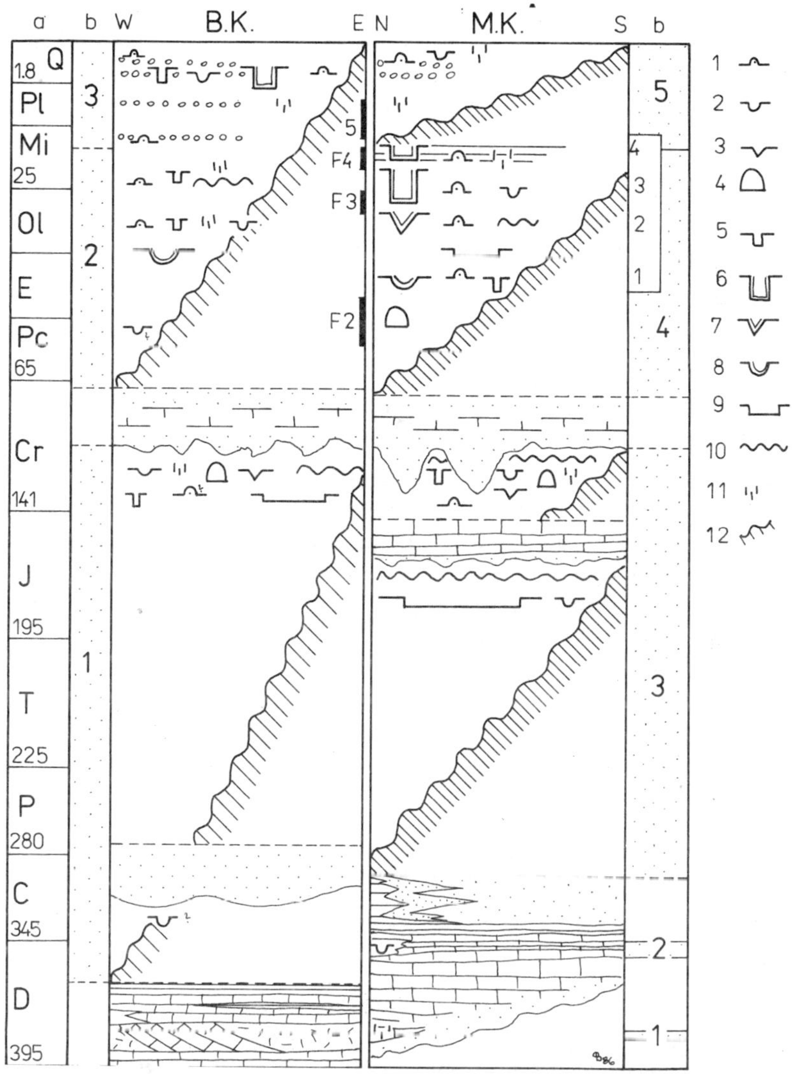

2–50. Distribution of paleokarst within the geological history of the Bohemian Karst (B. K.) and Moravian Karst (M. K.). a. periods; numbers indicate the age of the period base in Ma; b. karstification periods. 1. caves; 2. dolines; 3. geological organs; 4. mogotes, karst cones and towers, karst inselbergs; 5. shaft; 6.–8. valleys and karst canyons; 9. poljes and regional depressions; 10. corrosional surfaces; 11. karren; 12. periods of paleorelief evolution (F 1–F 5 = peneplanation phases in the Bohemian Massif, according STÖRR, KUŽVART and NEUŽIL, 1978).

A similar age of fossilization was determined for some caves in the **Tišnov** area near Lažánky (site 10). Lažánky 2 is a channel-like horizontal cave filled with a sequence of fluviatile sands and gravels with redeposited loesses and eolian sands in the upper part. The abundant assemblage of small mammals indicates a Biharian (Early Pleistocene) age for the site (HORÁČEK, unpubl. rep.).

The very spectacular karst of the **Horní Morava**, North Moravia, is situated near the Polish/Czechoslovak border in crystalline limestones at relatively high altitudes. Very deep and large karren and corrosional longitudinal depressions are filled with variegated and kaolinized deposits. The relief reaches up to 30 m. Some authors suggest a Tertiary (Neogene) age for this site, but others suggest a young, Quaternary, origin during interglacial periods. In the latter case the fill would therefore be redeposited from weathering profiles upslope.

West Carpathians

The West Carpathians represent an allochthonous cover folded and displaced by the Alpine Orogeny. The structure is characterized by the décollement of nappes from their basement. The latter comprises epi-Variscan platform with Proterozoic elements and Variscan consolidation. The Carpathians are divided, according to the age of the Alpine deformation, into the Inner Carpathians with a predominant Paleoalpine, Upper Cretaceous structure, and the Outer Carpathians with a predominant Neoalpine, Tertiary, structure (ROTH, 1980). The Outer Carpathians are composed mainly of flysch suites. The Inner zone consists of crystalline (Proterozoic-Paleozoic), Paleozoic and Mesozoic complexes. A common characteristic of both zones is the complex nappe structure. For more details see e.g. BYSTRICKÝ, MAZÚR and JAKÁL (1972).

Karst Rocks

Outcrops of karst rocks in the West Carpathians amount to 2 700 km^2. They occur almost exclusively in the Inner zone. There are some limited areas with Paleozoic carbonates, but a majority of karst rocks belong to the Mesozoic. Karst phenomena, with rare exceptions, are developed in the Middle Triassic Guttenstein Limestones, and Middle to Upper Triassic Wetterstein Limestones with their thicker dolostone sequences in places. The total thickness of Triassic carbonates exceeds 1 200 m (BYSTRICKÝ, MAZÚR and JAKÁL, 1972). Subcrops of Mesozoic rocks under Paleogene deposits in the Central Carpathian Basin and under very thick Neogene sediments in sub-Carpathian Basins are known but their extent has not been detected.

Paleokarst in the Inner Carpathians

This section presents only a short review of paleokarst forms plus some new ideas concerning paleokarst evolution in the Inner Carpathians, owing to the limited knowledge of paleokarst there.

The origin, fossilization and rejuvenation of the karst features were determined by the usual lithological and structural conditions, plus the following factors:(1) a prevailing "geosynclinal" regime from the Triassic until the end of the Early Cretaceous; (2) the different geotectonic position of local emerged areas of Triassic to Oligocene paleokarst when compared to their recent emplacement in the nappe structure of the Carpathians; (3) Eocene-Oligocene deposition in intramontane basins, and (4) increasing tectonic activity since the Miocene, accompanied by the uplift of mountain chains and subsidence of large basins.

Pre-Eocene Paleokarst

Karstification during this long period occurred mainly during the Cretaceous and Paleogene. Data on pre-Cretaceous paleokarst are rare. Paleokarst created in this period resulted from karstification either at the site of primary sedimentation (e.g. Triassic), or on emerged cordillera in the synorogenic phase (Cretaceous) and on emerged areas during the post-orogenic Paleogene.

Upper Triassic deposition in the area that is now **Muráň Plateau** (site 37) took place in lagoonal conditions. Cycles in the Dachstein Limestones (Rhaetian) were terminated by short emergences with subaerial dessication and karstification. Karstified fissures several centimetres deep and small cavities were formed in the freshwater lens (BORZA, 1977).

BORZA (1980) and BORZA et al. (1980) described breaks in the deposition of Lower Cretaceous limestones at Strážovské vrchy (site 34) and in Malé Karpaty Mountains (site 28). Valanginian to Early Hauterivian and Aptian/Albian emergence at **Strážovské vrchy Mountains** was accompanied by deposition of conglomerates and the formation of different types of neptunian dykes. In the **Malé Karpaty Mountains** there are breaks in the Upper Aptian to ?Lower Albian. Glauconitic limestones cover an undulating surface; subaerial corrosion processes cannot be excluded during the breaks and emergent periods, although karst features are not noted here.

Relicts of speleothems in the Upohlavské Conglomerates (post-Santonian) indicate the emergence of the **Pieniny** cordillera during Early to Middle Cretaceous times, with the development of karst and caves (site 36; MIŠÍK, 1978; MIŠÍK and SÝKORA, 1981). The cordillera emerged from the sea over a width of several tens of kilometres and with diversified relief. River basins were able to form. The karst was destroyed during the Late Cretaceous and relicts of speleothems and fresh water limestones were deposited in conglomerates.

The area between **Dobšinská Ice Cave** and **Nekészeny** (Bükk Mountains of northern Hungary) emerged at the beginning of the Cretaceous and remained so until the Senonian (MIŠÍK and SÝKORA, 1980). Upper Jurassic shallow marine strata were karstified during this period. Relicts of karst pockets filled with deposits containing Upper Cretaceous-Paleocene pollen grains prove the karstification in the Silica Unit (Gombasek Quarry, in area C; MELLO and SNOPKOVÁ, 1973; BÁRDOSSY and KORDOS, p. 143).

Karstification in the Late Cretaceous to Eocene period probably became more widespread after the Senonian phase of the Alpine Orogeny, when marine deposition ceased in nearly the entire Inner Carpathians. The continental regime was interrupted by deposition of the Central Carpathian Paleogene strata (Eocene to Oligocene), which covered a morphologically diversified relief in the Inner Carpathians. For example, in the **Liptov Basin** (site 32) near Ružomberok, the Upper Lutetian sea covered karstified and weathered Triassic dolostones in narrow to broad embayments up to 20 km long. The bays were originally valleys, i.e. rias (GROSS and KÖHLER, 1980).

Post-Oligocene Karstification

During the Neogene the northern elements of the Carpathians were generally uplifted. Simultaneously the southern edge of the mountains suffered considerable subsidence. Since the beginning of the Miocene, the Pannonian-Vienna Basin has substantially enlarged. In the Badenian, the Paratethys Sea transgression coverd large parts of karst massifs both in South Moravia and in South Slovakia. In the Malé Karpaty Mountains of West Slovakia, Badenian marine deposits infill caves, e.g. in **Děvínská Nová Ves** (Neudorf an der March; site 27) near Bratislava (see also HORÁČEK and KORDOS, p. 604). Here, marine deposits directly overlie cave sediments with a rich mammal fauna of the biozone, MN 6 (ZAPFE, 1960). Pre-Upper Badenian (i.e. older than 14 Ma) speleothems in small caverns in the **Malé Karpaty Mountains** have been described by MIŠÍK (1980); their age is based on finds of Lower Badenian rodents in fillings (FEJFAR, 1974).

In the **Ponický Karst** near Banská Bystrica (site 31) relicts of karst cones, deep karren and depressions filled with Badenian gravels and volcanoclastic rocks are reported (ŠILAR, 1968).

In South Slovakia there are more well dated karst deposits of Pliocene age, e.g. Ivanovce near Trenčín (site 30; FEJFAR, 1976); Koliňany near Nitra (site 29; FEJFAR and HORÁČEK, 1983) or Brekov in the East Slovakian Lowland (site 35; HORÁČEK and LOŽEK, 1984), etc. They are mainly narrow, tubular solution channels or vertical fissures filled with yellowish ochre clay or sandy loam, sometimes with nodular ores rich in Fe, Mn, Ti. A variety of Pliocene and Early Pleistocene karst and karst deposits are known from the whole of Slovakia; the extent of this chapter does not allow us to describe them in detail.

The **Slovak Karst** (Fig. 2–41 area C) consists of several karst plateaus that are separated by deep valleys or canyons. It is composed of highly folded and faulted Triassic to Upper Jurassic carbonates, with siliciclastic rocks at the base. Morphogenesis of the region was studied by LUKNIŠ (1962), MAZÚR (1965), SKŘIVÁNEK (1965, 1968), JAKÁL (1975) and many others. The pioneer work in this field was published by SAWICKI (1908, 1909); JAKÁL (1975) noted an even earlier study by SOBÁNYI (1896). The evolution of the Slovak Karst is newly discussed by JAKÁL (1975). The conclusions can be summarized as follows: (1) pre-Pannonian karstification was mostly vertical. The occurrence of mogote-like hills is a result of the lithology of the carbonate bedrock; (2) pediplanation during the Pannonian created distinct levelled surfaces; (3) tectonic activity at the end of the Pannonian caused river erosion, exhumation of weathering products and rejuvenation of the vertical karstification; (4) post-Pannonian uplift tilted the plateaus to the south and caused deep valley incision along structural lines; (5) Pontian subsidence led to the accumulation of gravels in the valleys; (6) Early Pleistocene tectonic movements caused river erosion of the gravel fills and differentiated uplift of some individual tectonic blocks. Mogote-like hills up to 150 m high, occurring on plateaus of the Slovak Karst were designated by JAKÁL (1975) as only lithology-controlled, but are clearly relicts of tropical karst landforms that originated in pre-Upper Miocene times (SKŘIVÁNEK, 1965, 1968; LUKNIŠ, 1962). The hills were transformed in the Pleistocene by periglacial weathering. The explanation presented by JAKÁL (1975) reflects only one aspect of the problem.

The Late Cenozoic history of the Slovak Karst has been affected to a considerable degree by its position between the uplifted northern zones of the Inner Carpathians (Slovenské rudohorie Mountains) and the subsided Pannonian Basin. Like other regions bordering this basin, the Slovak Karst was covered by sediments of the Badenian sea. Sands and glauconitic clays in the Včeláre Quarry are most probably a result of this transgression.

We propose the following post-regression development of the area (Fig. 2–51):

(1) post-Badenian erosion and exhumation of the Badenian cover was accompanied by rejuvenation of old surfaces, including the broad and deeply incised valleys of the Slaná and Bodva Rivers. There was increased surficial corrosion on the karst plateaus and probably also rejuvenation of some older subterranean routes. Because of the lack of a well dated record, direct evidence for the duration and sequence of these processes is missing.

(2) The next distinguishable phase is associated with the relative subsidence of the area. There was mass deposition of gravels, redeposited surface covers and

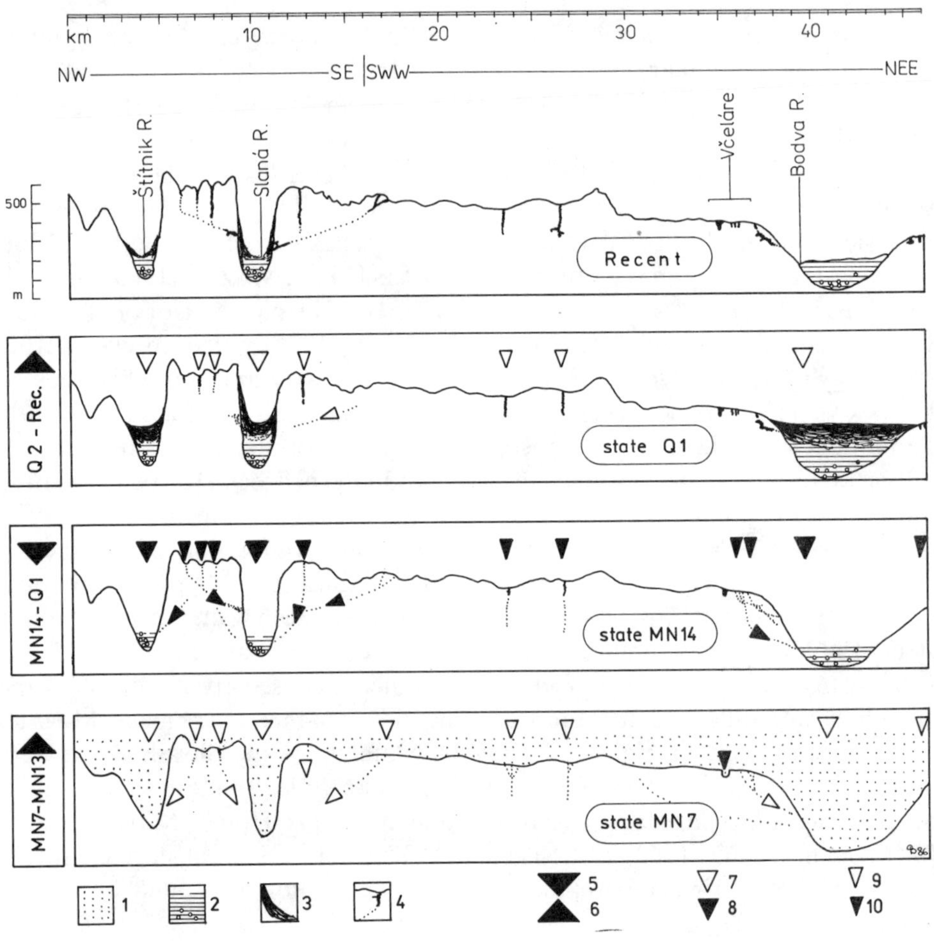

2–51. Schematic sketch of the evolution of the central part of the Slovak Karst (for details see the text). 1. Badenian sand ànd clay; 2. Upper Miocene to Lower Pliocene valley fill (MN 14–MN 16); 3. MN 17–Q 1 slope breccia and fluviatile deposits alternating laterally; 4. karst surface and underground waterways; 5. relative subsidence of the whole region; 6. relative uplift of the whole region; 7. river accumulation; 8. river erosion; 9. fossilization of karst cavities; 10. creative karstification or rejuvenation of underground karst phenomena.

weathered soils in the river valleys. At the same time, some vertical drainage routes became infilled and fossilized (cf. SKŘIVÁNEK, 1965) and subhorizontal karst passages became the main hydrologic system. Dating of these processes is possible by means of some fossiliferous fills, particularly at the southern edge of the area. We believe that many more such localities bearing the MN 15–16 faunal records from Osztramos Hill, Hungary (JÁNOSSY and KORDOS, 1977) or Včeláre Quarry (HORÁČEK 1980a, 1985; Fig. 2-52) belong to this phase. The same age is suggested by palynological analysis of the valley infills.

(3) The next phase was probably a continuation of the state achieved in phase (2). Filling and fossilization occurred more or less continuously. Mass debris accumulations characterize the plateau slopes. These detrital breccias form covers that are several tens of metres thick LOŽEK and ZÁRUBA (1965). It is

2–52. Včeláre 6 site. Narrow vertical cavities are filled with fossil-bearing Middle Villányan deposits, the somewhat broader karst form in the right half of the figure is filled with Q 1–2 sediments (Photo by I. Horáček).

believed that the breccias have never been covered by any younger deposits. The faunal record from the breccias and karst fissures, e.g. Honce (site 33) date the top of the breccia deposition to the end of the Early Biharian, i.e. to the period close to Q 1/Q 2 boundary sensu HORÁČEK (1979). There are other supporting data. A number of fossiliferous karst fills contain assemblages corresponding to biozones, MN 17-Q 1, but there are none younger than Q 2/1.

(4) From these observations, it can be presumed that since Q 2/1 the region experienced considerable uplift when compared with the level of the Pannonian Basin. We may also expect an increased vertical erosion rate on the karst plateaus, with enlargement of vertical subterranean shafts and the stabilization of drainage systems. Such a tendency can be favourable for the preservation of fossilized fissures. On the other hand, it does not provide conditions for new infilling phases, at least until the hydrological scheme of the region comes to be changed.

Practical Consequences of Paleokarst

Because of the small extent of karst areas in the Bohemian Massif, the engineering hazards are very limited. There are only a few mineral deposits exploited in the karst and associated with paleokarst. Mining of iron ores in the Moravian and Němčice Karsts has a long tradition. In the Moravian Karst, the mostly limonitic ores were mined from deep depressions in the environs of Rudice, Lažánky and Habrůvka (cf. WANKEL, 1882). In the Němčice Karst iron ores were mined in cave passages cutting old deep and narrow depressions. The mining lasting for centuries but finished at the end of the 19th century. The only deposits now mined are accumulations of sands of the Rudice Formation in the Moravian Karst, used in the neighbouring Blansko and Brno metallurgical plants as moulding sands. Some clay beds were utilized in the past for production of ceramics in Olomučany. Between Maršov and Lažánky in the Tišnov Karst kaolin was dug from extensive and deep karst depressions. Limited shows of oil and gas are associated with paleokarst traps in the upper parts of Devonian carbonate subcrops in the Carpathian Foredeep.

Engineering problems in the West Carpathians are somewhat more important. There have been many dam construction works in the karst areas. Problems with caves and caverns were particularly important in the construction of the Čierný Váh Dam, where several shafts and caves had to be filled with concrete and the entire floor of the reservoir covered with plastic foil and bitumen. No karst-related mineral deposits are exploited in the West Carpathians, only some minor occurrences of bauxite, iron and manganese ores being known.

The most important problems in Czechoslovakia as elsewhere are concerned with water supplies from karst areas. Intensive industrialization and agricultural production contributes to pollution of karst groundwater, making processing of the water necessary. Some pollution is a consequence of the rejuvenation of paleokarst horizons by large engineering works.

Conclusions

Paleokarst in the Bohemian Massif constitutes an important part of the present karst landscape. Even pre-Upper Cretaceous karst land forms survive in the relief down to the present. This has been caused by favourable sequence of burial and exhumation processes and of tectonic developments in the area. Most of the paleokarst forms are polycyclic and polygenetic, being the best examples of polycyclic karst of the Central European Type as defined by PANOŠ (1964a).

The study of particular paleokarst sites has contributed significantly to our knowledge of the geological history and development during the continental periods, which in the Bohemian Massif have lasted with only short interruptions since the end of the Paleozoic. The preservation of paleokarst fills (in some cases fossiliferous) has been important for numerous branches of geology, particularly paleontology, paleoecology and paleogeography.

Only two periods of karstification can be distinguished in the Bohemian Massif, the pre-Cenomanian and the post-Santonian; within them the occurrence of lesser karstification phases or events are more or less clearly evident. The oldest paleokarst is of Middle Devonian (Givetian) age, the youngest one is of Quaternary age.

The paleokarst of the West Carpathians has been studied less intensively. Traces of Mesozoic, Triassic and Upper Cretaceous karstification are rare, but evidence of Upper Cretaceous-Eocene karstification is more distinct. A majority of paleokarst sites have been correlated with a post-Oligocene karstification period when the most conspicuous karst forms, distinguishable even in the present karst landscape, originated and then became fossilized.

PALEOKARST OF HUNGARY

György Bárdossy and László Kordos

Introduction

The karst phenomena of Hungary were described by DARÁNYI (1972). As that paper concentrated on karst hydrology, we give more emphasis to morphological aspects of the karst and to its origin. Our main attention is focussed on karst phenomena generated by superficial processes; hydrothermal paleokarst is presented in a brief review by MÜLLER (p. 155–163).

The modern exposed karst areas of Hungary comprise 1 350 km^2, which is about 1.5 % of the total area of the country (JAKUCS, 1977 a, b, c). Even this small area is dispersed in isolated units in the Transdanubian Central Mountains and the North Hungarian Mountains. Very small, recent karsts also occur in the south in the Mecsek and the Villány Mountains and finally on the Austrian border near Lake Fertö (Fig. 4-26).

Drilling and mining activities during the past 100 years have revealed buried paleokarst surfaces over a much larger area. They formed during repeated emergence-karstification phases. They occur mainly in the vicinity of exposed karsts and at many places they are directly connected with them. In addition, a large, deeply buried paleokarst was found by oil drilling in southwest Hungary, near Nagylengyel. Altogether, paleokarst phenomena occur beneath 20–25 % of Hungarian territory. The other parts of the country consist of deep, relatively young basins filled with Neogene and Quaternary clastics and volcano-sedimentary rocks; e.g. Great Hungarian Plain, the Little Hungarian Plain and the deep basin of southwest Hungary.

The paleokarst areas will be decribed in the following sequence (Fig. 4–26): (1) the Bakony, Vértes, Gerecse and Buda-Pilis Mountains as parts of the Transdanubian Central Mountains, (2) the Nagylengyel area, (3) the Mecsek Mountains and the Villány Mountains, (4) the Bükk and Uppony Mountains and (5) the Aggtelek karst plateau.

Hungary is more rigid tectonically than the surrounding, strongly folded and overthrust mountain belts of the Eastern Alps, Dinarides and Carpathians. Paleokarst formation was strongly influenced by the fault zones because crush-

P. Bosák, D. C. Ford, J. Głazek and I. Horáček (Editors):
Paleokarst. A Systematic and Regional Review. – Elsevier and Academia. Amsterdam and Praha. 1989.

ing of the carbonate rocks facilitate their drainage and karst dissolution. In certain periods, ascending hydrothermal solutions also moved through these block-faulted systems to produce large-scale dolomite pulverization and, in some places, caves with distinctive forms.

History of Karst Investigation

Scientific speleological investigations were carried out as early as the 18th century. Consideration of contemporary factors influencing speleogenesis and the evaluation of fossil caves, however, began in the middle of our century (KEREKES, 1948).

Writing on the genesis of the terra rossa that covers the karsts, KRENNER (1876) was the first in Hungary to suggest that it must have developed during warmer periods in the past (the Tertiary). Following the discovery of bauxite deposits, more significant paleokarst research began when several Cretaceous-Paleogene buried karst surfaces were recognized (GYÖRGY, 1923; TELEGDI RÓTH, 1923,1927; VADÁSZ, 1946). An evaluation of the morphogenesis of these paleokarsts from a climatological point of view was undertaken by P. Z. SZABÓ (1956, 1957, 1960), after BULLA (1956) had introduced the necessary geomorphological concepts. Investigation of Neogene and Quaternary cave sediments started in 1847 when S. PETÉNYI found Pleistocene bone remains at Beremend. The excavation and dating of the fossil bearing cave deposits were carried out by means of paleovertebrate investigations (KORMOS, 1911; KRETZOI, 1956; JÁNOSSY and KORDOS, 1977, etc.).

Phases of Paleokarst Development

Carbonate rocks play a subordinate role among the Paleozoic strata of Hungary. Devonian and Permian limestones are present in the Bükk and Uppony Mountains and Carboniferous and Permian ones in the Transdanubian Central Mountains. No karst phases of Paleozoic age are observed in them.

A thick sequence of shallow marine limestones and dolostones was formed during the Triassic. The Upper Triassic Hauptdolomite and Dachstein Limestones have the largest extent and are several hundreds of metres thick. Marine sedimentation was continuous. No paleokarst processes operated during this period.

During the Jurassic there was continuous sedimentation in the Bakony, Gerecse and Mecsek Mountains with carbonates occurring at several levels. In contrast, stratigraphic gaps are observed in the Jurassic sequence of the Vértes and Villány Mountains. They are interpreted by GÉCZY (pers. comm., 1984) as submarine hiatuses in the deposition; consequently no superficial karstification occured during the Jurassic period.

Lower Cretaceous Paleokarst

An emergence began during the Tithonian in the **Villány Mountains.** The area most probably consisted of low-lying islands until the Middle Barremian. As the climate was wet-tropical, there was intensive karstification of the uplifted lime stones. The karst relief is relatively smooth, showing generally rounded forms. The depressions are 1–3 m, exceptionally 4–8 m deep and closely spaced. They were filled with bauxite during the Lower Barremian. Between the individual bauxite pockets, the paleokarst surface is less diversified. All the positive and negative karst forms are smoothed. The planation effect of later marine transgressions upon this paleokarst surface cannot be excluded. It is also probable, that the groundwater table was near the flat surface of the emerged land. The entire paleokarst was covered by the marine Nagyharsány Limestone of Upper Barremian-Aptian age.

Albian Paleokarst

The next karstification phase can be assigned to the Lower Albian when a narrow strip of land was uplifted along the SW-NE axis of the **Bakony Mountains**. The uplifted area was about 400 km^2. Under a wet-tropical climate, karstification started immediately on the surface of carbonates, mainly on the Rhaetic Dachstein Formation and on Lower and Middle Jurassic limestones.

Medium scale karstification occured, normally producing on the average 1–6 m deep karst depressions. Soon after their origin, bauxite clay filled them, mainly along the southern rim of the uplifted area. Most of the bauxite deposits are small lenses but near Alsóperepuszta a stratiform bauxite deposit covered all of the paleokarst over an area of nearly 3 km^2. Here, drilling and bauxite mining has revealed details of the paleokarst morphology, detecting sinkholes and domed surfaces at places (VADÁSZ, 1946).

Subsidence started soon after accumulation of the bauxite, and the marshy/brackish-water Tés Formation was deposited first. It was followed by the Upper Albian marine Zirc Limestones which covered the entire paleokarst (CSÁSZÁR, 1984).

Turonian-Lower Senonian Karstification Phase

At the end of the Cenomanian a large-scale uplift started in the Transdanubian Central Mountains and in its southwest prolongation in the North Zala Basin. There was tropical karstification on a surface of several thousands km^2 in extent during Turonian and Early Senonian times.

North Zala Basin

Oil drilling detected paleokarst phenomena at depths of 1 000 to 3 000 m in Karnian-Norian dolostones and in the Rhaetic Dachstein Limestones. Karst cavities, 1–6 m deep, were discovered as well as millimetre and centimetre size vugs. The paleokarst surface is covered by Upper Senonian sediments, but Eocene or Miocene sediments are developed at places (DEDINSZKY, 1972).

Transdanubian Central Mountains

Paleokarst phenomena were preserved in the SW Bakony Mountains, in the Halimba area, in the Nyirád area, between Sümeg and Csabpuszta in the Iharkut-Németbánya area. Highly diversified paleokarst was developed on Karnian-Norian dolostones and locally also on Dachstein Limestones. The paleokarst forms are mostly covered and filled with bauxite deposits. Some of karst depressions continued to develop below the bauxitic cover, producing complicated karst forms (BÁRDOSSY, 1982).

In the **Halimba area**, a stratiform bauxite deposit of more than 7 km^2 covers a karstified surface. Two erosional valleys 500–700 m long, 60–200 m wide and

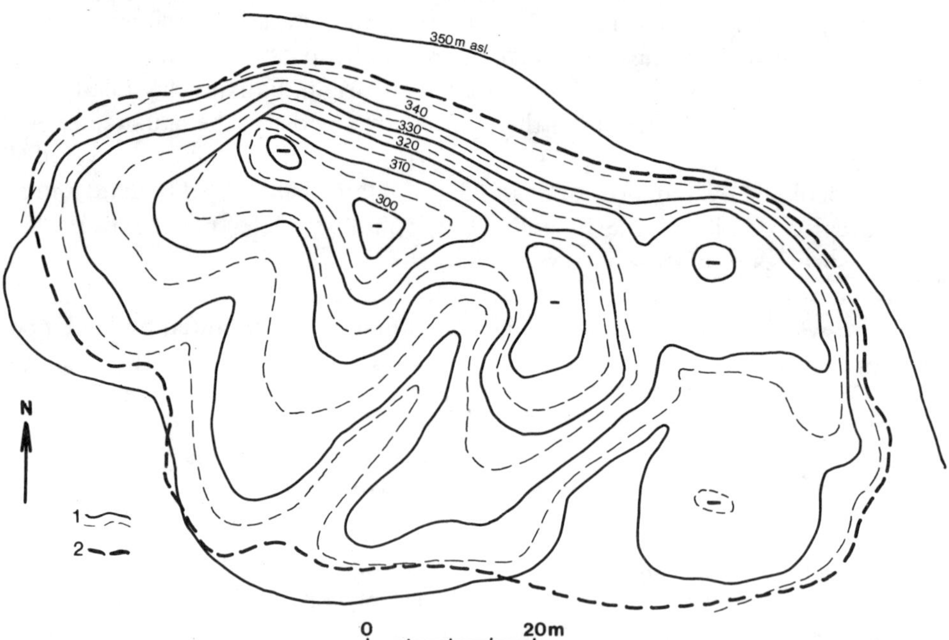

2–53. Paleokarst topography covered with bauxite in Iharkut. 1. isopachs; 2. contour of deposit.

15–30 m deep were revealed. Subsequent karstification modified them with low depressions, cones and domes with dimensions of 1–5 m vertically. A rugged paleokarst surface was found between and on both sides of these troughs,

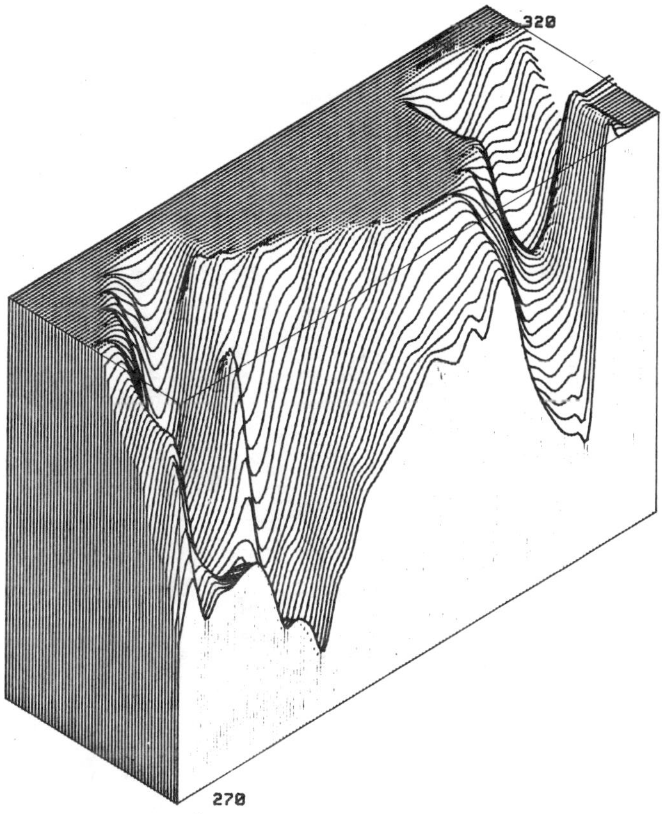

2–54. Three-dimensional representation of the karst topography at Iharkut bauxite site No. 5; constructed by computer program ELGI of the Hungarian State Geophysical Institute.

containing 10–20 m deep oval depressions and 1–5 m high domes. Small pinnacle forms, 1–2 m high, were observed as well as bauxite-filled fissures penetrating several metres into the footwall limestones. Cave galleries could be traced 10–20 m below the paleokarst surface, crossing several subvertical fissures and cavities filled with red bauxite clay. Drilling encountered similar bauxite-filled cavities as much as 50–100 m below the paleokarst surface.

Between **Sümeg** and **Csabpuszta,** a northwest dipping paleokarst surface is developed. The degree of karstification diminishes to the northwest towards the contact of the Karnian-Norian dolostones with the Rhaetic Kössen Formation.

In the **Ajka Basin**, the degree of karstification was weaker with less diversified paleorelief. The area most probably consisted of a low-lying land surface near the shoreline. It was covered by brown-coal beds of paralic character in the Upper Santonian. The karst environment had a strong influence on the coal deposits, which are considered to be distinctive *karst coal deposits* (SZÁDECZKY-KARDOSS, 1955).

2–55. Large paleokarst canyon-like depression at Iharkut bauxite mine (Photo by T. Lénárd).

In the **Iharkut-Németbánya area**, prospecting for bauxite and open pit mining revealed a peculiar paleokarst (Figs. 2–53 to 2–57). The area was an elevated plateau, high above the groundwater table, which allowed the formation of deep sinkholes. Most of them follow ancient tectonic zones. Some sinkholes were aligned, resulting in canyon-like formation, but the others are isolated. Generally 5–20 sinkholes are clustered together in a closely connected group separated from other groups by narrow, steeply walled dolostone ridges. At the junction of several sinkholes, karst cones or towers occur. Sinkholes are 10–30 m diameter and 30–60 deep, less frequently up to 100 m; the deepest having a depth of 114 m. Bauxite filled the sinkholes and also buried the ridges and cones. The paleokarst was finally covered by the Csehbánya Formation (Upper Santonian), consisting of fluviatile-lacustrine sandy marls.

The Turonian-Lower Santonian paleokarst probably extended over the entire axial zone of the Transdanubian Central Mountains. Several large areas were not submerged during the Upper Santonian transgression. Paleokarst develop-

2–56. Large paleokarst depression at Iharkut bauxite mine (Photo by T. Lénárd).

ment continued in these areas until the Middle Eocene transgression. In the Nyirád area an original Upper Senonian cover was eroded and karstification continued also during the next karstification phase.

Aggtelek Plateau

According to JAKUCS (1977b) karst development started here during the Cretaceous, continuing intermittently until the present time. His argument was corroborated by the findings of Upper Cretaceous pollen in a karst cave filling near Gombasek on the Czechoslovak side of the plateau. It indicates a continental period with wet-tropical climate (MELLO and SNOPKOVÁ, 1973).

2–57. Vertical walls of deep paleokarst depression at Iharkut bauxite mine (Photo by T. Lénárd).

Paleocene-Lower Eocene Paleokarst

General uplift took place at the end of the Cretaceous in the Transdanubian Central Mountains. Intensive tropical karstification continued in the regions described above and it began on the newly emerged Senonian limestones.

North Zala Basin

This paleokarst phase had a strong impact on the Upper Triassic carbonates and on the Campanian Ugod Limestones. Karstification was particularly strong in the latter, producing caverns, vugs and overall high porosity. It was later filled with oil and gas and this limestone is now an important reservoir rock. The entire area was covered by Eocene, Miocene and Pliocene sediments to a total thickness of 1 000–3 000 m (DEDINSZKY, 1972).

Transdanubian Central Mountains

At **Cserszegtomaj** on the southwest rim of the Bakony Mountains, 20–50 m deep sinkholes with almost vertical walls were found, filled with refractory grade bauxite clays. Because they are covered only with Upper Pliocene sediments, their age is much discussed. Based on mineralogical, geochemical and lithological analogies with the pre-Eocene bauxite deposits, BÁRDOSSY (1982) contended that they started to form during the Paleocene-Lower Eocene karstification phase. Two caves were also discovered in this area at the contact of Norian dolostone and Upper Pliocene sandstones. The origin of the more than 2 000 m long labyrinth they compose was first considered to be a product of ascending hot water (LEÉL-ÖSSY, 1953). Recently KÁRPÁT (1982) argued that they are a surface paleokarst feature whose development continued after accumulation of the Upper Pliocene sandstones.

Sümeg and Nyirád area. Here, the paleokarst surface, which extends over more than 100 km^2 and was developed on Norian dolostone, was covered by bauxite. Paleokarst formation probably started during the Turonian-Lower Senonian karstification phase but continued after erosion of a thin Upper Senonian cover. The karst depressions are 10–20 m, exceptionally up to 45 m deep. Well exposed by bauxite mining the paleokarst is very rugged with 1–2 m high cones and pinnacles alternating with narrow funnels. Below the bauxite the dolostone is weathered to a loose, floury material to depths of 0.5–2.0 m. We attribute this phenomenon to the enhanced leaching power of rainwater filtering down through the bauxites.

Karstification, as a rule, was strongest along fractured tectonic zones. Prospecting proved that most of the bauxite-filled karst depressions formed along such zones. Furthermore, tectonic fracturing and karst dissolution resulted in very high permeability in the dolostone. As a consequence it contains enormous quantities of karst water, forming a continuous groundwater system throughout the Transdanubian Central Mountains (see more details in BÖCKER and VIZY, p. 533–548).

The smoothed paleokarst surface is developed on Upper Senonian Ugod Limestones in the region of **Csabpuszta**. Depressions are less than 10 m deep and positive karst forms are rounded. Pulverization of the limestone below the bauxite is limited to a 5–20 cm zone. This paleokarst area is limited to the northwest where the Ugod Limestones are replaced by marls of the Polány Formation (Campanian–Maestrichtian).

The southern rim of the **Halimba Basin** (termed Malomvölgy) and the **Szöc Basin** is characterised by karst forms very similar to those mentioned in the Nyirád area. Karstification affected Norian dolostones here. A similar but smaller paleokarst area was detected by bauxite prospecting and mining near

Kislöd. A stratiform bauxite deposit covered and preserved paleokarst forms in Norian Dolostone. Karstification was of medium degree.

At **Csárdahegy** near Urkut, karst sinkholes were detected in Early Jurassic limestones by manganese ore mining (Fig. 2–58). The sinkholes are very narrow

2–58. Urkut-Csárdahegy manganese mine. Paleocene to Early Eocene paleokarst forms are developed in early Lower Jurassic limestones (Photo by T. Lénárd).

and 20–30 m deep, with almost vertical walls. This paleokarst was described as early as 1933 by FÖLDVÁRI. The sinkholes and the Mn-ore are covered by Middle Eocene sediments. A similar paleokarst was described by P. Z. SZABÓ (1964) from the **Eplény** manganese ore mine situated 25 km to the northwest. Recent laboratory investigations, as wells as observations made in new mining sites, led Z. SZABÓ, GRASSELY and CSEH-NÉMETH (1981) to argue that at Csárdahegy, ascending hydrothermal solutions dissolved the Lower Jurassic limestones and that even the manganese ore is a product of this hydrothermal activity. In our opinion this very complicated question needs further investigation.

On the northern rim of the Bakony Hills, beween **Fenyöfö** and **Bakonyoszlop**, a paleokarst extending over more than 50 km^2 was found by bauxite prospecting. Norian dolostones and Rhaetic Dachstein-type Limestones were karstified here. Karstification was strongest along fractured zones. Sinkholes and graben-like depressions 20–40 m deep developed, filled with bauxite. Middle Eocene sediments cover the entire territory. The paleokarst ends to the northwest along a line where the dolostones were replaced by marls of the Karnian Veszprém Formation. The bauxite also disappears along this line, proving the close connection of paleokarst and bauxite accumulation.

An important paleokarst is situated near **Iszkaszentgyörgy** and **Kincsesbánya** at the northeast end of the Bakony Mountains. Karstification occurred on Norian-Karnian dolostones over an area of more than 40 km^2. Large stratiform bauxite deposits filled karst depressions with a depth of 5–15 m, which is a karst relief of medium extent. The paleokarst ends to the southwest where dolostone is replaced by Middle and Lower Triassic marls.

In the **Gánt area**, in the Vértes Mountains, tectonically fractured Norian dolostones underwent strong karstification resulting in a very rugged karst relief over 24 km^2. Depressions 5–30 m deep are separated by steep walls, cones, and pyramids. The larger features are modified by smaller ones over 1–2 m in vertical extent. The entire paleokarst was covered by a large stratiform bauxite deposit. Open pit bauxite mining during the past 50 years has unearthed the paleokarst surface at many places. A dark-red iron crust several centimetres in thickness covers the surface of the strongly weathered, pulverised dolostone. Block-faulting continued after the accumulation of the bauxite deposits, and downfaulted areas were covered with marshy and freshwater Middle Eocene sediments. Erosion removed most of the bauxite from the uplifted areas and destroyed most of the paleokarst there as well.

In the Gerecse Mountains a paleokarst of about 15 km^2 occurs between **Nagyegyháza** and **Csordakut**. Coal and bauxite prospecting discovered this paleokarst, developed on Norian dolostones. Large stratiform and lenticular bauxite deposits cover the surface and fill its depressions. A particular feature of this area is a dolomite fanglomerate, interbedded with the bauxite and resting at places directly the paleokarst. Middle Eocene coal seams bury the entire occurrence.

The **Tatabánya Coal Basin**. Observations in mines revealed that the Eocene coal seams were deposited directly onto a paleokarst surface in Norian dolostones and Rhaetic Dachstein Limestones. At places mottled kaolinic clay and red bauxitic clay filled the karst depressions. Similar paleokarst occurs to the northeast near **Dorog** and to the southwest near **Oroszlány**, with Middle Eocene coal seams covering the paleokarst. In all these areas the vertical dimensions of the karst forms are more restricted than in the bauxite areas described above. This suggests a lower altitude, with a near-surface groundwater table during development of the paleokarst. Large, open karst cavities were encountered, mainly in the Dachstein Limestones, during the mining. They were filled with karst water and there were some catastrophic floods into the coal mines. Apparently the coal formation was influenced by the karst environment, which is why SZÁDECZKY-KARDOSS (1955) called them *karst coals*, attributing their high sulphur content and the particular mineralogy of their inorganic constituents to this influence.

At the northeast end of the Transdanubian Central Mountains, near **Pilisvörösvár, Pilisszántó** and **Budakeszi**, 10–20 m deep karst sinkholes were detected and mined on Norian dolostones. They are filled with low-iron refractory clays and bauxite clays. They are in most places covered only with thin Upper Pliocene or Quaternary sediments, but some remains of an original Middle Eocene cover have been discovered. We presume, on this basis, that this paleokarst also belongs to the pre-Eocene karstification phase, while not excluding the possibility of its further development during later periods and under near-surface conditions.

Bükk Mountains

In the southern rim of the Bükk Mountains Upper Eocene limestones cover Upper Triassic ones. Very probably a pre-Eocene karstification occured here (BALOGH, pers. comm., 1984). On the Bükk high plateau, Upper Eocene limestones that escaped erosion in a karst depression have been found recently. This proves that there was pre-Eocene karst activity on the Upper Triassic limestones of the high plateau itself.

Oligocene Paleokarst

In Hungary, Oligocene sediments were deposited in the NE-SW ranges, mostly covering the area of the present Mid-Hungarian Mountains. The sediments, of epicontinental-terrigenous and marine origin and whose deposition began in the Eocene, extended to cover most of the carbonate rocks. Surviving Oligocene emerged areas were unsuitable for karstification, which is why there are hardly any Oligocene karst forms in Hungary.

In the Triassic dolostone in **Kajmát quarry** at Bodajk (Bakony Mountains). 4–5 dolines filled with laminated terrigenous sediments were found. Their diameter is 3–13 m and their depth is 2–6 m. From some of them, vertical solution channels lead downwards. At a depth of about 20 m, there are man-sized cavities apparently created by phreatic processes that are today preserved as cave fragments. In one of them, supposedly the richest Middle Oligocene vertebrate assemblage of Central Europe was found (KRETZOI, 1971).

At **Csákberény** in the Vértes Mountains there is a fissure with solutional walls in Middle Eocene limestone. It is filled with a greyish and yellowish sandy clay in which Ruppelian bone remains (*Protaceratherium albigense*) were found (KRETZOI, 1971). Both of these karst deposits are parts of the Upper Oligocene gravels of the Csatka Formation. This sediment, of mainly fluvial origin, covers the surface over a large area, the total carbonate area of Transdanubia (1 600 km^2) apart from some more elevated plateau areas in the Keszthely and Bakony Mountains (KORPÁS, 1981).

Neogene Karstification Phase

Extensive Miocene transgression from the relict Oligocene basins with continuous sedimentation at several localities is observed. Generally, the Miocene sediments are deposited discordantly on broad areas (Little Hungarian Plain, SW Transdanubia, SE Hungary). The planation of the Mid-Hungarian Mountains in the Neogene took place under tropical-subtropical conditions (BULLA, 1956).

The surface of the **Bükk Mountains** is considered by PINCZÉS (1970) to be the Lower and Middle Miocene peneplane. However, in places Upper Eocene marine sediments are covered by Miocene volcanic ash. According to PÉCSI (1963), the planation of the karst mountains took place in the Sarmatian and was connected with the karstification. In spite of this, no clear Miocene paleokarst has been found in Hungary.

On the Bükk plateau partly degraded rhyolite tuffs were found (JÁMBOR, 1977). They fill surface depressions with both autochthonous and allochthonous material. Recently, under the Upper Pleistocene level in the so-called Kálmán-rét Shaft a thick rhyolite tuff, redeposited in the Pleistocene, was found. In the Slovak part of the Aggtelek karst ANDRUSOV et al. (1958) first indicated the presence of tuff in the terra rossa covering the surface. Later its presence was proved at several sites in Hungary too (BIDLÓ and MAUCHA, 1964; SZENTHE, 1969).

In the Pannonian there was geomorphological inversion in the Carpathian Basin, i.e. formerly elevated surfaces subsided and were covered with the various sediments of the Pannonian lake. The Mid-Hungarian Mountains, including Bakony, Vértes, Gerecse, Buda, Pilis, Bükk, Aggtelek karsts, and the Mecsek-Villány Mountains remained elevated as islands. Probably their low elevation and cover of earlier sediments limited their karstification. That is why it is hard to estimate the regional distribution and extent of the Pannonian karst.

In the southern part of the **Vértes Mountains** large fissure systems developed among Triassic dolostone cliffs of an escarpment, developed probably due to mechanical causes. Later, Lower Pannonian greyish clay filled them. The clay contains the rich vertebrate assemblage of the Csákvár Cave. Later, in the Early and Middle Pleistocene this cavity system was altered several times by various processes in differing conditions (KRETZOI, 1951).

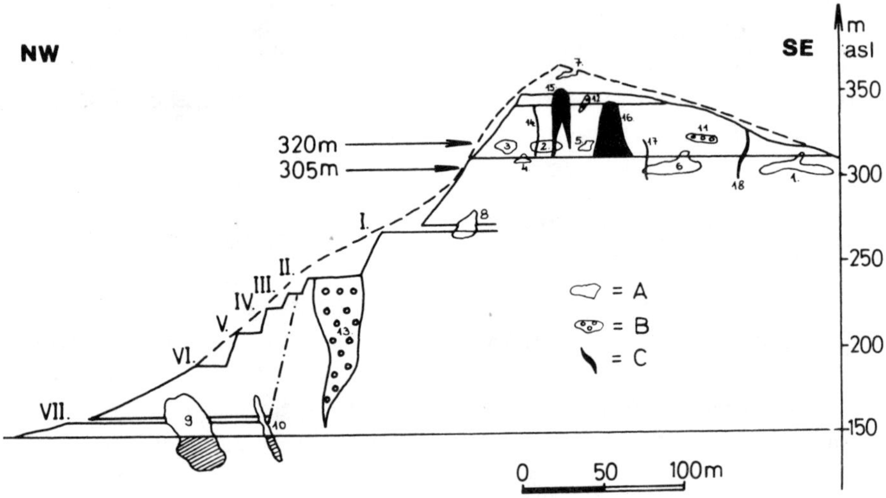

2–59. Cross-section of the Osztramos Hill. A = open caves; B = iron-clay filled karst fissures; C = fossil caves and fissures.

In the metasomatic dolostone surface of the **Rudabánya Range**, valleys can be observed which were filled first with pre-Pannonian reddish clay and then by Lower Pannonian swampy sediments containing prehominid remains (KORDOS, 1982).

Terrestrial karst formations known in the Cretaceous Ugod Limestone at **Sümeg** and in the Lower Jurassic Pisznice Limestone near **Tardosbánya** (Gerecse Mountains) are thought to have formed during younger stages of the Pannonian.

2–60. Cross-section of fossil cave at Polgárdi, an important paleovertebrate locality (after KORMOS, 1911). 1–4. layers with bone remains; 5. layer with rodent remains; pm = Paleozoic limestone; pah = Pannonian clay and sand; a = Pliocene loam; l = loess; h = humus.

The formation of abrasion terraces that developed in the highest regions of the **Gerecse-Pilis Mountains** occured in the earliest period of the Upper Pannonian. The strong karstification of the mountain was evidenced by huge travertine deposits near karst springs (KRETZOI and PÉCSI, 1982; SCHEUER and SCHWEITZER, 1980, 1981).

Since the Pliocene, certain karst areas already exposed were subject to strong karstification. Cavities in the Upper Carboniferous, crystalline Polgárdi Limestone (Transdanubian Central Mountains) are filled with sediments from the Miocene/Pliocene boundary period (Fig. 2–60). The largest cavity system, opened in 1910, began with a small surficial doline that could be followed downwards through a steeply inclined clastic cone for some tens of metres as a fossil cave. At the same time an open cave of similar character was described (KORMOS 1911). Although masked by speleothems, its original walls were solutional and showed the features of a sinkhole. As mining progressed in the region, other smaller fossil cavities of the same age and structure were found at **Polgárdi** (KORDOS, 1984; Fig. 2–60).

The most complete chronological sequence of karst cave development is known at **Osztramos Hill**, an isolated part of the Rudabánya Range. There are two fossil cave levels at 312 m and 305 m a.s.l. and an active cave at the modern water level (Figs. 2–59 and 2–61). On the basis of age determinations of rich vertebrate finds the cavity at 312 m was largely filled with Middle Pliocene sediments. The cave floor, originally formed at the local karst water level, was covered by 3–4 m thick, redeposited Pannonian sandy clay and then by several

2–61. Limestone quarry at Osztramos with the mark of fossil water level. Localities with vertebrate remains are numbered (Photo by L. Kordos).

metres of laminated calcite (Fig. 2–62). Its age (Middle Pliocene) was indicated by the included vertebrate remains. This means that open cavities can be continuously traced back to that period. The cave level at 305 m a.s.l. is filled by Lower Pleistocene deposits (JÁNOSSY and KORDOS, 1977).

2–62. Cross-section of fossil cave at Osztramos (320 m a.s.l.). 1. Pannonian yellow sand and clay; 2. calcite (Photo by L. Kordos).

Numerous data were collected in the **Villány Mountains** (Csárnota, Villány, Beremend Hill) about the more widespread karstification which took place at the Pliocene-Pleistocene boundary. The Csárnota 2 locality is the stratotype of Csarnotian (Upper Pliocene) while the Villány 3 locality is that of the Villányian (Uppermost Pliocene). Both are speleothem-filled, karstified fissures. In the same areas solution caves partly filled with terra rossa are also known. In the Cretaceous limestone at Beremend, caves sculptured both by hydrothermal and by cold karst waters can be found. They are filled with Lower Pleistocene sediments.

Transition from Paleokarst to Recent Karst

In Hungary the majority of karst rocks were buried until the end of Miocene times or even until the Pleistocene, i.e. the present karst surfaces are of pre-Neogene age. They have been buried and are now in the process of exhumation. Significant elevation of the mountain surfaces took place at the end of the Pliocene and in the Middle Pleistocene (KRETZOI and KROLOPP, 1972) when the subsidence of the basins dividing them became stronger. This process led to denudation of the overburden of the karst areas and the restoration of karst processes.

Several stages in the development of the most important Hungarian karst mountains (Bükk) can be differentiated (HEVESI, 1978, 1980). The older epigene doline and cave generations developed in the Upper Pliocene, the younger karst at the end of the Middle Pleistocene. In the case of the Bükk Mountains, the geomorphological effects of the cold periods in the Pleistocene must not be forgotten. During these periods the plateau at 800–900 m a.s.l. was subjected to periglacial conditions and the speleogenesis was of the high mountain type with sinkholes and shafts. This took place during the Riss and Würm glacials (JAKUCS, 1977c; KORDOS, 1976).

Conclusions

The continuous development of the present karst in Hungary can be traced from the Late Neogene. During this period, the modern scenery has developed as the product of different karstification processes operating amongst varied geological, geomorphological and climatic conditions.

On the basis of paleogeographic and paleomorphologic evidence, we consider that in all the karstification phases described, most of our paleokarst regions were low-lying, flat land emerging only a few metres above the ancient groundwater levels. In some bauxite regions, karst plateaus of 50–300 m altitude can be presumed. Higher karst relief was exceptional but there is one instance in the Iharkut-Németbánya plateau. Forms of real mountain paleokarst have not been detected in Hungary so far.

HYDROTHERMAL PALEOKARST OF HUNGARY

Pál Müller

Modern hydrothermal activity, signified by springs where the water temperature is more than 35 °C, is found in the Hungarian karst at the SW (Héviz) end and NE (Budapest) end of the Transdanubian Mountain Range and in Southern Transdanubia (Harkány; Fig. 2–63). Warm springs occur elsewhere, too, being common around the edges of the Hungarian karst regions.

Hydrothermal activity earlier in the geological history of the Hungarian karst regions is evidenced by various features such as caves (MÜLLER and SÁRVÁRY, 1977), mineral vein fillings (SCHAFARZIK, 1928; SCHERF, 1928), siliceous impregnations (BÁLDI and NAGYMAROSI, 1976), wide-spread deposits of freshwater limestones (SCHEUER and SCHWEITZER, 1974), metasomatic effects in limestones and dolostones (VITÁLIS and HEGYI-PATKÓ, 1974, 1977) etc. The most conspicuous features which can be attributed to these thermal waters occur in the **Buda Hills**. These are the best-studied and, perhaps, the best-developed examples in the country. This chapter is based mainly on studies made on them in the Budapest area.

A substantial number of the paleokarst features can be understood simply by extrapolation from the modern processes seen at work in the area. Study of some others suggests that they represent the effect of waters at higher temperature and pressure and of different chemistry from that observed now. SCHAFARZIK (1928) suggested a multiphase evolutionary model, and KOVÁCS and MÜLLER (1984) a two-phase model, for the mineralization and cave-forming processes of the Buda karst area. In the latter's model, the early phase occurred in rather deeply buried carbonate rocks (below 500 to 1 000 m of Oligocene clays and silts), probably with volcanic influence during one or several periods. There were influxes of fluids and heat of volcanic origin into the buried karst rocks, with large scale thermal free convection and material transport.

Ore-depleted mineral deposits are frequent in veins all around the Buda Hills. Several generations of calcite of various crystal habits and two generations of baryte, fluorite and cinnabar (SCHAFARZIK, 1928; NAGY and PELIKÁN, 1976)

P. Bosák, D. C. Ford, J. Głazek and I. Horáček (Editors):
Paleokarst. A Systematic and Regional Review. – Elsevier and Academia. Amsterdam and Praha. 1989.

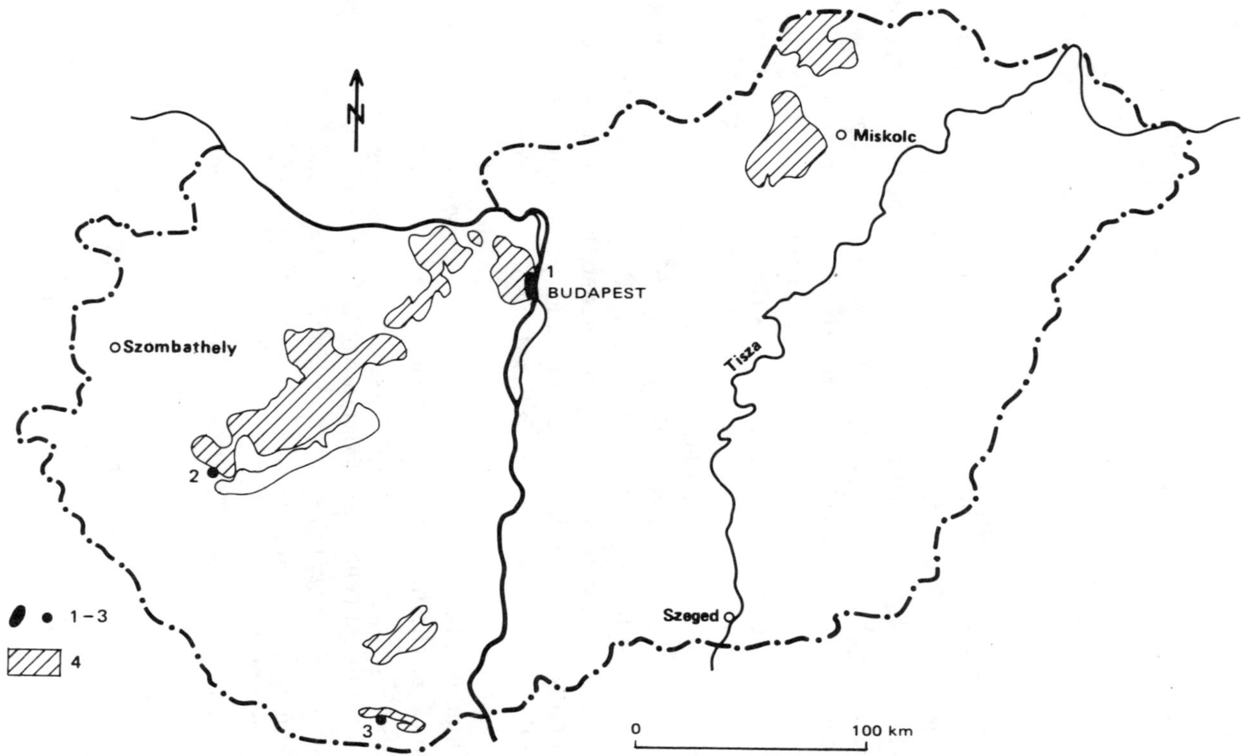

2–63. Hydrothermal karst springs in Hungary. 1–3. Main hot karst springs; 4. Main karst areas.

are interpreted as telethermal formations. Such veins are often intersected by caves that are obviously younger. Veins of baryte or coarse crystalline calcite were less soluble than the rock matrix in the cave-forming fluids, and consequently they stand out from the walls. One half of a symmetric baryte vein with a former central void occurs on the walls of Ferenc-hegyi Cave, appearing to be a final deposit from a cave-filling solution; further study revealed the remnants of the collapsed other half of the vein, however. Fluid-inclusion microthermometry of a calcite vein that is intersected by a gallery of Pál-völgyi Cave suggested a chloride-rich (4.5–8%) and hot (160–200 °C) source solution (GATTER, 1984).

Such features are most easily interpreted as the results of processes in deeply buried, volcanically affected karst as proposed by SCHAFARZIK (1928) and by KOVÁCS and MÜLLER (1984). BÁLDI and NAGYMAROSI (1976) attribute the siliceous cement of Early Oligocene sandstones (Hárshegy Sandstone) of the Buda Hills to such sources, as well. VITÁLIS and HEGYI-PATKÓ (1974, 1977) explain several features observed in Triassic dolostones and limestones by Early Badenian volcanism, which left huge masses of andesite in the vicinity of the Buda and Pilis Hills. Presumably a part of the previously mentioned mineralization, the interconnection of some fissures and some small cavities may be attributed to this event. However, evidence of earlier volcanic episodes is also present in the area. Triassic and Cretaceous events (HORVÁTH and ÓDOR, 1984) may be significant only in the case of some minerals restricted to Mesozoic rocks. The most complete mineral associations, however, are in fissures of Priabonian limestones. As these minerals must have been deposited when those of their host limestone were already diagenetically mature (buried at least deeply enough for the formation of the calcites studied by GATTER, 1984), any Priabonian volcanism can be at most only partially responsible for the observed mineralization.

During Middle Badenian (Middle Miocene) times, the bulk of the thick Oligocene cover was eroded and partly redeposited in troughs around the elevated parts of the Buda Hills (KÓKAY, pers. comm.). This process exposed Triassic and Priabonian carbonates to surface karstification. Slow evolution then led to the second (still active) cave-forming hydrothermal phase of KOVÁCS and MÜLLER (1984). The present-day thermal springs of Budapest are believed to be heated by the normal geothermal flux, which is higher here than the continental average. The water is mainly of meteoric origin, i.e. from infiltration in karstic and non-karstic areas, whilst fluids of other origin are thought to be negligible contributors to the solutions, with the exception of carbon dioxide (see below).

The thermal flux at Budapest is associated with the comparatively thin lithosphere of the inner-arc Pannonian Basin: this is believed to have been even thinner in Late Tertiary times (HORVÁTH and ROYDEN, 1981). Free (or open) convection currents in the karst waters convey thermal energy to the spring areas from deeply buried parts of the carbonate formations. The model for the behaviour of these currents was developed by VENDEL and KISHÁZI (1964) and by LIEBE and LORBERER (1978a,b). Their views are represented schematically in Figure 2–65, in which it is also made evident why springs of very different temperatures can exist close to each other, e.g. in the area of the Lukács and Császár Baths, along the Danube River.

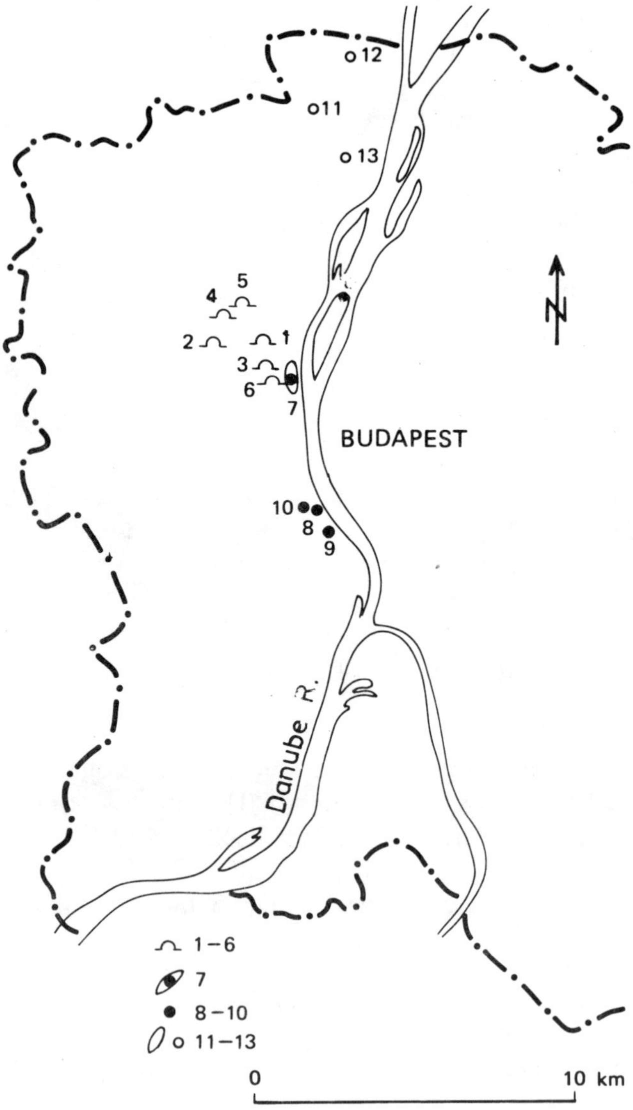

2–64. Location of karst springs and major caves in Budapest. 1–6. caves: 1. Szemlö-hegyi barlang, 2. Ferenc-hegyi barlang, 3. József-hegyi barlang, 4. Pál-völgyi barlang, 5. Mátyás-hegyi barlang, 6. Molnár János barlang; 7. warm and hot springs at Lukács and Császár baths; 8–10. hot springs; 11–13. warm springs.

While moderately warm waters (20 to 25 °C) have a chemical composition similar to that of regional cold karst waters, hotter waters contain much higher amounts of CO_2, HCO_3, Ca and Mg ions. It is proposed that the source of the increased CO_2 content (which is the cause of increased amounts of HCO_3, Ca

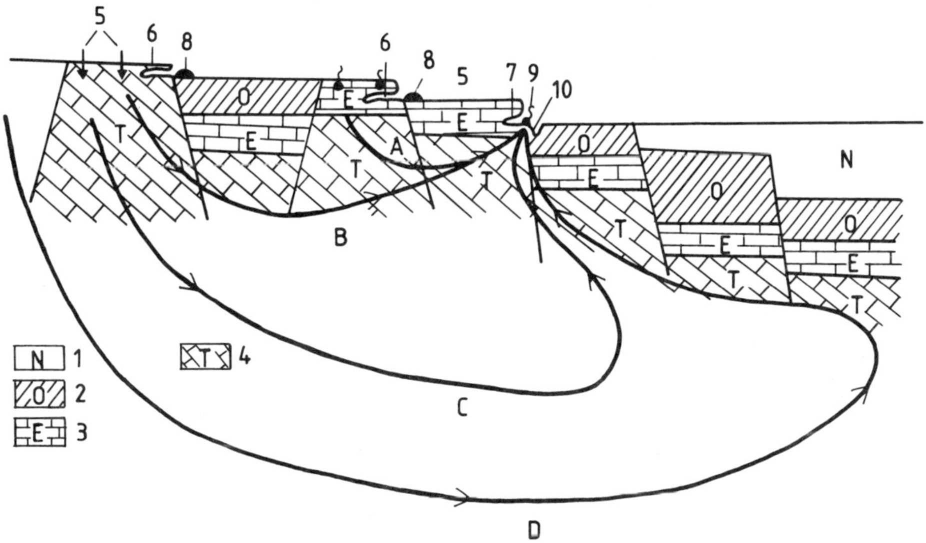

2–65. Schematic representation of the origin of hot and warm springs of Buda, as proposed by VENDEL and KISHÁZI (1964) and by LIEBE and LORBERER (1978). 1. Neogene; 2. Oligocene clays and silts; 3. Eocene limestones and marls; 4. Triassic dolostones and limestones; 5. infiltration; 6. fossil caves; 7. active hot-spring cave; 8. travertine deposits; 9. springs of different temperature; 10. Danube River. A–D = pathways of karst waters, B, C and D increasingly outside the profile.

2–66. Popcorn calcite (on the wall) and gypsum (on the floor) in the newly discovered József-hegyi Cave (Photo by T. Hazslinszky).

and Mg ions as well) is in metamorphic processes in the deep substratum (MÜLLER, 1971; ALFÖLDI et al., 1979). The ascending CO_2 is dissolved in the warm waters during their long-duration, circulating pathways, which extend over an area of approximately 1 000 km^2 in the case of the Buda springs. Areas of springs thus are often the zones of convergence of waters of different temperature and chemistry. The resulting mixing corrosion may be the main cause of the cave systems in the Buda Hills (MÜLLER and SÁRVÁRY, 1977). The most spacious caves, Szemlö-hegyi, Ferenc-hegyi and József-hegyi, are located closest to the Danube spring area noted above where waters emerging at different vents have temperatures differing as much as 45 °C, i.e. between 20 °C and 65 °C (Fig. 2–64).

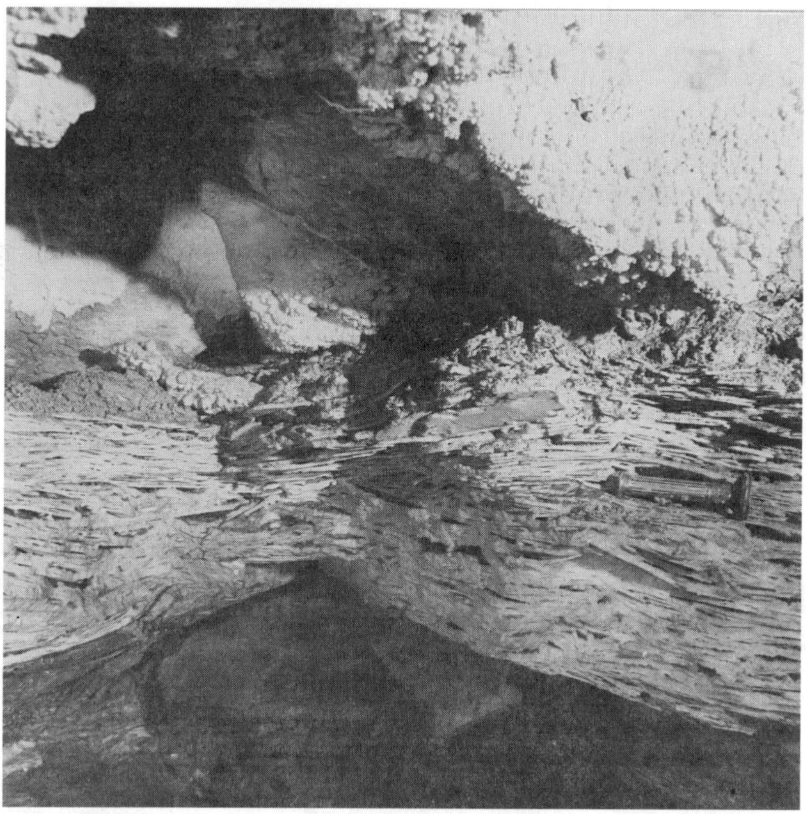

2–67. Accumulation of calcite laminae (calcite ice) in Pál-völgyi Cave (Photo by T. Hazslinszky).

The prominent horizontal extent of these caves may be connected with thermal-density stratification of the waters, causing mixing and corrosion to occur along subhorizontal planes. Several types of minerals are present in the caves. The most spectacular ones are *popcorn-like* calcite concretions (Fig. 2–66) covering cave walls in layers up to 50 cm or more in thickness. The concretions occupy the lower half or one-third of the height of the passages, while the uppermost walls are bare rock. Probably, after each lowering of the water level, a period of vigorous ventilation facilitated precipitation of the popcorn by removing CO_2 from the air immediately over the water. Thin films of calcite (*calcite ice*) precipitated on the water surface, too. Fragments of these laminae sank onto the bottom and were thickened there by further precipitation (KRAUSZ, pers. comm.; Fig. 2–67). The settling of such thin (fraction of a mm) sheets was controlled by flow patterns in the water and by wall geometry, resulting in local accumulations in forms likened to spruce or fir trees, which were subsequently covered by popcorn layers (Fig. 2–68). Similar formations were interpreted as geyser-cones by PANOŠ (1960). From fluid inclusion studies, popcorn concretions appear to have been deposited from waters very similar to those rising in the modern springs (GATTER, 1984).

Prominent, often very regular, spherical niches of one to five m in diameter are frequent in the upper, concretion free, parts of passages in the three caves noted above (Fig. 2–69). These niches (or solution pockets) often have only one

2–68. Szemlö-hegyi Cave. Spruce tree-like formation, interpreted as a geyser cone (geysermite) or as an accumulation of calcite laminae, covered by „popcorn" (Photo from the archives of Hungarian Institute for Environmental Protection).

entrance, at their lowermost part. MÜLLER (1974) and MÜLLER and SÁRVÁRY (1977) suggested that small-scale convection currents were responsible for their development, either in air or water. Physical properties of such convection currents were theoretically studied by RUDNICKI (1979), who proposed a similar

2–69. Cross-section of a spherical niche in a quarry. Sátorkö-pusztai Cave, 40 km NW of Budapest (Photo by T. Hazslinszky).

explanation. Spherical niches in general seem to be much more frequent in hydrothermal caves than elsewhere.

Thermal spring activity in the Buda Hills, probably very similar to the modern activity, can be traced back into the Early Pleistocene or even to Late Pliocene times, in the form of travertine deposits (SCHEUER and SCHWEITZER, 1974). The age of some of these deposits has been determined by contained mammalian faunas (JÁNOSSY in KROLOPP et al., 1976) and by molluscs. In the absence of fossils, the age of the caves must also be estimated by their elevation. In the case of Ferenc-hegyi and Szemlö-hegyi Caves one can assume a late Early Pleistocene or Middle Pleistocene age. Molnár János Cave at the Lukács Bath is the youngest member of this group and is still actively discharging water. Such estimates are based on the questionable assumption that, after the formation of the cavities, no major subsequent tectonic events have affected the hills containing them, i.e. the rate of uplift of the Hills has been uniform.

PALEOKARST OF ITALY. SELECTED EXAMPLES FROM CAMBRIAN TO MIOCENE

Maria Boni and Bruno D'Argenio

Introduction

We present here a review of selected examples of the paleokarst phenomena recorded in the Phanerozoic carbonate rocks of Italy. The country is characterized by a great variety of stratigraphic and tectonic phenomena, both in space and time. They compose a complex pattern of different orogenic systems or fragments of them, each showing peculiar and distinctive paleokarst features.

In the last 20 years the geological literature on Italian paleokarst has been quite abundant, although it has focused on a few topics that are mostly related to mineralization problems. Cretaceous bauxites, Triassic and Permo-Triassic base metals, baryte and fluorite ores have been interpreted in terms of a paleokarst origin. Only minor attention has been paid to paleokarst features not related to economic accumulations.

The different tectonic components of Italy have undergone several cycles of deformation and hence the conditions inducing emergence, karstification and karst fossilization in carbonate rocks have been controlled mostly by paleotectonic events. This is used as a main guideline in discussing the paleokarst occurrences. Of course, due attention has also been paid to the lithology and geometry of the sedimentary bodies and to the climatic and detrital fills, etc., which cause fossilization of both surface and underground karst cavities.

We have not surveyed the widespread Late Pliocene-Pleistocene paleokarst phenomena, which developed during the latest stages of uplifting and climatic change in Italy.

The Structural Grain of Italy

The General Setting

Lying in the centre of the Mediterranean, the land mass of Italy mostly coincides with a post-Paleozoic collisional mountain belt extending from Sicily and the Southern Apennines to the Northern Apennines and the Alps and then

P. Bosák, D. C. Ford, J. Głazek and I. Horáček (Editors):
Paleokarst. A Systematic and Regional Review. – Elsevier and Academia. Amsterdam and Praha. 1989.

and the Southern Apennines to the Northern Apennines and the Alps and then merging into the Dinarides and Hellenides. The convergence of these structures is generally towards the Adriatic Sea, which is considered the foreland to the Periadriatic Orogeny (D'ARGENIO, CHANNEL and HORVÁTH, 1980). Only Sardinia, which displays merely minor deformation in post-Paleozoic times, lies outside the belt, as it is on the western side of the Tyrrhenian Sea which is a Late Tertiary episutural feature (Fig. 2–70).

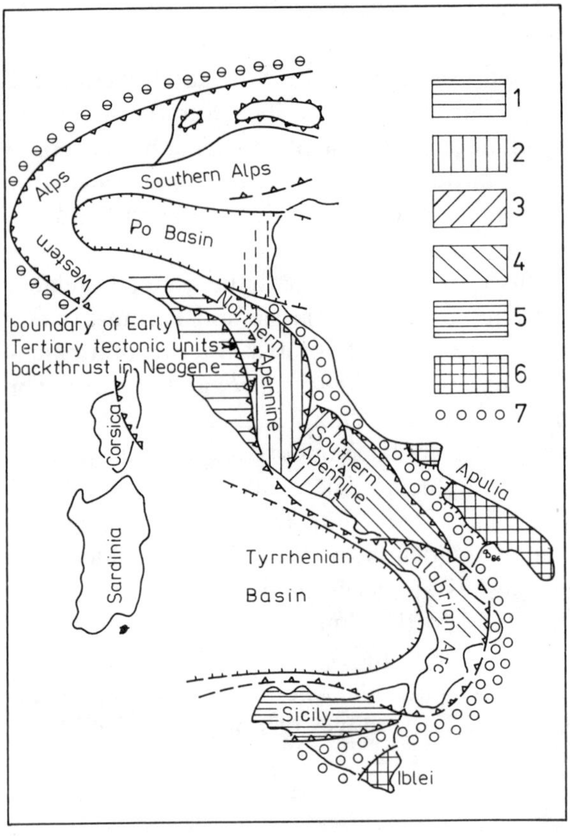

2–70. Main geologic provinces of Italy. 1. Liguria-Tuscany; 2. Umbria-Marche; 3. Latium-Abruzzi; 4. Campania-Lucania; 5. Sicily; 6. Apulia-Iblei foreland; 7. foredeeps.

Two Mesozoic continental margins of contrasting character, now deformed, may be recognized in this region. They originally faced the Jurassic-Cretaceous Tethys Ocean. The southern margin forms a belt extending from North Africa through Sicily, the Apennines, Southern Alps and Dinarides, to the Hellenides. The morphology of this margin began to evolve in the Triassic and continued during the opening of Tethys in the Jurassic and Early Cretaceous, when the sedimentary pattern was controlled by normal faults generally arranged sub-parallel to the junction between the oceanic and continental crust. This resulted in alternating shallow and deeper water sedimentary domains. The sediments were largely carbonate with little terrigenous influx.

In contrast, the northern margin, whose rocks (with the exception of Sardinia) are not discussed in this review, was characterized by a simpler continental shelf morphology. Between these two different margins there are various units of uncertain Mesozoic position, such as Calabria, Kabylias, and the Austro-Alpine units. However, these units were probably largely situated south of the main Jurassic oceanic spreading axis.

Deformation began on the eastern side of the Periadriatic Region in the Early Cretaceous and continued intermittently until the Late Tertiary, whereas the area on the western side was generally undeformed until the Late Cretaceous-Early Tertiary. The deformed areas of Italy may be divided into several sectors: Sicily, Calabrian Arc, Apennines, Western and Northern Alps, Southern Alps. Within the convex or concave curvature of the belt several basins, often strongly subsiding, developed during the Late Tertiary: Tyrrhenian Sea, Po Valley, Northern and Southern Adriatic Sea, the Ionian Sea.

Other peculiar elements of the Periadriatic Region are the transverse lines which cross the meandering orogenic belt. These juxtapose orogenic segments which display quite different structural characteristics or amounts of shortening or even of paleogeographic evolution.

The orogenic belts consist of thrust sheets piled up against stable forelands. Only in few cases does the foreland outcrop (Sicily, Southern Apennines), otherwise it is generally obscured by alluvial plains or covered by modern seas. The thrust sheet boundaries very often follow the main facies trends, with the exception of the Southern Alps which are less deformed and where the facies belts are discordant to the strike of the chain.

The Paleozoic evolution of peninsular Italy is less clear, because rocks of this time interval crop out mostly in the nappe structures of the Alps, Tuscany and Calabria.

The Paleozoic evolution of Sardinia is better known. Its rocks were deformed during Caledonian and Variscan times, then peneplaned and covered by Mesozoic and Tertiary deposits that are almost undeformed. The Paleozoic rocks range from Lower Cambrian to Carboniferous in age. These rocks suffered regional metamorphism which affected the southern part of the island only slightly. However to the North they are metamorphosed into phyllites, micaschists and gneisses, and grade to magmatic-anatectic granites even farther north (Cocozza et al., 1974). During Late Variscan times, many granitic stocks accompanied by dyke swarms were emplaced.

Sicily

Sicily is a cross-roads in the complex of Mediterranean geology, because the sedimentary and tectonic events that characterize its history have taken place (from Late Paleozoic to Pleistocene times) in an area of high mobility (Catalano and D'Argenio, 1982). The most stable part is the south eastern corner (Iblei Plateau) which displays foreland characteristics and African affinities.

The foredeep is filled with 1 000–1 500 m of Upper Miocene-Lower Pleistocene terrigenous, evaporitic and clastic carbonate sediments with intercalations of olistostromes. These sediments cover the front of the Sicilian chain. The chain has an East-West trend and consists of piled thrust sheets verging to the South. Most of this chain crops out on the Northern side of the island, extending towards the South along two transverse belts which connect the Palermo Mountains with the Sciacca Mountains and the Nebrodi Mountains with the Hyblean Plateau. In the northeastern corner of the island (Peloritani Mountains) a pile of thrust sheets formed of sedimentary and crystalline terrains crops out as part of the Calabrian Arc and overlies the Sicilian thrust belt with a complex tectonic contact (Taormina Line).

The different units of the chain were created during the Late Tertiary in several tectonic phases. Two main stages are recognized: Early-Middle Miocene and Tortonian-Pliocene. The original thrust planes have been progressively dissected by Plio-Pleistocene normal faulting that has produced different amounts of tilting and uplift. In terms of paleogeographic and tectonic restoration, there were several stages of evolution beginning with extension (Triassic-Early Tertiary), and followed by compression (Late Tertiary) which preceded and partly accompanied the Pleistocene uplift of the island.

Calabrian Arc

The Calabria-Peloritani Arc is a strongly arched orogen connecting the Apennines with Sicily and trending toward the African foreland. It consists of the northeastern corner of Sicily (Peloritani Mountains) and the southernmost part of peninsular Italy (Calabria). In the last 20 years several genetic models have been proposed for this segment. At present the Calabria-Peloritani Arc is interpreted as a fragment of an Early Tertiary Europe-verging "Alpine" chain, which was backthrust onto the Periadriatic continental margin during Late Tertiary times (AMODIO-MORELLI et al., 1976; BONARDI and GIUNTA, 1982).

The geologic history of the Calabrian Arc includes two main orogenic cycles: Variscan and Alpine. Rocks of the Variscan cycle are present in the Longobucco and Longi-Taormina units (Sila, Aspromonte and Peloritani area or Calabro-Austro-Alpine domains) and in the basement of the Mesozoic-Early Tertiary basins.

This area is thought to have been located originally on the southern side of Tethys, between the domains of the Apennines and those of the future Alpine chain.

Apennines

The Apennines display a nappe structure very similar to that of Sicily. There are four main sectors, Campania-Lucania and Latium-Abruzzi comprising the Southeastern Apennines, and Umbria-Marche and Liguria-Tuscany comprising the Northern Apennines. Several different structural units appear within them. Some of these belong to *internal* (oceanic or quasi-oceanic) domains and occupy the highest structural position in the nappe pile. Internal nappes are known both in the Southern and in the Northern Apennines (*Ligurid* and *Silicid* nappes) and, even though they originated as nappes during the Late Cretaceous-Early Tertiary, they were backthrust into their present tectonic positions only during the Late Tertiary.

The foreland crops out only in the Southern Apennines (Apulia). In the North it is buried under the Adriatic Sea. Towards the foreland both the topographic relief and the amount of tectonic thickening decreases. Only gravitational gliding (of Pleistocene age) occurs in the latest stage of deformation.

The foredeep has an almost continuous outcrop from the Abruzzi to Sicily; it consists of thick marine Pliocene and Quaternary clastic sediments derived almost entirely from the uplifted Apennine and Sicilid nappes. These marine sediments are covered unconformably by a generally thin blanket of continental Quaternary deposits.

The nappe system in this chain forms two main arcs separated by a transverse line (Ancona-Anzio Line). Thus, a partition of the Apennines into two segments is generally accepted. The Ancona-Anzio Line, an oblique overthrust (over-printing a transcurrent fault) which runs roughly from North to South, separates a western segment (Northern Apennines) from an eastern one (Southern Apennines).

On the concave side of the Apennine arcs extensive volcanic activity began in the Late Tertiary together with the main uplifting phases, which sometimes involved more than 5 km of vertical displacement. The deepening and oceanization of the Tyrrhenian Sea occurred during the same interval (SELLI, 1974).

Southern Alps

The Southern Alps, together with the Austro-Alpine structures of the Eastern Alps formed the northernmost fragment of the African continental margin, facing the Jurassic Tethys Ocean. Their paleogeographic domains are considered part of the African Promontory, or Adria, whose evolution has been recently reviewed (D'ARGENIO, CHANNEL and HORVÁTH, 1980).

While the Southern Alps appear only mildly deformed, the Eastern Alps have been detached and displaced to the North to create the complex Austro-Alpine nappe system. The Insubric Lineament separates these two areas.

The Southern Alps plus their southern (*external*) border were affected by compression during Late Cretaceous and Tertiary postcollisional events. The Alps *s.s.*, were already an uplifted chain and acted as an obstacle to further northward displacement of southern tectonic elements, forcing deformation towards more readily available southern areas. Some sectors underwent more intense compression because there underthrusting of the lower crust was easier, while others were only slightly deformed.

Up to Middle-Late Miocene times, although there are local differences in each sector, the tectonic activity displays a common timing. Probably the principal event in the Western sector of the Southern Alps developed during the Late Oligocene. Upper Miocene-Pliocene deformation is very weak or lacking there although there was more intense deformation in the eastern sector, still involved in the compression.

The Southern Alps are strongly differentiated along their E-W axis. South of the Insubric Lineament, a number of different structural sectors can be identified (CASTELLARIN, 1982): (a) Canavese and Ivrea-Verbano, (b) Lombardian Alps, (c) Judicarian zone, (d) Dolomites, and Lessini, (e) Eastern Dolomites, Cadore, Carnia, Friuli.

In contrast to the other Periadriatic sectors, deformation and shortening in the South Alpine region developed quite obliquely to main facies trends.

Pre-Alpine deformation (mostly Variscan) is known particularly in the eastern Southern Alps (Paleocarnic Chain; VAI, 1970) where a nappe pile of Carboniferous age forms the original basement of the Alpine sequences and was further deformed with them during Tertiary times.

Sardinia

West of the Tyrrhenian Sea, Sardinia and a large part of Corsica comprise one large block which has often been referred to as a microplate or microcontinent. However, Sardinia reached its present position in the Mediterranean only during the Miocene and, from a geologic stand-point, belongs in large part to the European foreland of the Alpine chain. Together with much of Corsica it is an extension of the Maures-Esterel and Montagne Noire of France (COCOZZA et al., 1974; COCOZZA and JACOBACCI, 1975). Sardinia may be considered part of the European foreland, detached from the mainland during the deformation of the European (*Alpine*) margin of Tethys in the Early Tertiary. ARGAND originally proposed, and it has recently been supported by paleomagnetic data, that Sardinia was rotated counterclockwise more than 30° (max 45°) away from the European mainland during the Oligocene and Early Miocene.

Paleozoic Paleokarst

There is little evidence of pre-Mesozoic paleokarst in Italy. The more interesting features have been studied extensively because of their association with ore deposits. They are known in SW Sardinia (Cambrian and Ordovician; BONI, 1982) and in the Carnian Alps at the contact between Devonian and Carboniferous sediments. For the latter, however, a non-karst model of genesis has also been proposed (SPALLETTA, VAI and VENTURINI, 1981).

Sardinia

Cambrian

The Iglesiente-Sulcis area (SW Sardinia) experienced several karst cycles during its long geological history, some of which are linked to significant concentrations of ore minerals. A first erosion cycle, which can occasionally be correlated with one kind of karst feature is thought to have occurred during deposition of the lower part of the Laminated Dolomite Member of the Gonnesa Formation (Lower Cambrian). Such episodes have been attributed to very gentle epeirogenic movements of the carbonate platform. The accompanying dissolution phenomena are also very slight. They are early diagenetic and, most probably, contemporaneous with the dolomitization itself. This is probably to be classified as *microkarst*, with small cavities and enlarged microfractures often filled with baryte or sparry carbonates. According to CARANNANTE et al. (1974), COCOZZA et al. (1974) and BONI (1979), at the end of some of the depositional cycles in the Laminated Dolomites, erosive, fairly oxidising phases of emergence must have taken place leading to the formation of iron-rich paleosoils. There are stratiform baryte and sulphides within some depositional cycles also. On the basis of further observations throughout the Cambrian district, we (the present authors) cannot support the proposition that the accumulation of metals was a result of Cambrian eluvial enrichment. The iron oxides and calamine ores in the Laminated Dolomites seem a later oxidation phenomenon of the earlier sulphides. Evidence of this origin includes the spongy boxwork structure of the ores (very similar to the classical iron-caps) with its highly porous and earthy-looking accumulations of calamine and lead-oxide minerals. Such diagnostic features may be observed quite clearly, e.g. in the Baueddu Mine and in the Malacalzetta-M. Cuccu Mannu outcrops. In addition, more evidence has been found in drill cores from the S. Benedetto and Candiazzus Mines which show many similarities to the well-known, unaltered massive sulphides of the Campo Pisano-Genna Luas and Seddas Moddizzis Mines.

Ordovician

The second major karst cycle of Sardinia also occurred in the Iglesiente-Sulcis. It can be attributed to the Sardic tectonic phase and the ensuing emergence of the folded Cambrian series. It is characterised by large cavities fossilized mainly by supergene mineralization. These cavities can be traced quite easily along almost all of the contact between the Cambrian limestone and dolostone and Ordovician sediments, from Candiazzus-Buggerru in the NW part of the Cambrian region right through Oridda (Arenas) to the numerous outcrops in Southern Iglesiente (S. Giovanni, Arco Sa Gruxi) and the Sulcis area (Concas de Sinui, Villamassargia, Barbusi; PADALINO et al., 1973). This is a classical example of a karst phase which developed over a folded basement as shown by LAGNY and ROUVIER (1976). The very strong climatic variations (from warm-humid to dry) indicated clearly by the hematite and red cements of the Ordovician transgressive puddingstone, may also explain the massive erosion and karstification of the Cambrian rocks. It includes deep lateritization and accompanying mineralization. Climatic influences should also have played a part in the epigenetic dolomitization and accumulation of some pre-Ordovician quartz crusts of varying thickness (PRETTI, SALVADORI and VIOLO, 1978).

The unconformity surface is characterized by oxidation minerals such as cerussite, hematite, calamine and some baryte, mostly in the Sulcis area. Predominantly sulphide ores with baryte gangue have been found occasionally (galena > pyrite-sphalerite). Fluorite and copper seem to be present also along the Ordovician contact, and in greater concentrations than in the older Cambrian ores, sometimes being of an economic grade.

In our opinion (BONI, 1982) it is unlikely that these varied mineralizations derive only from one or a few karst episodes that preceded the Ordovician transgression. Rather they indicate a range of depositional phenomena on the complex morphological surface that developed upon the folded Cambrian terrains. In fact, besides the features that quite clearly may be ascribed to a pre-Ordovician karst (i.e. Arenas, Monte Pira Roma), there are certain orebodies (chiefly consisting of sphalerite, pyrite and galena) which probably already show the characteristics of real transgressive sediments in the more depressed areas of Cambrian structures.

Carnian Alps

Devonian

Over a very large area extending for about 100 km from the Monte Croce Comelic pass to the Karawanke region, a complex of polymetallic deposits occur. This is genetically related to a Middle and Late Devonian paleorelief and contains Cu, Pb, Zn, Sb, Hg, $BaSO_4$ and CaF_2 (BRIGO and DICOLBERTALDO, 1972; OMENETTO and BRIGO, 1974). The ore-bearing horizon is a lithostratigraphic unit lying over the paleorelief and covered by transgressive Lower and Upper Carboniferous terrigenous deposits. The geometry of this horizon is related to the paleotopography of the Devonian relief, which was controlled by the synsedimentary tectonics and by a karst-type erosional cycle. The intensity of this erosion appears to have varied in different areas, depending upon the local time of emergence.

The strata-bound deposits are related to this evolution and consist of stratiform bodies at the base of the Carboniferous plus discordant vein and paleocavity fillings within the Upper Devonian limestones. The ore-bearing horizon shows different paragenetic associations and consists mainly of a more or less mineralized siliceous crust with the higher ore concentration occurring in the cavities and veins beneath it. The dominant ore minerals are sphalerite (Cd- and Ge-bearing), tetrahedrite, bournonite, galena, chalcopyrite and cinnabar, with accessory Ni (Co)-minerals. Locally, large quantities of fluorite and baryte predominate. The scarcity of pyrite and marcasite is remarkable. It has been suggested that the ore concentration is the result of progressive pedogenic enrichment during the Devonian emergence (VENERANDI-PIRRI, 1978), as in the case of the Ordovician ores in the quartzites of SW Sardinia.

The regional zoning seems to reflect the original distribution of primary anomalies, probably depending upon consistently discriminated metallic and non-metallic ion supplies from the continental areas (and from contemporaneous volcanic activity?).

A different view concerning the genesis of the mineralization between the Devonian carbonate facies and Variscan radiolarites and flysch has been expressed by SPALLETTA, VAI and VENTURINI (1981). With the exception of the Pontebba area these authors do not support a paleokarst origin of the cavity network in which most of the minerals are concentrated and instead favour deposition in neptunian dykes during a sudden deepening of the Devonian carbonates. The source of the metal ions is supposed to be from oceanic volcanism through upwelling processes, this being due to the purported intraoceanic and isolated position of the Paleocarnic (Devonian) carbonate platform.

Permotriassic and Triassic Paleokarst

Sardinia

In the Iglesiente-Sulcis region, in addition to the formation of karst in the Paleozoic, there was intense activity at the boundary between Paleozoic and

Mesozoic times. This is recorded in the so-called *Post-Hercynian Peneplane*. The related deposits have been investigated in detail because of their ore content (baryte and silver-rich galena). The mineralization has been exploited for a very long time; silver was known to the Phoenicians and Romans and in the Middle Ages was mined again by the Pisans. The geologic setting together with some genetic considerations is summarized here.

The last phase of the Variscan deformation in Sardinia was followed by a general *rejuvenation* of the Paleozoic relief that had already developed during a previous long period of emergence. A new, intensive erosion cycle caused deep denudation of the Paleozoic rocks, including weathering of the limestone and dolostone of the Cambrian Gonnesa Formation. The combined effect of weathering and percolating water, taking place under highly variable climatic conditions in this Formation with its high metal content (more than 1 000 ppm Pb+Zn in places), produced a peculiar karst system. As matter of fact, a very special climate is believed to have prevailed in the Iglesiente-Sulcis area (as in most of the rest of continental Europe) at this time, marked by strong arid periods during which lagoonal-to-evaporitic conditions predominated.

Such characteristics are typical of the Campomà area (COCOZZA and GANDIN, 1976), where the sequence is of Triassic age. However, it appears likely that the major karst formation was established as early as in Permian times, because of the peculiar climatic conditions. These led to the formation of thick paleosoils rich in iron and baryte which spread throughout the whole area, reflecting oscillations between arid climates and wet climates favouring karstification. During the Triassic the aridity increased and the rainfall became more sporadic. Evaporation increased also. These conditions could have produced high salt contents in the circulating waters and consequently pH and the solubility of silica rose constantly. Ba and other metal ions became exceptionally mobile in chloride-rich solutions. Large karst cavities formed and were then filled. Cave fills of chemical as well as of fragmental origin developed. Collapse breccias predominate, probably an indication of more arid climatic conditions.

Between the different depositional fabrics, textures associated with recrystallization are common. Calcite is present in many different types, i.e. as detrital fragments, as idiomorphic zoned grains of variable size, and as chemically precipitated *crystal meadows*.

Baryte occurs either as large and late tabular crystals which frequently replace carbonates or in the matrix as thin needles which are replaced by quartz and galena pseudomorphs. Excluding the fragments of country rocks in the breccia, the gangue minerals consist of coarse crystalline quartz and minor calcite. Quartz is also abundant as a microcrystalline constituent in the fine-grained internal sediments of the karst cavities, where it has been found as idiomorphic bipyramidal crystals with inclusions of carbonates. Quartz occurs also as cements in the conglomerates which at places rest unconformably on eroded internal sediments. Silicification is associated with those conditions also required for epigenetic dolomitization e.g. the well known *yellow dolomite* (BONI, 1982). Alteration occurred not only along fracture zones and cavity walls but also down almost vertical ore-bearing horizons, extending to considerable depth. Zones where this paleokarst is particularly well developed are the Iglesiente Valley (for instance, Mt. S. Giovanni and P. ta Candiazzus) and to the south, the Barega-Monte Arcau-Sa Bagattu hills and Barbusi-Monte Tasua and the hills around Nebida and Masua. They were explored for economic ore deposits in the past.

In the Buggerru area there was a different kind of karst development that caused intense fracturing along still active tectonic lines, with local circulation of metal-rich waters and the formation of a few large cavities containing Permo-Triassic ores. Of additional interest is the clear distinction in this kind of Permo-Triassic ore between real cave fillings that combine detrital and chemical sedimentation in small karstic networks, and the bigger veins that follow important tectonic lines for several hundred metres and in which chemical deposition prevails. In most cases the mineral and ore paragenesis are the same in both the karst feature and the veins.

From the similarity of paragenesis in the tectonic and karst cavity fillings (Reigraxius, Malacalzetta, etc.), one may wonder whether many of them should be regarded as mixed products of both pure karst leaching and deposition with normal meteoric waters (BONI, 1979; PADALINO et al., 1973), and of tardo-magmatic or tardo-tectonic thermal waters. Although the thermal karst model (DŻUŁYŃSKI, 1976) is not a favoured theory for the Sardic ores at the present time, it could help to elucidate many geochemical relationships apparent in the ore-forming processes (BONI, 1986). Also it may provide controls on the sedimentary environment which was similar to that where the modern formation of special kinds of calcareous tufa and crusts from hot waters in geothermal fields and strongly tectonized zones took place.

Southern Alps

There are important paleokarst features in the Triassic carbonate rocks of the Southern Alps. Lead-zinc-fluorite-baryte deposits are associated with some of them (ASSERETO et al., 1976). The paleokarst cavities have a different stratigraphic position and characteristics in the Lombardian Prealps when compared to the more eastern Cadore and Carnian regions due to a differing paleogeographic evolution. There are many similarities between the mineralization in the Southern Calcareous Alps and that of the same age in Austria (Bleiberg) and Yugoslavia (Mežica).

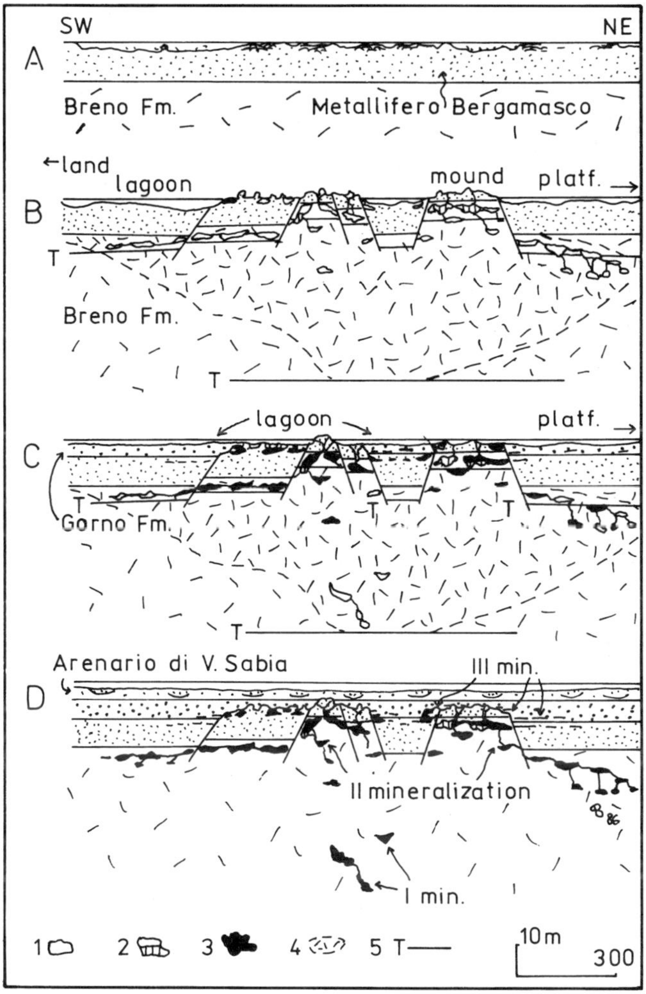

2–71. Reconstruction of the paleogeographic evolution and metallogenic phenomena during the Early Carnian at Paglio Pignolino (after ASSERETO, JADOUL and OMENETTO, 1977). A. Eastern epicontinental platform of the Bergamasco Metallifero Limestone; B. The crisis on the platform due to local emergences and terrigenous influx. Karstification started; C. Internal lagoon with brackish water and poor circulation. There is a high concentration of organic matter and syngenetic ore in the black shales. Subterranean karstification continued and karst cavities were filled with mineralized primary sediments; D. Fossilization of karst forms and of ore deposits by the expanding marine delta. Filling of the karst network finished, followed by the diagenetic changes of the mineralization. 1. karst cavities; 2. early internal sediment, mainly calcitic; 3. mineralized internal sediment; 4. hypothetical maximum extent of aquifers with fresh and brackish water; 5. main impermeable horizons (tuffites and marls).

Middle-Late Triassic of Lombardy

A thick Ladinian carbonate build-up (Esino Limestone) is known in Val-Brembana. At the top of the sequence (Upper Ladinian) an emergent phase is recognized. This may be correlated to the important tectonic activity accompanied by volcanism that occurred at the Ladinian-Carnian boundary. The subaerial exposure produced vadose diagenesis, karstification and dolomitization below the surface. Other sediments were deposited above the Esino Limestone during the Late Ladinian and Carnian; the discontinuous Calcare Rosso and above it the Breno Limestone, both displaying peritidal carbonate platform facies with well-marked cycles (OMENETTO and VAILATI, 1977).

Between Camerata Cornello and Monte Vaccareggio karst networks are particularly well developed in the Esino Limestone, with cavities up to 2–3 m in diameter filled with carbonate and clastic deposits. The whole rock was then extensively dolomitized.

At Paglio Pignolino there is minor mineralization (fluorite and Pb–Zn sulfides with quartz and sparry calcite gangue) in the karst cavities (ASSERETO, JADOUL and OMENETTO, 1977; Fig. 2–71). In the same area, three levels of elongated, interconnected paleokarst cavities of varying dimensions (generally some tens of m) occur in the upper part of the Breno Formation. The cavities are of two main types: (a) stratoconcordant, mainly horizontal, with roof collapse breccias; (b) discordant and filled with chaotic material. The filling is extremely varied, ranging from collapse breccias to laminated marly sediments, crystal silt, ore-bearing sediments and cements. During the Lower Carnian (upper part of the Breno Formation or Calcare Metallifero) a large stromatolitic coastal plain developed on the carbonate platform (Gorno). Synsedimentary tectonics produced a paleorelief which was probably associated with a more southerly emergent volcanic region (BRUSCA et al., 1981). This second emergence took place at the top of the Calcare Metallifero, which suffered erosion and karstification. A network of vadose and phreatic cavities was created, partly controlled by impermeable horizons in the underlying carbonate units. A thin layer of black clays was then deposited during the ensuing marine transgression at the top of the *Metallifero Bergamasco* (Gorno Formation; ASSERETO, JADOUL and OMENETTO, 1977).

This drowned paleorelief created a number of submarine highs which were covered by black shales that display early mineralization with fluorite, quartz, minor baryte and lead-zinc sulphides at their base. Evidence of seeping down of the ore muds (the mineralized black shales) into the underlying karst cavities is often observed in the Gorno district. Further, the persistence of karst voids provided additional opportunities for ore concentration by polyphase recrystallization, solution-collapse, migration and redeposition, all superimposed on the primary karst fabrics.

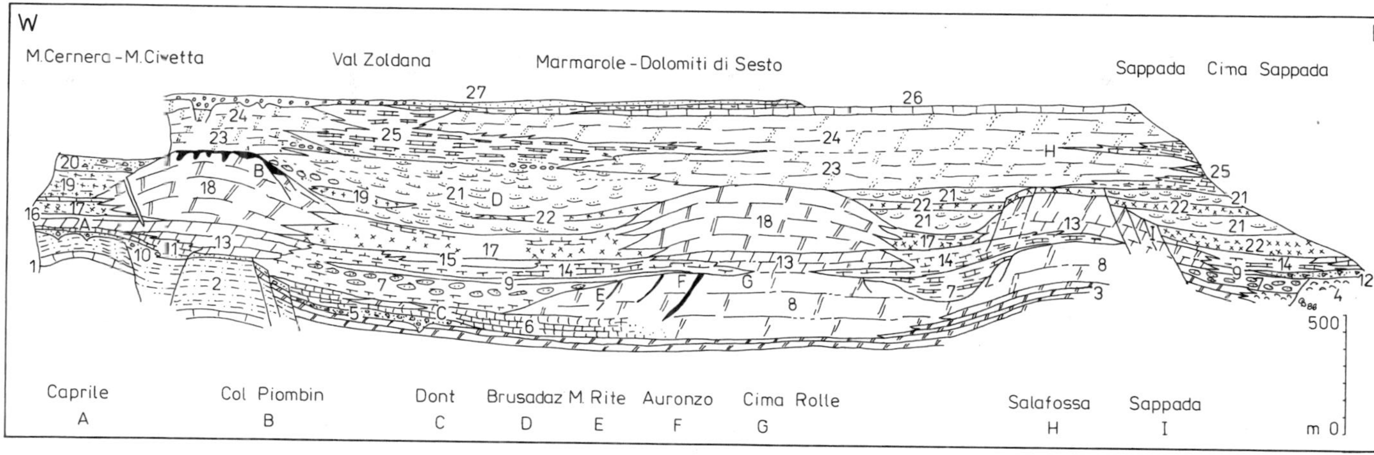

2–72. Stratigraphic and structural sketch of the Triassic in the Eastern Dolomites and the position of Pb-Zn mineralization (after ASSERETO et al., 1977). 1. Bellerophon Formation; 2. Werfen Formation; 3. Lower Serla Formation; 4. Lusnizza Formation; 5. Lower Conglomerate; 6. algal nodular limestone; 7. Dont Formation; 8. Upper Serla Formation; 9. Bivera Formation; 10. Richthofen Conglomerate; 11. Morbiac Limestone; 12. Ugovizza Breccia; 13. Contrin Formation; 14. Ambata Formation; 15. Daonella Marl; 16. Plattenkalk Formation; 17. Buchenstein Formation; 18. Sciliar Dolostone; 19. basic volcanics; 20. Marmolada Conglomerate; 21. Wengen Group; 22. Pseudobuchenstein Formation; 23. Lower Cassian Dolostone; 24. Upper Cassian Dolostone; 25. S. Cassiano Formation; 26. Dürrenstein Dolostone; 27. Raibl Group.

Middle-Late Triassic of Cadore and Carnia

The Triassic sequences of the Cadore and Carnia are characterized by carbonate platforms bordering deeper basins where clastic sediments accumulated. Four platform "episodes" have been recognized; Serla Dolomite (Lower Anisian), Contrin Formation (Upper Anisian), Sciliar Dolomite (Ladinian), Cassian Dolomite (uppermost Ladinian-Middle Carnian). From the Anisian to the Carnian there were three main erosional phases. The emerged areas were located mainly towards the west during the Anisian and towards the south during the Late Ladinian and the Carnian. Thick volcanoclastic sediments accumulated in the basins mainly during the Ladinian. Lead and zinc mineralization occurs mostly along the rims of the carbonate platforms, within breccias and/or cavities that are possibly related to synsedimentary faults. Therefore, only some of them may be associated with paleokarst phenomena (Fig. 2–72).

Lead and zinc deposits are known in three of the four carbonate buildups. In the Serla Dolomite the mineralization is within karst cavities according to LAGNY (1971). It is distributed over a wide area at the margins, where the carbonate platform grades into basinal facies. However, according to ASSERETO et al., (1977a), there is no clear evidence of emergence in the Serla Dolomite and the ore bodies are associated with synsedimentary faulting and the infilling of scarp breccias and neptunian dykes. At Monte Rite, however, the ore bodies (baryte and sulphides) do fill a real paleokarst network in the Serla Dolomite, later fossilized by deeper water sediments (ASSERETO et al., 1977). In the Sciliar buildup only one Pb-Zn deposit is known where the genetic interpretation is karstic (Col Piombin). In the Cassian Dolomite (Late Ladinian-Carnian) the Salafossa lead-zinc deposit, which is well known and of considerable economic interest, is located within the upper part of the carbonate body and penetrates some one hundred metres downwards. A karst genetic model was proposed for Salafossa by LAGNY (1971, 1974), and CROS and LAGNY (1969). A contrasting interpretation was offered by ASSERETO et al. (1977). In fact, there is disagreement on the age of the Salafossa host-rock, not to mention the age of the orebody. The former was interpreted by LAGNY (1975) as Late Anisian and by ASSERETO and BRUSCA (1976) as Late Ladinian-Carnian. The mineralization is considered by LAGNY (1974) to be the filling of paleokarst cavities associated with a Late Anisian unconformity, and by ASSERETO et al. (1977a) as an epigenetic impregnation of tectonic breccia along a fault of Alpidic age (probably a rejuvenated Triassic fault).

The main parts of the Salafossa orebody (90 %) consist of a network of fractures and breccias with sparry dolomitic and Zn-Pb sulphide cements. A minor portion of the orebody (10 %) consists of true karst cavities that are filled by small breccias and internal sediments, both being interpreted as deposited later than the main mineralization episode (ASSERETO, JADOUL and OMENETTO, 1977).

Middle-Late Triassic of Friuli

In the Tarvisio-Pontebba area a large Pb–Zn ore concentration is exploited at the Raibl-Cava del Predil Mine. This mineralization is either confined in the Late Ladinian-Early Carnian *Dolomia Cassiana* (KLAU and MOSTLER, 1983) or it is located along the lowermost part of the overlying Raibl Formation (dark grey bituminous marls and dolostones of Carnian age).

There are no clear relationships between the Raibl mineralization (mainly along faults) and unquestionable paleokarst. However, some internal sediments described by BRIGO and OMENETTO (1978) show similarities with the paleokarst ores of Permo-Triassic age of SW Sardinia (BONI and AMSTUTZ, 1982).

Sicily

Late Triassic of Sicily

In the Late Triassic (Norian) carbonate platform of Western Sicily (Trapani and Palermo Mountains), a sequence crops out which is formed by some tens of cyclothems deposited in environments ranging from subtidal dolostones with *Megalodon* to intertidal loferitic dolostones and dolostone breccias (CATALANO, D'ARGENIO and LO CICERO, 1974). At the end of the supratidal intervals, there is evidence of paleokarst solution resulting in the formation of small cavities (up to some dm in width) filled with vadose silt and many generations of cement. The dissolution may also affect the underlying subtidal interval.

Jurassic Paleokarst

Jurassic paleokarst is reported from Sicily, the Apennines and the Southern Alps. However, the infilling materials are often marine and the cavities themselves may be considered to have originated only in part by solution during subaerial exposure. Generally the features reported as paleokarst are located in the topmost part of Late Triassic-Jurassic shallow water sequences that suffered rapid drowning and remained as submarine highs covered by reduced pelagic sediments.

The Trapanese domain of Sicily is a facies belt extending from the westernmost coast eastwards for 100–200 km. It displays a number of seamount-type structures that developed during the Jurassic due to discontinuous drowning of a Triassic carbonate platform. Middle and Upper Jurassic basinal sediments were deposited around these submarine paleohighs and on top of condensed sedimentary sequences at the top of them. A network of cavities both of the neptunian dyke type, or parallel to the bedding and several metres long but seldom more than 1 m high, may be observed in the platform carbonates directly below the condensed sediments. The cavities may be related to paleokarst solution and partial filling (WENDT, 1971; JENKINS, 1970) during short emergent episodes immediately preceding the drowning. The last stages of cavity filling, however, are of pelagic sediments (WENDT, 1971; ABATE et al., 1982).

Reports of several Jurassic paleokarst features of minor importance in the Apennines and Southern Alps may be found scattered in the geologic literature of the last 20 years. COLACICCHI (1967), discussing the Central Apennines, reports occurrences of paleokarst in the shallow water Lower Jurassic carbonates of the Marsica area (Prato Rosso). COLACICCHI and PIALLI (1967) discuss some paleokarst in the Jurassic (late Lower Jurassic) of Umbria (Monte Cucco). FAZZUOLI and SGUAZZONI (1984) describe paleokarst cavities in the Lower Jurassic of the Alpi Apuane (Mte Pisanino). In the Dolomites CASTELLARIN (1972, 1982) describes Jurassic paleokarst cavities within shallow water carbonates. In this as in the previously mentioned occurrences, the distinction between true paleokarst and convergent forms is sometimes difficult.

Cretaceous Paleokarst

Southern Apennines and Apulia

Mid-Cretaceous

Widespread bauxite deposits of Cretaceous age rest on paleokarst unconformity surfaces in the shallow water carbonate sediments of Southern Italy. These bauxites, though mined until a few years ago, have never been of major economic importance except in Apulia. A comprehensive study of this subject was published by BÁRDOSSY et al. (1977).

The bauxites are emplaced within the sequence of two paleogeographic units, the Abruzzi-Campania carbonate platform and the Apulia carbonate platform. These were originally separated by the Molise-Marsica Basin. The bauxites mark a widespread stratigraphic hiatus created by an emergence and are disconformably overlain by Cretaceous rudistid limestones. They may also pass laterally to bauxitic breccias or even to clayey deposits (D'ARGENIO, 1969). In addition, in the Abruzzi-Campania carbonate deposits, two almost parallel belts with bauxite intercalations are known. In these belts, which run roughly along the strike of the Apennines, the width of the stratigraphic hiatus increases eastward. To the east only one horizon is known, representing a large interval, sometimes embracing the Albian (and even part of the Aptian) and the Cenomanian-Turonian. To west, two bauxitic horizons have been recognized, each representing a shorter emergent episode. The lower bauxitic horizon is the most conspicuous, the upper one being very thin or absent (in which case the hiatus may be documented biostratigraphically).

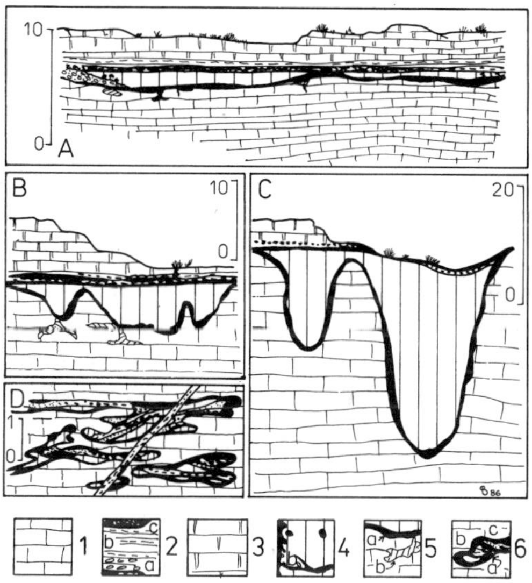

2–73. Forms of bauxite bodies. A. Flat, contiguous lenses; B. Thicker, scattered lenses; C. Vertically developed bodies, infill of sinkhole or canyon-like depression or cavity; D. Subterranean network of karst cavities, filled or mostly only partly filled with bauxite or reworked bauxitic material: 1. footwall limestones, generally white or pink micrites and biomicrites; 2a. calcareous conglomerates with clayey matrix; 2b. green or red clays; 2c. brown coal seams; 3. hanging wall limestone, white intra and biomicrites, sometimes stromatolites and tidal channel breccias; 4a. bauxite; 4b. calcareous breccia embedded in the bauxitic material; 5a. calcitic crusts at the bauxite/footwall contact; 5b. karst cavities in the uppermost footwall; 6. paleokarst cavity network contoured by calcite crusts and filled with laminae of reworked bauxites (6a), varicoloured micrites and calcisiltites (6b) and bioclastic calcarenites (6c).

Lower paleokarst horizon. The lower bauxite horizon is never continuous but consists of many small concentrations of different type, grouped in several outcrop areas. Its thickness is generally variable, from a few cm to more than 10 metres. As in most karst bauxites, the cavities filled with bauxite were generated by solution. The forms of the bauxite bodies (BÁRDOSSY et al., 1977) are various: (i) flat contiguous lenses, whose thickness ranges from 50 to 200 cm; (ii) thick scattered lenses, never thicker than 10 metres; (iii) vertically developed bodies filling canyon-like cavities with steep walls; thicknesses greater than 40 metres, and (iv) networks of karst cavities and veins rimmed by calcite crusts and filled with laminae of reworked bauxites, varicoloured micrites and calcisiltites and bioclastic calcarenites (Fig. 2–73). The latter type of cavities is generally restricted to the upper horizon in the more western belt.

Each different mode of occurrence predominates in one or more districts, although it is not entirely confined to them: (1) prevalence of type (i) forms: Campania, type area is Monte Maggiore; (2) prevalence of type (ii) forms: Abruzzi, type area is Monti d'Ocre; (3) prevalence of type (iii) forms: Apulia, type area is Northern Murge; (4) occurrence of type (i+ii) forms: Apulia, Gargano Peninsula; Campania, Monte del Matese; Abruzzi District, Monti della Marsica, and (5) occurrence of type (iv) forms: Campania and Abruzzi, type areas are Monte Camposauro and Monte Maggiore in Campania.

The bauxites of Campania and Abruzzi occur mostly as small deposits in flat pockets with diameters of 10 to 100 metres. The bauxite thickness varies from 0.5 to 3 metres with rare maxima of 6 metres. The basal surface is slightly karstified. The bauxites grade laterally into calcareous breccias and conglomerates cemented by reddish calcite and/or bauxitic clays. The bauxites of Apulia fill deep cavities opening onto the top of the Lower Cretaceous carbonate platform sediments (internal parts of the platform) and are disconformably covered by Lower Senonian carbonates deposited in similar environments. The cavities are connected by narrow ducts; in the Murge (Spinazzola) there are canyon-like features with very steep walls and filled with bauxite. These bauxite deposits are 100 to 400 m long, 20 to 80 m wide and 10 to 15 m thick. Some cavities are almost 40 m deep. To what extent karstification played a role in carving these latter, deep cavities is, however, not completely clear. Probably there was combined action by linear erosion processes and by karst solution.

Boehmite is the main alumina mineral; gibbsite and diaspore are less common. Almost all silica is present in the form of clay minerals, the most important being kaolinite. Most bauxite deposits have a massive, non-stratified structure, though very faint stratification could be observed in some parts of the Regia Piana deposits (Matese). Detrital bauxite grains are arranged parallel to the stratification.

Because the Cretaceous sediments of the carbonate platforms consisted mainly of limestones, one could be tempted to conclude that the bauxites were derived from their insoluble residues. Indeed, there was strong chemical weathering of the limestones and their insoluble residues certainly contributed to the material composing the bauxites. However, there was not enough to form the bauxite deposits. This opinion is supported by the very low insoluble residue content in footwall limestones. Furthermore, big differences between the Cu, Li and Pb content in the footwall limestones eliminate them as source rocks. The same is true for the monazite, xenotine, cheralite and baddeleyite detrital grains found in these bauxites.

D'ARGENIO (1969) and BÁRDOSSY et al. (1977) supposed that wind-blown, mainly pyroclastic, material contributed to the formation of the bauxites, plus a minor amount of mainland-derived dust (Fig. 2–74). Similar ideas were advanced by COMER (1974) for the formation of Jamaican bauxites, by TAYLOR and HUGHES (1975) for the bauxites found on Rennel Island (see BOURROUILH-LE JAN, p. 443–471, for a comprehensive discussion — Editors).

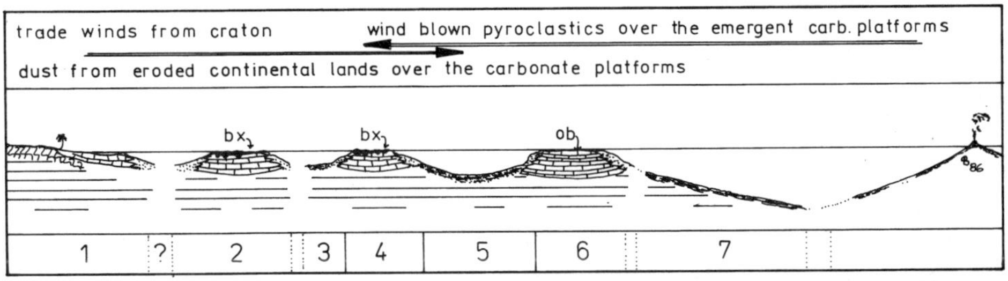

2–74. Palinspastic reconstruction of the Apenninic continental margin during the Cenomanian. The *Orbitolina* level (ob) as well as clays infilling Late Cenomanian paleokarst and bauxites (bx) probably derived, at least partly, from the volcanism active at this time (after BÁRDOSSY et al., 1977). 1. emergent continental margin, pre-Alpine basement and related structures; 2. Apulia carbonate platform and Gargano Basin emergent in this time; 3. Molise-Marsica Basin; 4. Abruzzi-Campanian carbonate platform, emergent; 5. Lagonegro Basin; 6. Campania-Lucania carbonate platform, submerged; 7. continental slope and rise.

Upper paleokarst horizon. The bauxites are good examples of deposits in surficial karst, though underground karst has also been reported. In several outcrops of the Abruzzi-Campania carbonate platform a short stratigraphic gap (uppermost Cenomanian-lowermost Turonian) has been discovered as well (SARTONI and COLALONGO, 1963; CARANNANTE, FERRERI and SIMONE, 1974).

While the lower stratigraphic gap is represented by the bauxite horizon already discussed, the upper gap is marked by a network of small paleokarst cavities. Their dimensions vary from few to some tens of centimetres. The fillings consist of sparry calcite and polychrome calcareous silt, sometimes with ostracods and fecal pellets (CARANNANTE, FERRERI and SIMONE, 1974). Many generations of internal sediments may be recognized, some of them with a reddish colour and oxide minerals. On the eastern side of M. Camposauro (Campania), there are no bauxites; instead, 30–40 m of Albian-Cenomanian lenses of poorly-sorted breccia crop out, composed of bauxite fragments and polychrome limestone elements, interbedded in shallow water grey limestones (D'ARGENIO, 1967; BERGOMI, MANFREDINI and MARTELLI, 1975). Upwards they pass abruptly to whitish rudistid limestone.

At both levels some large paleokarst cavities may be observed (*Marmi di Vitulano*). They have a prevailingly horizontal development. Four types may be distinguished: (i) interstratal; (ii) intrastratal; (iii) transverse fracture cavities (preceding the complete filling), and (iv) subvertical fracture cavities (i.e. perpendicular to the bedding; D'ARGENIO, 1967).

The episodes of marine emergence in the host carbonates are indicated by an increase of vadose diagenesis, with pervasive solution which caused redistribution downwards of loose sedimentary particles (CARANNANTE, FERRERI and SIMONE, 1974). Subsequently, lenticular cavities and cavity networks were developed by phreatic waters which enlarged the preexisting voids, often creating coalesced larger cavity systems. Large interstratal cavities (several metres long) are often localized at the transition between grey limestones and overlying rudistid limestones.

The filling of the cavities by alternating episodes of chemical and mechanical deposition appears quite complex. Among the chemical episodes is a first generation of radiaxial cements that are remarkably impure, followed by further generations of cements intercalated with detrital sediments. Among the latter there are clayey materials, grey and reddish to yellowish and reddish calcisiltites. In the grey calcisiltites, ostracods and small burrows may sometimes be observed. Clay materials, partly derived from reworked bauxite, were modified in their geochemical and crystallochemical characteristics during transport and diagenesis. Illite-montmorillonite mixed-layer clay minerals, illite and pM kaolinite predominate over more ordered kaolinite and over bauxite, with the exception of detrital sediments containing bauxite clasts (CARANNANTE et al., 1978; POZZUOLI et al., 1977).

Late Cretaceous

Minor paleokarst features are known in Southern Italy in the Upper Cretaceous limestones of the Monte Vesole-Trentinara which belong to a Mesozoic platform (Campania-Lucania carbonate platform). Two regressive episodes have been documented, probably connected with short gaps in deposition and resulting in paleokarst networks of decimetric size filled with pale pink kaolinitic sediments. The kaolinites show a high crystallinity (polytype T) like the so called *flint-clay* (BONI, STANZIONE and ZENONE, 1979).

Sardinia

Paleokarst features are also known in Sardinia where, as in Southern Italy they are signified by the presence of bauxites.

The bauxites cropping out in NW Sardinia (Nurra) are associated with an emergence and karstification which took place between the Middle and the Late Cretaceous (PECORINI, 1956, 1965; COCCO and PECORINI, 1959; Figs. 2–75 and 2–76). The bauxites form a quite irregular horizon with an average thickness of

1–2 metres, lying on dolostones, limestones and marls which range in age from Late Jurassic to Barremian-Aptian. The hanging wall consists of Turonian and Senonian hippuritic limestones. The bauxites cover a surface paleokarst similar

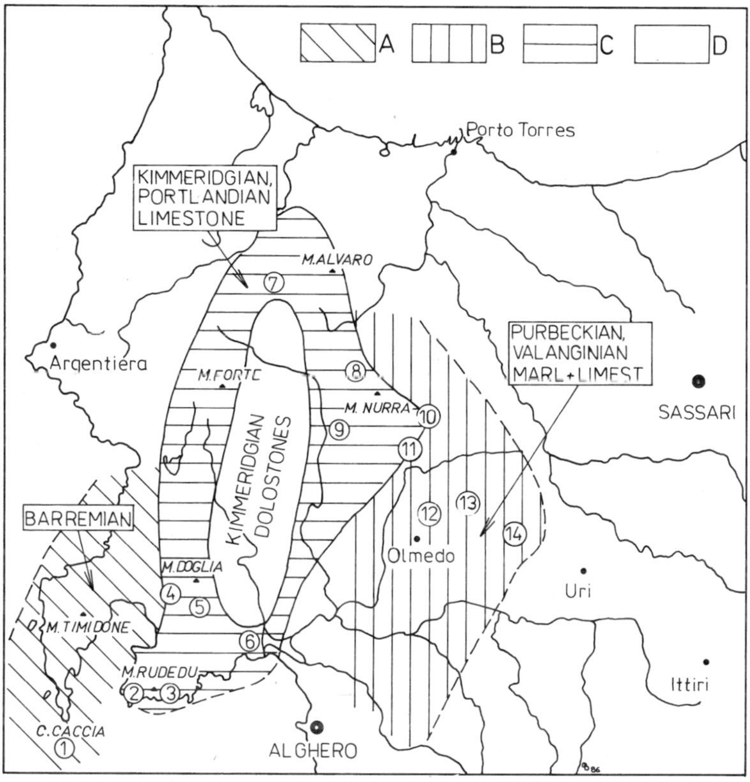

2–75. Erosion in mid-Cretaceous times in northwestern Sardinia. A. Minimum of erosion (< 100 m). Barremian folded limestones crop out below flat lying Turonian rudistid limestones. No bauxites; B. More than 100 m erosion. Valanginian limestones are the youngest deposits below the stratigraphic gap. The most important bauxite ores occur; C. More than 200 m erosion. Portlandian limestones occur below the Turonian ones. Rare bauxite occurrences; D. More than 300 m erosion. Kimmeridgian dolostones crop out in the narrow belt as the oldest rock below the gap. No bauxites. Encircled numbers refer to columnar sections in the fig. 2–76.

to that of the Matese-Mte Maggiore type in the Apennines. The form of the cavities is elongated and can be followed for some hundreds of metres. Sometimes a lateral passage to small breccias with bauxite matrix in the underlying carbonates can be observed (Pecorini, 1956).

The bauxites, mostly of boehmitic type, have been considered to derive from materials transported for a short distance over a low relief topography related to a tectonic phase developing during Mid-Cretaceous times (Cherchi and Tremolieres, 1984).

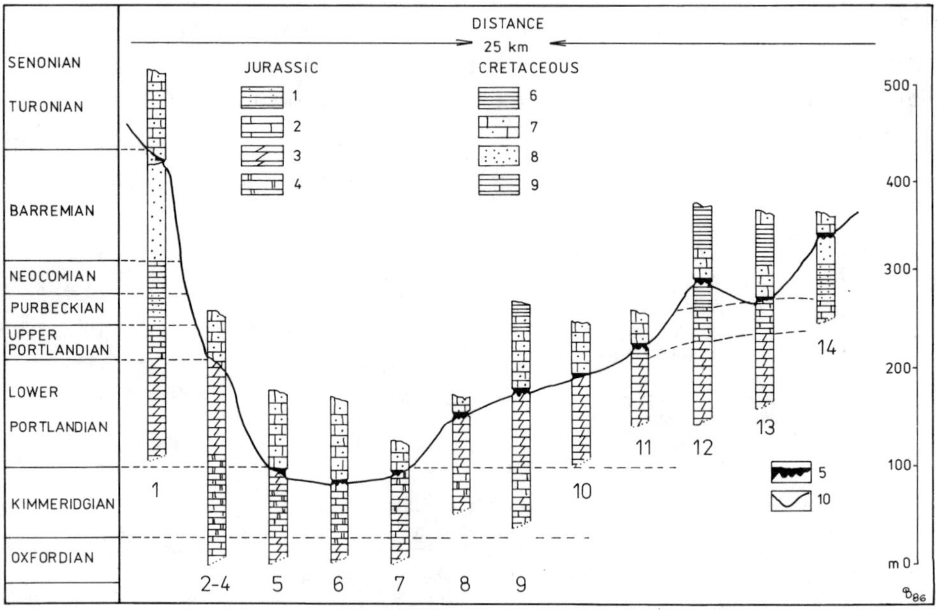

2–76. Mid-Cretaceous emergence in Sardinia. For location of columnar sections in northwestern Sardinia-Nurra see Fig. 2–75. (after PECORINI, 1956). 1. marls and limestones with Characeans; 2. limestones with *Clypeina jurassica;* 3. dolostone; 4. limestones with *Kurnubia* and dolostone; 5. bauxite; 6. marls with *Globotruncana;* 7. Hippuritic limestones; 8. limestones of the Urgonian facies; 9. marly-detritic limestones; 10. Mid-Cretaceous stratigraphic gap with deep erosion and bauxite formation covered by Upper Cretaceous deposits.

Tertiary Paleokarst

Tertiary paleokarst features are scattered over all Italy. The following is only a brief and most probably incomplete review of them.

"Terra Rossa" of Murge

In the Northern Murge (near Poggiorsini) a small paleokarst horizon with resedimented bauxite clasts in a terra rossa-type matrix may be observed between the Altamura Limestone (Santonian) and the *Planorbis* Limestone (Upper Paleocene-Eocene; IANNONE and PIERI, 1978). The hanging wall of the bauxitic horizon rests unconformably on the karstified footwall. Outcrops of this paleokarst extend for only a few metres.

Bauxite of the Salento Peninsula

In the Salento Peninsula (Southern Apulia), redeposited bauxites occur on Cretaceous limestones. They form small lenses filling karst depressions in the footwall and covered by Miocene to Pleistocene sediments (COTECCHIA and DELL'ANNA, 1959). The thickness of the ore bodies varies from 1 to 15 m. Bauxite pebbles and cobbles of 0.5 to 5 cm (from the erosion of the Cretaceous bauxites?) are embedded in a red bauxitic clay. They have a quite high gibbsite content (3%) when compared with the Mesozoic bauxites, probably as a result of redeposition.

Red Clays of Trentinara

In the tectonic Alburno-Cervati Unit created from deformation of the Campania-Lucania carbonate platform (D'ARGENIO, PESCATORE and SCANDONE, 1973), Cretaceous sediments grade into the Paleocene-Lower Eocene limestones, with green shale and marl intercalations (Trentinara Formation) between Capaccio-Trentinara villages and Mount Coccovello. The depositional environment is considered to have been brackish-marine to lagoonal. The upper part of this sequence is cut by an erosion surface with up to 10 metres deep pockets filled with red clays. This karst surface is disconformably covered by Early Miocene calcarenites. The red clays, deposited in a continental environment in a time interval preceding the Miocene transgression, are mostly kaolinitic (*bauxite* clay) and very rich in detrital minerals (BONI, 1974). Their source area is believed to have been outside of the carbonate platform. Paleokarst surfaces of the same type and age are also present further south in the Pollino Mountains.

Yellow Ochre of Verona

Underground karst known in the Lower Priabonian of the Lessini Mountains (Verona area, Northern Italy) formed during a Oligo-Miocene continental phase. During the Middle Miocene, a deep karst network was completely fossilized with glauconitic marine sediments very rich in fossils and with yellow ochre sediments that have been extensively exploited (CORRA, 1975). It is thought that at least part of the ochre may be related to the dissolution of clayey limestones, very rich in pyrite, in the Upper Priabonian. The karst networks, on the contrary, are much better developed in the pure limestones of the Lower Priabonian, while they also display depigmentation and alteration phenomena along the cavity walls and roofs.

Discussion

We have reported here the published evidences of pre-Pliocene paleokarst in Italy. From our field experience it appears that many other similar occurrences exist in the carbonate rocks of Italy, not to mention the evaporites. They wait to be described and interpreted.

Post-depositional cavities developed in carbonates and then filled with both chemical and/or detrital sediments are very common in the shallow water facies and sometimes also in deeper carbonate deposits. The main problem we have faced in our research and in the present review, has been to distinguish between what may be defined as paleokarst and what is the result of fossilization of cavities which show morphological convergence with paleokarst voids (e.g. marine bioerosional cavities of supratidal to subtidal regime). Even though true karst processes (and, thus, true paleokarst occurrences) involve surface and/or underground dissolution of carbonate and sulphate rocks by meteoric waters in continental conditions, we have often found it very difficult to determine whether some of these cases reported here and interpreted by the authors as paleokarst, may really be examples of it. At any rate, in all the occurrences that appear related to paleokarst, whatever the nature of the filling materials (including mineralization), an emergence period of the carbonate sediments can be documented.

Three types of *paleokarst* are noted in the Italian carbonate rocks.

(1) The first type is characterized by a more or less short stratigraphic gap (generally no more than two stages are missing) and lack of unconformity with the overlying deposits. No important removal of superficial materials by erosion is recorded and there is no evidence of deep underground karst solution. It may be supposed that the emergence was related to minor eustatic sea level lowering or even to non-deformational (epeirogenic) uplift. This type shows no direct relationships with tectonics. Examples include a minor part of the Cambrian karst of Sardinia, the Triassic karst of Sicily and the Cretaceous karst of Apennines and Apulia.

(2) The second type of paleokarst is firmly linked with tectonics and mostly displays synsedimentary characteristics. In this case, also, the stratigraphic gaps are not very long (a few stages) but there is clear evidence of discrete erosion and unconformities may be observed. Underground paleokarst is well developed but does not reach great depth beneath the original surfaces of emergence. Tectonic events (folding, faults) may be fossilized together with the karst features, by overlying marine deposits. Volcanism may also develop during deformation and emergence, so providing materials for the karst filling. The Cretaceous of Sardinia, the Devonian of the Carnian Alps and the Middle Triassic of the Dolomites belong to this type of paleokarst.

(3) The last type is related to the emergence and denudation of a deformed belt and has the characteristics of prolonged karst evolution. In this respect it is post-deformational, or post-tectonic, karst. The stratigraphic gaps are very wide, the karst is mature and there is deep erosion with removal of large amounts of material. Paleokarst of this type which is fossilized has developed on a peneplane. Good examples are the post-Variscan (Permo-Triassic) karst and the Ordovician karst of Sardinia.

PALEOKARST RELATED ORE DEPOSITS OF THE MAGHREB, NORTH AFRICA

Yves Fuchs and B. Touahri

Introduction

In North Africa a great number of ore deposits are located in karst features. Some of them are of economic interest, e.g. lead-zinc deposits like Oued Mekta (Morocco), El Abed and Djebel Gustar (Algeria), Djebel Trozza and Jebel Ajered (Tunisia), copper deposits at Djebel Khlak (Morocco), fluorite at Hamman Zriba (Tunisia), antimony deposits such as Koudiat Bou Diar, Ain Kerma (Algeria), vanadium deposits at Jebel Goraa (Tunisia), kaolinite-halloysite deposits of Djebel Debbagh (Algeria) or Ain Khamouda (Tunisia), etc. These deposits and a great number of smaller occurrences show a division into three groups whose genesis is related to particular phases of the evolution of the African carbonate platform. All these deposits are located in autochthonous or para-autochthonous units of the Alpidic system in the Maghreb. The three different groups are:

(1) ore deposits located in karst cavities underneath an unconformity or disconformity surface covered by pelitic and mainly carbonaceous sediments (black shale type). Examples are the lead-zinc deposits of Oued Mekta (Morocco), El Abed and Deglen (Algeria) and Jebel Ajered (Tunisia), as well as the fluorite deposits of north eastern Tunisia.

(2) ore deposits in karst cavities developed during a continental emergence. The karst is filled with argillaceous materials and oxidised ores. Examples include the halloysite deposits of Djebel Debbagh (Algeria) and of Ain Khamouda (Tunisia) as well as some karst occurrences of vanadium at Jebel Goraa (Mahjoub, 1975).

(3) some paleokarsts which developed during and after the major tectonic events of the Miocene. These show lead-zinc occurrences like Jebel Hallouf (Tunisia) and perhaps some parts of Oued Mekta (Morocco), or antimony at Bou Zitoun (Algeria).

P. Bosák, D. C. Ford, J. Głazek and I. Horáček (Editors):
Paleokarst. A Systematic and Regional Review. – Elsevier and Academia. Amsterdam and Praha. 1989.

Ore Occurrences and Deposits Located Underneath an Unconformity and Later Covered by Marine Sediments

At different times during the deposition of carbonate rocks on the epicontinental platform, epeirogenic movements induced local emergence. These emerged areas are restricted to a few square kilometers each. Their distribution is determined by the local relief of the platform reactivated by epeirogenic movements, or by diapiric structures. Underneath the emerged surfaces paleokarst can develop. This paleokarst has very limited depth; it ranges from a few metres to a maximum of 40 m. However, it shows mature characteristics. The horizontal part of the karst system is well developed and the host rocks are intensively dolomitized. All these characteristics indicate that the water table was not far from the surface. Such emergences occurred at different times during the Mesozoic and the Cenozoic in the epicontinental domain.

The Unconformity Between the Aalenian and the Bajocian in Morocco and Western Algeria and Related Paleokarst Ore Deposits

In this region, emergence was restricted to some shallow water areas characterized by tidal carbonate sediments. Karst developed along fracture zones, some of which were active during the sedimentation and the emergence (Fig. 2–77). Epigenetic dolomitization related to the karstification process results

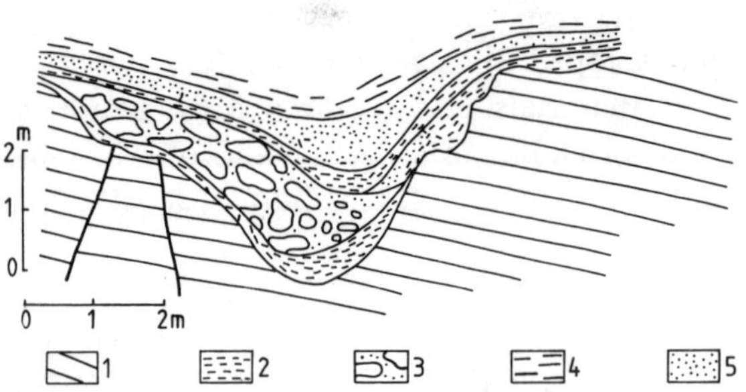

2–77. Relationship between paleokarst depression and post-Lower Jurassic fault structures (after TOUAHRI, 1983): 1. Lower Jurassic limestone; 2. black shales with sulphide ores; 3. breccia; 4. Upper Jurassic marl; 5. sand with pyrite.

from meteoric water circulation (TOUAHRI, 1983). After this emergence, deposits filled the karst cavities such as dolines, and paleovalleys of the unconformity surface. In these depressions euxinic conditions prevailed. Black organic-rich sediments with sulphides (pyrite, chalcopyrite, galena) were deposited. They are the main ore concentration in El Abed.

The Unconformity at the End of Jurassic Period in NE Tunisia and its Fluorite Deposits

This region was recently described by THIBIÉROZ (1974) and TOUHAMI (1978). Some local emergence with unconformity occurred during the Lower Jurassic (Domerian) but the main event took place after deposition of the Portlandian platform limestone. The exposed surface can be observed at Jebel Mecella, Hamman Zriba and Hamman Jedidi. Exposure was for a short period at Jebel Mecella (perhaps one ammonite zone or less) and for a long one at Hamman Zriba where the Abiod Limestone (Cenomanian) lies directly on the Upper Jurassic. The ore consists of fluorite, baryte, celesto-baryte, galena, sphalerite and pyrite. It fills depressions such as dolines in the unconformable surface. Between the dolines, marine deposits of Cretaceous age lie directly on the Portlandian limestone. Karst phenomena with substantial vertical extent are rare, except at Jebel Mecella, where some of them could even be recent developments of karst. Formation of the ore deposits is probably related to submergence of the karst unconformity surface. The deposition begins, on the surface, with black bituminous dolostone (Jebel Mecella) or black shales (Hamman Zriba,

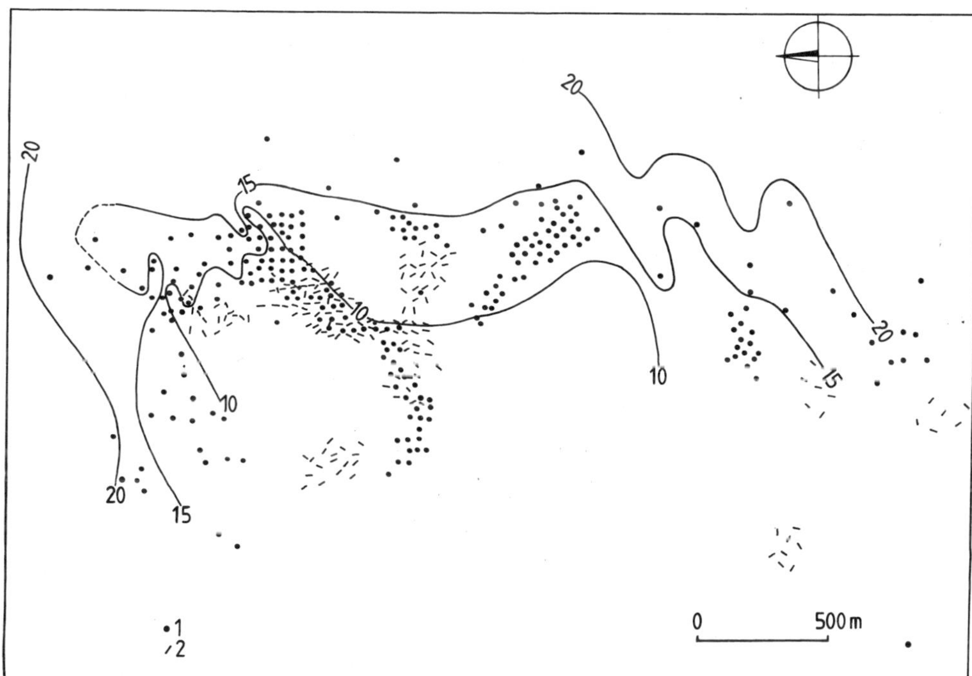

2–78. Isopachs of the highest uneroded level of the Aptian at Jebel Ajered, Tunisia (after FUCHS, 1973): 1. boreholes; 2. dug pits.

Oued Guebli). In these, horizons with Cretaceous microfauna many cm thick were found (THIBIÉROZ, 1974). This district is one of the most important in the world since it contains 7 million tons of CaF_2 (3.4 % of world reserves).

The Unconformity at the Aptian-Albian Boundary in Central Tunisia and Eastern Algeria

Lead-zinc and baryte deposits are located in karst structures related to this surface in the central part of Tunisia and the eastern part of Algeria (FUCHS, 1973; POPOV, 1968). The emergent areas are located at the top of diapiric structures which were active during the sedimentation phase (Fig. 2–78). This synsedimentary paleogeographical differentiation influenced the facies distribu-

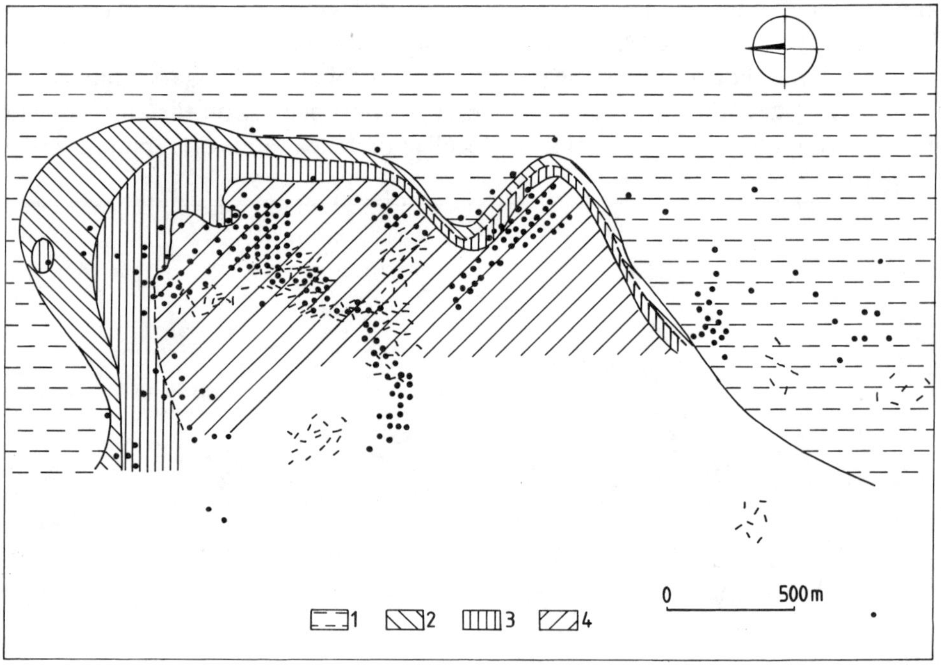

2–79. Aptian-Albian surface at Jebel Ajered, Tunisia (after FUCHS, 1973). 1. Albian (Fahdene Marls) lying conformably on Aptian limestone; no karstification; 2. zone with a conglomerate between the Aptian and Albian; 3. Albian on karstified Aptian limestones; 1–5 m of Upper Aptian are eroded; 4. Albian on karstified Aptian limestones; more than 5 m of Upper Aptian is eroded.

tion. On the main part of the epicontinental platform, argillaceous limestones were dominant during the Aptian while reef or parareef limestone was deposited on the diapiric zones. Development of erosional unconformities were sometimes repeated (twice at Jebel Ajered). The emerged areas are small, a few square kilometres (Fig. 2–79). The depth of karst structures ranges from a few metres to a maximum of forty metres. In the lowest levels of the karst, horizontal circulation is common. Brecciation, dolomitization of host-rock and, in some cases, silicification accompanied the karstification. The karst unconformity surfaces are overlain by black marls termed Fahdene Marls in Tunisia. These marls filled up the karst cavities e.g. at Jebel Hamra where the ore deposit consists of cavity fillings and breccia cement, including galena, sphalerite, pyrite, baryte.

The post-Burdigalian Unconformity and Related Fluorite and Sulphide Deposits of Oued M'Tak, Tunisia

This little, non-economic deposit shows some interesting characteristics. In the northeastern part of Tunisia, the Burdigalian consists of neritic limestones with pectens and sea urchins. Its upper limit is an unconformity that is particularly well seen at El Haouaria at the end of Cap Bon. There, a paleokarst surface was developed and then filled up with pyroclastic material. The diagenetic evolution of these ashes, in a reducing environment, resulted in glauconite-rich rock with some rhyolitic quartz, altered ferromagnesian minerals and fluorite, galena, cinnabar, sphalerite, jordanite, boulangerite (FUCHS, 1973). Black marls of Middle Miocene age overlie the surface.

Paleokarst Related to Regional Emergence

These paleokarsts are very different from the first type with respect to their morphology and content. They are represented in the central Tunisia district of Bled Zelfane with the Ain Khamouda deposit (AMOURI, 1977), and in eastern Algeria at Djebel Debbagh.

In central Tunisia during the Mesozoic, carbonate platform deposition prevailed. After the Campano-Maestrichtian there was emergence and the island of Kasserine-Tebessa was formed, with an area of about 6 000 km^2. The emergence lasted until the Early Miocene when continental and lagoonal sandstones covered the island. During this long exposure a paleokarst was formed, the best examples of which are at Ain Khamouda north of Kasserine. Here, the depth of the karst features reaches 70 to 80 m and the cavities are filled with: (1) white clay (halloysite), (2) a red deposit as crusts in fissures and filling caves, and (3) a sandy formation of Miocene age which represents the last stage of the karst filling and may be partially from collapse of overlying sediments.

The white clay consists mainly of halloysite plus minor trioctahedral smectite with appreciable amounts of ZnO (RIVIÈRE et al., 1985). At other nearby sites there are nearly amorphous Zn hydroxides and minute particles (3 to 10 nm) of pure silver. Some galena contains thyobacteria (AMOURI, DÉVIGNE and FUCHS, 1978). The red crust consists of hemimorphite, goethite, smithsonite plus minor calcite, quartz, galena and baryte. The morphology of some boxworks indicates possible presence of monheimite. Observations in the transmission electron microscope (T.E.M) shown that algobacteria type of aerobial chladmobacteria were present (AMOURI, DÉVIGNE and FUCHS, 1978). The Zn content of red deposits is 17 to 25 %. They are the main ore reserves of the deposit. The observation of the senior author confirmed the premise of M. AMOURI that the white clay was deposited prior to the red crust.

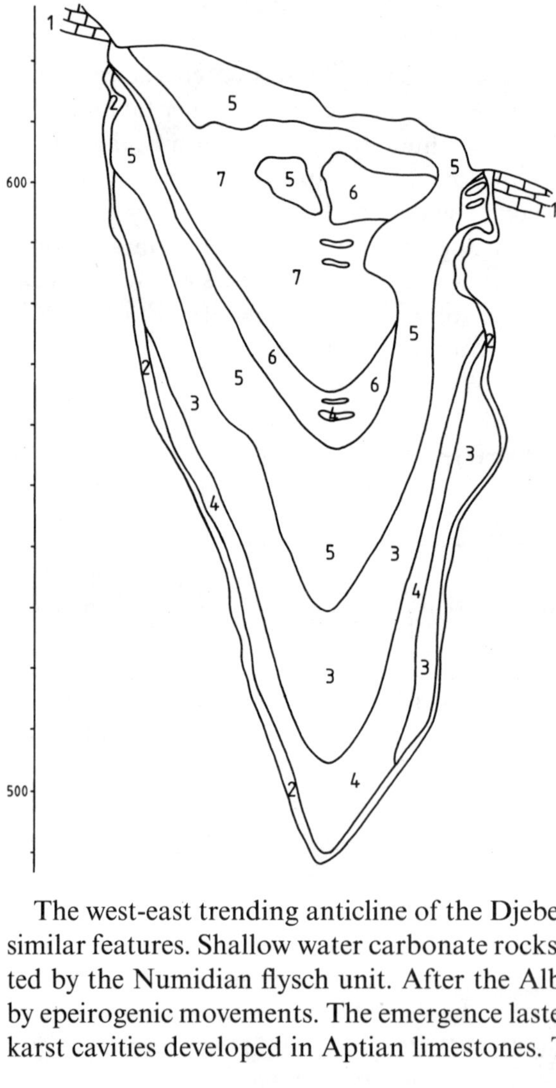

2–80. Cross-section through paleokarst cavity in Djebel Debbagh, Algeria constructed according boreholes and mining works. 1. Cretaceous limestones; 2. black clay layer with Mn oxides; 3. halloysite with alunite; 4. halloysite mixed with detritus; 5. iron crust with scorodite; 6. reworked halloysite with iron hydroxides; 7. montmorillonitic clay.

The west-east trending anticline of the Djebel Debbagh, Algeria, shows many similar features. Shallow water carbonate rocks of Cretaceous age are overthrusted by the Numidian flysch unit. After the Albian there was emergence caused by epeirogenic movements. The emergence lasted until the Maestrichtian. Paleokarst cavities developed in Aptian limestones. The diameter of vertical forms in their upper part is 15 to 50 m and the depth is up to 130 m (Fig. 2–80). The subvertical walls of cavities never show traces of the dolomitization. The fill of paleokarst consists of different zones: (1) halloysite plus minor alunite; (2) brecciated halloysite and alunite with some iron hydroxides; (3) montmorillonite; (4) sandstone breccias, and (5) so-called *gossan zone* which consists mainly of goethite and contains seams similar to the red formation of Ain Khamouda. The *gossan* contains minor amounts of sulphide (chalcopyrite), important quantities of scorodite and gold as traces of about 1 g per ton. The halloysite-alunite deposit fills up the deepest part of the paleokarst (Fig. 2–80). The *gossan* was formed in a later stage. Montmorillonite and the sandstone breccias could be the final deposits. Research is continuing (A. TALEB, University of Alger) and one can expect that the different phases of the filling of these cavities and the prevailing geochemical conditions will be more precisely described in the near future. Two types of phenomena could be responsible for the Djebel Debbagh deposits: (1) deposition of weathering products in a moderately acid leaching environment which was progressively transformed into a restricted one by an upward movement of the water table and (2) post-sedimentary evolution by partial leaching of an homogeneous deposit.

Paleokarst with Remobilized Ore Deposits

All over the Maghreb, tectonic activity during the Tertiary induced a general emergence. At the same time, the intensity of the tectonics favoured the formation of karst systems developed along faults or diaclases. Some of the sediments deposited in these quite recent paleokarsts contain ore. In the eastern part of Algeria an antimony-rich district was described by TOUBAL (1984). To this district belong different types of antimony deposits, from vein type to strata-bound and also different karst types. In one occurrence (Bou Zitoun) it is possible to observe fracture dissolution and filling with stibnite and calcite. Some stalagmitic deposits have shown (T.E.M.) 50 nm thick layers of senarmontite and aragonite.

The Oued Mekta deposit in Morocco is a lead-zinc deposit of the El Abed type. According to many geologists (BALCON, 1980; in DUPUY, 1984) an important part of this deposit was reworked during Late Tertiary times in a paleokarst system. Some karst remobilizations were described by ROUVIER (1971) at Jebel Aouane in Tunisia where a strata-bound lead-zinc deposit is located in a mollasse post-tectonic basin. In the underlying Senonian limestones there was an older paleokarst system. Remobilization of the strata-bound ore induced a partial filling of the paleokarst. The chemically deposited sediments have the form of stalactites, stalagmites and crusts and consist of calcite alternating with sulphide crusts of galena, sphalerite or pyrite.

Conclusions

Ore deposits located in paleokarst are of significant economic importance in North Africa. The ore forming process is quite different for each type of karst. However, the main factor underlying formation of these karst ore deposits remains the epeirogenic instability of the platform. This instability is the cause of emergence on a regional scale at the island of Kasserine-Tebessa. It is also clearly the factor inducing local emergence. The epeirogenic movements activated pre-existing structures of the platform or created diapirs. Because they are associated with positive structures, the emergence and related weathering occur on reef facies in many cases. These are particularly readily karstified.

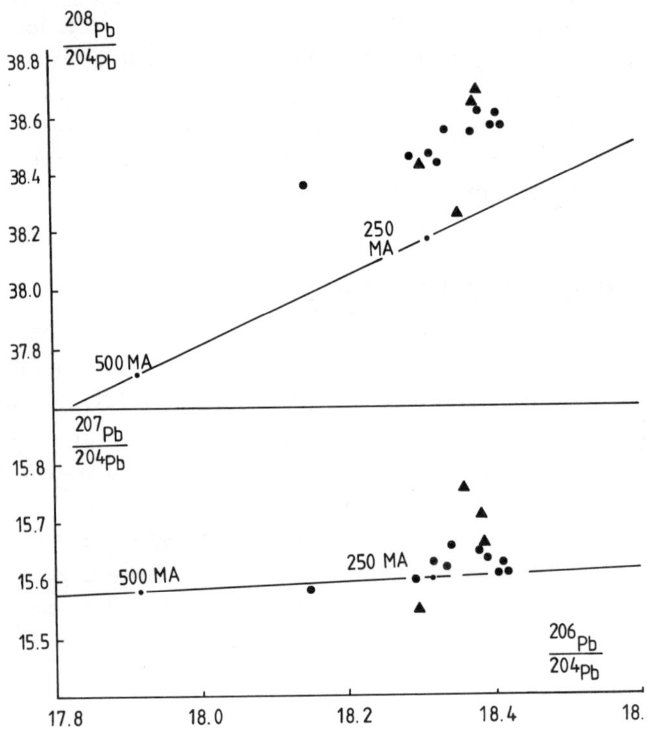

2–81. Distribution of $^{207}Pb/^{204}Pb$ ratios versus $^{206}Pb/^{204}Pb$ ratios and $^{208}Pb/^{204}Pb$ ratios versus $^{206}Pb/^{204}Pb$ ratios in galena of paleokarst-related ore deposits from Touissit region (RAJLICH, LEGIERSKI and ŠMEJKAL, 1983) (black triangles) and of the western part of Algeria (El Abed; black dots).

The assumption that the structural pattern plays an important role in the distribution of paleokarsts containing ore deposits is not sufficient to explain the origin of metals and the ore deposit formation processes. Regarding the origin of metals, it seems that a volcanogenic origin can be advanced for Oued M'Tak only. For all other deposits, the source may be the leaching of weathered profiles on the basement (TOUAHRI, 1983) or of sediments on the platform. The $^{208}Pb/^{204}Pb$ and $^{206}Pb/^{204}Pb$ isotope ratios in El Abed, Saida and Deglen in Algeria, and the results obtained by DUTHOU, EMBERGER and LASSERRE (1976) concerning

2–82. Stalactite sections, Jebel Hallouf, Bou Aouane, Tunisia. The central channel is filled with pyrite, smithsonite and galena (Photo by Y. Fuchs).

the Touissit orebody of Morocco are consistent with the expected ratios of the Variscan granites and Paleozoic sediments of the basement. The wide range in the ratios shows that the leads are coming from different sources. The prime source could be weathered profiles developed on the Paleozoic basement under favourable climatic conditions during a period of tectonic stability. An epeirogenic movement could have induced leaching of the elements present in the profiles and the formation of a strata-bound geochemical anomaly. Such an anomaly was found at the base of the black marls of Fahdene (FUCHS, 1973). The first type of paleokarst, located on restricted emerged areas, would be the traps where these metals would be more concentrated because of the euxinic conditions prevailing there. The metal supply was transported during the final stage of karst evolution, as the marine water was flooding the paleokarst structures. In this case, the karst-related ore deposits are characterized by a pervasive dolomitization of the host-rock. The $\delta^{13}C$ and $\delta^{18}O$ ratios of dolomite in the El Abed deposit show a major influence of meteoric water in the dolomitization process (TOUAHRI, 1983). At Jebel Hamra, however, the $\delta^{13}C$ and $\delta^{18}O$ of dolostones developed under the Aptian-Albian unconformity show that the dolomitization process occurred through mixing of meteoric water with seawater during the final stage of paleokarst evolution (M'RABET, 1981). The $\delta^{34}S$ ratios in sulphides of El Abed (TOUAHRI 1983) and Oued Mekta (RAJLICH, LEGIERSKI and ŠMEJKAL, 1983) do not show any evidence of bacterial sulphate reduction. They are, though, in agreement with the values occurring in organogenic coastal environments.

Concerning the second type of karst (Ain Khamouda and Djebel Debbagh) no isotopic measurements are available at present. The occurrence of strictly aerobic bacteria is related to a non-euxinic environment. The morphology of these paleokarsts shows evidence of circulation under pressure (pressure dumps and slabs). Dissolution phenomena and intrakarstic breccias are absent. The lack of dolomitization shows that there was no stagnant aquifer and no percolating water. The presence of minor amounts of sulphides (galena) in halloysite is related to a bacterial sulphuration process and not to a reducing environment. The saucinite is a co-syngenetic mineral of halloysite and zinc was introduced in the paleokarst system as it was an active karst (RIVIÈRE et al., 1985).

The great number of paleokarst related ore deposits in North Africa are related to the tectonic evolution of the platform. On this carbonate platform there are some quite varied types of karst, and the processes which induced formation of the ore deposits are of two main types related to the different environments of paleokarst evolution.

PALEOKARST OF YUGOSLAVIA

Dušan Gavrilović

Introduction

Carbonate rock terrains comprise approximately one third of the area of Yugoslavia. Given that the carbonate series is in some places over 5 000 m thick and tectonically very broken, highly favourable conditions for the development of the karst have been created. Traces of karst erosion are encountered on Triglav (2 863 m) and some other peaks of the Julian Alps, of Durmitor and Prokletije and also at a depth of 3 000 m below the level of the Adriatic. It is clear that all these forms could not have originated simultaneously. Some of them are undoubtedly of considerable age, i.e. they are the relicts from earlier periods and phases of karst erosion.

In the course of the past 300 years, more than 3 000 articles, treatises and studies have been published on geological, geomorphological and hydrological characteristics of the Yugoslav karst. The attention of researchers was chiefly focussed on recent and subrecent phenomena. Paleokarst was only of peripheral importance or it was totally neglected as a phenomenon. Very little still is known about the geomorphological characteristics of the paleokarst.

The Yugoslav terminology for paleokarst is poorly elaborated and mostly borrowed from the literature in foreign languages, i.e. it is international. The term *fossil karst* is used to indicate a karst where the process of corrosion has been interrupted. *Paleokarst* has much the same meaning but also that the geological age of the karstification has been determined. *Paleokarst* in a narrower sense is *fossil karst*, overlain and isolated by non-carbonate sediments and rocks and without any water circulation. Particular categories of paleokarst are *relict karst*, formed in different climatic conditions, and *exhumed karst* which can be at the same time also a relict karst. In both these cases the former essential characteristics of the relief have been preserved. The only specific term which refers to paleokarst forms is the *pyramid* — a miners' name for the cone of rock fragments in the base of bauxite deposits.

P. Bosák, D. C. Ford, J. Głazek and I. Horáček (Editors):
Paleokarst. A Systematic and Regional Review. – Elsevier and Academia. Amsterdam and Praha. 1989.

Geological Backround of the Karst

In Yugoslavia there are five distinct morphostructural and geotectonic units: the Alps, the Dinaric Mountains, the Carpatho-Balkanian Belt, the Serbo-Macedonian Belt of horsts and rift-valleys, and the Pannonian Basin. At the present day, carbonate rocks crop out extensively only in the three mountain systems. In the Serbo-Macedonian Belt and in the Pannonian Basin they crop out only sporadically or are to be found under a thick cover of Tertiary sediments (Fig. 2–83).

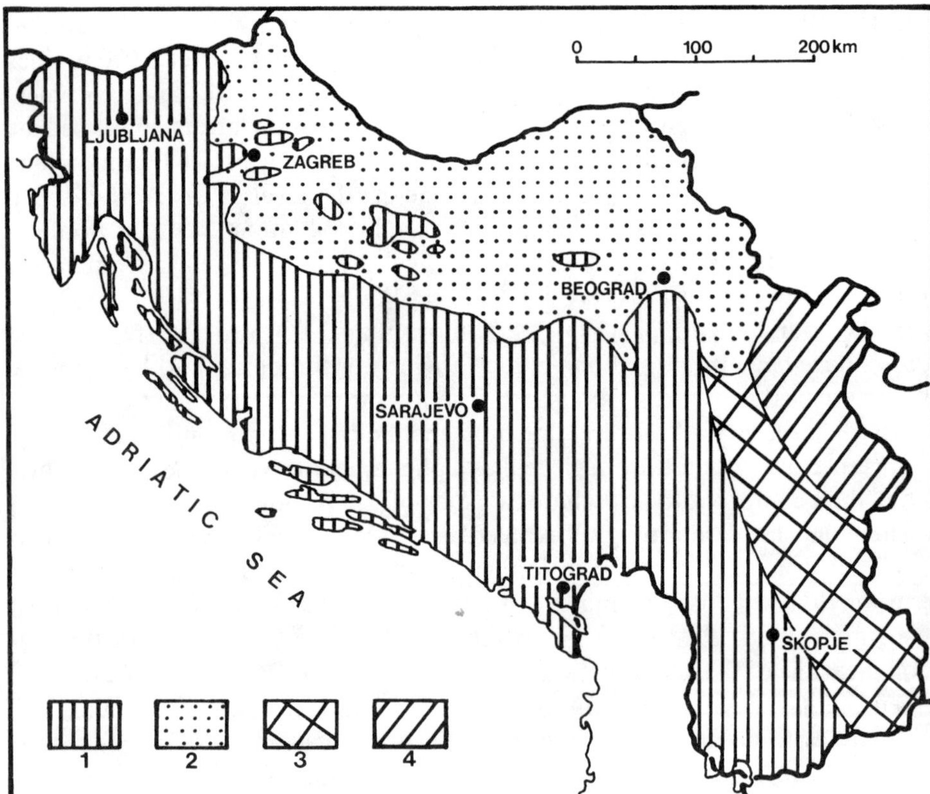

2–83. Major morphotectonic areas of Yugoslavia. 1. Alps and Dinarides; 2. Pannonian Basin; 3. Serbo-Macedonian Belt; 4. Carpatho-Balkanian Mountains.

The oldest carbonate rocks that are known occur in the Serbo-Macedonian Belt and in the base of the Carpatho-Balkanian Mountains. They are mostly in the form of interstratified layers (up to 20 m thick) and lenses of crystalline limestones (marbles) within gneisses and chlorite schists. Besides dolomite and calcite they often contain quartz and other minerals. According to micropaleontological research they are of Proterozoic age. The greatest complex of older Proterozoic marbles was discovered near Prokuplje (11 km^2). These marbles originated as marine limestones formed in shallow waters and they are now collectors of underground waters, i.e. karst is developing in them.

Carbonate rocks of the Cambrian, the Ordovician and the Silurian, represented by marbles, have been discovered in the base of the Carpatho-Balkanian Mountains, in the Serbo-Macedonian Belt and in several places in the Inner Dinarides. They occur as smaller lenses and intercalations in schists, are rather impure and display no karst phenomena.

The crystalline limestones and dolostones of Jakupica Mountain in Macedonia are the largest marble massif in Yugoslavia. They have a thickness of over 500 m and are intersected by many faults. A typical karst is developed in them with all its morphological and hydrological variety. To illustrate the great thickness of the carbonate mass and the degree of its karstification under the peak Solunska Glava (2 540 m) a pothole was explored to a depth of 450 m. Although the age of these marbles has not been precisely established yet, the majority of determinations place them in the Lower Paleozoic and some even in the Precambrian.

Dolostones and limestones of Devonian age have been discovered in the Savinja Alps, between Travnik and Fojnica, east of Sarajevo and northwest of Valjevo, where they are not fully differentiated from the Carboniferous. Little is known of this carbonate formation because it is for the most part covered by younger sediments. In the Devonian crystalline dolostones and dolomitic limestones of Sebešić, northwest of Fojnica, underground and superficial forms of karst have been described.

In sum, carbonate rocks of Precambrian and Lower Paleozoic age are local phenomena only. They are rather thin, impure, inserted into impermeable rocks and seldom karstified.

The Alpine-Dinaric karst rests upon the Lower Paleozoic basement. The Carboniferous formations are comprised mostly of clay and shales, sandstones and conglomerates with larger and smaller intercalations of limestone and dolostone in which karst phenomena are encountered. Permian deposits are also clastic to a great extent but in some places carbonate rocks prevail. In contrast to the Alps, where Lower Permian limestones are overlain with sandstones, the Upper Permian is represented by dolostones in Lika and west of Valjevo by bituminous limestones.

In the Alps and the Inner Dinarides, Triassic limestones are widespread. They are dolomitic for the most part. The Triassic carbonate formations were deposited discordantly upon the Paleozoic in most instances. It is rather heterogeneous and within it there are several erosional and tectonic disconformities, some of which are marked by the occurrence of bauxite and clastic rocks. In the central part of the Dinaric Mountains, the Triassic carbonate formations are overlain by vast covers of Jurassic diabase hornfels, serpentinites and amphibolites, tens to several hundred of metres thick. In most cases, over the serpentinites Cretaceous limestones have been deposited. From the position of the serpentinite cover within the Mesozoic carbonate rocks, it is certain that they have played a very important part in the evolution of the karst, both in the upper and lower parts of the limestone. In the Inner and Central Dinarides the Mesozoic carbonates are overlain by Senonian flysch. During the Paleogene the whole of this area was dry land.

The Outer Dinarides are the region of the classic Dinaric karst. Unlike the Inner and Central Dinarides, which have a very heterogeneous lithologic composition and where karst is only sporadically developed, in the Outer Dinarides Jurassic and Cretaceous limestones are very widespread and drilling has established that there are also Triassic carbonate deposits beneath them. In addition to being clean and homogeneous, without important inclusions of impermeable rocks, the Mesozoic carbonate series is over 5 000 m thick in some places. All this made it possible for karst to develop to very great depths, often considerably below the level of the Adriatic Sea. Paleocene limestones of Istria and Eocene flysch are deposited discordantly over the Cretaceous limestones. The sedimentologic properties of the carbonate rocks and their fossil contents indicate that throughout the entire Mesozoic in the Outer Dinarides, a shallow marine basin existed; its bottom was slowly subsiding but it was not exposed to stronger oscillations.

During the Laramide, Pyrenean and Sava Orogenies, the Mesozoic sediments of the Dinarides and the Alps were lifted and arched in the form of vast monoclines, anticlines and anticlinoria in which horsts and graben appeared as lesser structural units. Owing to strong lateral thrusting, at many places overthrusts occurred. The formation of deep graben, which later became basins and karst poljes, and the overthrusting of impermeable rocks onto the limestones and dolostones, had a great influence on the further morphological and hydrological evolution of the karst. The more the mountain chains rose, the deeper karst processes penetrated into the carbonate mass, so that the karst has developed continuously since the Late Cretaceous, i.e. for over 70 Ma.

The Carpatho-Balkanian karst is developed in Mesozoic carbonate rocks, primarily in Cretaceous and Jurassic limestones and dolostones. The Mesozoic carbonates rest upon the Paleozoic schists and sandstones. The lower parts of the series consist in most cases of Jurassic dolostones which pass upward first into stratified and then into massive limestones. The Mesozoic sedimentation was completed at the end of the Cretaceous by deposition of the Senonian flysch. After that came the Paleogene continental phase which was characterized by intensive tectonic movement, volcanism and erosion.

The karst is developed in rather small, isolated areas of carbonate rocks that are separated by basins, river valleys or terrains composed of impermeable rocks. Although the thickness of the carbonate series amounts in most cases to 100–300 m, in some places it is more than 700 m. In addition to being discontinuous in the horizontal, it is not uniform vertically either. Between units of pure limestones and dolostones there appear layers of marls and sandstones. The limestones are very often pierced by veins of magmatic rocks. Along almost the entire western border of the Carpatho-Balkanian Mountains there is overthrusting. Triassic and Jurassic limestones of the *western belt* are thrust over Permo-Triassic red sandstones and then, together with them, over Miocene lacustrine sediments and the Cretaceous limestones in the East. In this manner, a wide zone of impermeable rocks is squeezed between the western and eastern limestone belts. All this has contributed to giving the carbonate rock terrains in the Carpatho-Balkanian Mountains the characteristics of fluviokarst.

At the beginning of the Miocene there was a great marine-lacustrine transgression from the Pannonian and Pontian Basins which spread, via the tectonic depressions, far to the South. It inundated the lower parts of the relief in the Carpatho-Balkanian Mountains and in the Inner and Central Dinarides. During the course of the Miocene and the Early Pliocene marine-lacustrine transgressions occurred repeatedly, leaving thick sediments. At the bottom of individual basins the thickness of these Neogene sediments generally amounts to 100–400 m, but in the Morava depression it reaches 800–900 m and in the Sava depression (near Kutina) more than 4 000 m. The Neogene sediments cover considerable karst areas. In the Pannonian Basin and in the Metohija Basin the lakes disappeared at the end of the Levantian and in other parts sometimes as early as the Pontian. An unusual karst relief occurs sporadically in the Miocene marls and clayey limestones; its greatest extent is found south of Beograd.

Paleokarst Phenomena

There is no evidence for the existence of the karst phenomenon before the Carboniferous in Yugoslavia. Pre-Carboniferous karst, if it developed at all, has become difficult to recognize because of later mineralogic changes in the carbonate rocks.

The oldest karst form discovered to date could be a fossil speleological structure discovered north of Krupanj in West Serbia. It is a vertical shaft created by corrosion in Middle Carboniferous limestone, and it was entirely filled with antimony ore. The exploitation of the ore revealed a pothole, 5–10 m wide and about 30 m deep (Fig. 2–84). It is considered that this shaft was made by karst erosion in the course of the Late Carboniferous and the Early Permian (DJURIČKOVIĆ, 1982). However, considering that the mineralization is of Oligo-Miocene age, the shaft could be much younger.

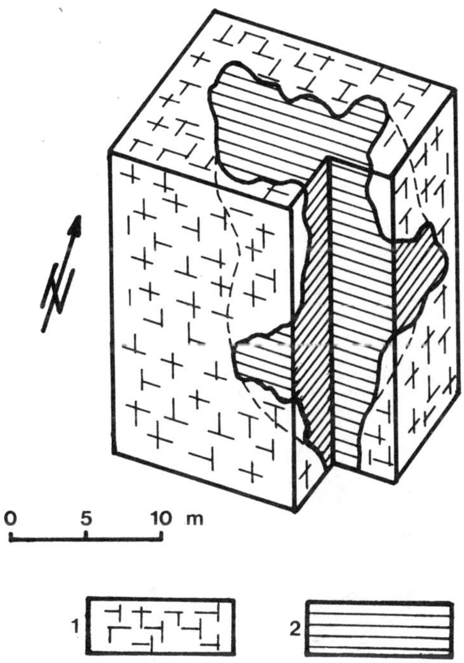

2–84. Stolice ore body near Krupanj (after DJURIČKOVIČ, 1982). 1. Middle Carboniferous limestone; 2. ore body – exhausted.

Triassic Paleokarst

Over vast areas in the Outer Dinarides, numerous bauxite deposits formed during the transition from the Middle to the Late Triassic. This Triassic paleokarst is best preserved and studied on Velebit Mountain, in the region of Lika, and in the environs of Nikšićko Polje. The karst is formed in Ladinian limestones and the cover is of Karnian sandstone, marls and clayey limestones.

The exhumed Triassic paleokarst of Debeljak on Velebit Mountain, west of Donje Pazarište village, has been described in detail by POLJAK (1952). According to him, in those places where the Karnian roof of basal breccia with hematite had been eroded, there appeared on the surface a paleorelief composed of dolines, karst passages, potholes and solution channels, essentially different from the recent karst relief. The Triassic dolines have a diameter of 3–5 m, are 1.5–2.5 m deep and have rather wide bottoms. They are connected with one another by shallower and deeper passages. On the sides of dolines are solution channels 6–16 cm wide and deep. The potholes are 2–3 m wide and 30–40 m deep. Their walls are covered with deep and wide solution channels which often reach almost to the very bottom.

Jurassic Paleokarst

Jurassic paleokarst is widespread in the Outer Dinarides of Montenegro and Herzegovina, where it is always associated with bauxite deposits. The floor of the bauxite deposits and of the paleokarst is formed of carbonate rocks of Late

2–85. Sub-ore karst relief in the Kutsko brdo bauxite deposit near Nikšić (Photo by A. Grubić).

Triassic to Early-Middle Jurassic and even Kimmeridgian age, whereas the cover or roof is almost always composed of Tithonian-Valanginian layers (exceptionally also of Kimmeridgian strata). Two different explanations have been proposed for this relationship; the first considers that during the Jurassic there were several periods of karst erosion (MILOVANOVIĆ, 1965), and the other suggests that one phase of karst erosion impacted simultaneously upon rocks of different ages (PAVIĆ, 1956).

When an ore deposit at Kutsko brdo (east of Nikšić) was exploited, on the floor of the bauxite a Jurassic paleokarst similar to tropical cone karst was discovered (GRUBIĆ, 1957). Limestone pyramids 5–10 m high are separated by dolines, potholes and passages. The pyramids are of irregular form, in most cases with almost vertical faces and furrowed with groovelike solution channels of various dimensions. Exhumed and exposed to modern weathering, the paleo-relief is quickly ruined and changes its aspect (Fig. 2–85).

Cretaceous Paleokarst

Most workers consider that during the Cretaceous there were at least two periods of karst erosion: one in the Early, and the other at the end of the Late Cretaceous. From the first period come bauxite deposits in Istria and on the peneplain of Banjani in western Montenegro. The floor of the bauxite consists of the Tithonian–Valanginian rocks and the cover is of Upper Cretaceous limestone. On the basis of discoveries of Lower Cretaceous continental flora in white bauxites at Banjani (PANTIĆ, 1958), it was established that these deposits had been formed in the course of the Hauterivian or the Barremian. Judging from the limited thickness of the bauxites, the underlying relief is smooth.

During the Late Cretaceous, karst was widespread in the Outer, and partly also in the Central Dinarides (Grmeč Mountain, Vrbas Valley and the Vlasenica region). In most cases the floor below the bauxite deposits is in Turonian limestones, and the cover is of Maestrichtian rocks. In the deposit at Kosturi, in East Bosnia, the maximum thickness of the bauxite layers is 80 m, but fault topography predominates in the floor instead of karst relief (FILIPOVIĆ, 1975).

Paleogene Paleokarst

At the end of the Cretaceous and in the beginning of the Paleogene, as a result of the Laramide Orogeny the greater part of modern Yugoslavia became land. To judge from the correlative sediments deposited in great tectonic troughs (synclines and rift-valleys) the intensity of subaeral erosion was significant. However, except for numerous bauxite deposits, traces of karst erosion from this period have not been preserved at the surface. The reasons for this absence are to be looked for primarily in the powerful tectonic and volcanic activities throughout the entire Paleogene, which also continued into the Early Miocene.

By means of electric sounding, bauxites have been discovered in several places in Istria, under Eocene sediments and hosted in Upper Cretaceous limestones. The bauxite deposits have the form of dolines, with diameters of 15–30 m and up to 25 m deep (Fig. 2–86). Northwest of Maršetići, under an Eocene cover 35 m thick, a bauxite pocket similar to a pothole has been discovered; in its upper part it is 15 m wide and is 25 m deep (KRULC and VIDOVIĆ, 1960). The discovery of a fossil doline near Nova Vas in the Mirna Valley is also interesting.

It is filled with bauxite in its lower part and with phosphate above. The doline is cut into Turonian-Santonian limestones and covered with Eocene limestones (ŠKERLJ and ŠKERLJ, 1973).

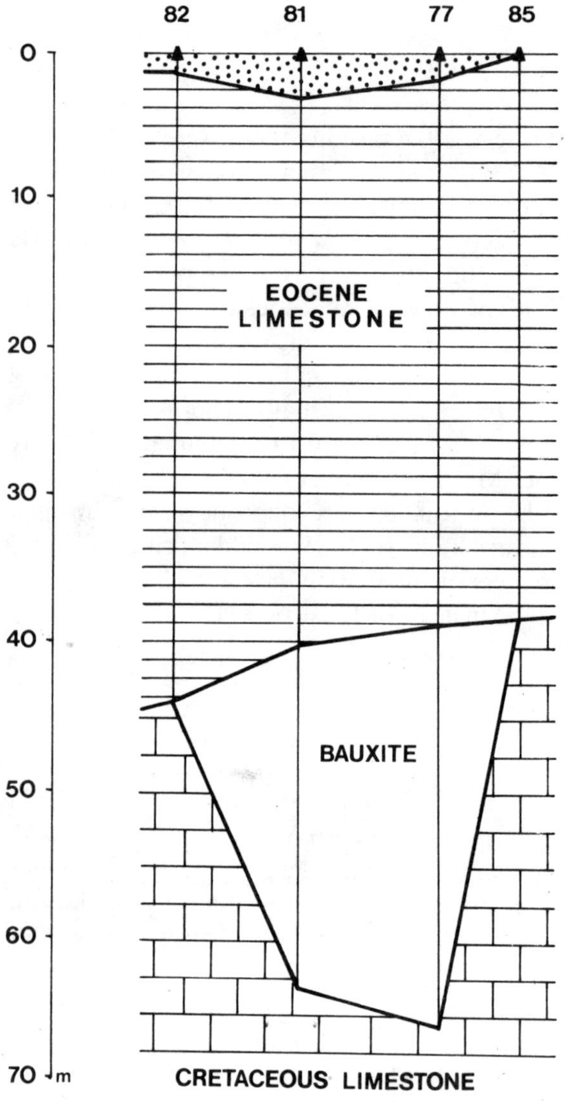

2–86. Bauxite deposit near Katunari, Istria, discovered by geoelectric soundings (after KRULC and VIDOVIĆ, 1960).

When exploratory drilling was made in the southern part of the Pannonian Basin, at two places large accumulations of underground waters in cavernous Triassic limestones were discovered. Both have the character of karst artesian aquifers. The first is situated south of the mountain chain of Fruška gora, in the floor of the Vrdnik Coal Basin. The drill-hole, in the middle of the basin, penetrated into Lower Triassic limestones beneath Lower Miocene coal-bearing sediments where, at a depth of 330 m, water gushed out. The limestones were drilled only some ten metres deeper, to approximately 340 m (120 m b.s.l.). The limestones are very cavernous, which caused some technical problems during drilling (MILOJEVIĆ, FILIPOVIĆ and DIMITRIJEVIĆ, 1968). In the Tertiary basin of Valjevo-Mionica, under the impermeable Miocene sediments, the Triassic cavernous limestones were also drilled. When the borehole reached the Triassic limestones at a depth of 138 m, water under pressure burst from them also (MIJATOVIĆ, 1984). In our opinion, these caverns in the Triassic limestones were formed during the Paleogene karstification phase.

At Ilidža spa near Sarajevo, an exploration borehole passed through Miocene sediments (150 m) and Eocene flysch (900 m) then, at a depth of approximately 930 m b.s.l., it penetrated cavernous Paleogene and Upper Cretaceous carbonate rocks in which there was thermo-mineral water under artesian pressure. On the basis of the analysis of Carbon–14, it was established that the water was 18 000 ± 350 years old (MIOŠIĆ, 1982). Despite the measured age of the water (which is considered to be rejuvenated) the widened fissures and channels in the Paleocene und Upper Cretaceous limestones were obviously formed during the transition between the Paleocene and Eocene.

In the Dinaric littoral and in Yugoslav territorial waters of the Adriatic there have been 58 deep prospecting boreholes, 14 of which are over 4 000 m and two over 6 000 m deep. Most of them passed, at different depths, through cavernous and karstified carbonate rocks in which oil and gas sometimes occurred or there was loss of the drilling fluid. In two boreholes in the northwestern part of the Dugi otok Island, karstified Lower Cretaceous rocks were discovered at the great depth of 2 960 m below sea level (AKSIN et al. 1981). Near Rovinj, in Istria, cavernous Triassic dolostones lie at a depth of 1 350–2 330 m and on the Vis Island, in Lower Cretaceous carbonates, in the interval 732–2 040 m there occurred a considerable loss of drilling fluid (MILOVANOVIĆ, 1965). The aggregate karst phenomena in the Mesozoic carbonates in this area could have originated only in the course of the Paleogene. However, the cavernous character of rock masses at the contact of the Adriatic platform and the Dinarides, where tectonic movements were very active throughout the entire Tertiary and Quaternary, need not always be a consequence of karst processes.

The presence of Paleogene karst relief was established along the western border of the Beljanica and Kučaj Mountains in the Carpatho-Balkanian system by means of prospecting boreholes. This paleorelief is developed in Cretaceous limestones and is covered with Miocene coal-bearing lacustrine sediments, whose thickness is sporadically more than 400 m. In the paleorelief there are dolines, uvalas and karst valleys. Two uvalas near Resavica Mine are about 1 km wide and 100 m deep (B. MAKSIMOVIĆ, 1956).

Probably one of the most uncommon paleokarst forms is the great system of

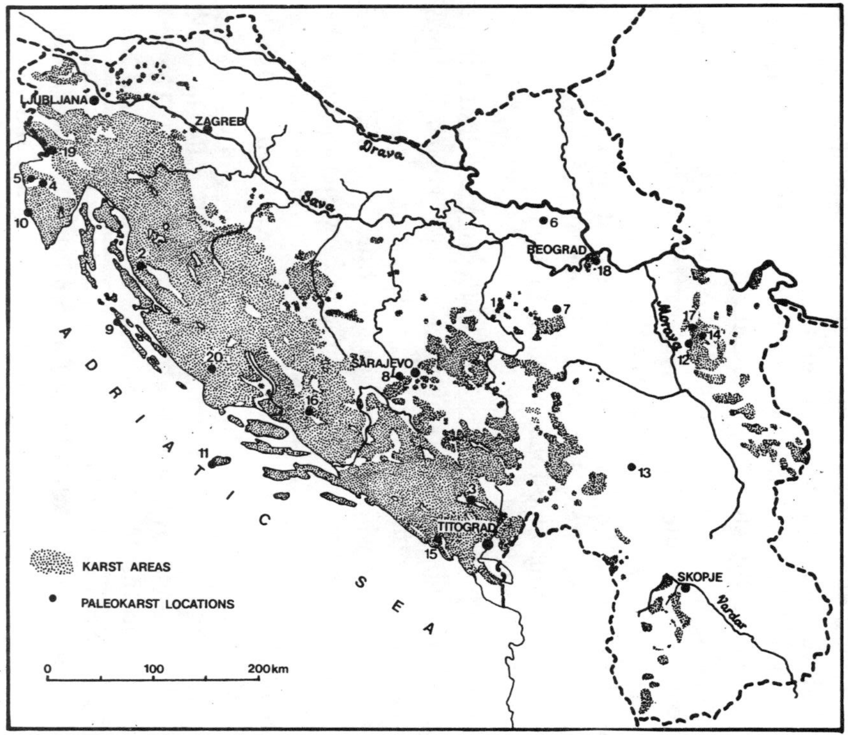

2–87. Extent of superficial karst relief and of some paleokarst phenomena in Yugoslavia. 1. Stolice ore body; 2. sub-ore relief at Debeljak; 3. sub-ore relief at Kutsko brdo; 4. bauxite deposit at Katunari; 5. fossil doline near Nova Vas; 6. artesian karst water table at Vrdnik; 7. artesian karst water table at Mionica; 8. artesian karst water table at Ilidža; 9. caverns in the prospecting borehole, Dugi otok; 10. caverns in the prospecting borehole, Rovinj; 11. caverns in the prospecting borehole, Vis; 12. paleokarst of the Resavica Coal Mine; 13. Stari Trg shaft system; 14. caves-ruins at Beljanica Mountain; 15. Peć shaft at Pestingrad; 16. Red and Blue Lakes; 17. exhumed cone karst of the Beljanica Mountain; 18. fossil dolines near Beograd; 19. fossil dolines near Divača; 20. Dočina bauxite deposit.

potholes discovered in the Stari Trg Mine, on the southern slopes of Kopaonik Mountain. In search of lead-zinc ores, the passages in the cave system were investigated between 730 m and 135 m a.s.l. They were formed in the marbles near their contact with schists, by the action of thermo-mineral waters. Within this system in addition to vertical channels, there are also horizontal passages and cave chambers several ten of metres in diameter (PETROVIĆ, 1969). Before the mining operations, all the channels below 545 m altitude, i.e. lower than the bottom of the Ibar Valley, were permanently filled with water. Because the age of the Kopaonik granites has been determined, by the K–Ar method, as Middle Eocene (DELEON et al., 1961), it may be stated with some certainty that this system of potholes originated in the younger Paleogene.

Neogene Paleokarst

Following the Sava orogenic phase of the Oligo-Miocene, when the initial trends of the present relief of Yugoslavia were created, karst developed more or less continuously to the present time. There were some interruptions of short duration only during the lake deposition which occupied the Pannonian and other major basins. We must think of relict phenomena contained in the contemporary relief, rather than of fossil forms (Fig. 2–87).

It has been established that cave remnants on the southern slope of Beljanica Mountain, Carpatho-Balkanian system, were formed at the end of the Miocene, together with the development of a karst peneplain at 700 m a.s.l. (GAVRILOVIĆ, 1976). The 225 m deep pothole, Peć, at Pestingrad above Kotor is approximately of the same age. Its mouth lies on the edge of a great karst slope overlooking Boka Kotorska (910 m a.s.l.). The great subterranean cavity near Imotski is also certainly of Neogene age. Above it a ceiling collapse formed the depressions of

2–88. Exhumed Upper Miocene cone karst on the western slope of Beljanica Mountain (Photo by D. Gavrilović).

the Red and Blue Lakes. Though these are obviously fossil forms, because they are nested in the recent karst relief and are not isolated by sediments they should not be included in the paleokarst. Such forms are very common, not only among the speleological remains but also in the surface relief. The entire recent karst of Yugoslavia is a complex of Tertiary and Quaternary forms.

Paleokarst of Neogene age so far has been positively established only on the west side of Beljanica Mountain, Carpatho-Balkanian system. There, it has been partly exhumed from beneath lacustrine coal-bearing sediments. The relief consists of dolines and conical limestone hills, similar to *tropical karst* (Fig. 2–88). This relief is considered to have originated at the end of the Miocene in a humid and warm climate (GAVRILOVIĆ, 1960).

Pleistocene Paleokarst

Paleokarst of Pleistocene age has been best preserved and it extends over the widest area. Along the Adriatic coast the submarine relief often includes typical karst landscapes, with numerous dolines, uvalas, potholes and caves. Considering the high mountainous hinterland from which water flows in large quantities via subterranean channels, many submerged coastal cave systems are still active as freshwater springs. Where caves have lost every hydrological connection with

2–89. Fossil doline filled with loess, in Upper Miocene limestone near Beograd (Photo by D. Gavrilović).

the land, the date of their fossilization depends directly on the depth at which they are situated, and on the rate of eustatic oscillations of sea level. The rise of global sea level in the Holocene has also fossilized some karst forms on dry land, such as the subsiding dolines of Dragon's Lake (Zmajevo jezero) near Rogoznica and the Dead Sea (Mrtvo more) on the Lokrum Island near Dubrovnik, Bačina Lake uvala near Kardeljevo, and many others.

The occurrence of paleokarst has been noted in several places also in the Sarmatian limestones south of Beograd. Karst topography of the younger Pliocene and the older Pleistocene there is covered by loess deposits whose original thickness exceeded 10 m. In the course of the Holocene the loess cover has been eroded to a considerable extent. It is now very much thinned so that the karst is reactivated. Nevertheless, there are numerous dolines and vertical channels which have lost their former hydrologic function and whose presence is no longer reflected by the topographic surface (Fig. 2–89).

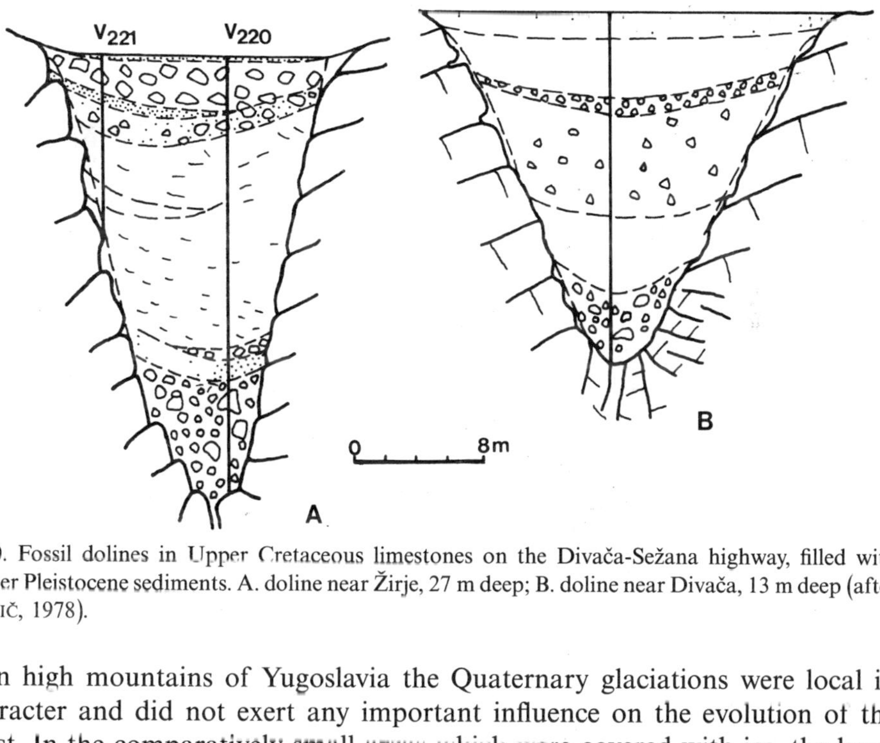

2–90. Fossil dolines in Upper Cretaceous limestones on the Divača-Sežana highway, filled with Upper Pleistocene sediments. A. doline near Žirje, 27 m deep; B. doline near Divača, 13 m deep (after HABIČ, 1978).

In high mountains of Yugoslavia the Quaternary glaciations were local in character and did not exert any important influence on the evolution of the karst. In the comparatively small areas which were covered with ice, the karst process was either decelerated or interrupted but, on the other hand, it was intensified on the periphery of the periglacial zone. There are many examples of karst forms covered with glacial, solifluction and fluvial depostis (Fig. 2–90). In most cases, however, karst processes were reactivated in the postglacial period, so that real fossil forms are very rare.

Conclusion

Various mining operations, prospecting boreholes and other geostudies discovered many examples of paleokarst forms from different periods of the Mesozoic, Tertiary and Quaternary. On the basis of known paleokarst phenomena, the existence of at least eight karstification phases have been established: (1) Carnian, (2) Upper Jurassic, (3) Barremian, (4) Senonian, (5) Early Paleogene, (6) Late Paleogene, (7) Late Miocene-Early Pleistocene, (8) Holocene. In contrast to the Mesozoic paleokarst phases, which were interrupted by transgressions, the younger phases are not everywhere clearly distinguished because frequently karstification continued over longer periods of time. In many places in the Dinarides and in Carpatho-Balkanian Mountains, karst has developed with more or less the same intensity throughout the Paleogene, during the Neogene and the Quaternary or continuously from the Late Cretaceous to the present day.

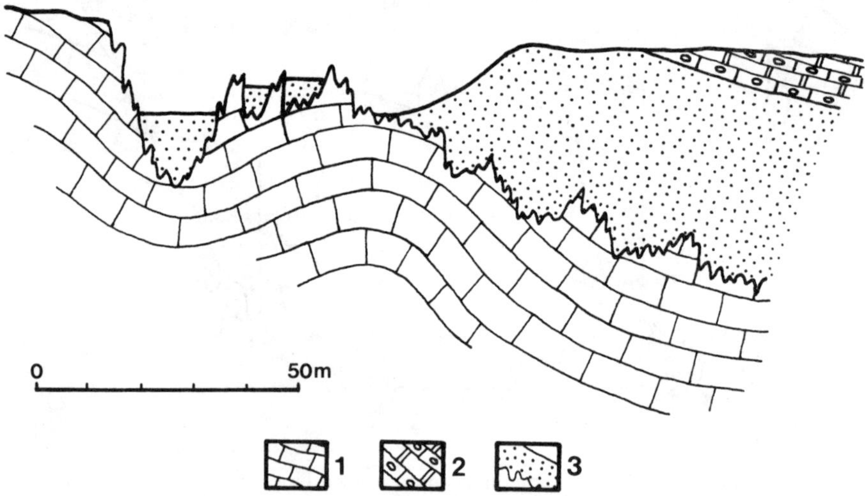

2–91. Profile through Dočina bauxite near Drniš with characteristic sub-ore relief (modified after SAKAČ, 1966) 1. Middle Eocene limestone; 2. Upper Eocene limestone; 3. bauxite.

The well known paleokarst phenomena are in most cases associated with bauxite deposits, the age of which varies from Triassic to Eocene. From the fact that in limestone floors beneath ore deposits depressions of the solution pipe (*geological organ* and *pyramid*) type prevail, it is obvious that the bauxites are a product of disintegrated weathering crusts (Fig. 2–91), formed under conditions of tropical climate and on very flat relief, i.e. they are connected with tropical peneplains. The sub-ore limestone relief can only have the character of subsoil karst. The bauxite deposits in true dolines and in smaller structural troughs are of allochthonous origin and formed chiefly by redeposition of the ore as a consequence of tectonic movements.

The existence of a tropical climate in the Yugoslav area during the Paleozoic and Mesozoic has been demonstrated by numerous sedimentological and biological paleoclimatic indicators and by numerous determinations of the $^{18}O/^{16}O$ isotope ratio (POLŠAK, 1976; POLŠAK and PEZDIČ, 1978). In addition to Early Paleogene bauxite pockets, some other forms of paleokarst are encountered but they are very rare. There is no bauxite from the Late Paleogene but the karst paleorelief is very varied, including dolines, uvalas, potholes, dry valleys, and other forms. Under the influence of strong differential tectonic movements which began at the end of the Cretaceous and were constantly intensifying, the style of the development of the karstification processes gradually changed. It became more and more the vertical dimension and penetrated deeper and deeper into the carbonate mass. Superficial and subsoil corrosion lost their former intensity and deep corrosion became predominant. The underground cavities gradually grew and the focus of karst corrosion was transferred into them. Consequently, the transition from the Mesozoic tropical peneplain through the Miocene cone karst to the Pliocene and Quaternary deep karst, though accompanied by climatic changes, was mainly determined by the tectonic uplift and subsequent drop of the outflow level.

Owing to this progressively deeper penetration by karst processes, there was some reactivation of paleokarst of different ages. At the same time, under the influence of tectonic movements and erosion processes, in many places the paleokarst was exhumed and incorporated into younger paleokarst or into the modern landscape. It is obvious that the modern karst landscape is a complex of forms and phenomena that have originated under different natural conditions and in different epochs.

The knowledge of paleokarst thus far acquired changes many of our notions not only of the morphological, but also of the hydrological, properties of the karst. It has been established that in addition to the common cave and fissure groundwater, there also exist artesian karst groundwater bodies with aquifer horizons that are situated in some cases up to one thousand metres below the Adriatic Sea level. A system of grikes and other fissures filled with water, has been discovered even at a depth of 3 000 m b.s.l. This means in practical terms that the base of karstification in the Dinaric Mountains is situated at the base of the carbonate series, which is thus karstified throughout its thickness.

Since there are several horizons of paleokarst of different age in the Dinaric Mountains and the Carpatho-Balkanian Mountains the question may be asked: "*In which direction did karstification and the corrosive widening of grike, etc. systems proceed?... from the surface downwards into the carbonate mass only or simultaneously from above and from below?*" Although our conclusions on the genesis and the essence of the karst are chiefly drawn on the basis of its superficial and subsurficial manifestations, one should not neglect the numerous occurrences of hydrothermal ore deposits and thermal waters. Sources of thermo-mineral waters of vadose origin are rather frequent in the Yugoslav karst and many of them function considerably longer than most ordinary (meteoric) springs. Thus, for instance, the springs at the spa of Niška Banja have emitted water without any interruption for at least 27 000 years (VUČIĆ and PAVLOVIĆ, 1958). During this time they have discharged together with water, more than 4.8 million m^3 of dissolved limestone, a mass sufficient to form a cube with edges 364 m long.

There is no doubt that the study of paleokarst is complex and poses many difficulties, but at the same time it offers new possibilities for visualizing the essence of karst phenomena in their entirety. Future studies will probably make essential modifications to our existing concepts of the genesis and age of forms in the karst relief.

PALEOKARST OF BULGARIA

Ivan Stanev and Stoyan Trashliev

Introduction

Paleokarst is widespread in Bulgaria. It is known mainly in boreholes, in carbonate sediments of Devonian, Carboniferous, Middle and Late Triassic, Bathonian, Late Jurassic, Early Cretaceous, Maestrichtian, Sarmatian and Neogene age. In most cases, recent karst is superimposed conformably or disconformably on the older paleokarst levels. Therefore, carbonate sediments exposed at the surface show mainly features of Neogene and Recent karst; only in rare cases can relicts of older paleokarst phases be observed.

Everywhere the paleokarst stages are parts of thick aquifers in the North Bulgarian (Moesian) artesian basins and in other smaller artesian basin in South Bulgaria. They contain fresh waters, saline waters, industrial waters, thermal waters and oil and gas. They are associated with accumulations of solid mineral deposits — gypsum, iron hydrothermal and metasomatic ores, lead-zinc metasomatic ores, bauxites, etc.

Paleokarst in Bulgaria is divided into two genetic groups: infiltrational and hydrothermal. Infiltrational paleokarst is further subdivided into platform (Moesian) and mountain (Balkanide) types (STANEV, 1970).

Infiltrational Paleokarst (Tab. 2–1)

Devonian Paleokarst

Devonian paleokarst occurs in borehole R–3 Gomotartsi (NW Bulgaria, Vidin) in Devonian dolostones, dolomitic limestones and limestones at a depth of 3 206–4 014 m (Fig. 2–92). Four paleokarst levels were formed (3 206–3 278 m; 3 443–3 680 m; 3 680–3 797 m; 3 797–4 014 m) separated from one another by unkarstified carbonates. The total thickness of the paleokarst and carbonates is 700 m. They form a single aquifer with thermal waters (t = 95 °C) of sodium-chloride type and mineralization of about 90 g · dm^{-3}

P. Bosák, D. C. Ford, J. Głazek and I. Horáček (Editors):
Paleokarst. A Systematic and Regional Review. – Elsevier and Academia. Amsterdam and Praha. 1989.

Table 2–1

Infiltrational Paleokarsts in Bulgaria

Age of Paleokarst	Age and Type of Host-rocks
Devonian	Devonian limestones and dolostones
Late Carboniferous	Lower Carboniferous limestones
Rhaetic	Middle Triasic limestones
Late Bathonian	Bathonian limestones
Berriasian-Early Hauterivian	Lower Cretaceous limestones
Berriasian-Late Hauterivian	Lower Cretaceous limestones
Berriasian-Barremian	Lower Cretaceous limestones
Early Hauterivian	Upper Jurassic-Lower Cretaceous limestones
Early Hauterivian-Aptian	Lower Cretaceous limestones
Early Hauterivian-Albian	Lower Cretaceous limestones
Aptian-Quaternary	Lower Cretaceous limestones
Early Sarmatian	Badenian gypsum
Neogene-Quaternary-Recent	Middle Triassic, Upper Jurassic, Maestrichtian and Sarmatian limestones

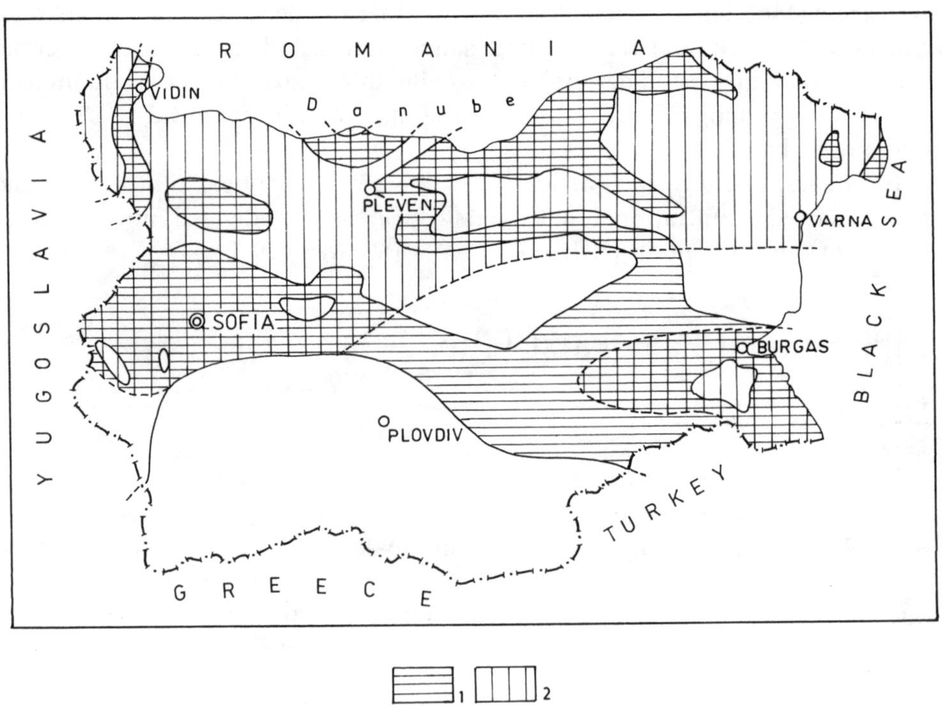

2–92. Paleokarst in Middle Triassic and Middle Jurassic carbonate sediments. 1. Rhaetic paleokarst in Middle Triassic carbonate rocks; 2. Upper Bathonian paleokarst in Bajocian-Bathonian rocks.

with a high hydrogen sulphide content. This platform type of paleokarst was formed as a result of breaks in sedimentation at the end of the Devonian and the beginning of the Early Carboniferous.

Upper Carboniferous Paleokarst

This paleokarst was also found in borehole R-3 Gomotartsi, in limestone and dolomitic limestones of Early Carboniferous age at a depth of 2 435–2 516 m. It also forms an aquifer with thermal waters (t = 90 °C), hydrogen sulphide or sodium-chloride type and mineralization of about 80 $g \cdot dm^{-3}$. The paleokarst resulted from a break in sedimentation during the Late Carboniferous.

Rhaetic Paleokarst

The Rhaetic paleokarst is distributed widely, mainly in North Bulgaria, while in South Bulgaria it is to a considerable extent re-worked by younger paleokarst phases and by folding (Fig. 2–92). It occurs at depths of from tens of metres down to 2 000–2 500 m. It is absent in areas where Upper Triassic sediments were deposited. From its morphology and thickness it is thought that this paleokarst was denuded. Two well-defined distinct hydrodynamic karst zones are distinguished — vertical and horizontal. The total thickness ranges up to 200–300 m which suggests a subdued or hilly paleorelief with altitudes less than 300–400 m above the erosional base. Lack of alluvial deposits of Rhaetic age indicates that there was an almost total underground paleokarst drainage (holokarst). In many places the Rhaetic paleokarst is filled with breccias of Anisian, Ladinian, Karnian and Norian rocks with clayey-sandy or carbonate cement. These breccias occur mostly in the upper parts of the paleokarst and are related to the upper levels of the vertical hydrodynamic zone. Sinkholes, filled with terrigeneous sediments of Early Jurassic age, are common. The lower levels of the paleokarst are not infilled. They contain fresh waters, highly mineralized thermal waters, oil and gas.

During the Tithonian, within the confines of the flysch trough of South and Central Bulgaria, the Rhaetic paleokarst was an excellent filter medium for hydrothermal-metasomatic fluids. Hydrothermal hematite was deposited in the karst voids while in the rocks between them metasomatic siderite accumulated, which was later replaced by limonite (e.g. the iron ore deposits at Kremikovtsi, Golyama Rakovitsa, etc.). The hydrothermal fluids were generated by a powerful expulsion stage under high temperature and pressure conditions (STANEV, 1970)

In the Balkanide tectonic zone the Rhaetic paleokarst was deformed by younger folding, being converted to a coarse limestone breccia. Relicts of cave deposits are preserved between the individual blocks and are a good indicator that distinguishes this paleokarst from simple tectonic or sedimentary breccias.

Upper Bathonian Paleokarst

The Upper Bathonian paleokarst formed in Bajocian-Bathonian terrigeneous carbonate formations. It is found in northern, western and southeastern Bulgaria (Fig. 2–92). The terrigeneous carbonates are not generally and regularly karstified, so the paleokarst is sporadic. Its vertical range varies from several metres to 40–50 m. In many places it is eroded or fractured and re-worked by later tectonic events. At the end of the Bathonian a shallow, low relief paleokarst basin developed. The resulting karst forms are small — pores and small caverns which cannot be related to any definite hydrodynamic zone. At present this paleokarst contains both fresh and sodium-chloride waters, locally subthermal or thermal.

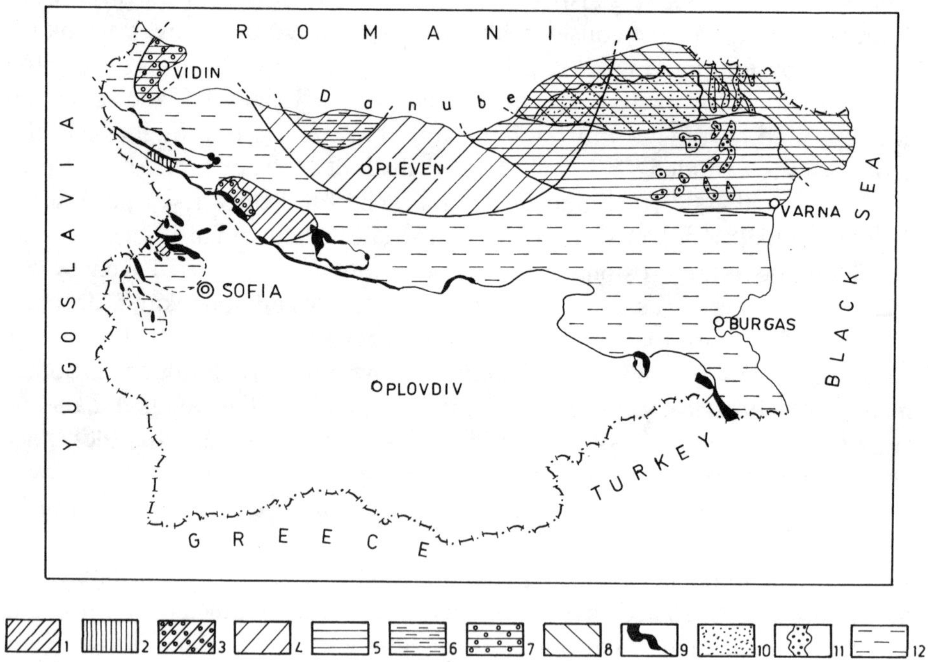

2–93. Paleokarst in Upper Jurassic-Lower Cretaceous carbonate sediments. 1. Berriasian-Lower Hauterivian paleokarst; 2. Berriasian-Upper Hauterivian paleokarst; 3. Berriasian-Barremian paleokarst; 4. Middle Berriasian paleokarst; 5. Lower Hauterivian paleokarst; 6. Lower Hauterivian-Aptian paleokarst; 7. Lower Hauterivian-Albian paleokarst; 8. Aptian-Quaternary paleokarst; 9. Neogene-Quaternary and Recent paleokarst; 10. exposures of Recent, rejuvenated Lower Hauterivian paleokarst; 12. Upper Jurassic-Lower Cretaceous carbonates which are unaffected by ancient or current karst process.

Middle Berriasian Paleokarst

The existence of Middle Berriasian paleokarst is established from boreholes in the north and northwest of Bulgaria (Fig. 2–93). It formed in limestones and dolostones of the Middle and Lower Berriasian and in the upper part of the Tithonian limestones, at depths of 1 000-2 000 m. The thickness of karstified carbonates varies from 200 to 300 m. Its origin and development are related to positive tectonic movements in the central and northwestern parts of the Moesian platform during the Middle Berriasian. In geological terms the karstification was brief because, during the upper half of the Berriasian, this paleokarst was covered by younger carbonate sediments which were not affected by any karst processes.

All the facts suggest that this dry land was relatively low and gentle, below 200–400 m a.s.l. The paleohydrodynamic karst zoning indicates a denuded, small-scale karst of platform type, typical of a subtropical climate. Today the waters in it are mineralized, of the sodium-chloride type.

Berriasian-Lower Hauterivian, Berriasian-Upper Hauterivian and Berriasian-Barremian Paleokarst

These paleokarst phases are of restricted occurrence, upon what were islands of Upper Jurassic to Lower Cretaceous limestones in west and northwest Bulgaria. These islands were low, relatively short-lived and with small surface drainage. As a result the paleokarst processes operated slowly and did not create any strongly expressed hydrodynamic zones. The thickness of karstified strata is variable. In places it is entirely eroded away. After it was formed, it suffered intensive tectonic deformation within the Balkanide fold zone and was reworked to a coarse breccia.

Lower Hauterivian Paleokarst

The Lower Hauterivian Paleokarst developed at a higher stratigraphic level and affected limestones and dolomitic limestones of the Upper Jurassic and Lower Hauterivian (*Radiatus* Ammonite Zone). The total thickness of karstified strata is about 200–300 m and its depth from 100 to 1 800 m. Here the paleokarst is again a result of positive tectonic movements (Fig. 2–93). Dry land was created over large parts of NE Bulgaria, Romania and the present Black Sea. Upon this new land an extremely well watered paleokarst basin with vigorous underground drainage was formed. In these conditions well expressed vertical and horizontal hydrodynamic zones developed, as proven by borehole data. The thickness of the karstified zone indicates that the relief was gentle, below 400 m a.s.l and without vegetation. In many places this paleokarst was rejuvenated by later karst processes.

The thick aquifer formed in this paleokarst now contains fresh and mineralized waters of bicarbonate, sulphate and chloride types. Thermal waters occur in its southern and, particularly, in its eastern parts. The Tyulenovo oil pool in East Dobrogea is also related to this paleokarst.

Lower Hauterivian-Aptian and Lower Hauterivian-Albian Paleokarst

These paleokarst phases are of restricted occurrence. They develop in two islands in NW and north-central Bulgaria (Fig. 2–93) and are a direct continuation of the Lower Hauterivian paleokarst that is geographically remote. These paleokarsts are totally preserved and covered by thick sediments of Aptian, Albian and Late Cretaceous age which form the Moesian platform. They display distinct vertical and horizontal hydrodynamic zoning. Waters are thermal, mainly of sodium-chloride type.

Aptian-Quaternary Paleokarst

The Aptian-Quaternary paleokarst formed in Hauterivian and Barremian limestones of NE Bulgaria (Fig. 2–93). It belongs to the platform type of karst in which vertical hydrodynamic zoning is dominant while horizontal zoning occurs in the form of numerous smaller karst levels one above the other and which mark the cyclic uplift of the area. The thickness of this karstified zone varies from 100 to 500 m. The lack of modern surface drainage in NE Bulgaria is related mainly to this paleokarst.

At the end of the Berriasian a new paleokarst period began which extended to the Quaternary. The most characteristic feature of this epoch is that the paleohydrological conditions, the evolution of the relief and the paleokarst are inter-related and are all determined by the tendency towards constant uplift in the North Bulgarian Swell.

Due to marine transgressions, development of this paleokarst was interrupted during the Cenomanian and the Maestrichtian. During the Paleogene and the Miocene there were partial breaks affecting eastern parts of the Dobrogea dry land area and during the Pleistocene, considerable parts of the same area were affected. During the Quaternary the North Bulgarian Swell was again uplifted and this caused a new activation of the karstification.

The kaolin deposits in NE Bulgaria are associated with this paleokarst. At present the aquifer formed in it contains mainly fresh bicarbonate waters.

Lower Sarmatian Paleokarst

This occurs in the northwesternmost part of Bulgaria where it is exposed by boreholes and mining works in the area of the Koshava gypsum mine, Vidin District. The stratigraphy consists of Upper Cretaceous marls overlain by Lower Badenian clastics and Middle Badenian evaporites which are represented by gypsum and a few limestones with a total thickness of 10–45 m. The evaporites are covered by Upper Badenian marls and clays 10–30 m thick. At the end of the Late Badenian, tectonic movements uplifted the strata as linear anticlines and synclines. Erosion led to the formation of a hilly relief.

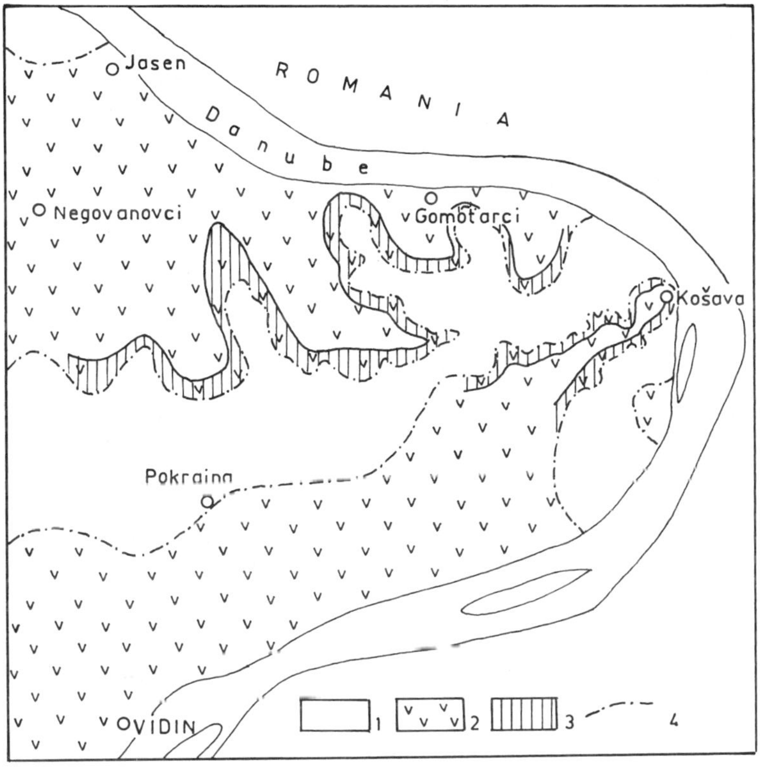

2–94. Distribution of karst phenomena in Badenian gypsum of the Vidin district. 1. area stripped of gypsum in Early Sarmatian time; 2. area with preserved gypsum; 3. zone of Early Sarmatian paleokarst; 4. boundary of recent distribution of gypsum rocks.

Upon the exposed gypsum (Fig. 2–94), shallow sinkholes formed which terminated with deep ponors and swallow holes. Horizontal karst is better-developed in the lower parts of the gypsum beds. Caverns and channels of varied size and morphology were created. The larger cavities are infilled with collapse material from overlying gypsum plus fragments of marl and limestone transported from outside. The fills are cemented by gypsum and clayey to marly material. The maximum vertical range of karstification was about 40 m.

Karst formed during the Early Sarmatian was buried by clay and sandy-clayey sediments of Sarmatian, Meotian and Lower Pontian age. The paleokarst in gypsum rocks is not in any hydraulic connection with the upper aquifers and is filled with syngenetic marine and mixed waters.

Neogene-Quaternary and Recent Karst

This karst is widespread in Bulgaria and is developed upon modern outcrops of carbonate sediments of Middle Triassic, Late Jurassic, Early and Late Cretaceous and Sarmatian age (Fig. 2–93). It has completely erased characteristic features of the older paleokarst phases and superimposed a new (and in most cases, unconformable) hydrodynamic karst zoning. Therefore, at present it is almost impossible to detect any paleokarst on the surface; only relicts may be found. The Neogene-Quaternary and Recent karst contains fresh bicarbonate groundwaters.

Hydrothermal Paleokarst

Hydrothermal karst of Tertiary age is known in the area of the Erma Reka deposit, close to the frontier with Greece. It is proved in boreholes at 500–600 m depth. Its thickness is more than 2 000 m. The rocks are marbles of the Philipovo metamorphic group (IVANOV, 1981), in which a thick deep-seated aquifer is formed (IVANOV and STANEV, 1982). They are overlain by gneisses of the Rhodope Massif (Fig. 2–95), which play the role of an upper aquiclude or aquitard.

In the conditions prevailing in this deep aquifer a typical hydrothermal karst has developed. Two distinct hydrodynamic paleokarst zones are distinguished; a lower with deep-seated vertical karst at a depth over 2 000 m; and an upper, with horizontal karst at a depth from 1 to 14 m. The vertical karst is associated mostly with normal faults affecting the deep-seated aquifer, while the horizontal karst is developed in marbles deformed and fractured as a result of tangential displacements.

Rich lead-zinc metasomatic ores at Erma Reka are related to the horizontal paleokarst (quartz-cavernous zone). They comprise mainly galena and sphalerite with subordinate pyrite, sulphosalts, magnetite, hematite, etc.

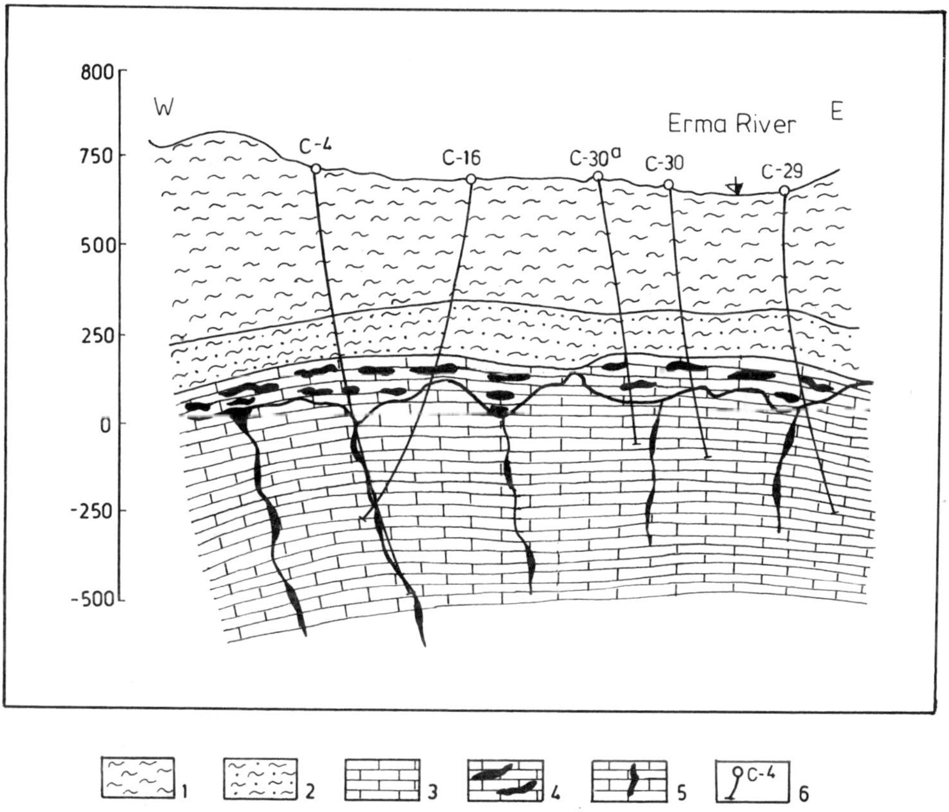

2–95. Cross-section through the Erma Reka deposit (after MANEV and IGNATOVSKI, 1981). 1. migmatites; 2. leucocratic rocks; 3. marbles; 4. horizontal Tertiary paleokarst with metasomatic lead-zinc mineralization; 5. vertical Tertiary paleokarst; 6. borehole.

The hydrothermal activity is a product of a Tertiary thermal expulsion system which originated as a result of the powerful temperature effects of a deeply intruded magma and high pressure. In this case, the vertical paleokarst was the conduit for hydrothermal solutions which, reaching the horizontal zone, deposited the rich lead-zinc mineral bodies.

Initially the hydrothermal fluids, which were very aggressive with respect to the marbles, created the paleokarst and the paleokarst hydrodynamic zones. Later they formed the metasomatic mineralization. The same solutions penetrated along transcurrent faults and normal faults into the gneiss formation overlying the marbles where they formed hydrothermal lead-zinc mineralization of the vein type.

Mineral Deposits Connected with Paleokarst

Bauxites

Bauxite deposits have been found only in the Trăn area, west of Sofia (Fig. 2–96). The bauxites are of karst type and infill negative forms developed on Tithonian limestones and dolostones during the formation of the Lower Cretaceous relief. They are covered by Turonian terrigeneous rocks: clays, siltstones, sandstones and a few conglomerates and argillaceous limestones (KARANOV, 1965).

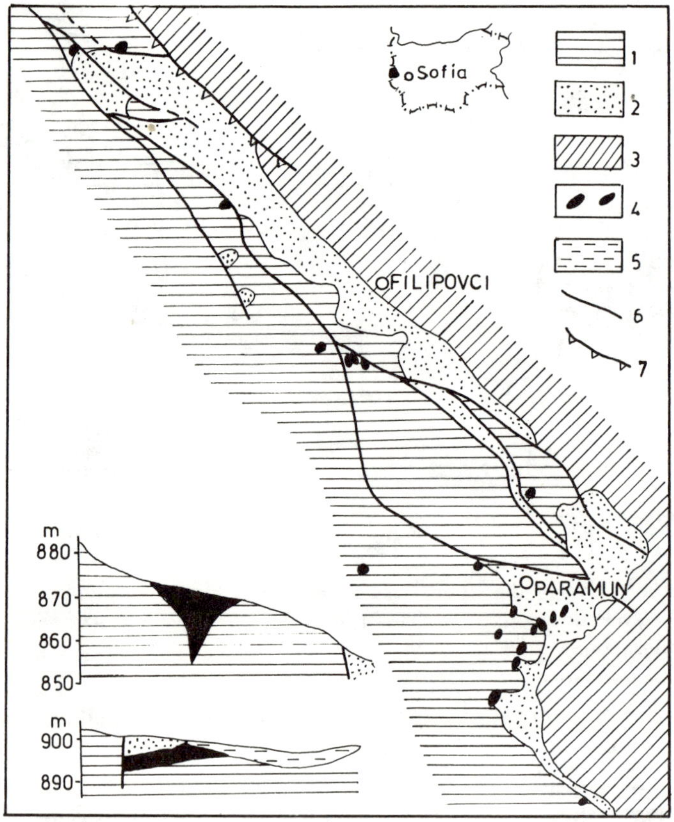

2–96. Geological sketch map of the bauxite deposit at Paramun and Filipovci villages and cross-sections of some bauxite bodies (after KARANOV, 1965). 1. Tithonian limestone and dolostone; 2. Turonian clay, siltstone, sandstone and conglomerate; 3. Senonian sandstone, marl and limestone; 4. bauxite; 5. red clay (Quaternary); 6. fault; 7. reverse fault.

The bauxites occur in irregular bodies, commonly of elongated form. They are 50–100 m long, 20–50 m wide and 1–15 m thick. Small bauxite lenses 1–10 m in size are also found. Parts of some bauxite bodies are directly exposed on the surface due to erosion of the Turonian sediments, while others are still under the Turonian cover. Some of the older and of the more recent karst depressions are filled with reddish Quaternary clay which comprises redeposited bauxites and Turonian clay with rare pieces of Turonian sandstones and hard bauxites.

Two types of bauxites are distinguished: (1) clayey, soft, with a lower Al_2O_3 content and (2) hard, compact, with a higher alumina content. Usually, the first type forms the base and the top of bauxite bodies. Deposits represented only by this type are probably relics of older, partially washed-out larger bauxite bodies.

Boehmite dominates in the bauxites while hydrargillite is rare and occurs in small amounts mainly in the upper parts of the bodies (TODOROVA and STEFANOV, 1973). Clay minerals are represented by kaolinite. Iron hydroxides include goethite and hydrogoethite.

The bauxites originated as redeposits of Lower Cretaceous weathering products of the lateritic type into the karst landforms. The weathering products were supplied from the south and southwest, close to the deposits (KARANOV, 1965). The infilled karst forms were then buried under the terrigenous deposits.

Continental formations of the terra rossa type in Lower Cretaceous karst have been found also in other places. They occur in the area of Belogradtshik in small karst cavities formed at the Tithonian-Barremian-Aptian boundary (TODOROVA and STEFANOV, 1975a); in the area of Trăn (outside the region of bauxite deposition) along the boundary between Upper Barremian-Lower Aptian limestones and Turonian conglomerates (TODOROVA and STEFANOV, 1975b); in the vicinity of Sofia (Slivnitsa), etc.

Kaolin

Kaolin deposits in Bulgaria are associated with the Aptian-Quaternary paleokarst in the northeast. The total area of paleokarst infilled with kaolin sands is about 3 000 km^2 (Fig. 2–97).

The paleokarst is developed in Hauterivian and Barremian limestones, which are extremely pure — between 92 % and 98 % $CaCO_3$ (KRASTEV, 1974). Formation of the paleokarst began during the Aptian. Processes were more intensive during the Early Albian when the range of karstification was up to 150 m (KRASTEV, 1975).

The karst landforms display complex and irregular outlines. There are limestone ridges with depressions between them. A majority of sinkholes are funnel-shaped with diametres of 20–150 m and depths up to 50 m, the deepest terminating with ponors. The largest depressions, some of which in scale and morphology correspond to poljes, also show a complex and irregular relief that is a result of the mutual connection of collapses, sinkholes and blind valleys.

During the Middle Albian the sea covered this area and buried the paleokarst under thick feldspar and quartz sands which entirely infilled the negative karst features. After the Middle Albian the area uplifted again but remained a peneplain during almost the entire Late Cretaceous, Paleogene and Neogene. During

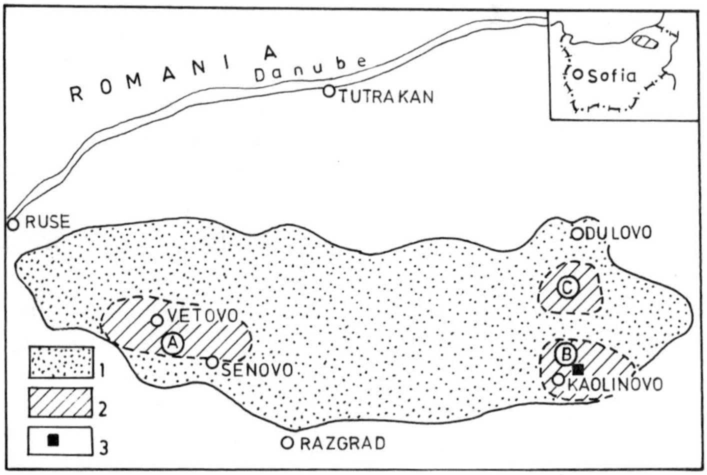

2–97. Distribution of Aptian-Quaternary paleokarst filled with kaolin sand. 1. paleokarst with kaolin sand; 2. areas of economic kaolin deposits: A. Vetovo-Senovo area, B. Kaolinovo area, C. Dulovo area; 3. location of the Todor Ikonomovo-south kaolin deposit (see Fig. 2–98).

individual periods of humid and hot climate this favoured deep chemical weathering of the arkosic sands which led to complete kaolinization of feldspars and other aluminosilicate minerals. A considerable part of the kaolin deposits and the buried karst were probably bevelled by the Sarmatian-Pontian peneplain. Now the kaolin sands are covered only by Quaternary deposits (soil, loess and clay) 10–40 m thick.

Although kaolin sands occur over almost the entire paleokarst area, economic kaolin deposits are concentrated mainly in three regions; Vetovo-Senovo, Kaolinovo and Dulovo (Fig. 2–97).

In the **Vetovo-Senovo area** surface karst forms are widespread and commonly have large dimensions, from 1–2 to 8–10 km^2. The deposits at Vetovo (11.2 km^2), Zlaten dol (8.2 km^2) and Esenitsite (5.0 km^2) are located in such karst poljes. Other deposits are smaller.

The **Kaolinovo region** is also intensively karstified and so numerous kaolin occurrences are found there. In comparison with the Vetovo-Senovo region, karst forms are deeper but of smaller horizontal extent. The largest features are less than 2 km^2 in size (Fig.2–98).

The **Dulovo region** is not so intensively karstified but, in terms of the size of the karst forms, it occupies an intermediate position. Together with smaller features there are poljes up to 3–4 km^2 (Kolober, Balaban). The depth of erosion is 90 m.

The kaolin raw material is a mixture of quartz sand and kaolin in the ratio 4 : 1. Dressing yields two industrial products — washed kaolin and quartz sand.

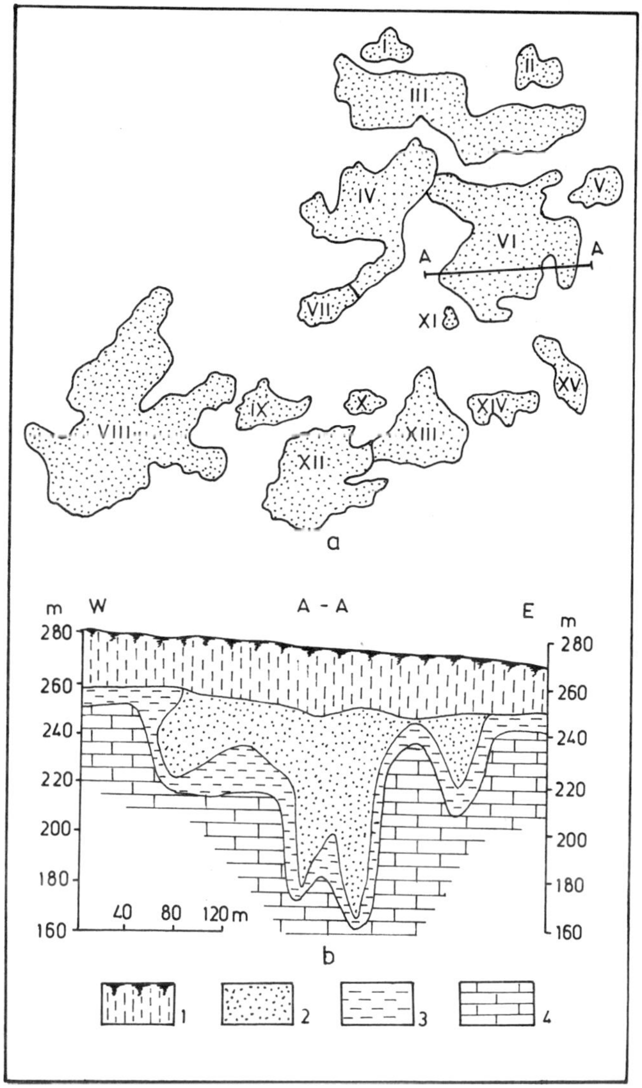

2–98. The Todor Ikonomovo-south kaolin deposit. a. paleokarst filled with kaolin sand (I–XV = numbers assigned to prospected and developed kaolin bodies); b. cross-section through kaolin body VI. 1. Quaternary (soil, loess, clay); 2. workable kaolin sand; 3. non-industrial kaolin sand containing layers of clay and sandy clay; 4. Lower Cretaceous limestone.

Iron Ores

Recently it was proved that many iron ore hydrothermal-metasomatic mineralizations in Bulgaria are not products of hydrothermal activity related to magmatism but to hydrothermal mixing activity located in a flysch trough of Tithonian age (STANEV and PANAYOTOV, 1976). These solutions were concentrated in a pre-existing natural reservoir (an aquifer) in the Middle Triassic limestones, with the Rhaetic paleokarst on them. There were linear drainage zones of regional extent to west and east.

When the hydrothermal solutions in the area of Kremikovtsi, Golyama Rakovitsa and other deposits reached the Rhaetic paleokarst, they acted initially upon pillars between the paleokarst forms, depositing manganese siderite ores in them. Later, the paleokarst forms themselves were infilled by hydrothermal hematite and baryte. In many places the paleokarst limestone breccia was converted into hematite-baryte breccia. Hydrothermal activity terminated with pyrite, sphalerite, galena and chalcopyrite mineralization.

After the ore deposition, the Tithonian flysch trough was intensively folded. During the Early Cretaceous it became dry land and there developed an intensive infiltration system which turned the hydrothermal-metasomatic ores into supergene ones — limonite, psilomelane, pyrolusite, manganite, cerussite, malachite, etc. After a hiatus during the Late Cretaceous, supergene processes continued during the Paleogene, Neogene and Quaternary.

Oil Accumulations

Oil accumulations in North Bulgaria are associated mainly with Rhaetic paleokarst (at Dolen Dabnik) and Lower Hauterivian paleokarst (at Tyulenovo). There are other small, non-economic accumulations in the other paleokarst phases.

Thermomineral Waters

In most of North Bulgaria thermo-mineral waters are associated with aquifers created during all the individual phases of paleokarst that have been described. The waters are highly mineralized (20–90 $g \cdot dm^{-3}$) and of sodium-chloride type. The temperature is from 30 to 90 °C. They are used as energy sources or for balneologic purposes.

Only NE Bulgaria is an exception in this respect. There, large parts of the Lower Hauterivian and the whole of the Aptian-Quaternary paleokarst bear fresh waters which are intensively used for domestic and industrial purposes.

PALEOKARST OF ROMANIA

Marcian Bleahu

Introduction

Carbonate rocks (limestones, dolostones, marbles) in Romania cover an area of 4 400 km^2, representing only 1.4 % of the country. This low percentage is due to the great extent of Pliocene and Quaternary deposits that overlie older formations in the plains. In addition to carbonate rocks, karst forms are found also on salt and gypsum in limited areas (BLEAHU, 1972).

Limestones are exposed in the Carpathian Mountains. They are of Paleozoic, Mesozoic and, less importantly, of Tertiary age (Fig. 2–99). The foreland zone, represented by the Moesian Platform, lies south of the South Carpathians and of North Dobrogea. It consists predominantly of Paleozoic and, especially, Mesozoic limestones overlain by thick Cenozoic and Quaternary detrital cover. Outcrops are present only in the South Dobrogea. Karstification here takes place at depth, being to a great extent of Mesozoic age.

Geological Backround

The Romanian Carpathians can be divided into three units: (1) the East Carpathians which extend from the northern border of the country (being the continuation of the West Carpathians) to the Ialomiţa Valley, (2) the South Carpathians extending from the Ialomiţa Valley to the Danube, and (3) the Apuseni Mountains in the western part of the country between the Mureş, and the Someş Valleys. Limestones are not uniformly distributed in these regions. They are less developed in the East Carpathians while the western part of the South Carpathians and the Apuseni Mountains are very rich in them, where karst forms are very prominent.

Precambrian and Paleozoic limestones and dolostones are present in some relatively restricted areas in the Carpathians and are mostly metamorphosed. Their phases of erosion and the existence of any karst relief is not established but karst cannot be excluded.

During the Mesozoic, limestone deposition started in the Anisian. Throughout the Triassic,

P. Bosák, D. C. Ford, J. Głazek and I. Horáček (Editors):
Paleokarst. A Systematic and Regional Review. – Elsevier and Academia. Amsterdam and Praha. 1989.

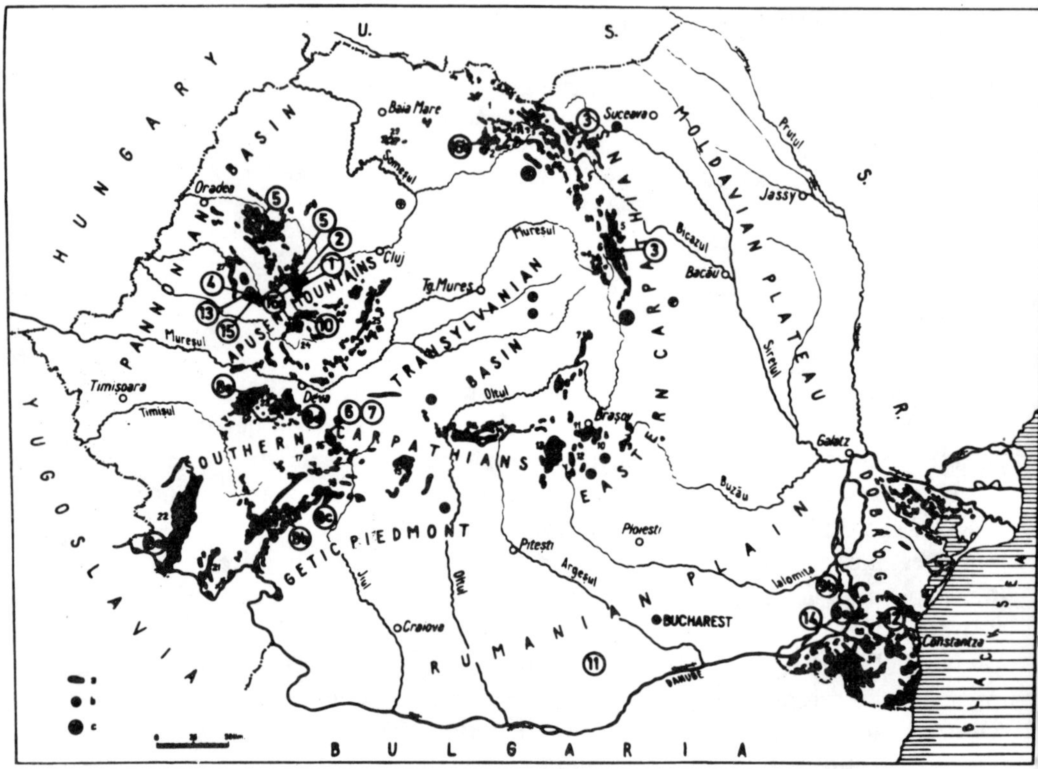

2–99. Outcrop of karst-forming rocks in Romania (after BLEAHU, 1972). a. limestones and dolostones; B. karst developed on salt and gypsum; c. volcanic rock karst; 1. Maramureş Mountains; 2. Rodna Mountains; 3. Rarău Massif; 4. Giurgeu Mountains; 5. Ceahlău Massif; 6. Hăghimaş Massif; 7. Virghiş Valley Massif; 8, Perşani Mountains; 9, Ciucaş Massif; 10. Piatra Mare Massif; 11. Postăvarul Massif; 12. Bucegi Mountains; 13. Piatra Craiului Massif; 14. Făgăraş; 15. Lotru and Căpăţîna Mountains; 16. Sebeş Mountains (Luncani Platform); 17. Retezat Mountains; 18. Vulcan Mountains; 19. Mehendiţi Plateau; 20. Mehedinţi Mountains; 21. Almaj Mountains; 22. western side of the Banat Mountains; 23. Poiana Ruscă Mountains; 24. Metaliferi Mountains; 25. Trascău Mountains; 26. Bihor Mountains; 27. Codru-Moma Mountains; 28. Pădurea Craiului Mountains; 29. Preluca Massif; 30. North Dobrogea; 31. South Dobrogea. Circled numbers are sites discussed in the text.

limestones and dolostones were deposited in great amounts, especially in the Apuseni Mountains where there are two great tectonic units composed of Triassic carbonates: the Bihor Autochthon and the Codru Nappe System. In the Autochthon there was a great break in the sedimentation that lasted almost throughout the Late Triassic, when an exokarst paleorelief formed. The same situation happened in the Codru Nappe System at the end of the Norian.

The second important period for limestone deposition was in the Late Jurassic. In all Carpathian units, Tithonian limestones covered large areas. Development of the present complex exo- and endokarst forms began. On their surface in the Bihor Autochthon, paleorelief was associated with bauxite deposits.

The Cretaceous was also a period of limestone formation, especially in the Carpathian inner units. Clastic formations prevailed in the outer ones (flysch, molasse). Generally, the Lower Cretaceous is richer in limestones (Urgonian limestones) on which, in the South Carpathian, an important relief formed.

The Tertiary formations in the Carpathians are only exceptionally of limestone, which occupies very limited areas and is without notable karst. During the Quaternary, various karst processes operated on the present relief, creating surface features filled with Pleistocene deposits, and developing endokarst networks which are also filled to a great extent with old deposits.

In the Moesian Platform, the second great structural unit with carbonate rocks, deposition of Paleozoic and Mesozoic limestones and dolostones alternated with periods of emergence that probably generated karst relief that was buried by subsequent transgressive formations. The most important karstification took place during the Paleogene. These fossil karsts are not visible as they now occur at great depth. They are known from boreholes. In South Dobrogea such ancient karstification is indicated by the modern circulation of water to great depths.

Terminology

So far few studies on the paleokarst in Romania have been made, so that the adoption of specific terminology has not been necessary. The fullest studies have dealt with bauxites but even in this case, only the term *paleokarst* was used. Under the general heading *paleokarst* we mean the following formations and processes which will be further discussed:

(1) Former exokarst (surficial) areas whose depressions were filled with younger formations and which were subsequently completely buried by transgressive formations. Thus, a former surface was fossilized and buried. The infilling depostits are generally continental and the transgressive cover is marine.

(2) Former endokarst networks, filled with later material. They must be correlated with continental periods when groundwater circulation was possible, but corresponding surface karst features cannot always be identified. These old networks can be recognized as fillings visible in outcrops and, especially, when intersected by the passages of modern caves.

(3) Former endokarst networks which were not infilled and that can be identified because of water circulation in them even today. If the paleoexokarsts are generally easy to define and identify, the situation is more complicated in the case of fossil endokarst because cave passages can contain very old deposits. Thus, the stream terrace deposits found in a modern river passage may be old but we do not include them in the paleokarst classification, restricting this term to cases where the fillings are older than the present karstification. Also, we shall not deal with karst processes acting under an impermeable cover which generate, for example, buried lapies.

Phases of Karst Development and Some Karst Phenomena

Triassic Paleokarsts

The question of fossil karst associated with Precambrian and Paleozoic limestones has not been studied, so that the earliest known instances of paleokarst date from the Triassic. These are all to be found in the Apuseni Mountains. In stratigraphic order they are:

(1) The Zugăi Formation in the **Bihor Mountains** (BLEAHU et al., 1981) is a detrital sequence situated close to the boundary between the Anisian and the Ladinian. It comprises: (i) quartzitic, partly conglomeratic, sandstones; (ii) predominantly maroon clay shales; (iii) limestones intercalated with millimetric red clays and (iv) breccias and megabreccias with olistoliths and paraconglomerates with flattened elements. Taking into account the fact that in addition to the Anisian clastics, some of the limestones within breccias are Wetterstein Limestones, the formation is likely to be situated a little above the Anisian/Ladinian boundary. It is a continental formation formed from the deposition of clastic material in torrential channels. On the whole the formation shows a

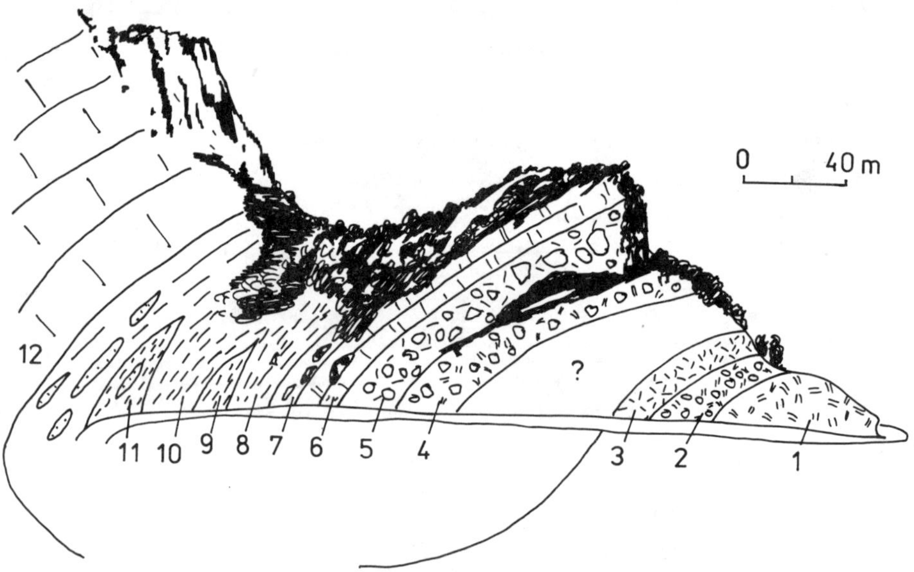

2–100. The Zugăi Formation in the Arieş Valley. 1. Massive dolostones; 2. dolomitic breccia; 3. black dolostones; 4. dolomitic breccia; 5. limestone breccia with red matrix; 6. massive limestone bearing dolostone breccia; 7. breccia with flat pebbles of white Wetterstein Limestones and black Anisian limestones; 8. clayey or calcareous red shales; 9. yellow shales; 10. violet-red shales with sandstone interbeds; 11. grey shales with sandstones; 12. massive Wetterstein Limestones; 4–10 are karst void fillings.

stratigraphic succession. It fills large channels that can only be of karst origin (Fig. 2–100). One must therefore suppose that there was at least partial exposure and erosion of the platform on which the Wetterstein Limestones were deposited, while in places solution processes and clastic filling of the depressions took place.

(2) The Ordîncuşa Formation in the **Bihor Mountains** (BLEAHU et al., 1981) which appears in the same unit of the Bihor Autchthon as the previous one but is situated higher in the geological column, in the upper part of the Ladinian and above the Wetterstein Limestones. Here, there are also (i) quartzose sandstones, (ii) red and violet clay shales, (iii) limestones with red argillaceous laminae and (iv) thin, platy limestones (as millimetric layers), breccias and conglomerates. In the Ordîncuşa Valley the formation crops out at several levels, displaying a clear stratigraphic sequence but here, too, it appears that this material comes from karst fillings. This fact is shown by the existence at numerous places in the Bihor Autochthon of fissures one metre or more wide, cut in Gutenstein Limestones and in Anisian dolostones as well as in the Ladinian Wetterstein Limestones, and filled with Early Lower Jurassic quartzitic sandstones. Considering that Upper Triassic deposits are lacking in the Bihor Autochthon, it is obvious that at the end of the Ladinian (possibly beginning only at the end of the Lower Carnian) the whole area emerged and stayed above sea-level until the "Rhaeto-Lias" transgression. During this continental phase, surface karstification processes were active, filling depressions, cavities, etc. with synchronous continental clastics and also, partially, with limey platform materials (intertidal or supratidal limestones). Much of the negative karst relief remained open for a long time, and was later filled with continental materials that mark the beginning of the Jurassic sedimentation cycle.

(3) Triassic paleokarsts in the **East Carpathians** are recognized in two stratigraphic positions, probably corresponding to the same continental phase. SĂNDULESCU (1975) describes fissures and solution depressions filled with red ferruginous material in massive Campilian-Anisian dolostones of the Bucovinian Formation. The filling sometimes becomes the matrix between angular dolostone fragments, showing that the basement solution was very intense. The author uses the term *siderolitic facies*, which points to a karstification process. Transgressive Middle Jurassic sandstones overlie this karst form. As Ladinian limestones are also known in other parts of this unit, it is thought that the most likely time for this continental phase is in fact the Late Triassic.

Farther north in the **Rarău Mountains** and within the same unit (the Bucovinian Formation), STĂNOIU (1967) also described an obvious paleokarst. Here, limestones with *Diplopora* (assigned to the Ladinian) are overlain by reddish, gritty spathic, oolitic limestones, within which there occur brachiopods, lamellibranchs (Pectinides) and ammonites (*Phylloceras* sp. and *Bullatimorphites* sp.), indicating a Late Bathonian-Early Callovian age. These limestones penetrate fissures in the Triassic limestones. Sometimes a breccia with subjacent Triassic limestones occurs between the two formations, and is cemented with gritty spathic limestones of the Jurassic type (Fig. 2–101). A post-Ladinian karstifica-

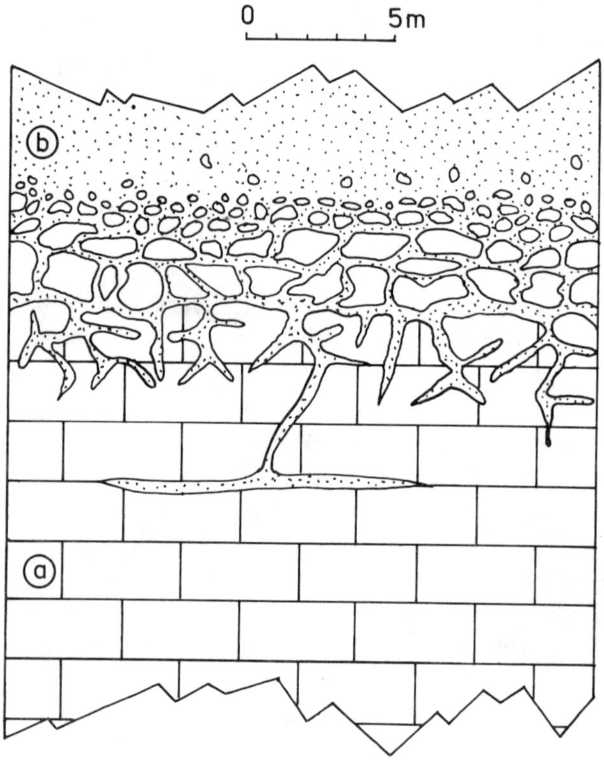

2–101. Contact between the Ladinian limestones and the transgressive cover (after STĂNOIU, 1967). a. Ladinian grey limestones; b. red gritty limestones of Late Bathonian-Early Callovian age.

tion is clearly indicated with the cavities having been filled by the first transgressive deposit. The same situation is found here as in the Bihor Mountains where the karstification and infilling took place simultaneously, but there the fillings were older (partly Carnian? or Hettangian).

(4) Post-Norian paleokarst in the **Codru-Moma Mountains** is represented by neptunian dykes in Carnian-Rhaetic Dachstein Limestones of the Coleşti Nappe; one of the upper nappes of the Codru System (PANIN et al., 1982). The deposits of the neptunian dykes lie chaotically in crevasses of a karst nature. The deposits include red and cherry-coloured micrites with *Involutina liassica*, cherry-coloured and green micrites with small, thin belemnites resembling those from the Pliensbachian, brachiopods, pentacrinids and red and greenish marls with Aalenian belemnites (*Brachybelus breviformis* and *Belemnopsis* sp.). It is interesting that the facies of the Jurassic deposits from karst fillings is not known anywhere other than in the Apuseni Mountains. This means that during a

continental period a large stratigraphic pile overlying the limestones was removed by erosion. In fact, it is surprising that there are only two sites where these paleokarst formations are known.[1]

Cretaceous Paleokarsts

Neocomian Paleokarst

(1) The Neocomian paleokarst of the **Pădurea Craiului** and the **Bihor Mountains** is represented by bauxites, deposits of this age being the most important ones in Romania. Bauxites are distributed over a large karst area developed on various Jurassic limestones including massive reef limestones, Farcu Limestones (Oxfordian-Kimmeridgian, possibly the basal Tithonian), Stramberg Lime-

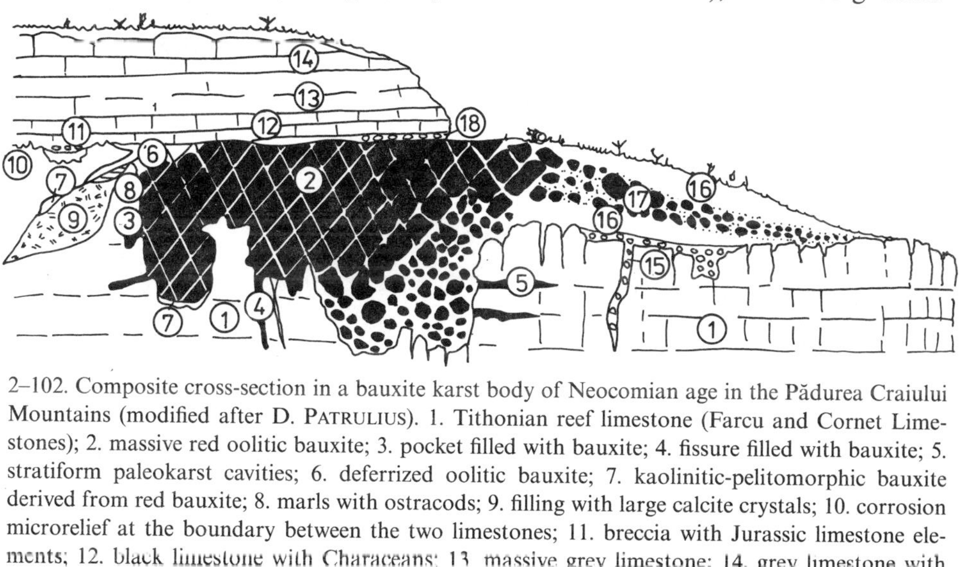

2–102. Composite cross-section in a bauxite karst body of Neocomian age in the Pădurea Craiului Mountains (modified after D. PATRULIUS). 1. Tithonian reef limestone (Farcu and Cornet Limestones); 2. massive red oolitic bauxite; 3. pocket filled with bauxite; 4. fissure filled with bauxite; 5. stratiform paleokarst cavities; 6. deferrized oolitic bauxite; 7. kaolinitic-pelitomorphic bauxite derived from red bauxite; 8. marls with ostracods; 9. filling with large calcite crystals; 10. corrosion microrelief at the boundary between the two limestones; 11. breccia with Jurassic limestone elements; 12. black limestone with Characeans; 13. massive grey limestone; 14. grey limestone with Pachiodonts; 15. Quaternary gravel infiltrated in Pleistocene karst cavities; 16. yellow loam; 17. bauxite gravel with red bauxite loam matrix; 18. yellow loam with limestone blocks, infiltrated at the contact between bauxites and the limestone with Characeans.

stones, Cornet Limestones and Albioara Limestones (black lagoonal oncolitic limestones), the last two types being of Tithonian age.

The fact that the karst erosion surface is cut on various limestones shows that there was a tectonic phase that led to dislocation of the limestones by block

[1] neptunian dykes are also of submarine origin formed during distinct breaks in deposition by infilling of cracks in the sea bottom (Editors)

faulting. In this area bauxites are deposited in large, irregularly shaped doline-like depressions several metres deep with pockets, in recesses, lapies and in some shafts several tens of metres deep (Fig. 2–102). The first deposit overlying the bauxites is a black lacustrine limestone with Characeans that is covered by a limestone bed with brackish water gastropods, after which marine conditions set in. Since the limestone with Characeans is of Barremian age, it follows that the bauxites are Neocomian (Berriassian-Hauterivian).

The bauxites are red, oolitic, ferric, or ferro-ferric, hematitic, diasporic, rarely diasporo-boehmitic, with a constant content of titanium minerals (anatase and secondarily, ilmenite or rutile). They are totally devoid of terrigenous detrital material (PAPIU et al., 1975). By iron leaching processes the red bauxites grade to white kaolinized pelitomorphic varieties infiltrated into fissures close to the primary deposits. Their massive aspect, devoid of stratification, and the lack of detrital material as well as their geochemical characteristics are arguments in favour of an autochthonous origin, that means they were formed by physico-chemical precipitation in the karst depressions (PAPIU and MÎNZATU, 1969).

Together with the forming of the Neocomian karst plain there were endokarst processes that created cave networks filled with detrital and bauxitic material. Geophysical investigations have pointed out numerous underground pockets, some of which are in the process of being investigated by exploration drifts. In 1977, in one of these exploited lenses, skeletal remains of Eocretaceous dinosaurs were found (JURCSAK and POPA, 1978) which were studied in detail. From these findings the mode of evolution of the Cretaceous karst has been established; it comprised several phases of cutting and filling of the karst cavities with various materials (PATRULIUS, MARINESCU and BALTRES, 1983).

During a first karstification phase following the Berriasian emergence, irregularly shaped shafts with vertical and even overhanging walls and lateral passages were generated. These cavities were compared by POP and MÎRZA (1977) with phreatic cavities of the *caguanes* type of Cuba. They were filled with massive diasporic and hematitic, rarely boehmitic, bauxite in their upper parts (Fig. 2–103A). As a result of the uplift of the platform, the water table was lowered generating a second phase of phreatic karst depressions that are considered to be of the *aston* type of Cuba. These are larger, bell-shaped or discoidal cavities (Fig. 2–103B) with the walls covered by spathite. That was the time when the dinosaurs that used to live near or in the lakes containing bauxite mud began to populate the Apuseni Mountains. During seasonal floods this material was carried into the underground cavities (together with the dinosaur bones) giving rise to a second bauxite generation which this time was arenaceous, pelitomorphic, layered, boehmitic and hematitic (Fig. 2–103C). With further platform uplift came a third generation of small karst holes (mostly fissures; Fig. 2–103D) that were subsequently filled with a third generation of karst deposits (Fig. 2–103E); these consist of kaolin and partly of boehmitic clays, and are emplaced at the periphery of the bauxite bodies from the first and second generations.

As already shown, the various bauxite types allow us to establish the complex evolution of a limestone platform where exo- and endokarst processes operated

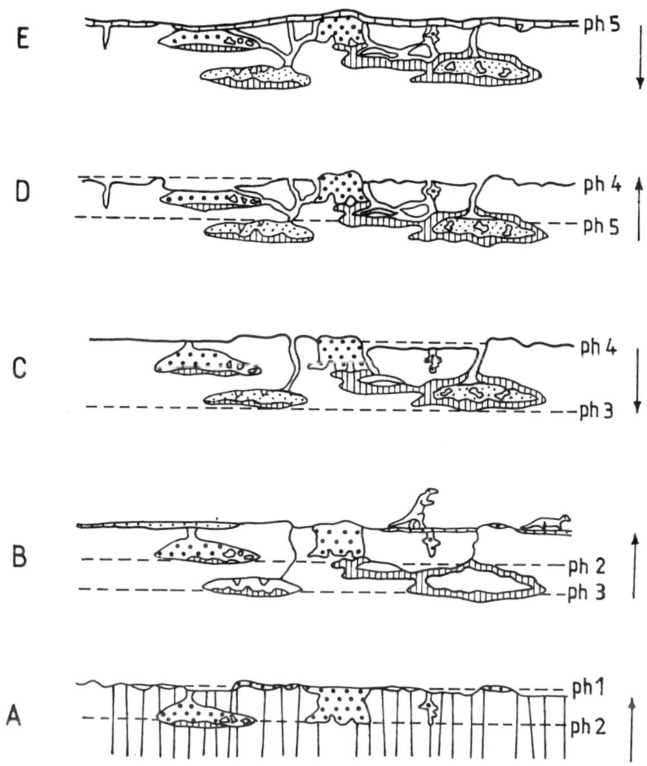

2–103. Evolution of the Neocomian karst system in the Padurea Craiului Mountains (after PĂTRULIUS, MARINESCU and BALTREŞ, 1983). A. the first phase of phreatic karstification which is filled by the first generation of diasporic and hematitic bauxites; B. formation of the second deep phreatic karst with discoidal caverns lined with spathites. Remains of dinosaurs living on the swamp banks penetrate the voids with bauxitic mud; C. the second generation of holes is filled with a second generation of boehmitic and hematitic, arenitic and pelitomorphic bauxites, mixed with dinosaur skeletal remains and limestone blocks; D. the third generation of karst cavities develops on fissures; E. filling with kaolinitic clays that also infiltrate the periphery of the bauxite bodies. The dashed lines indicate the relative position of the water table; the arrows indicate the platform movement; vertical lines represent spathite; large dots indicate the first generation of bauxite; small dots indicate the second generation of bauxite.

simultaneously. They resemble those of modern karstification but were entirely fossilized 130 Ma ago.

(2) There is a stratigraphic gap between Tithonian and Urgonian limestones in numerous places in the **Carpathians**. It is shown especially by the lack of some

paleontological horizons. In some other places there is only a tendency for a gradation from limy platform facies to littoral detrital ones. A great continental phase of Wealden facies (as in the Pădurea Craiului and the Bihor Mountains) is found nowhere else, however, there are some indications of some brief continental periods marked by deposition of some clastics over karst topography. Such a case is found in the western part of the Sebeş Mountains (a region mentioned later in connection with a younger paleokarst). In a cave near Dosul Lăcşorului there is a layer of continental sandstones several metres thick. The sandstones lie in depressions carved in the Tithonian limestones, and are transgressively overlain by Urgonian limestones (H. MITROFAN, pers. comm.).

(3) In the **East Carpathians** there is also a break of sedimentation at the same level, but there are only a few indications of paleokarst. In the Tithonian limestones on Postăvaru Mount, for instance, a decimetric channel filled with Neocomian marls was observed (information provided by M. SĂNDULESCU).

Albian Paleokarst

(1) The Albian paleokarst with bauxites occurs in the **Sebeş Mountains** in the central South Carpathians. It is on the western margin of the mountains at their contact with the Haţeg Basin; thus, the bauxite deposits are known as the Haţeg Bauxites. During the Late Aptian and the Albian, a large karst plain exhibiting

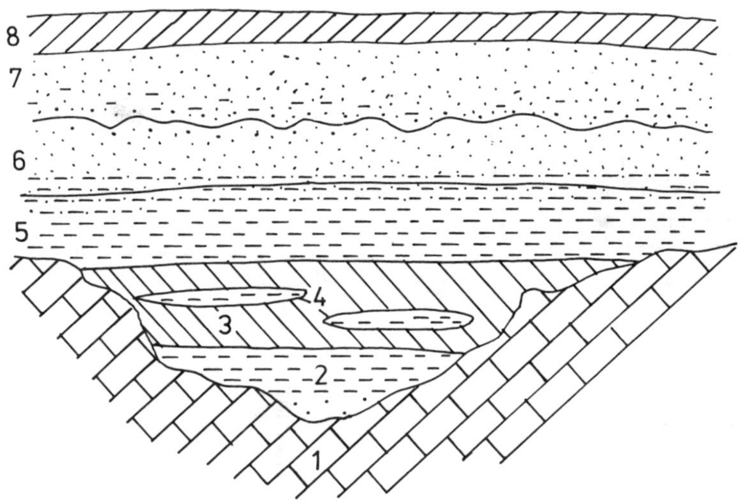

2–104. Paleokarst depression filled with Albian bauxite in the Sebeş Mountains (after PAPIU et al., 1971). 1. Urgonian limestones; 2. basal clayey-detrital horizon; 3. bauxite; 4. clayey-detrital intercalations; 5. upper detrital horizon; 6. sandstone and microconglomerate; 7. Cenomanian transgressive cover; 8. Recent alluvial deposits.

a mature, smooth relief with large closed depressions formed here on Jurassic and Urgonian limestones (paleontologically proved up to the Lower Aptian). The depressions were covered by lakes with still, shallow water in which were deposited lateritic muds and detrital material derived from crystalline schists. Sedimentation in the lakes was gravitational, a macroclastic layer being deposited at the base and overlain by a hematite-bauxite layer and a ferro-kaolinitic one (Fig. 2–104). The presence of fine micas points to the secondary, allochthonous nature of this material. As the gravitational separation was not perfect, the bauxite is to a great extent mixed with clays and fine clastic material which makes it difficult to dress. Reworking took place in this succession of sediments, sometimes causing inversion of the succession. Then it was covered by the sea, with deposition of the sandstones and detrital clays that mark the Cenomanian trasgression (PAPIU et al., 1971). Paleogeographic conditions differentiate the bauxite deposits of the Sebeş Mountains from those in the Pădurea Craiului Mountains by the following characteristics: (i) bauxites originate from allochthonous material that was formed elsewhere and was deposited here; (ii) the negative karst depressions were larger and not so deep, so that the deposits appear as continuous beds, not simple doline fillings; (iii) they are younger, the karstification being subsequent to the great Austrian Orogeny that played an important role in the formation of the Carpathian Chain.

(2) Other Albian paleokarst in the **South Carpathians**. The Austrian Orogeny was manifested throughout the Carpathians and was followed by a period of emergence in which some platforms formed. This was followed by the Cenomanian transgression. The continental phase is not always marked by continental deposits, but in some places where levelling took place on limestones, it can be recognized by some local karst deposits. We quote five such cases:

(i) In the western part of the great **Reşiţa-Moldova Nouă synclinorium** near the Danube, Upper Aptian limestones with *Orbitolina* are sometimes overlain by pockets or lenses of red continental deposits consisting of breccias and red clays. The clays sometimes show evidence of having been redeposited and sometimes consist only of red limonitic crusts. Such formations also infiltrate into clefts and cavities within the limestones. At Gîrnic the limonitic crust attains a few metres in thickness and there has even been an attempt to mine it. The first deposits that transgressed over the Aptian limestones consist of marine sandstones assigned to the Albian (Gura Columbului Sandstone) or Cenomanian (Radimna Sandstone), the exact period of the paleokarst formation being impossible to establish (S. NĂSTĂSEANU, pers. comm.).

(ii) Further east at Pecinişca on the **Mehedinţi Plateau**, (in the Danubian tectonic unit) Urgonian limestones are sporadically covered with alteration crusts, while lenses and pockets of iron oxide approximately 20–30 cm thick are found within them, marking a brief continental episode. They are transgressively overlain by the Nadanova Member, of Cenomanian age.

(iii) Still further east in the **Tismana zone**, lateritic soils occur in the same stratigraphic position, between the Urgonian limestones and the Nadanova Member while, in places, small pockets richer in aluminium are found. They provide economic bauxites. They are in an uncertain stratigraphic position as cover rocks are lacking.

(iv) The bauxites in the **Poiana Ruscă Mountains** are more interesting. They overlie Tithonian limestones and are themselves overlain by the argillaceous-detrital sediments of the Cenomanian transgression. These bauxites are massive, diasporic and hematitic and grade, towards their upper parts to schistose bauxites with detrital materials. In places they are metamorphosed at their contact with eruptive bodies, becoming pyrite-corundum-diasporic in nature. They are considered autochthonous bauxites of the same type as those in the Bihor and Pădurea Craiului Mountains, for which reason they are assigned to the Wealdian Neocomian facies (PAPIU et al., 1984). However the proximity and geological context of the South Carpathians point to their more likely origin being in the great Albian continental phase.

(v) The Poieni Formation of the **Poiana Ruscă Mountains** (MUREŞAN, 1970) is interesting as it partly reveals paleokarst processes. It is a continental deposit of chaotic accumulations of slightly ferruginous quartz blocks frequently displaying cavernous structure, associated with laterites and limonitic or manganese rocks. It overlies epimetamorphic rocks of the massif which are either non-karstifiable or limestones and dolostones. When it overlies the latter, the deposit fills negative forms of the surface karst (dolines, lapies) and also endokarst voids. This situation is encountered, for example, in a cave at Româneşti in the north-western part of the Poiana Ruscă Mountain, formed in Upper Devonian marbles. Some 200 m from the entrance, a polygenetic breccia occurs on the walls and ceiling, comprised of angular blocks of red, sometimes banded, quartzite with rare pebbles of white quartz or crystalline schist. The blocks vary up to 1 m in length. The breccia matrix is hard, also quartzitic, sometimes a hard compact argillite. A large part of this cave is carved in this breccia which is an obvious old filling. At the contact of the new breccia with the limestone cave wall, white and reddish millimetric quartzite beds occur, indicating a gradual, slow filling. The Poieni Formation is in places transgressively overlain by Cenomanian sandstones and clays pointing to its Albian age.

An important Albian continental phase is thus outlined in the Southern Carpathians. It led to the formation of a peneplain that extended over limestone and dolostone of various ages (Paleozoic, Jurassic and Cretaceous) generating a paleokarst characterized by bauxites, limonitic crusts, breccias and residual clays.

Aptian and Younger Paleokarsts in Dobrogea

(1) **South Dobrogea** lies in the eastern part of the Moesian Platforms between the Danube and the Black Sea. Like any stable platform, it comprises a sedimentary sequence of variable thickness. There is an alternation of lime and clastic formations that are separated by numerous breaks but there are no major unconformities because of the stable tectonic conditions. The most important calcareous formation consists of massive reefal dolomitic limestones reaching 250–900 m in thickness of Jurassic age. The limestones are generally below the surface, being reached by boreholes, but they crop out towards the Black Sea coast north of Constanţa and farther north in the Casimcea syncline of Central Dobrogea. They are succeeded, without a visible stratigraphic gap, by Berriasian marls and Berriasian to Valanginian limestones, both being 50 to 150 m thick. In South Dobrogea the marls are absent and the Jurassic-Urgonian limestones form a single mass.

The most important continental episode was in the Early Aptian. It is represented by continental formations with multicoloured sandstones alternating with variegated kaolin clays, all of which exhibit cross-bedding. This continental Aptian, present only towards the northern part of South Dobrogea, is found on an erosion platform which, on the limestones, exhibits a karst plain character. It shows depressions and a typical relief which was buried by the continental deposits. Thus, along the Cara-Su Valley at Mircea Vodă where Cretaceous limestones crop out, they contain solution pockets and metre–scale dolines covered by thin bed of fine, well-rounded quartz gravel, overlain by a rather chaotic filling with variegated kaolin clays. In places the gravels are absent and the filling consists only of kaolin clays (M. CHIRIAC, pers. comm.).

(2) The erosion platform extended farther north in **Central Dobrogea** and is there composed of Proterozoic greenschists overlain by Jurassic limestones (Bathonian-Kimmeridgian). These limestones also underwent karstification which is indicated by pits, dolines, etc. filled with Aptian variegated kaolin clays. It is visible in various places along the banks of the Danube (e.g. north of Hîrşova in Oxfordian-Kimmeridgian limestones, at Hîrşova in Callovian and Oxfordian limestones and at Topalu in Kimmeridgian limestones, etc.; BĂRBULESCU, 1974).

At almost all outcrops, the Aptian paleokarst is transgressively overlain by Upper Aptian and Albian marine deposits, represented by sands and glauconitic sandstones. Sometimes the karst excavations are so deep that the Albian deposits penetrate them too, and so have been protected against subsequent erosion.

We have already shown that, since it was a stable platform with frequent low amplitude oscillations, Dobrogea was subject to recurring marine transgressions and regressions which favoured some continental conditions at various times.

The Aptian episode described above was the most important but some other phases can be mentioned. They are indicated only by a few outcrops and so do not permit generalizations concerning their scale. Thus, for instance, in the Jurassic limestones at Topalu in addition to the Albian filling, there is a karst void filled with glauconitic sandstones bearing Santonian echinoids, indicating a pre-Santonian karstification (BĂRBULESCU,1973). Also in the Jurassic limestones, fissures filled with clays containing Badenian and Sarmatian faunas are observed, pointing to some pre-Badenian karstification (M. CHIRIAC, pers. comm.).

Santonian Paleokarst

There are Santonian bauxites in the Metaliferi Mountains which occupy a limited area. The floor is formed by Precambrian crystalline limestones on which a karst relief was sculpted during the Santonian. The depressions became filled with detrital material and lateritic mud formed at the expense of the crystalline schists and granites nearby. The material was carried here as pelitic

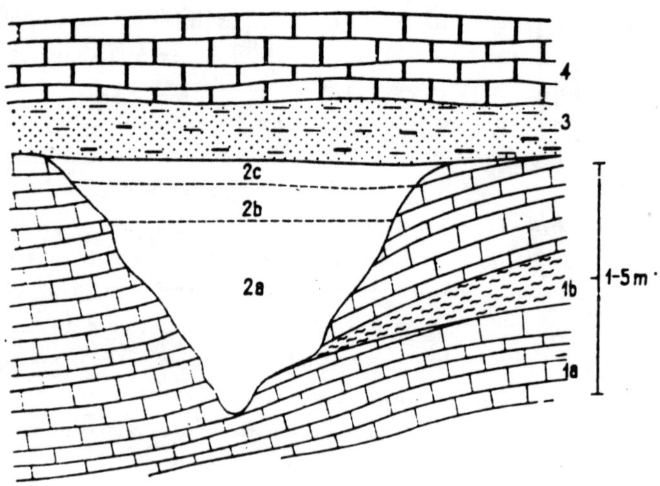

2–105. Paleokarst depression filled with Santonian bauxite in the Metaliferi Mountains (after PAPIU et al., 1975). 1a. metamorphic limestone; 1b. metamorphic shale; 2a. ferric bauxite; 2b. sideritic ferrous bauxite; 2c. ferrous bauxite with pyrite and coal; 3. clayey-gritty shales with coal; 4. limestones with Hippurites.

suspensions, grains or colloidal solutions, precipitation taking place concomitantly with gravitational deposition. This led to the formation of some bauxite lenses and bodies in the succession shown in Fig. 2–105. The capping bed of the paleokarst consists of gritty-clay shales with coal, overlain by hippurite-bearing limestones representing the Gosau Formation of Campanian-Maestrichtian age (Papiu et al., 1975).

As in the case of the Sebeş Mountains paleokarst, the bauxitic material is allochthonous, having been brought by torrential mud flows that filled the karst cavities.

Tertiary Paleokarsts

(1) The Paleogene paleokarst beneath the **Romanian Plain** is buried under a Neogene sedimentary cover 100–300 m in thickness. It has been penetrated by numerous oil exploration wells. This karst was generated at the end of the Cretaceous, when the Moesian Platform was uplifted and subjected to intense denudation. The erosion surface truncates various deposits, such as Tithonian, Urgonian and Senonian limestones, on which intense karstification took place.

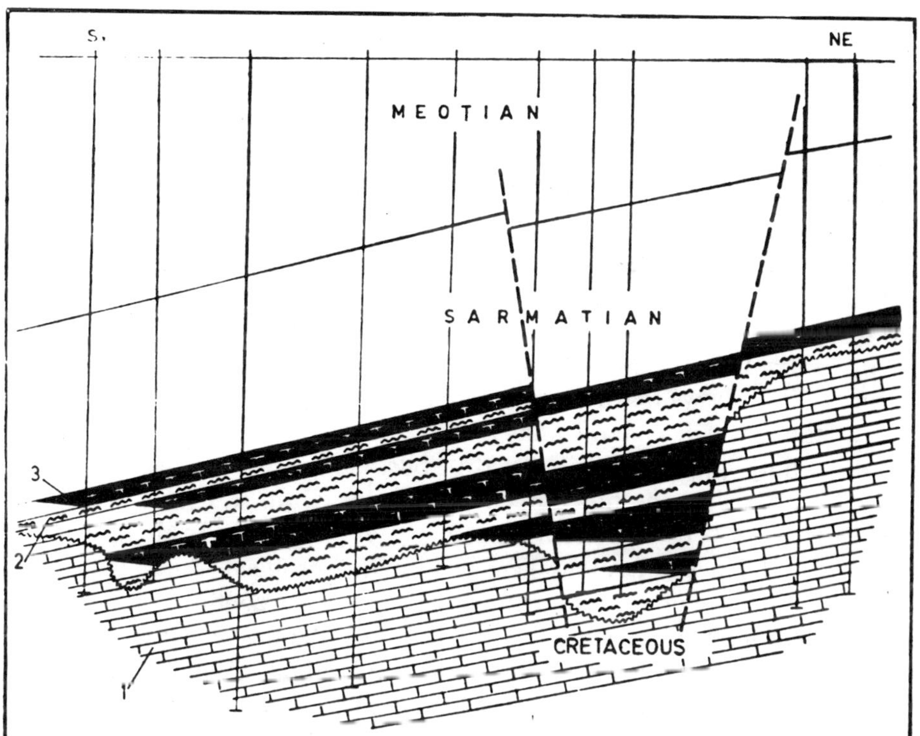

2–106. Section in the Romanian Plain west of Bucharest based on drillings (after Paraschiv, 1979). 1. Senonian limestones; 2. lower part of the Sarmatian accumulated in paleovalleys and karst depressions; 3. oil deposits; 4. upper part of the Sarmatian, transgressive over eroded areas; dashed line = faults; thin vertical lines = boreholes.

They were discovered by drilling and geophysical prospecting and appear as large paleovalleys and dolines. For example, a little to the west of Bucharest there are closed depressions of 7–8 km in length and 3–6 km in width with a maximum depth of 200 m (PARASCHIV and BUCUR, 1975) (Fig. 2–106). They were cut in Senonian limestones. West of the Olt River the structure rises so that the peneplain also truncates Urgonian, Jurassic, Triassic and Paleozoic limestones, with solution and the formation of karst relief. So far no specific karst deposits have been found (terra rossa, limonitic crusts, etc.), probably because they could not be observed in cores. The large denudation area, partly of a karst plain character, was covered transgressively by Sarmatian clastics which also filled the karst features (PARASCHIV, 1966). It is worth noting that these fillings offer good conditions for oil accumulation.

There is intense karst water circulation taking place now through the limestones on the Moesian Platform, especially through formations below the less-fissured Barremian limestones. There is no indication of the age of these waters but, considering the great depth at which they circulate (over 2 000 m), it is supposed that their channels were dug under former hydrodynamic conditions, when the limestones were in a much shallower position. Thus we suppose that these channels date back to the time of the Paleogene karst plain, being therefore paleo-endokarst of the same age.

(2) The karst conduit networks of **South Dobrogea** constitute one of the serious problems of karst hydrogeology in Romania. We have already mentioned the thick Jurassic-Urgonian limestones of South Dobrogea that may attain 1 000 m in thickness. These limestones are marked by an intense karst water circulation. Close to the seashore, at Caragea Derment, there are numerous water wells with a discharge of 500–13 000 m^3 per 24 hours for a drawdown of 4 m. Water is ascending since the hydrostatic level changes from +17 to +3 m. Drilling shows that the water circulates through the entire thickness of the limestones which contain numerous large voids, making drilling difficult (loss of drilling liquid, dropping of drill bits, etc.). Studies made by isotopic tracers have revealed the flow direction of water (from west to east) and the aquifer dynamics (ŢENU, DAVIDESCU and SLĂVESCU, 1983), so that we are not concerned with these questions here. The major problem is not the present circulation but the manner and time of formation of the conduits that carry this karst circulation. This question is important because the recharge zone has a maximum altitude of only 100 m, but water circulation takes place to a depth of 1 000 m, which fact cannot be explained even by the bathypreatic theory of karst conduit genesis. We must therefore suppose that these channel networks, which are still being used, formed in the geological past when a greater hydraulic head existed. It follows that the South Dobrogea territory was at a higher level than the zone where the waters left the limestone massif, even if this level used to be much lower. As the Black Sea has existed since the Cretaceous, the

difference in level that needs to be established is that between the mainland and the sea level. A low level of the Black Sea during the Miocene could be explained by the *Messinian crisis* that completely emptied the Mediterranean Basin (LETOUZY et al., 1977). HSÜ (1978) considered that the Black Sea was completely drained during the Upper Miocene (5.5 Ma), when gravels, followed by lacustrine deposits were deposited, indicating the period of dessication. Gravels were discovered near the Bulgarian coast at a depth of 865 m below the sea floor; the water being 2 000 m deep. Further investigation has proved the Messinian lowering of the Black Sea to the level of approximately -1 600 m b.s.l. (HSÜ and GIOVANOLI, 1979). These data indicate a lowering of almost 3 000 m. Even if subsequent tectonic sinking of the sea floor took place, there is still a lowering which indicates a sinking of 1 000 m of the hydrographic network in the carbonate massifs.

The second possibility, namely the rise of the continent with respect to the Black Sea level, can be seriously considered because the water circulation is similar to that underneath the Romanian Plain which, in fact, is a much lowered part of the same structural unit — the Moesian Platform. Thus, we can suppose that because formation of the Moesian channels that circulate water at present probably took place in the Paleogene, a similar age and process is valid for South Dobrogea.

This proposal is supported by the fact that a borehole piercing Cretaceous limestones at Ovidiu reached several one metre large pockets of Lutetian limestones with nummulites; it indicates that there existed a network of channels which was partly filled at that time.

Quaternary Paleokarsts

The exact identification and definition of such formations is often difficult. Thus, in the case of some buried lapies it is hard to determine whether they are old formations covered by subsequent burial (being therefore paleokarst forms) or are the result of present dissolution processes operating under a cover. That is why we mention here only cases where the fossilization process is certainly Pleistocene.

(1) The Pleistocene paleokarst in the **Vaşcău Plateau** is marked by the existence of some ferro-manganese deposits of a residual type in karst depressions and voids carved in the Anisian dolostones of the southern part of the Codru-Moma Mountains. The section (Fig. 2–107) shows that the relief of the karst forms is irregular, solution being not only of a doline type filled with allochthonous material. Moreover, numerous irregularities show that it was a very fine-scale corrosion that took place only under a cover, which entitles us to consider the filling as autochthonous and therefore residual.

The bottom of the dolostone relief is covered by a thin manganese ore comprising pyrolusite and ferri-ferrous manganomelane. There follows a thick bed consisting of concretionary iron ore agglomerations in which fragments of

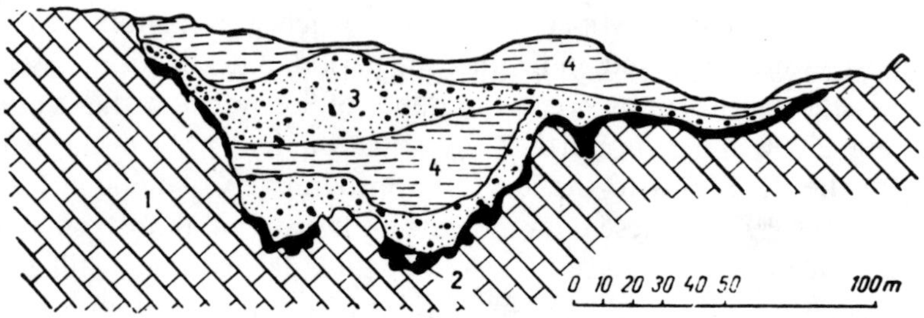

2–107. Cross-section through Pleistocene manganese and iron deposits in the Vaşcău Plateau (Codru Moma Mountains). 1. Anisian dolostones; 2. manganese ore; 3. iron ore; 4. red clay of terra rossa type.

manganese limonite ranging from grains to blocks of 0.5 m^3 in size are welded. This mass contains pockets with red or blackish clay, especially towards the upper part where the iron concretions occur in ever smaller inclusions. The top is usually formed only of the red clay.

There is no indication of the date of the filling. We assign it to the Pleistocene because there is a similar deposit of limonitic concretions which is older than the upper terraces of the rivers bordering the mountainous zone.[1]

The problem raised by this paleokarst is not that of its age but of its genesis. Initially there existed a clay filling resulting from limestone alteration (and therefore of residual nature) which was penetrated by meteoric waters which caused vertical separation and some concentration of the included metallic elements. While iron was deposited in the upper part of the filling, manganese was preferentially transferred to the lower part, forming a continuous ore there. Such an interpretation can be criticised because of the extremely low amount of iron and manganese in dolostones, so that the existing clay and ore accumulations require a huge amount of rock to be dissolved. However, this is the general problem of clays of terra rossa type in karst regions, accumulated in immense quantities from limestones containing tiny amounts of clay material (under 1 %).

(2) Pleistocene paleokarst in **South Dobrogea** developed in Jurassic and Urgonian limestones that underwent intense karstification where they outcropped at times during the Pleistocene. The karstification is visible in the walls of limestone quarries, where dolines and shafts are often noticed. For instance, in Cernavodā a doline about 80 m in diameter could be observed before

[1] Similar deposits are common, for example, in Hungary, Poland, etc. and their age is limited to the Pliocene (Editors)

exploitation for cement. It was situated in Urgonian limestones, exhibited a thin bed of red clay at the margins and was filled with loess which displayed bedding with red fossil soil bands. Such formations could be observed on a smaller scale along the Cara-Su Valley and in valleys situated farther south where limestones crop out on their banks.

(3) **Moneasa** Pleistocene paleokarst is similar to the previous one in the sense that it can only be examined in vertical section in a quarry. In this case there are Lower Jurassic limestones (Upper Sinemurian-Pliensbachian) with pits of 1–2 m diameter. They were exposed in the quarry to heights of over 15 m and were filled with a fine quartz sand, with a slight pelitic-arenitic matrix. Diagenesis is weak and a weak infilling bedding can be noted. There is no indication of age but the material exhibits the same red colour as that of the above-mentioned limonitic crusts. The possible age is therefore Pleistocene (?).

(4) **Breccias** in caves are deposits that may be assigned to paleokarst in some cases. There are various types of breccias that may be found as fillings of older channels which are intersected by the modern underground network. The fillings are difficult to date because of the lack of marked faunal elements or of calcareous crusts for radio-isotope dating. We shall present a few examples:

(i) In Magura Cave and in a pothole at Şesuri in the Bihor Mountains there are breccias of rounded limestone of local type within a matrix of terra rossa type which is quite highly altered by diagenetic processes. The present passages cross old channels that are completely filled with the breccias, or follow them for short distances.

(ii) In Tăuşoare Cave a gallery is filled with 10 m of layered alluvium, indicating phases of strong water flow alternating with others of weak flow that allowed only sand transport. As the whole fill mass is partly cemented it can be concluded that it is old, without specifying its age.

It would be interesting to discuss the age of fillings in modern active caves because some of them are obviously old. This is the case where speleothems cover Paleolithic settlements and where fillings with Pleistocene faunal successions have been found, e.g. in Adam Cave where a complete Riss-Würm succession was described with an alternation of warm and cold climate faunas (SAMSON, 1969).

Some fillings studied by palynological analyses are also old, such as the fossil ice mass of the Scărişoara glaciere, or of calcite deposits in speleothems datable by radioisotopes. Such fills cannot be regarded as true paleokarsts because they belong to the present karstification cycle even if they are Pleistocene or older. We did not deal either with the exokarst forms or with the traces of Pleistocene endokarst periglacial sculpture (such as those in the Bihor Mountains described by BLEAHU, 1966), as they do not correspond with the paleokarst definition used here.

Practical Consequences

The main metalliferous deposits connected with Romanian paleokarst have been described above. They are bauxites and ferro-manganese ores. Bauxites are associated with three phases: (1) the Neocomian continental phase with autochthonous bauxites, (2) the Albian and (3) Santonian phases with allochthonous bauxites. Accumulations of ferro-manganese ores are Pleistocene and autochthonous.

One of the richest raw materials associated with the paleokarst is the oil of the Moesian Platform, which has locally accumulated in karst fillings of Sarmatian age cut in the post-Cretaceous platform. It is worth noting that the accumulation did not occur only in the fillings of solution depressions but also in all types of platform entrenchments, such as paleovalleys, situated on the limestones.

Finally, waters circulating through the karst channels are important as they are intensely exploited. Some waters use a network of channels formed under paleogeographic conditions different from the present ones, therefore, paleokarst networks. They are found in the Moesian Platform (both in the western part which is now sunk beneath the Romanian Plain, and in the uplifted part in North Dobrogea) and elsewhere. For example, waters circulate beneath thick Neogene cover on the western margin of the Apuseni Mountains; these waters become thermal due to their very deep circulation and their rapid outflow along faults.

Conclusions

In the Romanian Carpathians and in the Moesian Platform facing them there are numerous examples of paleokarst relief, with which specific deposits are associated that indicate their presence (Fig. 2–108). The periods of emergence

		Apuseni Mts.	South Carpathians	East Carpathians	Romanian Plain	Dobrogea
Pleistocene		—				
Neogene						
Paleogene					—	—
Cretaceous	Senonian	—				
	Turonian					
	Cenomanian					—
	Albian		—			
	Aptian					
	Neocomian	—	—	—		—
Jurassic						
Up. Triassic		—		—		
M. Triassic		—				
L. Triassic						

2–108. Periods of paleokarst development in the Romanian territory.

that led to the formation of some of the paleokarsts are due either to some particular paleogeographic evolution of the carbonate marine platforms that generated the large limestone masses or to some general geotectonic conditions. The paleokarst forms at the Anisian/Ladinian boundary in the Bihor Mountains, where local supratidal processes accompanied by solution and deposition of continental deposits took place, belong to the first category.

The second category comprises the other paleokarst forms. They are derived from regional emergences that were consequences of geotectonic cycles. These are: (1) the Late Triassic emergence of the Bihor Autochthon and the Rarău Mountains; (2) the post-Norian emergence in the Codru Nappe System of the Apuseni and Rarău Mountains; (3) the Neocomian emergence in the Bihor Autochthon with the formation of bauxites, as a consequence of Late Cimmerian movements; (4) Late Aptian-Albian emergence throughout the South Carpathians with the formation of bauxites (Sebeş, Tismana, Poiana Ruscă) and crusts and fillings (Poiana Ruscă, Mehedinţi, Reşiţa zone) as a result of the Austrian Orogeny; (5) intra-Senonian emergence with the formation of bauxites in the Metaliferi Mountains and post-Senonian emergence in the Moesian Platform, as a result of the Laramide Orogeny; (6) Quaternary phase, of which only the older forms covered by younger deposits or deposits that were subjected to intense diagenesis are considered paleokarstic. This definition does not apply to all forms situated on the present karst surface or to caves that belong to the present cycle of karstification.

PALEOKARST OF THE UNION OF SOVIET SOCIALIST REPUBLICS

Rostislav Alekseevich Tsykin

Introduction

This chapter discusses the distribution of pre-Quaternary karst. As a consequence, most surface relief forms and explorable caves are not considered.

Karst is not confined to the uppermost zone of the lithosphere. Our knowledge about it has been enhanced by deeper crustal studies. There are now many examples of the discovery of paleokarst features in oil and gas regions where they have been intercepted at depths of 3–5 km or more.

The discontinuous patterns of karst development that occur may be said to constitute the *karstosphere* (MARUASHVILI, 1969). This is clearly developed on continental platforms. Its lower boundary there may be 500–1 000 m down in the igneous or metamorphic basement because this has bodies of marble, carbonatite, etc. The maximum thickness of the platform karstosphere is 10–15 km. In folded regions the karstosphere is developed in a more fragmentary manner. Its thickness in the Caledonides and Variscides exceeds 3–5 km.

The finding and dating of paleokarst are typical geological tasks. For these purposes paleogeographical and stratigraphical methods are used that are appropriate to its position in the section (e.g. underneath an unconformity) and to the geological age of the covering strata. Biostratigraphical methods allow us to determine the geological age of sedimentary fillings by study of their pollen and spores, the remains of stems and leaves of plants, the fauna, etc. Numeric dating methods may also be used, especially for hydrothermal karst. But there are often situations where there is no possibility of determining the precise age of evidently ancient, deep-seated karst that is filled with water and hydrocarbons.

Paleokarst Investigation in the U.S.S.R.

The first investigators of karst in Russia (VARSANOF'EVA, 1915, KRUBER, 1915) were limited in their historical analysis by a wish to recognize and depict

Bosák, D. C. Ford, J. Głazek and I. Horáček (Editors):
Paleokarst. A Systematic and Regional Review. – Elsevier and Academia. Amsterdam and Praha. 1989.

stages of the karst cycle. The first publications about pre-Jurassic karst were those of BARKOV (1932), discussing results of his work in the area of Samarskaya Luka in the Volga Valley. Shortly after, DUBYANSKY (1937) recognised Paleogene karst in the chalk of the Voronezh anteclize. In the 1940s there were reports of the occurrence of some epochs of karstification during the geological development of the Russian Platform (YAKUSHOVA, 1949) and the Ural (SIGOV, 1947; MALAKHOV, 1947).

G.A. MAKSIMOVICH divided karst into modern, ancient, inherited and rejuvenated types and gave the first summary about paleokarst in the U.S.S.R. (MAKSIMOVICH, 1963; MAKSIMOVICH and ARMISHEV, 1962). A special section is devoted to the paleokarst of Povolzh'e (the Volga Basin) in a monograph by STUPISHIN (1967). He distinguished Early Fammenian, pre-Visean, Early Visean, Sakmarian-Artinskian, Late Kungurian, Kazanian, and Triassic-Middle Jurassic epochs of karstification in this area. LUNSGERSGAUSEN (1966) described Late Paleozoic karst in the spurs of the Kharaulakh Mountains (N. E. Siberia) and proposed calling their fragmentary fillings "influvium".

KORZHUEV (1972) differentiated six karst cycles (pre-Riphean, Middle Cambrian, pre-Permian, pre-Jurassic, pre-Cretaceous, pre-Pliocene) on the Siberian Platform; they are related to stages of the continental evolution of this geostructure. KUCHERUK (1976) in his summary of the Podolian karst in the East European Platform described features of pre-Visean, pre-Middle Carboniferous, Bashkirian, Permo-Triassic, Early-Middle Jurassic, Cretaceous-Paleogene, post-Sarmatian, and Pliocene karstification phases. MAMATKULOV (1979) distinguished Middle Devonian, Early Carboniferous, Late Carboniferous, Early Cretaceous and Neogene karstification in the Tien Shan Mountains. In the Altay-Sayan Mountains TSYKIN and TSYKINA (1978) found relics of Paleozoic (Early Devonian, Permian) and Mesozoic-Cenozoic (Middle Jurassic, Early Cretaceous, Late Cretaceous, Paleogene and Neogene) karst.

In sum, investigations of both platform and folded regions have revealed evidence of many karstification periods within the different cycles of development that occurred in these structural provinces.

Geological Conditions of Karst Development

From the karst point of view the most interesting areas in the U.S.S.R. include some ancient platforms (East European and Siberian), young platforms (Turanian and West Siberian), the folded regions of the Baikalian, Caledonian and Variscan Orogenies and some Mesozoic and Alpine Orogenies, plus some marginal depressions which are considered with their adjoining platforms.

East European Platform (Craton)

The East European Platform includes the Baltic and Ukrainian Shields, the Timan Arch and the Russian Platform (Fig. 2–109). There are deep depressions (aulacogens), horst uplifts and domal anticlines in the Paleozoic cover. Positive geostructures are the more favourable for karst development.

Carbonate rocks of Precambrian age are widespread in the Lower Proterozoic (Krivoy Rog Group of the Ukrainian Shield, Kursk Group of Voronezh Massif), the Middle Proterozoic (Onega

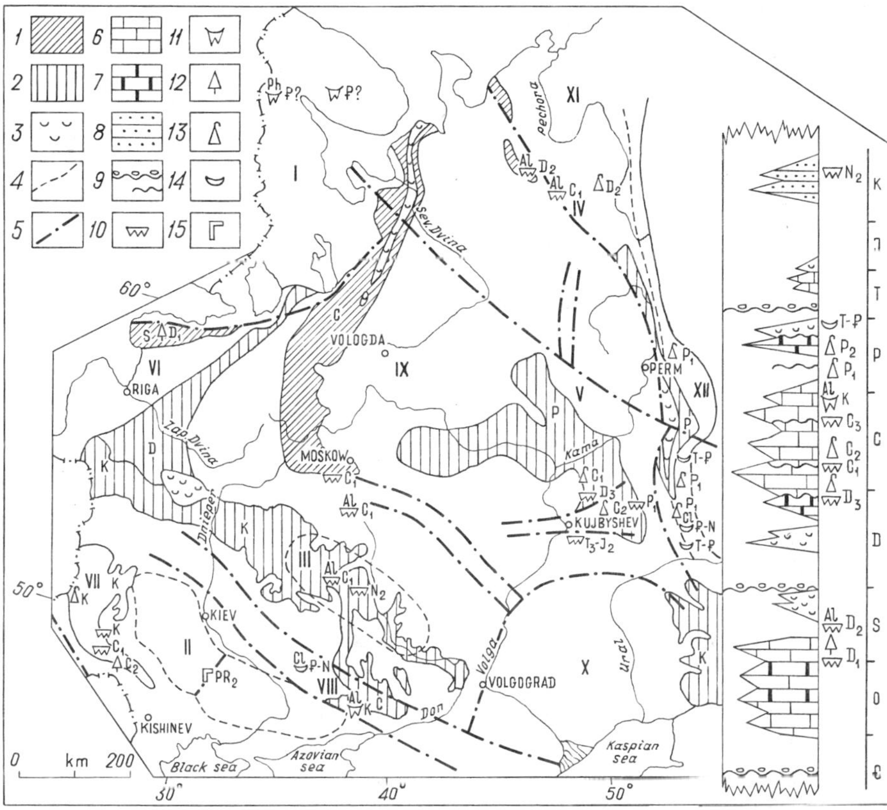

2–109. Distribution of paleokarst in area and in the stratigraphic column of the East European Platform (within the territory of the U.S.S.R.) 1–3. outcrops of karstified rocks: 1. carbonates, 2. terrigeneous-carbonate sequences, 3. sulphate and terrigeneous-carbonate sequences; 4. boundary of some geostructures; 5. basement faults; 6–8. stratigraphic position of karstified rocks: 6. limestones, 7. alternating limestone and dolostone, 8. chalk; 9. regional and local hiatuses, 10–14. types of hypergene karst: 10. naked (here and in the following Figures, above the symbol is the abbreviation of the mineral resource and to the right is the age), 11. covered, 12. water-bearing, 13. oil and gas-bearing, 14. buried subsidence depressions; 15. hydrothermal karst. Abbreviations: Al. bauxite, C. coal, Ph. phosphorites; Structures: I. Baltic Shield, II. Ukrainian Shield, III. Voronezh Massif, IV. Timan Arch, V. Volga-Ural anteclize, VI. Baltic syneclize, VII. Lvov depression, VIII. Dnieper-Donetz aulacogen, IX. Moscow syneclize, X. peri-Caspian syneclize, XI. Pechora depression, XII. Uralian foredeep.

Group of the Baltic Shield), the Riphean (Bystrinsk Formation of the Timan Arch, Kaltasinsk and Malokamyshinsk Formations of Kaltasinsk depression), Vendian (Osipovichsk Formation of the Orshansk depression, Serdobsk Group of the Pachelma aulacogen).

During the Cambrian, the main part of the Russian Platform was a region of denudation. At the end of the Early, the Middle and Late Paleozoic, karst rocks played an important role. Limestones and dolostones comprise the greater part of the Ordovician and Silurian systems. Silurian carbonates occur in the Dniester Basin and Timan Arch, too.

After the interruption during the Late Silurian and Early Devonian that marked the finish of the Caledonian Orogeny in central and eastern regions of the Russian Platform, terrigeneous-carbonate deposits of Middle and Late Devonian age accumulated. In the Volga-Ural region reefs have been found in the peri-Ural depression. Salt-bearing formations accumulated in the Dnieper-Don region.

Limestones and dolostones form extensive horizons in the Carboniferous system, which is widely distributed in the central and eastern regions of the Platform and in the peri-Ural depression. There were brief interruptions in the pre-Visean, Early Namurian and Bashkirian. The Permian is represented by large terrigeneous-carbonate and salt-bearing deposits. In the Late Permian (Ufimian) there was a hiatus.

Lower Triassic deposits are widespread but contain few karst rocks. In the Middle Triassic there was a long phase of continental exposure of the platform, marking the end of the Variscan cycle. Only in the peri-Caspian syneclize is there a complete section of Lower Mesozoic deposits, including sequences of carbonate rocks. The Jurassic is represented by late Middle and Upper Jurassic strata; these are karst rocks (carbonates and sulphates) in the Dnieper-Donetz aulacogen and in the Lvov Basin.

There is a stratigraphic gap at the base of the Cretaceous. The Lower Cretaceous is poorly developed, with continental facies. In the Late Cretaceous in the southern part of the Russian Platform great thicknesses of chalk accumulated. There are no karst rocks in the Paleogene, which is limited to the southern part of the platform. Neogene rocks occur in its southwest part, including coquina limestones and reefs.

The Siberian Platform (Craton

The Siberian Platform (Fig. 2–110) comprises the Aldan Shield, the Anabar Massif, Olenek Promontory and the Lena-Yenisei Platform. The latter consists of syneclizes, anteclizes and monoclines (MALICH, 1975). The Angara-Tasejeva syneclize and the peri-Baikal monocline form the "Irkutsk Amphitheatre", where karst was studied for a long time in connection with the design of the Angara Barrage hydroelectric power stations (VOLOGODSKY, 1975).

Metamorphic carbonate rocks and also carbonatite masses are known in Archean rocks of the Aldan Shield and Anabar Massif. Bedded crystalline limestones and dolostones occur amongst Lower Proterozoic deposits preserved in later depressions (Kodaro-Udokansk).

Limestones and dolostones are important in the Upper Proterozoic (Riphean, Vendian) outcrops at the margins of the platform (peri-Sayan, peri-Baikal, and Turukhan regions) and in the Yudoma-Maya foredeep. Karst rocks of varying composition predominate in the Cambrian and in the northwestern and eastern parts of the platform also occur in Ordovician, Silurian and Devonian deposits. Terrigeneous-carbonate, sulphate and salt formations of Early-to-Middle Cambrian are especially widespread. Their accumulation continued with interruptions into the Ordovician and Silurian in the northern part of Tunguska syneclize, and into the Devonian in the Vilui syneclize (Kempendyay uplift). There are huge areas (more than 0.5×10^6 km^2) of Lower Paleozoic karst

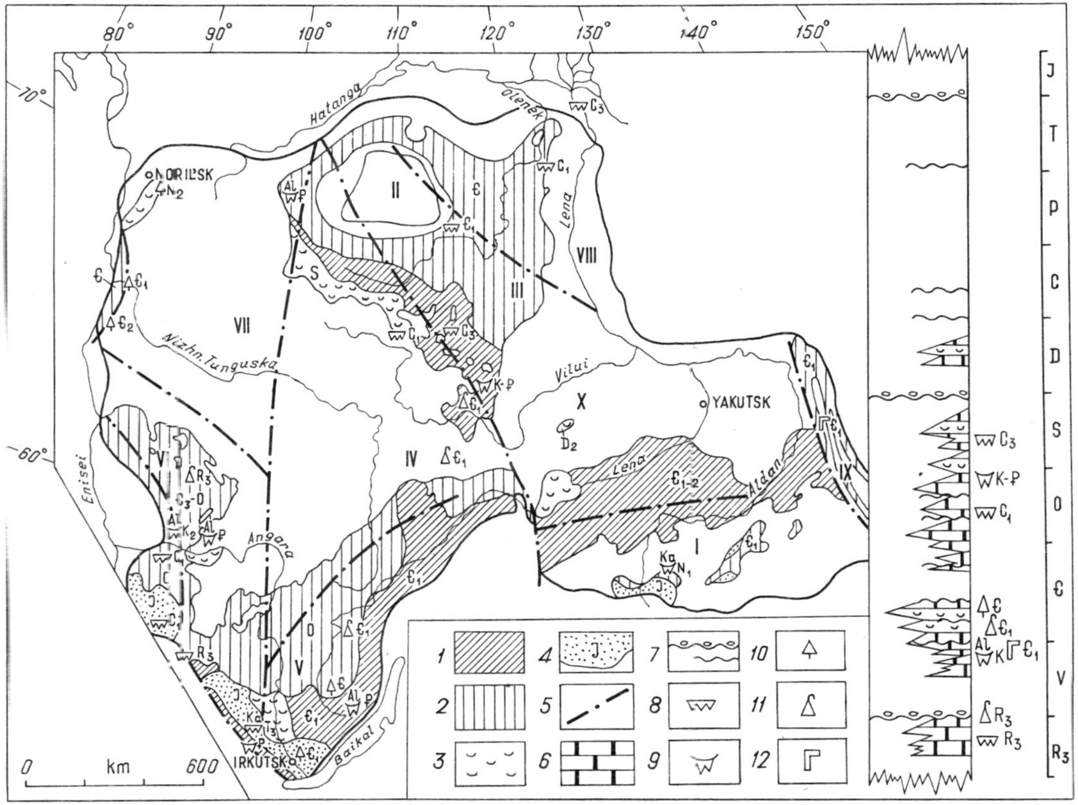

2–110. Paleokarst of the Siberian Platform. 1–3. outcrops of karstified rocks: 1. carbonates, 2. terrigeneous-carbonate sequences, 3. sulphate and terrigeneous-carbonate sequences; 4. superimposed depressions at the platform edge containing coal deposits; 5. basement faults; 6. alternation of limestone and dolostone; 7. regional and local hiatuses; 8–11. types of hypergene karst: 8. naked, 9. covered, 10. water-bearing, 11. oil and gas-bearing; 12. hydrothermal karst. Abbreviations: Al. bauxite, Ka. kaolin; Structures: I. Aldan Shield, II. Anabar Massif, III. Anabar-Olenek anteclize, IV. Nepsk anteclize, V. Irkutsk Amphitheatre, VI. peri-Yenisei monocline, VII. Tunguska syneclize, VIII. Verkhoyansk foredeep, IX. Yudoma-Maya foredeep, X. Verkhoyansk syneclize.

rocks on the margins of the platform and in the Anabar-Olenek anteclize. From the karst point of view the Angara folds, the Chadobets, Kempendyay and Khantaisk-Rybinsk uplifts are of greatest interest amongst the local positive structures.

In the course of accumulating the Upper Precambrian and Lower Paleozoic deposits there were significant periods of marine regressions and continental development during the Late Riphean, Middle Cambrian, Middle Ordovician and Early Silurian.

In the Upper Paleozoic (terrigenous and coal-bearing formations) and Mesozoic (volcanic, terrigenous and coal-bearing formations) karst rocks are almost entirely absent.

The Turanian Platform

The Turanian Platform (Fig. 2–111) has a predominantly Variscan basement which is complex in structure and topography. Riphean and Paleozoic deposits contain units or greater formations of carbonate rocks. They are exposed on the flanks of the Tien Shan, the uplifts of Mangyshlak, Sultanuizdag and Tuarkyr.

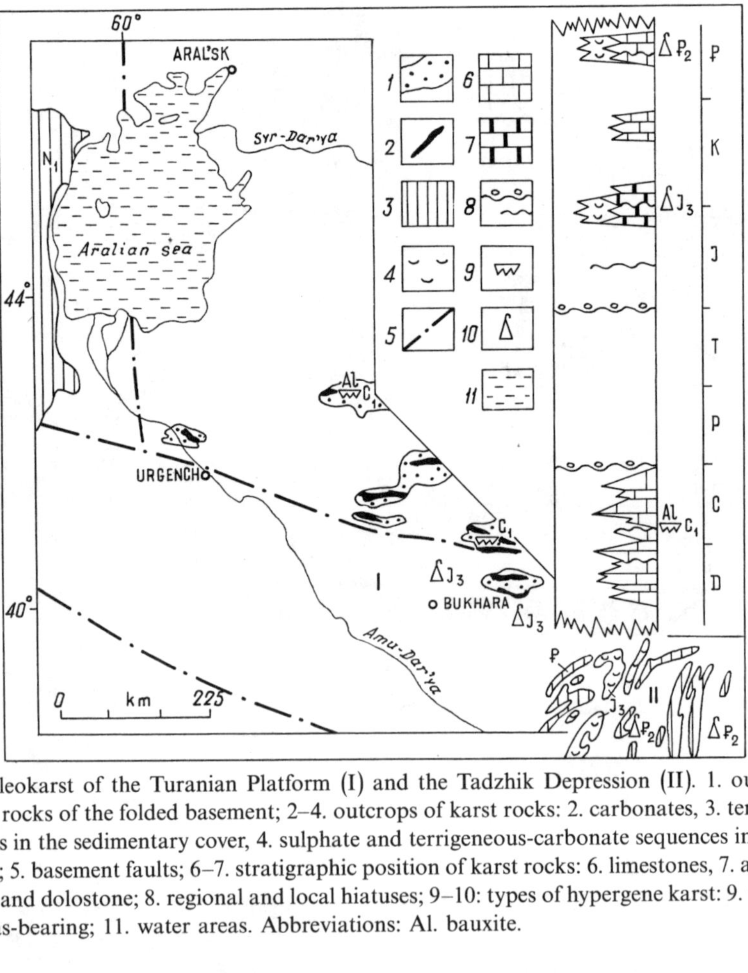

2–111. Paleokarst of the Turanian Platform (I) and the Tadzhik Depression (II). 1. outcrops of non-karst rocks of the folded basement; 2–4. outcrops of karst rocks: 2. carbonates, 3. terrigeneous carbonates in the sedimentary cover, 4. sulphate and terrigeneous-carbonate sequences in sedimentary cover; 5. basement faults; 6–7. stratigraphic position of karst rocks: 6. limestones, 7. alternating limestone and dolostone; 8. regional and local hiatuses; 9–10: types of hypergene karst: 9. naked, 10. oil and gas-bearing; 11. water areas. Abbreviations: Al. bauxite.

There are four depositional stages in the cover (GARETSKY, 1972): A Triassic-Lower Jurassic stage comprises terrigeneous-carbonate and coal-bearing formations. A Middle Jurassic-Lower Miocene sequence disconformably overlies it and includes terrigeneous-carbonate (Upper Jurassic-Neocomian), salt (Upper Jurassic) and carbonate (Oligocene) formations. Deposits of a Middle Miocene-Lower Pliocene stage are widely distributed and disconformable. They are represented particularly by terrigeneous-carbonate strata. There is a final disconformity at the base of Pliocene-Quaternary terrigeneous deposits.

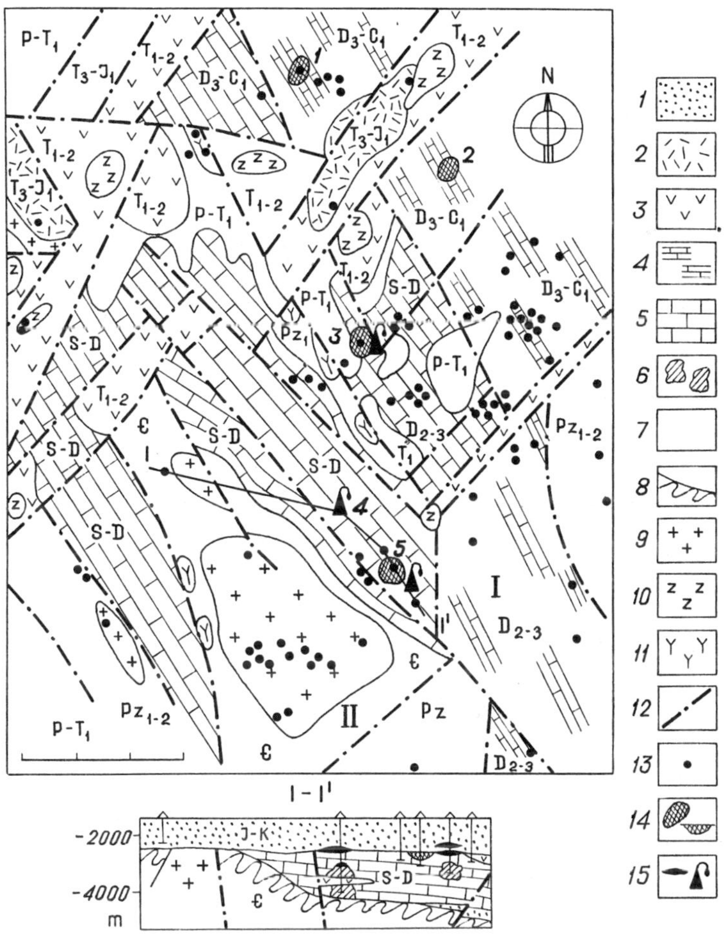

2–112. Paleokarst in the basement of the southern part of the West Siberian Platform (after ZHERO, SMIRNOV and SURKOV, 1977). 1. sedimentary cover, 2. persilicic effusives, 3. subsilicic effusives, 4. patches of limestones in terrigeneous strata. 5. limestones, 6. dolostone-limestone bioherms, 7. non-karst Paleozoic deposits, 8. Lower Paleozoic folded basement (in cross-section), 9, granites, 10. gabbros, 11. ultrabasics, 12. faults, 13. boreholes reaching the Paleozoic deposits, 14. bauxite bodies, 15. oil accumulations (on the map and in section). Structures in the basement: I. Nurolsk depression, II. Mezhovsk Massif.

West Siberian Platform

The West Siberian Platform has a heterogeneous basement. It is of Early Precambrian (Baikalian) age in the Eastern, fore-Yenisei part. There is Caledonian basement in the southeast part, the Salair Ridge, fold belt of Kuznetskiy Alatau and the Eastern Sayan. Variscan basement is developed in the remainder of the platform, Ural Variscides in the western part and the special Ob-Tazovsk system in the central region. There are numerous rift valleys in the basement which is comparatively evenly downwarped towards the centre of the platform. These rift valleys are filled with Triassic volcanic deposits.

Many boreholes have encountered carbonate and terrigeneous-carbonate deposits of Late Devonian and Early Carboniferous age, but Upper Silurian and Lower Devonian karst rocks are rare. They are distributed in the buried Ural-Kazakhstan foredeep (the Tobol Basin) and in the Om-Vasyugansk megablock, where the Nurolsk depression and the Mezhovsk Massif with a granite core occur (Fig. 2–112). Limestones and dolostones are found on the flanks of these structures, complicated by faulting. Karst-hosted bauxites and major oil deposits in fissure-karst aquifers were discovered there. The cover strata are 3.5–5 km thick but contain no karst rocks. The reason is that the paleokarst developed at the pre-platform (aulacogen) stage of evolution of this structure.

In summarising the geological conditions of paleokarst development on the platforms, we can assert that karst rocks are widely distributed in cover strata on the ancient platforms. There are sulphates and halite strata as well as carbonates. They are especially characteristic of foredeeps. There are very few karst rocks in the ancient basement; they are represented by marble, carbonatite bodies, and some quartzites which are a special case of karst.

On the young platforms, carbonate rocks occur widely in the basement. Their presence is smaller in the cover and sometimes they are absent entirely, as in the West Siberian Platform. Gypsum (anhydrite) and rock salt occur in deposits of varying age. There were numerous interruptions in the accumulation of the cover in the ancient and young platforms, marked by disconformities and small angular unconformities. Depending upon the structure of the karstosphere there may be several levels of karstification.

Fold Belts

The **Baikalian Orogeny** is important in South Siberia. Here the Yenisei, peri-Baikalian and Patoma fold belts are of interest.

In the Middle Proterozoic of the *Yenisei Range* (Fig. 2–113) crystalline limestones comprise most of the Penchengin Formation. Massive limestones and dolostones, with lenses of magnesite compose the Sosnovsk and upper part of the Potoskuisk Formations of the Middle Riphean. Bedded limestones are characteristic in the Upper Riphean Tokminsk and Dashkinsk Formations. Thick Vendian dolostones are developed in the Vorogov depression (the Pod'emsk Formation). Organodetrital dolostones of the Lower Cambrian Lebyazhinsk Formation are widespread. They are gently dipping and unconformable.

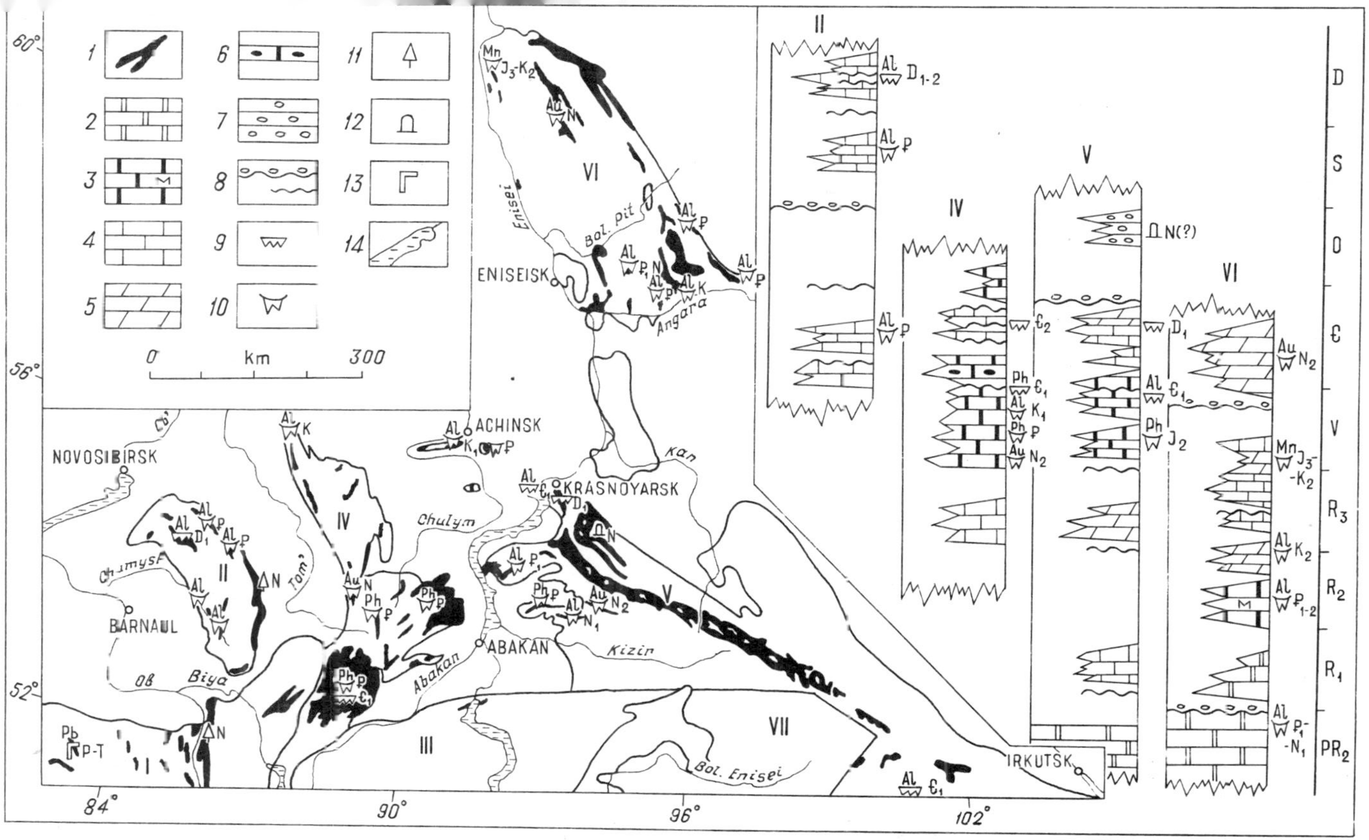

2–113. Paleokarst of the Baikalides and Caledonides of Southwest Siberia. 1. outcrops of carbonate rocks; 2–7. stratigraphic position of karst rocks: 2. marbles, 3. alternating limestone and dolostone, lenses of magnesite (m), 4. limestones, 5. dolostones, 6. lenses of sedimentary-diagenetic phosphorites in limestone-dolostone strata, 7. conglomerates; 8. regional and local hiatuses; 9–11. types of hypergene karst: 9. naked, 10. covered, 11. water-bearing; 12. hollow; 13. hydrothermal karst; 14. water areas. Abbreviations: Al. bauxites, Au. gold placers, Mn. manganese ores, Pb. polymetallic ores, Ph. phosphorites; Structures: I. Gorno-Altay, II. Salair, III. Western-Sayan, IV. Kuznetskiy Alatau, V. Eastern-Sayan, VI. Yenisei, VII. Southeast Tuva.

The *peri-Baikalian fold belt* (Fig. 2–114) consists of the Baikal anticlinorium and peri-Baikal foredeep. Karst occurs in Archean marble (Olkhon Island and peri-Olhkon region) and in Riphean dolostones and limestones of the Baikal Group (VOLOGODSKY, 1975). In the Patomskoe Nagorie belt carbonate deposits occur in the middle of the Patoma Group (Riphean) and are represented by marbles and limestones preserved in synforms.

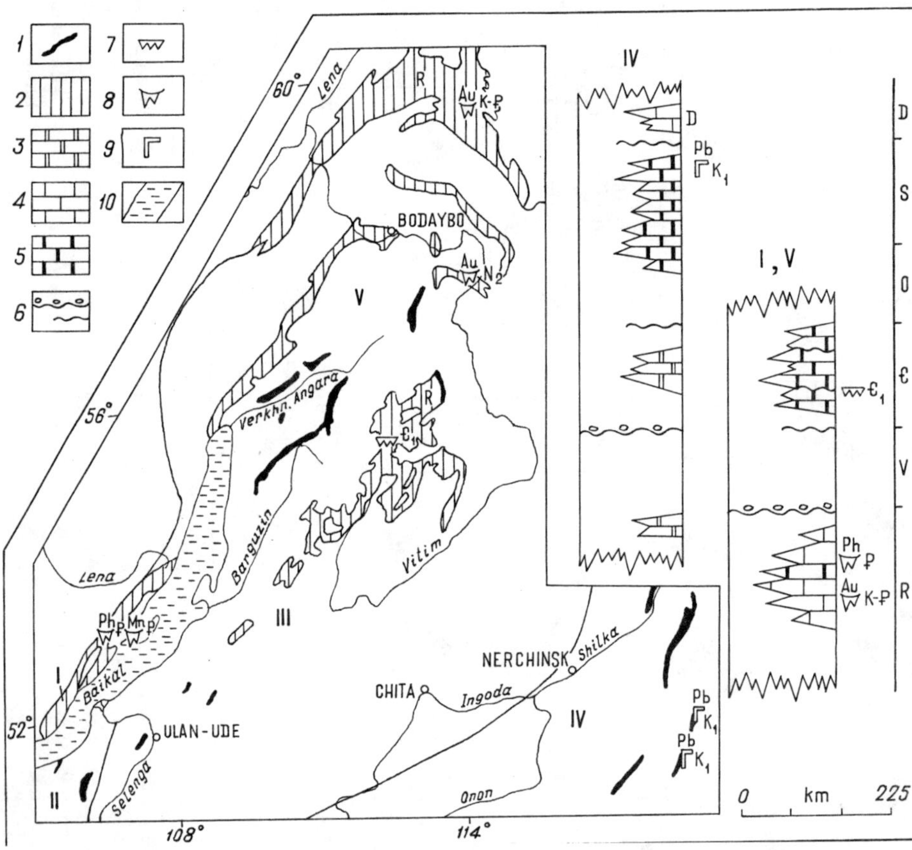

2–114. Paleokarst of the fold belts of Southeast Siberia. 1. outcrops of carbonate rocks; 2. terrigeneous-carbonate sequences; 3–5. stratigraphic position of karst rocks: 3. marbles, 4. limestones, 5. alternating limestone-dolostone; 6. regional and local hiatuses; 7–8. types of hypergene karst: 7. naked, 8. covered; 9. hydrothermal karst; 10. water areas. Abbreviations: Au. gold placers, Pb. polymetallic ores, Ph. phosphorites; Structures: I. peri-Baikalian of Baikalian age, II. Khamar-Daban of Caledonian age, III. Barguzin-Vitim of Caledonian age, IV. East Trans-Baikalian of Mesozoic age, V. Patoma of Baikalian age.

In the **Caledonide** regions the *Barguzino-Vitim area,* Eastern Sayan, Kuznetskiy Alatau and Salair Ridge are most significant. In Barguzino-Vitim there are widespread granitic intrusions. Carbonate deposits occur in separate fold-blocks (Kotero-Ulunsk, Tsipa-Amalat and Kurbo-Etantsin). They are contained in the basement complex (Lower Proterozoic marble) and are prominent in a geosynclinal complex of Riphean-Cambrian age (marble, bedded and massive limestones and dolostones).

The Lower Caledonides (Salairides) of *Eastern Sayan* contain abundant carbonate rocks amongst the Middle Proterozoic deposits (Derbinsk Formation), the Riphean (Mansk, Ungutsk Formations), Vendian (Ovsyankovsk Formation), Cambrian (Bokson Group, Kameshkovsk, Lodochnaya, Torgashinsk Formations). Especially interesting are the Ordovician conglomerates of the Mansk syneclize, which contain large caves (TSYKIN and TSYKINA, 1978).

Carbonate rocks are widely developed in *Kuznetskiy Alatau.* Upper Riphean and Vendian limestones and dolostones comprise the Turim, Zapadnosibir (Yenisei) and Martuhin Formations. The Cambrian rocks are the Belkin and Sagovaya, Usin and Bogoyul Formations. Local hiatuses occur between the Riphean-Vendian and the Lower Cambrian and inside the Cambrian sequence.

The *Salair Ridge* is a Late Caledonian-Early Variscan structure containing patches of massive and bedded limestones of the Vendian, Lower Cambrian, Silurian and Devonian. There are hiatuses within the Lower Paleozoic and in the Lower Devonian strata.

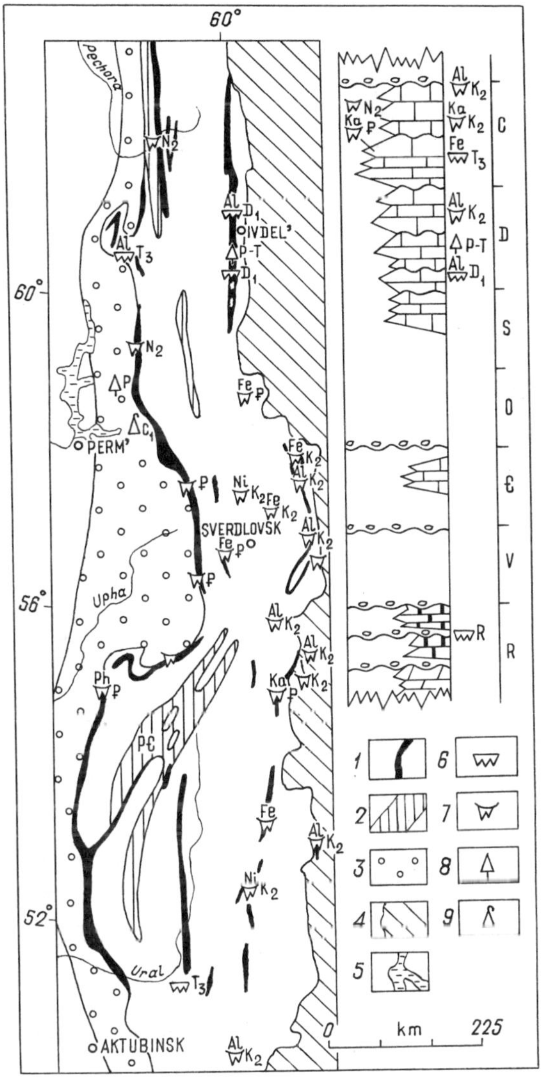

2–115. Paleokarst in the Ural Variscides. 1. outcrops of carbonate rocks; 2. outcrops of terrigeneous - carbonate sequences; 3. the Uralian foredeep; 4. the West Siberian Platform; 5. the Kama reservoir; 6–9. types of hypergene karst: 6. naked, 7. covered, 8. water-bearing, 9. oil and gas-bearing. Abbreviations: Al. bauxites, Fe. iron ores, Ka. kaolin, Ni. nickel ores, Ph. phosphorites.

The **Variscan** fold belts of the Ural, Kazakhstan and Tien Shan are rich in paleokarst. In the north-south extension of the *Ural* there is a sequence of compressed anticlinoria and synclinoria (Fig. 2–115). The stratotype for the Riphean occurs in the Bashkirian anticlinorium. In its basement group the Satkinsk Formation consists of dolostones with lenses of magnesite. Above it are patches of dolostones and limestones with siderite and limonite lenses (Bakalsk Group). The Middle Riphean Urmatinsk Group overlies it unconformably. There are horizons of limestones and dolostones in its upper part. The Upper Riphean is represented by the Karatau Group with a similar lithology.

Carbonate rocks of the Upper Ordovician, Silurian, Devonian and Carboniferous outcrop in the flanks and cores of the synclinoria. They are typical of the late stage of the Ural geosynclinal development. There are local breaks in the Paleozoic sediments.

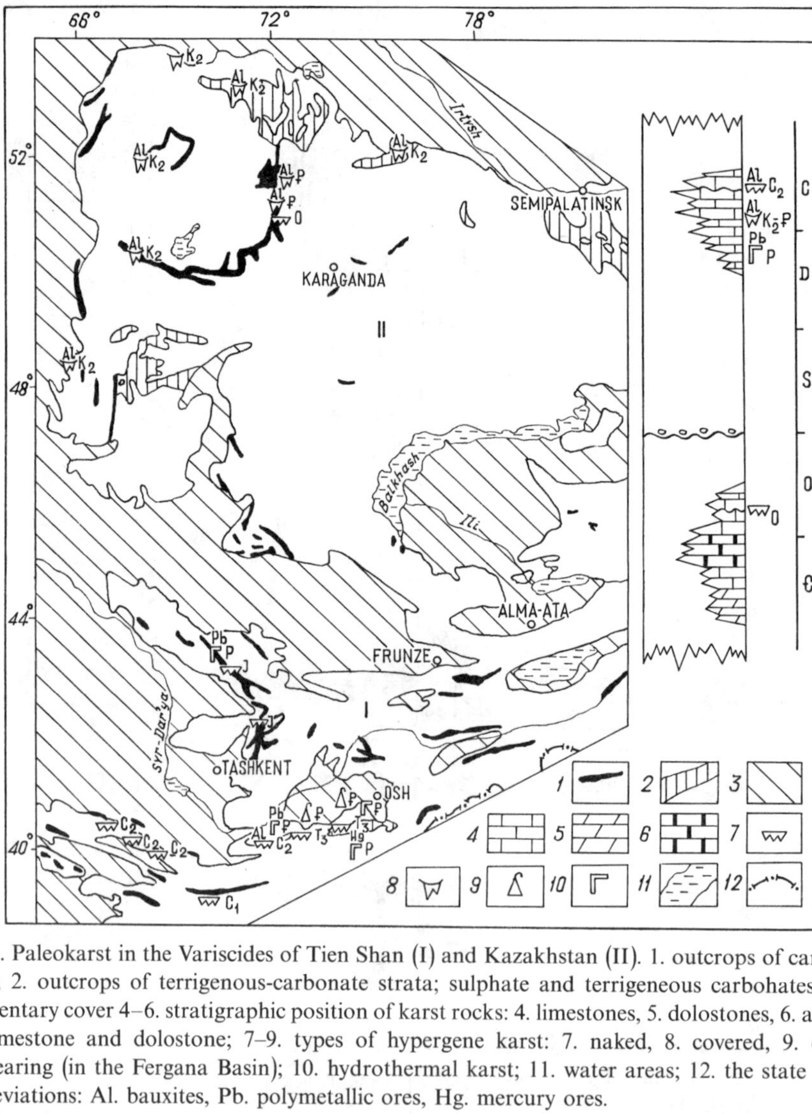

2–116. Paleokarst in the Variscides of Tien Shan (I) and Kazakhstan (II). 1. outcrops of carbonate rocks; 2. outcrops of terrigenous-carbonate strata; sulphate and terrigeneous carbohates of the sedimentary cover 4–6. stratigraphic position of karst rocks: 4. limestones, 5. dolostones, 6. alternating limestone and dolostone; 7–9. types of hypergene karst: 7. naked, 8. covered, 9. oil and gas-bearing (in the Fergana Basin); 10. hydrothermal karst; 11. water areas; 12. the state border. Abbreviations: Al. bauxites, Pb. polymetallic ores, Hg. mercury ores.

The Variscides of *Central Kazakhstan* have a complicated mozaic composition. Carbonate rocks occur in the Lower Paleozoic in mountain ranges of Karatau and peri-Balkhash, and are widespread in the Upper Devonian-Lower Carboniferous (Fig. 2–116).

Variscan folding is tight to the East (Alai Ridge) and more open to the West, in the South and West *Tien Shan*, around the large Fergana Basin. The Gissar, Zeravshan and Turkestan Ranges are fold-block uplifts between boundary faults. The Paleozoic in these ranges is represented by several different facies. The geological complexity of southwest Tien Shan is a result of superimposed folding and thrust nappes.

Carbonate rocks occur in the Lower and Middle Cambrian (Turkestan Range), and the Silurian (Zeravshan and Karatau Ranges). Limestones and dolostones are very prominent in the Devonian and Lower Carboniferous. Sulphate-carbonate deposits of Upper Jurassic, Cretaceous and Paleocene are developed in the southwest flanks of Gissar Range.

Paleokarst of varying genesis is found in local structures in areas of Mesozoic folding. There is paleokarst in the Nerchinsko-Zavodsk anticlinorium of the *East Trans-Baikalian* fold system. Here the limestone and dolostone are in the lower structural stage, in the Cambrian (Bystrinsk Group) and sometimes in the Devonian.

Riphean and Cambrian carbonates outcrop in the Kharaulakh uplift in the *Verkhoyansk-Chukotka* system. They formed a lower structural storey onto which the Upper Paleozoic Verkhoyansk Supergroup is deposited with distinct unconformity.

Dalnegorsk fold block is interesting in the southen margin of the Sikhote-Alin meganticlinorium (Primorsk anticlinorium). Thick Triassic limestones developed here and are divided into blocks by faults with which high-temperature hydrothermal paleokarst is associated.

The Crimea, parts of the Caucasus and Tadzhik fold systems are especially interesting in the Alpine orogenic belt. Weakly deformed limestones of Late Jurassic-Early Cretaceous age, with a thickness of about 600 m, are developed in the *Mountainous Crimea* (Fig. 2–117).

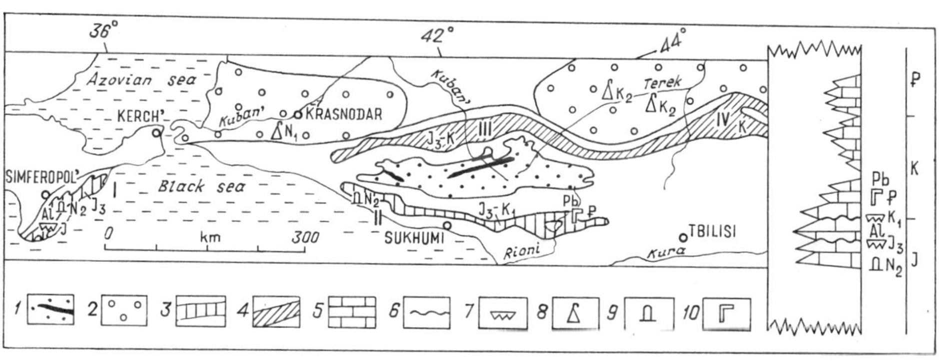

2–117. Paleokarst of the Alpine belt of the Mountainous Crimea and Causasus. 1. outcrops of the pre-geosynclinal complex with bodies of carbonate rocks (black); 2. foredeep; 3. outcrops of carbonate rocks of the geosynclinal complex; 4. the same, terrigeneous-carbonate sequences; 5. limestones; 6. regional and local hiatuses; 7–10. types of karst: 7. naked, 8. oil and gas-bearing, 9. hollow, 10. hydrothermal. Abbreviations: Al. bauxites, Pb. polymetallic ores; Structures: I. Mountainous Crimea, II. Sochinsk and Abkhazia-Rachinsk, III Labino-Malkinsk, IV. Dagestan.

Carbonate rocks occur in folded basement (in the Cambrian, Silurian, Devonian and Carboniferous systems) and in the geosynclinal complex within the meganticlinorium of the *Great Caucasus,* also. Its lower strata are Permian and Triassic, overlain unconformably by massive and bedded limestones and dolomitic limestones of the Upper Jurassic, Lower and Upper Cretaceous which are divided into local structural-facies zones (Laba-Malka and Dagestan to the north, Sochi and Abkhazia and West Georgia to the south). Especially interesting are Abkhazia and West Georgia (from the Psou River to Erco Lake) where weakly to moderately folded and faulted, thick limestones crop out. There are big reefs, distinguishable in the topography. Paleogene limestones and marls are known in this region, too. Neogene coquinoid limestones are widespread in the West fore-Caucasus and Dagestan.

The *Tadzhik* fold system is composed of the South Tien Shan and North Pamir mountain ranges. There are wide valleys. A Mesozoic foredeep deformed by orogenesis and the Alpine fold zone of distinctive anticlinoria and synclinoria can be recognised. Carbonate rocks in the basement belong to the Devonian and Carboniferous. They crop out in the cores of the Gissar anticlinorium. Unconformable Mesozoic deposits occurring between them are Callovian-Oxfordian limestones, Kimmeridgian-Tithonian gypsum, anhydrite and salt, which can be traced on the flanks of the folds. Sulphate and carbonate rocks also occur in the Cretaceous and Paleogene.

In summary, in folded areas karst rocks are rare in the Precambrian pre-geosynclinal basement; they are represented by some marbles, carbonatites, etc. The volume of karst rocks increases greatly in the Paleozoic and Mesozoic basement. Limestones and dolostones especially typify the middle and late stages of geosynclinal filling. There are carbonate-rich geosynclines in many orogenic areas, where great thicknesses of massive, often reefal, limestones and dolostones accumulated. Prominent breaks with unconformities are discovered at the tops of groups and formations. Sulphate and salt-bearing rocks are associated with intermontane depressions. Magnesite and conglomerate are special karst rocks. The products of different periods of karstification are distributed in complex ways within the karstosphere as consequences of the complex fold-block deformations and nappe tectonics that have occurred.

Paleokarst Genesis

The investigation of karst in the context of historical geology and paleogeography must not be separated from the analysis of the genesis of karst. In most places, paleokarst is a product of surface processes with meteoric waters (hypergene), *the child of the continental epoch* (STUPISHIN, 1967). Hypergene origin is undoubted for karst landforms exposed at the surface, even where they are filled with pre-Quaternary unconsolidated sediments. Buried karst forms discovered by drilling that occur directly under prominent stratigraphic unconformities, that are widespread and display sediments produced in weathering crusts are equally distinctive. Sometimes it is possible to interpret the environment of karstification that produced the ancient hypergene karst. The presence of towers, cones, dolines etc. can be interpreted from the characteristic roughness of the unconformity surface. We can also distinguish between naked, mantled and covered types of paleokarst from the manner in which the overlying deposits cover and seal (armour) the buried karst forms.

Sometimes cavities filled with deposits or water occur a few tens of metres below the unconformity. In these situations it is possible to speak about the cave environment or about the phreatic zone of that particular phase of paleokarstification. If insoluble strata (sandstone, effusive rocks etc.) are cut by the unconformity while karst forms are discoverd in underlying soluble strata, then the presence of screened paleokarst is established. It can be recognized by local collapse or sagging of insoluble rocks.

It is often found that there was simultaneous karst solution and accumulation of insoluble overlying deposits. Therefore, the solution in such cases took place under the sedimentary cover, the thickness of which increased at a rate determined by the rate of deep karst solution. We refer to this situation as *screened karst*[1] (TSYKIN, 1980).

Ancient hydrothermal karst features are found in many deposits of polymetals, antimony and mercury located within carbonate rocks. It is of subsurface genesis, cropping out at the surface in some places as the result of lengthy denudation. Hydrothermal karst is characterized by its close relationship with faults and fissure zones, its local distribution, the mineralization of karst voids by calcite, baryte, fluorite, chlorite, sulphides of iron, copper, lead and zinc.

In some places paleokarst of hypergene or hypogene origin has been drastically transformed and the primary genetic features are almost lost. For example, cavities of the phreatic zone when uplifted are drained, infilled with collapse and speleothem material (the relict cave stage) and perhaps intersected by the lowering erosion suface. Mantled or covered karst features exposed in eroding valley sides (and, rarely, in watershed areas) are altered by slope erosion processes with the result that remnant relief is formed.

In many karst regions there are such transformed ancient features. Their evolution may interpret following detailed investigation.

Periods and Phases of Karstification

The probability of finding exposed paleokarst and buried paleokarst decreases with increasing geological age. For this reason evidences of Precambrian and Early Paleozoic karst are not numerous.

Researchers in western Europe and North America would consider this to be one particular type of covered or interstratal karst. The term covered karst is taken to describe deep solution beneath insoluble cover strata, whether or not there is detrital accumulation associated with the solution-induced subsidence or collapse (Editors' note).

Proterozoic Paleokarst

No hypergene karst is known from before the Late Proterozoic in the U.S.S.R. Our most ancient karst is hydrothermal, examples of which are found in many places in the Krivoy Rog iron ore basin (the **Ukrainian Shield**). There is a major dolomitic marble in the upper part of the Krivoy Rog Group (Lower Proterozoic). Its thickness is about 500 m and it is observed in the Saksagan locality over an area of 50 km^2 and at depths extending to more than 1 000 m. Saline waters under pressure occur in well fissured dolostone patches and flooded mines. There have been cases of water bursting from large karst cavities, carrying a lot of sandy-clayey material and dolostone debris with it. Cavities were found by drilling at depths down to −700 m. In the northern part of the basin, where there was intensive alkaline metasomatism, large blind cavities were encountered in carbonatized quartzites. One of them had a length of more than 1 500 m, a width about 25 m and a depth about 350 m! Undisturbed deposits of Paleogene age were found in this cavity. In the author's view, quartzites are karst rocks because quartz is quite soluble in hot alkaline waters. At a temperature of 100 °C, for example, quartz dissolves faster than calcite.

The alkaline metasomatism in the Krivoy Rog iron ore basin is probably the result of hydrothermal activity during the stage of protoplatform development (Middle Proterozoic). The formation of the rich hematite ores is associated with that stage, as many investigators have supposed.

There is evidence of syngenetic karst in the Riphean limestones and dolostones of the Timan Ridge and the Ural (RODIONOV, 1963).

Vendian karst is found on the **Siberian Platform**. In the Kujmbinsk area (Tunguska syneclize) Upper Riphean dolostones at depths of 2 000–2 400 m are overlian paraconformably by Lower Cambrian salt-bearing deposits. The dolostones are light grey, dense and massive. In a drill core taken from the upper 70 m of them caverns are visible. There was an inflow of oil and gas from this zone.

In the Olenek uplift, dolostones with small sinkholes are unconformably overlain by Lower Cambrian sandstones plus calcareous clay or breccia. Sinkhole walls are usually covered with a thin iron carbonate crust.

Similar but more ancient (Upper Riphean) formations are found in the peri-Sayan region along the Uda River near the town of Maksut. The eroded surface in dolostones of the middle part of Baikal Group contains sinkholes 1–2.5 m wide and about 3 m deep. They are filled and overlain by phosphatic sandstones of the upper part of the Group. Thin phosphate crusts were deposited on the corroded surface of the dolostone (VOLOGODSKY, 1975).

There is allophane and gibbsite mineralization in the peri-Baikalian and Yudoma-Maya foredeeps in Riphean dolostones overlain paraconformably by sandstones and siltstones. This was interpreted by some investigators as evidence of a Riphean laterite weathering phase (SLUKIN and ZHIVTSOV, 1975). Others considered the aluminous deposits to be epigenetic infiltration products of the Cenozoic denudation period (CHERKASOV, 1978).

Cambrian Paleokarst

Ore-bearing Lower Cambrian karst has been studied in the Lower Caledonides of the **Altay-Sayan** region. On the Early Cambrian sea floor, reefs (now dolomitized) of the Belkin and Sagovaya Formations developed in the Gorno-Shorsk phosphorite-bearing basin. They were uplifted and drained periodically and island karst developed. This was filled with chemico-clastic deposits, which consisted of siliceous-phosphatic breccias and rare allites, and which filled steep walled depressions (sinkholes, trenches, fissures; compare recent phosphate deposits in High Carbonate Islands of the Pacific Ocean, BOURROUILH-LE JAN, p. 443–471, Editors' note). This Cambrian karst is masked by more widely developed Cretaceous-Paleogene karstification (FOMINSKY, 1974).

The bauxites of the Bokson deposit of Eastern Sayan occur in dolostones. They are of comparatively constant thickness and low quality (KIRPAL and TENIAKOV, 1978). There are karren and ore breccias at the base of the bauxite beds. The deposit accumulated during the Early Cambrian continental episode when weathering products were transported to a coastal karren zone. After deposition, they became bauxitized.

Lower Cambrian buried karst has been discovered over a large area at the head of the Vitim and Taloy-Maloamat watershed (**Bargusin-Vitim** fold belt). Conglomerates, mudstones and breccias occur on a rugged surface of algal dolostones of the Tilim Formation. There are small sinkholes in some places, with Al_2O_3 enrichment of the sericitic shales filling them.

There is Middle Cambrian karstification in the **Irkutsk Amphitheatre.** Small sinkholes formed in dolostones of the Angara Formation. They are filled with light green clay and dolomitic residuum and are overlain by variegated clays of the Osinsk Formation. These clays are thought to be products of weathering. During the karstification, gypsum and salt in the Lower Cambrian Angara Formation of this area was completely dissolved.

There are horizons of regional extent in the Lower-Middle Cambrian carbonates of the **Siberian Platform** with reported shows of brines, oil and gas. Some investigators explain their formation by breaks in sedimentation, with syngenetic corrosion. In the Irkutsk Amphitheatre five such permeable horizons are recognized. There are eight in the Turukhan uplift and four in the Nepsk anteclize (OVCHARENKO, 1978). The interruptions occur only in the Middle Cambrian and are not found everywhere. We can recognise, therefore, an important Middle Cambrian phase of karstification, but there is also a possibility of selective dissolution during uplift in the Lower Cambrian.

Ordovician and Silurian Paleokarst

Evidence of karstification in the Ordovician and Silurian is rare. Late Ordovician activity is seen in the **Kazakhstan** fold belt, in the eastern part of Teniz trough (Taskolsk area). Here, in crystalline limestone, a layer of bauxite breccia was found which consists of fragments of calcite, diasporic bauxite, illite and argillite. This rock fills fissures and karren in the limestone and is overlain by terrigeneous facies (KIRPAL, 1976).

In the Estonian combustible shale basin (**Baltic syneclize**) Silurian limestones and dolostones contain blind cavities; they are isometric, pipe-like forms. Vertical fissures widened by corrosion also occur. Some of this karst is filled with sands and clays containing Lower Devonian fauna. Middle Devonian deposits transgress the eroded Silurian rocks at the margins of the basin. The karstification supposedly began in the Late Silurian, and was repeated several times.

Devonian Paleokarst

There were not less than four phases of karstification in the Devonian. At th margins of the **Baltic syneclize** an Early Devonian phase is proposed (CHIKISH-EV, 1978). Karst depressions filled with sandy-clayey deposits of the Middl Devonian were formed under continental conditions upon the Silurian car bonate basement.

There is evidence of a Middle Devonian phase in the centre of the **Tima Ridge**. Peculiar bauxite strata (the Vezhau-Vorykvinsk deposit) occur on karstified surface of Riphean limestones; in age the bauxite is correlated with th North Ural bauxite basin (DEMINA and KUZMINA, 1975).

A Late Devonian phase is well displayed in the **Volga-Ural anteclize** (STUPISH-IN, 1967). Dolines 10–20 m deep and outliers of 50–60 m were created. Th karstified surface is overlain by limestones of Late Fammenian-Tournaisian ag

In the **Pechora Basin** the buried reef that is the Zapadno-Tebuck oil field h a karstified core of 0.2 × 2 km. This was discovered during drilling by t catastrophic loss of drilling fluid and break of the drilling tool (ENTSOV ar SHAITOV, 1973).

In the regions of Caledonian and Variscan folding the same features occur. the **Eastern Sayan** on the northern slopes of the Torgashino Range Midd

Devonian sandstones and mudstones are deposited unconformably onto a rugged surface of Cambrian limestone. There is a breccia of these in the basal part of the Devonian column. Clastic dykes of calcareous sandstones and blind bodies of breccia consisting of fragments and residua of grey limestones with a reddish-brown mudstone matrix are traced lower in the Cambrian limestones (TSYKIN and TSYKINA, 1978). The proportion of the karst inclusions in the limestones is 3.2 %.

In **Salair Ridge**, Lower-Middle Devonian strata display four phases of karstification (Early Emsian, Early and Late Eifelian, Givetian). Beds and lenses of bauxite correspond to each phase. Their bases are highly irregular, in contrast to the flat tops. Ore-bearing breccia occurs in the bases (KIRPAL and TENIAKOV, 1978). A reconstruction of the depositional environment is that there was a shallow Devonian sea in which limestone were deposited. The sea regressed periodically. On the exposed islands naked karst developed quickly in the

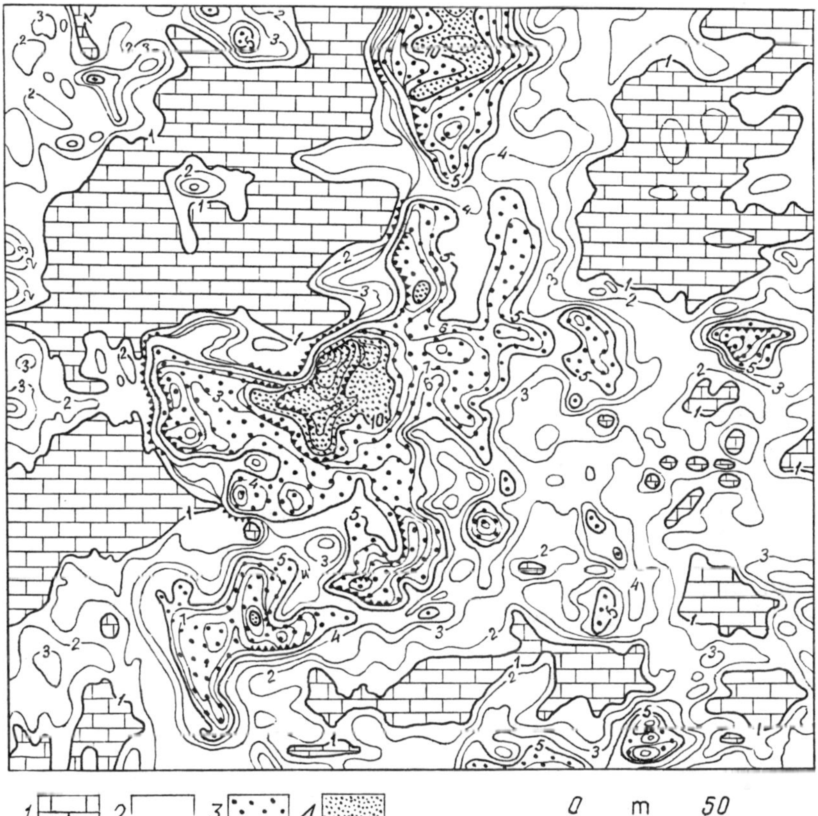

2–118. Changing thicknesses of ore horizons in a part of the North Ural bauxite area (after BUSHINSKY, 1975). 1. thicknesses of ore horizons less than 1 m, 2. 1–5 m, 3. 5–10 m, 4 more than 10 m.

tropical climate and bauxitic material accumulated. In times of transgression the bauxites were covered with further limestones.

In the **Ural** area the formation of bauxites deposits is associated with the Middle-Late Eifelian period and contains three phases. They are fully displayed in the North Ural bauxite basin on the surface on the Late Silurian-Lower Devonian limestones where in Early Eifelian, Late Eifelian and pre-Givetian times coastal karst belts were formed. These trapped the alumina minerals. The bauxite bodies have irregular bases and are often underlain by ore breccia; they are from 0 to 20 m thick (Fig. 2–118). There are networks of veins and pockets of bauxite in the limestones, created by the filling of fissures and blind cavities (BUSHINSKY, 1975). There are permeable fissures, caverns and smaller cavities in the Lower Devonian-Middle Carboniferous carbonates that enclose the bauxites. They are good aquifers, containing waters that are fresh to a depth of 1 000 m. This is the result of fissure karstification during later phases.

There are Eifelian and Frasnian bauxite deposits of similar type in the South Ural Basin (KIRPAL and TENIAKOV, 1978).

Carboniferous Paleokarst

Three periods of karstification are recognized in the Carboniferous of the Russian Platform, the pre-Visean, pre-Bashkyrian and post-Bashkyrian. Some of them have 2–3 phases.

Development of karst landscapes in the Lvov Basin, Moscow syneclize, Voronezh and Volga-Ural anteclizes and Timan Arch occured in the pre-Visean period. In the eastern part of the **Lvov Basin** (Podol'e) at the beginning of the Visean there was a phase of continental emergence when intense karst development took place. In limestones and dolostones of Late Silurian and Early Devonian age, dolines, ponors, bodies of breccia and dolostone grus were formed (KUCHERUK, 1976). The same features appearing in Middle-Upper Devonian mixed sulphate-dolostone strata are correlated by KUCHERUK with the post-Namurian and post-Bashkyrian periods. They indicate continental emergent periods with a warm wet climate. Cave systems were well developed beneath the surviving unconformity; voids constitute about 20–30 %. Solution cavities are observed to a depth of 3.5–4.8 km.

In the **Moscow syneclize** karst landforms are buried by the Tula coal measures, which extend unconformably across Tournaisian limestones and Upper Devonian gypsum and dolostones. The upper surface of the coal measures is highly irregular, with outliers and hollows. This morphology in the coal beds reflects the older karst relief on one hand, and subsidences associated with later epigenetic karstification on the other hand.

In the **Voronezh anteclize** during the pre-Visean denudation phase there was intense quartz dissolution in jaspilites and development of laterite and bauxite on exposed shales of the Lower Proterozoic Kursk Group (GERASIMOV and SIDORENKO, 1974). Linear depressions about 2 000 m long, 200–600 m wide and 100–120 m deep were created in the truncated exposures of steeply dipping beds of jaspilite in which the quartz is completely dissolved to great depth. Visean limestones then buried this basement topography.

In the **Southern Timan**, Visean bauxitic strata are found on karstified surfaces of Devonian limestones (Kedva-Vapovsk deposits). They are overlain by Namurian dolostones. The thicknesses of the bauxites are at a maximum in sinkholes and they pinch out above positive features in the buried relief (KOLOKOLTSEV, 1975).

The karst denudation surface that formed during the Late Tournaisian in the **Volga-Ural anteclize** was the same type as seen in modern tropical karsts, such as in Cuba and Jamaica (STUPISHIN, 1967). It is two phase karstisication, according to buried karst relief on limestone of the Tula Formation. We consider the Early Carboniferous to be the time of two phase karstification, according to buried karst relief on Tula limestones, which were then overlapped disconformably by Namurian carbonates. The Late Namurian and post-Bashkyrian periods of karstification explored in Bashkyria are associated with the formation of regionally persistent fissure karst aquifers now hosting oil and gas.

The Late Devonian-Early Carboniferous (pre-Visean) denudation is well marked in the **Siberian Platform** (GERASIMOV and SIDORENKO, 1974). Its products are the alumina-rich kaolins (flint clay) studied in the Kokuisk area of the Tunguska syneclize, where they fill sinkholes in Lower Cambrian limestones. The kaolins are overlain by Visean sandstones. Near Podayaisk at the western edge of the Angara-Tasejeva syneclize, kaolin and flints of the same phase fill sinkholes in Upper Devonian limestones, and are overlapped by Lower Carboniferous terrigeneous deposits.

Pre-Bashkyrian karst is known in the basins of the Nizhny Viljukan and Markha Rivers in the eastern Tunguska syneclize. There are grey-green clays containing fragments of native rocks at their base, resting on rugged surfaces of Ordovician and Silurian limestones. They are overlapped by Middle-Upper Carboniferous terrigeneous rocks (GERASIMOV and SIDORENKO, 1974).

The pre-Bashkyrian karst is found in the Variscides of the southern **Tien Shan**, with associated bauxites (KOCHNEV et al., 1975). There is both widespread and fault-located paleokarst. Small bauxite bodies occur in sinkholes and cavities connected with them. The deposits are aligned along the strike of the Lower Carboniferous limestones. Bashkyrian carbonate deposits overlap them. Deposits of fault-located karst are limited to Devonian limestones and occur as veins and columnar bodies of ore-bearing breccias. They are associated with contacts between limestones and shales. The cover deposits have been eroded, so the age of the bauxitic breccia is determined by analogy.

In the lower structural storey of the Mesozoic fold belt of the **Kharaulahk Mountains** there is pre-Permian (post-Bashkyrian) karstification (LUNSGERSGAUSEN, 1966; GOGINA and LEONOV, 1984). In the Lena Valley, 60–100 km south of the river mouth, relics of sinkholes and cavities are exposed in massive Riphean limestones. They are filled with various rocks — breccia, sandstones, mudstones. These deposits are overlain by Permian terrigeneous strata. Karst forms extend to depths of more than 300 m beneath the Permian base.

Some geologists contend that there are Lower Carboniferous karst bauxites in the **Ural** fold belt, the stratotype being the Zhuravlinsk deposit (KIRPAL and TENIAKOV, 1978). Investigators in the 1930s ascribed these deposits to Cenozoic sulphate weathering and infiltration of aluminium salts. In recent years further confirmation of that proposal has appeared (ENTSOV, 1982). Both the mode of ore occurrence (it formed as crusts on the limestones in the eroded cores of anticlines) and the peculiarities of its content (the presence of gibbsite, alunite, kaolinite) distinguish it from the typical geosynclinal bauxites described above.

Permian Paleokarst

Examples of Permian paleokarst are not as numerous. It is recognised in the **Volga-Ural anteclize** in the Russian Platform. MASLOV (1945) was the first to study the properties of the paleokarst in the Sakmarian-Artinskian reefs of the Ishimbay oil field. He distinguished four stages in the development of the porosity of the reefs. The first was reef growth, with primary cavities filled with organic debris. The second was a dewatering of the reef, and karstification. Karren, sinkholes and pinnacles formed on the surface and caverns and lesser cavities developed in the rock. The third stage was partial destruction of emerged areas by wave action, with the accumulation of clastic material, until the stage ended by submergence of the reef platform. The fourth stage was the deposition of sulphates and rock salt in the lagoon and also in the caverns of the reef. The karst of the Ishimbay field is considered by MASLOV to be pre-Upper Artinskian.

Sakmarian-Artinskian reefs are observed along the entire **Uralian foredeep** (Fig. 2–119). In many of them there is evidence of paleokarst and huge deposits of oil and gas have been discovered (MAKSIMOVICH and BYKOV, 1978).

There is evidence of Late Permian karstification in the **Volga-Ural anteclize**. It was caused by continental emergence in Ufimskian times. Sulfate and carbonate deposits of Early Permian age were subject to karstification, with later inflows of brine, oil and gas.

In **Central Povolzh'e**, STUPISHIN (1967) described a phase at the end of the Sakmarian-beginning of the Artinskian (Early Permian) when the eroded tops

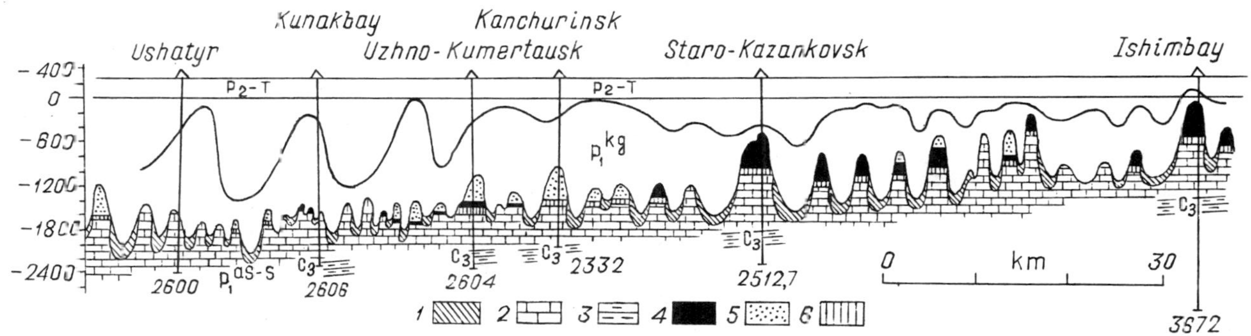

2–119. The Permian reefs of the Uralian foredeep which are reservoirs for oil and gas (after MAKSIMOVICH and BYKOV, 1978). 1. intereef facies (bituminous clayey limestones), 2. reef facies (massive limestones and dolomitic limestones), 3. sandstones and mudstones, 4. oil, 5. gas, 6. zone of oxidation of hydrocarbons.

of domes of Upper Carboniferous-Asselian carbonate deposits were attacked by karst processes. However, he considers the Late Kungurian phase to be more important. In this phase most of the Kungurian salt-bearing deposits were destroyed and karst relief formed which was then buried by Kazanian (Upper Permian) deposits.

Traces of karstification of Late-Carboniferous-Early Asselian are reported in the productive strata of the Vuktyl gas-bearing deposit of the **Pechora depression**(the upper horizons of the gas-bearing beds).

In the **Ural** fold belt Permian paleokarst is presumed in the Kiselov Coal Basin. Carbonate rocks accumulated there from Early Devonian until Early Permian times, to a total thickness of about 2 000 m. They are divided by minor terrigenous units into three main aquifer levels. The porosity (formed by the karst cavities) in the near surface parts of the upper aquifer (Upper Carboniferous-Lower Permian) is 5–6 %. With depth this declines to 1–2 %, but some cavities (in cross-section up to 3–4 m) are open down to depths of 1 100 m in the static water zone. The active karstification in this carbonate formation is supposed to have occurred as a result of the Late Permian orogeny. Karstification recurred in this area time and time again.

Karst deposits of a Late Permian phase were discovered by us in the Mazulsk limestone quarry near the town of Achinsk (TSYKIN and TSYKINA, 1978). Lithified, rhythmically bedded conglomerate, limestone, sandstone and carbonaceous mudstone bodies, crossed by veins of white calcite, occur within massive grey, irregularly jointed limestone of Precambrian age. They are complicated, curved vein-shaped or network bodies. Their Upper Permian age was determined by palynology.

Hydrothermal karst of the Variscan metallogenic cycle (Upper Carboniferous-Permian) is well seen in the **Kazakhstan** and **Tien Shan** fold belts at lead and zinc (Achesay, Karagach), antimony (Dzhizhikrut, Kadamzhai) and mercury deposits (Khaidarkhan etc.). For example, the main mined deposit at Achesaysk (in the Karatau Range) is a cave breccia. It is up to 30 m thick and is traceable down-dip for 320 m. Secondary argillo-ochre cerussite and smithsonite ores extend to depths of 200–250 m. In addition to the Paleozoic hypogene karst there is Paleogene surficial karst at this site (KUTYREV and LYAKHNITSKY, 1982).

Numerous cavities which have no connection with the surface are known in mines at the Kadamzhai and Khaidarkhan deposits. They are up to 100 m in length. The walls are covered by normal calcite speleothems overlying baryte, fluorite and Iceland spar. In the shallow subsurface the cavities are dry but deeper ones become filled with water. Cavities have been explored to −450 m (SULTANOV, 1972).

Triassic Paleokarst

In the Triassic, because of considerable erosion of Lower Triassic deposits, the preserved evidence is only of Middle Triassic-Early Jurassic karstification. This probably included two or three phases. There are traces of pre-Middle Jurassic karst in the **Moscow syneclize** and the **Volga-Ural anteclize**. In the former, Middle Jurassic deposits are widespread. They rest on overwashed, irregularly karstified surfaces of Lower-Middle Carboniferous carbonates. In buried caves there are clays and other clastic fills overlain by basal conglomerates or by sandstones of the Triassic sequence. There are pockets and veins of other deposits in the limestones and masses of grus with rock debris in the dolostones.

In the southwestern part of the **Siberian Platform**, many depressions developed in the Irkutsk superimposed foredeep (VOLOGODSKY, 1975). Here Lower Jurassic coal measures rest on Lower Cambrian dolostones in some places. The strata that survive, largely dolostones with zones of silicification, disintegration and brecciation, are the remains of the solution of intervening anhydrites and salt. This process extended to depths of 500–800 m below the Late Triassic paleosurface. The aggregate thickness of salt beds and anhydrite dissolved in this phase is about 560 m while their total thickness in the Cambrian deposits is about 800 m. The speed of underground chemical solution was 0.03–0.032 mm per year (ANTSYFEROV, 1979). Systems of depressions and antiforms are explained as consequences of the irregular patterns of subrosion.

Late Triassic paleokarst was detected at the Belsk uplift. Here, in poljes and other depressions which contain many sinkholes, deposits of the Rhaetian-Lower Jurassic Zilarinsk Formation occur. The surface of irregularly silicified dolostones is covered by siliceous breccia. Kaolinitic claystones 0–20 m thick are exploited as raw materials for making ceramics. The formation of these rocks is thought to be a result of sedimentary-diagenetic process connected with weathering of karst deposits (VOLOGODSKY, 1975)

In the east of the Tunguska syneclize (Botuobya, Vilui, Markha River Basins) in combined erosional and karstic depressions in Lower Paleozoic carbonates, there are grey clays 0.5–4 m in thickness that are overlapped by Lower Jurassic conglomerates and sandstones. The depth of some filled dolines is 100–250 m.

On the **West Siberian Platform**, Middle-Late Triassic karst (as noted above) is found in the Nurolsk area. The paleosurface on carbonate rocks of Silurian and Upper Devonian-Lower Carboniferous age has relief up to 200 m that proves the presence of karst forms. Lenses of boehmitic and diasporic bauxite 4-6 m thick were discovered by a few boreholes that passed through dolines or a polje (ZHERO, SMIRNOV and SURKOV, 1977). There was loss of drilling fluid in many boreholes and intervals of disintegrated rocks were recorded. A flow of brine and oil was obtained from the top of these rocks.

Middle-Late Triassic karst with bauxite is known in the basement of Volochaevsk and Bogoslovsk Basins in the **Ural** fold belt, beneath Rhaetian-Early Jurassic coal measures. Drilling indicates complicated morphology on the Silurian limestones. They are overlain by clay in which the lenses of bauxite are formed, surrounded by allitic layers (KIRPAL and TENIAKOV, 1978).

Jurassic Paleokarst

In **Povolzh'e** an Early Jurassic paleokarst morphology is well studied (BARKOV, 1932; STUPISHIN, 1967). Permian carbonates contain sinkholes, shafts, trenches, potholes and hums or pinnacles. The karst paleorelief is about 50 m. Caves are filled with clays and sands containing dolostone fragments.

There were many subsidence depressions formed as a result of interstratal solution of the rock salt and part of the sulphate rocks that comprise the major part of the Kungurian strata in the **Uralian foredeep** (GARYAINOV, 1980). These structures are 6–20 km in length and 2–8 km in width. The depth of subsidence that was compensated by infilling with Mesozoic and Cenozoic strata was 400–500 m. In the centre of the foredeep, active synclinal subsidence began in the Middle-Late Triassic and continued (with interruptions) to the Paleocene-Upper Pleistocene. The synsedimentary development of these structures is identified by the increasing thickness and depth of marker horizons and by facies changes from the flanks to the centre.

The origin of some nickel and iron ore is associated with Late Triassic-Early Jurassic hypergenesis in the **Ural** (GERASIMOV and SIDORENKO, 1974). In the Tagil synclinorium, stratiform intrusions of serpentinite occur in Silurian-Lower

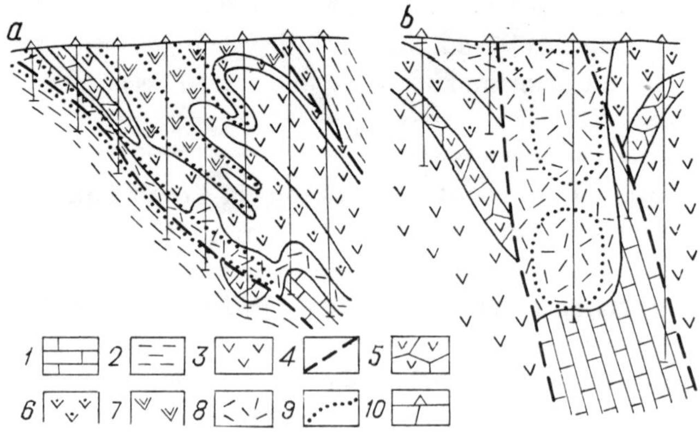

2–120. Geological cross-section of the silicate nickel deposits of the Ural (a. Sinarsk, b. Lipovsk; after GLAZKOVSKY, GORBUNOV and SYSOEV, 1978). 1. limestones; 2. schists; 3. serpentinites; 4. faults; 5–7. zones of serpentinite weathering crust: 5. disintegration, 6. leaching; 7. ferrugination and silicification; 8. karst deposits; 9. contours of ore bodies; 10. boreholes.

Devonian greenstone strata which contain lenses of marble. The marble is dissolved along the contacts with the serpentinites and schists. Dissolution breccias and sandy-clayey deposits accumulate in these places (Fig. 2–120). The contact-karst processes extended to depths of 250–300 m (GLAZKOVSKY, GORBUNOV and SYSOEV, 1978).

In the Akkermanovsk iron ore deposits, the lower ore (Lower-Middle Jurassic) lies unconformably on a karstified surface of Visean-Namurian limestones. The deposits are of pebbles, sand and clay with lenses of siderite. They fill a polje containing sinkholes. An upper deposit of limonite which fills higher karst depressions is mined. Evidence of three phases of karstification (Late Triassic, Oligo-Miocene and Quaternary) has been revealed by detailed research there (GARYAINOV, VASIL'EVA and AGANIN, 1978).

Bauxites associated with buried karst of Late Triassic-Early Jurassic age occur in the **Tien Shan** (Kshut, Maylisu, Shing etc.). They fill depressions in Paleozoic carbonates. They display breccia textures and form small lenses with rugged bases that are the result of filled karren, sinkholes and shallow shafts.

Fissured and porous Upper Jurassic limestones forming a karst monadnock have good porosities in the Kokhanovsk oil field in the **Lvov Basin**. The monadnock is buried by Cretaceous rocks. It is a product of Late Jurassic karstification.

Sinkholes up to 120–150 m deep were discovered by drilling in dolostones and limestones of Anabar-Udzha and Vilui-Kotuy districts in the **Anabar anteclize**. In some of them Upper Jurassic deposits were detected, in others there were Cretaceous and Cenozoic deposits (PROKOPCHUK, LEVIN and METELKINA,

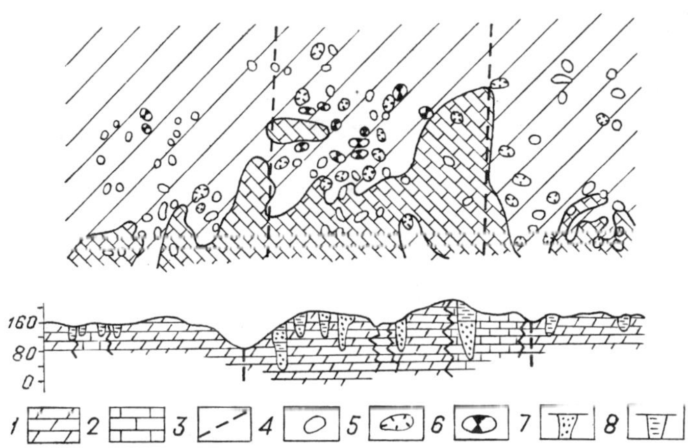

2–121. Sinkhole-shaped depressions of the Anabar-Udzha watershed (after PROKOPCHUK, LEVIN and METELKINA, 1982). 1. dolostones; 2. limestones (on the coross-section), Paleozoic carbonate strata (on the map); 3. faults; 4–6. karst depressions: 4. shallow (less than 30 m), 5. intermediate (30–60 m), 6. deep (more than 60 m); 7. silts and sands, sometimes with pebbles; 8. clays and loams, sometimes with debris.

1982). This is evidence of several karst phases, beginning in the Middle Jurassic (Fig. 2–121).

There are Callovian-Oxfordian carbonate rocks in the **Turanian Platform**, with numerous deposits of oil and gas. There was karstification of reefs and carbonate mounds (CHERNOV, 1973) in a few stages in the Late Jurassic-Early Cretaceous. In the roofs of such structures, disintegrated and cavernous rock was found, with high losses of drilling mud and dropping of drillstrings.

Relics of hypergene and hydrothermal karst of Jurassic age are known in the orogenic belts. A Middle Jurassic age was determined for the Seybinsk phosphorite deposit in **Eastern Sayan.** Small dolines and linear contact depressions in Vendian limestones are filled with phosphorite-bearing sediments. They are siliceous and clayey, brown or grey-white. The phosphorites are friable and rocky, with a brecciated texture in the grey-white units.

In conditions of a stable tectonic regime and peneplanation there was kaolinitic weathering and karstification in the **Ural** fold belt. Numerous deposits of kaolin and brown iron ore are associated with covered karst (SIGOV, 1969). In the Alapaevsk iron deposits, for example, there are depressions up to 2 000 m long, 300–500 m wide and 150 m deep in Middle Carboniferous limestone. Their bases are irregular and there are chert pebbles and other debris at the base of the Jurassic filling. The thickness of these fillings and cover amounts to 200 m. There are pockets and lenses of brown iron ores and sometimes siderites (CHIKISHEV, 1964).

Cavities of varied shape, filled with Jurassic sandy clay are known in the Chatkal Range in the **Tien Shan**. They are in Lower Carboniferous limestone, have lengths up to 25 m, base areas up to 12 m^2 and extend to 60 m below the modern erosion base (MAMATKULOV, 1979).

In the **Mountainous Crimea**, karst bauxites are associated with a Late Jurassic phase which has been prospected around Basman Mountain. The bauxite deposits rest on rugged surfaces in Oxfordian limestones. As usual in such cases, there is ore breccia at the base. It is succeeded by bauxitic clays, then lenses of bauxite breccia and layered bauxites alternate. Cover strata are Tithonian limestones and sandstones.

Cretaceous Paleokarst

Cretaceous paleokarst has been investigated in many regions and we can recognise with confidence prolonged karstification divided into three or four phases. In some places the karst development extended into the Paleogene.

In the Lvov Basin of the **Russian Platform** various underground and surface karsts developed in limestones and dolostones of Silurian, Devonian and Jurassic age under the humid climatic conditions of the Neocomian. Surface landforms

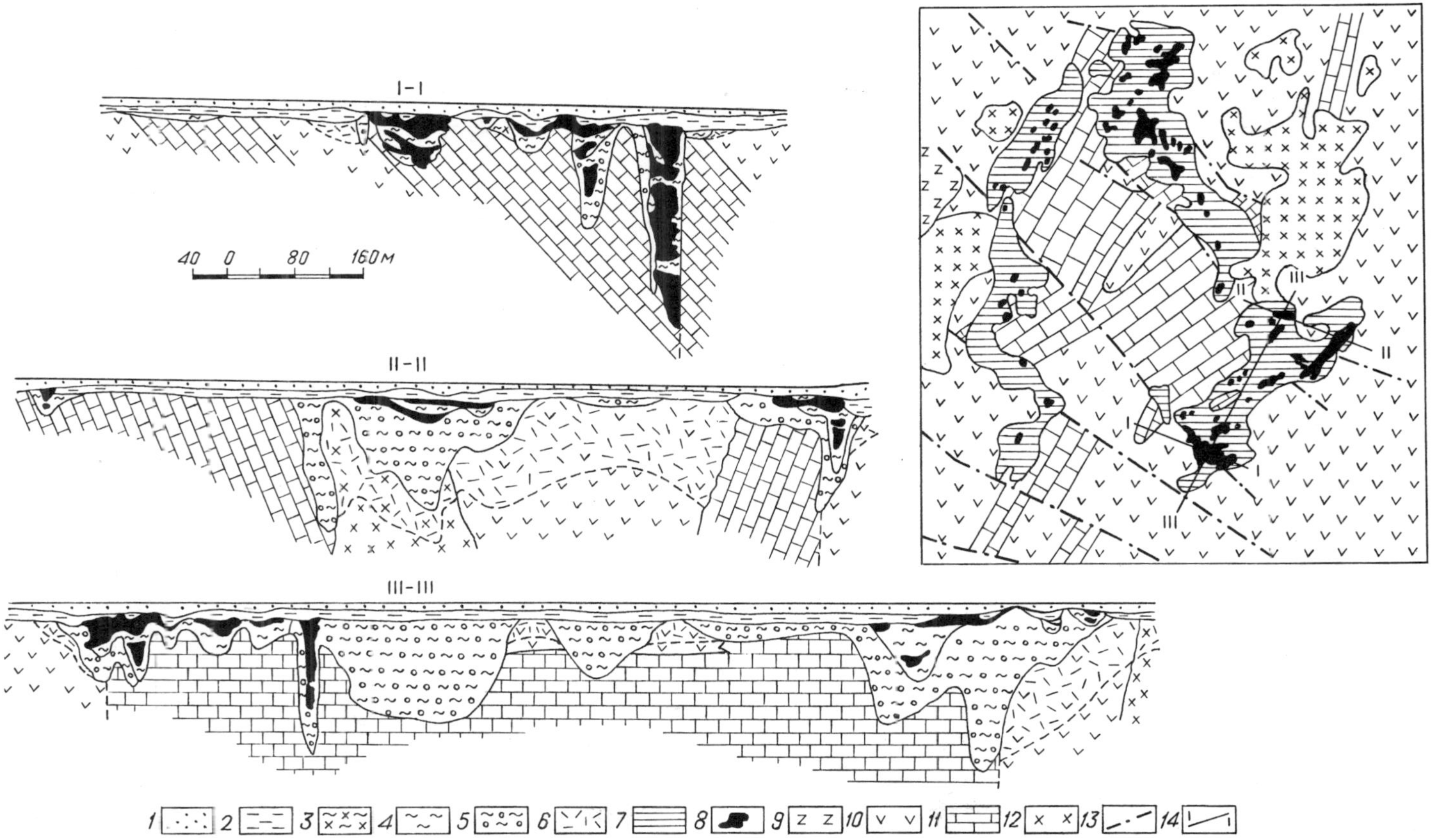

2–122. Geological map and cross-sections of the Belinsk bauxite deposit (after ZHUKOV, 1974). 1. Quaternary deposits; 2. Lower Paleogene, marine clays; 3. Eocene opoka clay and sands; 4. Upper Cretaceous clays of bauxite-bearing strata; 5. clays underlying bauxite horizons; 6. Mesozoic weathering crust; 7. bauxite-bearing deposits on the map; 8. bauxite bodies; 9. Triassic basalts; 10. Lower Carboniferous tuffs and porphyrites; 11. Visean limestones; 12. Lower Carboniferous diorites; 13. faults; 14. lines of sections.

containing calcareous sands are buried by Upper Cretaceous marine deposits (KUCHERUK, 1976). In the peri-Azov section of the Platform, where the Ukrainian Shield subsided under the Carboniferous deposits of the Dnieper-Donetz aulacogen, the Pokrovo-Kireyevo bauxite deposits occur. The productive strata are Lower Cretaceous fills of dolines in limestones. There are also cavities filled with clay. The base of the karst deposits is kaolin overlain by bauxites alternating with allites. The thickness of bauxite lenses varies from 1 to 75 m (BASS et al., 1971).

On the **Siberian Platform** the peri-Baikalian and peri-Yenisei monoclines contain bauxites at Korsakov, Punsk and Endinsk. Small karst depressions are partially preserved in Lower Cambrian limestones and dolostones. They are filled with grey clays containing bauxite lenses whose thicknesses are to 40 m.

There are small dolines filled with gold-bearing terrigeneous sediments in the Bolshaya and Malaya Valukhta Basins in the Patoma Highland (SHOFMAN, 1973). Their depth is 40–60 m and four horizons dating from Upper Cretaceous to Neogene are distinguished within their filling.

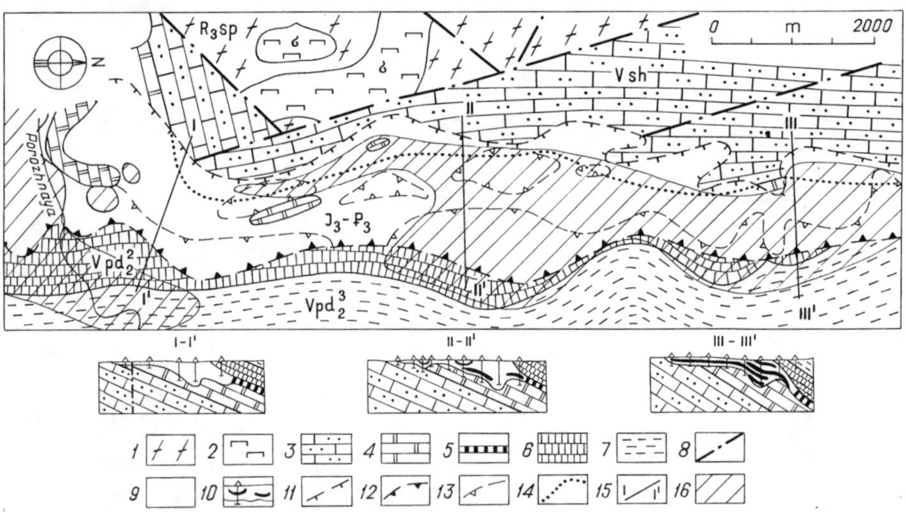

2–123. Geological maps and cross-sections of Porozhinsk manganese deposit, Yenisei Range. 1. Upper Riphean metamorphic schists of the Suchopitsk Group, 2. ultrabasics, 3. Vendian sandy limestones of the Suchorechensk Formation, 4. Vendian dolostones of the lower member of Pod'emsk Formation, 5. the first patch of the upper member of the Pod'emsk Formation (tuffs and tuffites with manganese ore lenses), 6. silicates of the second patch of the upper member of Pod'emsk Formation, 7. mudstones of the third patch, same member, 8. faults; 9. Mesozoic-Cenozoic deposits of the karst depression (on the map), 10. the same, in cross-section (black — lenses of manganese ore), 11. rearward shallow flank of karst depression, 12. frontal steep flank passing under original rock flank of karst depression, 13. deep (more than 100 m) parts of the depression, 14. base of the Pod'emsk Formation under Mesozoic-Cenozoic deposits, 15. lines of section, 16. area of manganese ores.

In the **Turgay Basin** there are karst bauxite deposits in two regions. Most occur in the west (Ayatsk, Batalinsk, Belinsk, Karasorsk, Krasnooktyabrsk, Temirsk etc.). In Central Turgay there are the Zapadno-Ubagansk deposits, (KIRPAL, 1976). Karst depressions range from 40–60 to 150–200 m in depth (Fig. 2–122), with clays and loams with clastic structure usually at the base. Grey clays with bauxite lenses to 50 m thick overlie them. The fill is completed by lignitic clays. There are two rhythmic sequences (Aptian-Albian and Cenomanian-Turonian) in some deposits; in others just the lower or upper part of sequence is present.

In the Baikalides of the **Yenisei Range** we recognize a Late Jurassic-Cretaceous phase in the Vorogovo-Tissk synclinorium in the Porozhinsk ore region. The covered karst of this region is distinctive because its morphological peculiarities and the content of its deposits depend on the direction of dip of the Vendian dolostones (TSYKIN and KOSTENENKO, 1984). The deepest parts of the depressions occur along the contact between the dolostones and the overlying tufaceous sedimentary rocks (Fig. 2–123). The depressions are filled with proluvial, alluvial and lake deposits totalling 350 m. In the 200 m of rhythmically bedded lake deposits, spore-pollen complexes of Late Jurassic, Neocomian and Senonian age recognized.

In the dolostones and underlying sandy limestones, a system of valleys and dolines developed with depths to 230 m. They then amalgamated into one great depression which is traceable for 15 km, and wide of 300 to 1 200 m. The principal concentrations of carbonate manganese occur in the top of the dolostone strata (0.5–2.5 % Mn) and in overlying tuffites, tufaceous mudstones and sandstones (to 12.5 % Mn). With weathering and karstification, the residual manganese was concentrated and partially dissolved; it migrated and accumulated in the karst deposits at alkaline and alkaline-oxidizing geochemical barriers. About 50 lenses of carbonate manganese and oxide manganese ores were discovered by drilling. They are detrital-clayey material saturated with manganite, psilomelane, todorokite, vernadite, rhodochrosite, goethite with ore concretions and veinlets. The structure and distribution of these karst ores indicate their genesis by infiltration.

In the peri-Angara part of the Yenisei Range, an Upper Cretaceous karstification phase was discovered in the Verkhoturovsk bauxite deposits. There are two ore horizons in it, Senonian and Eocene. The first occurs in the lower part of the deposits which fill karst depressions in Riphean limestones, dolostones and magnesites. They are particularly located along the contact with the shales. Clarification and argillitization characterise the bauxites. There are also deposits of kaolin in the Senonian productive horizons.

In the northern part of **Kuznetskiy Alatau** the bauxites of Barzassk region have an Aptian-Albian age. Lower Cambrian limestones and dolostones form large blocks among effusive rocks and shales, and certain sinkholes and poljes occur

2–124. Sinkhole-shaped depression filled with horizontally bedded Neocomian sand, clay and marl deposits in Mazulsk limestone quarry. Width of depression about 25 m. Illustrated part from about −10 m to −20 m (Photo by R. A. Tsykin).

2–125. Tower-shaped outlier of limestone exhumed from clay deposits in the first stripping levels of Mazulsk Quarry. General height of the outlier is about 30 m (Photo by R. A. Tsykin).

filled with sand and clay deposits with lenses of ferruginous bauxite, brown iron ore and kaolin. The thickness of productive strata is as much as 100 m.

We have studied Cretaceous covered karst deposits of flux limestone at Mazulsk in the **Arga Range** (TSYKIN and TSYKINA, 1978). In the massive Vendian limestones a polje-shaped depression of about 30 m was formed, in the floor of which there are sinkhole subsidences (Fig. 2–124) and pinnacles protruding through the fill (Fig. 2–125). Karstic deposits consist of variegated clays with inclusions of metasomatic quartzites and bauxite pockets.

In the **Ural** fold belt, many of the deposits of kaolin and bauxite are associated with the Cretaceous phase of karstification. The morphology of the bauxite-bearing covered karst was studied in Alapaevsk, Rezhsk, Sukhoy Log mining regions and in the Mugodzharsky uplift. In the Paleozoic limestones and dolostones and along their contacts with aluminosilicate rocks small valley-shaped and sinkhole-shaped depressions were discovered. Their deposits are dated as Aptian-Albian (Sinara Formation). They consist of clays, loams and sands, with lenses of bauxite and limonite in the clays. There are some overlapping deposits of kaolin (e.g. at Troitsk-Bainovsk).

In the **East Trans-Baikalian** fold belt there are hydrothermal karst forms and deposits. They accompany skarn-polymetallic ore mineralization of an Early Cretaceous metallogenic epoch. At the deposits of Blagodatsk, Kadainsk, Klichkinsk and Mikhailovsk (RZHECHITSKY, 1966) in Upper Ordovician-Lower Silurian dolostones, cavities of 40–100 m^3 filled with groundwater were found, together with cave breccias and ores deposited in the karst. On the walls of cavities penetrated by adits there are crusts of fluorite, zeolite and calcite. Hydrothermal karst was explored to -360 m and is believed to extend deeper.

Paleogene Paleokarst

The Late Cretaceous karstification continued into the Paleogene in some regions. In addition, new karst phases began. They have been investigated in many platforms and orogenic belts.

There is great interest in a chalk paleokarst discovered by DUBYANSKY (1937) on the **Russian Platform**. It is a characteristic feature of the Belorussian, Voronezh and Ulyanovsk-Saratov regions. As was shown by later research (RODIONOV, 1963), during the Danian and Paleocene the sea withdrew and sinkholes, valleys and open fissures developed. Surface features were filled with sands and loams, and underground features with green clay.

In the **Lvov Basin**, karst developed throughout the Paleogene. There are systems of valleys, caves and karrens under Tortonian beds (KUCHERUK, 1976).

In the **Baltic Shield** there is a supposed Cretaceous-Paleogene karst in the Lake Imandra area. Middle Proterozoic limestones there contain caves filled with sand and clay deposits. They are overlain by a Quaternary moraine (AFANASYEV, 1977). In the Kovdorsk pluton filled cavities and covered sinkholes occur in carbonatites whose depth is about 150 m. They are filled with fragments of alkaline ultrabasic rocks and magnetite ore, cemented by staffelite.

In the eastern part of the **Uralian foredeep,** many subsidence basins developed in the Paleogene as a result of the marine regression and associated activation of salt tectonics. They are 1–4 km long and 0.5–1 km wide, and depths to the top of the remaining sulphate and halite deposits (Kungurian) reach 150 m (GARYAINOV, 1980).

On the **Siberian Platform** there is Paleogene karst in three regions. In the Irkutsk Amphitheatre on the watersheds of Irkut-Olkha, Shinikhta-Olkha, Belaya-Malaya Eret', covered relics of sinkholes in Lower Cambrian limestones and dolostones are filled with variegated clays and pigment minerals. At Bereleisk in dolostones, two 10 m deep sinkholes display grus at the base of their fillings, and contain pockets of white stony brecciform bauxites. They are overlain by variegated clays.

In the **Chadobets uplift** the Ibdzhibdek bauxite deposit (Paleocene-Eocene) is in a polje-shaped depression with a floor complicated by small rises and dolines (cf. Fig. 3–35). It occurs in Lower Cambrian dolostones, has a length of 6 km, a width of 1.0–1.5 km and is about 160 m deep. Bauxites occur as lenses in variegated clays.

In the western part of **Anabar-Olenek anteclize** (Maimecha-Kotui region) in Lower Cambrian dolostones, sinkhole-like forms with a depth of 30 m are preserved. They are filled with bauxitic clays, which contain Eocene spore-pollen spectrum (CHEKHA, 1982), or homogenous friable bauxites (Alysardakh occurrence). There are also subterranean interbedded bodies of allites and bauxites at Mooustakhsk.

In the eastern part of the **Aldan Shield** (Uchur-Maya region) a strata of Cambrian carbonates contains the large karst depression of Mar-Kuel. In its floor, sinkhole-shaped depressions filled with Oligocene sand and clay deposits were found, suggesting a long-duration, multi-phase history for Mar-Kuel (KARAVANOV, 1973).

In the eastern part of the **Turanian Platform** (Tadzhik depression) and also in the Fergana intermontane basin, in Paleogene carbonates (Turkestan, Alay, Suzak Members) permeable horizons are persistent along strike and confine the oil and gas deposits of many multistoried fields such as Auval, Zapadny Palvantash, Kichikbel, Kokaitu, Selkhoro, Khodzhiabad, Changyrtash etc. These reservoir beds are fissure-karsts created during brief emergent interruptions in the sedimentation (MAKSIMOVICH and BYKOV, 1978).

Paleogene bauxite-bearing karst occurs in the peri-Angara part of the **Yenisei Range.** Depressions there have developed in different karst rocks — crystalline limestones (Tatarsk bauxite deposits), dolostones and magnesites (Kirgiteysk and Verkhoturovsk) and limestones and dolostones (Porozhninsk and Sukhoe). Bauxite-bearing depressions vary from 40–60 m to 7 500 m in length and are about 120–150 m deep, except at Tatarsk where depths reach 240 m. The deposits are the result of three karstification phases (Danian-Paleocene, Eocene, Oligo-Miocene).

The Algashtyksk bauxite deposits of the Sesimsk synclinorium of **Eastern Sayan** are Eocene. They are contained in kettle-shaped depressions up to 150 m deep, developed in Vendian dolostones (TSYKIN and TSYKINA, 1978). The Teleksk karst phosphorites whose Paleogene age is presumed by analogy are located nearby. The polje-shaped depressions are complicated by lesser depressions and filled with phosphatic clays, friable and stony phosphorites that are mostly infiltration-metasomatic breccias.

In the southern part of the **Kuznetskiy Alatau** (Mrassk central massif) a Paleogene age is supposed for karst phosphorites in the Gornoshorsk Basin (ZANIN, 1969). Some depressions have been prospected, the greatest of which

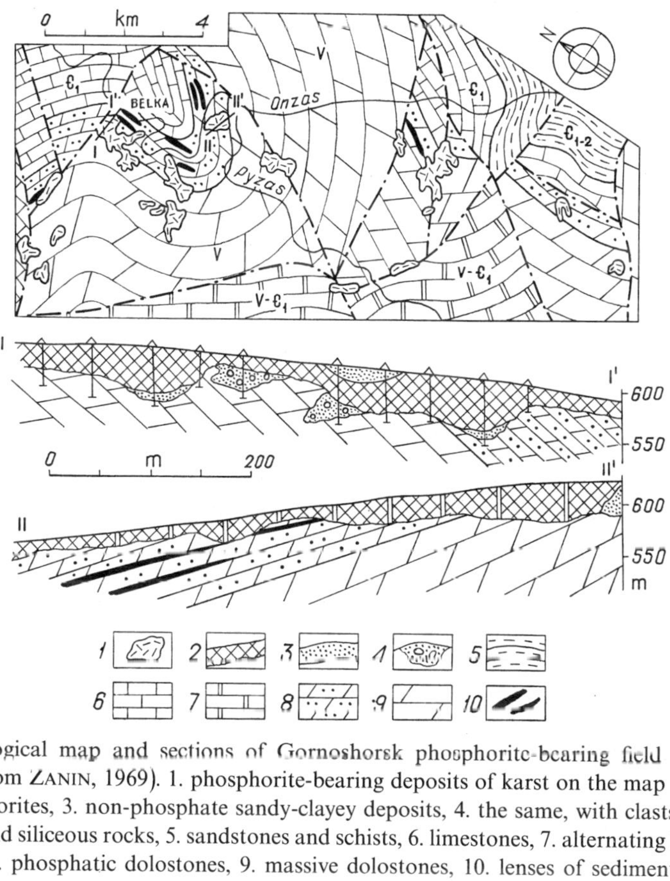

2–126. Geological map and sections of Gornoshorsk phosphorite-bearing field (modified and completed from ZANIN, 1969). 1. phosphorite-bearing deposits of karst on the map (Paleogene), 2. karst phosphorites, 3. non-phosphate sandy-clayey deposits, 4. the same, with clasts and debris of dolostones and siliceous rocks, 5. sandstones and schists, 6. limestones, 7. alternating limestones and dolostones, 8. phosphatic dolostones, 9. massive dolostones, 10. lenses of sedimentary-diagenetic phosphorites.

is the Belka deposit (Fig. 2–126). Its phosporites are situated above phosphorite-bearing cherty dolostones of Upper Riphean-Lower Cambrian age that contain polje forms extending to 1 500 m with depths to 80 m. The Obladzhansk deposit in the Batenevsk Massif is similar. Rich phosphorites fill small (cross-section to 400 m) but deep (to 150 m) depressions.

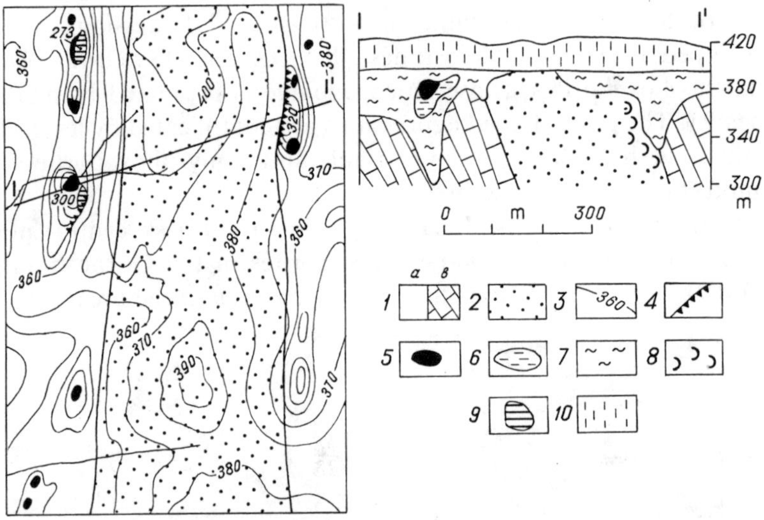

2–127. Geological plan and cross-section of the karst area of Salair Ridge (after A. M. MALOLETKO, unpubl.). 1. limestones (a. on plan, b. in section), 2. schists, 3. contour lines on the surface of the original rocks, 4. subvertical walls of karst depressions, 5. bauxites, 6. kaolins, 7. variegated clays, 8. weathering crust, 9. Paleogene carbonaceous clays, 10. Pleistocene loams.

Small covered karst depressions are preserved in the **Salair Ridge**. They are located in Paleozoic limestones and along their contacts with shales and effusive rocks (Fig. 2–127). They are filled with variegated clays containing lenses of bauxite and kaolin. A Paleogene age was determined for the Aprelsk, Vaganovsk and Samzhnevsk deposits (GERASIMOV and SIDORENKO, 1974).

In **Central Kazakhstan** within Paleozoic limestones and especially along their contact with shales, there are chains of depressions containing bauxite deposits (Amalgendy, Tselinograd and Ekibastuz ore regions), whose bases contain sand and clays. They are overlain by Paleocene clays with bauxite lenses, above which are Oligocene sands and clays at some sites.

In the northern spur of the **Alai Range** of the Tien Shan the derelict mine, Tyuya–Muyun, and nearby Baritovaya Cave are of karst interest. They occur in a narrow ridge of Carboniferous limestones and are associated with hydrothermal karst of Oligo-Miocene age (FERSMAN, 1924, 1930). Vertical and horizontal mineralized cylindrical forms extend to depths of 220 m (cf. Fig 3–51) and concentric sheets of calcite, baryte and red clays were deposited in sequence in them (cf. Fig. 3–52). At their centres an open cavity is usually preserved. In Baritovaya Cave the walls were covered with crystals of baryte, which were mined [Editors' note: the Tyuya-Muyun site is well known thanks to interesting mineralization of copper and uranium vanadates most probably of infiltration origin. The main vanadate minerals are: tyuyamunite, turanite, tangeite, some ferganite and vanadinite (CHIRVINSKY, 1925; ALEKSANDROV, 1927; FERSMAN 1930; SMOLYANINOVA, 1970). The temperature of calcite precipitation was determined, by fluid inclusion studies, to be 40–65 °C. Baryte was precipitated from cold water (Y.V. DUBLYANSKY, pers. comm., 1985). The leaching of heavy metals (Cu, U, V) from Silurian and Devonian shales and slates is connected with Jurassic or Tertiary intense weathering (BOSÁK, 1986)].

In **Primorye** (Pribrezhny anticlinorium of Sikhote-Alin) the formation of skarn-polymetallic deposits in the Dalnegorsk ore zone is attributed to the Paleogene metallogenic epoch. Hydrothermal karst cavities of several cubic metres capacity were opened in Triassic limestones. The Verkhnee deposit is interesting because of the various forms of calcite druse covering the cavity walls.

The Kvaysinsk lead-zinc deposit in the Gagra-Dzhavsk zone of the **Great Caucasus** developed in a Middle Paleogene hydrothermal karst phase. Cavities filled with ore and breccia occur in Upper Jurassic limestones. According to ZHABIN, SAMSONOVA and ASTAKHOV (1984) there were three phases of karstification — Late Jurassic (small cavities below an erosion surface and filled with Lower Cretaceous clays), Eocene (hydrothermal karst) and Pliocene-Pleistocene (development of the well known Kudarskie Caves).

Neogene Paleokarst

Neogene karst is everywhere. Elements of the modern karst relief began to appear and most large cave systems known in the U.S.S.R. were created. Also many relict caves are known; they are openings mostly in the higher parts of valleys and in karst outliers.

Paleokarst filled with Neogene deposits is also widespread. There appear to have been distinct Miocene and Pliocene karstification phases according to available data.

On the **Russian Platform** there is mainly Pliocene karst in many regions (KUCHERUK, 1976; RODIONOV, 1963; CHIKISHEV, 1978). In some areas of the Baltic and Moscow syneclizes and Voronezh and Volga-Ural anteclizes, sinkholes and cavities filled with sandy-clayey deposits are encountered in boreholes and quarries. Their age varies from Pliocene to Middle Pleistocene.

On the Siberian Platform, fragments of sedimentary cover are preserved in the northern part of the **Aldan Shield**. There are horizontally bedded Lower Cambrian dolostones and limestones, which are sometimes overlain disconformably by Jurassic conglomerates and sandstones. A peculiar karst was formed here, including barren gold-bearing deposits. Its age is considered to be Neogene.

The primary gold concentrations, which are contained in eluvial-karstic placers, occur in Late Jurassic feldspathic and quartzose metasomatites and

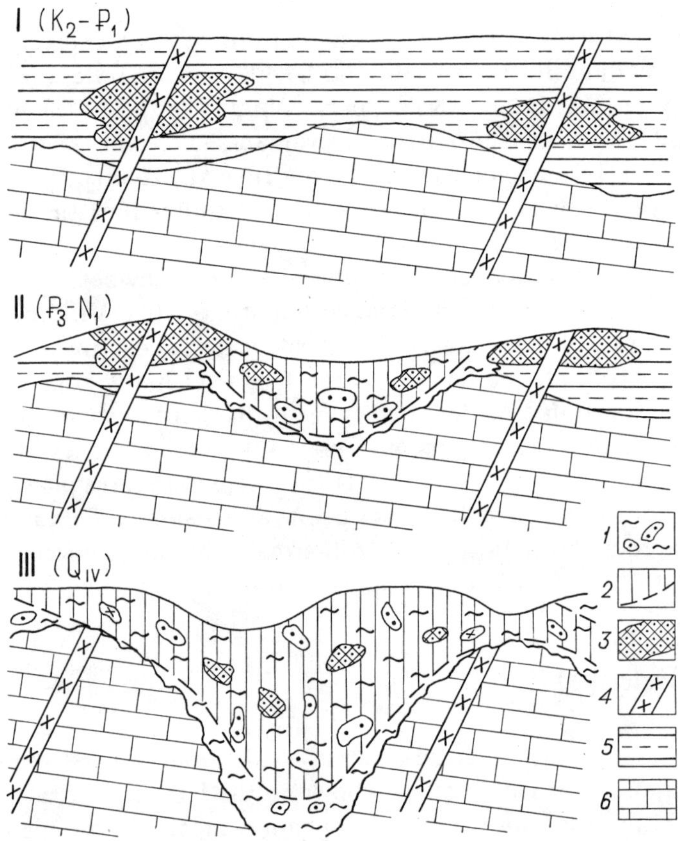

2–128. Scheme of formation of the eluvial gold placers in the Aldan Shield (I–III. stages of hypergenesis). 1. sandy-clayey deposits with clasts, 2. contours of placers, 3. gold-bearing quartzose metasomatites, 4. porphyry dykes, 5. Jurassic sandstones and conglomerates, 6. Cambrian bituminous limestones.

porphyry dykes (BORODAEVSKAYA and ROZHKOV, 1978). As a result of the permeability of the coarse-grained, fractured Jurassic rocks and the favourable geomorphological situation of a low plateau, conditions for covered karst development in the underlying dolostones and limestones were created. Solution took place in the phreatic zone, causing destruction and subsidence of conglomerate, sandstone and metasomatite blocks. These gradually weathered and altered into siliceous-clayey masses. The karstification front descended, while the weathering front moved up through the blocks. As a result, strata of weathering products of the gold-bearing rocks accumulated in karst depressions (Fig. 2–128). The gold grains in the sandy-clayey material are very small and friable, with traces of chemical regeneration of the peripheral parts. Due to erosion of the eluvial-karst placers some modern eluvial gold placers of karst type have formed in this region (ZHELNIN, 1977).

The Krasnozavodsk kaolin deposit in the basin of the **Khatymi River** is similarly situated. It is in a kettle-shaped depression 1 500 m long, 400–600 m wide and about 50 m deep. This is filled with clays, some homogeneous and others containing boulders of sandstones and dolostone debris. Pliocene variegated clay and Quaternary talus cover the clays.

In the **Patoma Highlands** and **Altay-Sayan Mountains** karst gold placers that formed in the Pliocene-Early Pleistocene have been prospected. For example, in the Sredne-Vitim gold region a buried placer rests on Riphean metamorphic limestones. The thickness of the placer alluvium varies considerably, because the bedrock floor contains many irregular depressions (KAZAKEVICH, 1972).

In the **Yenisei Range** partly buried karst gold placers were formed on Middle Proterozoic crystalline limestones (Centralny anticlinorium) and Cambrian dolostones (Kordo-Lebyazhinsk synclinorium), located at headwaters and under

2–129. Strongly dissected surface of grey, bedded crystalline limestone of the Lower Cambrian exposed from beneath a Neogene sedimentary cover as the result of the exploitation of a karst gold placer near Chibizhek. The height of some outliers reaches 6 m (Photo by R. A. Tsykin).

valley floors (FELDBARG and RYBAKOVA, 1974). Red sand and clay deposits with fragments of quartz and shales are gold-bearing. The placers occur on carbonate rocks or on silts which fills dolines.

In **Eastern Sayan** gold placers have been worked in the basin of the Jeb and Chibizhek and the headwaters of the Sisim Rivers. The bedrock floor that is exposed is very rough, with numerous outliers of limestone (Fig. 2–129). Gold-bearing Miocene red loams contain blocks and pebbles of different rocks (TSYKIN and TSYKINA, 1978).

In the **Kuznetskiy Alatau** Neogene karst gold placers were worked on the Bol'shaya Siya, Severny Kozhukh, Taidon, Federovka Rivers etc. These deposits also varied rapidly in thickness (KAZAKEVICH, 1972).

At the **Ural** iron ore deposits, gold and diamond placers are associated with Pliocene karstification. In the Akkermanovsk iron ore deposit, the polje-shaped depression is filled with weathering products of ultrabasic rocks, plus brown iron ores in the upper horizons (GARYAINOV, VASIL'EVA and AGANIN, 1978).

On the western slope of the Ural a series of contact depressions of great length occurs. The depressions are filled with pebble-sandy-clayey deposits of different genetic types. The thickness of accumulated alluvium of Pliocene-Lower Pleistocene age is 30–50 m (STEPANOV, 1967).

The Practical Value of Paleokarst Studies

Many kinds of ore and non-ore mineral resources are associated with paleokarst. Karst aquifers may play an important role in some oil and gas fields and may yield fresh and mineral water as well. These are all beneficial features of paleokarst. It may have deleterious effects, such as the flooding of mines in carbonate strata. In deposits of flux limestone and dolostone and also in gypsum, sand and clay inclusions are undesirable. However, they are almost everywhere, constituting between 1.5–2 % and 10–15 % of the mined masses.

In the U.S.S.R. *karst bauxites* are the principal bauxites. Soviet geologists distinguish two groups: geosynclinal, the main representatives of which are the Devonian deposits of the North Ural; and platform bauxites which include the many deposits of the Timan uplift, the Turgay depression and other regions (BUSHINSKY, 1975).

Geosynclinal deposits are characterized by stratiform relationships and the persistent quality of the bauxite. The distribution of beds is determined by block-folding structure of the ore-bearing region. Platform deposits are subdivided by morphological features. The bauxites are in multiple lenses within karst deposits and bauxite quality is very variable. Lenses are often surrounded by veneers of allites and red clays. There may be several modes of genesis in one deposit, but probably the chief mode is laterite-lite formation connected with desilicification of clays.

In the Ural there are hypergene deposits of silicate *nickel ores* (cf. GINZBURG, 1966). At some, a type of karst can be distinguished which contains from 30 to 70 % of the ore reserves (Lipovsk, Rogozhinsk, Sinara, Cheremshanka, Sheleinsk). In such ores the nickel content is 1.2–2 times higher than in the eluvium of ultrabasic rocks. Nickel minerals in karst formations are infiltrates.

Where there are hydrothermal *lead-zinc, antimony,* and *mercury* deposits in carbonate rock there is usually associated karst. There are pre-, syn-, and post-ore forms of hydrothermal karst. In the pre-ore forms the mineral was precipitated into voids and extends in zones from the walls to the centre (or from the bottom to the roof) of the cavity (e.g. the deposits at Kvaysa, Kadainsk, Tyuya-Muyun). Collapse breccias with cementation of the clasts by sulphides are characteristic of syn-ore forms. Post-ore karst is preserved as voids; however, the walls are often covered with crystal coatings. When deep-seated hydrothermal karst cavities are breached there may be inrushes of water into mine workings.

Paleokarst depressions often contain deposits of brown *iron ores* and siderite. Karst iron ores are important for the metallurgical industry of the Ural (Akkermanovsk, Alapaevsk deposits and others). Their genesis is disputed. Some geologists consider them to be the product of mechanical redeposition of iron accumulated in weathering crusts of basic or ultrabasic intrusive rocks. Others suppose the siderite to be deposited in karst lakes, and then oxidized and partially redeposited. A third opinion (supported by the present author) is that there is precipitation of ores from groundwaters at geochemical barriers (for a discussion of origin on the iron ores see BÁRDOSSY , FUCHS and GŁAZEK, p. 419–429 ; Editors' note).

Manganese ore is common in paleokarst in the U.S.S.R. but not in economic accumulations. An exception to this rule is the Porozhinsk deposit in the Yenisei Range, noted above (p. 283). The ores are a detrital-earthy type with an average manganese content of 18.0 %, iron 9.5 %, phosphorus 0.5 %. The deposit is complex, which is a common peculiarity of covered karst, and contains bodies of ceramic clays, pigment minerals, dolostone grus and silicate debris.

Placers of *gold* associated with paleokarst are known in the Aldan Shield, Altay-Sayan Mountains, the Yenisei Range, Patoma Highlands and the Urals. Gold-bearing deposits are developed in riverbeds of small rivers or in dry valleys. They contain a high content of clay.

Phosphorites accumulate in covered karst. Phosphorus is derived from its original source in the form of minerals and partially in solution (ZANIN, 1969) during the weathering process. The ore quality is variable and depends on the intensity of phosphate metasomatose and the composition of gangue minerals. High iron content (in the form of hydroxides) is unfavourable while the presence of dispersed quartz favours ore processing because the quartz is easily separated. P_2O_5 can be dissolved in citrates to a concentration of 20–30 % and this concentration can be increased 3–4 times by mechanical activation. Therefore a phosporite can be used as mineral fertilizer.

Many *clay* deposits are located in paleokarst. They are especially abundant in the Ural and in the Uralian foredeep. About 200 deposits occur in these regions. Great deposits are known also in the Salair Range, Irkutsk Amphitheatre, Turgay depression etc. Kaolinitic and illite-kaolinitic clays are thought to be redeposited products of weathering crusts. However, in situ kaolinitization can contribute to the evolution of the deposits. Kaolin is also a product of the silicification of bauxites or the desilicification of loams and other deposits.

Brown coal deposits occur in many depressions in the Uralian foredeep, peri-Caspian syneclize and Irkutsk Amphitheatre, etc. The thickness of the coal measures reaches up to 100 m and the reserves constitute at least a hundred million tons (Voroshilovsk deposit of the South Ural Basin, Novodmitrievsk of the Ukraine, etc.).

Numerous *oil and gas* deposits are connected with paleokarst reservoirs; 8 % of the oil reserves of the U.S.S.R. are located within paleokarst traps (MAKSIMOVICH and BYKOV, 1978). They are concentrated in Nurolsk depression (West Siberian Platform), Nepsk anteclize (Siberian Platform), Uralian foredeep and Turanian Platform. Mineral waters and salt brines are secondary products of karst reservoirs.

Conclusions

The sedimentary mantle of the Earth's crust developed with interruptions. This is characteristic of both platform and geosynclinal regions. The interruptions were times of development of continental geostructures, with phases of hypergene karstification.

In the U.S.S.R., at least 20 such phases have now been established. There are two phases in the Late Precambrian, eleven in the Paleozoic, four in the Mesozoic and three in Cenozoic. Various types of surface and underground karst developed during each phase. Directly underneath the unconformities are the buried (armoured) forms. Development of naked types of karst was characteristic of the Precambrian and Paleozoic periods. In the Mesozoic and Cenozoic periods, covered and partly subterranean karsts predominate. In the first few hundred metres below the unconformities are cavities filled with clastic deposits and waters. In addition, during the Mesozoic and Cenozoic periods, screened (subjacent) karst developed actively in rock salt and gypsum. It is expressed at the surface by subsidence depressions. Hypergene karst of one or another phase has (as a rule) a regional extent, sometimes with several levels of karst development.

Paleohydrothermal karst is characteristic of folded regions and the basements of ancient and young platforms. Its features are associated with fault zones and impermeable rock barriers. They display localised distribution patterns.

Paleokarst deposits are polygenetic. They are formed partly in subaerial environments of sedimentation and partly in subterranean ones. Subaerial deposits are subjected to weathering and may be converted into bauxites and kaolins. Those deposits of Meozoic and Cenozoic covered karst having tremendous thicknesses and a compound distribution of facies are peculiar. They developed by simultaneous sedimentation, chemical transformation, and mechanical migration of material due to the deepening of the karst depressions. The accumulation of bauxites, iron, manganese and phosphate ores and placers of valuable minerals are a characteristic feature of covered karst (TSYKIN, 1980, 1985).

The environment of surface and subjacent karst belongs to a special domain of continental lithogenesis. There is a paragenetic association of continental sedimentary deposits which is peculiar to the karstosphere. However, a part of all paleokarst remains as voids. That is why underground waters and (in deeply buried paleokarst) hydrocarbons are collected.

PALEOKARST OF CHINA

Zhang Shouyue

Introduction

Carbonate rocks are widespread in China. They cover over one fifth of the country and crop out over more than 1 250 000 km^2. They contain some of the most important and complex karst terrains in the world. China is one of the classical karst regions.

Carbonate outcrops range from metamorphosed limestones and dolostones of Archean age to Cenozoic reef limestones. Karstification of these formations produced a variety of surface and subsurface features. Some were created during several different phases of karstification from Sinian to Quaternary. Numerous paleokarst localities are known. The Cenozoic Era is the most important period of karst development in China.

Methods of Investigation

Geologic methods are used successfully to investigate karst hydrogeology and hydrology as well as the petrology of carbonate rocks and karst development. Studies of the paleohydrologic and sedimentologic conditions are needed in carbonate rocks. Paleohydrodynamic maps have been drawn for the Permian karst gas reservoir in the south of the Sichuan Basin.

Conventional geophysical methods and drilling are also used, with differing degrees of success. In general, borehole geophysics and drilling are the most successfull and give more information than the other methods.

Studies of sedimentary fillings on planes of disconformity or unconformity and in caves are especially valuable in karst terrain. Speleothems have been dated by several radiometric methods, including ^{14}C, U-series methods and thermoluminescence. Over 50 $^{230}Th/^{234}U$ ages have been obtained for 40 speleothems in 18 cave sites from the eastern mainland of China.

From stable isotope studies of speleothems, the paleoenvironment from several cave systems for last 350 000 years B.P. has been discussed.

P. Bosák, D. C. Ford, J. Głazek and I. Horáček (Editors):
Paleokarst. A Systematic and Regional Review. – Elsevier and Academia. Amsterdam and Praha. 1989.

Historical Review of Paleokarst Research

As early as the time of Han dynasty (206 B.C. — A.D. 220), mammalian fossil bones and teeth in caves were already well-known to the Chinese; they used them as a kind of medicine. However, Chinese geologic sciences did not really begin until 1913. Scientific investigation of paleokarst is very young.

Many caves contain very precious paleontologic, paleoanthropologic and archeologic remains preserved in Pleistocene and Holocene deposits (TEILHARD de CHARDIN et al., 1935). Beginning in 1916, scientists started to collect and excavate mammalian fossils in a number of them. During the 1920s, a Cenozoic Research Laboratory was established for the purpose excavating in the well-known Zhoukoudien (Choukoutien) site and collecting mammalian fossils in other regions of China. We knew that the caves or pits of South China are rich in fossils. They have yielded a mammalian fauna headed by *Ailuropoda* and *Stegodon*. Its age is determined to be Middle Pleistocene, or contemporary to the *Sinanthropus* remains known at Zhoukoudien in North China (PEI, 1965). The discovery of Beijing (Peking) Man, of course, created a stir in academic circles worldwide.

Before 1949, also, some geologists and geographers spread PENCK's and CVIJIĆ's theories for karst development and applied them in discussions of the physiographical stages of karst terrain in the Cenozoic in China.

Since the founding of the People's Republic of China in 1949, scientific research into karst has made rapid progress. A large number of localities of Quaternary mammalian fauna have been discovered (HUANG, 1982). It is now possible for us to understand the origin and evolution of karst and caves. Paleokarst forms have been investigated by many authors and a great number of papers deal with them. The main interest is in the occurrences of mineral resources such as bauxite, karst oil and gas reservoirs and karst water resources. Foundations of dams, leakage of storage reservoirs and engineering hazards in mines are discussed in some papers (ZHANG and YUAN, 1982).

General Description

Carbonate rocks occur in all the main tectonic regions of China (Fig. 2–130). Karst is developed in rocks which differ widely in composition, texture and structure, and which belong to formations ranging from the Archean to the Pleistocene in age. Karstification of these carbonate formations in various climatic zones produced a variety of surface and subsurface features in eastern China.

In western China, the karst regions belong to the Qinghai-Xizang Plateau climatic type, an arid region. The landscapes of carbonate terrains there are the

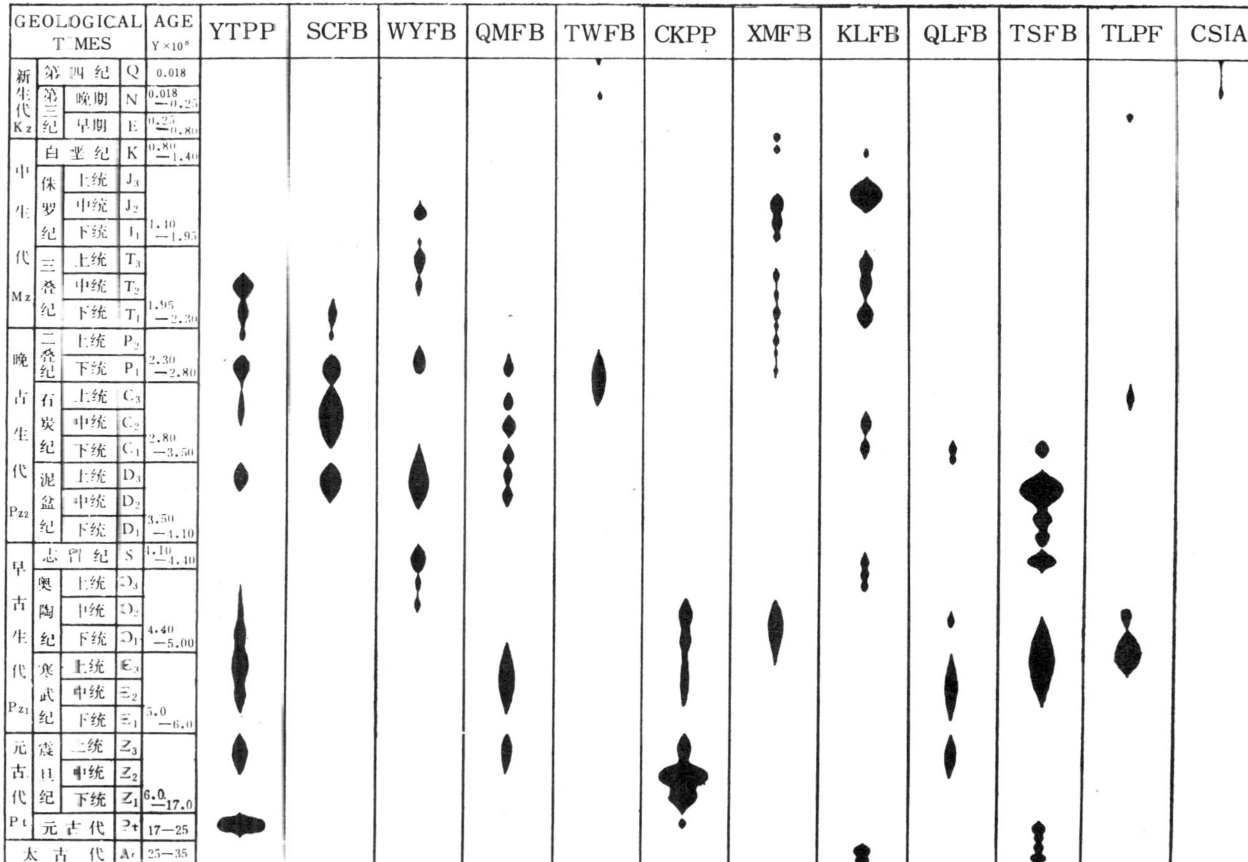

2–130. Ages of carbonate rock formations in various tectonic units in China. Paraplatforms: YTPP = Yangtze; CKPP = China-Korea. Platforms: TLPF = Talimu. Geosynclinal fold belts: SCFB = South China; WYFB = West Yunnan; QMFB = Quinling Mountains; TWFB = Taiwan; XMFB = Ximalaya (Himalaya); KLFB = Tien Shan Mountains. Non-tectonic units: CSIA = coast, shelf, South China islands.

same as in the non-carbonate terrains. Frost and arid action are the dominant exogenic processes and corrosion is weak. This paper, therefore, is chiefly concerned with the situation in eastern China.

Geologic Background

The crust of China has experienced multiple tectonic movements, each of which has its own characteristics. The tectonic movements closely controlled the development and evolution of the carbonate formations.

The eastern part of China comprises mainly the Yangtze paraplatform, the geosynclinal fold belt of South China and the China-Korea paraplatform.

The **China-Korea paraplatform** is triangular in shape. It includes all of North China and the southern part of Northeast China. The thickness of carbonate rocks is generally 1 000–2 000 m, ranging from Sinian (Precambrian 600–1 700 Ma) to Ordovician in age. However, in specific areas, such as the Yanshan Mountains near Beijing, carbonates may be as much as 6 000–7 000 m thick. This paraplatform is unconformably overlain by well-developed Cambrian to Middle Ordovician carbonate rocks, Middle Carboniferous to Lower Permian coal-bearing strata of littoral and terrestrial facies, and Upper Permian to Middle Triassic red beds. The absence of the Upper Ordovician to Lower Carboniferous is characteritic. Jurassic and Cretaceous times were marked by intrusions and extrusions of intermediate to acid igneous rocks and by faulting and folding of the cover strata. This crustal deformation is called the Yanshanian Orogeny. In the Mesozoic and Cenozoic, block-like vertical differential movement led to the formation of large-scale basins and graben where oil- and gas-bearing sediments of mainly continental facies accumulated.

The **Yangtze paraplatform** covers nearly the whole territory of the Yangtze River Basin. The sedimentary cover on the basement of the Yangtze paraplatform is rather thick. It began with molasse, tillite and carbonate of latest Proterozoic age (Sinian System). The carbonate rocks are several thousands meters thick and range in age from Sinian to Triassic. The Paleozoic-Mesozoic cover consists mainly of marine carbonates, clastics, and continental basalt flows. The carbonate facies belts in the major sedimentary environments (deep continental shelf and basin, carbonate platform margin and carbonate platform) are all generally well-developed. During Late Mesozic times, the eastern part of the paraplatform was the site of large-scale magmatism. The Cenozoic strata are red beds and oil- and gas-bearing sedimentary formations. The Yanshanian Orogeny created the latest tectonic framework and morphological features.

All of the Upper Proterozoic and Lower Paleozoic strata were folded at the end of the Early Paleozoic and unconformably overlain by terrestrial Lower Devonian clastics in the geosynclinal fold belt of South China. This Late Caledonian fold belt is a part of the Marginal-Pacific geosynclinal region. After the Caledonian Orogeny the belt was transformed into platform and combined with the Yangtze paraplatform.

In the **geosynclinal fold belt of South China**, the thickness of Upper Paleozoic carbonate rocks is 2 000–5 000 m. Middle and Upper Devonian and Lower Carboniferous stata are widespread and thicker than on the Yangtze paraplatform.

Indosinian movements took place during the Triassic and the Devonian to Triassic cover was folded. From the Late Precambrian to the Late Paleozoic (Variscan Orogeny) or Triassic (Indosinian Orogeny), the platform area was subjected to oscillatory movements. During the Indosinian and Yanshanian periods, the modern tectonic framework and morphological features were developed in the eastern part of China. Since the Himalayan (Ximalaya) movements in the Cenozoic there has been strong, large-scale uplift and block-like vertical differential movements associated with fault blocks. Pre-existing landforms of the Mesozoic were mostly disturbed or reformed.

A comparison of the carbonates deposited during the same periods in different tectonic units will now be made for Eastern China (ZHANG, 1984). In the Sinian carbonates, dolostones and impurities

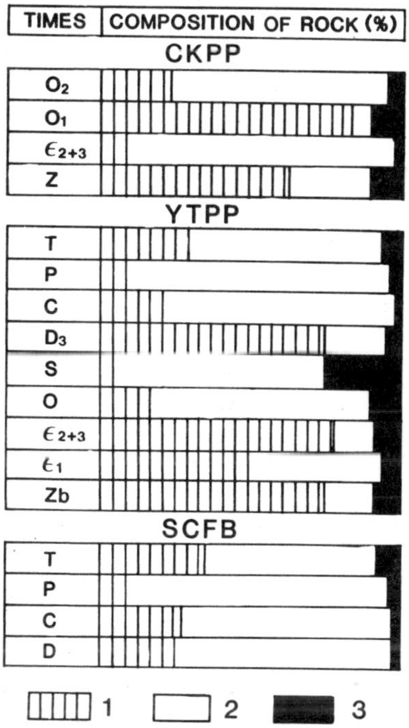

2–131. Generalized composition of carbonate rock formations in Eastern China. 1. dolomite; 2. calcite; 3. impurities. For abbreviations see Fig. 2–130.

are relatively more abundant than in other formations. In the Cambrian System of the Yangtze paraplatform, dolostones are dominant and impurities are higher, but on the China-Korea paraplatforms the opposite is true. During the Ordovician, the composition of carbonate rocks on the Yangtze and China-Korea paraplatforms was similar. Since the Devonian period, the composition of carbonate rocks on the Yangtze paraplatform and in the geosynclinal fold belt of South China have been similar. In both areas, calcite is dominant and impurities are low (Fig. 2–131).

Karst Zonation in China

In principle, karst zones vary depending on which characteristics are emphasized. At a first order of karst zonation, China may be divided into four karst regions based on climate. Each region is a climatic zone with a distinctive water balance (ZHANG, 1980). At the second order, karst areas are divided by tectonic units. A given karst area displays similar geologic conditions, contains carbonate rocks of similar age, has a similar geologic history and characteristics of crustal movement. That is, there are broad similarities in how the karst is formed and how the karst and pore space developed. The third order of karst zonation

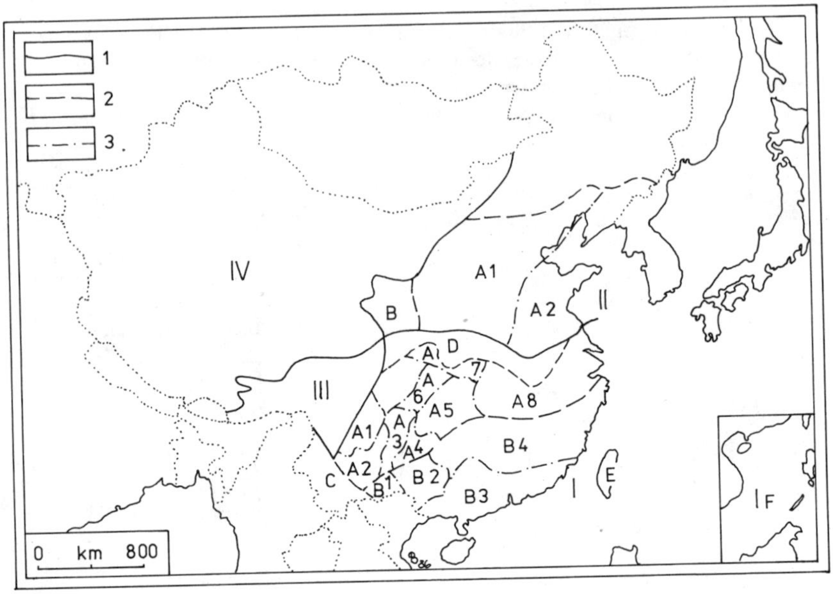

2–132. Karst zonation of China. 1. boundaries of karst regions; 2. boundaries of karst areas; 3. boundaries of karst subareas. I. *Corrosion and erosion-corrosion region of tropical and semi-tropical humid climatic type:* IA. Yangtze paraplatform (karst in Proterozoic to Mesozoic carbonates): IA1. Southwest Sichuan (canyon-mountain karst); IA2. East Yunnan (plain-hilly plateau karst); IA3. West Guizhou (uvala-quifung[1] mounatin plateau karst); IA4. Central Guizhou (plain – quifung and funglin[2] mountain plateau karst); IA5. Guizhou and Hubei (uvala-quifung mountain karst); IA6. East Sichuan (uvala-quifung mountain karst); IA7. Hubei and Sichuan (uvala-quifung mountain karst); IA8. middle course of Yangtze River (plain-quifung and hilly low mountain and hilly plain karst); IB. Geosynclinal fold belt of South China (karst in Upper Paleozoic and Mesozoic carbonates): IB1. Southeast Yunnan (plain-funglin plateau karst); IB2. Guizhou and Guangxi (uvala-funglin mountain karst); IB3. Guangxi and Guangdong (plain-funglin plain karst); IB4. Hunan nad Jiangxi (polje-quifung mountain and hilly karst); IC. Geosynclinal fold belt of West Yunnan (karst in Paleozoic carbonates); ID. Geosynclinal fold belt of Qinling Mountains (karst in Upper Paleozoic carbonates); IE. Geosynclinal fold belt of Taiwan (karst in Upper Paleozoic metamorphic carbonates); IF. Coast, shelf and islands (karst in Cenozoic coral reef limestones). II. *Corrosion-erosion region of mid- and warm-temperature zone, semi-arid and sub-humid climatic type:* IIA. China-Korea paraplatform (karst in Archean to Ordovician carbonate rocks): IIA1. Shanxi, Hebei and Liaoning (arroyo-mountain karst); IIA2. Shandong and Liaoning (arroyo-mountain and hilly karst); IIB Geosynclinal fold belt of Qilian Mountains (karst in Proterozoic and Paleozoic metamorphic carbonates). III. *Corrosion-denudation region of humid climatic type of the Qinghai-Xizang plateau.* IV. *Denudation region of arid climatic type of the Qinghai-Xizang plateau and temperate zone.*

[1] Quifung: A type of cone karst that is characterized by gently sloping, hemispherical limestone hills, the diameter being several times the height. Quifung is a transliteration from Chinese.

[2] Funglin: Tower karst and cone karst. A transliteration from Chinese.

produces karst subareas. The results of various agencies produce different types of karst landscapes which are related to phases of cave development and to hydrogeological characteristic of the phreatic zone. Geologic structure is considered a defining characteristics in a few karst areas.

Karst in China may be divided into four regions, eight areas, and fourteen subareas (Fig. 2–132). This zonation provides a scientific basis for evaluation, modification, and utilization of karst for national economic growth.

Terminology

Types of karst can be classified according to the presence or absence of cover strata, the burial of carbonate rocks, the age of karstification and the relation of the cover to the karstic strata (QUINLAN, 1972). The main types described in Chinese literature may be summarized: (1) *naked* or *exposed karst* — which is locally covered by loose or unconsolidated sediments; (2) *covered karst* — mostly covered by loose or unconsolidated sediments; (3) *buried karst* — which is covered by post-karst rock or sediment and there is normally a disconformity or unconformity; (4) *intrastratal karst* — which is covered by less soluble or non-carbonate strata where any karstification is younger than in those strata (*subjacent karst* is a synonym); (5) *modern karst*[1] refers to karst that developed during Pliocene to Holocene; (6) *recent karst*[1] refers to karst that developed from the Mesozoic to the end of the Miocene; (7) *ancient karst*[1] developed before the Mesozic; (8) *paleokarst* and *fossil karst* developed mainly during past geologic time, but this latter term also refers to any inactive karst forms that are covered by later rock or sediment. Some karst scientists have defined paleokarst in a narrow sense as that developed before Cenozoic time. In a broad sense, paleokarst is that developed before the Quaternary (REN, 1982). We propose to prefix a paleokarst with its time of development based on dating or on lithostratigraphic correlation with other dated profiles.

Time of Karstification

In order to understand the laws of development pertaining to karst in time and space, various criteria of geologic survey and exploration are used. More than half of the surface karst terrain occurs in Eastern China where the tectonic framework and gross morphological features were fixed mainly during the

[1] This distinction in Chinese papers is quite different from that comonly used in Anglo-Saxon publications (see T. D. FORD'S and PALMERS' papers in this volume; cf. QUINLAN (1972) and many others). However, this explanation given by ZHANG may be very useful for readers utilizing the Chinese publications (Eds.).

Yanshanian movement times. Absence of Cenozoic carbonate formations is characteristic, therefore, except in the South China Sea. For these reasons, the term *paleokarst* in Eastern China is commonly applied in a narrow sense to that developed bcfore the Cenozoic. It can be subdivided into several karstification periods (KARST RESEARCH GROUP, 1979).

Karst development was possible during the remote geologic past because there were carbonate rocks on the continent and precipitation occurred. The amount of atmospheric oxygen in the Early Phanerozoic is thought to have been about one third of the present amount. Throughout the Phanerozoic the proportion of oxygen in the atmosphere has been increasing but the process has been uneven. A curve for carbon dioxide fluctuations shows that its concentration varied within the range 0.1 to 0.4 % during the most of the Phanerozoic (Fig. 2–133; BUDYKO, 1982). The present concentration has reached an unprecedented low level, several times lower than the mean concentration during the Phanerozoic. The present atmosphere contains a relatively small amount of carbon dioxide ($2.33 \cdot 10^{18}$g or 0.033 % of the atmosphere by volume).

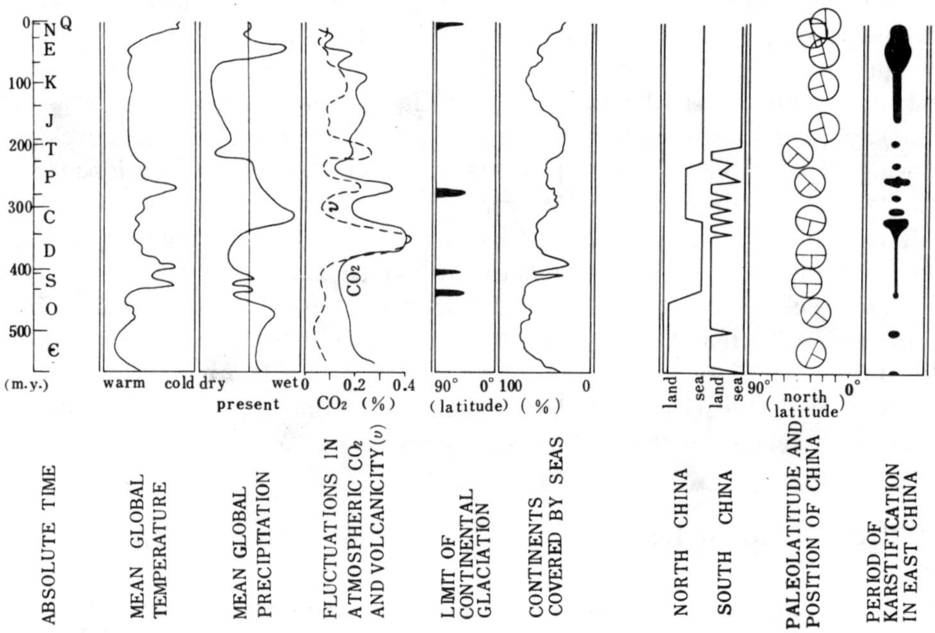

2–133. Generalized global fluctuations in some influential factors of karstification and the paleolatitude, karst periods in China since the Paleozoic (partly based on data of SEYFERT and SIRKIN, 1979; FRAKES, 1979; BUDYKO, 1980).

Comparison of a curve of volcanic activity (v) and the carbon dioxide curve shows that fluctuations in volcanic activity undoubtedly have influenced the carbon dioxide content of the atmosphere. For this reason CO_2 is available, too, for karstification during past geologic time. Karst features in the geologic column have been confirmed by the presence of karst-related mineral deposits.

As shown in Fig. 2–133, karstification in Eastern China occurred mainly during five periods. They are the Precambrian, Early Paleozoic, Late Paleozoic, Mesozoic and Cenozoic karst periods, corresponding to various tectonic cycles or sub-cycles. Each karst period displays some karst phases associated with the interruption of sedimentation and corresponding to tectonic episodes.

In this report, we discuss only the most important phases of the first four karst periods. The last one will then be discussed in more detail.

Precambrian Karst Period

This is the oldest karst period. There are some sedimentary gaps between the Sinian and Cambrian Systems and within the Sinian System itself in a variety of terrains. The diversity of relief on the disconformity surface is about 10 m in 50 m distance in Jinsha County (Guizhou Province). In Kaiyang County (Guizhou Province), a bore-hole drilled Cambrian mudstones lying on a disconformity surface below which were Sinian carbonate rocks containing a cavity big enough to require several tons of clay and cement to be injected to seal it.

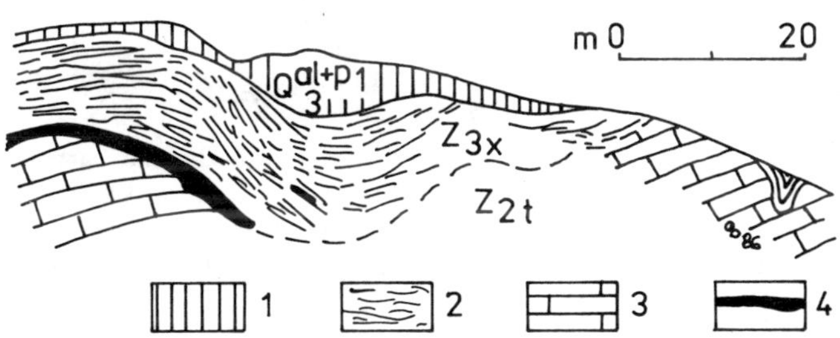

2–134. Cross-section of the Sinian karst at Xushiu County in North China (after Ministry of Geology, 1965). 1. Quaternary (Q_3^{al+pl}); 2. Xiamaling Formation (Z_{3x}) ferruginous slates; 3. Tieling Formation (Z_{2t}) dolostones; 4. ferrolites.

In North China the thickness of the Sinian System is reduced because of denudation during various Precambrian periods. In Xushui County (Hebai Province), Middle Sinian dolostones are disconformably overlain by Upper Sinian ferruginous sediments. The karst landforms are depressions ranging from 10 to 20 m in depth and 20 m width (Fig. 2–134).

Early Paleozoic Karst Period

The paleokarst features of the Middle Ordovician carbonates are typical of the last tectonic episode of the Caledonian cycle in the China-Korea paraplatform. The Ordovician carbonates are disconformably overlain by Middle Carboniferous clastics. On the top of the Middle Ordovician there are a series of closed depressions (ponors and sinkholes) and caves. On the karst denudation surface, bauxite, fire clay and pockets of iron ore are deposited.

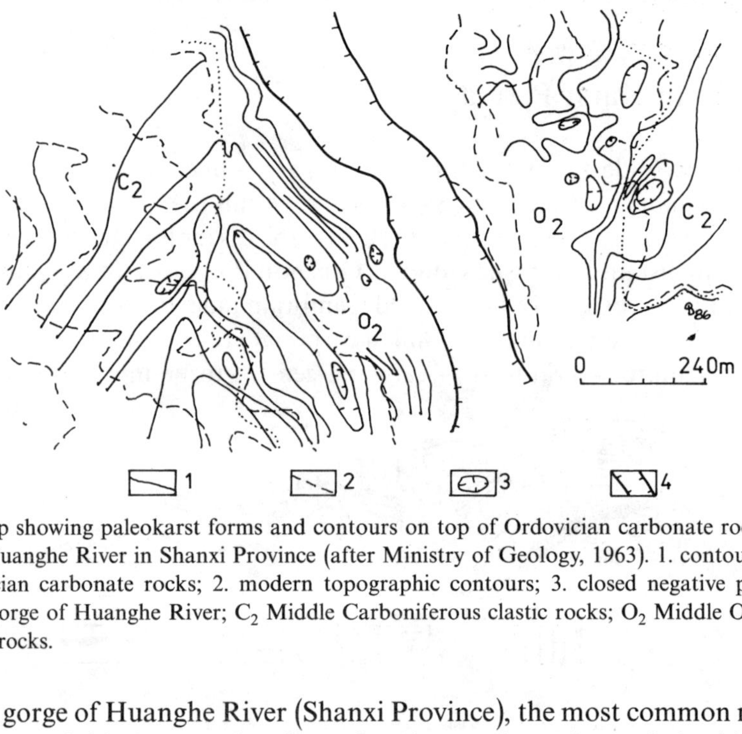

2–135. Map showing paleokarst forms and contours on top of Ordovician carbonate rocks in the gorge of Huanghe River in Shanxi Province (after Ministry of Geology, 1963). 1. contours on top of Ordovician carbonate rocks; 2. modern topographic contours; 3. closed negative paleokarst forms; 4. gorge of Huanghe River; C_2 Middle Carboniferous clastic rocks; O_2 Middle Ordovician carbonate rocks.

At the gorge of Huanghe River (Shanxi Province), the most common negative features are sinkholes on the disconformity surface. One of the biggest is 60 × 120 m, by 20 m deep. The sinkholes are filled with bauxite and clastic rocks (Fig. 2–135). The major subsurface paleokarst features are caves with a filling of Carboniferous clastics, which occur at a certain depth under the disconformity surface.

In the central part of Guizhou Province, bauxite and iron ore accumulated on the top of Cambrian carbonates, and are disconformably covered by Middle Carboniferous dolostones.

In the north margin of the Talimu platform in West China, the Middle Ordovician carbonates about 200 m thick were completely denuded in numerous areas. This paleoplain of denudation preserves some undulating karst landforms.

Late Paleozoic Karst Period

One of the most important karst phases occurred in the Dongwu tectonic episode of the Variscan cycle, between the Early and Late Permian. Karst features of this phase are widespread in various terrains of South China. The relict thickness of late Lower Permian carbonates, deposits of bauxite and fire clay on the undulating disconformity plane, and caves with gas reservoirs below the paleokarst surface confirm the formation of karst at that time.

According to incomplete statistics for the southern part of the Sichuan Basin, about 25 % of the holes drilled into Early Permian gas reservoirs struck caves, there being a total of 40 cavities. In one high-yield gas well, the maximum height of the open cavern was more than 4.45 m; this was 3 000 m below the modern surface. The maximum loss of drilling fluid was 5 895 m^3.

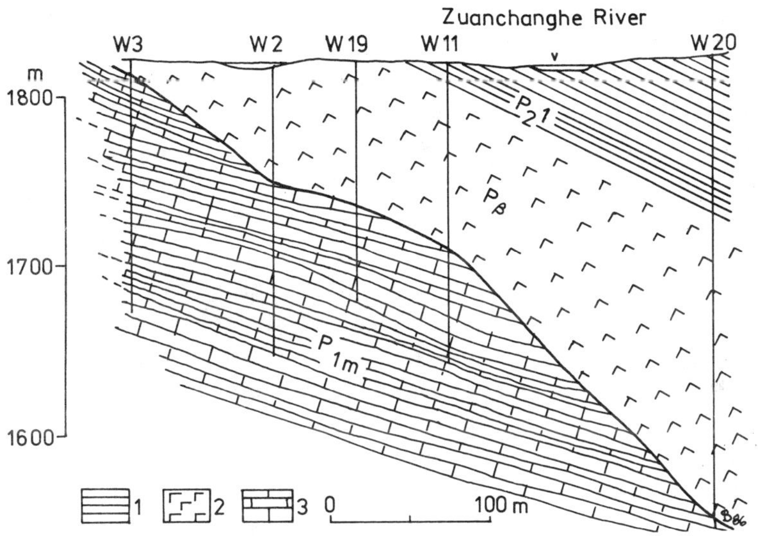

2–136. Geologic section of the paleolandforms of the Permian karst at Luoping County in South China (after Bureau of Power Industry of Yunnan Province, 1973) 1. Longtan Formation (P_21) clastic rocks; 2. Omeishan basalts (P_β); 3. Maokou Formation (P_1m) carbonate rocks.

At the famous *stone forest* area of Lunan County (eastern Yunnan Province), various late Lower Permian strata are overlain by Upper Permian basalts. The strata have been eroded to a difference in thickness of about 230 m over a horizontal distance of 5 km, which reflects the height difference of the paleo-relief. In Luoping County (the same Province), the paleorelief differences reach 110 m within a distance of 400 m horizontally (Fig. 2–136).

There are some other karst phases of local importance, such as at the end of the Permian, between the Devonian and the Carboniferous, and at the end of the Carboniferous.

Mesozoic Karst Period

This karst period corresponds to the Indosinian tectonic movements. Karst features of the Middle Triassic are preserved at a disconformity overlain by Jurassic clastics. In Sichuan Basin, filled caves were encountered when drilling 40 m below the ancient land surface. Sometimes, where Upper Triassic clastics were not completely denuded, gypsum beds below the land surface of that time are relatively well preserved. Solution breccia formed near the denudation plane where the gypsum was dissolved.

Cenozoic Karst Period

Carbonate rocks ranging in age from Precambrian to Triassic were exposed at outcrop by faulting, folding and denudation to begin the latest karst period in the Cenozoic. The effects of strong karstification that are currently seen on the surface are products principally of the Yanshanian and Himalayan tectonic movements since the Late Cretaceous. Based on the altitude of the denudation planes and corresponding sediments, three karst phases can be recognized. They correspond with the three physiographic stages; their name and altitude vary from area to area. They are products of the Eocene, Neogene, and Quaternary respectively, but the problems of more precise dating have not been resolved.

Rich paleontologic, paleoanthropologic and archeologic remains preserved in the Quaternary sediments permit the determination of the upper age limit of cave development. Over 60 important cave sites have been discovered and excavated in East China until 1982.

Based on 53 ^{230}Th/^{234}U ages for different speleothems in 18 caves from the eastern Mainland of China and on stable isotope studies, the paleoenvironment of speleothems has been established for the past 350 000 years (Fig. 2–137; ZHANG and ZHAO, 1985). By attributing epochs of deposition to interglacials and nondeposition to glacial periods, interglacials are recognized at: (1) present to 20 ka; (2) 75–120ka; (3) 170–230 ka; (4) 250–310 ka. These interglacials correlate with periods of high sea level and warm-wet climate observed on the marine foraminifera isotope record (SHACKLETON, 1976).

Paleoenvironments of Karst Development and Plate Tectonics

Karst is a kind of exogenic geologic process closely related to climatic zones. The paleoclimate is controlled by the changing paleolatitude as a result of continental drift and is recorded by the sediments in the geological column.

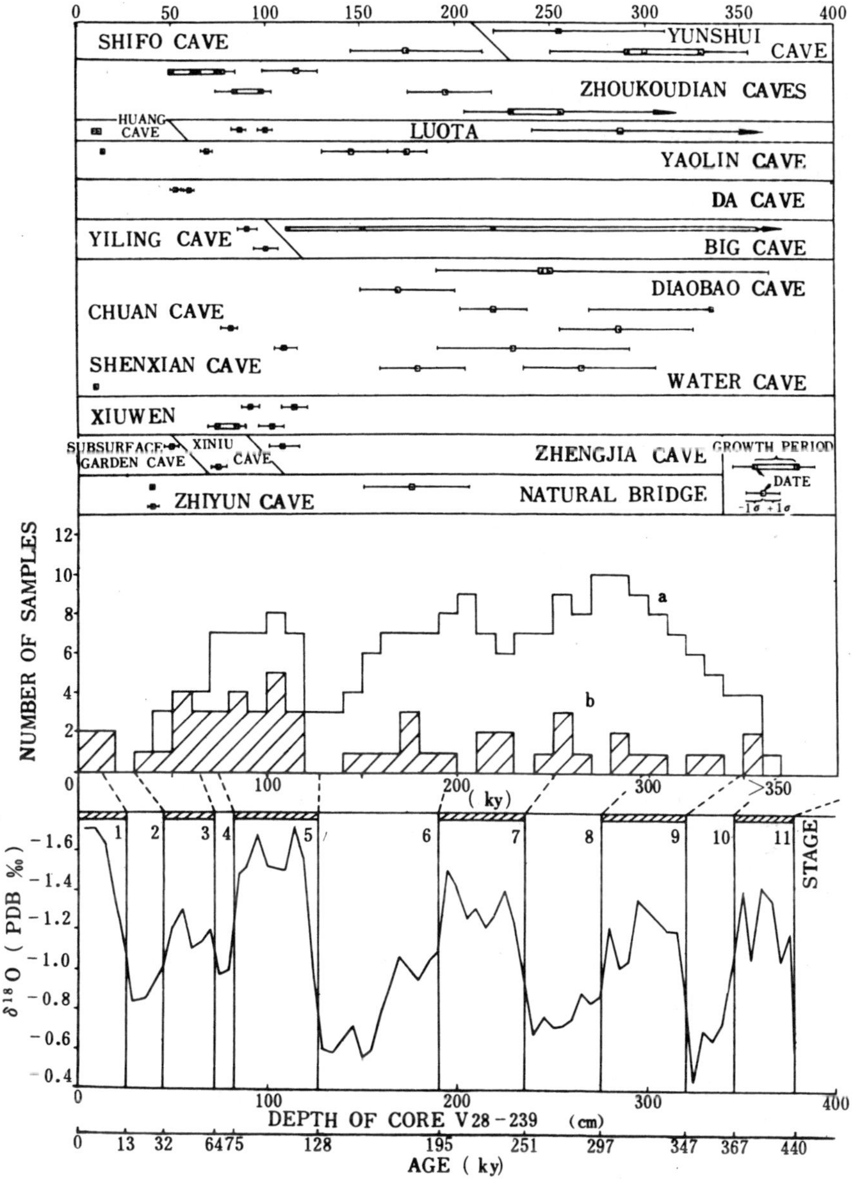

2–137. Distribution of speleothem ages in Eastern Mainland of China. Comparison between frequency distribution of speleothem dates and the isotopic record of marine foraminifera from core V 28-239 (Shackleton, 1976). The histograms represent (a) taking into account errors of 1σ standard deviation in the dates; and (b) neglecting these errors in ages. The numbers above the $\delta^{18}O$ curve are the isotopic stages.

Taking into account reconstructions of past continents (FRAKES, 1979; SEYFERT and SIRKIN, 1979) and the position of the China plate as shown in Fig. 2–133, paleoenvironmental conditions favourable for karstification have existed since the Paleozoic (ZHANG, 1986). The Middle Ordovician to Early Carboniferous in North China, between the Early and Late Permian in South China and since Middle and Late Cenozoic times are the most important periods of karstification.

Practical Consequences

Various practical aspects associated with paleokarst in China were investigated in the course of geologic survey and exploration of different mineral resources, especially for oil and gas, water supply, and for the evaluation of environmental geology and the use of caves for commercial tourist purposes.

Mineral Resources Associated with Pre-Holocene Karst

As mentioned above, *bauxite* and *fire clay* as residual deposits are formed on the karst plane of denudation, which is mainly of Ordovician-Carboniferous age in North China and Guizhou Province in South China, and of Permian age elsewhere in South China. The secondary enrichment of bauxite by modern karst has been also found.

Caves in Permian paleokarst are filled with *kaolin* deposits in Jiangsu Province. Their genesis is ascribed to hydrothermal alteration-secondary enrichment.

Different Quaternary *placer deposits* preserved in sinkholes, uvalas and karren have been exploited; e.g. tin in Guangxi Zhuang Autonomous Region and Yunnan Province, diamonds in Shandong Province and mercury ores. Bat *guano* is exploited in relict caves.

Karst Reservoirs of Oil and Gas; Water Supply

Gas in karst reservoirs of Permian and Triassic age occur in the Sichuan Basin. The Northern oil and gas area includes the Cenozoic subsidence zone in the Liaoning, Shandong, Hebei, Henan Provinces, and Bohai and Huanghai Seas (ZHANG ZHIGAN, 1980). In these buried fault-block mountains, important oil reservoirs occur. They are mainly in the Sinian and Early Paleozoic carbonate strata. There, the karstification periods are mainly of Sinian or Early Paleozoic, and of Paleogene age.

Waters from the paleokarst in Ordovician carbonates are exploited for the water supply of the larger towns and industrial districts in North China.

Evaluation of Environmental Geology

The Yilihe River storage reservoir (Yunnan Province) was built in a karst area composed mainly of Paleozoic carbonates. The impounded water discharges along a paleokarst erosion surface below Late Permian basalts and then a fault, to reach the Longtan Valley which is 900 m lower than the Yilihe River and only 16 km distant.

Karst waters in paleokarst areas often cause serious obstacles to mining. There have been a number of coal mines flooded in North and South China. Paleokarst rejuvenation has caused some damage to buildings.

At present, the most important economic role of caves in China is tourism. Over 50 caverns have been equipped with electric lighting and made accessible to tourists.

Conclusions

The data obtained by many karstologists permit the characterization of karst processes during past geologic periods in the territory of China:

(1) Relics of karst forms originated in older periods and phases of karstification are in various degree replaced by Quaternary karst phenomena, which prevail in the present-day scenery.

(2) In East China, times of karstification mainly comprise five periods, corresponding with the main tectonic cycles or subcycles. Each karst period involved several phases associated with sedimentary gaps and corresponding to particular tectonic events.

(3) The Middle Ordovician to Early Carboniferous in North China, the Early to Late Permian in South China, and the Middle to Late Cenozoic times are the most important periods of karstification.

(4) The development of karst during the past was influenced by the paleoclimate which was controlled by the changing paleolatitude as a result of continental drift. Climate and tectonic mobility of plates are the important factors for karst development and evolution.

PALEOKARST OF CANADA

Derek C. Ford

Introduction

Canada is a large country where much of the geology is known only in a general manner, not in detail. In the past, Canadian geologists have shown little interest in karst or paleokarst phenomena. It is only in the last ten years or so that the potential economic significance of paleokarst has begun to be quite widely appreciated. To compile a nationwide and comprehensive review of what is known of Canadian paleokarst would require much interpretation of unpublished maps, notebooks, etc. It is too big an undertaking for this short chapter.

As a consequence, the chapter will consider two subjects. The first is a summary of some striking but conventional paleokarst phenomena that are known in the Precambrian and Phanerozoic rocks, plus an assessment of the potential for other paleokarst there. PALMER and PALMER (p. 337–363) give a similar but more thorough review for the adjoining United States of America. The broad phases of rock deposition and karstification that they define there also apply to much of Canada.

The second is a discussion of the effects of Quaternary glaciations upon karst and paleokarst. This recognises the fact that Canada differs from most countries because nearly all of its karst rock outcrops have been severely and repeatedly glaciated. In addition, permafrost prevails in the arctic regions: its significance is briefly considered at the end of the chapter.

It follows that a broad definition of *paleokarst* will be used. It includes not only subjacent and buried karst, but also surface karstic features that are *fossil* (or *relict*) in the sense that they were created under environmental conditions quite different from those occurring today.

Karst Rocks in Canada

In outcrop there are approximately 570 000 km^2 of limestone and marble, 600 000 km^2 of dolostone and 80 000 km^2 of gypsum and anhydrite (Fig. 2–138). This is very similar to the extent of karst rocks in the People's Republic of China

P. Bosák, D. C. Ford, J. Głazek and I. Horáček (Editors):
Paleokarst. A Systematic and Regional Review. – Elsevier and Academia. Amsterdam and Praha. 1989.

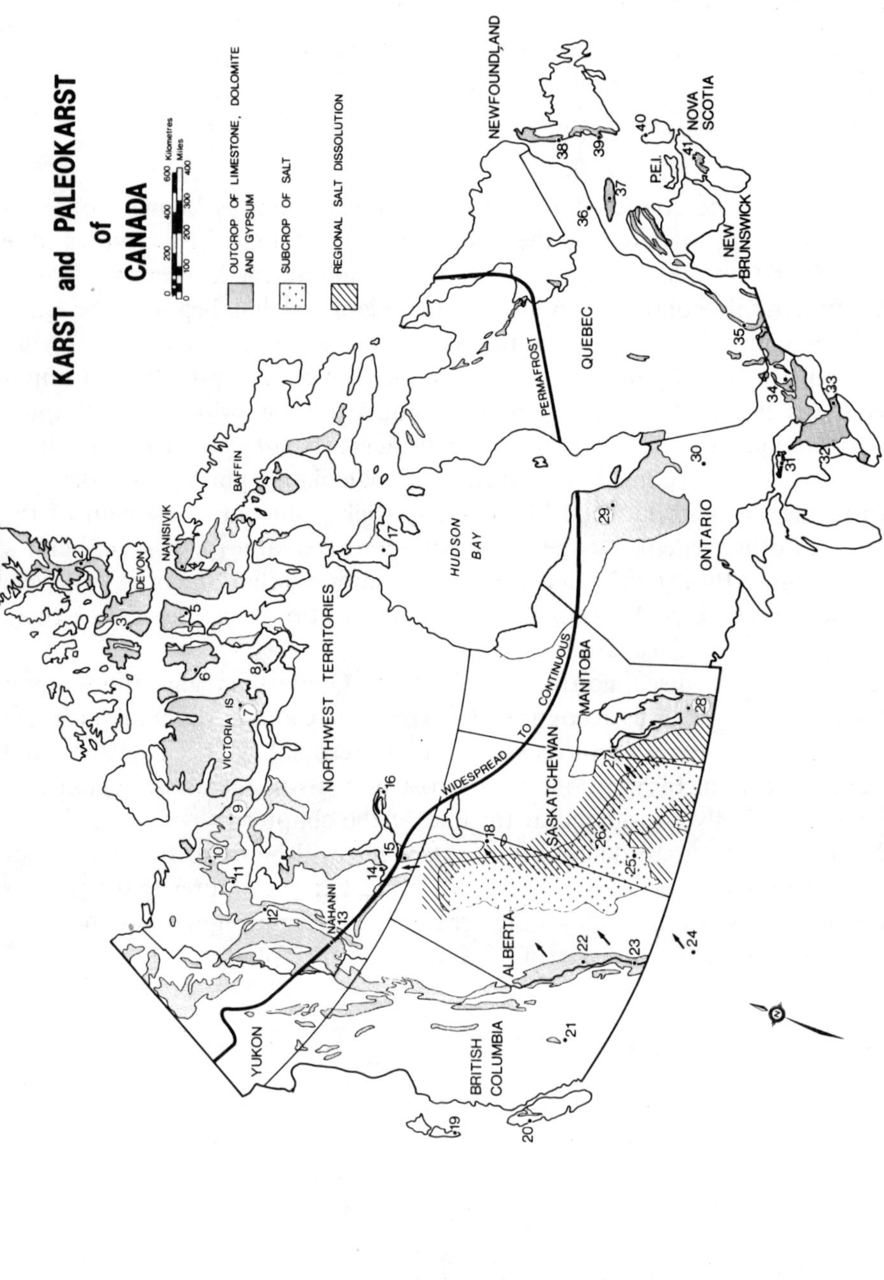

2–138. The major outcrops of karst rocks in Canada, plus the subcrop of the Elk Point Group salt deposits beneath the southern Interior Platform (Prairie Province). The numbers refer to karst sites discussed in the text.

(ZHANG, p. 297–311). More than 1 000 000 km^2 of other carbonates are present beneath only 50 to 2 500 m of cover strata. Most of these are in the Interior Platform that lies between the Canadian Shield and the Western Cordillera. There are also 500 00 km^2 of salt beneath the southern Platforms (Fig. 2–138).

The Platform units are well-bedded or reefal rocks displaying little tectonic deformation. The same conditions extend through the southern Arctic islands (sites 5–8, Fig. 2–138), the Hudson Bay Lowland (site 29), southern Ontario (31–33) and the St. Lawrence Lowlands and Anticosti Island (35–37). Strata in the Appalachian geological province (38–41) display Alpine-style deformation and displacement. There is comparatively simple folding in the northern Arctic islands (1–3) and the northwestern Cordillera (11–13). Overthrusting predominates in the Rocky Mountains (22, 23). Within the Canadian Shield there is a range from highly deformed and metamorphosed carbonates in some greenstone belts (e.g. 34) to merely tilted platformal blocks in horst and aulacogen situations (4 and 16).

In all the geological provinces there are extensive exposures of bare carbonate rocks. However, in the platform terrains most karst *outcrops* in fact are concealed by a mantle of 1–50 m of Quaternary glacial and proglacial sediments.

Archean and Proterozoic Paleokarsts

The oldest karst rocks are Archean limestones and dolostones, known in small outcrops. There are no reports yet of paleokarst features in them. Dolostone, limestone and marble are quite widespread in the Aphebian (Lower Proterozoic, > 1 700 Ma) and are up to 2 500 m thick. In the Belcher Islands there are some probable minor paleokarst surfaces in thick stromatolitic sequences, representing brief and shallow vadose emergent periods (M. BEAUPRÉ, pers. comm., 1981).

The Grenville Supergroup

The Grenville Supergroup is prominent in central Ontario and Quebec (Fig. 2–138, sites 34, 35). It is mid-Helikian in age (approx. 1 400 Ma) and comprises up to 4 000 m of marbles, dolomitic marble, quartzite, amphibolite and paragneiss formations. Grade of metamorphism varies widely. There are some very pure dolostones preserving recognizable stromatolite units and minor paleokarst surfaces that indicate periods of shallow emergence during deposition (M. S. BOURQUE, pers. comm., 1983).

Near Madoc, Ontario (site 34) pure dolostones are tightly folded and block faulted (Grenville Orogeny, ~ 975 Ma). Today they display a hilly terrain that is being exhumed from Late Cambrian-Lower Ordovician cover rocks (BOURQUE, 1982). SPRINGER (1983; pers. comm., 1985) recognises cone karst features and cockpits in the hills, plus lesser dolines and grikes. There are terra rossa soils and possible caliche profiles preserved. The basal Paleozoic sediments are sharply unconformable upon them. SPRINGER (1983) considers this mature dolostone karst to be Precambrian in age, but it is possibly diachronous and could be as late as the post-Sauk phases (p. 317–318). Quaternary glacial scour has uncovered and retouched the resistant cone karst to create a remarkable composite modern landscape. Exploration continues; many more discoveries of Grenvillian paleokarst may be expected.

Neohelikian and Hadrynian Paleokarst of Borden Peninsula, Baffin Island (site 4)

Neohelikian ($\geqslant$ 1 000 Ma) dolostones, limestones, gypsum and anhydrite are widely preserved in the southern Arctic islands. They are block faulted and locally folded but not metamorphosed. Detailed studies have been made on the Borden Peninsula where there are Mississippi Valley type (MVT) zinc/lead deposits > 675 Ma in age. These are exploited at the Nanisivik Mine.

The host-rock is the Society Cliffs Formation, 260–900 m of dolostone with evaporite beds up to 3 m thick. OLSON (1984) recognises four distinct paleokarst phases: (I) evaporite solution and dolostone brecciation with local sinkhole development. This phase includes early to mesogenetic features and was concluded before deposition of disconformably overlying shales of the Victor Bay Formation (Neohelikian). (II) in the Hadrynian, a major phase of subaerial karstification and cave genesis in rocks exposed by local erosion. Massive sulphides and white sparry dolomite were later deposited in the karst cavities. (III) a minor phase when pyrites were locally oxidised, probably Late Proterozoic. (IV) a Pleistocene phase of solution corridor formation that is quite independent of the earlier phenomena: this is discussed with other glacial relict features later in the chapter.

The main ore body at Nanisivik is a sinuous, filled passage extending 3 km along the strike of a karst block. It is truncated by Quaternary erosion at both ends. It is horizontal, oriented to strike in medium to thin-bedded dolostone dipping 15°. It is 80–100 m wide and 8 m deep. The roof is flat and completely discordant to the stratal dip, as are fins of wall-rock projecting up to 20 m into the ore body. Horizontal sheets of layered pyrite transgress as much as 400 m laterally into the walls.

OLSON (1984) interprets this remarkable feature as a Mammoth Cave-type normal phreatic passage with vadose entrenchment (a keel) below. It was buried intact to -3 km or more by later sedimentation and filled with hydrocarbons.

The massive sulphides were then emplaced. D. C. FORD (1986) reinterprets the ore body and host cavity as being largely simultaneous in their formation. Small caves only of Phase II served as targets to divert ore fluids from boundary faults into the unfaulted body of the horst, where the great cavity was then excavated and the sulphides deposited at a slowly rising piezometric or gas/liquid surface. This model also permits a more economical interpretation of the oxidising paleokarst phase, III. It is an example of the hydrothermal dissolution-sulphide emplacement model advocated by DŻUŁYŃSKI and SASS-GUSTKIEWICZ (p. 377–397).

Lower Paleozoic Paleokarst

The Cambrian and lowest Ordovician (Tremadocian) saw widespread limestone and dolostone deposition in southeastern Canada, where it is truncated by the post-Sauk events. There are no carbonate deposits of these ages in the southern Interior Platform. There are important salt and gypsum deposits in the Mackenzie Valley (sites 11, 14). Some very thick limestones accumulated in the southern Rocky Mountains and in Ellesmere Island; no major paleokarst has yet been reported in them.

Post-Sauk (Beekmantown or Knox) Paleokarst

This important period of erosion in the Lower to Middle Ordovician (Canadian-Chazyan or Arenigian-Llanvirnian) was induced by the Taconic Orogenies and so is most prominent in the Appalachian region and its fringes (see PALMER and PALMER, p. 341–345). DESROCHERS and JAMES (1985, and pers. comm.) have described its form in a regional platform carbonate sequence in the Mingan Islands (site 36). It was an extensive karst plain there, exhibiting karren, small dolines and shallow caves with collapse breccias. The substrate was often well lithified and dolomitised before karstification began. There was a veneer of eolian sediments. A few metres stratigraphically above this surface, Chazyan (Llanvirnian-Llandeilian) limestones display a similar paleokarst with a relief up to 10 m. In between these two major surfaces and succeeding the latter, shallowing-upward calcarenite cycles of one-two metres each are terminated by local paleokarst horizons that are marked by lithification and the formation of sharp free karren. In this region, therefore, there is a long record of repeated shallow emergences with vadose diagenesis and shallow karstification.

At Daniel's Harbour in western Newfoundland (site 38) an important MVT deposit in Lower Ordovician limestones and dolostones is emplaced in intrastratal dissolution and collapse structures attributed to this phase (COLLINS and

SMITH, 1975). Strata are considered to have been elevated at least 90 m above the contemporary sea level, but more modern interpretations would also consider the possibility of salt/freshwater mixing zone dissolution at depth below sea level (see discussion of the Pine Point MVT, p. 321–323).

In the Montreal and Ottawa areas dolostone of Canadian (Tremadocian-Arenigian) age are unconformably overlain by limestones of Wilderness (lowest Caradocian) age. No major karst features have been described at the unconformity.

In the high Arctic islands the Bauman Fiord Formation is a gypsum and anhydrite deposit that is highly brecciated. It is of Whiterock (Llanvirnian) age. In northwest Baffin Island the dolostone Ship Point Formation (Canadian-Chazyan) displays upper collapse breccias and an accidented erosion surface with filled sinkholes and a relief of 10–20 m (TRETTIN, 1969). It, too, is buried by Wilderness deposits. In summary, throughout the eastern half of Canada there is abundant evidence of emergence with erosion, including karstification, in the lower part of the Middle Ordovician. This may be represented by a single unconformity with karst to a depth of several tens of metres or by multiple, but shallow, epikarst surfaces.

Middle Ordovician to Upper Silurian

Several minor episodes of emergence with vadose karstification are reported during this period. There are two at the Ordovician/Silurian boundary on Manitoulin Island (31) and two or three in southern Manitoba and the Hudson Bay Lowlands (28 and 29). More notable is a strong unconformity in the Middle Niagaran (Llandoverian-Wenlockian) but no major paleokarst has been described here.

In southwestern Ontario (32) and adjoining Michigan, U.S.A., the Silurian closed with the deposition of salt that locally buried pinnacle reefs. Basinal waters passing through the reefs during the Devonian sapped the salt to produce interstratal cavities filled with the breccia of higher formations (SANDFORD, 1964). 150–200 km to the East the Silurian/Devonian boundary is marked by a strong disconformity. There were two phases of coastal and pavement karren development upon dolostones (kluftkarren to −10 m). Each was succeeded by a filling of beach sands. Then platform limestone deposition (Gedinnian) was renewed (KOBLUK et al., 1977).

Devonian and Upper Paleozoic Paleokarst

This is the most important period of conventional paleokarst preserved in Canadian rocks. The Middle to Upper Devonian is of great significance in western Canada; the most important carbonate hosts of petroleum, natural gas and MVTD were deposited and partly eroded during this interval. For much of the time very shallow platform-lagoonal conditions existed over the Prairie region, in which some of the world's largest salt deposits (Elk Point Group) accumulated. The Pine Point barrier complex south of Great Slave Lake (site 15) impounded the lagoon for a while. Further north, limestone, dolostone and gypsum accumulations extended to the Arctic shelf. To the west the principal karstic carbonates of the present day in the Rocky Mountains and Mackenzie Mountains were deposited in greater thicknesses. Carbonate deposition continued in southeastern Canada as well, but most of it has been removed by erosion. There were major gypsum deposits associated with red beds in western Newfoundland and Nova Scotia.

The Bear Rock Formation

The Bear Rock Formation is a remarkable deposit that is present in outcrop or at shallow depth beneath approx. 200 000 km^2 of the northern Interior Platform (sites 9–12, 14). It is Eifelian in age. In the south it is composed predominantly of gypsum with lesser dolostone. In the north there is a thin, basal breccia beneath a thick cover of dolomitic limestones. The central, most extensive, part comprises 150–250 m of polyphase breccias of dolostone and minor limestone with surviving gypsum and anhydrite beds that are as much as 3 m in thickness. In outcrop this breccia is chaotic, with high vuggy porosity (Fig. 2–139).

2–139. An outcrop of the Bear Rock Formation (Eifelian) at Bear Rock, a few km from the Mackenzie River. This pinnacle is 15 m in height; the larger cave is 3 m in diameter. The Bear Rock Formation is a polyphase dolostone breccia with gypsum, anhydrite and minor limestone, totalling 150–200 m in thickness (Photo by D. C. Ford).

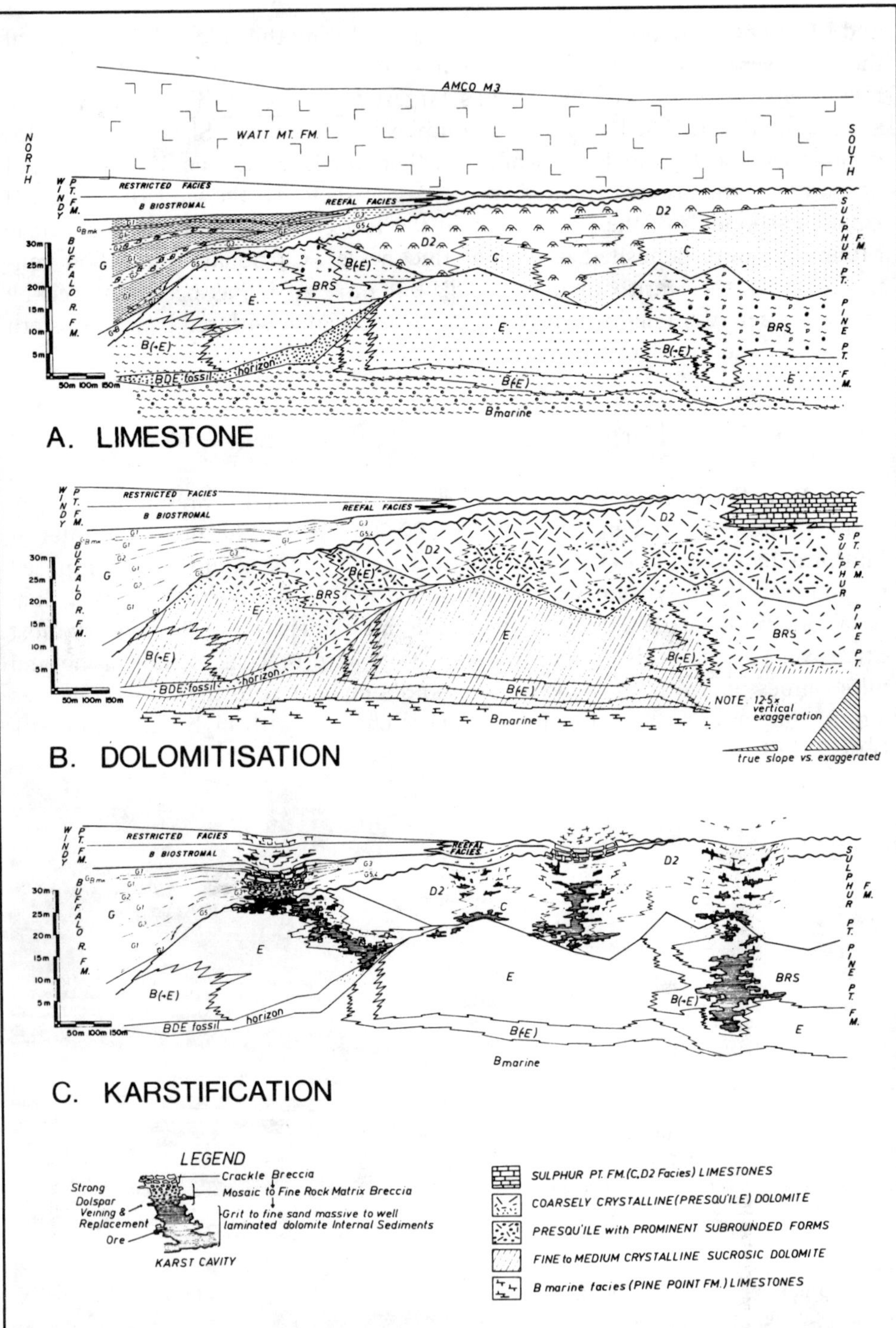
AMCO M3
WATT MT. FM.
NORTH
SOUTH
WINDY PT. FM.
RESTRICTED FACIES
B BIOSTROMAL
REEFAL FACIES
BUFFALO R. FM.
SULPHUR PT. FM.
PINE PT. FM.
D2
C
B(-E)
BRS
E
G
BDE fossil horizon
B(E)
B marine
30m
25m
20m
15m
10m
5m
50m 100m 150m
A. LIMESTONE
B. DOLOMITISATION
NOTE 12·5x vertical exaggeration
true slope vs. exaggerated
C. KARSTIFICATION
LEGEND
Crackle Breccia
Mosaic to Fine Rock Matrix Breccia
Grit to fine sand massive to well laminated dolomite Internal Sediments
Strong Dolspar Veining & Replacement
Ore
KARST CAVITY
SULPHUR PT. FM.(C,D2 Facies) LIMESTONES
COARSELY CRYSTALLINE(PRESQU'ILE) DOLOMITE
PRESQU'ILE with PROMINENT SUBROUNDED FORMS
FINE to MEDIUM CRYSTALLINE SUCROSIC DOLOMITE
B marine facies (PINE POINT FM.) LIMESTONES

The earliest brecciation evidently occurred during diagenesis (Eifelian). Brecciation may have been continuous since that time. Certainly there are episodes of intrastratal brecciation (or of accelerated brecciation) during the Givetian and the Cretaceous. A wide variety of unusual karst landforms has developed upon outcrops of the Formation during the Quaternary and intrastratal dissolution was probably accelerated during de-glacial phases. It is the foremost Canadian example of a solution breccia.

Pine Point and Horn Plateau

Pine Point is an Erian (Middle Givetian) carbonate barrier complex that has been traced from Great Slave Lake 500 km southwest to the cordilleran overthrusts. 50 km at the eastern end of the barrier are exposed in outcrops; here, the greatest concentration of paleokarst-hosted MVT ore bodies that are known in Canada have been found. There are 36 open pit mines. An excellent description and interpretation is presented by RHODES et al. (1984).

The carbonate barrier is a lensatic accumulation of fragmental carbonate sands and muds (Pine Point Formation) overlain by biohermal and bioclastic mounds (Sulphur Point Formation). Lagoonal gypsum is in lateral contact to the south. The barrier is extensively dolomitised, with fine-grained dolostone below and coarse, sucrosic dolostone replacing, largely, the Sulphur Point strata (Fig. 2-140). All deposits rest on an underlying marine limestone with gypsum (Chinchaga Formation) beneath it. They are buried by Late Givetian-Frasnian shales, marls and argillaceous limestones (Watt Mountain Formation and higher formations).

The karst and ore bodies follow two parallel trends along the strike, fixed at or close to the base of the sucrosic dolostone. There are two main types of features. What the mine geologists term *tabular karst* are sinuous and anastomosing solution channels 30–300 m in width and 1–20 m in height. They were filled with clastic sediments of local derivation. The ores have occupied the cavities, partly replaced the sediments and penetrated laterally into inter-channel rock (Fig. 2–141). These solution channels are interpreted as strike-draining, salt/freshwater mixing zone solution cavities and are likened to the modern *Boulder Zone* cavities of southern Florida (RHODES et al., 1984 and see PALMER and PALMER, p. 355–356).

The second type of karst features are the more familiar collapse cavities, created by stoping upwards from the channels (Fig. 2–140). They are also important hosts of ores. Deposition of the ores is attributed to sulphate reduction of basinal brines expelled due to overburden pressures (ANDERSON and MAC-

2–140. Middle Devonian paleokarst at the Pine Point zinc/lead depostits, Northwest Territories. Figure modifed from RHODES et al. (1984). (Reproduced with permission of Economic Geology).

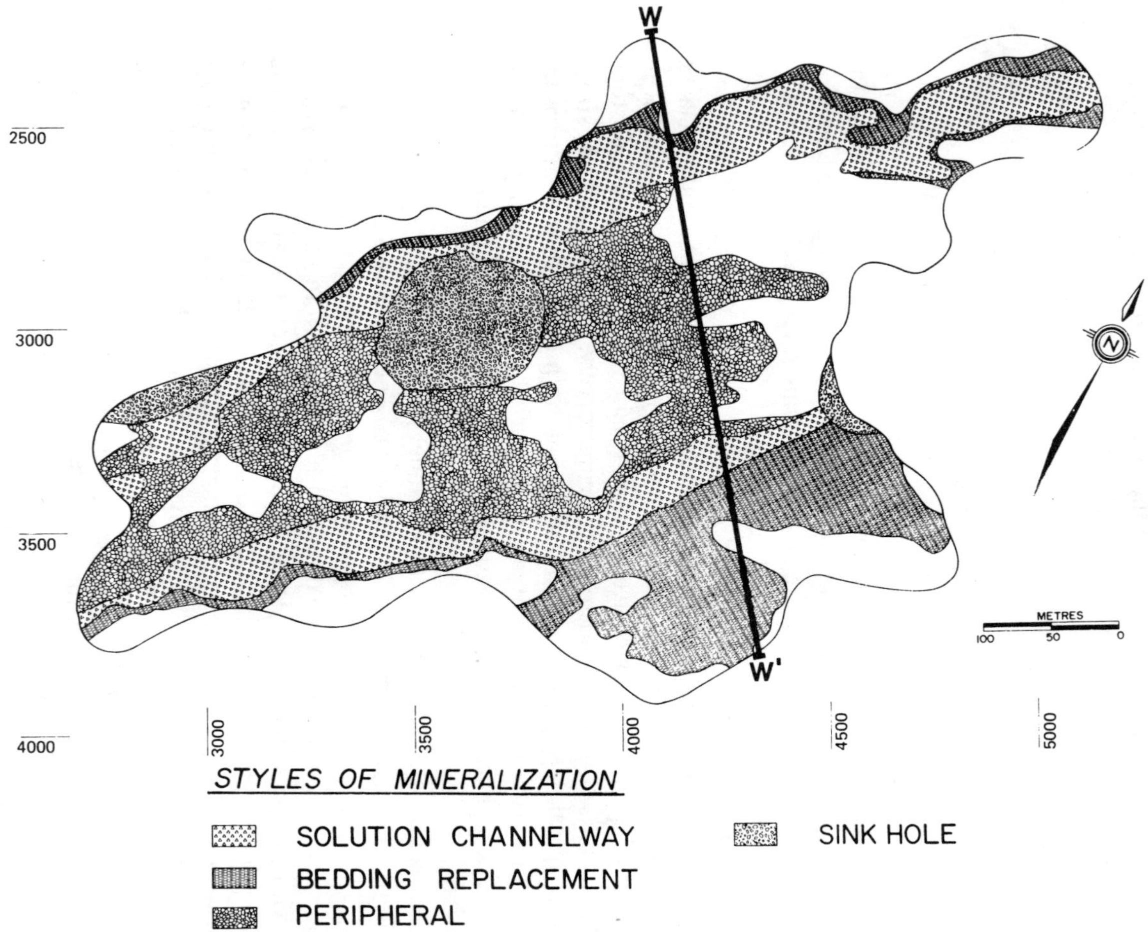

2–141. Pattern of solution channels, collapse sinkholes and replacement features containing zinc/lead ore at Pine Point, Northwest Territories (after RHODES et al., 1984; reproduced with permission of Economic Geology).

QUEEN, 1982). Precipitation temperatures were only 50–100 °C. There are several episodes of ore formation. Their ages are not established.

Some collapse cavities at the eastern (most eroded) end of the barrier complex have been rejuvenated. These are discussed in the glacial section of this chapter.

Horn Plateau (site 14) is 170 km NW of Pine Point. It is composed of pinnacle reefs that are also of Givetian age. They display fissure karstification to a depth of 100 m. This was terminated by deposition of the Frasnian shales.

In the Upper Devonian, carbonate deposition was generally resumed over the Interior Platform. At Lesser Slave Lake (near site 18) Fammenian limestones are terminated by a major erosional unconformity that is reported to display some large karstic features (OZORAY, 1977). There have been no detailed studies.

Eastern and Arctic Canada

Limestones, dolostones and minor gypsum from the Middle Devonian are preserved in southwestern Ontario. Disconformities within them display minor paleokarst. As noted above, there was dissolution of Silurian salt above pinnacle reefs at this time so that Givetian rocks are preserved as breccia pipe fillings within lower strata.

At Port-au-Port Peninsula in Newfoundland (site 39) a karst of small dolines, trough-like valleys and shallow caves that developed in limestones of Middle Ordovician age is being exhumed from a cover of Visean carbonates. Caves are filled with Visean sediments.

In Arctic Canada the Ellesmerian Orogenies (Upper Devonian) raised extensive Ordovician-Middle Devonian carbonate terrains above the sea. This probably induced much karstification. An important MVT deposit on Little Cornwallis Island (site 3) occupies a karst collapse cavity attributable to this episode (W. A. GIBBINS, pers. comm., 1982).

It may be noted that the post-Kaskaskia (Mississippian-Pennsylvanian or Namurian) paleokarst surface that is so prominent in the west-central United States (PALMER and PALMER, p. 345–355) does not seem to be strongly developed in the main carbonate sequences of Canada. It may appear in the lesser limestone formations of interior British Columbia, where strike-slip movement against the Pacific Plate has translocated strata from hundreds of kilometres to the south.

Elk Point Group Interstratal Karst

The Elk Point Group comprises one main and several subsidiary salt formations with minor anhydrite, red beds and dolostone. They are Upper Devonian (Frasnian-Fammenian) in age. They occupy approx. 500 000 km^2 in Canada

(Fig. 2–138). There is a minor extension into the United States. The aggregate thickness of the surviving salts ranges 10—400 m.

The salts are underlain by a regional platform carbonate that is dolomitised and serves as an effective aquifer. Upon this platform are local pinnacle and barrier carbonate mounds with relief up to 100 m. They are places where the overlying salt may be preferentially sapped by subrosion (WARDLAW and REINSON, 1971).

The top of the salt is now at depths ranging from 200 to 2 400 m below the land surface. The cover rocks are Paleozoic carbonates succeeded by Mesozoic (predominantly Cretaceous) shales and clays with minor sandstones, etc. The Mesozoic rocks are locally broken by open jointing but normally serve as aquicludes.

Commercial deposits of sylvite and carnallite occur at the top of the salt in southern Saskatchewan. This has led to intensive exploration by geophysical means, drilling and mining in this region. These have revealed many interstratal karst features. It may be presumed that there are similar phenomena elsewhere in the salt that have yet to be discovered.

Four distinct categories of karst features can be recognised:

(1) The margin (*zero sub-edge* or *salzhange*) of the salt has receded an average distance of 130 km along the eastern (or updip) side of the deposit. This is where cover strata are normally thinnest. There is a recession of 100 km at the distal (northern) end of the deposit, towards Pine Point. These recessions are signified by brecciation or anomalous dip (draping) in overlying strata. They commenced immediately after salt deposition. The presence of brine springs (discussed in the Mesozoic section of this chapter) suggest that some recession continues today.

(2) Southwest of Regina (site 25) there is a deep re-entrant where all salt has been removed along a frontage of 250 km. DE MILLE, SHOULDICE and NELSON (1964) suggest that this is located upon a basement ridge where the underlying carbonate is highly fractured, permitting more widespread basal solution. The maximum recession of the salt edge here (as attested by brecciation, etc.) is 100 km. The salzhange is steep. This feature appears to be inert, of Paleozoic age.

(3) The **Rosetown Low** (Fig. 2–142 A) is a closed depression where all salt has been removed over an area of 360 km^2 (i.e. it has the scale of a large polje). It is 110 m deep. Its development may be related to an underlying reef trend or, as the linearity of the northeast wall suggests, it may be fault controlled. It has been inert since at least the Tertiary. It is completely filled and without topographic expression today.

(4) Breccia pipes (or geological organs) appear to be abundant above the salt. Potash miners report many in their workings. A majority probably stabilised during the Paleozoic and have been inert since that time. However, at least 40 breccia pipes in southern Saskatchewan extend far up into the Cretaceous cover (Albian to Maestrichtian). Some have been activated during the Quaternary.

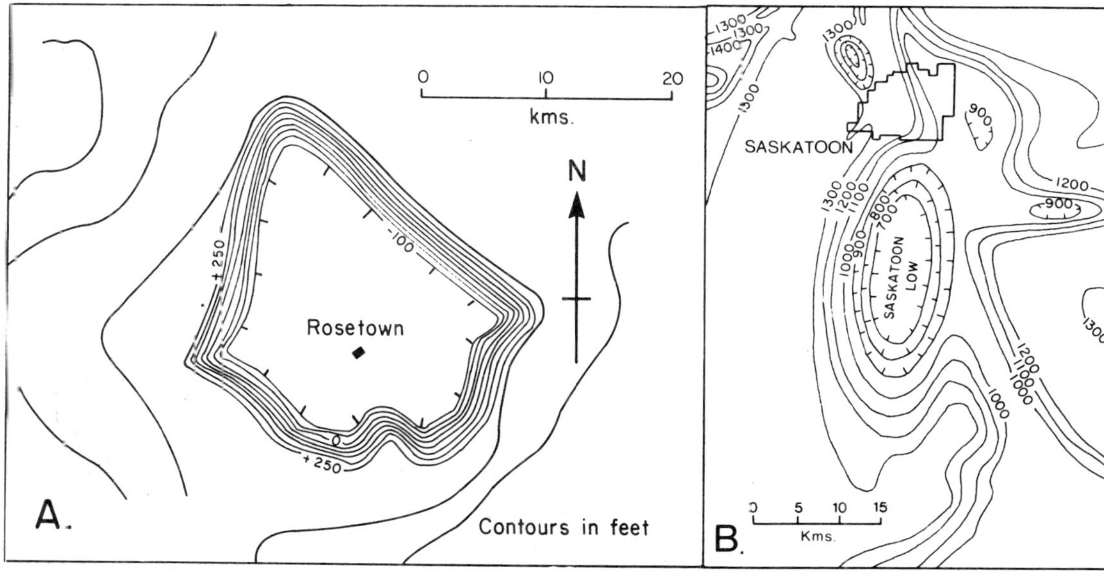

2–142. A. The Rosetown Low interstratal karst solution feature developed in the Elk Point salt. Contours (in feet) are on the top of the salt. The closed depression is approximately 100 m deep; salt is entirely removed from the central area of 360 km^2 (after DE MILLE, SHOULDICE and NELSON, 1964; with permission of the Geological Society of America). B. The Saskatoon Low and other interstratal karst depressions in the vicinity of Saskatoon, Saskatchewan. Contours (in feet) denote the stratigraphic top of the Cretaceous strata (after E. A. CHRISTIANSEN, 1967; with permission of the Canadian Journal of Earth Sciences).

Figure 2–142 B illustrates the **Saskatoon Low** (E. A. CHRISTIANSEN, 1967). This is a large depression that was last active at the close of the Wisconsinan (Würm) Glaciation. It has propagated upwards through 460 m of carbonate and 650 m of siliciclastic rocks plus 50 m of glacial till, a total of 1 160 m of foundering from the top of the salt (or 1 350 m of displacement to the base of the removed salt). Where explored by drilling (in the upper clastic cover) the collapse takes the form of cylindrical block faulting. Faults dip 70°. From an outer cylinder, 25 × 40 km in dimension, the total downdrop is 190 m. This outer limit is preserved in Maestrichtian strata truncated by Quaternary glacial erosion i.e. age of the outer collapse lies between Maestrichtian and Wisconsinan. An inner, distinctive collapse cylinder contains a downdrop of 70 m within Wisconsinan till. This 70 m depression, 8 × 16 km in dimension, is infilled by proglacial lake sediments so that it has no topographic expression today. The final collapse evidently occurred just before the glacial ice receded from the locality, ~12 000 y B.P.

At least four shallow closed depressions appearing in the modern prairie surface of southern Saskatchewan represent the tops of breccia pipes. They are stoped from salt at depths greater than 700 m. **Crater Lake** has been investigated by drilling (E. A. CHRISTIANSEN, 1971). It is a collapse structure of two nested cylinders with an aggregate downward displacement of 78 m. This created a doline 300 m in diameter. Infilling of the latest collapse commenced at c. 13 600 y B.P. and the topographic depression is now only 10 m in depth.

The Elk Point salt would appear to be one of the world's great interstratal karsts. There is much more to be discovered here.

Mesozoic and Tertiary Paleokarst

During the Permian and Triassic deposition of thin limestone units interbedded with siliciclastic rocks was common in northwestern Canada. A widespread erosional unconformity of upper Triassic (Norian) age there probably exhibits local paleokarst. The principal carbonate rocks of western coastal Canada (sites 19, 20) are also of this age but are predominantly deeper water facies and so are less likely to display paleokarst features.

Interbedded limestones, anhydrites and sandstones were also being deposited in the high Arctic islands during these periods. There is extensive solution brecciation. Outcrops of breccia include the most northerly Canadian karst rocks (site 1 at 82°30′N).

In Manitoba a karst surface with small solution features and mottling is known on Lower Mississippian (Visean) limestones. The surface is unconformably overlain by Lower Jurassic (Hettangian) red beds and anhydrite. The basal red bed is interpreted as a reworked terra rossa.

Elsewhere in Canada little new karst rock was deposited during the Permian to Jurassic interval and much was destroyed during periods of platform emergence associated with early Cordilleran Orogeny, the breakup of Pangea, etc. There are no important karst strata of Cretaceous or Tertiary age surviving in the country. With trivial exceptions, there are no Mesozoic or Tertiary strata of any kind in southern and eastern Canada. Any karst from those periods has been destroyed.

Middle Cretaceous Paleokarst

Mountain building in the Western Cordillera during the Late Jurassic and Early Cretaceous elevated most of the Interior Platform and induced new phases of karstification there. They were terminated in the Albian by the beginning of Upper Cretaceous-Tertiary clay, shale and siltstone deposition that, ultimately, buried all of the platform carbonates.

Karst that has been conserved and recognised in the south takes the form of the rejuvenation of Paleozoic collapse sinkholes and breccia pipes, or the creation of new collapses near to them. This is true of breccia pipes above the salt in Saskatchewan (DE MILLE, SHOULDICE and NELSON, 1964).

Far to the North, in the Anderson and Horton plains and Colville Hills (sites 9–11), sinkholes to 60 m deep are exposed in Ordovician dolostones overlying anhydrite. They are filled with Albian terrigeneous clays and shales (VAN EVERDINGEN, 1981; YORATH, BALKWILL and KLASSEN, 1968). This appears to be the earliest preserved karst in these regions. It was rejuvenated and extended during the Quaternary, following stripping of much of the cover rocks.

In the Athabasca Oilsands region of northeastern Alberta (site 18), the principal host-rock is the MacMurray Formation, a bituminous sand that is also of Albian age. This buries a quite rugged topography of paleovalleys and sinkholes developed in Upper Devonian limestones and dolostones overlying gypsum (OZORAY, 1977). The bitumen migrated via the carbonate rocks.

The Cargill Phosphate Deposit

A very unusual paleokarst is reported in the Shield rocks of northern Ontario (site 30) by SANDVIK and ERDOSH (1977). It is developed upon carbonatites within a carbonatite-pyroxenite-amphibolite intrusion of Aphebian age ($\sim$1 800 Ma) that is 7 × 1–3 km in outcrop. There are steep-sided solutional troughs and sinkholes as much as 1 000 m in length that preserve residual fillings of apatite crystals with minor goethite, magnetite, etc. Some fillings exceed 170 m in depth.

The deposits are unconformably overlain by sands and organic clays considered to be preglacial in age, followed by glacial lake clays of Wisconsinan or Würm III age. This karst could have originated in the Late Tertiary or even earlier, or it could be of Quaternary interglacial age.

Regional groundwater flow systems of southern Interior Platforms (Prairie Provinces)

Three long distance groundwater systems are recognised beneath the Prairies (Fig. 2–138). All of them are conveyed principally within the Devonian carbonate rocks. They are discharged as brine springs. The waters must be flowing through combinations of older paleokarst porosity with Cenozoic karstic interstratal or intrastratal porosity.

The largest system is believed to extend from a recharge zone in the Rocky Mountain foothills of Montana, U.S.A. (site 24) to springs along the northwest shore of Lake Winnipegosis (site 27). This is a straightline distance of 1 400 km. A second system extends from the Canadian foothills near site 22 to springs in the Athabasca Oilsands region (site 18). The third system is more local, extending 80–200 km from Cretaceous caprock hills in northern Alberta to springs along the south shore of Great Slave Lake near Pine Point (15).

The systems are driven by hydrostatic heads of 80 to 1 200 m today. They gain water from some areas of clastic cover rocks en route. The long southern systems could have been initiated as early as the Cenomanian and have more certainly been active since the late Rocky Mountains overthrusting during the Oligocene. Groundwater residence times are estimated to be greater than 50 000 years. Details are given in HITCHON (1969).

Quaternary Glaciations, Karst and Paleokarstification

During the Quaternary, ice caps grew and coalesced to create a multicentred icesheet, the *Laurentide Icesheet,* that covered all of Canada east of the Western Cordillera. The cordillera themselves were largely buried by valley glaciers and local ice caps except in the northwest extremities (e.g. Yukon Valley), which were too arid. The number of major expansions is not known, although it is certainly more than the four glaciations of older research (e.g. D. C. FORD et al., 1984). The earliest glaciation probably occurred before 2.2 Ma. Maximum ice thickness are believed to have exceeded 4 km. Zones of warm basal ice (pressure meltwater present at the glacier sole) and cold basal ice (too cold for melting) migrated with the expansion and recession of the icesheets (SUGDEN, 1978). There were protracted non-glacial intervals (interglacials or interstadials), some of them probably being warmer than the Holocene (HARMON et al., 1977). The aggregate depth of Quaternary erosion in southeastern Canada is thought to be ~100 m; this may be an overestimate: the depth is unknown elsewhere.

The effects of glacier action upon karst range between the extremes of complete destruction and of new creation. In the context of paleokarst, the effects may be categorised as follows:

(1) Destructive: (a) to erode karst features entirely, or (b) to make them fossil (relict) by diverting or destroying their water catchment e.g. caves drained and fossilised by glacial entrenchment of a valley; sinkholes or caves infilled by glacial debris.

(2) Transforming: altering the form and function of a karst so that it continues to evolve, but in a different style. Such karsts are polygenetic, with relict features.

(3) Rejuvenating paleokarst that has been buried and inert.

(4) Creating new karst landforms that will become relict (paleokarst) when the ice is gone.

Glacial Destructive or Fossilising Effects upon Karst

Destructive or fossilising effects are predominant (D. C. FORD, 1983a). Karren and shallow epikarst aquifers may be completely erased by the basal scour of warm ice. On resistant Ontario dolostones it is common to find the upper beds of clint-and-grike topography removed and surviving grike bottoms infilled with glacial clay (Fig. 2–144 A). This renders them buried and inert i.e. *paleokarst* in the strictest definition of the term.

Larger features such as dolines or poljes are rarely erased, even in soft rocks such as gypsum. But they may be partly eroded and then infilled and buried entirely beneath glacial detritus. This may render them inert (there is no groundwater solution and circulation through them), so that they are also paleokarst in the strictest sense. Quarrying, well drilling, etc. often exposes such features. There are probably tens of thousands of them concealed beneath the glacial drift plains of interior Canada. Good examples in gypsum that have been sectioned by Holocene sea cliff retreat occur in Newfoundland and Nova Scotia (sites 39–41).

In some karst terrains the response to glacial erosion and infilling has been much more complex. Some infilled sinkholes are partly re-excavated and reactivated by postglacial groundwaters. Other remain infilled (they lack any topographic expressions) but are able to renew their function as drains feeding in to a karst aquifer. Still others remain inert; fresh postglacial sinkholes may develop nearby to take over their function. As a result, the terrain is a multiphase, polygenetic and polymorphic, fluviokarst with relict glacial and karst elements.

Examples of regional karst aquifers being deranged, drained and fossilised by glacial entrenchment below their elevation are common in the Western Cordillera (e.g. D. C. FORD, 1983b) and display the same sorts of relationships as ZÖTL (p. 484–489) describes in the Limestone Alps of Austria. The oldest relict caves of this kind in Canada have been estimated at 6–10 Ma in age (D. C. FORD et al., 1981). The Castleguard alpine karst (site 22) is a modern subglacial, combined active and relict, karst system that has been intensively studied (D. C. FORD, 1983 c).

Glacial Transforming Effects upon Karst

Two radically contrasting examples will be mentioned here, to emphasise that a great variety of transformations are possible. The first is a terrain of very low relief, the upper carbonate aquifer at Winnipeg (RENDER, 1970; D. C. FORD, 1983d; and Fig. 2–143). This was an interglacial clint-and-grike solutional

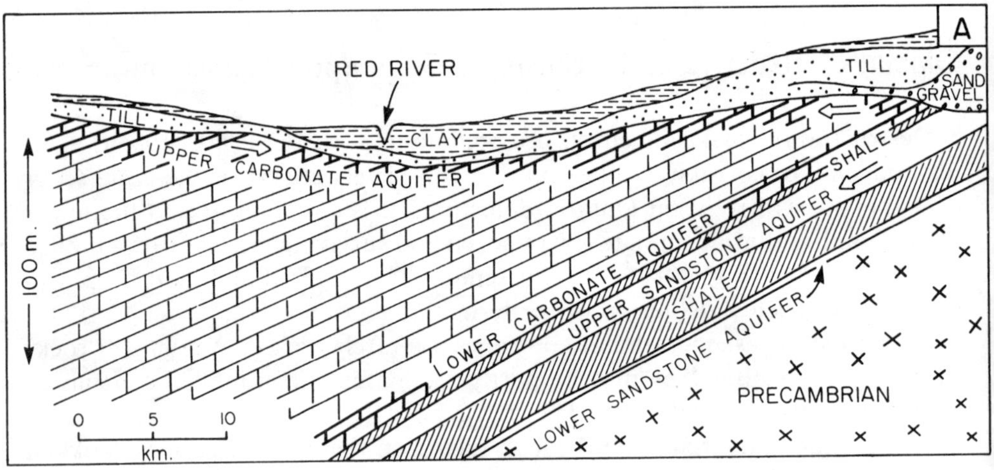

2–143. The Upper Carbonate Aquifer at Winnipeg, Manitoba. The aquifer is interpreted as an interglacial or interstadial solution pavement in dolostone, sealed by glacial deposits (from D. C. FORD, 1983d, adapted from RENDER, 1970; with permission of the Canadian Geotechnical Journal).

pavement in dolostone, extending ~4 000 km^2. During the last glaciation (Wisconsinan) it was buried by cold-based ice. Apart from some small rip-up patches, there was negligible glacial erosion. During glacial recession the pavement was buried (but not infilled or clogged) by till, and then sealed off by a cover of lake clays. The karst is no longer developing as a pavement but, instead, serves as an excellent aquifer with recharge occurring at the margins of the clays.

The second example is the Goose Arm karst of Newfoundland (site 38; KAROLYI and FORD, 1983). This is an ice-moulded, rugged terrain (local relief up to 350 m) in deformed Ordovician carbonates. There are more than 50 large sinkholes, as much as 1 000 m in diameter. Glacial scour by warm-based ice has eroded their rims. Their floors are filled with clay-rich till with the result that most now contain permanent ponds.

The large sinkholes will have developed in association with an efficient conduit aquifer linked to a few springs. Today this aquifer is infilled by clay injected subglacially, and so is inert. It is replaced by a mixture of local, short and postglacial, karst drains through sinkhole rims, plus surface overspill channels. The transforming effect has been the opposite of that which occurred at the Winnipeg karst.

Glacial Rejuvenation of Karst

In the course of a regional glaciation the crust is first depressed by a thickening ice cover, and then rebounds as this is removed. Rebound occurs in the presence of abundant subglacial and proglacial meltwater. Old fissures may be opened by rebound flexure and new ones created; water may be injected into them. This may effect some spectacular hydrologic rejuvenation of ancient paleokarst features that had been inert for long periods preceding the glaciation.

The principal examples known in Canada include some of the breccia pipes above salt in southern Saskatchewan. The Saskatoon Low (p. 326) was a Cretaceous collapse structure that was rejuvenated during the Late Wisconsinan, when 70 m further downthrow occurred. At Pine Point (Fig. 2–140) there are a set of collapse sinkhole-hosted MVT ore bodies termed anomalous (RHODES et al., 1984). Some of these display cores of Quaternary till that pass through the centres of the Paleozoic breccia and ore bodies. Evidently, episodic Quaternary dissolution has occurred at and below the Paleozoic karst solution base, perhaps extending into the Chinchaga anhydrite. Further examples of such sinkhole rejuvenation occur in the southern Arctic terrains such as Colville Hills, Anderson Plain, etc. PALMER and PALMER (p. 356–357) report similar features in northern Michigan (near site 32).

The deep injection of glacial meltwaters in this fashion is episodic and so can be considered to be a form of one-time or slug injection. It is suggested here that the big regional groundwater systems described above (Rocky Mountains to Lake Winnipegosis and to the Athabasca Oilsands area) may be maintained, at least in part, by such glacial slug injections rather than by continuous recharge that quantitatively matches the discharge when averaged over time. This also opens up the possibility that many buried paleokarst surfaces have been mildly rejuvenated by slug dissolution during the Quaternary.

Subglacial Creation of Karst Landforms

Certain types of karst landforms can be created or maintained only in the presence of glacier ice. A common class are caves carved by marginal meltwaters. These lose their catchment as soon as the ice withdraws. There are many examples in the Canadian mountains; LAURITZEN (p. 71–76) gives other examples in Norway.

A very different case is that of the potential karst terrain that is permafrozen when it is without glacier ice but that thaws when covered by an insulating ice sheet. The geothermal flux plus any frictional heat of basal ice motion is partly trapped at the glacier base because of the low thermal diffusivity of ice. The glacier base becomes warm and supplied with local pressure meltwater that is available for deep dissolution.

The southern Arctic islands (sites 5–8, 17) are carbonate platforms of low relief that are permafrozen today. Depth of the seasonally thawed layer is less than one metre. This is usually in till or felsenmeere; thus, modern thaw waters have very little access to the bedrock. Over extensive areas there are patterns of solutionally widened joint corridors. They are 100–1 000 m in length, 5–30 m wide (FRASER and HENOCH, 1959). Most are filled with till, permafrozen and relict today. They are interpreted as subglacial in origin (D. C. FORD, 1984). Solutional corridors in Proterozoic dolostone at Nanisivik (paleokarst phase IV of OLSON, 1984) are an example that are exceptionally deep because the terrain is more rugged.

Nanisivik Mine provides a further example of subglacial karstification that became relict upon deglaciation. A portion of the mine was a collapse structure at least 200 m in diameter, into underlying dolostones. Much of the collapse was supported by a cement of ground ice with an δ^{18}O isotopic ratio typical of the glacial-interglacial transition in the region. The modern temperature in the mine is -11 °C. Evidently, the rock thawed and basal meltwater was injected during periods of cover beneath deep glacier ice. This created the collapse (D. C. FORD, 1984).

Finally, some of the most fragile of all karst precipitates can develop only in the subglacial environment. These are subglacial regelation calcite precipitates (e.g. HALLET, 1976; LEMMENS, LORRAIN and HAREN, 1982). Around Castleguard Mountains (site 22; Fig. 2–144 B) they are more extensively developed and diversified in their form than at any other reported site (D. C. FORD, 1983c). When the ice recedes they become true relict karst. Rainwater and frost then quickly destroy them.

2–144. Illustrating the two extremes of glacier impact upon karst. A. The top beds of a solutional clint-and-grike pavement have been sheared off by Wisconsinan erosion and the base infilled with clay till. The till has been dug out to expose the contacts. Glacial action has rendered this a completely inert *ideal* paleokarst. Scene on Niagaran dolostones in Ontario. B. Subglacial calcite precipitates onto scoured limestone, exposed by glacier recession since 1940 A. D. These very delicate subglacial karst deposits are now being destroyed by rain and frost. Scene on Middle Cambrian limestones at Castleguard Mountain, Rocky Mountains (Photos by D. C. Ford).

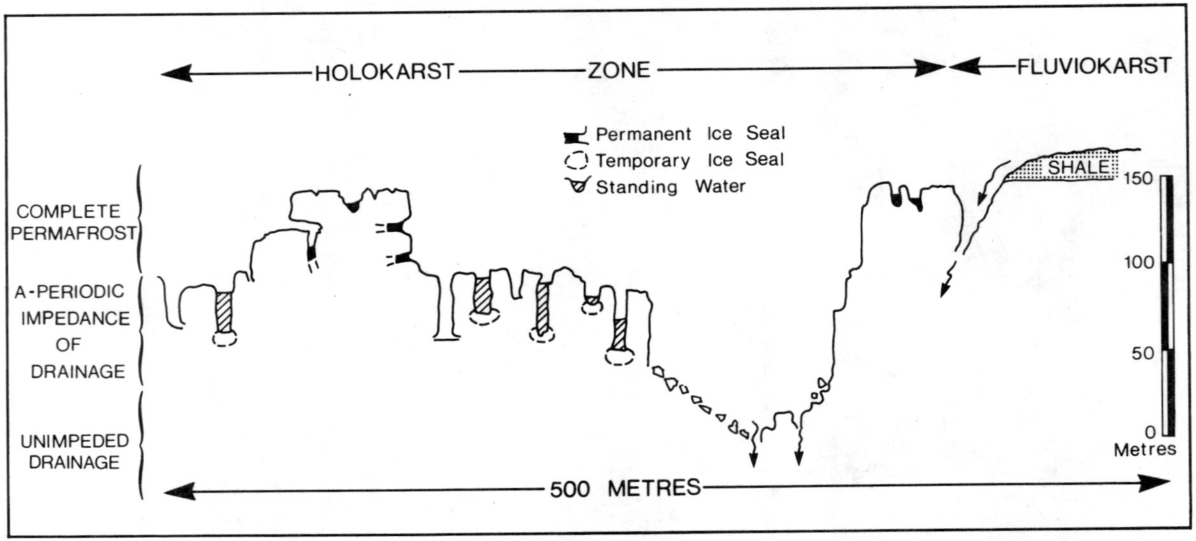

2–145. A model to depict the relationships between permafrost and karst activity in the mountainous arctic and sub-arctic regions of Canada, based on examples in the Mackenzie Mountains, Northwest Territories. Higher, isolated holokarst areas function as paleokarst. Lowest holokarst areas are normal (active) influent karst. Intermediate areas display mixed conditions.

Permafrost and Karst

Approximately 50 % of the outcrop of karst rocks in Canada is technically permafrozen (Fig. 2–138). In principle, this should restrict any karst solutional attack to the shallow, seasonally thawed layer (*karst sous-cutanée* of CIRY, 1962). Any deeper karst that was created before the growth of the permafrost becomes an ice-filled paleokarst.

These simple conditions appear to apply on carbonate rocks in the high Arctic islands (sites 2 and 3; e.g. WOO and MARSH, 1977). In the southern islands and mainland, where the permafrost may still be hundreds of metres in depth, the situation is more complex. In lowland karsts (e.g. 9–11) large dolines remain open and discharge their sinking waters through thawed corridors (talik) in the permafrost. Such dolines can be considered to have the same permafrost-inhibiting effect as small lakes. Some Holocene sinkholes have developed beneath postglacial lakes, draining them.

A model for permafrost-karst relationships in mountainous terrain is given in Figure 2–145. It is based on observations in the Nahanni karst (site 13; BROOK, 1976; BROOK and FORD, 1978a). This has a local relief of 50–150 m. Mean annual precipitaion is 300–400 mm and mean temperatures are −6 to −8 °C.

The karst displays three distinct hydrologic zones. The highest zone, where karst water catchments are smallest, is an ice-filled, inert paleokarst (Fig. 2–145). The lowest zone in an active holokarst with major dolines and small poljes; there is no significant permafrost impedance to groundwater circulation. The intermediate zone is one of a-periodic blockage. Ice accumulates in a drained sinkhole, sealing it. Over a period of one to several years water builds up over the seal until, eventually, the latter is ruptured by hydrostatic pressure and the pond drains in a few hours or days. This is a periglacial slug injection mechanism that maintains an active karst that, otherwise, would become relict. In this respect, it probably serves the same role as the glacial rebound slug injection discussed above.

Conclusions

Paleokarst is a rich and complex phenomenon in Canada. Its study is only beginning. The oldest surviving karsts are probably Archean. The youngest examples belong to the year when you read this chapter — subglacial calcite precipitates become exposed and, thus, relict each summer in the Rocky Mountains. Major paleokarsts are being recognised in the Proterozoic, including a cockpit-and-mogote topography in dolostone that certainly antedates the spread of vascular plants (normal forests, etc.). In the Phanerozoic the most striking paleokarst developed in the mid-Silurian, mid-Devonian and mid-Cretaceous. Repeated glaciations during the Quaternary, plus local and regional growth and contraction of permafrost, complicates the phenomena in all geologic provinces of the country because paleokarst can be both created or destroyed, conserved or rejuvenated by glacial effects.

PALEOKARST OF THE UNITED STATES

Margaret V. Palmer and Arthur N. Palmer

Introduction

Literature on paleokarst of the United States is widely scattered and consists mainly of passing reference in studies of local stratigraphy and economic deposits. Much of this work is cited by QUINLAN (1972, 1978), who is the only previous author to consider the subject on a continent-wide basis. A recent compilation of papers on paleokarst, edited by JAMES and CHOQUETTE (1988) covers many regional aspects of paleokarst, mainly from a stratigraphic and economic viewpoint, and has greatly increased awareness of this previously neglected subject among North American geologists.

This chapter is the first attempt to provide a comprehensive view of the origin and geologic relationships of the paleokarst of the United States. For that reason we have been conservative in our use of terminology in favor of detailed description of features and their origin. Terms used here are compatible with those of QUINLAN (1978), who distinguished two types of paleokarst: (1) *buried karst*, which is formed at or near the land surface by meteoric water and buried by younger rocks, and (2) *intrastratal karst*, which is formed by deep-seated processes within rocks already buried by younger strata. We favor the term "intrastratal" over the more common "interstratal", recognizing that most solutional processes at depth are not limited to boundaries between strata. We further differentiate intrastratal karst into *subjacent karst* (QUINLAN, 1978), which forms at depth only because of the relative insolubility of overlying strata, and *hypogenetic karst*, which forms by solutional aggressiveness produced by deep-seated reactions, not including those in which mixing causes a rejuvenation of aggressiveness derived originally from surface sources. A common example is the formation of hydrogen sulphide by the reduction of sulphates. Intrastratal features are considered paleokarst only if they are out of adjustment with the present geologic setting. This criterion is often difficult to apply, for intrastratal processes tend to operate over a long time span, and many features that qualify as intrastratal paleokarst were formed by processes still operating at a diminished rate today.

P. Bosák, D. C. Ford, J. Głazek and I. Horáček (Editors):
Paleokarst. A Systematic and Regional Review. – Elsevier and Academia. Amsterdam and Praha. 1989.

The major paleokarst zones of the United States include two buried surfaces of nearly continent-wide extent, which are of Early/Middle Ordovician age and mid-Carboniferous age. In addition, there are many lesser examples of paleokarst, only two of which are described here because of their regional importance: the buried and submerged karst surfaces of Florida, which relate to Cenozoic sea-level changes, and the intrastratal karst of Michigan, formed by the solution of sulphates and carbonate rocks at depth, partly by hypogenetic processes. Owing to differences in extent of exposure at the surface (and therefore available literature), as well as uneven geographic coverage in our own field work, this chapter gives the various paleokarst zones rather uneven treatment. Space for references is limited, so previous papers cited in this chapter are mainly general articles, which should be consulted for a more detailed review of the literature.

Geologic Background

SLOSS (1963) has identified and named six major Phanerozoic sequences of marine rocks that cover most of the low-elevation areas of North America, including most of the United States (Fig. 2–146). Deposition of these rock sequences was interrupted by periods of erosion when sea level was relatively low. Changes in sea level have been caused both by glacial episodes and by tectonic events, the latter also determining the pattern of sedimentary basins and platform deposits. Because the transgressions and regressions responsible for this stratigraphic record were gradual and incomplete, the unconformities between sequences are diachronous and are not present everywhere on the continent.

The two major paleokarst surfaces of the United States are related to the unconformities at the top of the Sauk sequence (Early/Middle Ordovician) and the Kaskaskia sequence (mid-Carboniferous). Although minor karst developed locally at the tops of other sequences, these two were the only ones to experience the deep dissection of carbonate rocks necessary to form widespread karst surfaces. In this paper these two paleokarst zones are referred to as the post-Sauk and post-Kaskaskia surfaces. The post-Sauk paleokarst is best developed throughout the southeastern United States and can be traced intermittently as far northeast as Newfoundland, Canada. The post-Kaskaskia paleokarst extends throughout the western part of the United States and southwestern Canada, gradually diminishing in intensity toward the east-central United States, where clastic rocks are exposed beneath the unconformity. The post-Kaskaskia surface is of particular importance, as it is the basis for subdividing the Carboniferous of North America into two separate periods, the Mississippian and Pennsylvanian.

The nature and distribution of these paleokarst surfaces is closely linked to the tectonic history of the continent during the Paleozoic Era. From the Ordovician Period to the end of the Paleozoic Era, North America underwent convergence toward its neighboring continents, Eurasia, Africa, South America, and associated microcontinents, forming the Appalachian Mountains along the east coast and the Ouachita Mountains in the southeast (Fig. 2–147). Meanwhile, the ancestral Rocky Mountains were forming along the west coast through convergence with an oceanic plate. The

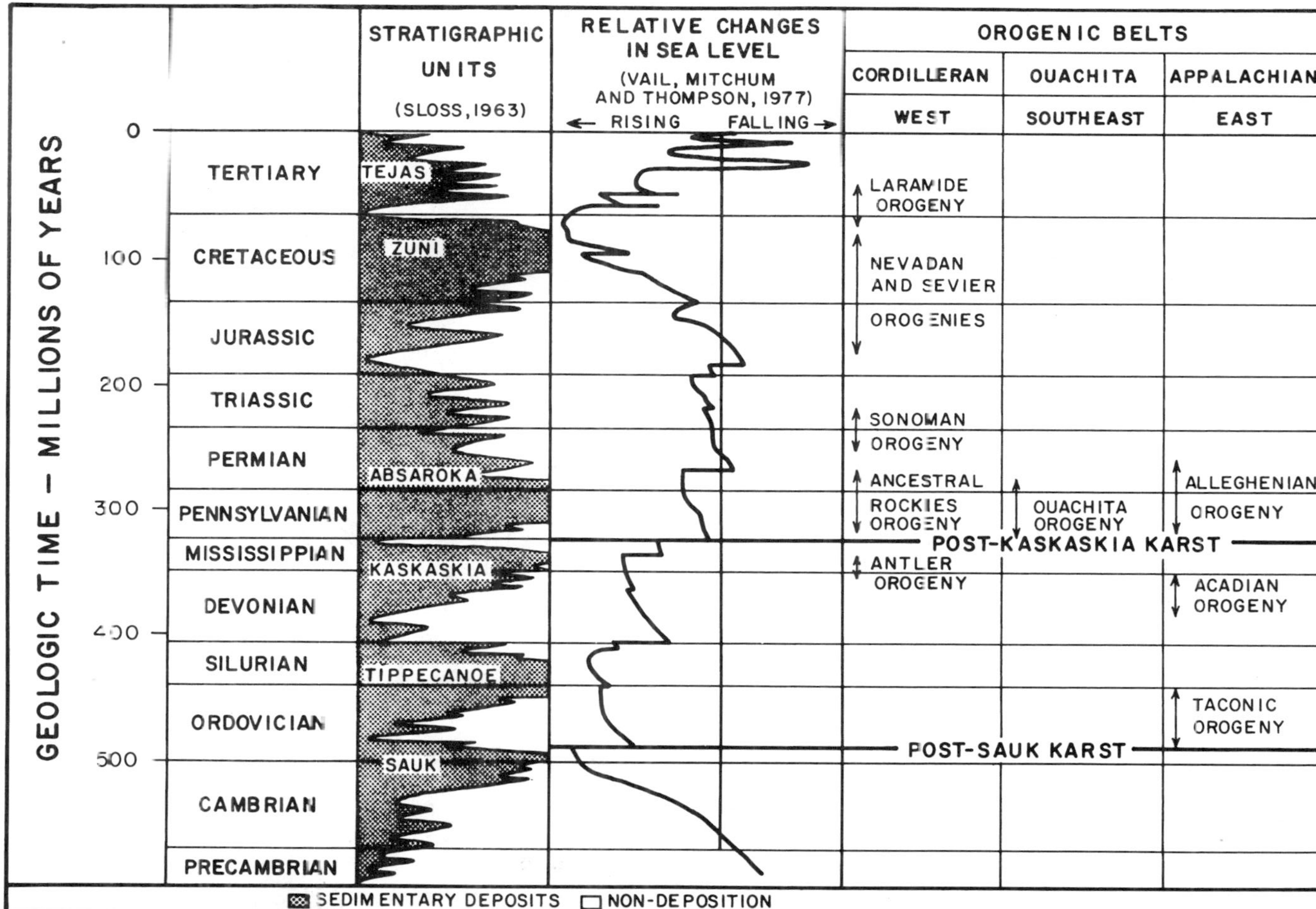

2–146. Geologic time scale, relating major paleokarst surfaces to stratigraphic sequences, relative elevations of sea level, and tectonic events.

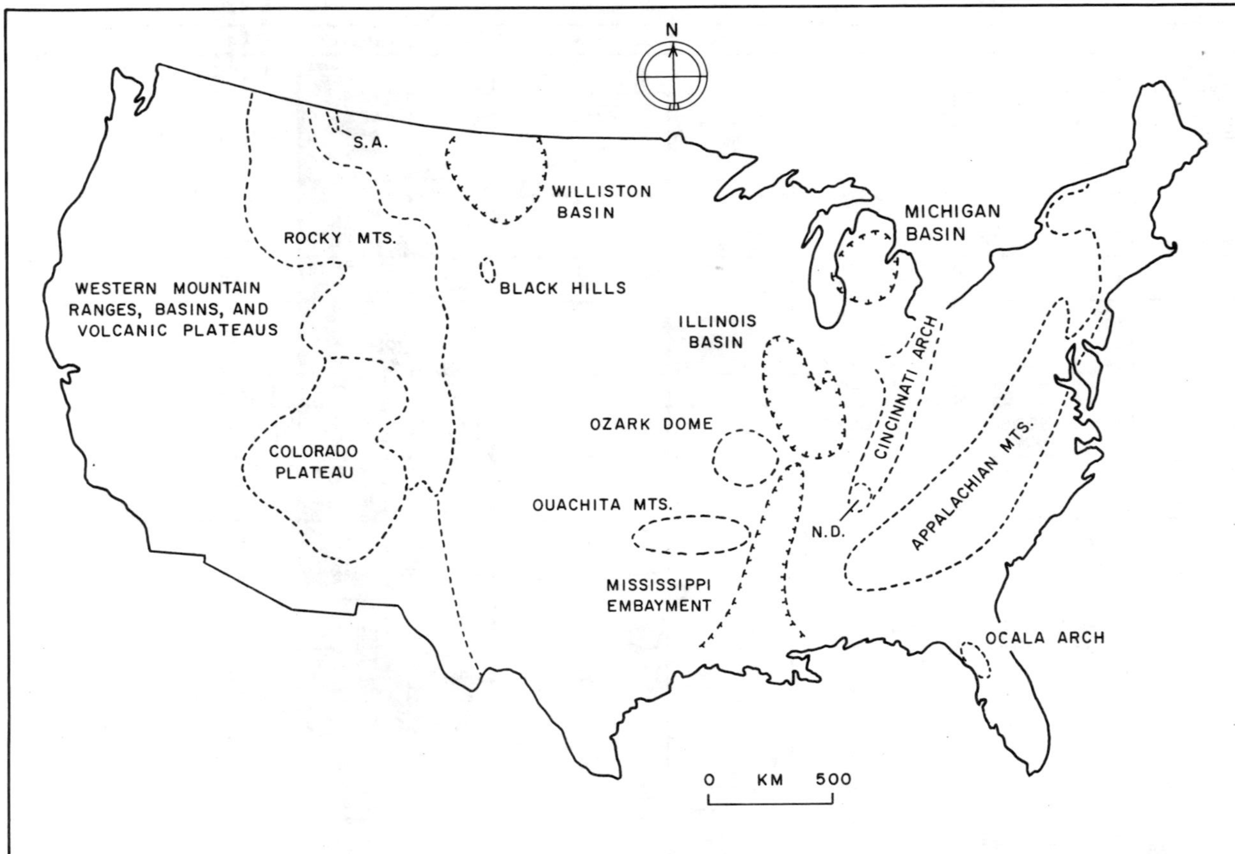

2–147. Location of structural features in the United States described in the text with reference to paleokarst. S. A. Sweetgrass Arch; N. D. Nashville Dome. The Ouachita Mountains today represent only the northern edge of the Ouachita orogenic belt, the early stages of which are shown in Figs. 2–148 and 2–149. Most of this belt is now buried beneath Cretaceous and Cenozoic rocks.

continental interior developed gentle domes and basins, largely inherited from older Precambrian structures. By the end of the Paleozoic Era, all the major continents had merged to form a single continent, Pangea. During the Triassic Period this supercontinent began to split apart into separate continents once again through crustal tension (Palisade Disturbance). Westward migration of North America since then has caused convergence with plates in that direction, forming the Rocky Mountains and ranges farther west (Fig. 2–148).

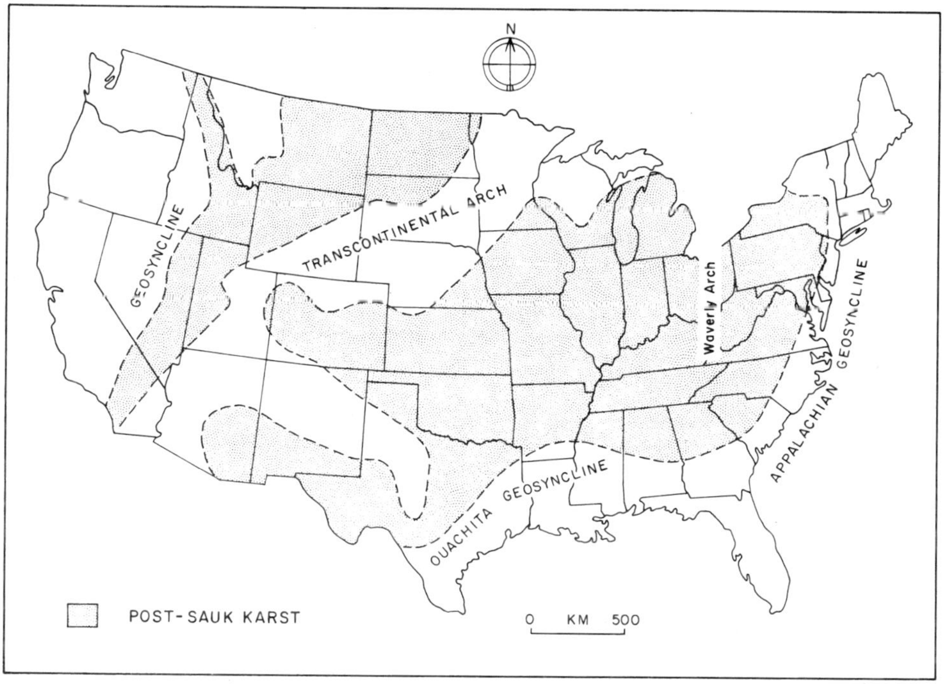

2–148. Approximate extent of the post-Sauk paleokarst surface at the end of the Early Ordovician.

Most of the exposures of post-Sauk paleokarst are located in the Appalachian mountain belt, where post-Ordovician folding brought it to the surface. The post-Kaskaskia paleokarst is best exposed in the western states, along the flanks of the numerous uplifts of the Rocky Mountains.

Buried Paleokarst Surfaces

Post-Sauk Paleokarst

From the Late Cambrian through the Early Ordovician all parts of the continent but the Canadian Shield and the Transcontinental Arch were below sea level. After a basal sandstone, most of the marine rocks deposited at this time were carbonates, youngest of which is the Early Ordovician Knox Group.

Owing to tectonic stability at that time, the carbonates have great lateral persistence. The Knox Group consists of supratidal algal limestone and dolomitic mudstone in shoreline areas, grading through shoal deposits to intraclastic, intertidal rocks in the miogeosynclinal areas. Chert is common in the lower beds but diminishes upward. The beds are increasingly arenaceous toward the top. The Knox Group reaches a maximum thickness of about 1 200 m in Tennessee and generally thins to less than a third of that in surounding states.

Continent-wide regression at the end of the Early Ordovician allowed erosion to occur on the exposed carbonates (Fig. 2–146). This surface was buried by the Tippecanoe rock sequence, whose basal rocks are of Middle Ordovician age. The resulting unconformity includes one of the two major paleokarst zones in the country. It extends more or less continuously throughout the southeastern part of the continent at least as far west as Missouri and includes scattered areas farther west.

Post-Sauk erosion and the beginning of Tippecanoe deposition occurred during the earliest phases of Appalachian deformation, the Taconic Orogeny, when large-scale thrusting and folding took place along the eastern edge of the continent. Meanwhile, thick geosynclinal sediments accumulated in the southeast and in the far west as the forerunners of later mountain belts. The onset of tectonic activity during the Ordovician may have been at least partly responsible for the extensive post-Sauk erosion, because adjustments in relative elevation between continents and sea level are common during major changes in crustal morphology.

Local relief on the post-Sauk unconformity is approximately 15–60 m and consists of large karst depressions and valleys (possibly of karstic origin), with intervening low hills. Paleokarst is best developed in the east, but because it is buried nearly everywhere, its character is known largely from mining and from well logs. The post-Sauk surface is exposed around the Ozark Dome in Missouri, but later weathering of the carbonates has obscured the paleokarst with thick blankets of cherty residuum.

During post-Sauk erosion, the Waverly Arch formed a relatively high area extending north-south through what is now Ohio and West Virginia (HARRIS, 1971). Karst depressions are deeper and more closely spaced on the paleo-arch than in the adjacent lowlands. Some are as much as 200 m deep, but it is questionable whether the largest of these represent true surface features, as they may have been enlarged or even initiated by stoping from below. On the paleo-arch the largest depressions are filled with collapse breccia overlain by gray-red mudstone and shale and are capped by a basal Tippecanoe shale. Breccia fragments are found as much as 200 m below the unconformity along fractures. Depressions in the former lowland contain sand and scattered chert fragments and are coverd by limestone rather than shale.

Most of the literature on the post-Sauk paleokarst concerns its relationship to the lead-zinc ores of central and eastern Tennessee (KYLE, 1983). Two distinct breccias of different age are commonly associated with these deposits, but they appear to be only partly related to the buried karst surface. The early breccia is best developed within the upper 150 m of the Knox Group, beneath the unconformity. The breccia consists of finely crystalline, angular dolostone fragments in a finely crystalline, green-gray dolomitic matrix (crystal diameter less than 50 microns in both the fragments and the matrix). In these respects it is similar to the breccia fill in the deep depressions. It is therefore probable that the early breccia was formed at or about the same time as the post-Sauk erosion by surface or near-surface processes. Breccia fragments in central Tennessee are bleached, as are those in the fill in the karst depressions of the post-Sauk surface. The bleaching is generally considered to be due to oxidation below the erosion surface. In eastern Tennessee the early brecciation event also produced a distinctive breccia with a red matrix, apparently about the same age but not identical to that with the green-gray matrix. The red-matrix breccia contains more silica, clay and iron oxide. Adjacent bedrock is stained red.

The late breccia is of unknown age and in places extends upward across the post-Sauk unconformity. It is therefore not related to the paleokarst. It is similar to the earlier one, except that the fragments are larger, up to several metres in diameter, and the crystal size in both the bedrock fragments and the matrix is larger (more than 200 microns in diameter). In general the late breccia formed around the early breccia and its pattern seems to have been controlled by the early breccia. Lead-zinc ore occurs only within the late breccia, whose larger crystal size is apparently due to hydrothermal alteration. Brecciation was probably caused by collapse into solution voids, as most of the blocks are located within a metre below their original positions. Both breccias typically form wide, tabular bodies or prisms with irregular tops and bottoms. In cross section the early breccia forms continuous zig-zag patterns extending upward from the ore zone to the unconformity, which typically lies about 200 m higher. This characteristic may cast doubt on whether the early breccia was formed at the same time as the paleokarst. Unfortunately, stratigraphic relationships are unclear because of intense folding and faulting where the breccias are exposed at the surface. Exposures in mines are extensive but limited mainly to ore zones, so comprehensive information on paleokarst is difficult to obtain. Most field work so far has been concerned with delineating the extent of the ore, rather than understanding the origin of the paleokarst, although knowledge of the latter would certainly aid in predicting ore distribution.

The Knox Group is a highly permeable aquifer of regional extent. The most permeable zones are within the top 60 m of the Knox at and beneath the post-Sauk unconformity, in solution zones along bedding-plane partings and within fractured, vuggy or coarse-grained dolostone and breccias. These rocks are used as a water supply where they are at shallow depth and for waste

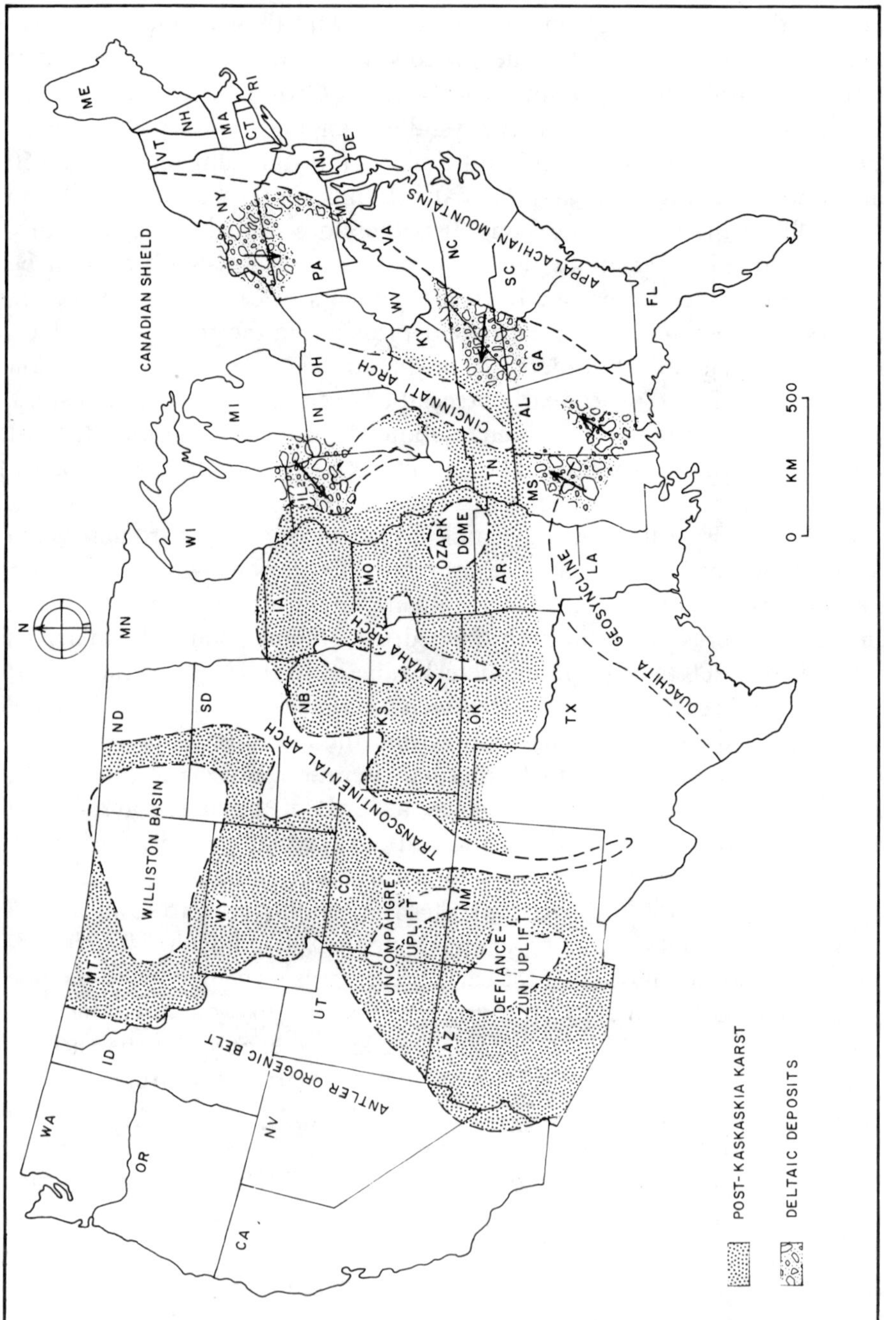

2–149. Paleogeographic setting during the post-Kaskaskia erosional episode, showing location of karst surfaces.

injection where they are deep beneath the surface and relatively isolated from nearby aquifers. In some places the static water level in wells lies above the land surface. Water in much of the aquifer is considered to be partly connate, but diluted by recharge from domal areas where the aquifer lies near the surface. Where Ordovician rocks are exposed along the Cincinnati Arch in central Tennessee and Kentucky, the post-Sauk unconformity comes within a few tens of metres of the surface. Flow through the aquifer is moderate in Tennessee, but it diminishes northward, with an accompanying increase in salinity. Westward, in Missouri, rocks of equivalent age crop out around the Ozark Dome and transmit a great deal of freshwater even where they lie at considerable depth.

Post-Kaskaskia Paleokarst

General Setting

The Tippecanoe sequence of deposition ended with widespread regression and erosion early in the Devonian Period, but with little karst development. These rocks were later covered by the Kaskaskia sequence, which extends in age from Middle Devonian to the end of the Mississippian Period. Most of these rocks are carbonates, particularly near the top. Erosion at the end of the Mississippian produced a widespread karst surface that was preserved by deltaic and transgressive marine sediments of Pennsylvanian age (Fig. 2–149).

Tectonic activity was widespread around the margins of the continent during the Mississippian Period. However, the central part of the continent was rather stable, allowing a broad variety of shallow-water carbonates to extend over large areas (Fig. 2–150). The Appalachians underwent renewed deformation and uplift during the Acadian Orogeny (beginning in the Middle Devonian), and throughout Kaskaskia deposition these mountains provided alluvial fans of clastic rocks that encroached into the epicontinental sea from the east, while additional sediment was carried in from the Canadian Shield. To the south, late-Mississippian through Pennsylvanian orogeny produced the Ouachita Mountains, which provided a source for alluvial deposits extending northward. In the west, convergence of crustal plates in Late Devonian through Permian time caused local uplifts from Idaho to Nevada, which were interspersed with periods of quiescence during the Mississippian that allowed carbonate rocks to be deposited over widespread areas. Erosion surfaces and alluvial deposits resulting from the western uplifts were of rather limited extent.

Karst features in the post-Kaskaskia surface are best developed in the western United States and diminish in number and clarity eastward, where the uppermost Mississippian rocks become increasingly clastic. Preserved erosional relief is rather uniformly 10–100 m, which is sufficient to have fostered extensive karst wherever carbonate rocks were exposed at or near the surface. Paleokarst is exposed in the eroded flanks of the numerous Rocky Mountain uplifts but is

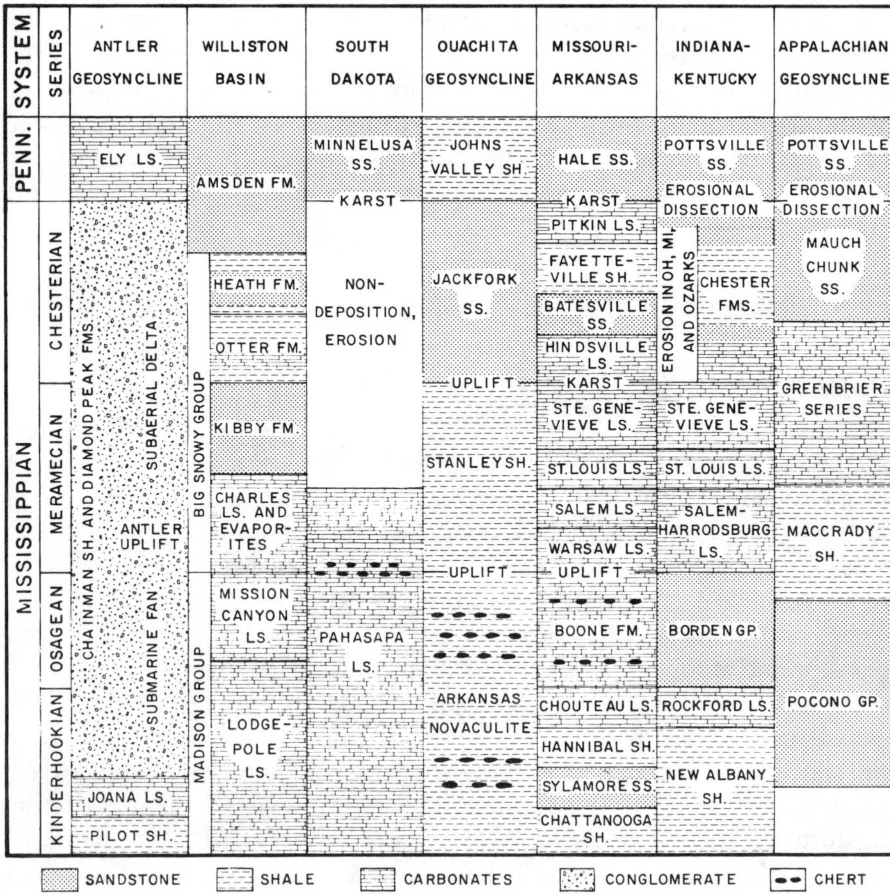

2–150. Representative stratigraphic sections through Mississippian and Lower Pennsylvanian rocks containing the post-Kaskaskia paleokarst.

deeply buried by a Tertiary alluvial cover in the central states. Where the erosion surface is exposed in Kentucky, Indiana, and Tennessee, almost the entire relief is developed in Chesterian clastic rocks, with only scattered karst features in the few carbonates that were exposed.

In the central United States, where most of the post-Kaskaskia surface is still buried, broad uplifts such as the Transcontinental Arch and Ozark Dome were reactivated just prior to or during the post-Kaskaskia erosional period. Karst depressions are abundant in these areas, although their date and origin are obscured where the post-Kaskaskia and post-Sauk surfaces intersect.

The great extent of the post-Kaskaskia surface and the uniformity of its dissection indicate widespread erosion under surprisingly similar conditions throughout much of the continent. Furthermore, the erosion surface is capped by rocks deposited under rather uniform conditions at or near sea level. It is unlikely that tectonic activity alone could have produced such extensive uplift followed shortly by a return to nearly the original elevation. A temporary drop in sea level is a more likely cause, and although evidence for such an event is scant, a low sea-level stand of about 1–2 million years is indicated by VAIL et al. (1977) on the basis of world-wide seismic profiles, micropaleontology, and radiometric age determinations (Fig. 2–146). A tectonic cause is more likely than polar glaciation in causing this drop in sea level. For instance, a decline in sea level as much as 300 m could have been caused by the changes in ocean-basin volume during the decay of an oceanic rise system (ARMSTRONG, 1969). Both the post-Sauk and the post-Kaskaskia erosion surfaces immediately pre-date large-scale thrusting and folding in the orogenic belts, which may support the view that the drop in sea level was tectonically controlled.

Depositional/Erosional History of the post-Kaskaskia Paleokarst

Exposures of post-Kaskaskia paleokarst are extensive in the western United States, where outcrops of Carboniferous rocks are clearly visible and well preserved in the rather arid and poorly vegetated setting, so its morphology and stratigraphic relations are much better known than those of the post-Sauk paleokarst. The following description traces the evolution of the karstic rocks of the Carboniferous in more detail than is necessary to the understanding of the post-Kaskaskia surface as such, but this informtion may be of use in future attempts to distinguish the various episodes of karst development.

The Mississippian Period in North America is divided into four epochs: from earliest to latest, the Kinderhookian, Osagean, Meramecian, and Chesterian (Fig. 2–150). From Late Devonian through the Kinderhookian, transgressive dark-gray shales were deposited in an epicontinental sea over much of the United States. The Chattanooga Shale in the east-central part of the country is an example. Deposits were thickest in the several broad structural basins of the continental interior, particularly the Illinois Basin and Michigan Basin of the central United States and the Williston Basin of North Dakota and eastern Montana. These basins originated early in the Paleozoic Era and periodically underwent renewed downwarping at later times. The Transcontinental Arch, of similar history, formed a relatively elevated, broad area of land or shallow water extending northeast-southwest through South Dakota, Nebraska, and Colorado, separating the major regions of sedimentation to either side. These structures appear to have been reactivated and modified Precambrian features.

During the Kinderhookian and Osagean Epochs, the eastern basins were filled to depths of several hundred metres by deltas of detrital silt (e.g., the Borden and Pocono deltas) derived from the Appalachian Mountains to the east and the Canadian Shield to the north. Siliceous rocks such as the Arkansas Novaculite were deposited in the Ouachita geosyncline north of a volcanic belt that bordered the continent on the southeast. Meanwhile, limestones were deposited in the west (Madison Formation and equivalents) in a shallow-water shelf environment extending from Montana and North Dakota to New Mexico and Arizona.

The earliest episode of Carboniferous karst development appears to be the result of minor subaerial exposure at the end of the Osagean Epoch as the result of uplift of limited extent. A cherty zone in undifferentiated carbonates surrounding the Williston Basin may represent the same exposure episode. An Osagean/Meramecian unconformity developed in relatively high areas in the east and southwest, along with minor dolines and caves now filled with Meramecian sediment. Deposition continued without interruption in basinal areas, while uplifted and tilted rocks of the Appalachians and Ouachitas were truncated by erosion.

Widespread breccias are common in the early Meramecian rocks of the western states near the top of the Madison Formation and its equivalents. These breccia zones are laterally continuous with evaporite deposits in the Williston Basin (Charles Formation) and probably represent collapse caused by intrastratal solution of the evaporites in exposed areas of the shelf (SANDO, 1974). These consist mainly of angular, unstratified, and unsorted carbonate and chert fragments in a red, clay-rich, siltstone matrix. Clasts vary greatly in size and shape. The floors of the breccia bodies are sharply defined, with small, unsorted fragments near the bottom grading upward to only slightly fractured roof bedrock with very little displacement. Illite is the main clay constituent in the matrix, as is true for the Pennsylvanian Amsden Formation, which caps the post-Kaskaskia paleokarst surface. Whether or not this similarity indicates that the breccia was formed during the post-Kaskaskia karst episode is uncertain. The breccias occur only in present domal areas, indicating that the solution of evaporites must have taken place in topographic highs during or after uplift. The post-Kaskaskia surface truncates most of the breccias, so it is likely that these areas were relatively high long before their major uplift during the Laramide Orogeny of Late Cretaceous and Early Tertiary age. Farther down the flanks of domal areas, however, strata were exposed to the circulation of meteoric water only after large-scale Laramide tectonism. Rocks as young as Cretaceous have slumped into some of the brecciated zones in these areas, indicating that at least some of the intrastratal solution greatly post-dates the Carboniferous.

The earliest phases of post-Kaskaskia erosion took place late in the Meramecian Epoch in areas of relatively high elevation in the west. A short erosional break in exposed areas of the eastern United States was followed by deposition of mainly clastic Chesterian rocks, forming a minor unconformity with a few minor paleokarst features. Paleo-dolines filled with Chesterian rocks are scattered throughout many of the central and eastern states, particularly in Missouri. By the end of the Mississippian, however, regression was widespread enough that most of the Carboniferous rocks throughout North America were exposed to erosion. At this time, renewed uplift of domal areas such as the Ozark Dome and Transcontinental Arch caused these areas to be deeply eroded, in many places through the entire thickness of Mississippian rocks.

The resulting erosion surface typically has about 20 m of local relief, with as much as 100 m in places. Dolines, fissures, and caves are common karst features, and a red paleosol covers parts of the surface. In areas of domal uplift in the central United States, karst features formed not only in the Mississippian carbonates but also in underlying rocks exposed by erosion. Around the Ozark Dome the post-Kaskaskia karst truncates carbonate rocks as old as Cambrian age. Deeply entrenched, meandering valleys are preserved in the surface in many areas, showing the presence of well-integrated surface drainage (McKEE and GUTSCHICK, 1969). Such valleys are most common in the east but also occur in Wyoming and Arizona.

The largely clastic Chesterian rocks are thickest and most continuous in the eastern United States, and therefore the post-Kaskaskia erosion surface becomes progressively less karstic toward the east. Where the Chesterian rocks reach thicknesses of several hundred metres, the paleo-topography is composed mainly of deeply entrenched, steep-walled valleys with very little paleokarst. Thin Chesterian limestones interbedded with the clastic rocks appear to have had limited exposure with only minor amounts of karst development of the subjacent variety.

The post-Kaskaskia surface is covered by clastic Pennsylvanian rocks, which filled valleys, surface depressions and nearly all of the caves that were formed at the end of the Mississippian. Some caves show evidence of only partial filling or none at all (CAMPBELL, 1977). The paleokarst fill is most commonly sand and clay colored deep red by hematite. The overlying Pennsylvanian deposits were generally cyclothemic, deltaic or estuarine sediments that accumulated near sea level. In the west they are typically red sandstones with lesser amounts of shale, siltstone, limestone, and conglomerate. In much of the east the basal Pennsylvanian includes a widespread quartzose conglomerate containing well-rounded quartz pebbles averaging 1–2 cm in diameter with a sandstone matrix. These deposits grade upward into sandstones and shales.

Orogenic uplift in the east apparently did not correlate exactly with the various phases of the orogenies in the west (Fig. 2–146), and it is unlikely that synchronicity of tectonic events throughout the continent can be invoked as an explanation for the post-Kaskaskia erosion surface. Uplift and erosion at the Osagean/Meramecian and Meramecian/Chesterian intervals in the Appalachians and Ouachitas contrast in time with the major phase of uplift in the Antler belt, which occurred during the early Meramecian (Fig. 2–146; SANDO, 1974).

A Detailed Example: Paleokarst of the Black Hills, South Dakota

There is no better exposure of post-Kaskaskia paleokarst, or indeed any paleokarst in the United States, than in the Black Hills (Fig. 2–147). There the paleokarst has been intersected by some of the largest caves in the world, including Wind Cave and Jewel Cave, totalling more than 180 km of passages in the Pahasapa Limestone (Madison equivalent). These caves provide a three-dimensional view of the deeper paleokarst features, and some of the upper-level passages are exhumed post-Kaskaskia caves (Fig. 2–151).

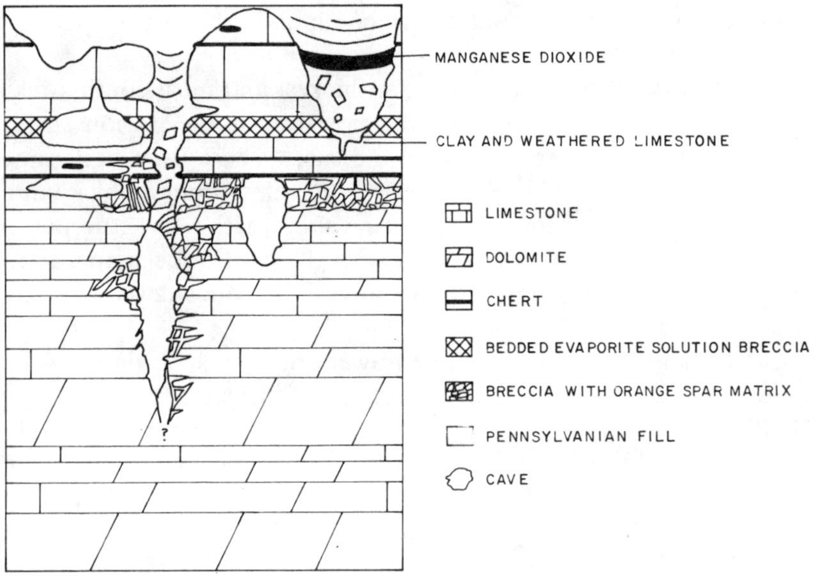

2–151. Relationship of the post-Kaskaskia paleokarst to strata and Tertiary caves in the Black Hills, South Dakota.

The largest caves are located in the southern Black Hills. There the Pahasapa is 100–150 m thick, dolomitic toward the base, and almost pure limestone in the upper third. At least one and commonly two zones of bedded chert separate the dolomitic limestone from the overlying purer limestone. The main phase of cave development took place in the Tertiary Period and involved a combination of artesian and hypogenetic processes (A. N. PALMER, 1984). Current opinion

2–152. Pennsylvanian fill in a post-Kaskaskia fissure intersected by a Tertiary passage in Wind Cave, South Dakota. The fill consists of fragments of sandstone, carbonate rock, and chert in a red matrix of sand and clay (Photo by A. N. Palmer).

favors a deep-seated origin by rising waters driven by thermal gradients and acquiring solutional aggressiveness by cooling.

Lower levels in the caves are mainly irregular fissures intersecting in a complex network pattern. Their pattern is strongly controlled by joints and faults of Laramide age (Late Cretaceous and Early Tertiary). Most of these passages are located in the middle third of the Pahasapa, but a few extend downward almost to the base of the formation. In contrast, the upper levels above the chert beds are broad, arched rooms and passages of irregular cross section. Deep fissures filled with Pennsylvanian sediment also extend downward from the paleo-surface as much as 50 m (Fig. 2–152). Detrital sediment is also visible in the floors of some upper levels that are intersected from below by collapse into underlying passages. This material is almost identical in petrology and mineralogy to the fill in paleodolines in the overlying post-Kaskaskia unconformity, and therefore represents partial or complete filling of Late Mississippian caves with basal Pennsylvanian sediment.

The fill in post-Kaskaskia dolines and fissures is well exposed in canyons, road cuts, and quarries throughout the Black Hills (Fig. 2–153). It typically consists of fine to medium subrounded quartz sand (60 %), undeformed grains of

2–153. Doline of post-Kaskaskia age near the top of the Pahasapa Limestone, filled with red Pennsylvanian sand and clay, exposed in a quarry near Jewel Cave, South Dakota (Photo by A. N. Palmer).

muscovite and biotite (20 %), feldspar (10 %) and chlorite (10 %), with trace amounts of clay, hornblende and tourmaline. The sand grains are surrounded by hematite-rich clay, which gives the fill its red color. In deep fissures and caves of post-Kaskaskia age the fill includes abundant sub-rounded to angular fragments of sandstone, carbonate rock, and chert (Fig. 2–153). The source of the clasts was local, but the matrix was derived from a rather remote igneous or metamorphic terrain. Layers of manganese dioxide up to 50 cm thick are interbedded with the fill in places and may have originated in organic-rich ponds or swamps.

In many places the Tertiary caves intersect filled fissures that may originally have connected directly with the overlying post-Kaskaskia surface. In the few fissures in which the bottom is visible, the lower part of the fill consists of yellow clay and weathered limestone fragments and is overlain by the more abundant red fill described above. The interpretation of this color change is not certain, although the red fill appears to be mainly sand of the Pennsylvanian Minnelusa

Formation. Fossil pollen or spores in the red fill indicate that it was closely linked with the surface.

The fill in post-Kaskaskia fissures and caves is bedded in most places, indicating transport by flowing water (Fig. 2–154), but is chaotic where the material

2–154. Bedded detrital sediment of Pennsylvanian age in a paleocave intersected by a Tertiary passage in Jewel Cave. The white bodies are remnants of calcite spar layers (Photo by A. N. Palmer).

subsided into voids, particularly in fissures (Fig. 2–152). Some of the fill is smeared and slickensided along fissures produced by later faulting. Many paleokarst features have been exhumed where their fill has been sapped into underlying Tertiary cave passages. Cones of detrital sediment spread out into the lower passages from the points of intersection (Fig. 2–153). In places the walls of paleokarst features are bleached by oxidation.

Two well-defined breccias occur in the Pahasapa Limestone of the Black Hills, one just below a major chert zone, roughly 20 m below the top of the formation, and another above the chert zone. The two are probably not contemporaneous, because they show subtle differences in mineralogy and structure. The lower breccia (Fig. 2–155) has a distinctive yellow-orange calcite matrix and in many places cuts discordantly across the limestone beds, whereas the upper breccia has a matrix mainly of green to yellow clay and is more highly concordant with the surrounding beds. Stylolites both pre-date and post-date the spar matrix in the upper breccia but appear only to pre-date the paleokarst fill. Both breccias show several stages of fracturing. These are most clearly distinguished in the lower one, in which the first stage of fracturing produced narrow cracks lined

2-155. Mississippian breccia with yellow-orange calcite matrix in Jewel Cave, below the major chert zone in the Pahasapa Limestone. The spar coating is of Tertiary age. Long dimension of photo is approximately one metre (Photo by A. N. Palmer).

with dolomite and pyrite and containing organic filaments, and a second involved imbricate shearing that displaced the fragments. Void space was later filled with the orange calcite matrix, which is rich in iron oxide and organic material. In places the spar seems to have assimilated the weathered edges of the bedrock fragments, producing a ragged contact. Post-Kaskaskia paleokarst features cut across both breccias, indicating that the breccias pre-date the Pennsylvanian and may be partly or entirely of karstic origin. The upper breccia is probably correlative with that described elsewhere by SANDO (1974) as the product of intrastratal solution of evaporites. The lower breccia shows evidence that fragments were wedged apart, probably by recrystallization of gypsum.

The developmental history of the Black Hills caves and related paleokarst features appears to be as follows: (1) Breccia originated in the Pahasapa Limestone near the end of the Mississippian Period, probably by intrastratal solution of evaporites; (2) Some of the upper levels now present in the caves formed during post-Kaskaskia erosion. Their position near the top of the limestone and mainly above the chert (which may have acted as a partial barrier to solution) suggests an origin by meteoric water; (2) Pennsylvanian sediments filled these early caves almost completely; (3) A single-layer coating of white dogtooth spar several centimetres thick coated the walls of post-Kaskaskia caves that had not been filled with sediment. This event took place sometime between Pennsylvanian burial and Laramide uplift of the region; (4) Post-Kaskaskia karst features were intersected by later caves formed by the main phase of solution during the Tertiary. Paleo-caves were partially exhumed of their Pennsylvanian fill and enlarged by solution. The dogtooth spar was intersected by the Tertiary passages, which in turn were later coated with a thicker, multi-layer coating of spar of varied crystalline form. Remnants of the dogtooth spar layer delineate the outlines of the original post-Kaskaskia voids that were not filled with Pennsylvanian sediment.

Tertiary and Quaternary Paleokarst of Florida

Florida is the recently emergent third of a flat-topped, steep-sided platform whose surface now lies only slightly above and below sea level. The platform has been periodically exposed during low sea-level stands during the Tertiary and Quaternary Periods, resulting in buried karst surfaces, karsted terraces and submarine springs (DAVIES and LeGRAND, 1972).

Many of the buried exposure surfaces are now below the present water table. Well drillers frequently encounter caves at depths as great as 90 m below sea level, and occasionally much deeper. Freshwater rises from springs located below sea level off the coast. Dating of the paleokarst surfaces is based on paleontological and stratigraphic evidence. Some cave and depression fill contains the bones of terrestrial vertebrates as old as Late Oligocene-Early Miocene. The earliest glacial-eustatic drop in sea level is thought to be of Late Miocene age and is represented by a karst terrace covered by phosphatic pebbles 200–400 m below present sea level on the southern Florida platform. A Late Pliocene sea-level stand produced a karst surface now located 35–42 m above sea level in the arched central portions of Florida. Evidence for Pleistocene sea-level fluctuations spans the range from 80 m above to 150 m below sea level, and exposure surfaces are represented by laminated crusts and corrosion zones in the Pleistocene limestone. In southern Florida dolines 75 m wide and 4 m deep formed on Pleistocene limestone during the latest (Wisconsinan) sea-level drop and are now filled with peat or by mangrove swamps whose circular patterns reveal the doline shapes.

An uplift in the north-central part of the state, the Ocala Arch, exposes the present-day bedrock aquifer, the Eocene Ocala Limestone (Fig. 2–147). Where it is exposed, rivers sink into flooded paleokarst systems of dolines and water filled dendritic cave systems and discharge upward at deep springs farther down the flanks of the arch. Downdip the Ocala becomes covered with the silty Miocene Hawthorn Formation, which confines the aquifer.

In southern Florida, wells have penetrated strata-bound horizons of cavernous permeability, called the boulder zone, as much as 900 m below sea level. PLUMMER (1977) suggests that this zone may have formed in areas of freshwater-seawater mixing when sea level was lower. However, it may instead be the result of intrastratal solution of evaporites, for PLUMMER also states that today water acquires sulfate as it flows seaward through the aquifer.

Paleokarst Formed by Evaporite Solution in Michigan

Although thick deposits of Pleistocene glacial drift cover most of the Michigan Basin, some unusual karst features are found at the northern edge of the basin (Fig. 2–156). These include large, water-filled dolines partly choked with

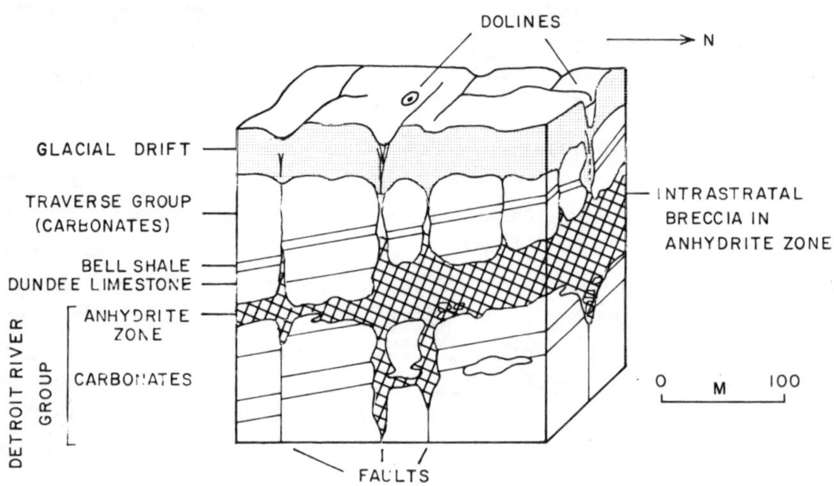

2–156. Generalized diagram of Mackinac Breccia and its relationship to surface karst in northern Michigan (after BLACK, 1983).

glacial drift, as well as submerged springs that discharge northward into Lake Huron. The area is underlain by the Silurian Salina Formation and the Devonian Detroit River Group, which contain large amounts of evaporites. The Mackinac Breccia of northern Michigan is the product of collapse due to solution of evaporites and lesser amounts of carbonate in the peripheral areas of the basin. The dolines are the surface expressions of vertical fractures filled with breccia. They formed by stoping upward along faults from the zones of evaporite solution. A deep regional flow system allows solution to occur to depths of at least 450 m. Drillers commonly encounter interconnected voids below the surface, and water levels in wells are usually close to the level of Lake Huron. Water in dolines, springs, and bedrock wells is high in sulfates (about 1 350 mg/liter). In one quarry, shafts filled with quartz sand are aligned along faults, with the bedrock locally dipping steeply toward them. Similar features were noted by BRETZ (1950) in mineralized slump pipes in the Missouri lead-zinc district. Since some springs and drift-filled dolines are located below the level of Lake Huron, it is likely that these features pre-date the last (Wisconsinan) glaciation, although solution, brecciation, and modification of paleokarst features probably continue today.

Economic Deposits in Paleokarst

Oil and Gas

The post-Sauk and post-Kaskaskia erosion surfaces are traps for oil and gas in certain areas, particularly where the circulation of groundwater is limited. Leached dolomite mounds along the post-Sauk surface are traps for oil and gas on the eastern flank of the Cincinnati arch. Geophysical mapping of these mounds is currently used to locate oil and gas wells of small but reliable production. The Oklahoma City Field, one of the ten largest oil fields in the United States, has yielded some of its oil from the Arbuckle Dolomite (Knox equivalent) about 120–150 m below the post-Sauk surface. This horizon contains irregular zones of medium-to-coarsely crystalline dolomite honeycombed with solution cavities as much as 2 m in diameter. Large amounts of oil were removed from this highly permeable reservoir, but the wells were quickly abandoned because the interconnected cavities rapidly filled with water.

The post-Kaskaskia paleokarst is not particularly noted as an oil trap. In the Williston Basin oil field in Montana and North Dakota, most of the oil in the Mississippian Madison Limestone comes from dolostones whose porosity is not related to the post-Kaskaskia surface. In the Kevin-Sunburst Field on the Sweetgrass Arch in Montana, some oil occurs in the paleokarst zone in joint controlled fissures. Elsewhere in the west the Madison paleokarst is a highly permeable aquifer and does not normally trap oil.

The post-Kaskaskia unconformity serves as a trap for about 10 % of the oil in the central United States, although much of it is not associated with karst. However, the Kraft-Prusa Field in central Kansas produces oil from caves and solutionally enlarged joints. Well logs show that the producing zone is located in what appears to have been cockpit karst in the Ordovician Arbuckle Dolomite. The karst surrounded Precambrian hills which probably acted as the aggressive water source for the karst when it was forming. One-third of the oil from the post-Kaskaskia paleokarst has come from the Sooner Trend in central Oklahoma. Oil and gas are trapped in highly fractured Osagean-Meramecian limestone, which is highly siliceous and has been shattered nearly to a breccia.

Hydrothermal Ore

Economically valuable lead-zinc sulphide deposits in carbonate rocks are common in two major areas of the United States: the upper Mississippi Valley, and central and eastern Tennessee (Fig 2–157). In addition, some economic deposits in carbonate rocks are associated with igneous intrusions in the western states. In all of these areas the ore post-dates the host strata.

Mississippi Valley Ore Deposits

Most of the lead in the United States has been mined from galena deposits in Paleozoic carbonate rocks surrounding the Cretaceous and Cenozoic deposits of the Mississippi Embayment. The ore-bearing rocks are nearly horizontal and are laterally continuous over wide areas. The Tri-State area of Missouri, Oklahoma, and Kansas is one of the best-known ore producers (McKnight and Fischer, 1970). The ore is located in the cherty Mississippian Boone Formation, which is truncated by the post-Kaskaskia karst surface. It is concentrated in bioherm mounds at the intersection between a series of gentle northwest-trending folds and basement faults along graben structures. The ore fluid tended to alter limestone to coarsely crystalline dolostone. Where chert is common, the surrounding carbonate bedrock is silicified. Solutional voids formed in most of the carbonate rocks before the ores were deposited, and these two processes may have been performed by the same fluid.

Ore bodies appear to have formed at shallow depths, generally within 150 m of an overlying erosion surface, and many end upward at an unconformity. This relationship implies that much of the ore and solutional bodies formed soon after the bedrock was lithified. However, in many areas, solutionally induced collapse and brecciation in ore-bearing zones have been reactivated, and some solution is still active today. The precise age of the solution features and ore deposits is therefore difficult to determine.

The ore zones have many similarities with buried karst and with zones of intrastratal evaporite solution, and it is probable that paleokarst has acted as a

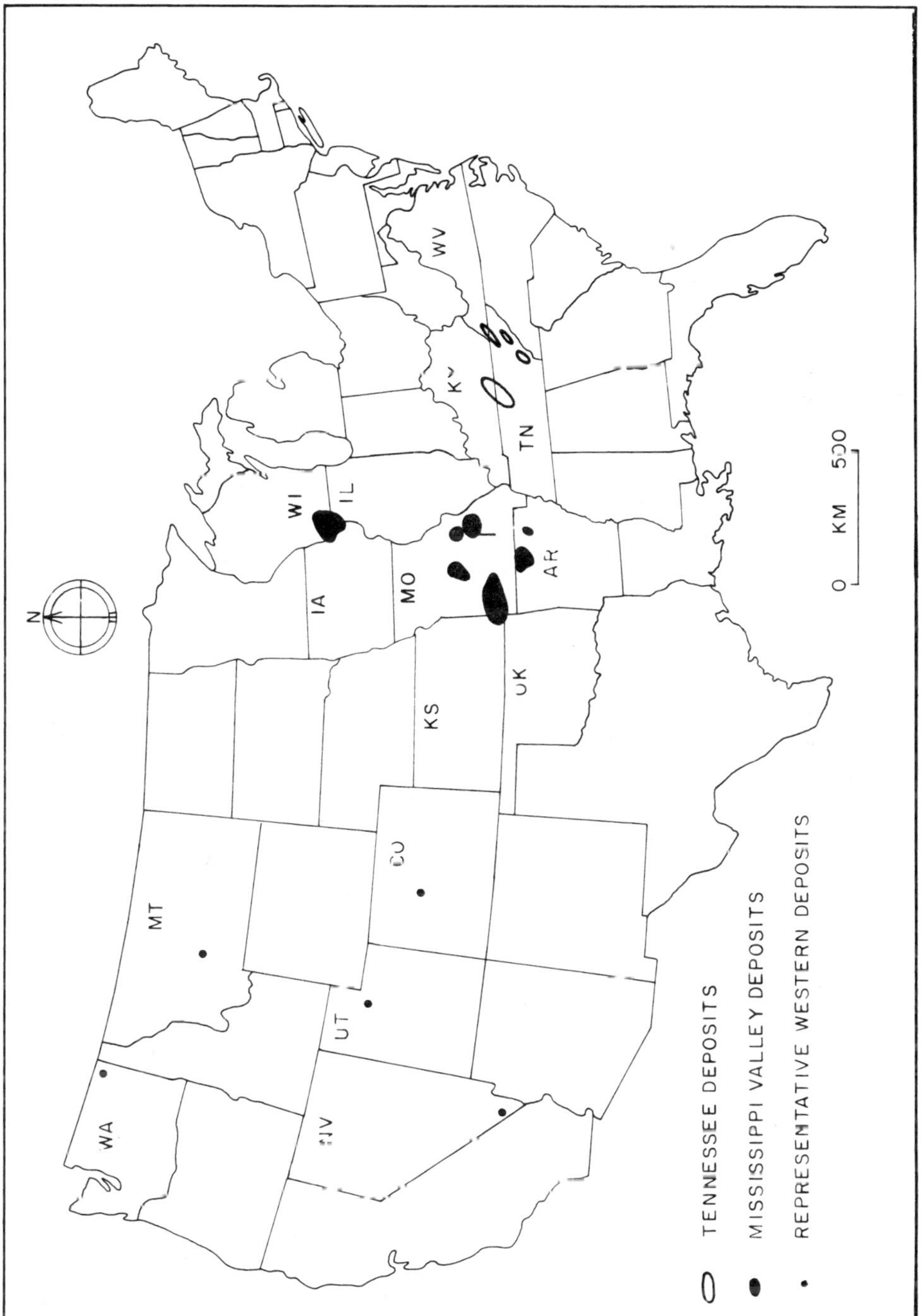

2–157. Major lead-zinc deposits in carbonate rocks of the United States associated with solutional features.

host for at least some of the ore. Because the ore appears to be concentrated around unconformities, it is thought that caves and dolines formed by meteoric water have provided pathways for the ore-bearing fluids. Solutional voids are partly filled with breccia, and overlying beds have slumped into them from above. Most of the ore is located in the matrix of the breccias. The breccias have been explained in a variety of ways, including explosive hydraulic fracturing, karst processes, and disruption by the recrystallization and later solution of evaporites. Breccia bodies exhibit four main shapes: cylindrical dome-shaped deposits, tabular curvilinear deposits, sheet-like deposits, and rectilinear fracture networks. Some ore also occupies isolated vugs partly lined with saddle dolomite crystals, which are indicative of hydrothermal solutions. Solutional thinning of limestone beds has caused slumping of overlying beds and left residual clay seams. Solutional voids contain no contemporaneous detrital sediments, so apparently the groundwater flow that formed them was slow and there were no surface recharge points for solutionally aggressive water.

The cylindrical breccia bodies are sometimes referred to as *circles* or *sinkholes* and have been described in detail by BRETZ (1950). Most are 30–90 m wide and about 50 m in vertical extent, but these dimensions can be as great as 2.5 km and 200 m respectively. The sides of the circles are usually along faults and the circle walls are lined with displaced but unrotated fragments of wall rock (called *crackle breccia* by mining geologists) that grade laterally into fractured bedrock. This gradation argues against a tectonic origin for the breccia. Solution along bedding planes around each circle causes the surrounding beds to dip toward the circle. The top of each circle is commonly sediment fill from the overlying formation which is tilted and deformed. Exposure in mines shows that most circles are located on small structural domes or bioherm mounds recrystallized to dolostone, which suggests a hypogenetic origin by rising fluids.

Strata-bound breccias are common along chert beds in the Mississippian Boone Formation in the upper Mississippi Valley. These are areas of intrastratal solution called *sheet deposits*. Cherty limestone is dissolved, forming brecciated beds up to 5 m thick and a square kilometre in area. Breccia that remains unmineralized forms so-called *boulder zones* that are major shallow aquifers in Missouri and Tennessee. Breccias along faults, joints and bedding planes form rectilinear maze patterns with many dead ends that resemble hypogenetic caves like Wind and Jewel Caves in South Dakota. These features are more common in the Ordovician Knox Group in Tennessee.

The Viburnum Trend of southeastern Missouri (shown as a north-south linear zone on Fig. 2–157) is a source of lead-zinc ore discovered only in the late 1960s. It is a series of wide, sinuous, tabular breccia bodies in the Cambrian Bonneterre Formation that parallel former shelf-edge algal reefs and which are about 90 m wide and 25 m thick. These may have originally been tidal channels. Slightly to the northeast, the Old Lead Belt, known since 1720, is the single greatest lead district in the United States. Ore occurs as disseminated replacement of bedrock. In a few places tabular breccia bodies merge with the bottoms of *circles*.

The upper Mississippi Valley lead-zinc district includes the contiguous edges of Wisconsin, Iowa, and Illinois. Here the host rocks date from Late Cambrian through Late Silurian. Ore is most common in Middle Ordovician dolostones, which are bracketed by two unconformities, the post-Sauk surface below and a Middle/Upper Ordovician surface above. The most highly mineralized formation is the Middle Ordovician Galena Formation. Ore occurs along fractures and bedding-plane partings (called *gash-vein deposits* and *pitch-and-flat deposits* respectively) in arcuate collapse zones around structural highs. In Iowa, small maze caves and widened joints also contain ore.

Several explanations have been given for the ore in the upper Mississippi Valley (OHLE, 1980). The valley is a periodically reactivated, graben-shaped rift zone. Alkaline igneous rocks, crypto-explosion structures caused by the escape of gas and hot water, and former volcanoes ranging in age from Late Cambrian to Cretaceous lie close to the mineralized areas. These features may have been the source of heat that initiated ore movement and heated the brines to ore-forming temperatures. Lead and zinc ores also occur in the oil-field brines along the Gulf coast, and basinal fluid may have moved upward and northward through the semi-lithified sediments in the Mississippi Embayment and discharged along fracture zones and structural highs in the surrounding bedrock. Zoning of ore minerals in Missouri from high to low temperature away from the Mississippi Valley suggests a single episode of ore genesis.

Fluid inclusions show that the ore-bearing fluids were saline and therefore probably of deep basinal origin, and that their original temperature was about 80–150 degrees Celsius. In a few places oil is associated with hydrothermal ore, which suggests that water rich in hydrogen sulphide, typical of carbonate oil fields, migrated from nearby basins and mixed with the metal-bearing brines. The bedded breccias and collapse domes of the Mississippi Valley are similar to those in the evaporite karst of Michigan. Gypsum inclusions have been found in the bedrock, which is otherwise now devoid of evaporites. It is possible that reduction of former gypsum beds provided sulfides to form a chemical trap for the metals.

Ore Deposits in Central and Eastern Tennessee

The other major ore area in carbonates in the United States is in the Lower Ordovician Knox Group around the Nashville Dome in central Tennessee and in the folded and faulted rocks of the Appalachians in eastern Tennessee (KYLE, 1983). These areas produce only zinc ore (sphalerite) in economic concentrations. The ore is found in shallow-water carbonates of the Kingsport Formation and overlying Mascot Dolomite at the top of the Knox Group, which are truncated by the post-Sauk paleokarst surface.

The major host for the ore is the late breccia of the upper Knox Group, described earlier, which surrounds and follows the trend of the unmineralized earlier breccia. Ore-bearing breccias also form network patterns in which the individual fracture zones are normally 2–5 m high and 3–6 m wide and which are restricted to limestone beds underlying confining beds of primary dolostone. Ore is mined down dipping beds to a depth of about 300 m and has been found in Cambrian rocks as much as 1 500 m below the main ore zone in the Knox Group. *Break-through* breccias in bodies as much as 60 m high and 300 m long extend above the Kingsport/Mascot contact. Some bodies of the early breccia have been enlarged by upward stoping to the post-Sauk unconformity, where they merge with the deep dolines along the Waverly Arch. These are similar to the *circle* deposits of Missouri. Ore occurs along the edges of tabular, reef-like bodies and inside sinuous, anastomotic breccias, like those of the Viburnum Trend of Missouri, which originally may have been tidal channels. Ore was deposited in the late breccia at the boundary between the limestone bedrock and the coarsely crystalline dolostone, mainly at the tops of the breccia bodies where previous solution was most intense. Stratified dolomite sand from the disintegration of the coarsely crystallized bedrock fills large open spaces and contains fragments of ore. In eastern Tennessee these deposits parallel the dipping beds of the rock, which suggests that the ore in that region pre-dates the Appalachian deformation and may be as early as Middle Ordovician. The ore-bearing fluids were probably derived from compaction of sediments in the Appalachian geosyncline and transmitted through permeable zones, including paleokarst. In central Tennessee, however, ore extends upward as far as Mississippian rocks, which suggests a younger and perhaps entirely different episode of ore deposition. The source area may have been sediments of the Mississippian Embayment. In both central and eastern Tennessee, most of the breccias do not contain ore and are either filled with calcite cement or retain large amounts of open space. The open zones in central Tennessee and southern Kentucky are significant aquifers, as described earlier.

Western Ore Deposits

Lead-zinc deposits also occur in carbonate rocks in the western United States in areas of domal uplifts caused by Laramide intrusions of monzonite (quartz-deficient granodiorite). Hydrothermal solutions have altered the Leadville Limestone in parts of Colorado, replacing limestone with coarsely crystalline dolostone or with silica. Exhumed, fracture-controlled maze caves in the area ar probably of hypogenetic origin, formed either by rising sulphide-rich water or by cooling thermal waters.

Paleokarst and related mineralization in the Leadville Limestone have been described by MASLYN (1976) and by a recent group of papers edited by De VOTO (1985). Mississippian caves in the area are filled with breccia and Pennsylvanian sediment, both of which contain lead, zinc, and silver ore. The caves appear to have drained to large paleo-valleys in the limestone that are now filled with Pennsylvanian detrital rocks. Typical cave fill consists of clay and dolomitic sand covered by a breccia of collapsed carbonate roof material having a matrix of dolomitic sand or black clay. Breccia clasts are chaotic at the bottom but nearly in place above. Any remaining voids were filled with black shale of the Pennsylvanian Molas Formation. Mineralization is also common in Mississippian soil rubble zones. The ore is considered to be partly late Paleozoic and partly Tertiary in age and is associated with a regional zone of dolomitization.

Copper, silver, lead, zinc, and manganese are mined from the Mission Canyon Limestone at Castle Mountain in central Montana, where an igneous intrusion has domed the overlying sedimentary rocks. The highest-grade ore is at the edge of the intrusion where the limestone is dolomitic and friable. The ore is found along bedding-plane partings, in brecciated and fractured rocks, and in solution cavities.

In the Colorado Plateau in the vicinity of the Grand Canyon, Arizona, breccia pipes formed by upward stoping of overlying sedimentary rocks from zones of solution in the Mississippian Redwall Limestone. Some are mined for copper and uranium. These breccia pipes may have originated in zones of post-Kaskaskia paleokarst, but their exact age and mode of origin is uncertain.

Conclusions

It is clear that the major paleokarst zones of the United States are significant to the erosional, depositional, and tectonic history of the continent, but much more research is needed before a detailed account can be made of the environments and large-scale processes that they represent. The study of paleokarst in North America is in its infancy.

The paleokarst features described here are host to extensive ore deposits and serve as regional aquifers and traps for petroleum. Efficient use of these resources depends on a careful assessment of their distribution. To do this, it is important to determine whether the host features (mainly solutional voids and breccias) are the product of surficial karst processes, mixing of fluids at depth, or hydrothermal activity. More detailed mapping of paleokarst and comparison with present-day karst will be necessary in making this distinction.

PART III

MINERAL DEPOSITS CONNECTED WITH KARST

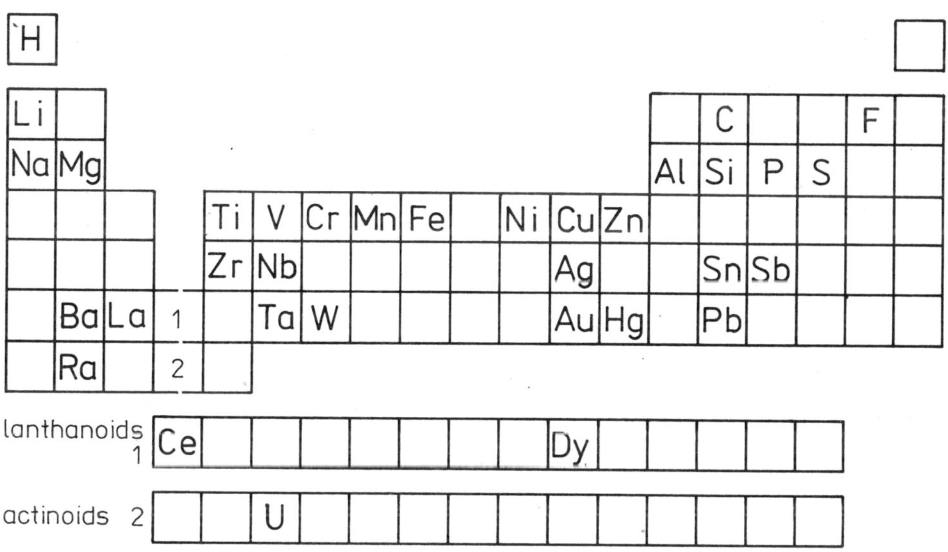

AN INTRODUCTION TO KARST-RELATED MINERAL DEPOSITS

Pavel Bosák

Much of modern industry relies upon the search for new deposits of raw materials. Not only new, non-traditional raw materials are of interest but also new unconventional host terrains are being explored. Although karst lands have long been recognized as loci in which certain natural resources can accumulate rpreferentially (ZUFFARDI, 1976, p. 175), karst-type ore deposits have been a major new discovery for many mining geologists in many countries during the past decade or so (BONI and AMSTUTZ, 1982, p. 73).

Numerous types of mineral deposits occurring at the karst surface and in the subsurface have been known for a long time. Some of them were being mined in Classical Greek times or even earlier, e.g. Chinerse exploitation of karst ores in the Alai Range of central Asia. In studies of such deposits, attention has been focussed rather too narrowly upon the deposits themselves, i.e. the delineation of deposits, structure, ore distribution and reserves, rather than upon the role of karst features and karst processes affecting the ore genesis, distribution and other characteristics.

The abundance of karst-associated mineral deposits contrasts sharply with the virtual absence of observable ore deposition occurring in modern karst terrains. In addition, some processes involved in mineral deposition from vadose and phreatic meteoric waters appear to bear little resemblance to modern observable and shallow karst groundwater activity (LAZNICKA, 1985, p. 727). Increasing attention is therefore being attached to paleokarst as a favourable environment for ore deposition (ZUFFARDI, 1976, p.175). Certain deposits at or near the present surface are explained by processes of exhumation and/or rejuvenation or that they belong to relict karst landforms, e.g. Vietnamese iron and bauxite deposits. Some phosphates and bauxites in the Pacific (see BOURROUILH-LE JAN, p. 443–471) can be cited as examples of modern karst deposits.

There are only a few review articles and studies concerning karst-associated mineral deposits, e.g. GINZBURG (1966), QUINLAN (1972), KROPACHEV (1972),

Bosák, D. C. Ford, J. Głazek and I. Horáček (Editors):
Paleokarst. A Systematic and Regional Review. – Elsevier and Academia. Amsterdam and Praha. 1989.

PERNA (1975), ZUFFARDI (1976), BOSÁK (1980c; in press). Many useful discussions and illustrations can be found also in books edited by AMSTUTZ and BERNARD (1972), KLEMM and SCHNEIDER (1977), RIDGE (1980), AMSTUTZ et al. (1982), SCHNEIDER (1983) and others as well as in The Handbook of Strata-Bound and Stratiform Deposits edited by K. H. WOLF and the very extensive compendium of LAZNICKA (1985).

Some Specific Characteristics of Deposits in Karst

Mineral deposits connected with karst show some common characteristics which have been noted by a majority of the above mentioned authors. They are as follows:

1. The mineral deposits occur mostly in carbonate host-rocks; other soluble rocks, such as rock salt, gypsum and anhydrite are far less frequent as hosts (ZUFFARDI, 1976, p. 176);

2. They occur beneath, and not necessarily close to, an unconformity or disconformable surface. The carbonate host-rock may be cut by the surface directly or separated from it by other rocks (ZUFFARDI, 1976, p. 176). Mineral enrichment is linked especially to erosion surfaces associated with major weathering horizons (SCHULTZ, 1982, p. 108);

3. The karst features in a majority of cases provide nothing more than open space for ore deposition, i.e. they serve as traps and filters. Filling of such voids can, and usually does, take place long after the karst developed and perhaps occurs at much greater depths and by ore crystallization from heated solutions (LAZNICKA, 1985, p.727–728).

The paleokarst origin of ore mineralization is also suggested by the following features (SCHULTZ, 1982, p. 108): (i) discordant layers exposed by erosion; (ii) the presence of weathering horizons; (iii) rather shallow erosion features; (iv) redeposited and allochthonous detritus accompanying chemically precipitated ore and gangue minerals, and (v) the likelihood of their volume or economic concentrations being derived from a previous enrichment.

In addition to the rugged topography of many unconformities on carbonate rocks and the preservation of remnants of soil profiles, QUINLAN (1972, p. 156) suggests that the following are the most common karst and related features in which mineral deposits are found in the rock record: caves, cave fills and related speleothem deposits, various types of breccias, breccia pipes, other solution-subsidence structures, and lag deposits of chert gravels. Other common karst landforms, such as sinkholes for example, are surprisingly rare in the rock record according to QUINLAN. Cave breccias and mineralized cave fill are, on the contrary, rare in modern caves. The regional and topical review we present here shows that surficial karst forms play an important role in the preservation of many paleokarst fills, not only various doline types and geological organs but also such large landforms as poljes, etc. Their role can be traced in bauxite, iron ore, clay, sand and kaolin, phosphate deposits, in coal deposition, etc.

The characteristic shapes of karst-related mineral deposits are summarised by ZUFFARDI (1976) and some textures also by SCHULTZ (1982). The form of deposits originated from thermal solutions has been studied particularly by DŻUŁYŃSKI (see DŻUŁYŃSKI and SASS-GUSTKIEWICZ, p. 377–397).

Hydrothermal Versus Recrystallization Theories

The substantial role of *hydrothermal* or *thermomineral* karst (KUNSKÝ, 1957) processes in the formation of some ore deposits in karst is becoming apparent, although there has been a long lasting controversy on the subject (cf. ZUFFARDI, 1976; DŻUŁYŃSKI and SASS-GUSTKIEWICZ, p. 377–397). The typical *karst ore*, i.e. MVT lead-zinc deposits, requires 70–150 °C connate or other brines acting at depth beneath the surface. However, the problem is rather odd (SCHULTZ, 1982, p. 109): "what is typical karst mineralization for one school of thought, is characteristic proof of hydrothermal replacement by minerals to other scientists". This statement is clearly illustrated by ZUFFARDI (1976, p. 195): "Rather high homogenisation temperatures in paleokarst ore mineral deposits are stressed by the supporters of hydrothermal genesis. Those temperatures, however, can be easily and reasonably explained by ... regional metamorphism of deposits. In fact, burial to a depth of 4 000 m under conditions of a normal thermal gradient would establish a sufficient temperature ... with consequent possible recrystallization and modification of the fluid inclusions".

It is difficult to evaluate this debate at present. But readers should bear in mind that heated waters without *any* associations with deep magmatic sources are thermal waters according to the standard definition. They may be deeply circulating meteoric waters (see MÜLLER, p.155–163; GŁAZEK, p. 77–105) or connate fossil waters expelled during diagenesis or deformation. At the same time it seems that many enrichment models, which explain the ore concentrations only in terms of karstification, leaching and the accumulation of trace elements from carbonate host-rocks or by the effect of hypersaline environments concentrating and trapping elements from the immediate neighbourhood of the karst, are inadequate because of the large amounts of ore that are deposited. It is probable that none of the individual models is universal. Some deposits can be formed exclusively by hydrothermal processes, some by karst enrichment with subsequent thermal ore recrystallization, some deposits may represent a combination of both sets of processes. Deep hypogene inter- and intrastratal karst processes and others of karst rejuvenation probably play a major role.

Classifications of Karst-Associated Mineral Deposits

Classification or typologies of karst-associated mineral deposits are based either upon the relationships between the mineral accumulation and the host-rock or upon the depositional processes believed to be operating at the time of mineral accumulation.

PERNA (1975) suggested the following classes: (1) mineral deposits of the *karst cycle*, or autochthonous and parautochthonous mineral deposits; where the mineral is the residuum of the dissolution of the carbonate rock and the place of deposition is the karst produced by that dissolution; (2) mineral deposits *in karst*, or allochthonous mineral deposits; where the minerals come from crystalline rocks or in no manner can be derived from the carbonate rocks hosting the karst that is the site of mineral deposition; (3) mineral deposits of *karst enrichment*; where the carbonate rock is weakly mineralized and undergoes superficial dissolution of the carbonate substance with (i) detrital enrichment in economic minerals i.e. *residual mineral deposits*, or (ii) dissolution and reprecipitation of economic minerals i.e. *supergene mineral deposits*, or (iii) alteration and concentration of minerals i.e. *alteration deposits*, and finally (iv) comparative enrichment due to circulating phreatic waters removing the carbonate i.e. *phreatic mineral deposits*.

ZUFFARDI (1976) similarly divided mineral deposits into: (1) *autochthonous* deposits, where the composition is qualitatively made up of the same components as the host-rock; (2) *allochthonous* deposits where the composition is made up of constituents foreign both to the host-rock and to its immediate surroundings (3) *parautochthonous* deposits where the composition is primarily of constituents which are foreign to the host-rock, but the same as its close surroundings. This process of accumulation may also be considered as a residual concentration (which is not the standard explanation of residual deposits — Eds.).

Minerogenetic and depositional factors led QUINLAN (1972) to distinguish: (1) *residual deposits* from which predominantly carbonate bedrock has been leached and which may be further altered by weathering; (2) *subsurface precipitates* from meteoric and/or stratifuge waters, some of which may also be hydrothermal; (3) *subsurface deposits* related to bacterial reactions; (4) deposits preserved in solution-subsidence structures, and (5) surface or near-surface deposits by *waters* from karst aquifers or allogenic streams.

KROPACHEV (1972) classified karst deposits according to the classical typology of SMIRNOV (in English 1975). In a similar classification BOSÁK (1980c) recognized: (1) *sedimentary deposits*; (2) *weathering deposits* (i.e. residual, infiltrational and mixed), and (3) *hydrothermal deposits*. All deposits can be contemporary with the karstification and can be originated by the karst processes themselves, i.e. *syngenetic deposits*, or they can fill pre-existing karst, i.e. *epigenetic deposits*. Syngenetic deposits are the equivalents of the autochthonous deposits of ZUFFARDI (1976) or deposits of the karst cycle and deposits of karst enrichment of PERNA (1975). They also include part of the allochthonous deposits of both authors.

Types of Minerals Deposited in Karst

Nearly 30 different types of mineral accumulations have been described in karst-associated deposits. QUINLAN (1972), ZUFFARDI (1976) and LAZNICKA (1985), have presented reviews although these are not complete. From the point of view of their economic importance and the amount of mining, the most important deposits are: (1) lead-zinc ores with silver, accompanied by copper, pyrite, baryte, fluorite; (2) bauxites; (3) oil and gas; (4) phosphates; (5) gold; (6) diamonds; (7) cassiterite and wolframite; (8) baryte; (9) uranium, and (10) antimony. Some of these deposits, namely lead-zinc ores, bauxites, phosphates and uranium ores are discussed in their own special chapters, thanks to specialists who were willing to contribute to this book. We complete the review with summaries of such traditional deposits as iron ores, clays, sands and kaolin which have been mined and studied for a long time. Each special contribution is arranged to present an explanation and review of ore-forming processes, and of the role of karst in the origin and distribution of the mineral accumulation, rather than as a review of world deposits.

We should not forget that the most important raw material of all occurs in karstwater. It is estimated that 25 % of the global population depends on karst water for domestic use (MILANOVIĆ, 1981). The role of paleokarst in groundwater distribution and circulation is summarized in this volume by ZÖTL (p. 483–509). Some interesting data can be found also in BÖCKER and VIZY (p. 533–548), WILK (p. 513–531) and MÜLLER (p. 155–163).

Oil and Gas

Carbonate rocks are one of the most important collectors of oil and gas accumulations. High porosity results from a complex sequence of early to late, vadose and/or phreatic diagenetic processes, dissolution, leaching, mineral in version, dolomitization and many others (cf. ELF-AQUITAINE, REECKMANN and FRIEDMAN, 1982). The primary porosity, and the secondary porosity to which karst belongs according to carbonate sedimentologists (e.g. ESTEBAN and KLAPPA, 1983; LOHMANN, 1985) serve as excellent traps for migrating hydrocarbons. Oil and gas fill not only small pores or vugs but also large caverns as has been proven by speleothems brought up by the oil pumped from deposits in the Gulf Coast Province of the U.S.A. Important oil and gas reservoirs in carbonate rock containing major paleokarst are known from the Near and Middle East and the Arabian Peninsula, from Europe (Hungary, Romania, etc.) and the U.S.A. (Williston Basin, Texas, Indiana, Ohio, Oklahoma, etc.). Rich oil and gas accumulations have also been detected in the U.S.S.R. (Volga-Ural and Dnieper-Donetz Oil Provinces, Dagestan, Fergana Basin, etc.; VYSOTSKY, 1976; WILSON, 1985).

Coal

Coal deposits are relatively common in karst. They develop in subsiding tectono-karst and subrosion basins and troughs. Less frequently they appear in the paludal and swampy zones of low lying karst plains. There is an interesting association between some bauxite and coal accumulations: black coal developed above bauxite horizons in Hunan Province (China); the Ajka Coal Basin (see BÁRDOSSY and KORDOS, p. 137–153; BÖCKER and VIZY, p. 533–548) in Hungary rests on bauxites in continental deposits. Brown coal is commonly found in the karst of Poland (GŁAZEK, p. 77–105), the greatest deposit being in the subrosion graben near Bełchatów. In the U.S.S.R., the largest karst-related coal deposits accumulated in the Kiselov Basin of the West Ural and in southeastern Bashkiria (both black coal), plus the Ufima Plateau, the southwestern part of Donetzbas and in the Dieper-Don Basin (all brown coals). Coal deposits in karst are known in Missouri (U.S.A.). See BOSÁK (1980c), GVOZDETSKY (1982).

Placer Deposits

Placer deposits in alluvial, lacustrine, deluvial and eluvial environments are often associated with rugged relief in the underlying bedrock. Karstified carbonate rocks may serve as excellent traps for heavy mineral deposits. Gold placers are of great economic importance in Siberia, e.g. Yakutia, Yenisei Range, Kuznetskiy Alatau, Salair Mountains, East Sayan Mountains, Patoma Highlands, etc., where they fill morphologically diversified paleokarst depressions of different sizes and ages (TSYKIN, 1985).

Diamond placers are very important in Africa. Bakwanga (Zaire) and the Lichtenburg-Ventesdorp area (Republic of South Africa) are the best known examples of diamond-bearing fossil (Neogene?) alluvial deposits emplaced in karst relief. Diamond placers in sinkholes are known also from the U.S.A. and U.S.S.R. (Koyva Basin and other placers in North and Central Ural Mountains, Vishera River Basin).

Cassiterite and *wolframite* placers also containing columbite are widely distributed in Southeast Asia, namely in Malaysia (Kinta Valley), Laos (Nam-Phatene River) and in North Vietnam (Tin-Tuc polje). A special kind of *lead* placer has been described from South China.

Placers containing the *rare earths* and other economically important minerals (leucoxene, tourmaline, monazite, rutile, zircon, ilmenite, almandine) have been intensively studied recently, mainly in North Siberia (between the Anabar and Udzhina Rivers) and in the Trans-Yenisei area, in the North and Central Ural Mountains.. *Niobium* and *tantalum*-rich minerals occur in karst placers in the U.S.S.R. as well, and in karst depressions in carbonatites in Tanzania where there are also Ce, La and Dy-bearing minerals like bastnaesite.

Precious stones are mined in Myanma (Burma), e.g. the Mogok deposit (ruby, spinel from alluvial and deluvial sediments) and in Sri Lanka, e.g. the Pelmadulla deposit (sapphire and ruby from karst pockets up to 40 m deep). See: BATEMAN (1950), GINZBURG (1966), GŁAZEK (1966), BARDET (1974), SMIRNOV (1975), PROKOPCHUK et al. (1985); CÍLEK Jr. (pers. comm., 1980).

Antimony

Antimony deposits in karst and paleokarst are of growing importance and interest. Fragments of Sb minerals (oxides) occur in residual clays in the Kwangsi and Honan Provinces of China. Cinnabar and stibnite associated with fluorite are contained within thin jasperoid lenses superimposed on paleokarst in Transbaikalia (Kelyansk deposit). Antimony together with gold occurs in the Bau deposit of Sarawak. The Bou Zitoun deposit in Algeria is described by FUCHS and TOUAHRI (p. 191–200). See: ZUFFARDI (1976) and LAZNICKA (1985).

Vanadium

Vanadates are commonly associated with the oxidation and cementation zones of certain Pb, Zn, Cu deposits in carbonate rocks, e.g. Berg Aukas (Namibia), Kabwe (Zambia), Akenab (SE Africa), Itacarambí-Januaria (Brazil). The Tyuya Muyun deposit in south Fergana (U.S.S.R.) contains a rich assemblage of Cu and U vanadates and of uranium minerals filling small vugs and forming impregnations in older calcite-baryte low temperature cave fills. The vanadates originated from infiltrating weathering solutions which leached Cu, U and V from black shales and slates and volcanics in the neighbourhood of the karst massif. Similar deposits also occur in La Rioja Province, Argentina. Vanadium deposits are reported also from Jebel Goraa in Tunisia (see FUCHS and TOUAHRI, p. 191–200; also BATEMAN, 1950; PERNA, 1975; DE BRODTKORB, 1982; LAZNICKA, 1985; BOSÁK, 1986).

Fluorite

Fluorite is a common gangue mineral in many MVT lead-zinc deposits. Its occurrence is widely noted in this book (e.g. FUCHS and TOUAHRI, p. 191–200; T. D. FORD, p. 51–70). One of the world's largest fluorite deposits occurs in the Transvaal, South Africa. An original cave was infilled with clayey sediments which were later altered hydrothermally with the precipitation of fluorite, baryte and galena (see BOSÁK, 1980c).

Apatite

Apatite deposits can be classified as an infiltrational-metasomatic type. They fill depressions and lesser irregularities of karst surfaces. The source of the phosphorus or of the apatite mineral itself can be found in the carbonate rock or in the vicinity of the karst. Deposits in Tennessee and Florida originated from percolating weathering fluids acting on the limestones of fissure walls by selective metasomatism. Apatite in karst is also known from Belgium (Liège) and the U.S.S.R. (Telek deposit in the East Sayan Mountains). A large apatite siltstone deposit accompanied by uranium ores has been noted by FUCHS (p. 475–478) in Bakouma, Central African Republic. Residual apatite fills karst depressions in carbonatites at Cargill, Canada (D. C. FORD, p. 327–328; see also BATEMAN, 1950; SMIRNOV, 1975).

Baryte

Baryte, as a common mineral in MVT lead-zinc deposits, forms in complex hydrothermal and low temperature mineral successions (see DŻUŁYŃSKI and SASS-GUSTKIEWICZ, p. 377–397; BONI and D'ARGENIO, p. 165–189; D. C. FORD,p. 313–336 and others in this book). Baryte is known also as a residual mineral accumulated in karst depressions in Strawczynek, in the Holy Cross

Mountains in Poland and in Washington County, Missouri in the U.S.A. Residual baryte derives from leached limestones in the latter deposit and from weathered baryte veins in the former deposit. A complex metasomatic origin of baryte is reported from Sardinia, Italy; its genesis there is associated with Ba^{2+} substitution for Ca^{2+} in gypsum via infiltration in caves. Low temperature baryte crystallized in paleokarst cavities in the Tyuya Muyun deposit of south Fergana, U.S.S.R. (see BATEMAN, 1950; PADALINO, TOCCO and VIOLO, 1980; BOSÁK, 1980c, 1986; LAZNICKA, 1985).

Nickel Ores

Nickel ores are associated especially with the contact type of karst developed along the boundary between carbonate and non-carbonate rocks. Weathering of ultrabasic rocks liberates Ni, Fe, Mn and Cr. These elements migrate in aqueous solutions into karst cavities where they may form complex precipitates. Deposits of this type are reported at Veschnii, Ufalei and Rezh in the Central Ural Mountains (GINZBURG, 1966).

Conclusions

Paleokarsts serve as significant loci for economically important accumulations of many types of mineral deposits. This wide variety of metallic and non-metallic deposits has been created by numerous ore-forming processes, some of which are directly connected with the karstification process, others taking place in younger fill in the karst. The diverse geological and geochemical paleoenvironments occurring in karst have permitted and supported this variety of associated mineral deposits. There are numerous special types such as nickel ores, vanadium and mercury deposits, zinc carbonates and silicates (cf. also LAZNICKA, 1985). The importance of paleokarst to ore production is particularly illustrated in MVT lead-zinc deposits, oil and gas accumulations, bauxites, placers of gold and precious stones, and phosphorites. Water resources are also very significant.

Pb-Zn ORES

Stanisław Dżułyński and Maria Sass-Gustkiewicz

Introduction

This article provides a brief treatment of underground karst structures associated with carbonate-hosted sulphide ores. Discussion is confined to the so-called Mississippi Valley-type Deposits (abbreviated MVTD) which are among the world's largest productive lead-zinc resources e.g. in Upper Silesia, the Southern Appalachians, the Tri-State District and the Eastern Alps. Examination of these karst structures has led to the interpretation of MVTD as a hydrothermal karst phenomenon. The present paper focusses attention on this interpretation and is based upon earlier publications by the authors (see references). The paper deals with some selected problems of MVTD which are of interest to karst scientists. The mineralogy and geochemistry of these deposits as well as the question of metasomatic ores lies outside of the main themes considered here. References to these subjects are kept to the minimum that is necessary for clarity of discussion. The meaning of terms which are not usually used in karst research and are not explained in the text are given in the **Glossary** at the end of this article.

Underground Karst

The term *underground karst* as used in this article applies to dissolution cavities, cavity fillings and rock disturbances by the action of aqueous solutions of any origin, composition and temperature. The disturbances mentioned, termed *karst tectonics* (BALWIERZ and DŻUŁYŃSKI, 1976), result from stress redistribution consequent upon solutional removal of rocks and include collapse breccias, minor cavity faults and tilting of strata towards regions of maximum dissolution (subsidence). Karst tectonic structures are products of brittle failure and involve dilatancy which is manifested as deformation openings such as interfragmental voids in breccias, open cracks and sag fissures developed by

P. Bosák, D. C. Ford, J. Głazek and I. Horáček (Editors):
Paleokarst. A Systematic and Regional Review. – Elsevier and Academia. Amsterdam and Praha. 1989.

separation of beds above karst caverns. Such openings, too, are included in the category of karst cavities.

Karst structures produced by the action of hot solutions are termed *hydrothermal karst* (KUNSKÝ, 1957; G. A. MAKSIMOVICH, 1969). Some comments on modern examples of such structures are pertinent to the following considerations. For references and details see JAKUCS (1977b).

The morphologies of hydrothermal caves are similar to those in caverns resulting from the action of cold meteoric waters. However, in massive carbonates spherical chambers produced by convection currents (RUDNICKI, 1979) are considered to be diagnostic of hydrothermal karst phenomena (JAKUCS, 1977b). In bedded carbonates, large bodies of collapse breccia are of common occurence. While ordinary meteoric karst conduits show little wall-rock alteration, this is much more marked in the case of many hydrothermal karst passageways. The rock alterations consequent upon the transfer of hot solutions include dolomitization, recrystallization, silicification and granular disaggregation of the rocks involved. At the least, stable isotope ratios of the rocks (oxygen and carbon ratios) are altered. Hydrothermal caves are invariably associated with deep tectonic fractures which serve as passageways for ascending fluids. Where the upward flow is unobstructed, the hot fluids may reach the Earth's surface and/or may gain access to pre-existing systems of meteoric caves. However, if barred by flat-lying impermeable strata, the rising hot solutions tend to spread laterally and give rise to horizontally-disposed karst conduits. From the economic point of view, the known examples of modern hydrothermal cavities are barren. The hot waters responsible for the formation of such features are not the ore-forming fluids proper although small amounts of iron sulphides, malachite, baryte and fluorite have been reported from the Hungarian occurrences (OZORAY, 1961). We have, as yet, no analogues of mineralized karst structures in MVTD among the present-day manifestations of hydrothermal karst processes although this may be a question of recognition rather than of their non-existence.

Explanations of Karst Ores in MVTD

No agreement has been reached concerning the genesis of karst ores in MVTD. In general, four possibilities seem feasible (DŻUŁYŃSKI and SASS-GUSTKIEWICZ, 1985): (1) the karst structures and the ore in them are products of processes involving cold surficial waters; (2) the karst structures were produced by the action of cold surficial waters and acted as traps for ores introduced by hot ascending solutions; (3) the karst cavities and the ores in them resulted penecontemporaneously from the action of hydrothermal solutions; and (4) the karst receptacles were produced by hot solutions before the introduction of sulphide ores by cold surface waters. With the exception of the last mentioned possibility (which seems the least probable), each of these explanations has its advocates and opponents.

The first interpretation is chiefly favoured by exponents of the sedimentary origin of MVTD (for references see ZUFFARDI, 1976). The hypotheses advanced may differ in detail but all have it in common that any interference by hydrothermal solutions is excluded. The base metals are thought to have been leached from the host strata or to be derived from rocks located outside the carbonate aquifer and which were subject to prolonged weathering. While supergene groundwaters are invoked by most of the authors opposed to hydrothermal interpretations, some seek an explanation in terms of marine transgressions over the karstified carbonate rocks and the sealing of the cave openings by marine sediments (e.g. LAGNY, 1975)

The second possibility is favoured by a number of investigators supporting the epigenetic and hydrothermal genesis of MVTD (e.g. R. T. WALKER, 1928; HILL, MORRIS and HAGEGEORGE, 1971; HAGNI, 1976). According to this interpretation, the formation of karst cavities and the emplacement of ores co-exist in space but are essentially unrelated events. The hydrothermal, ore-forming solutions gain access into pre-existing meteoric karst systems and there deposit the sulphide ores. This concept has recently been supplemented by the reasonable suggestion that the karst cavities serving as ore receptacles might have developed along the interface dividing fresh and saline groundwaters as is the case in low-lying coastal regions of carbonate rocks (RUDNICKI, 1980).

The third possibility was invoked, implicitly or explicitly, by many early workers (see later in this article) although the term *karst* has seldom if ever been used in this connection. This possibility is the essence of the concept of hydrothermal karst as the dominant factor in the formation of cavity-filling sulphide ores in MVTD. The remainder of this article is devoted to a discussion of the factual material upon which this concept is based.

Geologic Setting of Karst Ores in MVTD

In order to place the problem of karst ores in MVTD in perspective, it is necessary to summarize some principal features of these deposits. These are as follows:

(1) mineralogically, MVTD consist chiefly of sphalerite, galena, pyrite, marcasite and associated *gangue minerals* such as fluorite, baryte, or secondary crystalline dolomite;

(2) the ores tend to occur in the form of bedding-oriented tabular bodies. The formation of such bodies and the development of the mineralized karst structures took place in nearly flat-lying carbonate rocks prior to any significant tectonic disturbances;

(3) the ores may occur at different levels within their host-strata. However, the uppermost ore bodies are commonly located relatively close to disconformities reflecting the emergence of the host-strata above sea level;

(4) the host-strata may be cut by the disconformities which (in such situations), may display subaerial karst relief, or the ore-bearing carbonates are separated from the disconformity surfaces by impermeable argillaceous sediments;

(5) MVTD are as a rule enclosed in secondary crystalline dolostone or, to a lesser extent, in recrystallized limestones. These carbonates constitute the host-rock proper of the ores and represent neosomes showing cross-cutting metasomatic contacts with the surrounding non-mineralized host-strata i.e. the paleosome. Rock alterations associated with MVTD karst ores also include granular disaggregation and, occasionally, silicification;

(6) the ore bodies are very restricted in space, in contrast to the non-mineralized parts of their host-strata. The profusion of mineralized karst structures is limited to and coextensive with the ore bodies and their host-rocks. Outside the mineralized portions, the host-strata are virtually devoid of karst structures corresponding in age and stratigraphic position to those which serve as ore receptacles;

(7) although viewed as a whole MVTD are bedding oriented and generally referred to as *strata-bound,* they are genetically associated with ore veins in tectonic fractures perpendicular to the bedding. Such fractures have a considerable vertical extent (it is measured in hundreds of metres) but are devoid of any visible displacement. BAIN (in BAIN, VAN HISE and ADAMS, 1901)described the fractures in question as *crevices* and the ores in them as *crevice ores*. The mineralized crevices are seen to connect directly to the bedding-oriented ores and, what deserves emphasis, the bedding-oriented mineralized karst structures. Any rational explanation of mineralized karst structures in MVTD must give a convincing account of these characteristics.

Morphologies and Patterns of Karst Structures in MVTD

Dissolution Cavities

The host-rocks of MVTD are riddled with countless, generally small and low-ceilinged, karst conduits which are partly or entirely filled with sulphide precipitates or, to a lesser extent, with mineralized clastic cave sediments. The morphologies of such conduits are similar to those displayed by ordinary meteoric karst cavities and are indicative of phreatic conditions. In three dimensions, the conduits under consideration are tabular, tubular, irregularly domed or entirely irregular. Occasionally they may show triangular and roughly plano-concave cross-sections. The small-diameter tubular conduits may show anastomosing or network patterns but, most commonly, are entangled to form irr-

3–1. Sphalerite ribbon ore with sheet cavities, Trzebionka Mine, Upper Silesia (after DŻUŁYŃSKI and SASS-GUSTKIEWICZ, 1985).

egular spongeworks. However, the narrow sheet-cavities which tend to occur in closely spaced parallel sets (Fig. 3–1) are characteristic of MVTD. Such cavities are rarely encountered in meteoric karst domains and, as we shall see later in this article, their origin in MVTD is genetically related to the emplacement of metasomatic sulphide ores.

Most of the cavities under consideration are bedding controlled as is the case with karst structures developing in horizontally bedded carbonate rocks. Among other factors controlling the development of caverns in MVTD are the metasomatic boundaries of host dolostones and the upper boundaries of disaggregated rock bodies (Figs. 3–2 and 3–13).

High caverns with roofs that have not collapsed are very rare in MVTD. This is due to the fractured character of the host-rocks and to the fact that the mineralized karst passageways developed in stratified and flat-lying carbonate rocks. In such a situation, the enlargement of caves must tend to lead to cave breakdown. The height of caverns prior to their collapse is given by the vertical extent of preserved side-walls and the thickness of any fine-grained clastic cave sediments. In Upper Silesia, the largest caves may have been up to a few metres in height.

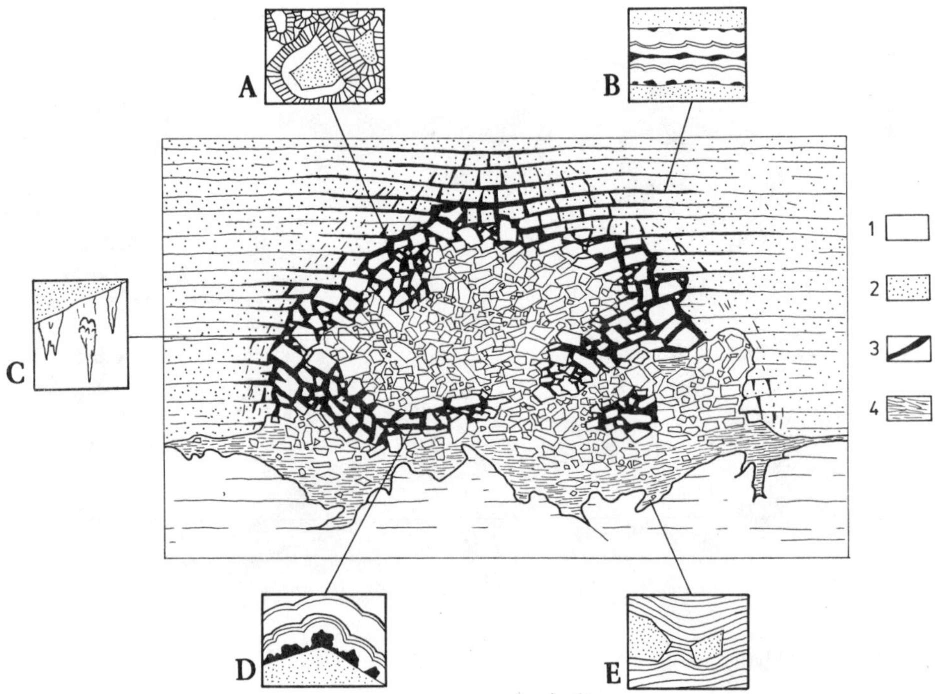

3–2. Schematic representation of transverse cross-section through karst collapse breccia developed along metasomatic contact of ore-bearing dolomite and limestone. Not to scale. Insets show details of sulphide mineralization. A. dolostone fragments enveloped and cemented by sulphide minerals; B. sag fissure filled with sulphide precipitates; C. sulphide dripstones in empty interfragmental voids; D. banded microcrystalline sulphide crustifing on dolostone blocks; E. detail of internal clastic sediment with dropstones. 1. limestones; 2. ore-bearing dolostone; 3. ore minerals; 4. internal sediment.

Solution Collapse Breccias

As noted, the fractured character of host-rocks and the vestiges of primary stratification promote the development of collapse breccias which are the most important receptacle of the cavity-filling type of ores in MVTD. The breccias in question occur in the form of tabular, irregular and commonly branching bodies. The horizontal extent of such bodies amounts to several hundred metres and their height may attain several tens of metres. The breccias are variously distributed within the host-rocks, but commonly they tend to develop along the lower metasomatic boundaries of these neosomes (Fig. 3–2). The lowermost boun-

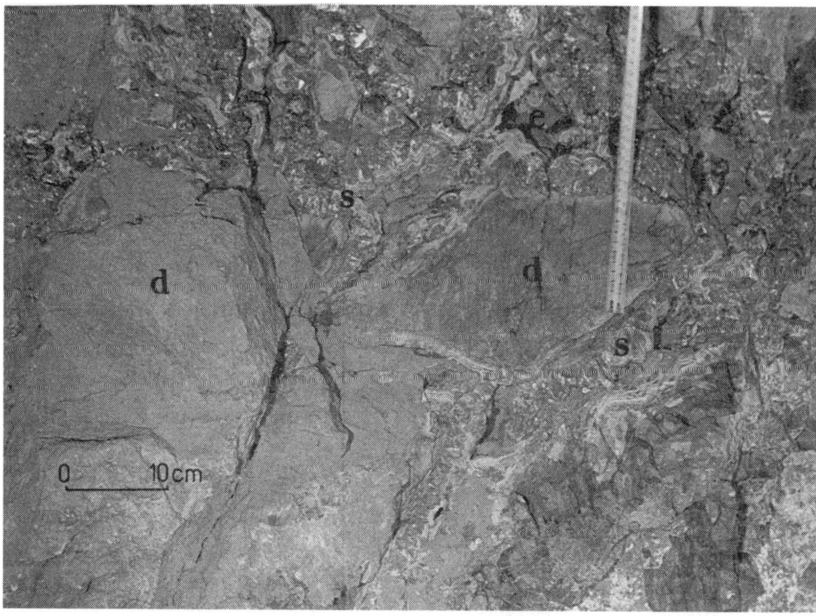

3–3. Detail of mineralized solution collapse breccia, Olkusz Mine, Upper Silesia (Photo by M. Sass-Gustkiewicz). d. dolostone blocks; s. sulphide incrustations; e. empty void.

daries of the breccias are sharply defined solution-etched surfaces representing the former floors of caves. Such surfaces show a relief of irregular depressions and ridges and are usually covered with fine-grained cave sediments (see below). These basal sediments are overlain by a chaotic mass of angular blocks (Fig. 3–3) which constitutes the main body of breccia. The interblock spaces are partly or entirely filled with sulphide precipitates or with a rock-matrix of small fragments and particles from the enclosing host-rocks. The upper boundaries of breccias are gradational and the chaotic rubble of angular blocks is seen to pass through crackle breccias into a network of cracks and fissures cutting the overlying strata in various directions. In transverse sections through the breccia bodies, such fissures may approximate the pattern of tension domes (McCormick et. al., 1971; Sass-Gustkiewicz, 1974). The breccias are also associated with extensive mineralized sag crevices in overlying carbonate layers.

Karst Fillings

As noted, the karst cavities in MVTD are filled with: (1) precipitates of sulphides and other ore minerals and (2) internal clastic sediments.

Precipitates

The morphologies of sulphide precipitates are similar to those known from non-mineralized hydrothermal and cold-water meteoric speleothems. Of common occurrence are drusy encrustations of empty voids, banded microcrystal-

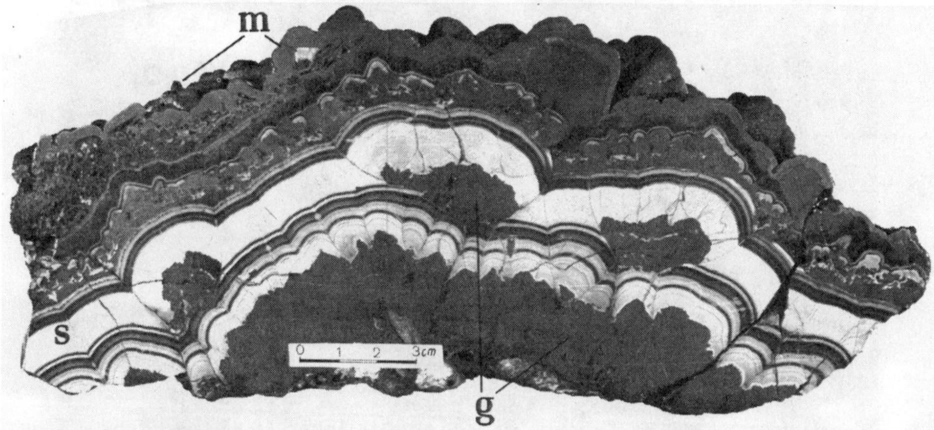

3–4. Incrustation of empty void, Olkusz Mine, Upper Silesia. s. banded microcrystalline sphalerite; g. galena; m. marcasite (after SASS-GUSTKIEWICZ et, al., 1982).

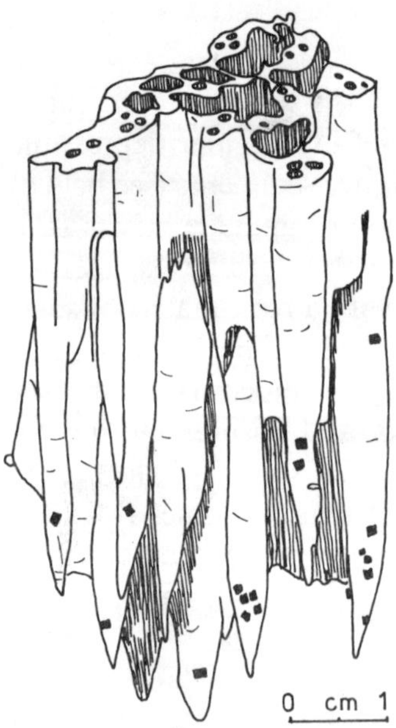

3–5. Complex of sphalerite stalactites. Central portions empty or porous. Note idiomorphic galena crystals on surfaces of stalactites (black), Pomorzany Mine, Upper Silesia (after SASS-GUSTKIEWICZ, 1985).

line crusting (Fig. 3–4) or *corraloid* forms and box-work structures. There are also flat ledges similar to cave *shelfstones* but produced under subaqueous conditions on surface of fine cave sediments. Most of these precipitates crystallized directly onto the walls of dissolution caverns or open fractures and onto

3–6. Marcasite dripstones, fragment of drapery, Pomorzany Mine, Upper Silesia.

the surface of rock fragments in breccias. Some of them, however, nucleated in suspension and settled under gravity to form caps and overhangs on the upper horizontal or sub-horizontal rock surfaces (OHLE, 1951).

The overwhelming majority of the sulphide precipitates show morphologies indicative of phreatic conditions. The observed variations in these morphologies bear a record of changes in composition and hydrodynamic properties of the ore fluids (see below). Sulphide precipitates indicative of vadose conditions are very rare in MVTD although sulphide dripstones such as stalagmites, stalactites (Fig. 3–5) and draperies (Fig. 3–6) have been reported from several districts (e.g. POŠEPNÝ, 1894; SOBCZYŃSKI and SZUWARZYŃSKI, 1975). Some of these drip-

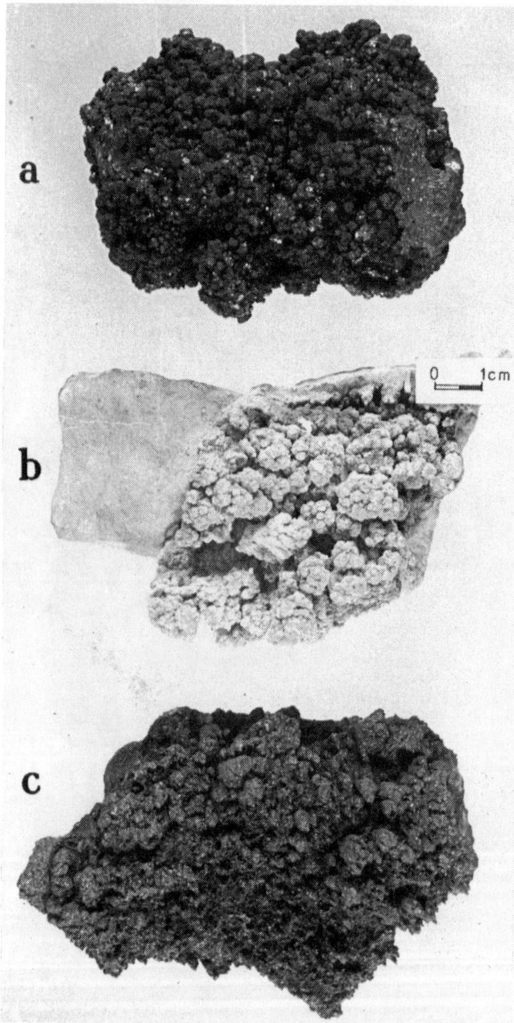

3–7. Comparision of *knobby* galena (a) and marcasite (c) forms from Pomorzany Mine with calcite speleothems (b) from meteoric Niedźwiedzia cave, Lower Silesia.

stones may be the secondary products of post-ore supergene remobilization processes effected by groundwaters enriched in H_2S. However, recent investigations by SASS-GUSTKIEWICZ (1985) indicate that the sulphide dripstones may also belong to the main paragenetic sequence of ore emplacement and be the products of the same ore-forming processes which are responsible for the bulk of MVTD. Indeed, between the unquestionably *phreatic* and *vadose* sulphide

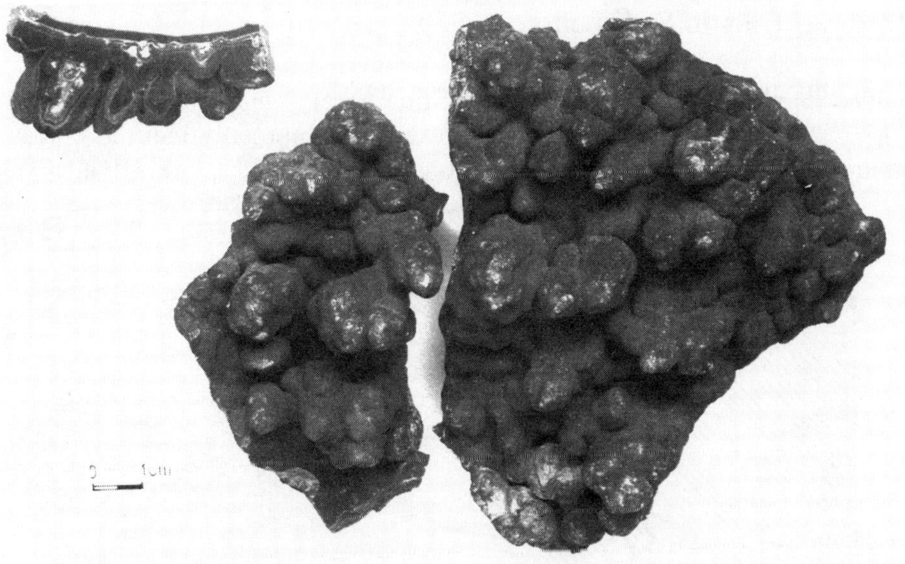

3–8. Mammilated sphalerite speleothem. Note successive banding in cross-section (left), Pomorzany Mine, Upper Silesia (Photo by S. Michta).

3–9. Examples of coralloidal sulphide speleothems, Pomorzany Mine, Upper Silesia. A. hemispherical sphalerite incrustation covered with drusy marcasite; B. grape-like marcasite speleothem.

precipitates, there are intermediate forms which correspond closely to the so-called *cave coral* or *popcorn* speleothems that are known from ordinary meteoric caves and , depending on morphological differences, variously named (see HILL, 1976). Some typical examples of such speleothems are shown in Figs. 3–7, 3–8 and 3–9.

Internal Clastic Sediments

The internal clastic sediments constitute a very small fraction of MVTD. These sediments consist of disaggregated and solutionally reduced dolostone grains, solutional residues, clasts of the enclosing rocks, ore fragments and

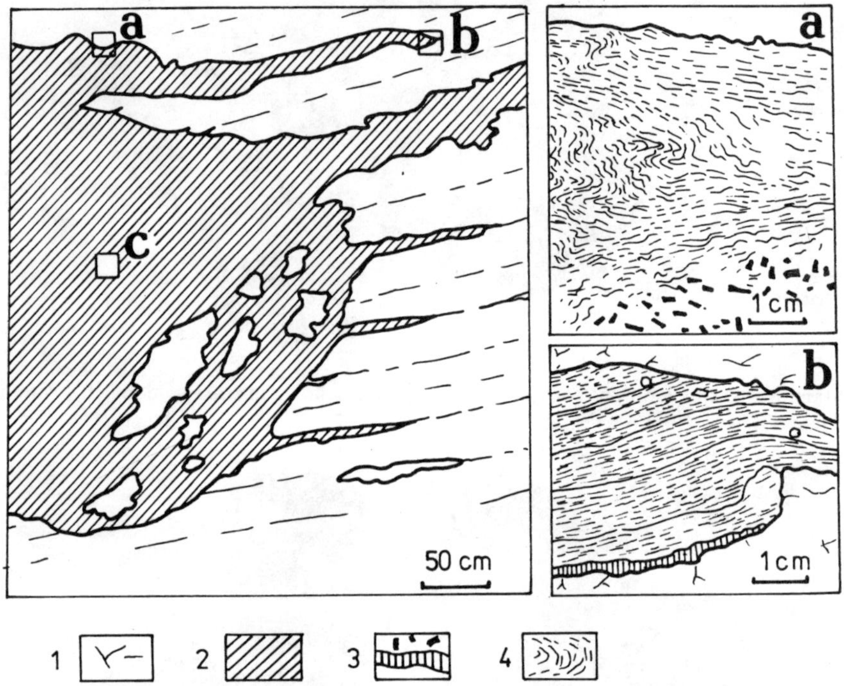

3–10. Mineralized fine-grained cave sediments. Insets (a) and (b) mark position of enlarged details (right). Inset (c) marks the position of sample shown in Figure 3–11, Bleiberg, Eastern Alps (after Dżułyński and Sass-Gustkiewicz, 1977). 1. host dolostone; 2. mineralized cave sediment; 3. sphalerite incrustations; 4. lamination in cave sediments.

authigenic sulphide crystals (for references see Dżułyński and Sass-Gustkiewicz, 1985). In places, the proportion of ore fragments and authigenic sulphide crystals is so high that the cave sediment may be referred to as an internal sedimentary ore (cf. Figs. 3–10 and 3–11; Sass-Gustkiewicz, 1975a). The sediments display structures that occur in external sediments laid down in marine or other aqueous basins. Such structures include horizontal and cross-lamination, graded bedding, scour-and-fill forms and soft-rock deformations. The last mentioned are chiefly produced by piecemeal and/or by massive roof failure of caverns.

Sedimentary structures indicating lateral redeposition by subterranean currents are limited to the very fine-grained sediments. The rock and ore fragments enclosed in these sediments are dropstones i.e. pieces of roof that dropped into the finely stratified mud. Where the mud was soft, the laminae are seen to bend down beneath the fragments or to flow out from under the dropstones giving rise

3–11. Detail of mineralized internal sediment from inset (c) on Fig. 3–10, Bleiberg, Eastern Alps (after DŻUŁYŃSKI and SASS-GUSTKIEWICZ, 1977).

to a series of minor folds and thrusts. Occasionally the soft cave sediment is subject to liquefaction (Fig. 3–12). We may add that a part of the graded bedding revealed by laminae may also be due to the action of fine suspension currents that are generated by piecemeal or massive roof failures of caves. On impact with the fine residues covering the cave bottoms, the falling rock fragments stir the muddy sediment and produce clouds of particles in suspension which may spred laterally to create the graded bedding (DŻUŁYŃSKI and SASS-GUSTKIEWICZ, 1982).

Although it is not of immediate consequence here, we add that such internal sediments have often been confused with those laid down on the sea floor (external sediments) and were used as a chief argument in support of the synsedimentary origin of MVTD (PARK and AMSTUTZ, 1968). The confusion is

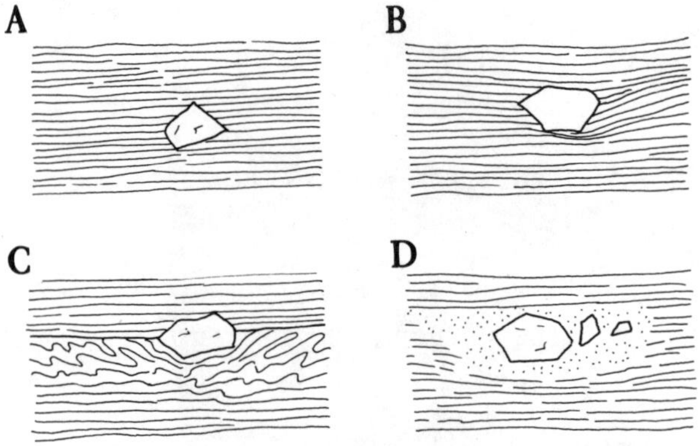

3–12. Examples of dropstones in internal sediments. A. dropstones in non-deformed laminated sediment; B. downbending of laminae beneath dropstone; C. soft rock deformation produced by impact of falling rock fragments; D. liquefaction aureole around dropstones.

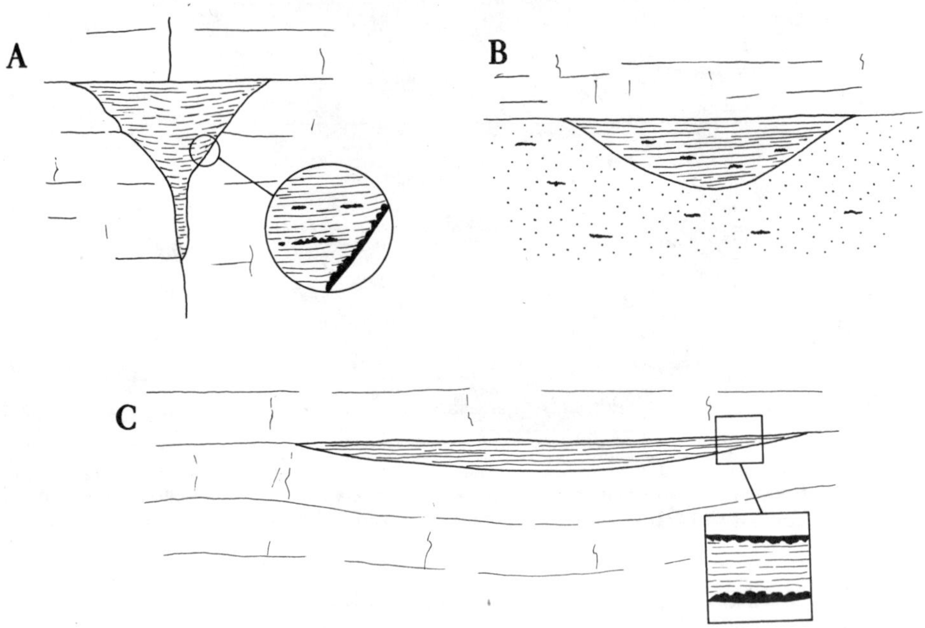

3–13. Examples of sediment-filled caverns showing confusing morphologies. A. cavern with triangular section; B. cavern showing planoconcave cross-section developed along boundary between disaggregated (dotted) and non-disaggregated dolostone; C. sediment-filled sag fissure developed along bedding interface. Sulphide ores are black.

augmented by the fact that the geometry of karst openings may impose upon clastic cave sediments shapes that approximate those of external sediments. We are referring here to such features as fine-grained ore-bearing clastic sediments in sag fissures, triangular, and plano-concave cavities (Fig. 3-13). Viewed in cross-sections such openings, if tightly filled with disaggregated dolostone grains, bear a superficial resemblance to the infillings of erosion channels or to normal sedimentary layers. This similarity is accentuated when the cavities in question were produced in flat-lying carbonates and the sedimentary interfaces within the cave sediments are parallel to those of the enclosing rocks (BOGACZ et al., 1973; DŻUŁYŃSKI and SASS-GUSTKIEWICZ, 1977).

Temperature of Sulphide Crystallization

The crystallization temperatures of ore minerals in MVTD, estimated on the basis of gas-fluid inclusions, range from 75° to 200 °C. The later-formed minerals tend to display the lower temperatures of formation (ROEDDER, 1971). The sulphide ores in the karst receptacles discussed are, thus, of hydrothermal origin. This conclusion is compatible with and supported by all the field evidence derived from the study of MVTD (DŻUŁYŃSKI and SASS-GUSTKIEWICZ, 1985). Particular reference is made here to the above-mentioned ore veins in tectonic fractures which acted as passageways for metal-bearing solutions ascending from deep-seated sources. Such solutions, by definition, are *hydrothermal.*

Relationship Between the Sulphide Ores and Karst Structures

The question that arises is whether the hydrothermal ores and their karst receptacles were produced by the same formative processes or whether the formation of karst structures and the emplacement of ores resulted from a coincidental superposition of genetically unrelated events. The answer to this question is provided by the mineralized collapse structures in MVTD. Such structures resulted from a succession of brecciation and mineraliziaton stages whereby the sulphides belonging to different stages have the same composition but may differ in texture, mode of occurrence and quantitative associations. Of importance however, is the fact that the successively younger sulphide minerals enclose and/or envelope the clastic products of earlier mineralization and brecciation episodes. Furthermore, the breccia bodies when seen in plan view tend to show a concentric and zonal arrangement of sulphides, with younger stages covering progressively more extensive areas (SASS-GUSTKIEWICZ, 1975b).

Such relationships indicate that the formation of the karst breccias and the emplacement of the sulphide ores were overlapping and penecontemporaneous phenomena and were part of the same formative process. This conclusion can be extended to cover the remaining karst stuctures in MVTD. All the karst structures in these deposits are interconnected and the sulphide ores are seen to continue uninterruptedly from one structure to another.

Hydrothermal Karst as a Factor in the Formation of MVTD

In summary, we come to the following conclusions: (1) the rock openings serving as ore receptacles in MVTD are karst openings; (2) the formation of such cavities and the emplacement of sulphide ores were not genetically unrelated phenomena but were parts of the same formative process and were essentially simultaneous, and (3) the ores in the cavities in question were emplaced by hydrothermal solutions.

These three conclusions constitute the principal premises upon which the concept of hydrothermal karst as a factor in the formation of MVTD is based (BOGACZ, DŻUŁYŃSKI and HARAŃCZYK, 1970; DŻUŁYŃSKI,1976; DŻUŁYŃSKI and SASS-GUSTKIEWICZ, 1985). This concept is not new insofar as general opinion is concerned. In many publications devoted to carbonate-hosted sulphide ores, the mineralized dissolution cavities are ascribed to the action of hydrothermal ore-forming solutions (for references see DŻUŁYŃSKI, 1976; SASS-GUSTKIEWICZ, DŻUŁYŃSKI and RIDGE, 1982). Viewed in this way, the terms *hydrothermal* or *thermomineral* karst (KUNSKÝ, 1957) may be regarded as modern expressions of old ideas inherent in the hydrothermal interpretation of MVTD.

Processes Involved in the Formation of Mineralized Karst Structures in MVTD

Having established the fundamental facts and relationships concerning karst ores in MVTD let us examine the question of mineralized karst structures from a process point of view. Where do the karst-forming solutions come from? What is known of the development of such structures and finally what was the environment in which the mineralized karst structures originated?

The development of extensive karst conduits concurrently with deposition of vast quantities of sulphides requires a copious flow of ore-forming solutions under persistently reducing or at least non-oxidizing conditions. The derivation of such solutions is a matter of conjecture. However, the association of bedding-oriented karst ores with the ore veins in vertical tectonic fractures indicates that the mineralizing and karst-forming solutions might have been available from deep-seated fluid reservoirs. It is feasible to envisage such reservoirs as *water sills* which are thought to develop below impermeable strata in deeper parts of the Earth's crust (FYFE, PRICE and THOMPSON, 1978). The sills in question may contain huge amounts of ore fluids under high pressures and temperatures which are abruptly released upon the appearance of tectonic fractures. The fluids will force their way upwards and, if barred by impermeable layers or on reaching the karst water table, may spread horizontally giving rise to mineralized karst structures.

Observation of modern occurrences of hydrothermal ore-bearing solutions indicates that the upward flow of such solutions commonly takes place in the form of episodic surges (LEBEDEV, 1974). Here earthquakes are responsible for moving quantities of ore fluids towards the surface (SIBSON, MOORE and RANKIN, 1975). The episodic surges may be among the causes of successive brecciation and precipitation phases. It is realized, however, that in natural fluids a delicate balance exists between dissolution and precipitation and even minor changes in conditions of fluid flow may shift the reaction from one side to another. Such changes in the flow conditions may also result from transformations within the evolving karst systems (see later).

The problem of the persistent capability of ore solutions to dissolve huge amounts of carbonate rocks is one of geochemistry and lies outside our present consideration. We mention only that apart from the ubiquitous carbon dioxide which is among the common constituents of hydrothermal solutions, the precipitation of sulphides itself is an acid-generating process (ANDERSON, 1983).

In general, the formation of mineralized karst structures in MVTD is preceded by a slow, non-integrated (dispersed) circulation of ore-forming fluids through the pore space of the host-rocks (Fig. 3–14 A). Such circulation affects large volumes of the rock involved and is conducive to the emplacement of metasomatic ores. The term *metasomatism* applies to processes of replacement in which the substitution of one mineral by another is essentially simultaneous with the dissolution of the former and deposition of the latter. If, however, the velocity of replacing solutions is relatively great, the dissolution runs ahead of the deposition giving rise to dissolution vugs (LINDGREN, 1918). Such vugs tend to develop along primary or secondary discontinuities (e. g. sedimentary laminations, sets of fractures and discontinuities consequent upon the spread of rhythmic diffusion banding) where the flow velocity is greater and the deposition of the metasome lags behind the dissolution of the paleosome. The confluence of vugs gives rise to the appearance of the sheet cavities noted above, which constitute an essential part of the so called *ribbon ores* (Fig. 3–1). Such ores are regarded as of intermediate type between the metasomatic and the karst ores proper.

Among the effects of non-integrated solution transfer is also granular or solutional disaggregation, i. e. the transformation of hard crystalline carbonates,

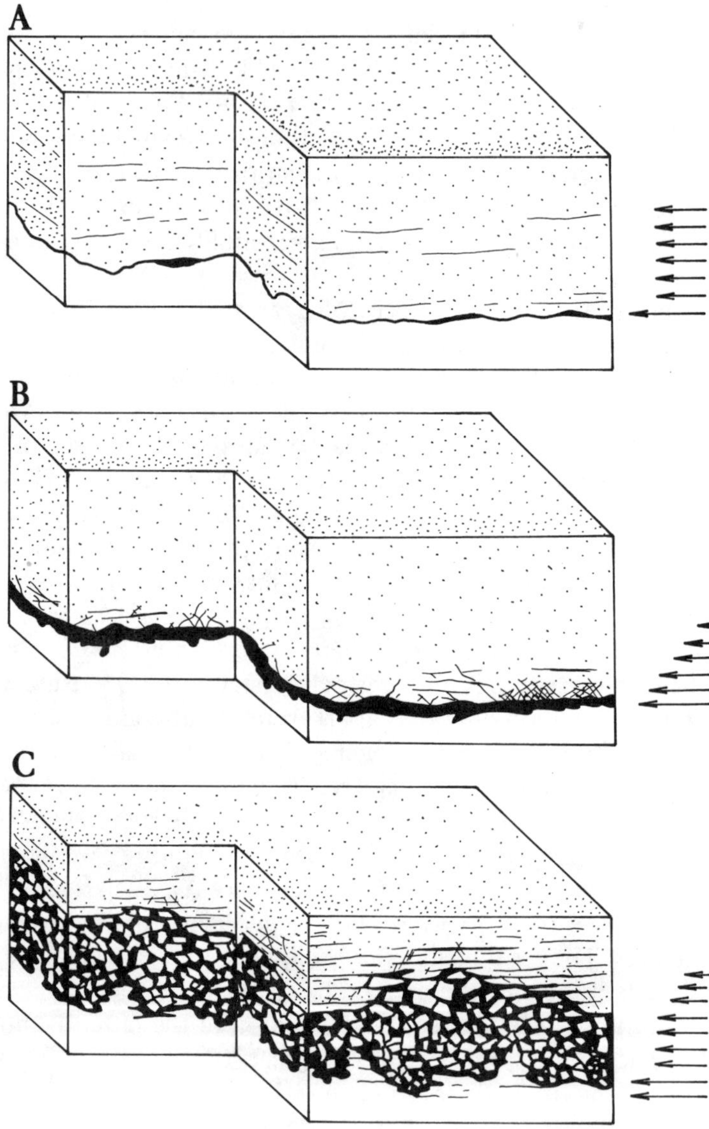

3–14. Diagram showing dissolution motivated transformations in Ore-bearing Dolomite (dotted) (after SASS-GUSTKIEWICZ, 1985). A. emplacement of metasomatic ores by non-integrated transfer. Note incipient karst conduits along metasomatic contact of Ore-bearing Dolomite with limestones (thick line); B. enlarged incipient karst conduits with cracks and sag fissures in overlying dolostone produced by subsidence; C. mature karst structures with collapse breccias. Arrows indicate direction of solution flow. Length of arrows corresponds to flow intensity.

mostly dolostones, into a porous, friable or incoherent mass of grains which, by dissolution of crystal edges, are sometimes reduced to silt-particles. Disaggregated carbonates, commonly referred to as *sanded* or *pulverized,* are of wide-spread occurrence in MVTD, though they are recemented and thus again transformed into hard crystalline rocks.

As already noted, the process of disaggregation is a factor controlling the development of karst cavities which, to a considerable extent, are produced by corrasion (attrition) of the disaggregated particles. In early stages of disaggregation, the pulverized parts of the rock may appear as *sanded branchworks* that resemble the pattern of spongework cavities in their complexity, close spacing and the manner in which disaggregated bodies are mutually interlocked. If washed clean by underground solutions such branchworks are transformed into spongework cavities proper. The disaggregation is also associated with solutional thinning which is due to the settling of grains reduced in size by dissolution or entirely dissolved by that process. This results in the development of voids along the upper boundaries of large sanded bodies or disaggregated layers (DŻUŁYŃSKI and KUBICZ, 1971; BOGACZ et al., 1973). There is a good reason to suppose that a number of mineralized karst cavities in MVTD originated in this manner (for details see DŻUŁYŃSKI and SASS-GUSTKIEWICZ, 1985).

With the appearance of karst cavities, the ore solutions become constrained to flow through them and the poorly integrated circulation gives way to a more rapid, integrated solution transfer (Fig. 3–14 B; SASS-GUSTKIEWICZ, 1983). This type of transfer promotes dissolution and leads to a further enlargement of the conduits. However, in more spacious caverns, the flow velocity is again reduced so that the conduits may become sites of sulphide precipitation and the accumulation of residues (Fig. 3–14 C).

As noted, the enlargement of caves provides conditions for massive roof failures and the development of collapse breccias. The formative mechanism of such breccias is known from observations (DAVIES 1951; WHITE and WHITE, 1969) and experiments (BALWIERZ and DŻUŁYŃSKI, 1976). However, a few comments are needed here on this subject. Extensive breccia bodies may also be created in the absence of large caverns. For instance, the previously mentioned solutional thinning of carbonate layers affected by disaggregation is commonly followed by stress redistribution which leads to the formation of cracks and brecciation. There are also breccias that originate from derangement of carbonate layers which are subject to simultaneous dissolution along a great number of transmissive fractures (*polycentric derangement breccias* – BALWIERZ and DŻUŁYŃSKI, 1976). Finally, the breccia bodies develop under phreatic and/or vadose conditions through progressive piecemeal breaking of the rocks involved and differential settling of detached blocks. This occurs concurrently with removal of the rock material in the lower portions of the breccia bodies. Admittedly, in MVTD also the lower parts of breccias consist exclusively of angular dolostone fragments. It should be borne in mind, however, that at the time of brecciation the crystalline host-dolostones were saturated by mineralizing solutions and, presumably, softened by incipient disaggregation. Such dolostones tend to split into smaller angular fragments and disintegrate rather than become rounded by dissolution.

The opening of new voids which is a natural consequence of karst tectonics brings about a sudden drop of pressure. It is also accompanied by a decrease in temperature of ore fluids when such fluids come into contact with newly exposed and cooler rock surfaces. In such situations, the underground karst structures which now have reached the mature stage of their development (Fig. 3–14C) rapidly become filled with sulphide precipitates. The cessation of hydrothermal ore-forming activity and the draining of ore fluids from the karst cavities provides opportunities for deposition of sulphide dripstones and related *vadose* speleothems.

The geological domain of hydrothermal activity that results in the formation of karst ores in MVTD can be envisioned as a nearly horizontal sequence of lithified carbonate sediments emerged from the sea to form a low-lying carbonate plateau. There is no indication whatsoever that the karst receptacles of ores developed concurrently with the deposition of the carbonate (*syngenetic karst* in the meaning of JENNINGS, 1968, 1971). On the contrary, even the uppermost sulphide-bearing karst structures in MVTD are separated from the disconformities overlying their host-strata by a thick sequence of other strata. In addition, there is good reason to suppose that much if not all of MVTD were emplaced after the uplift of the host-strata above sea level.

The relative proximity of the disconformity surfaces makes possible invasion by ore-forming hydrothermal solutions into a pre-existing system of meteoric caverns. However, the direct genetic relationships between the sulphide ores and their karst receptacles mentioned above rules out this possibility as a general explanation of the karst ores in MVTD.

MVTD are highly unstable when acted upon oxigenated meteoric groundwaters in near-surface regions. The sulphides are transformed into carbonatesand sulphates which are generally much more soluble than the corresponding sulphides. In such a situation, meteoric karst processes may start to operate, creating new underground karst structures superimposed upon the older hydrothermal ones. The meteoric karst processes are assisted by sulphuric acids and carbon dioxide which tend to form whenever the sulphide-bearing carbonate rocks are invaded by oxidizing solutions (MOREHOUSE, 1968). In the course of these activities, small amounts of sulphides, chiefly galena, may be remobilized, i. e. reprecipitated by the action of H_2S of organic or inorganic origin in the newly produced karst cavities. In view of what has been said earlier in this article, such supergene remobilization processes are totally inadequate to explain the huge accumulations of pure sulphide ores in karst receptacles of MVTD deposits.

Glossary

Crackle breccia: one whose fragments are slightly disjointed but with little or no displacement

Dilatancy: the property of expanding in bulk with deformation due to increase in shape between individually rigid particles or parts as they change their relative position

Host-rock: rock containing ore minerals and directly enclosing ore bodies

Host-strata: rock sequences containing the host-rock of ores

Metasomatism: replacement of an older parent rock (paleosome) by newly formed minerals (metasome)

Neosome: newly formed part of rock

Solutional disaggregation: transformation of hard crystalline carbonates into weakly coherent or incoherent masses of grains through the action of slow non-integrated fluid circulation

BAUXITES

György Bárdossy

Introduction

Bauxite is defined and discussed here in its petrographic context rather than in economic terms. The former is constant, the latter are changed continuously by demands of the world economy and by the development of processing technology. For economic purposes we suggest the term *aluminium ore*. Bauxite deposits can be divided into lateritic, karst and Tikhvin types. About 89 % of global bauxite tonnage belongs to the lateritic type, 10 % to the karst type and less than 1 % to the Tikhvin type.

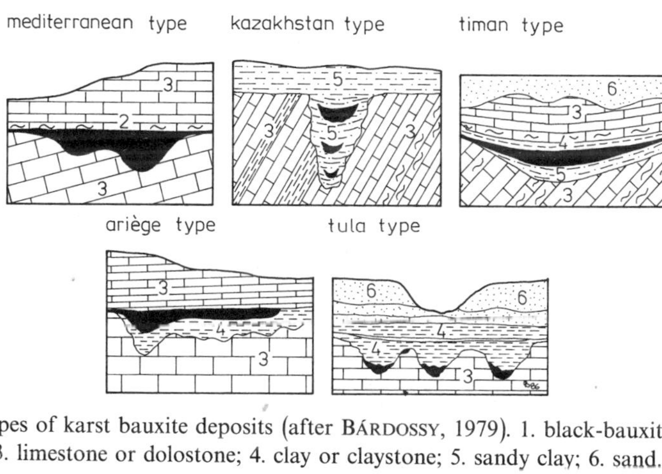

3–15. Main types of karst bauxite deposits (after BÁRDOSSY, 1979). 1. black-bauxite and bauxitic clay; 2. marl; 3. limestone or dolostone; 4. clay or claystone; 5. sandy clay; 6. sand.

Karst bauxite deposits have been divided by the author into six main types (BÁRDOSSY, 1982) as shown in Fig. 3–15: (1) Mediterranean type deposits; (2) Kazakhstan type; (3) Timan type; (4) Ariège type; (5) Salento type (not shown in Fig. 3–15 – Eds.) and (6) Tula type deposits.

The *Mediterranean type* of deposit is composed entirely of bauxite and bauxitic rocks. In the *Kazakhstan type* deep karst depressions are filled with

P. Bosák, D. C. Ford, J. Głazek and I. Horáček (Editors):
Paleokarst. A Systematic and Regional Review. – Elsevier and Academia. Amsterdam and Praha. 1989.

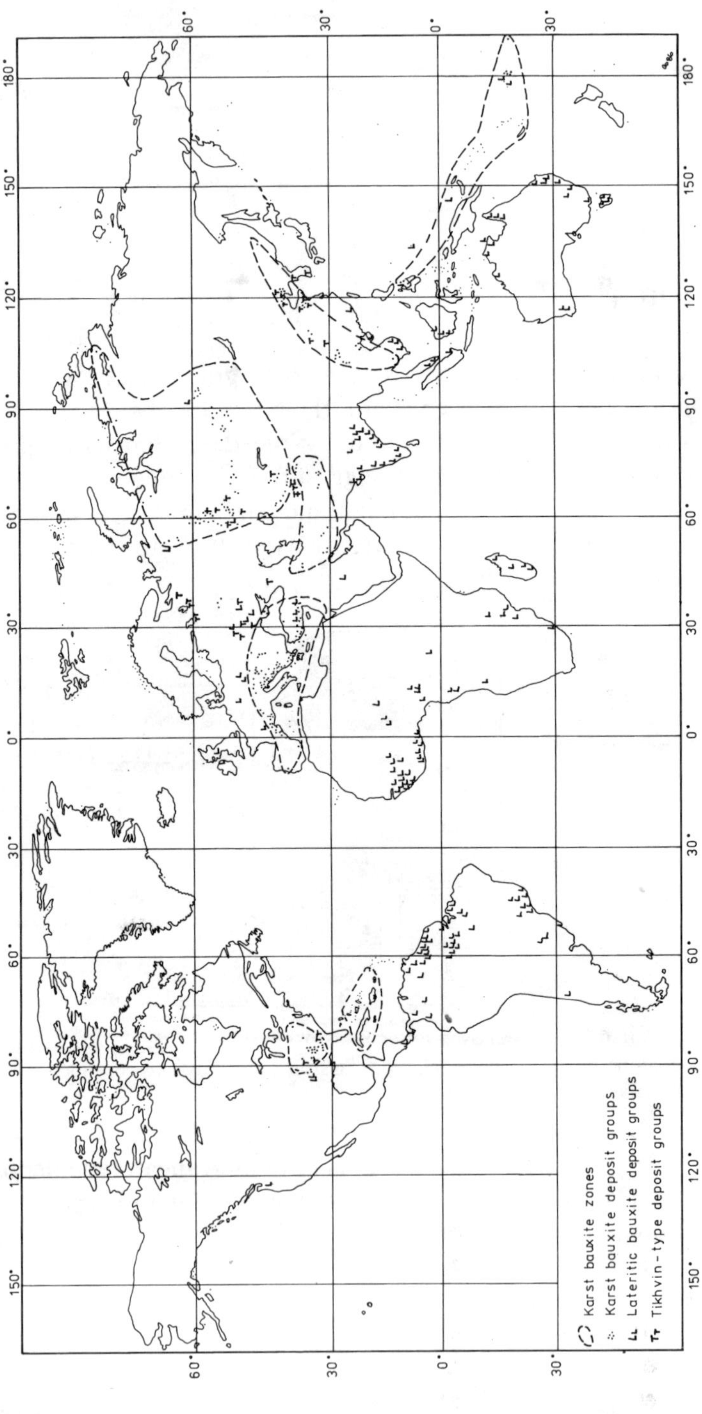

3–16. The world distribution of bauxites (after BÁRDOSSY, 1982).

various continental sediments, the bauxite occurring as intercalated bodies at several levels. *Timan type deposits* fill large, shallow karst depressions, their upper part consisting of bauxite, the lower one of continental sediments. The *Ariège type* is a peculiar transition between karst and lateritic deposits. Redeposited bauxite pebbles imbedded in red clay form the *Salento type* deposits. Finally, the *Tula type* is an accumulation of gibbsite precipitates in karst depressions as a product of pyrite oxidation.

76 % of the global karst bauxite tonnage belongs to the Mediterranean type, 14 % to the Timan type, 7 % to the Kazakhstan type, 2 % to the Ariège type, 1 % to the Salento type and less than 0.1 % is of the Tula type.

The world distribution of bauxites is shown in Fig. 3–16. The bulk of the lateritic bauxites occur in the tropical and subtropical belts – on the Guyana shield, in West Africa, India, Southeast Asia and Australia. The karst bauxite deposits can be grouped into seven belts as indicated in the figure: (1) Mediterranean belt with 36 % of the total karst bauxite tonnage; (2) Caribbean belt, 32 %; (3) Uralo-Siberian belt, 14 %; (4) East Asian belt, 14 %; (5) Iranian-Himalaya belt, 2 %; (6) Pacific belt, 1 %; and (7) North American belt, 1 %. All these belts are situated north of the main lateritic bauxite regions. Mediterranean type bauxites occur mainly in the southern karst bauxite belts, the Timan and *Kazakhstan type* deposits to the north of them.

The oldest karst bauxite deposits are of Late Proterozoic to Early Cambrian age (at Bokson, U.S.S.R), the youngest are sub-Recent to Recent (on the islands of the Southwest Pacific). The main periods of karst bauxite formation were during the Middle to Late Devonian and Early Carboniferous; the Late Cretaceous; the Miocene and Pliocene.

Review of the Bauxite Regions

The largest karst bauxite region is on the island of **Jamaica,** where the total tonnage of bauxite surpasses 2 billion tons (LYEW-AYEE, 1984). It covers the karstified surface of the White Limestone complex, of Eocene to Miocene age. The main bauxite areas are shown in Fig. 3–17. The area of highest altitude is in the Cockpit Country. Here karstification was strongest, producing a very rugged surface (Fig. 3–18). The bauxite fills 10–30 m deep sinkholes.

The plateaus of St. Ann, Manchester and Clarendon are covered by a great number of lenticular surface deposits that are 5–15 m thick (HILL and OSTOJIĆ, 1982). The bauxites are very porous, earthy, of red colour. They are called *plateau type bauxites*. In contrast, *valley type bauxites* occur in the lower valleys and in plains, e. g. Porus and Essex valleys where large, stratiform deposits are found. They are partly exposed, partly covered by alluvial gravels or lacustrine marls of some metres thickness (ANDERSON, 1971). The valley type bauxites are

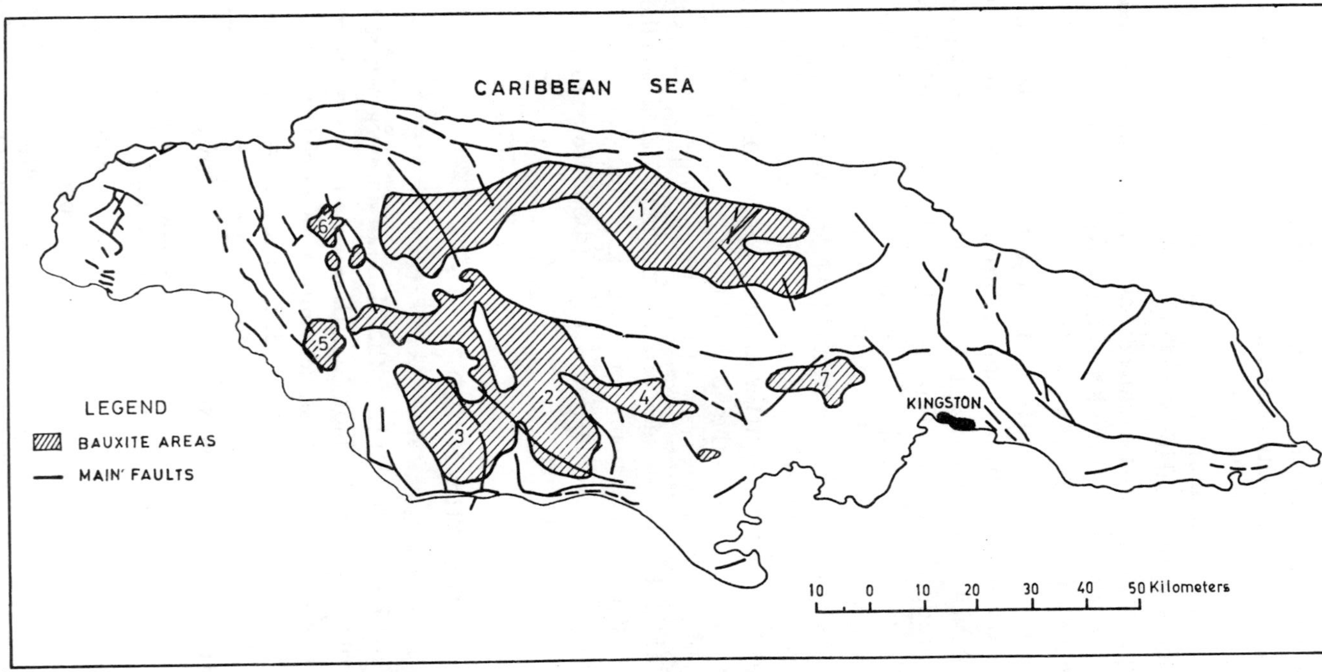

3–17. The main bauxite areas of Jamaica. 1. St. Ann Plateau; 2. Manchester Plateau; 3. Essex Valley; 4. Clarendon Plateau; 5. Nassau Mountains; 6. Cockpit Country; 7. St. Catherine Plateau.

characterized by an ochre colour. They contain more silica and less alumina than the plateau type bauxites (LYEW-AYEE, 1984; Tab. 3–1).

The bauxite is essentially gibbsitic with a subordinate boehmite content. Kaolinite is the main silicate mineral. The red colour of the plateau type bauxites

3–18. Aerial view of Cockpit Country (photo Tyndale-Biscoe).

is due to their major hematite content whereas the ochre colour of the valley type reflects their goethitic composition.

As a rule, karstification of the footwall limestone was stronger on the plateaus than in the lower valleys and plains.

Much smaller bauxite deposits of similar type occur in the southern part of **Haiti** and the **Dominican Republic.** They cover medium to strongly karstified

Table 3–1

Chemical composition of the plateau type and valley type bauxites of Jamaica (from LYEW-AYEE, 1984)

Average content in %	Plateau type	Valley type
Al_2O_3	48.8 ± 1.2	45.4 ± 1.0
SiO_2	0.8 ± 0.5	7.5 ± 3.3
Fe_2O_3	18.0 ± 0.8	18.8 ± 1.8
TiO_2	2.5 ± 0.2	2.2 ± 0.2

surfaces of Eocene to Oligocene limestones and have no overburden. The bauxite is red, soft, and essentially gibbsitic with subordinate boehmite.

Strongly tectonised, red, hard diasporic bauxites were found recently in the Pinar del Rio Province of **Cuba.** The deposits are in a highly folded Cretaceous limestone sequence.

Relatively small karst bauxite deposits occur also in the southeast part of the **United States.** In Gasconade County (Missouri) karst sinkholes in Cambrian to Ordovician dolostones are filled with grey to white diasporic bauxites, locally called *diaspore* (KELLER, WESTCOTT and BLEDSOE, 1954). The sinkholes are generally 5–10 m deep, but exceptionally they reach 70 m depth. Upper Carboniferous claystones and limestones are the overburden. The age of the bauxite is Middle Carboniferous.

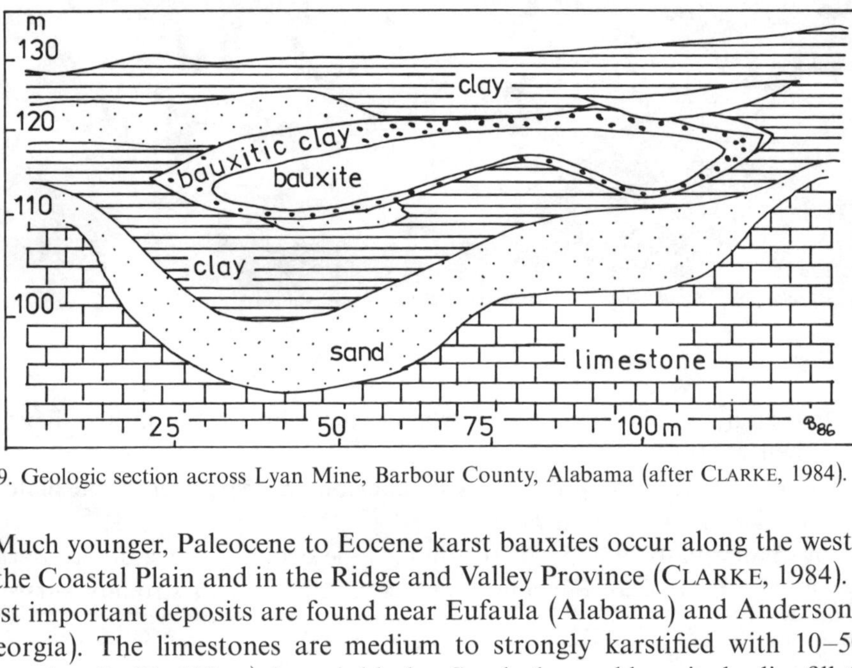

3–19. Geologic section across Lyan Mine, Barbour County, Alabama (after CLARKE, 1984).

Much younger, Paleocene to Eocene karst bauxites occur along the west rim of the Coastal Plain and in the Ridge and Valley Province (CLARKE, 1984). The most important deposits are found near Eufaula (Alabama) and Andersonville (Georgia). The limestones are medium to strongly karstified with 10–50 m (exceptionally 80–130 m) deep sinkholes. Sand, clay and bauxite bodies fill them (Figs. 3–19, 3–33). These are similar to the *Kazakhstan* bauxite deposits. The bauxite is gibbsitic, accompanied by kaolinite and hematite.

In **Europe** the main karst bauxite regions are in Greece, Hungary, Yugoslavia and France, those in **Greece** being the richest. Its principal bauxite areas are in the Parnassos, Ghiona and Helicon Mts. of central Greece. The main bauxite horizons are of Early and Late Upper Jurassic, Albian and Turonian ages. They are situated in relatively brief stratigraphic gaps in a shallow marine carbonate sequence. Most deposits are lenticular and 5–20 m in thickness. The footwall limestones are moderately to strongly karstified. The bauxite is generally of high quality with 50–60 % Al_2O_3 and 3–6 % SiO_2 content. Most bauxites are dark red, very hard, of diasporic or diasporo-boehmitic composition accompanied by kaolinite, hematite plus low amounts of goethite and anatase.

In **Hungary** the main bauxite region is in the Transdanubian Hills. It covers about 3 400 km². Three bauxite horizons can be distinguished, of Aptian-Albian, Turonian-Early Senonian and Paleocene-Early Eocene ages. 57 % of the total bauxite tonnage occurs in large stratiform deposits, 23 % in lenticular deposits, 10 % in sinkholes, 8 % in graben and the remaining 2 % in canyon-like, blanket and pocket deposits. The main groups of deposits occur near Nyirád, Halimba, Irhakut, Fenyöfö, Bakonyoszlop, Iszkaszentgyörgy, Gánt and Nagyegyháza.

The bauxite thickness is 5 to 30 m in the stratiform and lenticular deposits, which display moderate karstification. In the strongly karstified, sinkhole deposits, thickness of 60 to 114 m are quite frequent (BÁRDOSSY, 1982). For a detailed discussion of the karstification of the footwall surface see chapter of BÁRDOSSY and KORDOS (p. 000–000).

Bauxite quality varies from very high to low. The most frequent Al_2O_3 content is 48–55 %, the SiO_2 ranging 2–11 %. The mineral composition is mainly boehmitic, with some deposits being boehmito-gibbsitic. Kaolinite, goethite and anatase are the main accessory minerals. Most bauxites are brick-red, but ochre and mottled bauxite are also frequent. At some localities the bauxite was covered by marshy sediments. Under their influence, reduction and partial iron leaching occurred in the upper parts of deposits, producing grey pyritic and locally sideritic bauxites.

A second, small bauxite region occurs in South Hungary near the Yugoslav border at Nagyharsány. It consists of small pocket deposits situated between Lower Tithonian and Lower Barremian limestones. The footwall limestone is only slightly karstified.

In **Yugoslavia** the main bauxite areas are in Dalmatia, Bosnia - Herzegovina and Montenegro. Deposits occur in several stratigraphic horizons that are characterized by relatively short breaks in sedimentation. The oldest deposits are of Middle Triassic, the youngest of Oligocene age. The most important deposits were formed during the Late Jurassic, Cretaceous and Paleocene (GRUBIĆ, 1975). Most deposits are lenticular, and their footwall surface is moderately to strongly karstified. Bauxite thicknesses vary from 5 to 25 m. On the Istrian Peninsula deep sinkholes are filled with bauxite of 20–50 m thickness. The walls of sinkholes are often subvertical.

According to KORMOS (1943) bauxite-filled sinkholes occur on the surface of Senonian limestones in Knezopolje, near Mostar. During the mining of these deposits bauxite-filled caves were found in the limestone. These were also excavated and a system of ancient caves was discovered this way, entirely filled with red bauxite. Remains of ancient underground cascades and stalagmites prove that the caves were open and vadose before being filled with bauxite. The forms of some of these caves are shown in Fig. 3–20.

Important stratiform deposits occur at Vlasenica (Bosnia) and at Nikšićka Župa (Montenegro) in moderately karstified limestones.

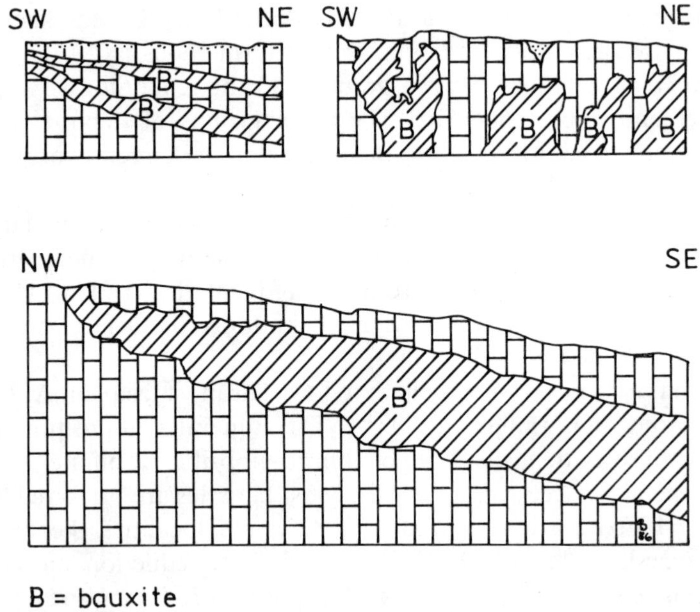

3–20. Geologic section of bauxite filled caves at Knezopolje, near Mostar, Yugoslavia (after KORMOS, 1943).

Most Yugoslav bauxites are of high to medium quality with 50–55 % Al_2O_3 and 1–8 % SiO_2 content. Generally they are boehmitic but the Paleocene and Eocene bauxites are mainly gibbsito-boehmitic and gibbsitic. Only a few deposits have a diasporic composition, e.g. Grebnićka Planina in Kosovo. Most Yugoslav bauxites are red; low iron, white bauxites occur in the Cenomanian of Montenegro.

The **French** bauxites are of a particular historical interest, having been discovered in 1821 by the French professor, BERTHIER. Their name, given by DUFRENOY in 1837, refers to the medieval town of Les Beaux in Southern France where BERTHIER made his discovery. World bauxite mining started in France, in the 1860's.

The main bauxite regions are in Southern France at Bédarieux, Villeveyrac, the Alpille Hills and around Brignoles. Smaller ones occur in the Haut-Var, at Revest and in Ariège, on the north slopes of the Pyrenees. Karstification of the footwall carbonates is moderate to strong. The strongest karstification was observed at Bédarieux and in the Haut-Var where 20–40 m deep, steep-walled karst depressions are filled with bauxite. In the other regions large stratiform and elongated strip-like deposits are surrounded by lenticular ones. The bedrocks are Jurassic to Lower Cretaceous limestones, less frequently dolostones. The hanging wall consists of Cenomanian to Senonian, locally Paleocene sediments.

According to a statistical study by LAJOINIE and LAVILLE (1979) most deposits can be divided into three zones, the lower and the upper being siliceous (SiO_2, 15–35 %) and the central having a low silica content (4–9 %). This latter zone contains 45–55 % Al_2O_3. The bauxite is mainly boehmitic, accompanied by kaolinite, hematite, goethite and anatase. The Haut-Var bauxites are gibbsito-boehmitic.

Ariège is the type region of the *Ariège-type* bauxite deposits. They were studied and described in detail by COMBES (1969). Upper Jurassic and Lower Cretaceous limestones were covered by Aptian marine clays and marls. After an emergence, lateritic weathering transformed their upper part into bauxite. Weathering was strongest in karst depressions in the limestone footwall, where drainage was best. Thus, there was a combined effect of normal lateritic weathering and of karst dissolution. The deposits are covered by Albian marine limestones and marls. The bauxite is red, hard, of medium quality, diasporic and diasporo-boehmitic. Effects of post-tectonic karstification on an overturned bauxite deposit were described also by COMBES (1969).

In **Spain** only small lenticular and sinkhole deposits are known. They are found on the southern slopes of the Pyrenees, in the Ebro massif, in the Maestrazgo and in the Betic Cordillera. Their age ranges from Middle Jurassic to Paleocene. The bauxite is generally of medium quality and boehmitic to boehmito-gibbsitic in composition.

In **Italy** small lenticular and pocket deposits are found in Abruzzi Mts, in Campania and on the Gargano Peninsula. Deep canyon-like deposits were mined in Apulia, near Spinazzola. The bauxite is of Cenomanian-Turonian age and of boehmitic composition (BÁRDOSSY et al., 1977). The type region of *Salento-type* deposits occurs on the Salento Peninsula, where a large number of lenticular deposits are scattered on a karstified surface of Turonian limestones. The bauxite probably comes from eroded Turonian deposits. Their transportation and redeposition may have occurred during the Oligocene and Miocene.

Large bauxite deposits have been discovered in recent years near Olmedo in northwest Sardinia. The footwall limestones and marls are only slightly karstified. The deposits are uniform 2–5 m thick and are covered by Turonian and Senonian marine limestones. The bauxite has a high alumina and low silica content and is essentially boehmitic (CECCARINI, OGGIANO and SALVADORI, 1984).

Small lenticular and pocket deposits occur in western **Romania** at Pădurea Craiului. They are of Early Cretaceous age and diasporo-boehmitic composition. Very small deposits are recorded also from **Switzerland, Austria, Czechoslovakia, Bulgaria** and **Albania.**

Important karst bauxite deposits occur in the **U.S.S.R,** the main region being in the North Ural Mts. The deposits are of Eifelian age, and rest on moderately karstified Gedinnian reef limestones. They form large stratiform and lenticular deposits. The bauxite thickness is generally 2–6 m; its lower part is red, and its upper part is green and grey. The latter contains fossils indicating a shallow marine, littoral environment. The bauxite generally has 50–55 % Al_2O_3, 3–15 % SiO_2 and 22–25 % Fe_2O_3. It is mainly diasporic, less frequently diasporo-boehmitic or boehmitic. The red bauxite also contains kaolinite and hematite, the green and grey contains pyrite, berthierine, leptochlorite and locally siderite (KIRPAL and TENIAKOV, 1978).

Similar but much smaller Devonian deposits are reported from the northernmost and from the South Ural Mts.

The type region of *Timan-type* bauxite deposits occurs in the South Timan Mts. The footwall limestone is of Late Devonian age. Its slightly karstified surface contains large, flat depressions filled with continental clays and sand with a maximum thickness of 40–60 m. The bauxite is situated on top of them with 4–6 m average thickness. The overburden begins with Lower Carboniferous marshy sediments. Under their influence the bauxite was reduced at most places to a grey pyritic material. The bauxite contains 40–65 % Al_2O_3, 12–28 % SiO_2 and relatively little iron. It is boehmitic with subordinate gibbsite.

The type region of *Kazakhstan-type* deposits occurs in Soviet Kazakhstan near Arkalyk and in the West Turgay area. Bauxites of the former region are Paleocene-Early Eocene, while those of the latter area are Albian to Turonian in age. The deposits fill depressions in strongly karstified Devonian to Carboniferous limestones and dolostones. Most frequently they display elongated, strip-like forms and follow the contact between carbonate rocks and shales. Sinkholes are also frequent. Most are 40–100 m deep but some reach 200–300 m depth. The lower parts of the karst depressions are filled with mottled clays,

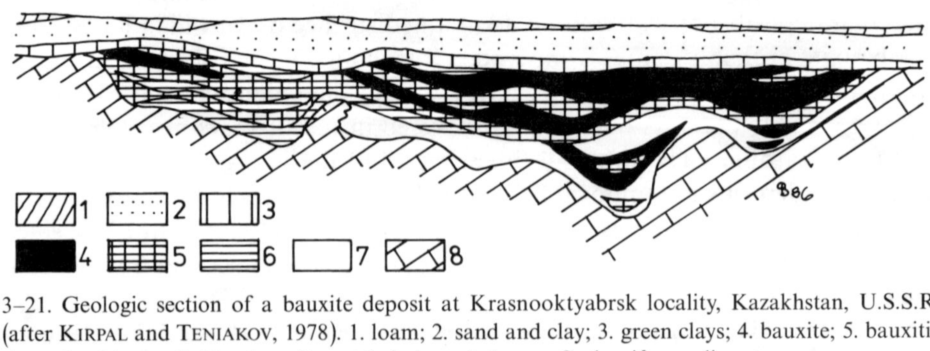

3–21. Geologic section of a bauxite deposit at Krasnooktyabrsk locality, Kazakhstan, U.S.S.R. (after KIRPAL and TENIAKOV, 1978). 1. loam; 2. sand and clay; 3. green clays; 4. bauxite; 5. bauxitic clays; 6. white kaolinitic clays; 7. mottled clays; 8. Lower Carboniferous limestones.

sandy clays and rare intercalations of thin bauxite. The upper parts consist mainly of bauxite with clayey intercalations. The deposits are covered chiefly with Lower Tertiary deposits (Fig. 3–21; cf. KIRPAL and TENIAKOV, 1978).

The bauxite is red; it contains 42–52 % Al_2O_3 and 5–15 % SiO_2. It is gibbsitic, accompanied by kaolinite, hematite and some goethite. In several deposits of the West Turgay area secondary processes have led to a strong sideritization accompanied by formation of calcite, pyrite and marcasite.

Similar but much smaller *Kazakhstan-type* deposits also occur in Siberia in the Salair region, along the Yenisei River and in the Chadobets area. Particularly deep sinkholes (down to 300 m depth) were discovered by drilling in the Yenisei River area (KIRPAL and TENIAKOV, 1978).

The oldest known bauxite deposits in the world are situated along the Bokson River in the Eastern Sayan Ranges. Karstification of their bedrock, Upper Proterozoic dolostone, was only very weak. The stratiform bauxite bodies are 4–5 m thick and are covered with Lower Cambrian sediments. The bauxite is dark red in its lower parts and greenish grey in the upper parts of the deposits. It contains generally 40–45 % Al_2O_3 and about 20 % SiO_2. It is mainly diasporic, accompanied by less boehmite (KIRPAL and TENIAKOV, 1978).

Large karst bauxite deposits are known in **China.** Their form of deposition is most like that of the *Timan-type* occurrences discussed before. The most important deposits are in the Liaoning, Shandong, Kweichow, Yunnan and Kwangsi-Chuang Provinces (WANG CHUNG-HSING, 1965; IKONNIKOV, 1984). They rest on slightly karstified surfaces of Upper Cambrian and Ordovician limestones and are covered by transgressive Middle and Upper Carboniferous sediments. The bauxite is mainly grey, hard and compact. Diaspore is the principal aluminium mineral, accompanied by subordinate boehmite. The bauxites generally contain 45–60 % Al_2O_3, 5–20 % SiO_2 and they are low in iron.

Karst bauxites are found in the northern part of **Vietnam** in the Lang-Son and Cao-Bang Provinces. *Mediterranean-type* karst bauxite deposits were formed here on slightly karstified surfaces of Upper Permian limestones. The 4–9 m thick lenticular deposits were covered after a short stratigraphic break by Upper Permian transgressive sediments (Fig. 3–22). After several tectonic events the entire area emerged in the Neogene and very strong karstification started under wet tropical climate. As a result one of the most beautiful tower karst regions of our Earth was formed, very similar to the neighbouring South China tower karst areas. The strong dissolution not only removed the Permian cover of the deposits but attacked the bedrock limestones at many places as well. The bauxite gradually broke up into blocks that started to creep down slopes. Secondary, *Salento-type,* deposits were formed this way, comprising hard bauxite fragments and blocks mixed with soft red clayey material. After washing, even deposits such as these may be of economic interest (BÁRDOSSY, 1982). The bauxite is diasporic, accompanied by hematite, boehmite, berthierine and kaolinite.

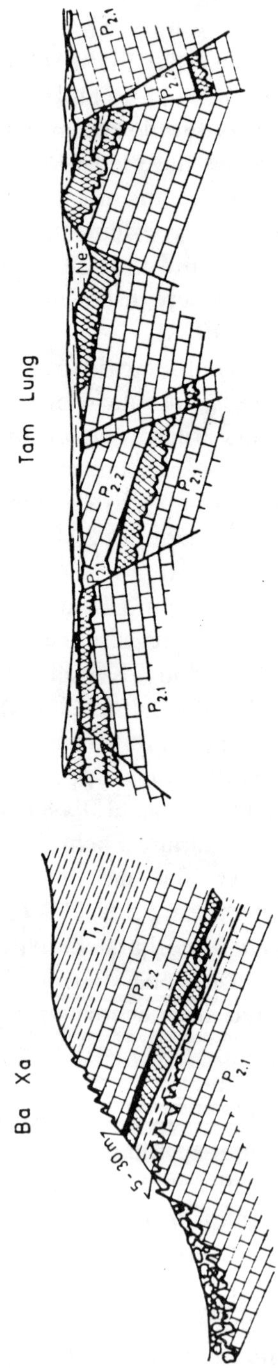

3–22. Geologic section of Permian bauxite deposits in Lang-Son Province, Vietnam (after Komlóssy, 1976).

A distinctive karst bauxite belt can be traced in the **Southwest Pacific,** beginning in the Philippines (Samar Island) and continuing through the Solomon Islands (Rennel) to the Loyalty Islands (Lifou, Maré), the Lau Islands and finishing in the southeast on Niué Island (see also chapter of BOURROUILH-LE JAN, p. 443–471). The youngest known karst bauxites are found here. Their bedrocks are Plio-Pleistocene limestones and they have no cover. Their formation started during the Pleistocene and continues at present. The limestone surface is weakly to moderately karstified, with depressions being less than 2 m deep. The bauxite is red, loose, earthy material with low silica and relatively high iron content. Gibbsite is the main aluminium mineral, accompanied by hematite, goethite and kaolinite.

Karst bauxites occur along the Taurus Mts in southern **Turkey,** in two stratigraphic horizons. The older ones are of Upper Permian age and are found upon weakly karstified Lower Permian limestones. The younger are Cenomanian and cover a moderately to strongly karstified surface of Lower Cenomanian limestones.

Comparatively small bauxite deposits form a belt from **Iran** through **Afghanistan, Pakistan** to **India** (Kashmir). They are found in several stratigraphic horizons, the lowest being in the Middle Devonian and higher in the Permo-Triassic, Upper Triassic, Middle Jurassic, Senonian and Paleocene. They correspond to relatively short periods of emergence. Subsequent compressional tectonism has strongly dislocated these deposits. The bauxite is generally red, hard and very compact. Diaspore is the main alumina mineral; less frequently, boehmite is also enriched.

It is significant that no karst bauxites have been found yet in the southern continents of Australia, Africa and South America, or in other parts of the Indian and Arabian peninsulas.

The Origin of the Bauxite Deposits and Their Relationship to Paleokarst Phenomena

The origin of the karst bauxites is strongly related to the underlying carbonates and their karstification. Pre-existing, syngenetic and subsequent karstification can be distinguished.

Most of the karst bauxite deposits are underlain by limestones. Marbles are rare. Dolomitic limestones and dolostones also have a limited association with bauxite. As a general rule, the host carbonate rocks are very pure, containing only a little clayey contamination. In those areas where pure carbonates pass laterally into marls, karstification quickly diminishes and the bauxite deposits also disappear. The one single exception is the deposit near Olmedo (Sardinia) where a large stratiform bauxite deposit extends over a slightly karstified surface of clayey limestones and marls of Neocomian age. Where this deposit covers pure limestones, karstification is more pronounced and the desilicification of the bauxite was more efficient due to better drainage conditions (CECCARINI, OGGIANO and SALVADORI, 1984).

The author has found no essential differences between the karst relief on pure limestones and on dolostones. Paleomorphologic and hydrogeologic conditions had a stronger influence on bauxite development than these differences in rock composition.

With respect to the role of *pre-existing karst,* it is established that karst depressions served as traps for the accumulation of initial bauxitic materials. It must be stressed that most probably the depressions were not yet fully developed when the first bauxite arrived. Their evolution went hand in hand with the accumulation of the bauxitic material. This syngenetic association can be explained by the effect of vegetation and of the bauxite itself upon infiltrating rainwaters. There was increased dissolution of the carbonate rocks below the bauxite deposits compared with surrounding bauxite-free areas. The reason must be the higher CO_2 content of these waters and perhaps the slower downward movement of rainwater in the bauxite. At many places in Jamaica the author has observed outcrops of very hard limestones that contain large-scale dissolution vugs and fissures around the deposits. On the other hand, below the deposits dissolution has operated on a finer scale resulting in a pulverised, soft, porous limestone beneath the contact.

The deepening of bauxite karst depressions can continue until their bases reach the groundwater table, where it stops. On highly elevated carbonate plateaus, sinkholes could reach 50–100 m depth, or even 200–260 m some *Kazakhstan-type* deposits (KRIVTSOV, 1969). Repeated alternation of arenaceous continental sediments with bauxite and lignite beds in these deposits was interpreted by PELIETEK (1971) as due to repeated uplifts with karst deepening, followed by a gradual rise of the groundwater level that governed the speed and nature of the sedimentation in the karst depressions.

In most bauxite regions the deposits follow ancient fractured tectonic zones of the footwall. Karstification could progress best in such zones because fracturing of the carbonate rocks facilitated their drainage and dissolution. Often the biggest and deepest bauxite filled depressions are located at the intersection of two fractured tectonic zones.

In Kazakhstan and Siberia several bauxite deposits are found on tectonic contacts between carbonate and aluminosilicate rocks. Deep sinkholes and karst canyons were formed along these fractured zones, reaching 100–200 m depth in places. In a deposit near Sokhatino (Siberia) karstification of the footwall limestone continued along tectonic contact even below neighbouring amphibolite, as seen in Fig. 3–23 (BUSHINSKY, 1975).

Tectonic grabens also serve as traps for bauxite accumulation. In these cases the initial tectonic surfaces were modified by later karstification. Small depressions and cone karst were found on the floors of the graben (below the bauxite) and surrounding fault-planes also underwent karstification.

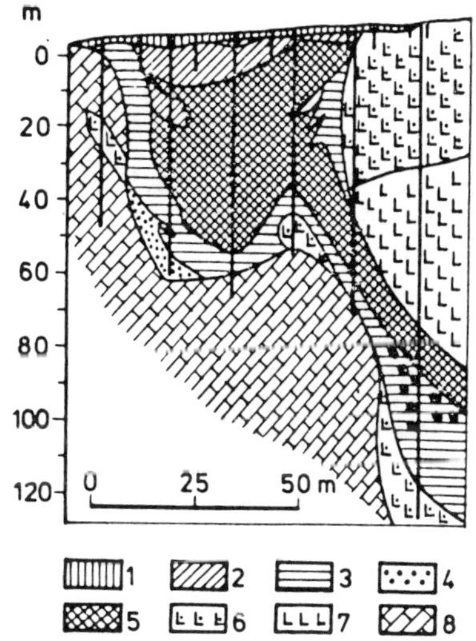

3–23. Bauxite deposits along a petrographic contact, Sokhatino deposit group, near the Yenisei River, U.S.S.R. (after BOBROV in BUSHINSKY, 1975). 1. sand, clay; 2. red bauxitic clay; 3. red clay; 4. sand; 5. bauxite; 6. clayey regolith; 7. amphibolite, Proterozoic; 8. limestone, Proterozoic.

In a few cases erosional valleys and canyons have served as places of bauxite accumulation. Syngenetic karstification can be observed below these deposits also.

It is significant that open caves have acted only exceptionally as places of bauxite accumulation (KORMOS, 1943).

In earlier literature on bauxites it was a widely held view that the karst relief did not develop further after the accumulation of the bauxite deposits. VADÁSZ (1951) was the first who stressed that karstification could continue after bauxite accumulation. In the tectonically tilted bauxite deposits of Yugoslavia and

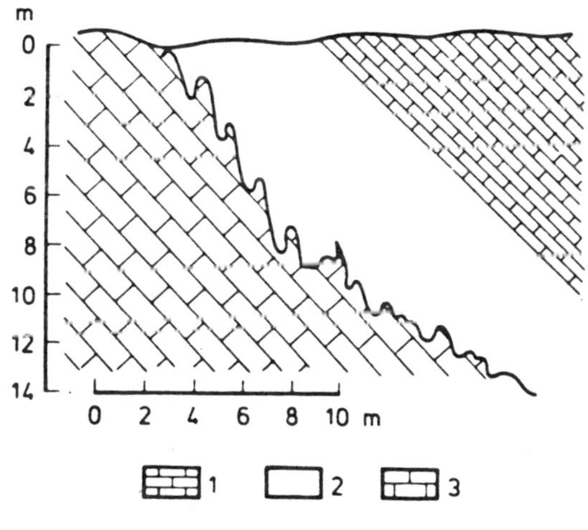

3–24. Vertical karst forms beneath a steeply dipping bauxite deposit at Anthimos open pit, Distomon, Greece. 1. Tithonian limestone; 2. bauxite; 3. early Upper Jurassic limestone.

Greece, vertically oriented karst forms are often observed (Fig. 3–24). It is obvious that they could be formed only after the tilting i.e. long after deposition of the bauxite and of its cover.

SAKAĆ (1966) found at Radinović, near Bosanska Krupa (Yugoslavia) steep karst features below a tectonically overturned bauxite deposit. The original hanging wall of Campanian-Maastrichtian age is now in the footwall position. The original footwall (of Turonian age) showed a much more gentle karst surface. COMBES (1969) described a similar overturned bauxite deposit from Cadarcet in the French Pyrenees (Fig. 3–25). The strong secondary karstification of the original hanging wall is obvious.

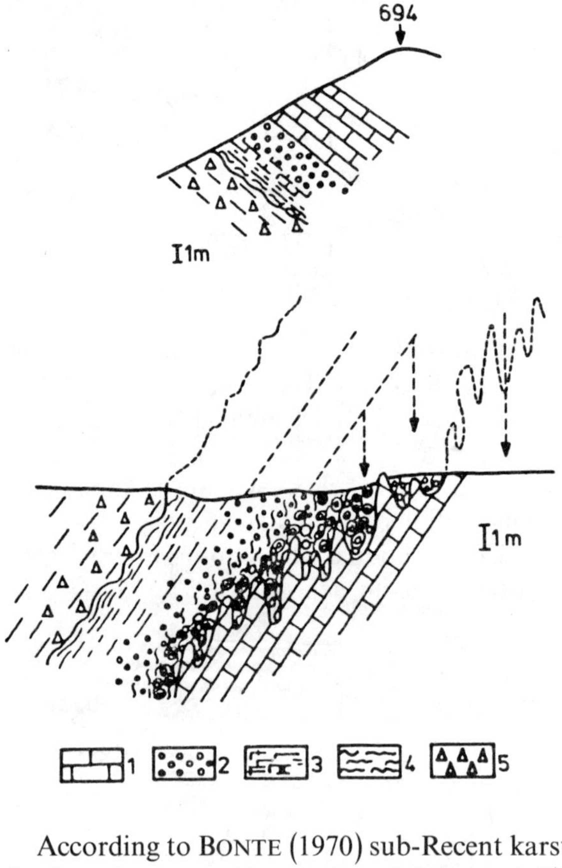

3–25. Secondary karstification below an overturned bauxite deposit near Cadarcet, Ariège, France (after COMBES, 1969). 1. Cretaceous limestone; 2. bauxite; 3. bauxitic clay; 4. laminated clay; 5. Jurassic limestone.

According to BONTE (1970) sub-Recent karstification took place under bauxite deposits exposed by erosion of their cover. This process resulted in particularly rugged karst forms, as observed in the Combecave-Pins-Batard open pit near Brignoles (France; Fig. 3–26). The karst surface gradually becomes gentler as the thickness of the overburden increases.

Recent continuation of karstification has resulted in the formation of open caves, cavities and fissures below several deposits, as the writer has observed in

Jamaica, Greece and Hungary. Generally, they are 0.5–2.0 m size. An open cave about 25 m diameter and 30 m height was noted in central Greece. Big blocks of bauxite fallen from the top of the cave covered its floor.

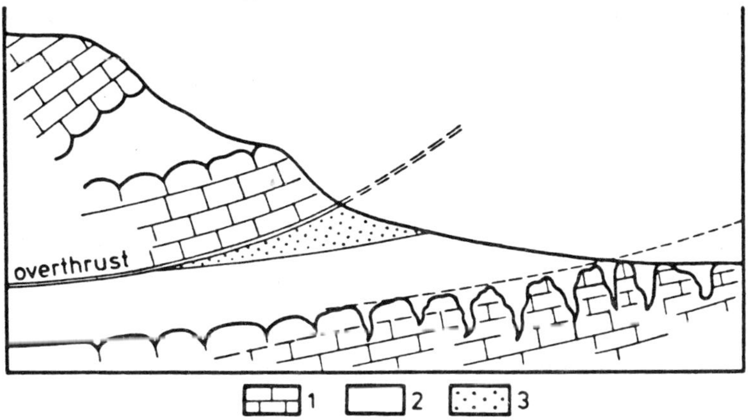

3–26. Relationship between karst relief and overburden thickness, Combecave-Pins-Batards open pit near Brignoles, France (after BONTE, 1970). 1. bedrock: Bathonian limestone; 2. bauxite; 3. Senonian sandstone.

Another important role of karst has been its influence on the *in situ* desilicification of bauxite. Material, more or less initially bauxitised, continues to desilicify in the karst environment. The good permeability of the carbonate footwalls has ensured that rainwater could easily migrate downward through the bauxites. Observations by the writer in Jamaica showed that even water from big tropical rainstorms could disappear through bauxite without any surface runoff being generated.

The descending waters dissolve part of the remaining silica content and take it away through the footwall rocks. This process has been most efficient where the drainage was best, generally over the deepest part of karst depressions. As a rule the lowest silica content is found in these parts of the deposits. It must be mentioned, however, that in a small number of deposits little, if any, relationship could be found between bauxite quality and the footwall karst surface, e.g. at Bitó II Mine, Hungary. In these cases desilicification did not continue after deposition of the bauxite, probably because of a lower permeability of the footwall and/or a higher groundwater level.

Karst depressions have protected bauxite deposits from subsequent erosion. In several regions of intense erosion, bauxite is preserved exclusively in the karst depressions. It has been totally eroded from the upstanding areas.

Several authors have attributed a crucial role to karst dissolution in furnishing the parent material of the bauxite. This is the so-called *terra rossa theory*. According to it, bauxite is a weathering product of the insoluble residues of footwall carbonate rocks. The name terra rossa has been used since the last century to denote the reddish coloured soils widespread upon karstified rocks in the Mediterranean area.

DOLFUSS (1904) was the first to propose this theory to explain the origin of French bauxites. It was elaborated in detail by TUĆAN (1912) and KIŠPATIĆ (1912) for bauxites of the Dinaric Ranges. This theory was applied by DE LAPPARENT (1930) to the bauxites of France, by HOSE (1963) to the Caribbean bauxites and by KOMLÓSSY (1967) to the bauxites of Hungary. According to DE WEISSE (1964) all of the Earth's karst bauxites have had this origin.

However, the theory encountered repeated criticism from others who pointed out that at most places the carbonate bedrock is very pure, containing only some tenths or hundreds of one percent of Al_20_3. Accumulation of the actual bauxite deposits would have required the dissolution of several hundred or even thousands of metres of carbonate rocks. Direct evidence was found that the *Kazakhstan-* and *Timan-type* bauxites were the products of lateritic weathering occurring on neighbouring aluminosilicate rocks. The weathering products were transported by surface waters into near-by karst depressions (BUSHINSKY, 1975).

For *Mediterranean-type* deposits the explanation of their origin by lateritic weathering is more difficult because, in most cases, they are relatively far from aluminosilicate rocks. During recent years evidence has been found at many localities (Jamaica, Italy, the North Ural, ets.) to suggest that volcanic tuffs carried by wind could also serve as the parent material (COMER, 1974). Floating pumice was also considered by some authors (TERCINIER, 1971).

Clays, marls and even hydrothermal springs have been suggested by other authors to explain the origin of karst bauxites. Finally, different chemogenic theories should be mentioned which suggest dissolution of alumina in the parent rocks, its transportation in aqueous solution and its precipitation in karst depressions. For example, the *sulphuric-acid theory* suggested that pyrite oxidation was the dissolving agent (ARKHANGELSKY, 1937). RUTTNER (1970) drew analogies with the acidic *black waters* of the Rio Negro, Brazil, and stressed the dissolving role of humic acids produced by decay of the rich tropical vegetation.

After visiting a great number of karst bauxite deposits around the world, the author has concluded that no one of the above models or theories has an overall validity. Parent materials and modes of bauxite origin may differ from one bauxite region to another depending upon its regional geologic evolution. There are bauxite regions where one single parent material can be presumed but in most places contributions in different proportions from several different parent materials seems to be the most likely explanation.

Global tectonism has had a great influence on bauxite formation. This is illustrated by the fact that 93 % of karst bauxites are situated in orogenic belts, whereas 96 % of the lateritic bauxites occur on stable platform areas. The author has found also that the *plate tectonic* theory offers the most satisfactory explanation for the distribution of bauxite deposits in Phanerozoic time. Using the continental motion reconstructions of SMITH, BRIDEN and DREWRY (1973) he showed that there is an almost total coincidence between the world bauxite

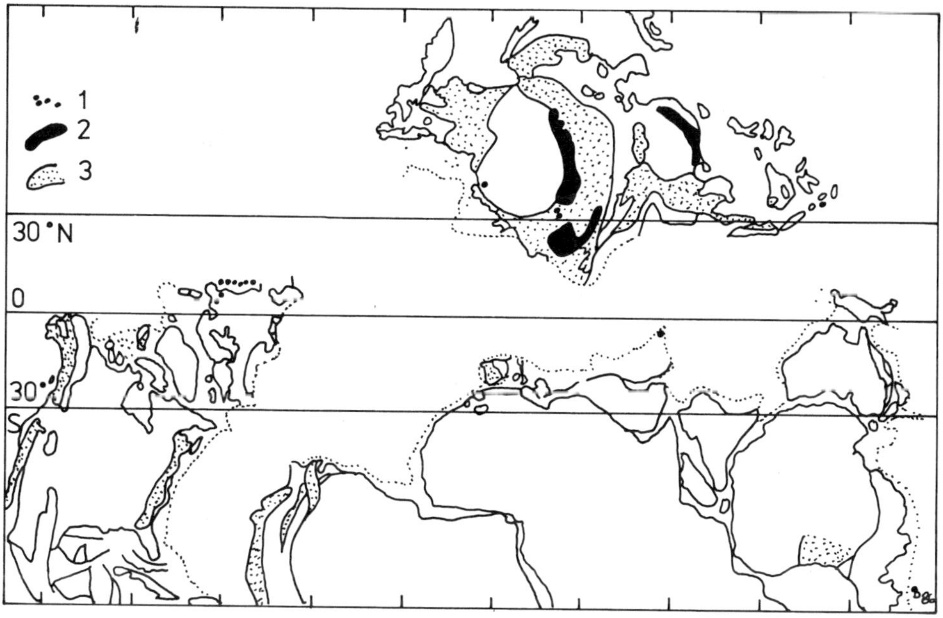

3–27. Devonian continent reconstruction after SMITH, BRIDEN and DREWRY (1973) with contemporary bauxite deposits added (after BÁRDOSSY, 1982). 1. bauxite deposits; 2. Caledonian orogenic belts; 3. Variscan orogenic belts.

distribution and the position of favourable climatic belts in any given epoch (BÁRDOSSY, 1982). As an example, the Devonian continental reconstruction with its corresponding bauxite deposits is shown in Figure 3–27. It can be seen that the deposits of the North Ural and Timan Mts (now near the Arctic Circle) were situated in the equatorial belt during the time of their formation.

Conclusions

About 10 % of the total world bauxite tonnage is directly connected with paleokarst. These deposits form a separate group called *karst bauxites* because they rest on the karstified surfaces of pure carbonate rocks. The degree of karstification ranges from very weak to very strong. Pre-existing, syngenetic and subsequent karstification can be distinguished below different deposits.

The role of paleokarst in bauxite formation can be summarized as following:

(1) karst depressions served as traps collecting the initial bauxite material from surrounding areas; (2) karst depressions provided places of optimum drainage, thus promoting further desilicification of the bauxites; (3) bauxites were protected in karst depressions from later surface erosion. This is prime reason why they are preserved at many places.

IRON ORE DEPOSITS IN PALEOKARST

György Bárdossy, Yves Fuchs
and Jerzy Głazek

Introduction

Iron ore deposits in paleokarst structures are frequent but only a few of them are of economic interest at the present time. During past centuries, however, such deposits were intensively mined all over Europe. High iron contents, occurring frequently in this type of deposit, were welcome for metallurgical reasons during Middle Ages and until the second half of the 19th century. There are some exceptions, such as North-East Bavaria in Federal Republik of Germany and Alapaevsk and Akkermanovsk in the U.S.S.R., where paleokarst iron ore deposits with important reserves are still mined.

Probably the best prospects for the discovery of new, rich iron deposits in paleokarst structures are in ancient and recent tropical domains. The latter have been poorly investigated up to the present. The Qui-xa deposit in North Vietnam may serve as an example.

Examples of Deposits

A great number of karst deposits that are classified as lead-zinc or baryte deposits also contain important quantities of iron even if they are uneconomic. At **Bled Zelfane,** Tunisia (AMOURI, 1977), paleokarst structures are filled with mixed ore containing halloysite, saucconite clays, goethite, smithsonite, etc. The zinc content ranges from 2 to 10 %, the Fe from 15 to 25 %.

In Belgium, Luxembourg and France a great number of small iron ore deposits in paleokarst structures are known because they were mined during the past. In the region of **Nismes** and **Couvin** in Belgium (SWYSEN, 1971) paleokarst forms (locally called *abannets*) developed in Givetian reefal limestones. These are large depressions. Their diameter can be as great as 100 m and their depth, 50 m. They are filled with limonite lenses and Tertiary sands. In Luxembourg and in the southern part of Belgium (LUCIUS, 1952; DEJONGHE, 1985)

P. Bosák, D. C. Ford, J. Głazek and I. Horáček (Editors):
Paleokarst. A Systematic and Regional Review. – Elsevier and Academia. Amsterdam and Praha. 1989.

paleokarst is developed in Bajocian limestones (Longwy Limestone). The karst depressions are filled with reworked lateritic products (locally called *bolus*) and iron-rich pisolitic concretions. The age of this filling is Eocene according to LUCIUS (1952), or Oligocene following DEJONGHE (1985). The pedological alteration of limestone and marls of Early Jurassic age formed residual clay deposits which then experienced lateritization under hot climatic conditions. Mobilisation of iron in these particular conditions induced the formation of pisolitic concretions in the paleokarst. These ores were mined from the protohistoric period onwards and were rich in iron (35 to 48 %) while depleted in phosphorus. The deposits were small in size. Mining activity ceased in 1852.

Similar deposits are known from France. STEINBERG (1970) reports pisolitic and detrital ores, locally called *siderolithique,* frequently filling karst depressions in Mesozoic limestones of **Poitou** in the West of France.

Such karst depressions can also be observed in the southern part of the Massif Central (Causses). They are filled with clays, detrital quartz and iron pisoids. Caddis worm (phryganeid) tubes indicate a fluviatile environment. The most important deposits of this type were described by ROUSSET (1967) in the Haut-Var at **Beausoleil** (SE France). Karst sinkholes filled with iron ore were mined here at the end of the 19th and the beginning of the 20th century. The total quantity of extracted ore was 350 000 tons with an average Fe content of 55 %. The karst is located within Bathonian and Bajocian limestones. The vertical depth of the sinkholes reached 200 m! The material filling the sinkholes consists mainly of goethite accompanied by lesser amounts of hematite and dolomite. Marcasite and pyrite nodules were observed in the lower part of the sinkholes, their quantity increasing with depth. Concerning the source of the iron, ROUSSET (1967) emphasizes the importance of pedological alteration of glauconitic marl of Early Cretaceous age during an emergence as the most plausible origin.

All of the deposits described above show some characteristics in common. The karst depressions are filled with goethitic or hematitic ores mixed with continental sediments. Sulphides can be present in the deeper part of the paleokarst, but minerals like siderite are not present and the phosphorus content is very low. Their vertical extension below the surface can be important. The ore reserves are in the range of tens of thousands of tons. These deposits were formed in continental environments beneath high plains surfaces.

A distinctly different type of iron-filled karst structure is known in the Federal Republic of Germany and in the U.S.S.R. Sedimentary iron ores filling karst depressions accumulated in a continental brackish or marine environment. Such karsts are known at Saint-Dizier (France), in Oberpfalz (Bavaria, F.R.G.) and in the U.S.S.R.

Near **Saint-Dizier** the karstification occurred at the top of the Jurassic limestones. The karstified surface is overlain by sediments of Neocomian age deposited in a brackish environment. The deposit consists of iron hydroxides and celestite.

In the North-Eastern part of **Bavaria** important iron ore deposits are located stratigraphically at the boundary between the Jurassic and Cretaceous. Sixty five

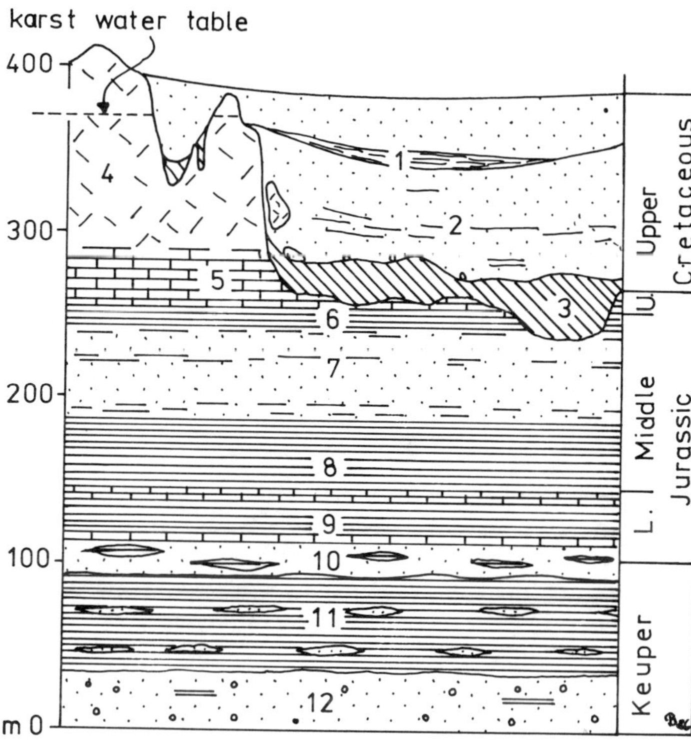

3–28. Stratigraphy in the iron deposit zone of Northeast Bavaria, Federal Republic of Germany (after GUDDEN, 1984). 1. *Cardium* clay, 0–30 m; 2. Cretaceous, Cenomanian-Coniacian-? Santonian, 0–150 m; 3. ore-bearing horizon, 0–60 m; 4. Upper Jurassic dolostone and 5. Upper Jurassic limestones, both more than 150 m thick; 6. Dogger γ–ζ, 5–15 m; 7. Dogger Sandstones, 55–100 m; 8. *Opalinus* clay, 50–60 m; 9. Lower Jurassic, 25 m; 10. Rhaetian to Lower Jurassic sandstones, 20 m; 11. refractory claystone, 45–55 m; 12. Upper Burgsandstone, 30–35 m

million tons of ore with an average Fe content of about 45 % have been mined since exploitation began. The ore fills troughs of a karst surface formed on Upper Jurassic limestones (Fig. 3–28). Their depth may be as great as 150 m and they can be several kilometres wide. The earliest ore-containing layers within them are of Early Cretaceous age (GUDDEN, 1984). The ore district forms a NNW extending tongue (Fig. 3–29) that was a marine gulf during the Late Cretaceous (Cenomanian-Coniacian). This paleogeography is probably related to a reactivated feature of the basement. The existence of NNW/SSE-trending structures in the deposits is related to some reactivation during the Early

Jurassic-Cretaceous period. Flexure was active during the sedimentation and faulting occurred after deposition of the ore and before the deposition of Coniacian marls. This tectonism induced the formation of a trough in which previously developed karst features were preserved from erosion and where the supply of dissolved iron in surface waters was important.

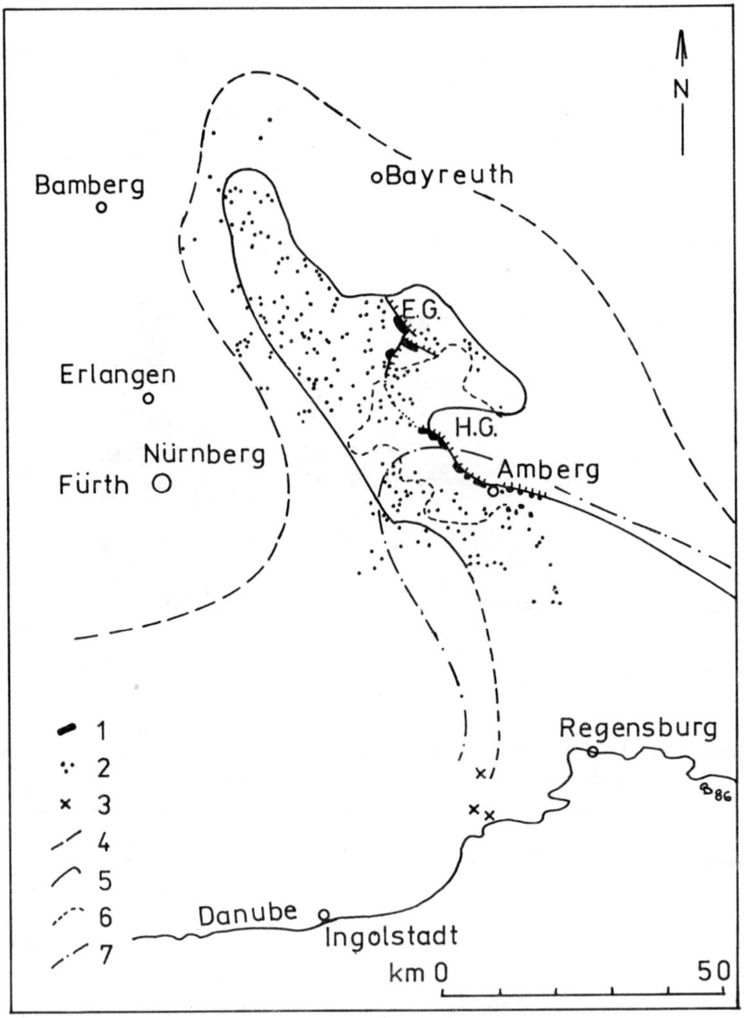

3–29. Paleogeographic situation of iron ore deposits, Northeast Bavaria, F.R.G. (after GUDDEN 1984). 1. main ore deposits of the Amberg Ore Unit; 2. occurrences of Cretaceous iron ores, partly reworked; 3. ancient iron ore mining near Painten, Neuessing and Kelheim; 4. approximate outer boundary of the Upper Cretaceous (Cenomanian to Santonian) depositional area; 5. approximate boundary of the depositional area of the Amberg Ore Unit; 6. extension of Cretaceous Neukirchen Ochres; 7. approximate outer boundary of Late Cenomanian glauconitic sandstone; H. G. Hahnbach uplift; E. G. Eibenstock uplift.

There is some controversy concerning the nature of the environment during deposition of the iron ore beds (continental, brackish or marine?). It seems obvious to us that this area was a trough with submergence of the karst surface.

The ore deposits are interbedded with clastic layers. There are three main types of ore: (1) white ore (Weisserz) containing mainly siderite with some accessory chamosite; (2) brown ore (Braunerz) with goethite and limonite; and (3) limonitic ochre ore (Ockererz). During the deposition of the ore, variations in the Eh caused the alternation of siderite and limonite. In many cases, erosion planes can be observed and might be related to post-depositional oxidation of the sideritic ore? Other changes, particularly limonitization, are locally associated with late circulation of oxidising water.

There are two paleokarst iron ore deposits in the U.S.S.R., described by SOKOLOV and GRIGOREV (1977). The first is near **Alapaevsk,** Sverdlovsk district, in the Northern part of the Ural Mountains. A paleokarst surface of Car-

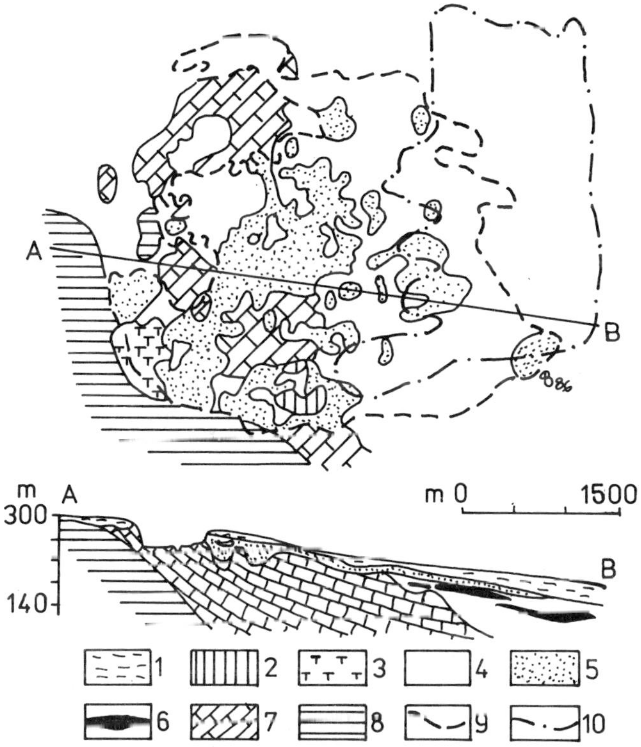

3–30. Geological sketch-map and cross-section (A–B) of the Akkermanovsk deposit, U.S.S.R. (after KISELEV, 1963 *fide* SOKOLOV and GRIGOREV, 1977). 1. brown and grey loams; 2. rubbly-pebbly redeposited iron ores; 3. yellowish-grey sandy clays, frequently with gypsum; 4. mottled clays, ferruginized sands and pebble beds; 5. brown ironstone, upper horizon; 6. sideritic clay ores, lower horizon; 7. Visean limestones; 8. siliceous shales; 9. outlines of brown ironstone, upper horizon; 10. outlines of sideritic ores, lower horizon.

boniferous limestones is overlain by a stratiform ore-bearing sequence of Mesozoic age. The overburden consists of Cenozoic sands and Quaternary loam. The thickness of the ore sequence varies from 0.5 to 70 m, being largest in the karst depressions. The deposit has a length of 5 to 10 km and width of 0.5 to 1 km. The upper part of the ore-bearing sequence consists essentially of goethite and the lower part (located below the groundwater table) of strigovite and hydrogoethite accompanied by siderite. The ore has a nodular and concretionary structure and is associated with ochreous clays. The average Fe_2O_3 content of high quality ore is 55 % (38.5 % Fe). The reserves of the deposit are estimated to be 42 million tons.

The other deposit is located in SE Ural 20 km west of Orsk at **Akkermanovsk.** A paleokarst surface of Visean limestones is covered by Jurassic continental sediments. At their base are stratiform iron ore deposits. The maximum thickness of the ore sequence is 35 m. Laterally, the ore passes into a sequence of mottled clays and ferruginous sands underlain by sideritic ore and clay. The ore consists of a mixture of goethite and hematite within a clayey groundmass. It contains 45–50 % Fe_2O_3 (31.5 to 35 % Fe). The structure of the ore is partially oolitic-pisolitic, partly blocky-rubbly. The reserves of the entire deposit are estimated at 280 million tons; about the half of them belong to the paleokarst type (Fig. 3–30).

Important iron ore deposits were discovered in the northeastern part of Vietnam, in the Phang-xi-pang (Fan-si-pan) Mountains (HOFFET, 1939). There, between two belts of Proterozoic (?) gneisses, there is a narrow synclinorium of crystalline schistose rocks (OSIKA, GŁAZEK and JUSKOWIAK, 1967) among which dolomitic marbles classified as Middle Devonian occur (GŁAZEK and JUSKOWIAK, 1964). All rocks have been metamorphosed to greenschist facies and were strongly folded during the Indosinian Orogeny (Late Triassic) (Fig. 3–31).

The planated mountain ridges are covered by thick laterite enriched with limonite and containing martite blocks that are without economic importance. However, in some places on the karstified marbles huge accumulations of limonite ore were found. The weathering and erosion processes that produced and concentrated them probably started in the Early Jurassic (GŁAZEK,1966). They were accelerated by the Himalayan uplift which caused the deep incision of the present meandering river pattern.

The planated **Qui-xa Hill** (Fig. 3–32) in the axial part of the synclinorium probably represents a fragment of pre-Late Tertiary surface. The ore-bearing sequence consists of porous limonite with intercalations of laterite, weathered gravels and detrital blocks from surrounding rocks. The ores compose 85 % of it. The thickness of these deposits reaches 150 m in karst depressions. The length of the deposit is 2 km and the width is a little less than 1 km; the average thickness is about 100 m. The ore consists of goethite, hydrogoethite and hydrohematite, with some manganese hydroxides. Relicts in the ore sequence in clude oxidized pebbles and blocks of magnetite quartzites and pyrite-bearing rocks. Particularly interesting are intercalations and lenses of post-carbonate, rhombohedral limonite enriched with manganese oxides, called *turgite*. Average

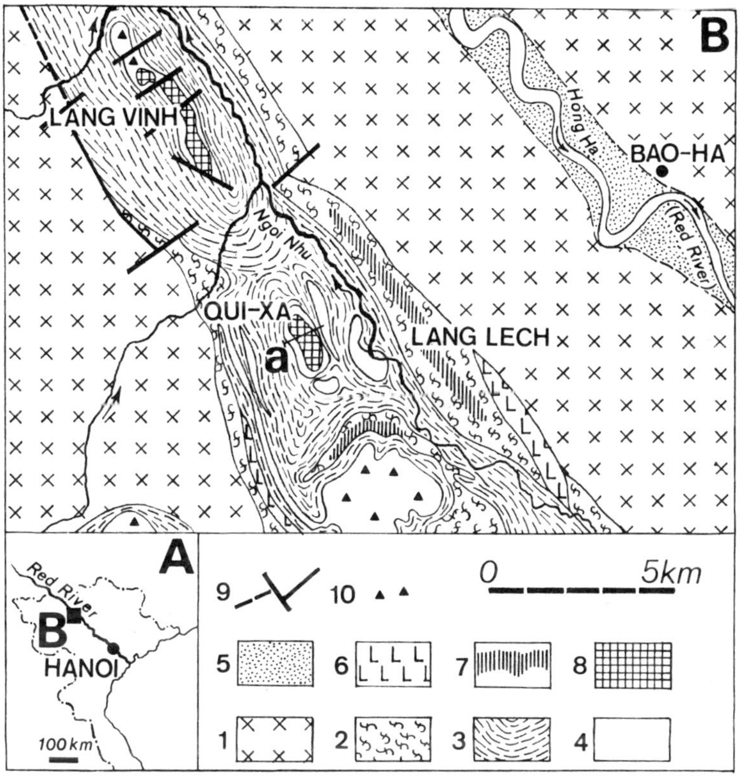

3–31. Geologic sketch-map of the ore bearing area in Vietnam (inset shows the location), modified after OSIKA, GŁAZEK and JUSKOWIAK (1967). Proterozoic (?) : 1. gneisses, 2. Lang Lech Group of mica schists; Paleozoic (?) : 3. Coc-xan Group of phyllites, 4. Devonian (?) marbles; Upper Tertiary: 5. conglomerates and brown coal deposits; Plutonic rocks: 6. metamorphosed gabbroids, Iron ore deposits: 7. primary magnetite-bearing quartzites, 8. karst limonitic ores; 9. faults and overthrusts; 10. mogotes.

ore composition is: Fe 54.8 %, Mn 3.0 %, SiO_2 3.75 %, Al_2O_3 2.0 %. P 0.08 %, S 0.025 % (KUBICKI in OSIKA, GŁAZEK and JUSKOWIAK, 1967). The reserves of these deposits are estimated at 106 million tons.

Similar but lesser deposits were found in **Lang Vinh** (HOFFET 1939). There, the thickness is up to 40 m (in sinkholes), the average iron content is 40 %, and the $SiO_2 + Al_2O_3$ content is 19.2 %. The reserves were estimated to be 10 million tons (GRZEGORSKI and SAWICKI in OSIKA, GŁAZEK and JUSKOWIAK, 1967). This deposit is much more dissected and eroded by the post-Neogene erosion. It is remarkable that both the Qui-xa and Lang Vinh deposits were originally treated as iron caps upon supposedly primary iron-bearing rocks (HOFFET, 1939; OSIKA, 1961); however, boreholes revealed only barren rocks (mostly marbles) beneath the ores.

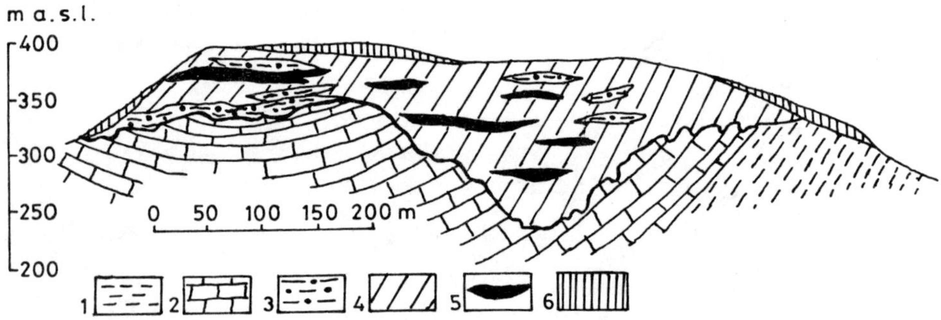

3–32. Cross-section of the Qui-xa deposit (after KUBICKI in OSIKA, GŁAZEK and JUSKOWIAK, 1967). Paleozoic substratum: 1. phyllites, 2. marbles; Ore-bearing sequence: 3. laterite with weathered gravels and blocks of older rocks, 4. limonites, 5. turgite intercalations, 6. laterite and limonite weathering waste.

The Role of Physico-Chemical Processes in the Genesis of the Paleokarst Type of Iron Ore Deposits

The existence of monometallic or mainly monometallic, exogene concentrations like the paleokarst iron ore deposits calls for processes which induce selective mobilization, transportation and deposition of iron.

The weathering of continental areas under hot and wet climatic conditions is particularly important. The abundance of lignitic fragments in the deposits (Saint-Dizier, Oberpfalz), and the kaolinization of the crystalline basement of the Bohemian Massif during the Early Cretaceous period seems to establish such climatic conditions for the formation of the Saint-Dizier and Bavarian iron ore deposits. The selective mobilization of Fe with respect to other elements, particularly Al, is one of the most important conditions for the supply of iron in the depositional areas (as defined by RUPPERT, 1984).

The iron is present in the soils mainly as goethite and can be taken into solution according to the following equilibrium reaction:

$$\alpha\,\text{FeOOH} + 3\,\text{H}^+ + e^- \rightleftharpoons \text{Fe}^{2+} + 2\,\text{H}_2\text{O} \qquad (1)$$

where

$$\text{Eh} = 0.826 + 0.059 \log \frac{[\text{H}^+]^3}{[\text{Fe}^{2+}]}$$

and

$$\log \left[Fe^{2+}\right] = \frac{0.826}{0.059} \, Eh - 3 \, pH$$

The solutions which are percolating through soils have pH values close to 4.0 and Eh values about 0.4 V (HERMS, 1982; RUPPERT, 1984). The equilibrium state, solution/goethite, is reached at a value of $10^{4.8}$ mol · l^{-1} (= 0.9 ppm Fe). Decrease of pH to a value of 3 means an increase of 1 000 times in the solubility of Fe^{2+}. A decrease of the Eh value of 0.1 V increases the solubility of Fe^{2+} some 50 times. Therefore the solubility of Fe^{3+} in a domain where the pH is higher than 3 never lies beyond 10^{-7} mol · l^{-1}. In soils, however, organic complexing agents are present. They can form chelate boundings with iron. The existence of organic substances resulting from the decomposition processes in soils induces very low pH values if no buffers (alkaline or alkaline-earth elements) are present. Soluble complexes containing Fe^{3+} or even Fe^{2+} are then formed. At a pH of 4 these Fe^{III} complexes are more important in solution than the free Fe^{2+} ions. Therefore, the Fe^{3+} ions are more easily complexed. The relation to other elements, principally to Al, shows that the $[Fe^{III}]$ complexes are selectively mobilized in solution. This process is essential for the selective mobilization of iron with respect to aluminium.

The concentration of Al in solution can be determined as follows:

$$2 \, Al^{3+} + 2 \, H_4SiO_4^0 + H_2O \rightleftharpoons Al_2Si_2O_5(OH)_4 + 6 \, H^+ \qquad (2)$$

with

$$\log \frac{[H^+]^6}{[Al^{3+}]^2 \cdot [H_4SiO_4^0]^2} = -6.25$$

$$Al^{3+} + 3 \, H_2O \rightleftharpoons \gamma\text{-}Al(OH)_3 + 3 \, H^+ \qquad (3)$$

with

$$\log \frac{[H^+]^3}{[Al^{3+}]} = -8.44$$

$$Si(OH)_4^0 \rightleftharpoons SiO_2 + 2 \, H_2O \qquad (4)$$

$\log Si(OH)_4^0 = -2.74$ (amorphous silica), -3.10 (soil silica), -3.94 (cristobalite), -4.00 (quartz). In equilibrium with kaolinite $[Al_2Si_2O_5(OH)_4]$, the Al^{3+} content of the solutions will be 0.4 ppm, with gibbsite $[\gamma\text{-}Al(OH)_3]$, 7 ppm. Reactions (2) and (4) are the most frequent and the Al^{3+} in the solutions is normally less than 1 ppm (RUPPERT, 1984).

Iron can be mobilized selectively. Fe^{2+} ions in solution do not oxidise or do so very slowly at $pH = 4$ and $pO_2 = 0.2$, so that the transportation of iron as ionic Fe^{2+} in solution is very easy if the bacterial activity in this relatively low pH domain does not induce an increase of the oxidation rate (to 10^6 times).

The karst environment of this type of iron deposit includes water that is saturated in the deepest part of karst, particularly with respect to $CaCO_3$. The water has a neutral or basic pH. In the equilibrium state, with a pCO_2 = 0.000 29 MPa, the pH of a solution saturated with $CaCO_3$ is 8.3. The CO_2 content of karst water can increase in response to the oxidation of organic matter and the pH for a CO_2 content of 0.098 MPa will be 6.6. The precipitation reactions of siderite, $FeCO_3$ (PLUMMER and BUSENBERG, 1982; SMITH and MARTELL, 1976; quoted by RUPPERT, 1984) are:

$$CaCO_3 \rightleftharpoons Ca^{2+} + CO_3^{2-} \qquad (5)$$

$$\log\,[Ca^{2+}]\,[CO_3^{2-}] = -8.12$$

$$Fe^{2+} + CO_3^{2-} \rightleftharpoons FeCO_3 \qquad (6)$$

$$\log \frac{1}{[Fe^{2+}]\,[CO_3^{2-}]} = 10.32$$

$$Fe^{2+} + CaCO_3 \rightleftharpoons FeCO_3 + Ca^{2+} \qquad (7)$$

$$\log\,[Ca^{2+}]\,[Fe^{2+}] = 2.20$$

The precipitation of siderite can begin as soon as Fe^{2+} is higher than 1/160 of the molar concentration of calcium.

The main part of the iron is mobilized as organic iron comlexes or is bound as colloids to humic compounds. The oxidation of organic matter during the transportation and the rapid change in pH as the solutions enter the karst system have an important influence on the stability of the complexes. Calcium can replace the iron bounded in the organic ligands. Iron hydroxides are then deposited. Such reactions occur in the first type of deposits described (i.e. Luxembourg, South-East France, etc.).

On the other hand, in the second type of iron deposits (N. E. Bavaria, U.S.S.R., etc.) some areas of flooded karst were characterized by euxinic conditions. In these areas aqueous oxygen was completely exhausted by the oxidation of organic matter. Eh was very low. Iron hydroxides and the Fe^{3+} organic complexes could be reduced and solutions with Fe^{2+} ions could appear. These solutions can precipitate siderite (5), (6), (7). All these conditions also favour the deposition of P and Mn, which often accompany siderite in this second type of deposit (RUPPERT, 1984).

The Qui-xa deposit seems to be a result of both processes: highly oxidizing conditions prevailed during the initial erosion and accumulation of ore, then euxinic conditions within the deposits caused precipitation of siderite and finally, Late Tertiary uplift caused a second oxidizing and enriching of the deposit.

Both types are characterized by: (1) high iron concentration; (2) predominant hydroxide mineral composition; (3) low silica admixture which enables cheaper metallurgical processing, and (4) very varied thickness and superficial distribution caused by the buried karst surface as well as by distribution of carbonate rocks within the substratum.

As a conclusion, it seems important to emphasize that these two types of iron deposits in paleokarst developed in quite equivalent environments with regard to the conditions for the mobilization and transportation of iron. The main difference seems to be the morphology of the karst and the tectonic environment. The first type evolved on stable continental platforms whereas the second type occurred at the borders of more active areas and were inundated because the iron ores were deposited shortly before marine transgression (Saint-Dizier, North-East Bavaria, Akkermanovsk).

CLAYS AND SANDS IN PALEOKARST

Pavel Bosák

Introduction

Clays and sands, together with kaolin, are the classical mineral deposits of paleokarst landforms. Usually, they are associated with smaller iron (limonitic) and manganese ores and large bauxite deposits. The clays, sands and kaolin have been mined in karst for a long time, at first mainly for their iron content (see BÁRDOSSY, FUCHS and GŁAZEK, p. 419–429), but later also as refractory and ceramic clays, moulding sands, etc.

These deposits have some common characteristics: (1) they are continental deposits; (2) they represent deeply altered products of weathering in tropical and subtropical conditions; (3) they cover highly diversified karst relief. *In situ* weathering of freshly deposited sediments is quite common (e.g. kaolin and bauxite deposits in the U.S.A., kaolin deposits in Bulgaria, etc.). However, they are more usually the transported and redeposited weathering products of magmatic, metamorphic and sedimentary rocks.

The regional distribution of examples of these deposits is briefly discussed by other authors in the Regional part of the book. Classical deposits of this type have been described from Czechoslovakia (the Rudice Formation; BURKHARDT, 1974; BOSÁK, 1980b), Poland (GRADZIŃSKI, 1977; BŁASZAK, 1970, 1976; i.e. moulding sands) and Belgium (CALEMBERT, 1945, 1947, 1948). Similar deposits are known from England (the Brassington Formation; T. D. FORD, 1972; WALSH et al., 1972) and Wales (the Pocket Deposits; WALSH and BROWN, 1971; see also T. D. FORD, p. 58–59). Some data are available from the description of the bauxite deposits of the Coastal Plain and the Valley and Ridge provinces of the U.S.A. (e.g. OVERSTREET, 1964). Large clay deposits are associated with the Kazakhstan-type of bauxite deposits in the U.S.S.R. (the Krasnoyarsk region, Turgay Depression, the Ural Mountains, Ukraine, etc., KRIVTSOV, 1968–1969). Other deposits have been noted e.g. by STANEV and TRASHLIEV (p. 227–229) in Bulgaria and by ZHANG SHOUYE (p. 297–311) in China or by TSYKIN (p. 253–295) in the U.S.S.R.

P. Bosák, D. C. Ford, J. Głazek and I. Horáček (Editors):
Paleokarst. A Systematic and Regional Review. – Elsevier and Academia. Amsterdam and Praha. 1989.

The common mode of occurrence, mineralogy, geochemistry and processes of origin led BOSÁK et al. (1979) to classify them as sediments of the *Rudice type*.

Selected Case Studies

Some typical examples of sand, clay and kaolin deposits are presented here. Moulding sands are represented by examples from Czechoslovakia and Poland, refractory clays by deposits in the Chadobets region (U.S.S.R.), and kaolin deposits by sites in the U.S.A. Detailed descriptions of in situ kaolinized sands in Bulgaria are also reported in the Regional part (STANEV and TRASHLIEV, p. 227–229) and other refractory clays are described in the Belgian paleokarst by QUINIF (p. 46–47).

Kaolin in the Southeastern United States

Kaolin and bauxite deposits in the southeastern United States occur chiefly in the Coastal Plain and the Valley and Ridge provinces. The deposits are a result of Paleocene-Eocene weathering on a continental surface.

In the **Coastal Plain province,** the Eufaula district of Alabama and the Springvale and Andersonville areas of Georgia are rich in kaolin and bauxite karst-affected deposits (CLARK, 1965; WARREN and CLARK, 1965; ZAPP and CLARK, 1965). They are restricted to the lower part of the Nanafalia Formation (Lower Eocene). The Nanafalia Formation lies unconformably over the Clayton Formation (Paleocene) of impure, partly fossiliferous and chalky limestones. The Nanafalia Formation consists of continental clastics which gradually pass into marine deposits toward the south and west. It is divided into a lower unit containing kaolin and bauxite lenses and an upper unit of sands and clays. The upper part of the formation is, in places, cut by channels filled with thin alternations of sand and kaolin, sometimes with carbonaceous sand with kaolin blocks at the base. The Formation is unconformably overlain by marine clastics of the Tuscahoma Formation (Eocene; Fig. 3–33).

Sands of the Nanafalia Formation are coarse- to fine-grained and commonly micaceous; around thicker kaolin bodies they become carbonaceous with coalified plant remains. Pure quartzose sands are developed in the lower unit. Quartz-pebble conglomerates are common also in places at the base of the Formation. The sands are cross-bedded, brown or light red to white, and generally unconsolidated. Locally, they are indurated by iron oxides. The heavy mineral association consists of magnetite, ilmenite, tourmaline, rutile, kyanite,

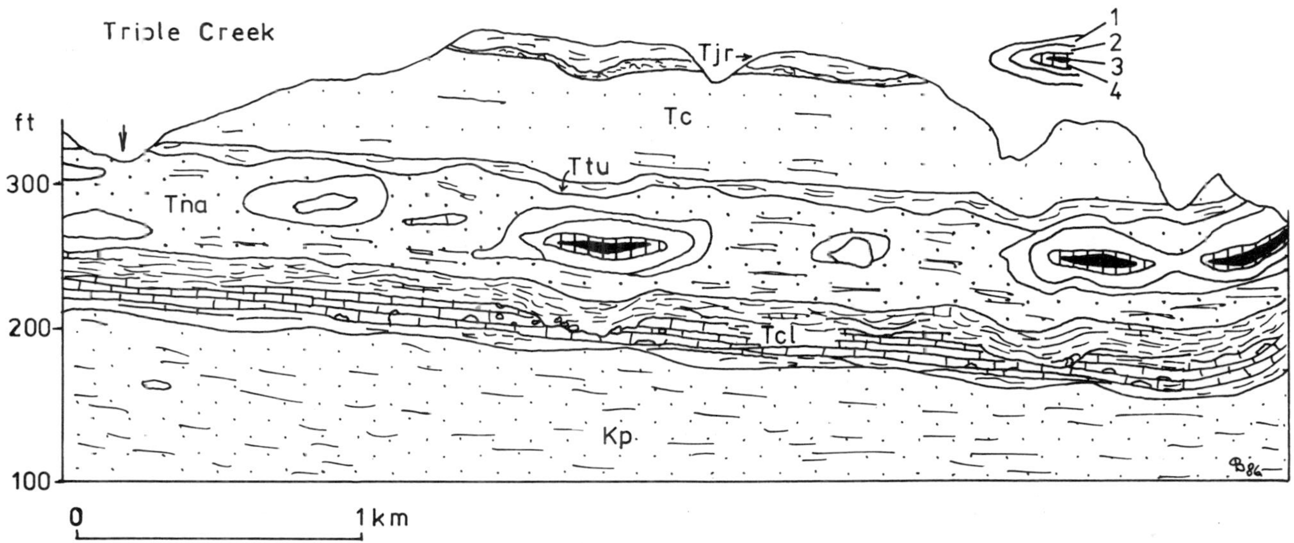

3–33. Cross-section through a part of the Andersonville bauxite district, Georgia, U.S.A. (after ZAPP, 1965). Eocene: Tjr. clay units of the Jackson Group; includes residuum from limestones of Oligocene age; Tc. sands of Clayborn Group; Ttu. Tuscahoma Formation; Tna. Nanafalia Formation; Paleocene: Tcl. Clayton Formation; Upper Cretaceous: Kp. Providence sand; 1. sandy kaolin; 2. kaolin; 3. bauxite; 4. bauxitic clay.

hypersthene, zircon, glaucophane (?) and ceylonite (?), muscovite being common to very abundant.

Kaolin forms irregular masses, bodies and lenses within the sands. Thicker examples (more than 10 m) pass gradually into bauxite clay and bauxite at their centres and grade downwards, laterally and upwards into sandy kaolin, kaolinitic sand and sand, often micaceous. The bauxite forms thin, tabular bodies (Fig. 3–33). Lenses of sand or sandy kaolin rarely occur within the sand-free kaolin masses. Some kaolin envelopes of bauxite are enriched in carbonaceous material.

The kaolin in the lower part of the unit is rather massive, while that in the younger channel fills is interlaminated with sand. Soft structure with pisoids is locally present, especially in the kaolin which grades laterally into pisoidic bauxite. The kaolin consists almost entirely of the mineral, kaolinite, locally with a small halloysite content. Muscovite, although abundant in the surrounding sands, is scarce in the kaolin and bauxite. Siderite forms spheroidal pellets in the kaolin enveloping bauxite. In its non-oxidized parts, the iron occurs as pyrite or marcasite; in peripheral, oxidized parts of the kaolin, iron oxides prevail. Compounds of iron, titanium and locally also manganese total less than 3 %.

The kaolin deposits thicken and thin over irregularities in the underlying Clayton Formation (Fig. 3–33). Its contact with the Nanafalia Formation is very undulating, with dome-like cones and depressions. The relief is as much as 30 m. In places, the surface of the Clayton Formation is indurated by silicification, and elsewhere it may be intensively leached. The residuum is composed of crumpled, decalcified clays and sands, with or without chert fragments. An underground drainage network developed here with a high frequency of solution cavities. Some sinkholes originated on the limestone surfaces. Karstification lasted until the transgression of the Tuscahoma Formation. The Nanafalia Formation infilled the sinkholes or was dumped into depressions originated by the collapse of subterranean cavities. This is proven by the brecciated and chaotic structure and by the faulting displayed by the kaolin and bauxite bodies. In the areas adjacent to and between Springvale and Andersonville the Nanafalia Formation has survived later erosion only in the sinkholes. The kaolin beds attain 17 m in thickness above the bedrock depression floors, the Nanafalia Formation being a maximum of 30 m thick.

Kaolin deposits extend from Virginia into Tennessee, Georgia and Alabama in the **Valley and Ridge province** (CLOUD, 1966, 1967; DENSON and WAAGE, 1966; DUNLAP et al., 1965; WARREN, BRUDGE and OVERSTREET, 1965; W. S. WHITE et al., 1966). They are denudational relicts filling karst depressions in Upper Cambrian and Lower Ordovician limestones and dolostones of the Knox Group or in the Mississippian Warsaw Limestone. In Northeastern Alabama, the sinkholes represent the contact-type of karst developed along the junction of sandstone beds with limestone or dolostone (CLOUD, 1967).

Kaolin bodies and bauxite lenses are associated with yellow to red residual clay and silt. The clays are gritty to silty and commonly contain chert inclusions or drusy, siliceous aggregates. The boundary between the residual clays and the kaolin-bearing clayey sequence is sharp. The centres of karst depressions are occupied by variegated clays, silty, sandy or lignitic clays, kaolinitic clays, kaolin and locally also bauxite. Sands and gravels locally intertongue with the clayey series or clays pass gradually into sands and gravels. In some places there are also lignite seams. The deposits are unconsolidated and are commonly coloured by iron compounds attributed to water-table fluctuations or groundwater diffusion.

Kaolin occurs in the form of irregular masses, lenses, downward-tapering cones or wedges, and some cylinders. Some kaolin bodies pass into bauxite toward the centre.

The residual clay is composed of illite, montmorillonite, mixed-layer minerals and kaolinite, plus abundant quartz. In some places there are rich impregnations of manganese minerals (wad, psilomelane, pyrolusite) and iron compounds

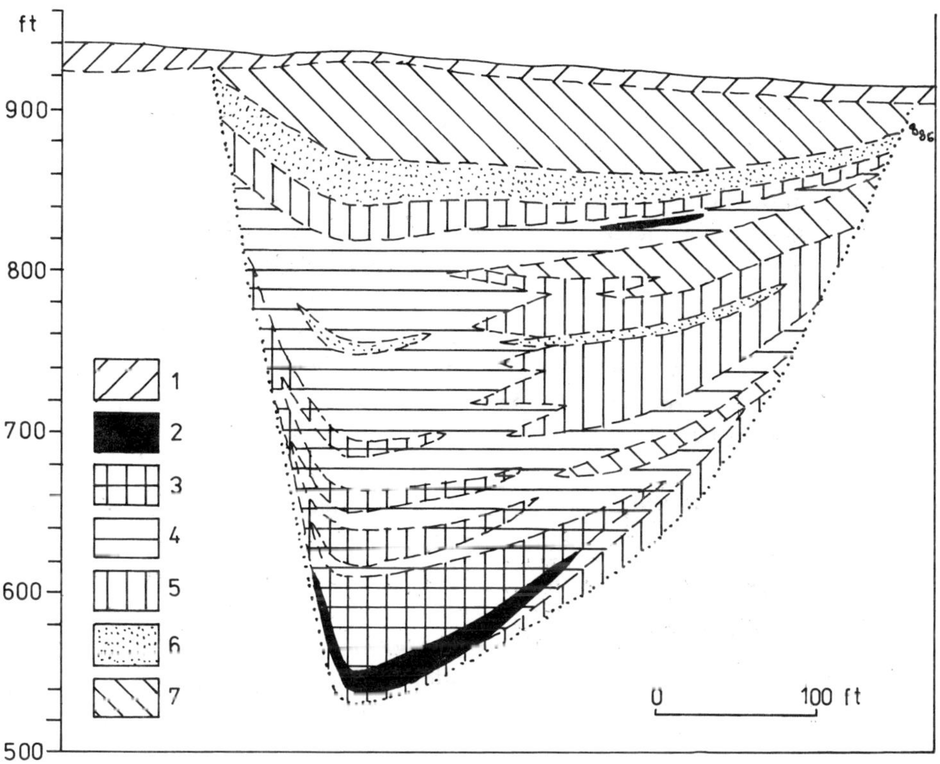

3–34. Bauxite deposit in the Rock Run Area, Northeast Alabama, U.S.A. (after CLOUD, 1967). 1. sandy yellow to brown soil; 2. grade A and grade B bauxite; 3. grade C bauxite; 4. grade D bauxite; 5. kaolin; 6. sandy kaolinitic clay; 7. lignite and lignitic or carbonaceous clay.

(mostly limonite and hematite). Kaolinitic deposits consist almost entirely of kaolinite with minor halloysite and iron oxides.

The karst topography on the Knox Group or Warsaw Limestone was formed during a long period of chemical weathering that was followed by the deposition of continental, probably lacustrine, deposits. The weathering created the residual clays. The continental sediments were deposited onto well-drained surfaces of carbonate rocks. Many collapse sinkholes developed, engulfing the overlying deposits in chaotic disarray. Evidence of slumping after the bauxite and kaolin was formed is indicated by small slickensides and by the mixture of displaced masses of different sediment types. Karst forms are irregular, having diameters up to some hundreds of metres and depths commonly 30–50 m, with a maximum of up to 120 m (Fig. 3–34).

Local finds of flora indicate a pre-Miocene age for the kaolin and bauxite but the distinction between Paleocene (Midway) and Eocene (Wilcox) age is uncertain (CLOUD, 1966).

Kaolins and bauxites in both provinces are believed to have formed by the *in situ* weathering in warm wet climates of transported detritus derived from crystalline rocks of the Ridge and Piedmont province.

Refractory Clays of the Chadobets Area, the U.S.S.R.

The Chadobets Plateau is located along the Chadobets River, a right bank tributary of the Angara River on the West Siberian Platform. Karst depressions with continental fill are developed in Sinian and Lower Cambrian limestones and dolostones, locally also in carbonatites. There developed a thick weathering crust on these Precambrian and Paleozoic basement rocks, as well as on Triassic basalt trap (KRIVTSOV, 1968–1969). Development of the weathering crust is correlated with the Uppermost Jurassic and Lower Cretaceous (SLUKIN, 1971; LEIPTSIG and LEVINA, 1975).

The continental fill of the karst depressions is of Albian to Cenomanian age. The sediments commonly begin with unconsolidated sandy-clayey beds, often with lenses of alumina-rich ironstones and of kaolinitic clays. They grade upward into kaolinitic clays and/or carbonaceous and lignitic clays with rare lignite seams. They are overlain by bauxite or bauxitic clays. The upper part of the fill is of variegated clays and kaolinitic clays. The lithology of the fillings is highly variable. The beds are often lenticular, showing traces of synsedimentary subsidence into karst depressions. The fills lie on preserved weathering crust or directly on the carbonate rocks. The thickness of the fill is commonly 10–80 m, locally increasing to 170 m.

The filled karst depressions are mostly dolines 10–100 m in diameter with depths up to 90 m (Fig. 3–35). Larger complex forms, including polje-like depressions and karst troughs having lengths of over 1 km, widths up to 800 m and depths of over 100 m are known, but are rare.

In addition to bauxite, high temperature refractory clays (1 710 °C) have been discovered. Their reserves in just one small part of the entire area are 13 million

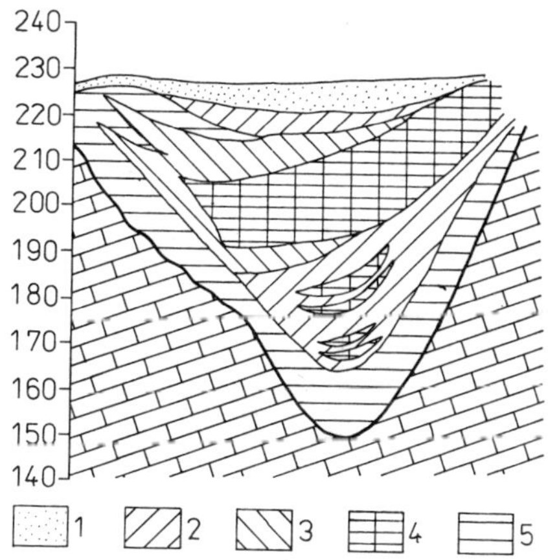

3–35. Ibdzhibdek bauxite deposit, Chadobets area, U.S.S.R. (after G. T. BAKLAENKO in KRIVTSOV, 1968–1969). 1. deluvial loam with rock fragments; 2. light grey to white kaolinitic clays; 3. allites; 4. bauxites; 5. yellow sandy loams.

tons (KRIVTSOV, 1968–69). Because of the similarity of all deposits in the area it can be expected that more of the refractory clays will be discovered.

The clastic-bauxite filling of the karst relief represents redeposited weathering crust. This is proven by fragments of lateritic ironstones, different bauxite types etc. in the fill. Deposition took place in lacustrine, swampy and paludal conditions intertonguing with slope (deluvial) deposits. It is possible that weathering continued also during the sedimentation.

The Lower Cretaceous bauxite-bearing sequence is unconformably overlain by Oligo-Miocene deposits, in places rich in ilmenite.

Moulding Sands of the Cracow-Wieluń Upland, Poland

Moulding sands and similar deposits occur in the Cracow-Wieluń Upland, the Holy Cross Mountains, in Upper Silesia and the North Sudetic Depression (BOSÁK et al., 1979). They are most extensive in the Upland; for example, 568 sand-filled karst depression have been explored in an area of 260 km^2 close to Częstochowa (BŁASZAK, 1970, 1973, 1976).

The karst depressions are filled with complicated sequences of clayey sands, often ferruginous, accompanied by subordinate muds, clays, pure quartzose sands and gravels of local origin (weathered Upper Jurassic limestones and cherts). Residual clays, muds and clayey sands sometimes contain manganese and iron impregnations. The sediments are vari-coloured – yellow, orange, white, greenish, etc.

The lithology of the deposits is highly variable in both the horizontal and vertical direction. Distinct horizontal bedding is rare. Some deposits display large stratal dips in various directions, resulting from differential compaction and/or due to subsidence from continuing karstification. The deposits are often structureless. Planar discontinuous lamination and small-scale cross-bedding are infrequent. Dessication cracks are common (GRADZIŃSKI, 1977).

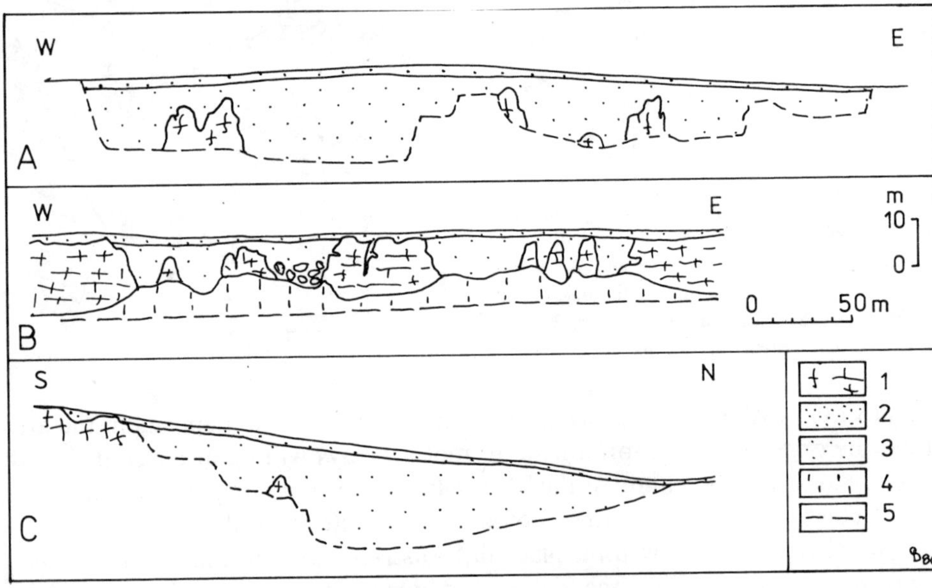

3–36. Examples of relation between moulding sands and limestone relief, Cracow-Wieluń Upland, Poland (after GRADZIŃSKI, 1977). A. Czatachowa sand pit; B. Wysoka Quarry; C. Hucisko sand pit; 1. Upper Jurassic limestones; 2. moulding sands; 3. Quaternary deposits; 4. debris; 5. bottom of excavations.

The moulding sands cover a morphologically diversified relief on the karstified Upper Jurassic limestones (Fig. 3–36). The principal karst depressions are highly variable in morphology and size. Smaller sinkholes and funnel-shaped depressions are also present. The area of the largest forms reaches several square kilometres; their depth is commonly 25–45 m, and exceeds 100 m at maximum (BŁASZAK, 1970; GRADZIŃSKI, 1977; GŁAZEK, p. 560).

The clay fraction consists largely of kaolinite and halloysite accompanied by lesser amounts of smectite and mixed-layer clays (montmorillonite-illite; BŁASZAK, 1970, 1976). The heavy minerals include disthene, tourmaline, rutile, garnets, less frequently epidote, magnetite and leucoxene (KRYSOWSKA-IWASZKIEWICZ, 1974).

The investigations of GRADZIŃSKI (1977) indicate that the moulding sands were deposited after the Turonian and before the Miocene and are probably of

Paleogene (Eocene?) age. The clastic material was derived from sandy Albian and Cenomanian deposits. Sedimentation took place directly in the karst depressions. These were occasionally filled with water heavily loaded with clastic material. In part the sands were deposited as local alluvial fans and associated debris flows.

Moulding Sands and Ceramic Clays of the Moravian Karst, Czechoslovakia

Moulding sands and ceramic clays, known as the Rudice Formation, occur in the central and northern part of the Moravian Karst in the vicinity of the villages Babice and Rudice. In addition, they are redeposited in some caves (e.g. Punkevní Caves) and in the Valchov Graben (BOSÁK, 1981a).

The Rudice Formation was the earliest reported karst-related deposit in Czechoslovakia and possibly elsewhere. Detailed descriptions were presented by ANDRÉ (1821), REICHENBACH (1834), REUSS (1854). UHLIG (1881) and WANKEL (1882). Due to very old underground mining for iron ores and ceramic clays, their descriptions are quite precise, in many sites being better than results from current observations. Recent open-pit mining for moulding sands reaches only several tens of metres, in contrast to the more than 100 m deep ancient mines and shafts.

The association of sands and clays with karst topography was first noted by REICHENBACH (1834) who stated that they fill karst depressions and shafts in the Upper Devonian limestones. UHLIG (1881) supposed that the Rudice Formation filled pre-existing karst forms. The earliest description of the Rudice Formation (ANDRÉ, 1821) stated only that it covers a „deeply rugged" surface of Devonian Limestones.

The bases of the karst depressions are covered by residual clays with quartz and chert gravel and fragments. Sometimes there are also thin layers of paleosoil and thicker, coarse block limestone screes. The basal part is rich in limonitic ores, developed at the contact between the Devonian limestones and the Rudice Formation, and in residual clays. The ores also fill irregularly shaped cavities in the limestones (WANKEL, 1882). The overlying sequence consists of residua from weathered Upper Jurassic rocks, represented by clayey-sandy deposits and white sandy clays.

The majority of the Rudice Formation is composed of an irregular alternation of clayey sands, local kaolin beds, muds, pure quartzose sands, and gravel with quartz and chert pebbles (Fig. 3–37). Some parts are indurated by siliceous cement and local ferruginous impregnations are also developed. The bulk of the sediments are unconsolidated. The deposits are vari-coloured. Sedimentary structures are very rare, the deposits being mostly structureless. Sometimes centripetal dips and small faults indicate synsedimentary or post-depositional subsidence. The Rudice Formation is overlain by paleontologically dated Cenomanian carbonaceous clays of the Peruc Member (BOSÁK, 1980b).

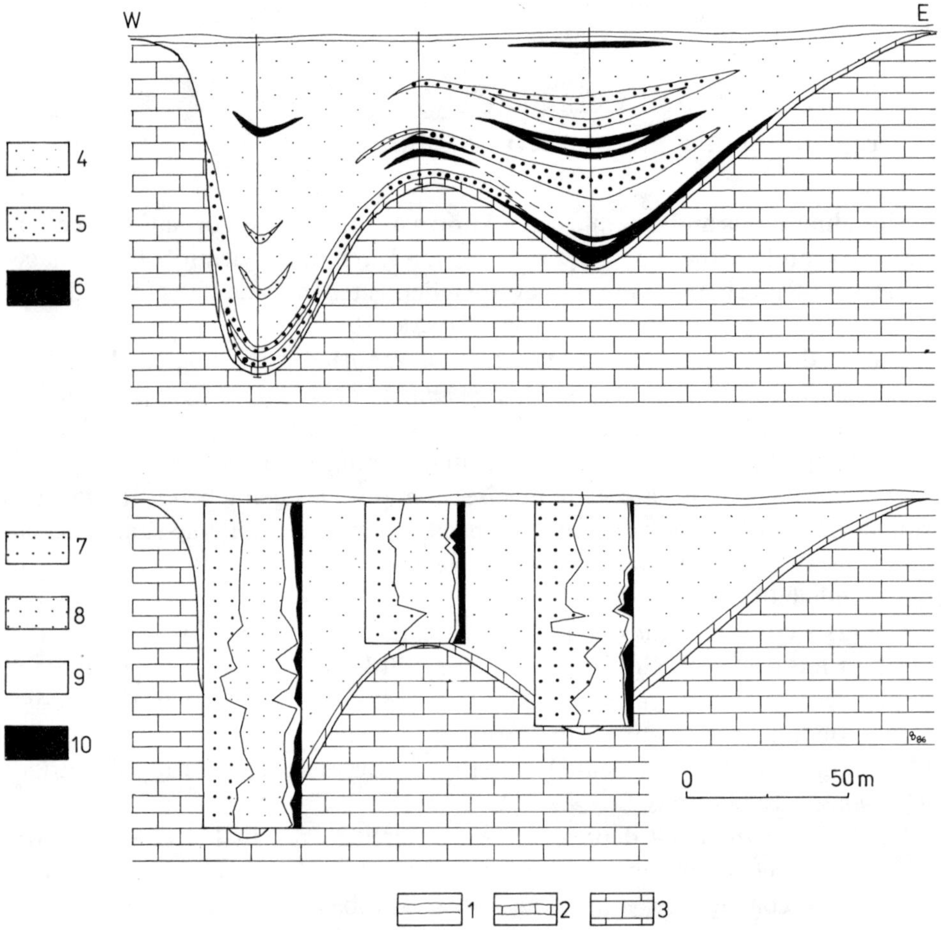

3–37. The Rudice Formation in the Novinky depression near Rudice, Moravian Karst, Czechoslovakia; the geology (above) and the grain-size (below) (after BOSÁK, 1980b, modified). 1. Quaternary cover; 2. clayey chert gravel; 3. Devonian limestones; 4. sands; 5. gravels; 6. clays; 7. grain-size under 0.02 mm; 8. grain-size 0.02–0.6 mm; 9. grain-size 0.6–1.0 mm.; 10. grain-size above 1.0 mm.

Mineralogical determination showed that the clay minerals are composed nearly entirely of kaolinite (polytype T prevails over the pM polytype) with some halloysite, ferrihalloysite, and montmorillonite. Feldspars (orthoclase, microcline, albite, oligoclase) as well as muscovite represent a common admixture. Boehmite and hydrargyllite have been detected among the Al compounds and goethite, hydrogoethite, lepidocrocite, hematite and hydrohematite are the Fe minerals (Bosák, 1980b). The heavy minerals consist mostly of a disthene-tourmaline-zircon assemblage with rutile, staurolite and lesser amounts of magnetite, andalusite, leucoxene, hematite and pyroxenes (Krystek, 1966).

The Rudice Formation fills karst depressions of highly variable morphology (Fig. 2–45); some can be classified as sand pipes (geological organ-pipes) and cockpits, some resemble sinkholes, complex dolines or chasms. The ancient mining also proved the presence of caves near the bottom of depressions filled with iron ores, sand and clay (cf. Reichenbach, 1834, plate I). Some of the depressions were formed prior to the clastic deposition and so contain paleosoil beds and fossil scree. The majority of the depressions were synsedimentary in origin, being sand pipes and cockpits. Deepening of these depressions halted only after the Cenomanian or even extended into the Paleogene, as is proven by sunken Cenomanian clays in some of the features.

The Rudice Formation, without any doubt, is of Early Cretaceous age and was accumulated in the interval between the Late Jurassic (Kimmeridgian regression) and the Cenomanian (Late Cenomanian transgression). The Formation represents redeposited weathering products of the kaolin-laterite type, from various rocks in the surroundings of the carbonate host-rocks. Deposition took place in a complex continental environment of alluvial plains with river channels, alluvial fans and deltas. Intermittent lakes supplied with clastic material by mass mudflows and outwash were also present.

Conclusions

The deposits of sand, clay and kaolin, associated in places with bauxite, iron and manganese, undoubtedly originated during long duration, intense weathering episodes on various rocks, under conditions favourable to the decomposition of less resistant minerals, accompanied by removal of alkalis, alkaline earths and carbonates (decalcification), and suitable for the concentration of amphoteric elements (Fe, Mn, Al) and SiO_2.

The high lithologic variability in both horizontal and vertical directions and the rare sedimentary structures in these deposits suggest their rapid transport and deposition in an environment unfavourable for the sorting of detrital material. The processes that contributed to the origin of these deposits were particularly rainwash and mudflows. The weathered material was washed into karst depressions that were deepening simultaneously. Due to continued drainage down into the karstified bedrock, the depressions were deepened even after the sedimentation had come to an end. The role of collapse subsidence is important in the preservation of some deposits (e.g. Brassington Formation, England).

The weathering, transportation and deposition occurred in warm and humid climatic conditions with clearly defined humid and dry seasons, i.e. in highlands under tropical to subtropical climates. The favourable climatic conditions contributed to the extensive *in situ* kaolinization and even bauxitization in places.

The mineralogical composition of all the described deposits and of other ones noted in this book is very similar. Clay minerals are represented mainly by kaolinite, accompanied by lesser halloysite with small amounts of smectite and mixed-layer minerals. Feldspars are mostly kaolinized; fresh minerals are rare. Muscovite, in places, is a common addition. The heavy mineral assemblage consists mostly of chemostable minerals resistant to weathering and partly also to transportation (disthene, tourmaline, zircon, rutile, staurolite, garnets). Less frequent are other minerals which represent, in part, local material (magnetite, andalusite, leucoxene, pyroxenes and many others).

The formation of deposits of this type require the co-existence of: (1) particular karst topography, (2) of rocks whose weathered residua provide sources of clayey-sandy material, and (3) adequate climatic conditions persisting for a sufficiently long time span.

THE OCEANIC KARSTS: MODERN BAUXITE AND PHOSPHATE ORE DEPOSITS ON THE HIGH CARBONATE ISLANDS (SO-CALLED "UPLIFTED ATOLLS") OF THE PACIFIC OCEAN

Françoise G. Bourrouilh-Le Jan

Introduction

At the end of the 19th and the beginning of the 20th centuries, the Pacific Ocean was the scene of great economic rivalry and the battleground for maritime competition between local and/or European powers. Some small islands, only a few kilometres across, were found to be covered with phosphates or phosphate products, generally by chance discovery. Their exploitation proved to be very easy due to the absence of cover rocks. A century after the discovery of phosphates, some other islands were found to be rich in aluminium and bauxite deposits (Fig. 3-38).

These Pacific Islands are termed here the *High Carbonate Islands* instead of "uplifted" atolls and in contrast to the *High Volcanic Islands*. DARWIN (1842), AGASSIZ (1903) and DAVIS (1928) supposed them to be tectonically uplifted atolls, but they are not. They are pieces of carbonate platform with Pleistocene or Holocene coral rim around a lagoon on a submerged karst geomorphology (MacNEIL, 1954; PURDY, 1974; BOURROUILH-LE JAN, 1977).

The exposed carbonate platforms of the southern Pacific may have undergone alternate submergence and emergence over several tens of millions of years. Emergence brought not only the development of the karst morphology, but also red soils (terra rossa) originating from volcanically transported products that, for a long time, were interpreted as residues of the dissolution of the carbonates on which they rest. It has been proved that the terra rossa cannot be the result of carbonate dissolution (TERCINIER, 1972; BOURROUILH-LE JAN, 1980). Instead these soils, which are rich in aluminous and phosphate products, originate in equatorial to tropical climatic conditions from the decomposition of volcanic products (ash, tephra and pumice) transported *via* aerial or marine paths. They have remained trapped on the karstified surfaces.

P. Bosák, D. C. Ford, J. Głazek and I. Horáček (Editors):
Paleokarst. A Systematic and Regional Review. – Elsevier and Academia. Amsterdam and Praha. 1989.

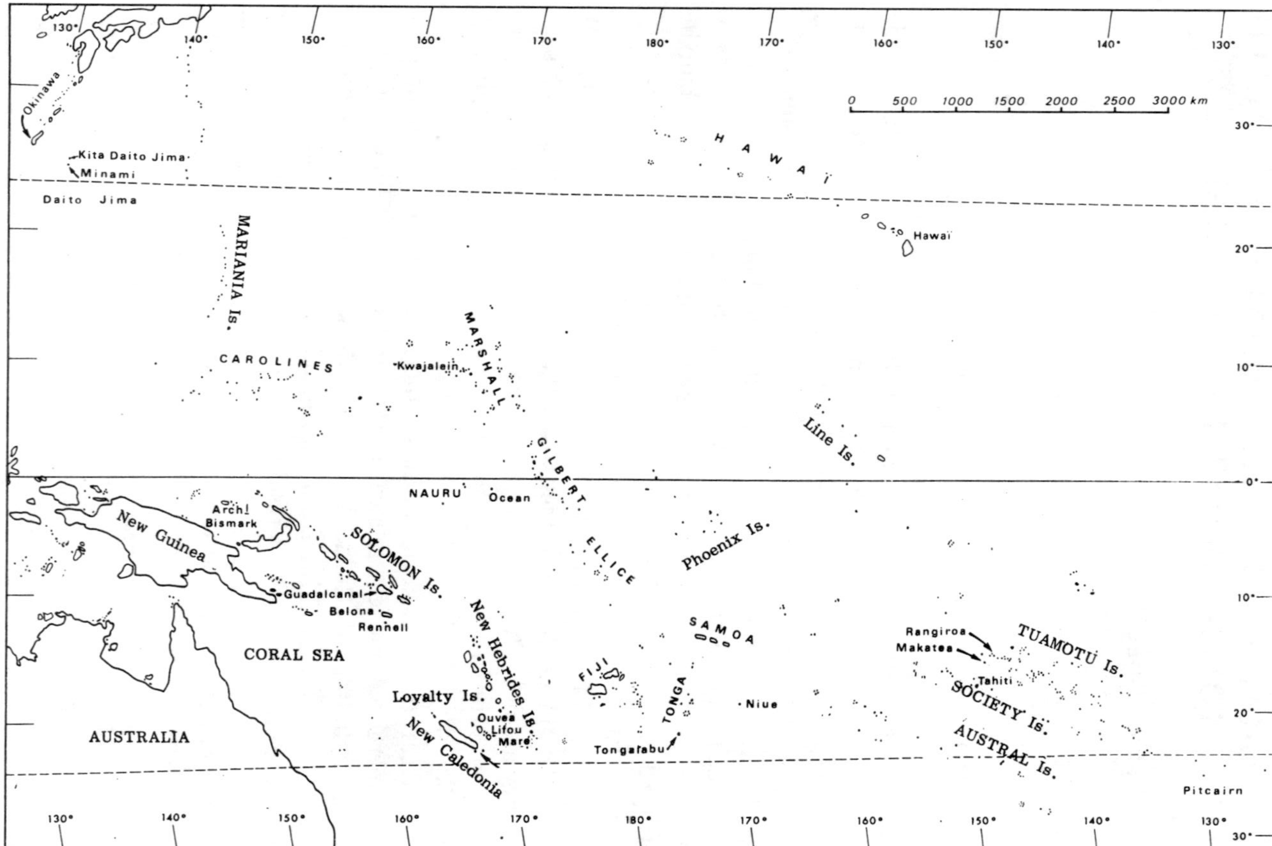

3–38. Location of some High Carbonate Islands: Rennell (Solomon Islands), Maré and Lifou (Loyalty Islands), Nauru, Ocean, Niue, Minami and Kita Daito Jima and location of some active volcanic centres: Tonga, Vanuatu (New Hebrides), Fiji, Société Islands, etc.

Systematic analysis of a number of these carbonate platforms has allowed us to show evidence of: (1) the karstic origin of the morphology, which was said by previous authors to be that of an "uplifted atoll"; (2) the systematic presence of modern soils of greater or lesser thickness that are red, loose, rich in alumina (bauxite) and phosphate, and which fill into and cover the karstified surfaces; (3) the link between certain soils (rich in both alumina and phosphate) and the influx of volcanic materials such as ash, pumice and tephra on and around these islands, and the later pedologic maturation of this volcanic material to the status of a soil that had previously been described by pedologists (TERCINIER, 1972).

The Holocene marine transgression since 15 000A B.P. raised the level of the surface of the mass of freshwater contained within the highly porous carbonates of these islands (Ghyben-Herzberg lens; CASTANY, 1967). The rise of the lens gradually submerged the karst relief of the High Carbonate Islands, bringing eventually about the formation of marshes, pools or lakes.

Where the karst morphology of the island was such that the sea was not stopped by remnant periphere walls, it penetrated extensively to create protected bays. There are preserved carbonate peaks with vertical cliffs that are sometimes stacks (bio-constructions corresponding to higher sea levels) but more often are residual karst towers resulting from dissection of older limestones; their loose, peripheral debris having been cleared away (e.g. Bay of Oupi, Island of Pines, New Caledonia).

Modern Bauxite of the Oceanian Karsts

In the Western Pacific, the High Carbonate Islands of Daito (Japan), Rennell (Solomon), Loyalty (New Caledonia), Tongatabu (Tonga) and Niue are covered by thick terra rossa containing granules or pisoids of phosphate and composed of poorly crystallised aluminium minerals.

Kita and Minami Daito Jima

Kita Daito Jima, in the East of the Ryukyu Archipelago (Japan), is 13.3 km^2, 4.8 km in length and rises 71 m a.s.l. An internally depressed plateau is surrounded by a circle of centre facing cliffs. A submerged karst network allows the development of marshes and lakes. The presence of phosphate within a thick, rich, red soil associated with volcanic ash and tephra has long been recognized (YAMANARI, 1935; Fig. 3–39). The soil grain size never exceeds 2 mm. The soil is very permeable, transmitting water to the underlying dolostone. The depth of buried karst pits (Fig. 3–39) is 5–40 m.

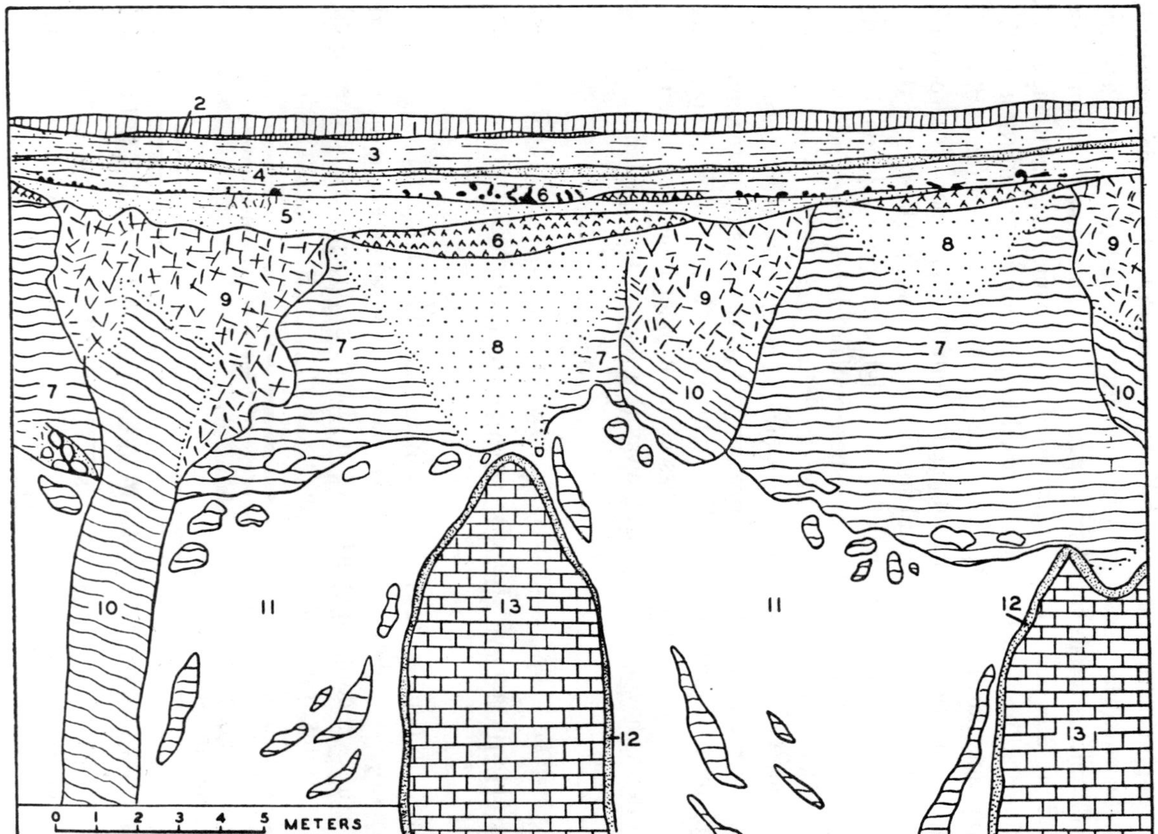

3–39. Cross-section of the soil of Kita Daito Jima (after YAMANARI, 1935). 1. soil; 2. thin layer of ferric aluminous phosphate; 3. upper layer; 4. middle layer; 5. lower layer; 6. phosphate with roots of concretional aluminium phosphates; 7. clay and pumices; 8. brown phosphatic clay; 9. reticulated ferric aluminous phosphate; 10. indurated, high content phosphate aluminium and iron; 11. loose aluminous phosphate; 12. black

Two complete soil analyses show the presence of alumina, silica, iron and phosphate together with traces of vegetal roots. For YAMANARI (1935), the presence of colloidal matter suggested that precipitation, coagulation and dissolution played an important role in the history of the deposit, and facilitated the silica separation at the bottom of a previous ancient lake. In the inner depression an iron pan has been found.

Fiji

In the Fijian Archipelago, there are several High Carbonate Islands in the Lau island group. LADD and HOFFMEISTER (1945), ROE, BURNETT and LEE (1983) provide brief descriptions of them: high carbonate plateau with peripheral rim and cliffs, i.e. classical Pacific karst morphology.

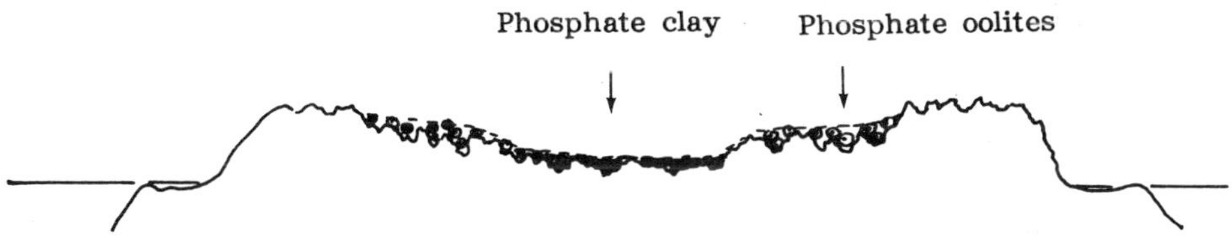

3–40. Schematic cross-section through Tuvuca Island (modified from ROE, BURNETT and LEE, 1983).

The soil consists of clay, pebbles and phosphatic ooids accumulated in hollows and pits on the very irregular carbonate surface. Ooids and pebbles are generally mixed with the terra rossa but on Tuvuca Island a clear granulometric progression exists, with phosphatic ooids on the higher rim of the island gradually fining to clay in its centre. Ponds and marshes in the peripheral area dry up during the dry season, indicating the lowering of a phreatic lens (Fig. 3–40). The mineral composition of the phosphatic soil is apatite, crandallite and goethite. Ooids are rich in Ca, P and F, with a higher peripheral concentration in Al, probably in the form of crandallite. The cement of the gravels and ooids consists of goethite, crandallite or secondary calcite. There are small quantities of manganese oxide.

The uranium series ages are: Vanua Vatu = 110 000 ± 15 000 a, and Tuvuca = 300 000 a, but the real meaning of those ages have to be discussed: age of soil content and/or contamination by the sea level rise?

Rennell, Solomon Archipelago

The huge central basin of Rennell is surrounded by high dolostone cliffs. It represents the results of intense karst erosion in an equatorial climate with high rainfall. It is covered by a bauxitic soil described by the Mining Journal in 1968 and by DE WEISSE in 1970. The soil of the western part and the lake deposits of the eastern part of the island were compared by TAYLOR and HUGHES in 1975.

The bauxite reserves are evaluated at $25 \cdot 10^6$ tons in the western part, with an additional 800 tons scattered in pockets. Soil composition averages $Al_2O_3 = 47.68$ %; $SiO_2 = 0.25$ %; $Fe_2O_3 = 18.36$ %. A cross section shows, above the bauxite, the presence of an upper surface rich in humus that contains phosphatic black pisoids.

The principal minerals are gibbsite ($Al_2O_3.3H_2O$) and goethite (FeO.OH), with 2.09 % TiO_2; 1.65 % P_2O_5, Mn and organic carbon 0.50 % (BOURROUILH-LE JAN, 1982). A radioactive anomaly is directly connected with a large accumulation of bauxite (12 times the background of $1.806 \cdot 10^{-7}$ C $\cdot$ kg^{-1} per hour). Analyses indicate that the radioactivity is associated with the K present in illite (only 0.005 % U and Th; MITSUI Co. 1970).

This soil occurs at +6 m in the western part of the island but in the eastern part it is covered by a lake created by the Holocene rise of the phreatic lens. The old tower karst topography forms numerous small islands. S. CHRISTIANSEN (1964) distinguished two types of lake floor: a hard, rocky, uneven one and a second type, flat and soft, covered by a brownish jelly of organic matter and trapped particularly in drowned avens. TAYLOR and HUGHES (1975) analysed the jelly: it is a gel of aluminous hydroxide, with a fluorescent coating when dry, due to a possible bacterial origin. Inside the gel are arenaceous and agglutinated forams and ostracods. Under the gel there is a grey clayey layer, rich in pyrite and of unknown thickness.

TAYLOR and HUGHES (1975) attributed the bauxitic soil to eolian volcanic ash with reference to WALKER'S work (1972). They therefore connect the presence of bauxite on Rennell with eruptions of Guadalcanal and New Georgia (andesite with hornblende) during Pliocene to Recent times. The nearest volcanic centre is situated 180 km from Rennell.

Lifou and Maré, Loyalty Archipelago

TERCINIER (1971, 1972) distinguished very clearly between the two different topographic zones: (1) the littoral platform made up of Pleistocene fringing reefs (200 000 and 125 000 a, GAVEN and BOURROUILH-LE JAN, 1980, 1981) and (2) the inner plateau, Miocene to Pliocene (BOURROUILH-LE JAN, 1979). On the coastal terraces of Lifou (TERCINIER, 1972), the soils are composed of rendzinas formed essentially from aragonitic bioclasts, of poorly evolved soils with allophane and lime-saturated andosoil formed from stranded volcanic pumices mixed with coral fragments. The analysis reveal amorphous products but boehmite can be recognized. Neither phyllosilicates from neomorphism nor gibbsite have been detected.

On the high inner plateau of Lifou, TERCINIER pointed out that bauxitic soils on karst occupy 1 000 km^2, more than 90 % of the surface of the island. Organic matter is abundant, more than 10 % in A horizons beneath forests and 3–5 % in B horizons. The P_2O_5 content is approximately 2 % at the surface and remains greater than 1 % in the B horizons.

The composition is as follows: alumina 54 ± 4 % (well crystallized boehmite, not so well crystallized gibbsite); 8–16 % of total alumina is amorphous; iron oxides 31 ± 3 % (poorly crystallized goethite and a few amorphous hydroxides); 7 % silica.

TERCINIER (1972) concluded that these soils formed from volcanic products, principally the pumices that can be found, for modern example, all along the low coastal terraces of Lifou and Ouvéa.

The Phosphate-Covered High Carbonate Islands

Oceanic phosphate deposits are numerous. Some are currently in production, others are exhausted, and still others are under exploration. In the Central Pacific the following examples may be noted: Ocean Island: exploitation finished in 1982; Nauru: 90 % of the deposit already extracted; Christmas Island (NE Indian Ocean): still being worked. In the Eastern Pacific there is Clipperton (abandoned in 1917) and Makatea (Tuamotu, exploitation stopped in 1966). Some atolls have also been recognized as being able to contain phosphates (Line Islands and Tuamotu).

From studies of several large phosphate deposits, their common characteristics are:

- location on the surface of a High Carbonate Island,
- the ore is trapped on and within karst,
- often located laterally in the inner basin of the island,
- the granulometry of the ore varies greatly, from powder to huge blocks,
- the position of the different granulometric classes is the same in the different deposits; the powders are usually in contact with the substrate, as ELSCHNER first noticed in 1913 in the case of Nauru. The carbonate substrate seems to have acted as a filter with very tiny meshes,
- the contact between phosphate and carbonate is always a surface of karst discontinuity, the phosphate often penetrating inside the karst cavities, reproducing on a small scale the distribution of the ore itself in the karst,
- the thickness of the ore is more or less constant on any given island but differs from one island to another,
- the phosphate facies are very different from the carbonate facies,
- the phosphate facies vary from subtidal to supratidal and even continental

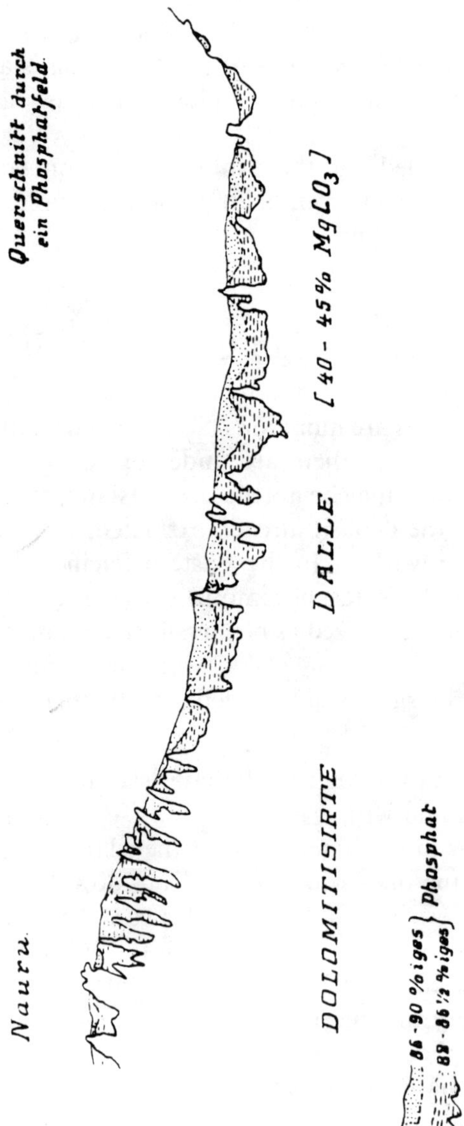

3–41. ELSCHNER'S cross-section of the general phosphate exposure of Nauru, (1913).

facies with caliche and pisoids, or they can be formed from direct precipitation with cements and concretions,
– the phosphate mineral is apatite, rich in elements such as Zn, Cu, Ba, V, Cr, Ni, Sr, La, Mn.

The Karst, a Morpho-climatic Trap

The Nauru karst has been very well described by ELSCHNER (1913; Figs. 3–41 and 3–42). The island is at latitude 0°25′ S and longitude 166°57′ E. It rises 65 m a.s.l. It is intensively karstified on its surface but the karst appears only after removal of the phosphate. Before mining began, the topography was slightly hummocky or smooth. The island was entirely covered by phosphate with a black phosphatic soil on the top and with only a few sharp dolostone peaks emerging through it. After removal of the phosphate, the morphology is seen to consist of joint-segregated peaks that are several metres in height.

The presence of phosphatic layers inside the carbonate has been observed by ELSCHNER (1913), BOHNE (1926) and ELLIS (1936) but their interpretation differs from one author to another. Since AGASSIZ (1903), geologists have agreed on a Tertiary age for the Nauruan carbonates although no micropaleontological evidence has been provided.

Makatea (Tuamotu) has been studied by DANA (1849), HUNT (1855), AGASSIZ (1903), ELSCHNER (1913), STUTZER and WETZEL (1932), WILDER (1934) and OBELLIANNE (1963). A High Carbonate Island located in the SW side of the Tuamotu chain of atolls, Makatea looks like Nauru but is higher. The phosphates rest on dolostone and are spread on different terraces around the inner depression of the island. They fill pockets of the karst surfaces, which are less well developed than those of Nauru (Fig. 3–43). Geomorphological studies in the field and in the laboratory using aerial photographs show a plateau 20 m high displaying two different lithologies and karst solution styles (Fig. 3–44).

Removal of the phosphate revealed broad, vertical shafts or avens in several parts of the island. They are aligned. Some are 20 m deep, others pass below sea level. The northern inner depression has not been a part of the mining. Several dolostone karst towers emerge from its smooth topographic surface. They are aligned, and can form real walls emerging from a reddish to brownish soil.

The Bedrock of the Phosphate Deposits

In Nauru, the karst surface is fossilized by lithified calcitic or phosphatic laminated deposits. Dissolution has attacked a dolomitized shallow water and coral-influenced wackestone to packstone facies enriched at the top in red algae, rhodolites and corals (BOURROUILH-LE JAN and HOTTINGER, 1988).

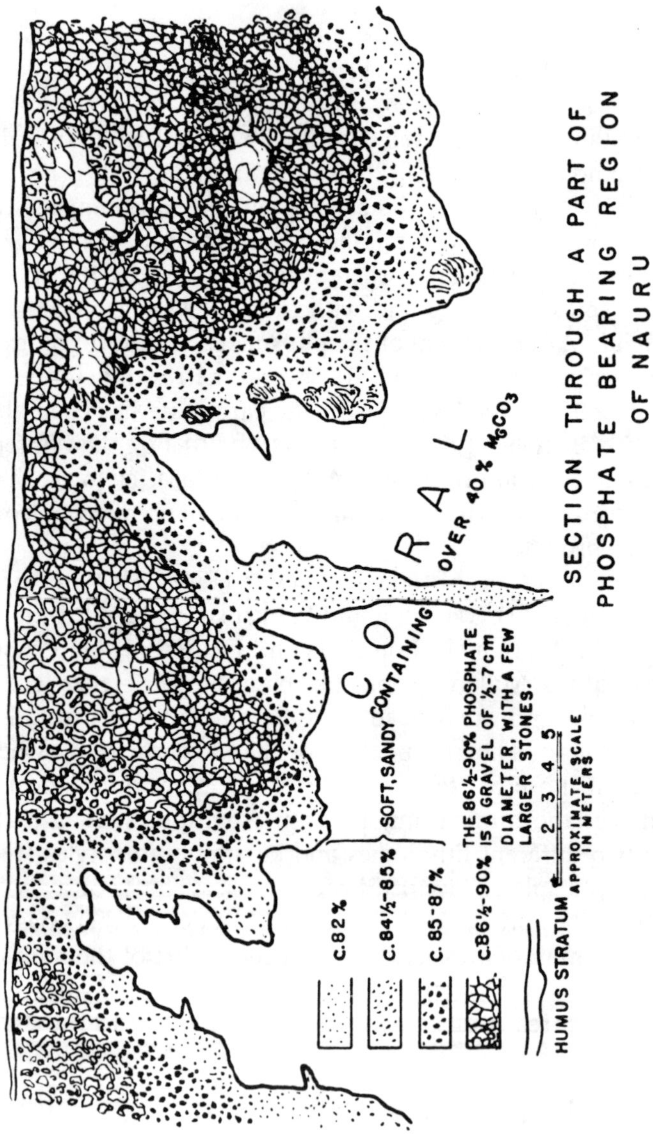

3–42. Cross-section of the phosphate ore and bedrock in Nauru (after ELSCHNER, 1913).

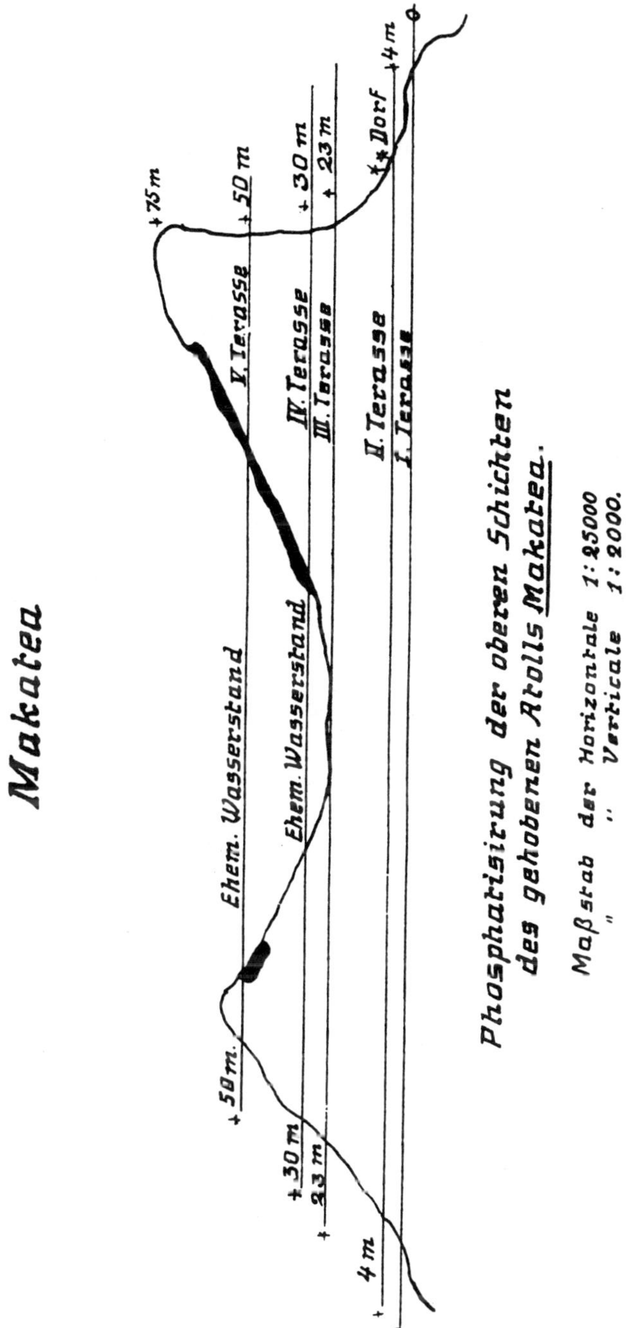

3–43. Schematic cross-section of Makatea (Tuamctu Islands) after ELSCHNER (1913), showing the location of the phosphate deposits.

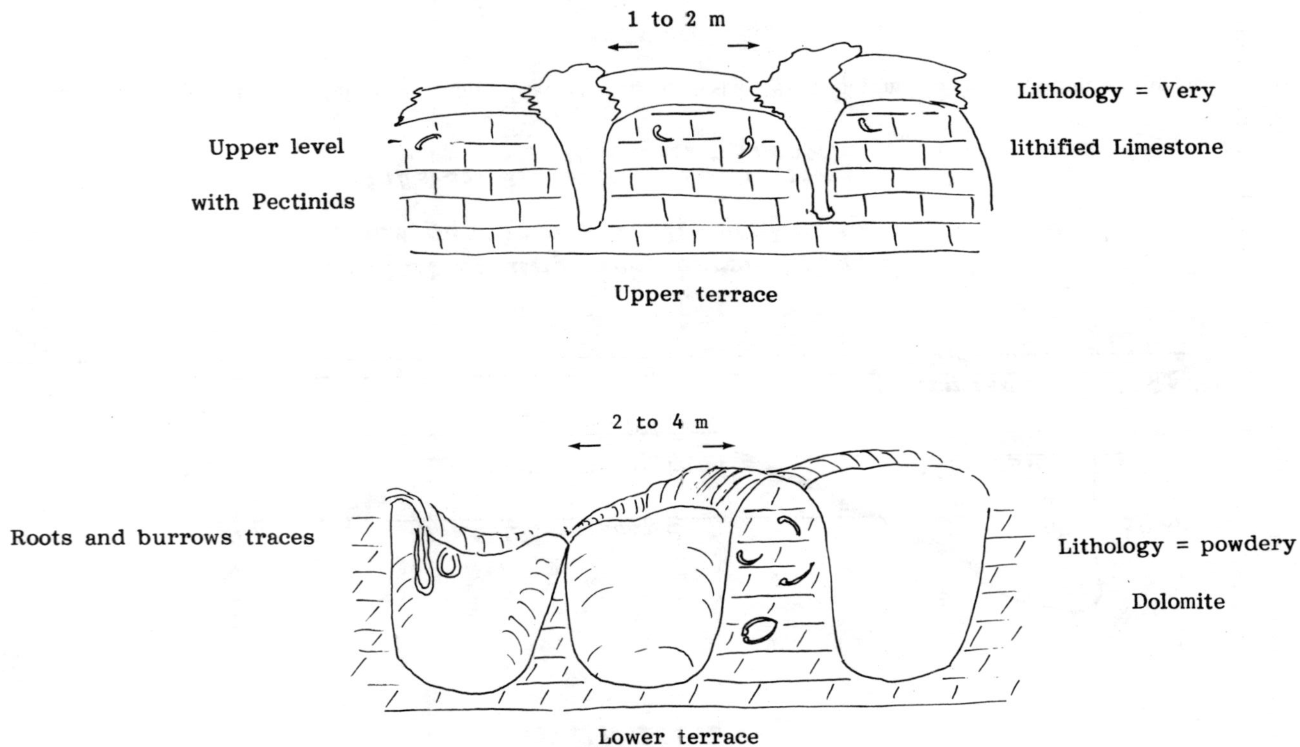

3–44. The karst surfaces of Makatea (Tuamotu Islands) as exposed by the removal of the phosphates. The differences in morphology are due to the different lithology, hard limestone and powdery and soft dolostone.

For AGASSIZ (1903) and ELSCHNER (1913), the dolomitized carbonates of Nauru date from the Miocene despite the lack of microfossils. In Makatea the carbonates are dated Lower Miocene (HOTTINGER 1985) and reveal a great diversity of hardness, mineralogy and facies (Figs. 3–45 and 3–46). On the upper surface of the island, a rough distinction can be drawn between dolomite and calcite facies. OBELLIANNE (1963) had previously noted the stratified and horizontal layered aspects of the Early Miocene formation, with giant cross-bedding and bafflestone structures of corals. Our studies show that the summit of the island (110 m) is composed of a soft, crumbly dolomitized wackestone with pellets and branched *Lithothamniae* which in certain places reveal some phosphatic elements (a very fine wackestone with traces of roots and phosphatic elements). At around +50 m, there is an occurrence of limestone – packstone rich in echinoids and encrusting forams, grainstone containing pelecypods, and mudstone rich in *Amphistegina*. Dolostones are composed of subtidal packstone with fine pellets or miliolids containing fairly thin shells of large pelecypods, and supratidal packstones with pellets and miliolids with burrows. The occurrence of dolomitized facies covering the phosphates may indicate a marine transgression after the phosphate formation.

Makatea therefore reveals various classical features of shallow water carbonate platform sedimentation that was active during the Early Miocene. These have been subject to karst erosion, and now the karst surface is fossilized by other facies which can either be calcareous or dolomitized. These younger (post-Early Miocene) deposits are: (1) stalactitic and stalagmitic coatings, bearing witness to a long emergence (Fig. 3–46); (2) less visible dolomitized fissure filling; mudstone with *Amphistegina* and mudstone with supratidal intraclasts; (3) calcitic crusts or caliches, separated by an erosional surface. Some display stromatolitic laminations and vegetal mats; (4) white caliches with detrital dolomite rhombs and dolomitic intraclasts, with bioturbation and roots; (5) calcitic mudstone with ostracods and phosphatic ooids (15 %), reworking detrital dolomite rhombs and dolomitic lithoclasts. It is possible to observe the dolomitization of the mudstone and the correlative dissolution of the ostracod carapaces from the bottom to the top of the sample (Fig. 3–46); finally (6) packstone containing calcitic pisoids showing supratidal, and even continental conditions of formation. Thus, sedimentological analyses of bedrock proved its carbonate platform origin and slow evolutionary control by rythms of emergence and karstification, i.e. by global sea level fluctuations.

The Ore and its Sedimentology

In Nauru, the phosphate displays two depositional modes: (i) there is a stratified mode where deposits are distributed between karst peaks and (ii) a non-stratified mix with sand, gravel and blocks which covers the first. The

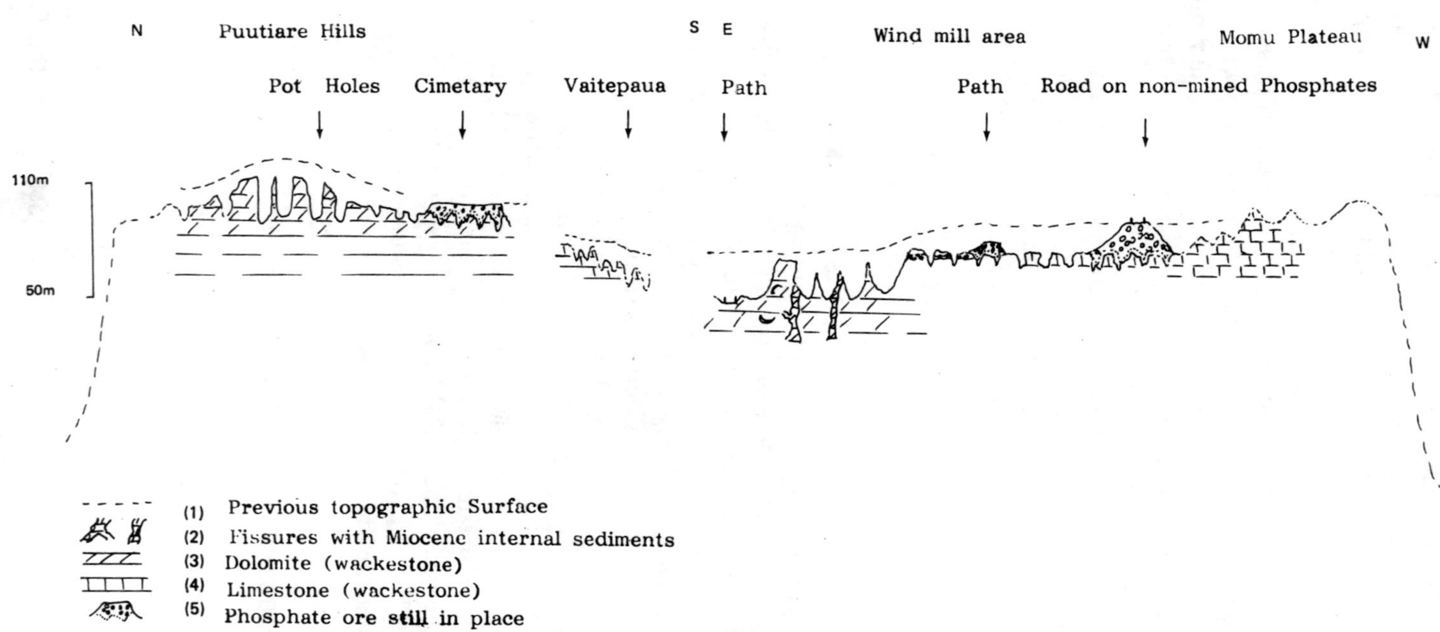

3–45. NS and EW cross-sections of Makatea (Tuamotu Islands), showing the former topographic surface and lithologic variations of the Lower Miocene rocks, variations of the topography under the phosphate, and finally, the synsedimentary fissures and their filling with dolomitized inter- to supratidal shallow water facies.

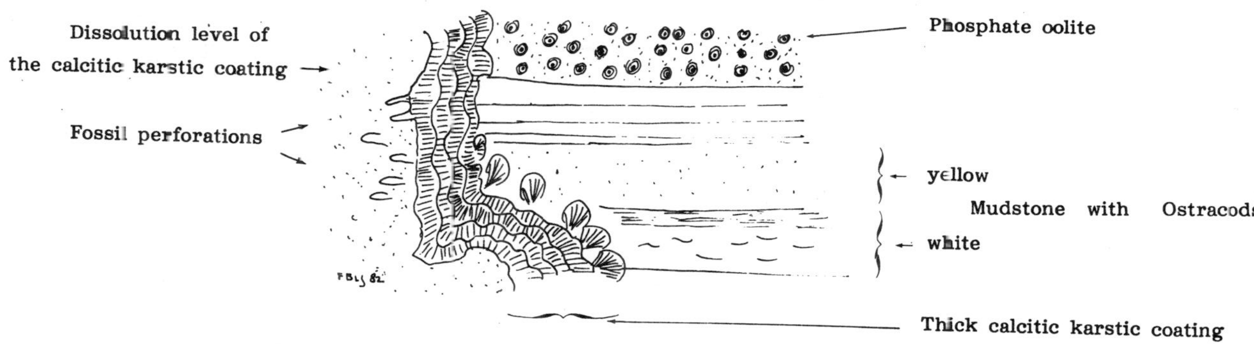

3–46. Details of the karst contact between the Lower Miocene bedrock of Makatea (Tuamotu Islands) and its phosphatic cover. The contact is complex and begins with an erosional and biologically perforated surface. Then several calcite stalagmitic coatings can be seen which are themselves fossilized by brackish to lacustrine, thinly laminated deposits of an ostracod-rich micrite with reworked detrital dolomite rhombs and containing phosphatic ooids towards the top.

phosphate deposit attains 15 m in thickness. It can begin with a calcitic, laminated facies lying in a discontinuity on the underlying dolostone, into which it can also penetrate. The loose phosphate has a very varied granulometry (lutites to rudites) and very varied facies in the reworked blocks (laminae, pisoids, corals). Some blocks and gravels are covered by a phosphatic varnish. All these characteristics bear witness to: (1) phosphatization of very different facies ranging from the subtidal to the continental; (2) erosional break and subdigging of the phosphatized facies; (3) trapping inside karst depressions where the low energy, calcitic facies had previously been deposited.

In Makatea, the phosphates have the same characteristics as in Nauru. The sediment is loose and variable. Phosphate ore fills a karst with about ten metres of relief generally and scattered avens which are at times more than 100 m deep. An exceptional feature of Makatea is that phosphate has also been found on the exterior terraces of the island. This is because some karst pockets on its summit have been breached and have discharged naturally their phosphate towards the outer parts of the island. As in Nauru, the phosphate facies are numerous and varied; some of them have been observed still in place at the contact with the carbonate.

Although they are entirely pure phosphate, we still use the DUNHAM (1962) textural classification. They are composed of phosphatic grainstone with ooids and stalactitic cement showing dissolution and precipitation; grainstone with phosphatic pisoids; packstone with phosphatic mudstone intraclasts; phosphatic caliches with traces of roots and rootlets. As in Nauru, phosphate pockets can be found inside the carbonates (described as ore No 5 by OBELLIANNE, 1963).

Comparision Between the Geochemistry of Bauxite, Phosphate and Soils

To understand the genesis of the observed phosphate sedimentary facies, chemical analyses were carried out on various formations of the studied area: the bauxite and its bedrock from Rennell; the phosphato-bauxitic soil and underlying rocks from the Loyalty Islands; a basalt outcrop at the bottom of a deep sinkhole in Maré (Loyalty Islands); floating pumice found on Ouvéa beaches (Loyalty Islands); and the red soils from Tongatabu (Tonga Archipelago).

A few general points may be established before a detailed analysis of the results is presented: (1) the studied soils come from different islands, differing because of geography and geological age; (2) their filling with volcanic detritus has thus occurred at different times for the different examples. The process is still continuing; (3) the volcanic debris have been or are emitted by volcanic complexes whose geographical location differs for each sample island (Rennell = Solomon and Vanuatu volcanism; Loyalty Islands = Vanuatu and Fijian volcanism; Tongatabu = Tongan, Fijian volcanism; Makatea = Tongan and Fijian, and Society volcanism); (4) lastly, the chemical composition of the magmas of the different volcanic massifs themselves may have changed with time.

Global Analyses of Major Elements

Four groups of samples can be differentiated; carbonates, soils, phosphates and volcanic products. The carbonates are characterized by the total absence of Fe, Si and Ti, the soils by the presence of Fe, Si, Al, Ca, Mg and Ti, the phosphates by P, Ca and traces of Si, Mg and Ti and volcanic products by Fe, Si, Al, Ca, Mg and Ti (Table 3–2).

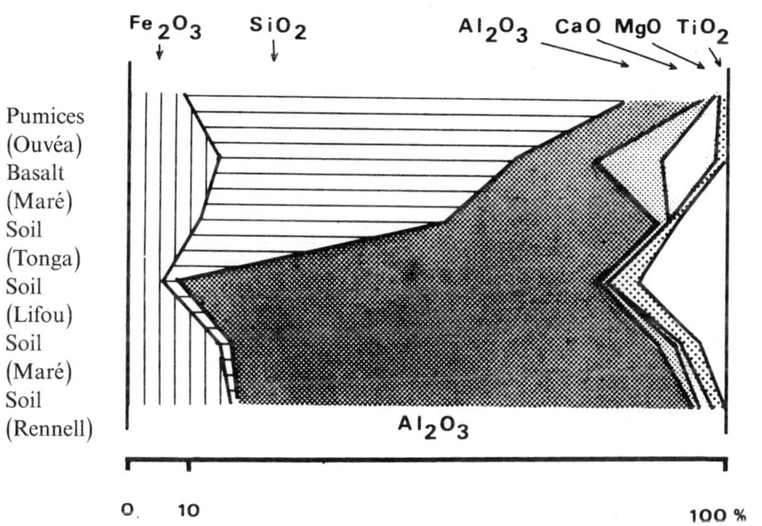

3–47. Major components of several soils from High Carbonate Islands, and of a few volcanic products from the central and SW Pacific.

There is major variation in the intensity of the alteration between, at the one extreme, soils from Tonga and the other those from Rennell (Fig. 3–47). This confirms the geological age of these islands, ranging from Pliocene to Holocene at Tongatabu up to the Middle Miocene at Rennell. Alteration of the volcanic products continues up to the disappearance of silica, with an amazing increase in alumina which is usually in the form of a jelly where the trihydrate gibbsite and the monohydrate boehmite have difficulty in rising. Progressive removal of Ca and Mg ions can be observed as well as the conservation of the same

Table 3–2

Mineralogy and geochemistry of several bauxitic; phosphato-bauxitic and phosphate covers of some High Carbonate Islands. Comparison with volcanic products (from BOURROUILH-LE JAN 1980)

Island	NAURU		RENNELL		LIFOU	MARÉ			OUVEA	TONGA
Rock	Phosphate Ore	Bedrock	Phosphato-Bauxitic Soils	Bedrock	Phosphato-Bauxitic Soils	Phosphato-Bauxitic Soils	Basalt	Bedrock	Modern Floated Pumice	Soil
Age	Late Miocene-Pliocene?	Early to Middle Miocene	Modern	Middle Miocene	Modern	Modern	Oligo-cene	Middle Miocene	Modern	Modern
Mineralogy	100 % Hydroxy-apatite	100 % Dolomite	Gibbsite Boehmite	99 % Dolomite	Gibbsite with Cranda-llite	Gibbsite or Boeh-mite Amorpho-us Matter	Plagio-clase +Diop-side	Calci-an Dolomi-te		Amorph. Matter
Number of analyses	3	2	3	8	1	9	13	12	1	1
ppm Pb		.	153	.	30	124	39	.	20	10
Zn	1 167	.	60	2	55	165	140	14	115	95
Cu	15	.	82	3	50	165	95	14	40	195
Mo	.	.	.	.	.	4.7	.	.	.	.
Ba	13	.	85	.	15	167	219	.	180	230
Ag	.	.	.	.	.	1	.	.	.	.
Bi	.	.	.	.	.	1.3	.	.	.	.
W	.	.	1 123	.	.	894	33	.	.	.
P	>9 999	.	>9 999	.	–	>9 999	2 442	.	–	–
B	.	.	77	.	.	51	11	.	.	.

V	20	·	420	·	320	392	305	·	55	590
Cr	80	5	1 643	10	1 300	1 565	228	17	·	175
Ni	12	·	224	1	·	204	132	·	·	·
Co	·	·	49	·	·	73	53	·	·	30
Sr	303	340	2 914	172	370	1 378	407	144	·	410
Sc	·	·	107	·	·	135	2	·	·	·
As	·	·	·	·	·	91	·	·	·	·
Sb	·	·	975	·	·	167	·	·	·	·
Ga	·	·	64	·	84	63	23	·	·	36
Ge	·	·	·	·	·	9	·	·	·	·
Be	·	·	32	·	·	79	160	·	·	·
Zr	·	·	1 561	·	560	1 174	·	·	·	·
La	62	48	–	·	660	345	–	·	45	200
Li	·	·	–	·	20	32	·	·	?	?
Mn	74	·	5 778	·	700	1 105	1 447	·	1 400	530
% Fe_2O_3	·	0.5	17 3	·	6	15	15	·	10	12
SiO_2	0.6	·	1	·	2	2	49	·	73	41
Al_2O_3	·	·	79	·	71	72	14	·	13	36
CaO	22.4	30.1	1	22.58	0.8	2.5	10.5	32	6	0.8
MgO	0.03	16.4	2	27.06	·	1.1	9.5	18.5	1.4	·
TiO_2	0.07	·	2.2	·	4.76	2.6	2.2	·	0.5	1.5
K_2O	1.4	0.25	–	·	0.2	–	–	·	1.2	0.2

– = Element not analysed, due to a change in the device programme
· = Element not reported, because the percentage is under the detectable limit of the instruments

Table 3–3

Mineralogy and geochemistry of soils from the western part of Rennell (Solomon Islands) (from BOURROUILH-LE JAN 1982)

Analysis	A	B	Pb ppm	Zn ppm	Cu ppm	Ba ppm	W ppm	P ppm	B ppm	V ppm	Cr ppm	Ni ppm	Co ppm
p. 1275	+		139	65	78	102	1 348	>9 999	76	455	2 367	148	39
p. 1276		+	163	47	61	47	880	6 986	67	453	319	362	66
p. 1311	+	+	156	69	106	107	1 142	9 999	88	532	2 242	162	43
Average			152	60	81	85	1 123	>9 000	77	480	1 642	224	49

The following elements have not been reported in the table because they are below the limits of detection of the equipment:
Mo < 4 ppm; Bi < 5 ppm; Sn < 10 ppm; As < 50 ppm; Ge <30 ppm; Ag < 3 ppm.
A = gibbsite; B = boehmite

percentage of Ti. In the Rennell bauxitic soils, aluminium minerals are poorly crystallized: 79 % gibbsite and boehmite, with 17 % Fe_2O_3, 2.23 % TiO_2, 2 % MgO and 1 % CaO (Table 3–3). Iron slowly increases in soils, where it appears in goethite.

In conclusion, from all these global chemical analyses it is difficult to maintain that the bauxitic and phosphato-bauxitic soils can derive from the simple attack, alteration, dissolution and residual concentration of the underlying carbonates (as has been claimed for a long time) because all of the carbonates are completely free of bauxite or phosphate.

The Trace Elements in the Soils, the Volcanic Products, and the Phosphates

All carbonates from the different islands are very poor in trace elements. In the Loyalty Islands, the soils of Lifou and Maré contain roughly the same elements except that in Maré they are richer in Pb, Zn, Cu, Ba, W, B, V, Cr, Ni, Co, Sc, As, Sb, Ge, Zr, Li, Mn. This suggests that Maré emerged earlier than Lifou. Maré is particularly interesting because of its geological setting and the altered basalt outcrops at the bottom of deep dolostone aven in the centre of the island. Chemical analyses have been carried out on the volcanic substrate of Oligocene age, the dolomitic bedrock (Middle to Upper Miocene) and the modern soils (Figs. 3–47 and 3–48).

In the Rennell soils, elemental analysis reveals Pb, Zn, Cu, Ba, W, B, V, Cr, Ni, Co, Sc, Sb, Ga and Be. Some of the elements are present in high proportions,

Sr ppm	Sc ppm	Sb ppm	Ga ppm	Be ppm	Zr ppm	Mn ppm	Fe_2O_3 %	SiO_2 %	Al_2O_3 %	CaO %	MgO %	TiO_2 %
6 008	103	954	58	31	1 567	2 780	15	1	81	2.0	1.9	2.50
138	106	1 013	69	32	1 449	9 333	18	1	82	0.5	2.1	2.26
2 595	111	957	66	34	1 668	5 221	19	1	74	0.7	2.0	1.94
2 913	106	974	64	32	1 561	5 778	17	1	79	1	2	2.23

notably P, Mn, Sr and Zn. Compared to the phosphato-bauxite soils of Rennell and Maré, the phosphates of Nauru are very similar but impoverished of trace elements. Only the following persist: Cu, Ba, V, Cr, Ni, La, and Mn.

Comparison between Tongan soils and modern floating pumices reveals that the two are very similar. The Tongan soils are not as rich in trace elements as the soils of other High Carbonate Islands, which are usually older.

In summary, the same minor elements can be found in the phosphates, in the soils and in the basaltic rocks but there are higher concentrations (especially of P) in the soils. Soils, basalts and pumices appear to be closely associated in terms of their composition and evolution. However, the dolomitic bedrock does not display any volcanic or bauxitic elements, suggesting its complete independence from the phosphatization or bauxitization processes.

TERCINIER (1971) interpreted the Lifou soils as being derived from altered pumices and ash. TAYLOR and HUGHES (1975) readopted this hypothesis for Rennell, as well as for Rennell and other islands we studied (BOURROUILH-LE JAN, 1980, 1982).

Geochemistry of Rurutuan Formations (Austral Archipelago): Intermediate Stage Between a High Volcanic Island and a High Carbonate Island

Rurutu is an emerged volcanic complex of Miocene to Recent age (BOURROUILH-LE JAN, 1985). The different carbonates (limestones and dolostones), the overlying soils and the internal sediments have been analysed. The Rurutuan dolostones are characterized by V = 10 ppm, La = 153 ppm, K_2O = 16 % and TiO_2 = 0.04 %. The Late Miocene carbonates above the dolostones also show the presence of Ba, Cr, Ni, Ga, Be, Li, and Mn. The volcanic rocks have as

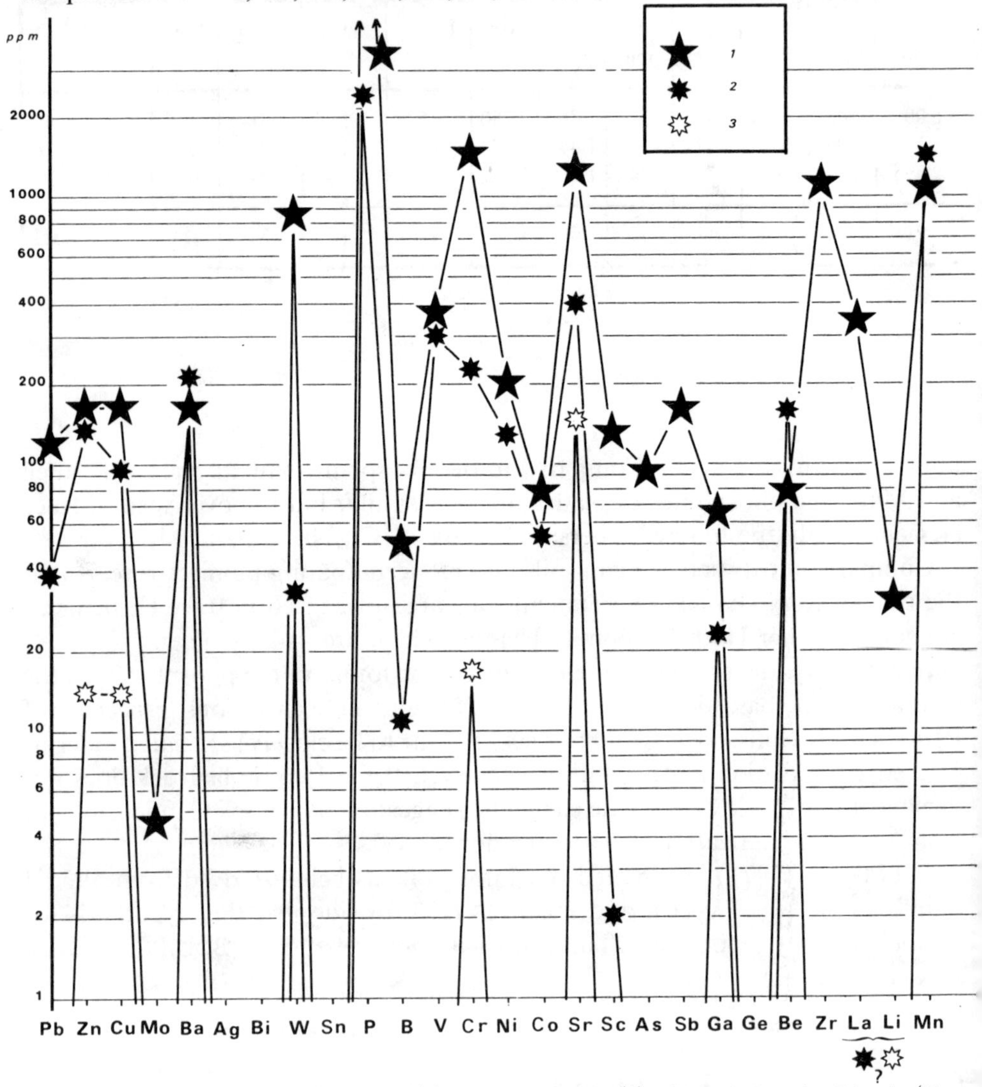

3–48. Comparison of soils (1), basalts (2), and carbonates (3) of Maré, Loyalty Islands (New Caledonia).

geochemical indicators: Zn, Cu, Ba, V, Ni, Co, Sr, Ga, Zr, La, Li, Mn, with 6 % MgO, 1 % K_2O, 3.5 % TiO_2.

These different geochemical indicators can be found again in the different soils located on the karst surface and inside karst features in the carbonate plateau of Rurutu, including modern soil formations and older fossilized formations which now fill the karst cavities. Because of their probably different magmatic origin, differences exist in their percentages. However, the soil presents the same fundamental geochemical assemblage as the volcanic rocks but with higher iron, silica and alumina, a higher content of Cr, Zr, La, Li, and above all of Mn.

These soils have also been sampled for their phosphate content. The intra-karst clays contain 6.6 % and the brownish to greenish soils only up to 0.6 % phosphates. Intra-karst sediment with continental gastropods and caliches shows the same geochemical composition but with contents reduced by 20–25 %.

In conclusion, chemical analyses show that the Rurutu soils are formed from volcanic products which are located here close to the carbonates. This indirectly confirms the volcanic origin of soils in the High Carbonate Islands.

Tectono-Karstic Circulation and Fissure Filling in the Carbonate Platforms

Karst groundwater circulation and sedimentary infilling of fissures and cavities are observed frequently in numerous High Carbonate Islands, including Minami Daito, Rennell, Nauru and Makatea. The intra-fissure sediments are finely laminated, contain ostracods and are rich in trace elements: Mn = 340 ppm, Cr = 74 ppm, Cu = 12 ppm, with some Ni, Sb, V, and 1–2 % of Fe_2O_3, Al_2O_3, TiO_2, and traces of clay and boehmite.

Underground karst flow connections are indicated also by the presence of phosphate ore in caves and marine-related notches and are located at an altitude of +50 m on the peripheral cliffs of Makatea.

Thus the presence of Mn and TiO_2 shows a relationship between the beginning of the deposition of internal sediment in the tectono-karstic fissures and the bauxite alteration with ionic migrations of Mn and Mg. Mg ions served for dolomitization during carbonate diagenesis.

Modern Phosphate Genesis on Clipperton Island

A field study organized by the Cousteau Foundation and supported also by the Singer-Polignac Foundation has shown that modern phosphate deposits are in the process of being laid down along the lagoonal leeward embayments on the windward side of the almost-atoll of Clipperton (BOURROUILH-LE JAN, 1983). These deposits are thin and are intensively burrowed by terrestrial crabs (*Geocarcinus*). The phosphatic minerals are poorly crystallized. Double carbonate mineral is formed (kutnahorite). At the same time, initial diagenesis acts on the old coral deposits of the atoll, i.e. mainly the dissolution of aragonite. The carbonate dissolution and the modern phosphate genesis are both linked to the very peculiar chemistry of the lagoon; this is now transformed into a meromictic lake with stratified waters, a lack of oxygen below 5 m and a very high H_2S content in the deeper waters.

A comparision among the modern phosphate deposits, the volcanic rocks of Clipperton, and modern bird guano deposited on the top of large coral boulders has been carried out. The same elements were found in the modern phosphate

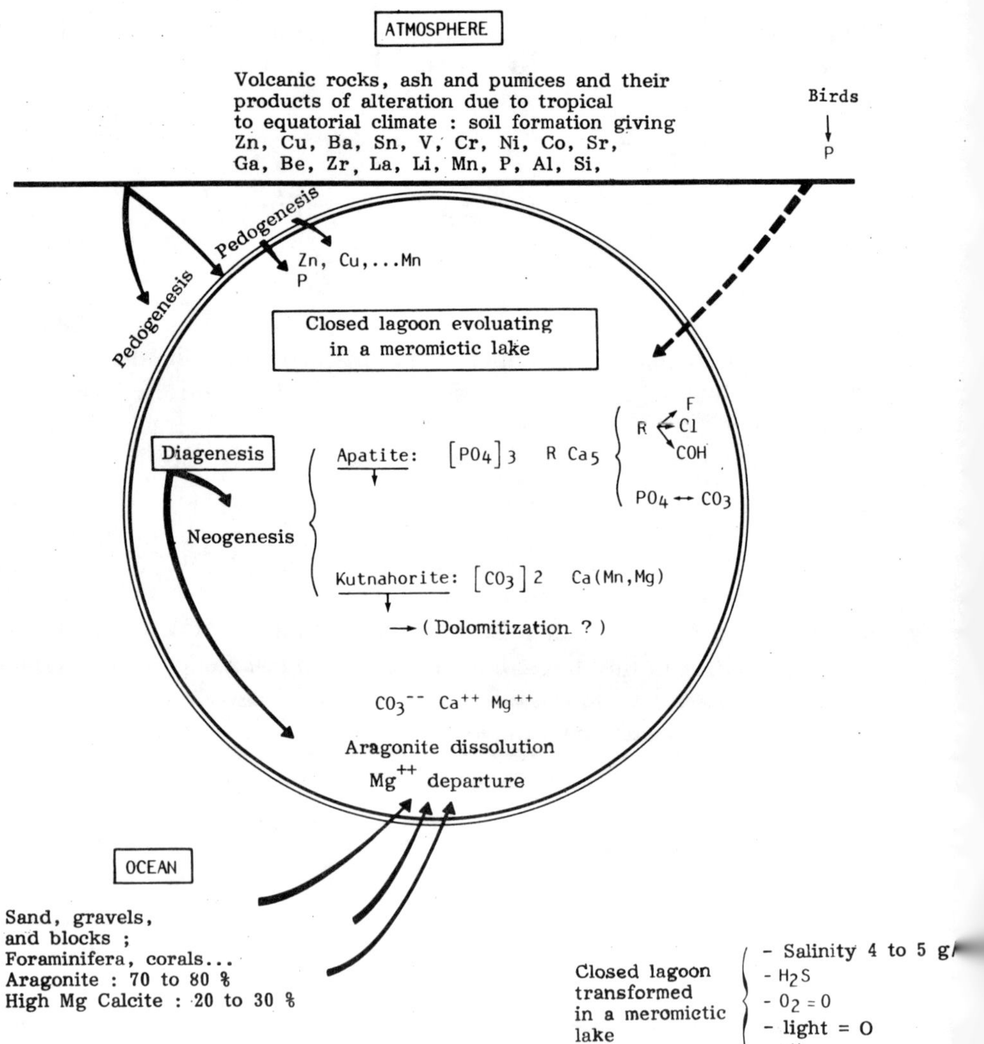

3–49. Model of phosphate genesis in the almost-atoll of Clipperton, ENE Pacific (after BOURROUILH-LE JAN et al., 1985).

deposits and the volcanic rocks, and none in the modern guano. This proves the indirect relationship between them; a pedogenic and also sedimentologic relationship (*via* a soil) in the restricted environment of the brackish lagoon (BOURROUILH-LE JAN et al., 1985; Fig. 3–49); and the hypothesis of guano source for oceanic phosphates seems not to be valid.

Discussion

The classical hypothesis for the Oceanic phosphates is based on the presumption that there is phosphatization from top to bottom (*per descensum*), by guano leachate acting on carbonate bedrock. This hypothesis can no longer be accepted because of the karst character of the ore/bedrock contact. Internal sediments and phosphates or phosphato-bauxitic soils are geochemically linked together, but are sedimentologically and geochemically independent of the carbonate substrate on which they rest. The karst contact between the phosphates and the carbonate bedrock (WARIN, 1967; OBELLIANNE, 1963) confirms our hypothesis. This surface is fossilized by the younger deposits (BRAITHWAITE, 1968). The karst topography is buried by deposits from a completely different sedimentary environment from that of the carbonate bedrock.

The morphology of the karstified bedrock surfaces of Nauru, Makatea and other islands is similar to some karst landscapes in China (KARST IN CHINA, 1976; ZHIYI, 1969) that formed under tropical conditions (annual rainfall = 1 500–2 000 mm; mean annual temperature = 20–22 °C; mean annual erosion rate = 0.12–0.3 mm).

There are no detailed sedimentological analyses of most phosphate deposits in the Pacific and Indian Oceans. This prevents paleogeographical reconstruction. Former researchers differentiated oceanic phosphates by their mineralogy, by their simple physical structure (loose or coherent, shield or carapace form) or by their supposed derivation from either ammonia-rich guano (*real guano* of WARIN, 1967) or phosphatic guano (*ancient guano*). The latter type was said to have lost its ammonia, oxalates and urates, all the components that unmistakably would indicate its origin.

A pedological origin for the phosphates has been advanced by FLICOTEAUX et al. (1977). The modern soils of the High Carbonate Islands are rich in iron and aluminium. GASTUCHE, FRIPIAT and SOKOLSKI (1963) showed that when Fe and Al are in competition, the aluminous phosphates are always formed first. Crandallite has been found in many of soils in the High Carbonate Islands (TERCINIER, 1972). Crandallite itself can be dissolved (PENEAUD, 1968) with an ensuing precipitation of a phase richer in Ca.

The effects of phosphate in seawater are not very well known but ARPER (1964) showed that its presence can stop the precipitation of Ca carbonates by the formation of complexes and colloids and by the inhibition of nucleus growth. This can therefore explain the bulk phosphatization of masses of coral deposits when the soils are eroded and dispersed on the High Carbonate Islands as a result of a marine transgression, as shown by the study of different ores.

The basic conclusions are as follows: (1) the carbonate rocks of these widely dispersed High Carbonate Islands are always poor in trace elements; (2) the carbonate substrate is independent of its cover (phosphate, phosphato-bauxitic soil or bauxite); (3) the phosphate, phosphato-bauxitic soil or bauxite cover derives, at least *pro parte,* from the degradation of volcanic rocks or ash *via* soil formation, and (4) the presence of phosphates or phosphato-bauxites in ancient geological records can be evidence of former volcanic episodes and their pedologic reworking. They may then indicate episodes of oceanic rifting, and they imply particular morphoclimatic and/or eustatic conditions. Therefore, they can be tied to curves of global sea level variation and, in particular, to phases of emergence and karstification.

Reconstruction of the Development of the Ore Cover on a Karstified Carbonate Bedrock

The example of Makatea has been chosed because some new facies of phosphatic ore have been discovered there, adding more details to the geological history obtained earlier from the Nauruan deposits.

Shallow water carbonate sedimentation began during the Early Miocene on top of a presumed volcanic summit. Seismic tremors caused the opening of gaping fissures which were progressively filled up with muddy sediments of low energy facies as well as by subtidal to supratidal facies with bird eye structures. The final sequence of the carbonate platform shows that it was completed with confined facies, miliolid facies and supratidal deposits. After this, dolomitization of the supratidal facies began. Due either to global eustatism or to local tectonics, the platform emerged slightly and there was intense dissolution of the exposed carbonates plus some formation of stalactitic and stalagmitic calcite which may be of the same age as the dolomitization affecting the supratidal facies. Among the karst features were freshwater ponds where ostracods were deposited in small horizontal beds. These deposits fossilized any previous stalactitic coatings and also reworked detrital dolomite rhombs or entire dolomitic lithoclasts; this establishes that dolomitization processes had already started. Phosphatic ooids are trapped in this lacustrine ostracod layer showing that the phosphatization processes had also begun.

We also note that the presence of continental facies lying upon carbonate marine bedrock indicates that the bedrock was slightly emerged and had developed large level surfaces on which caliches developed and where pisoids could grow larger through rolling.

At the same time there was trapping of volcanic ash and tephra. This was facilitated by the low elevation of the island. Weathering from the volcanic products which collected on the emerged areas was marked by the destruction of silicates, the formation of iron hydroxides, and the concentration of phosphates in the form of ooids within phosphato-bauxitic soils. At that time (Middle and Upper Miocene) a palm began to establish itself and became the now botanical puzzling endemic Makatea palm tree known as *Eupritchardia vuylstekiana* (WILDER, 1934).

The separation between iron, alumina and phosphates could have occurred as a result of marine transgressions reworking the phosphato-bauxitic soils (e.g. as on Christmas Island in the Indian Ocean). Following regression only the phosphates remained; these could be reworked several times. Phosphate may have started to dissolve and so to phosphatize other older facies such as the pisoids and caliche. It is possible that Makatea and perhaps also Nauru have gone through the Clipperton stage where a meromictic lake is present in which modern phosphates are deposited as a coastal and accretionary prism along the lagoon coast. A fall of the sea level, and therefore of the phreatic lens, leaves such phosphate deposits as flank deposits around a karst dissolution basin or polje.

The low solubility of the phosphate deposits is emphasized by their persisting through the up-and-down movements of the sea during the Quaternary.

Karst and Ore Deposits on Oceanic Islands: New Ideas

The phosphatic and bauxitic cover deposits of High Carbonate Islands such as Nauru, Makatea, but also of karst platform derived island as Aldabra, modern atolls etc., and the terra rossa of the Loyalty Islands, Rennell, Tongatabu, etc. are developed from the complex weathering of volcanic detritus transported by air (ash) or sea (pumice). The phosphates are concentrated in reworked soils after one or more phases of marine transgression has preferentially removed iron and silica liberated by humid tropical subaerial weathering. Phosphate may be precipitated as apatite, phosphatizing previous carbonate phases. Liberated Mg is concentrated in freshwater phreatic lenses beneath the soil and superficial karst where it may cause wide scale dolomitization of what were originally subtidal coral reefs. This is illustrated in Figure 3–50 which shows several Oceanic islands; each of them represents a different stage in the development of these processes because of their differing altitude with respect to the differing high sea levels reached during the Pliocene, Pleistocene and Hol-

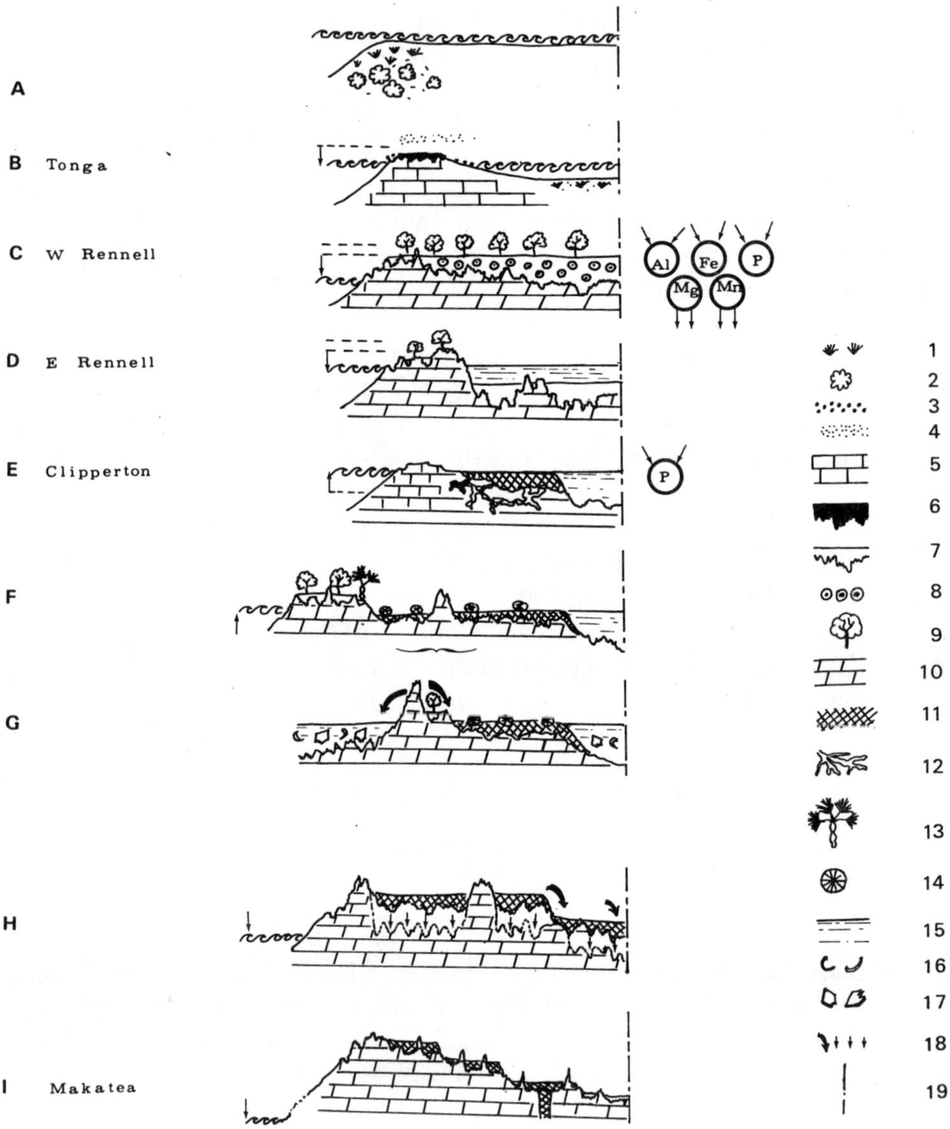

3–50. Reconstruction of the evolution of the High Carbonate Islands; from different examples studied: A. Shallow water carbonate platform with corals (1) and rodolites (2); B. Example of Tonga (Central S Pacific). Arrival of pumice (3) and ash (4) on the emerged part of the platform (Tongatabu) which has been a few metres high since the Pleistocene, with calcite diagenesis (5) and little evolved soil that is still rich in silica (6). Beginning of karstification. C. Western part of Rennell (W Pacific); bauxitic soil (7) with phosphate ooids (8), covered by equatorial forest (9), characterized by high rainfall, and dolomitized substrate (10) dated to Middle Miocene; D. Eastern part of Rennell (W Pacific); bauxitic gel from the reworking of bauxitic soil, covered by a meromictic lake and filling karst holes and submerged avens (S. Christiansen, 1964); E. Almost-atoll of Clipperton (ENE Pacific); modern formation of phosphate (11) deposits with inter- to supratidal sedimentary prism ocene, and also because of the difference in the location of the various moving plates of the Pacific Ocean.

Acknowledgments: the research has been supported by several organizations which the author wishes to thank, especially the Société Nationale Elf-Aquitaine (Production), the Singer-Polignac and Cousteau Foundations and the Muséum d'Histoire Naturelle de Paris and Ecole Pratique des Hautes Etudes, Paris. I am gratefull to Drs. Gy. Bárdossy and P. Bosák, and to Professeur G. Millot for their encouragement and interest in these studies.

---

with apatite and kutnahorite neogenesis from a preexisting soil (BOURROUILH-LE JAN et al., 1985). Holocene to Pleistocene carbonate rim beginning to be dissolved (12). Possible dolomitization; F. Makatea (SE Pacific) during the Late Miocene-Pliocene. Dolomitized platform (DANA, 1849; HUNT, 1855) of Lower Miocene (BOURROUILH-LE JAN, 1985), covered on its emerged zone by a tropical *Eupritchardia* forest (13) modern endemic palm tree of Makatea, reworking of the soil and the phosphate ooids, with phosphate deposits (11) in inter- to supratidal sedimentary prisms as in the modern example of Clipperton; formation of calcite pisoids (14) on the banks of meromictic lakes (15); G. Detail of F, Makatea (SE Pacific), meromictic lakes with ostracods (16), reworked dolomitic lithoclasts and isolated and broken crystals of dolomite (17); H. The termination of platform evolution on High Carbonate Island (the aspect of "uplifted atoll") with subjacent karstification, reworking and subdigging of phosphates (18), reworking and transport I. Schematic cross-section of modern Makatea (SE Pacific;). Diagrams not to scale. Half of the platform has been represented (19: axis of vertical symmetry).

PALEOKARST-RELATED URANIUM DEPOSITS

Yves Fuchs

The Problem of Uranium Accumulation in Karst

The presence of economic or sub-economic uranium concentrations in paleokarst is uncommon although some occurrences are reported e.g. Le Vigan (France), Bakouma (Central African Republic), Tyuya Muyun (Kirghizia, USSR), Orphan Mine (Arizona, U.S.A.) or Visé region (Belgium)...

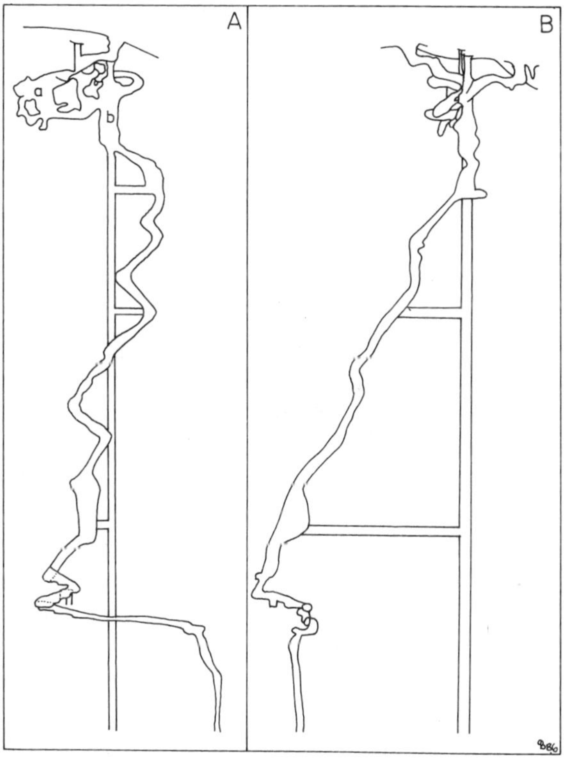

3–51. Tyuya Muyun uranium deposit in south Fergana, U.S.S.R. (modified after FERSMAN, 1930). a. Yellow Cave; b. Green Cave; A. section N–S; B. section W–E.

P. Bosák, D. C. Ford, J. Głazek and I. Horáček (Editors):
Paleokarst. A Systematic and Regional Review. – Elsevier and Academia. Amsterdam and Praha. 1989.

The paleokarst environment is not propitious for the mobilization and redeposition of important amounts of uranium. The carbonate rocks which are weathered during karstification and the regional environment of a normal carbonate platform are not rich in uranium. Most such carbonates contain of

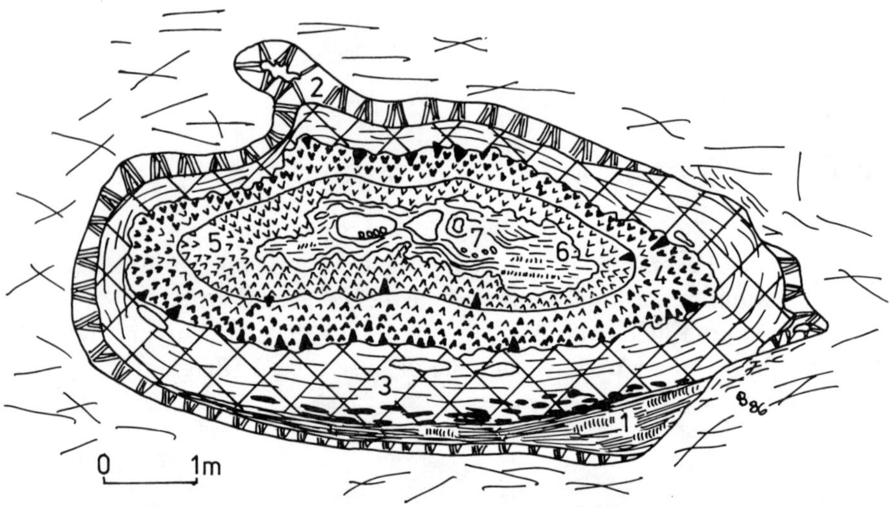

3–52. Schematic cross-section of the main vein in the Tyuya Muyun deposit (after FERSMAN, 1930). 1. lime crust; 2. columnar flowstone with stalactites and stalagmites; 3. coarse-crystalline calcite with tyuyamunite (CaO, $2UO_3$, V_2O_5, H_2O), touranite ($5CuO \cdot V_2O_5 \cdot 2H_2O$), tangeite ($2\ CuO \cdot 2CaO \cdot V_2O_5 \cdot H_2O$), radiobaryte, radiocalcite; 4. red baryte; 5. translucid baryte; 6. crust and terra rossa (later karst); 7. hole.

the order of 0.0X to 0.X ppm except for reef facies where the uranium content can be as high as 2 to 3 ppm (TATSUMOTO and GOLDBERG, 1959). In these coral limestones it seems that uranium is closely related to the organic matter content. Some other measurements of the sediments of the Gulf of Mexico show a strong relationship between uranium and organic carbon (TIN MO, SUTTLE and SACKETT, 1973). As a general rule carbonate rocks deposited on neritic platforms have a higher uranium content (up to 3 ppm) than do those deposited in oceanic environments. Platform carbonate rocks more frequently undergo karstification, as well, particularly when epeirogenesis is active. Nevertheless, it appears that they cannot be a consistent and adequate source of uranium for occurrences known in karst. For the main examples, some other origin must be put forward. At Tyuya Muyun, for instance, the presence of black shales near the karst ore deposit (copper and uranium vanadates) is known (FERSMAN, 1930). The vanadium, copper and uranium contents of these Silurian-Devonian black shales are obviously higher than those of the carbonate rocks where the karst is hosted. At Le Vigan (France) where there are some uranium occurrences in paleokarst in Cambrian limestones, ORGEVAL and CAPUS (1978) see in an uranium-rich sodic or sodi-potassic cineritic layer interbedded in the limestones the most probable source of the uranium. In the basin of Bakouma the products of weathering of the old Precambrian basement were transported toward the centre of the basin where a karst was developed. In this case, the most probable source would be the uranium-rich basement.

As a general rule we can say that the quantity of uranium which reaches karsts is unimportant. Particular geological conditions are necessary to explain an exceptional supply of uranium such as those at Bakouma or Le Vigan. Furthermore, the chemical conditions which prevail in karst do not favour deposition of uranium which is in solution in the water. The two main agents of karstification are CO_2 and H_2O. Solution of CO_2 from the air is faster than dissolution of the carbonate rock. In a karst structure, if the circulation of water is fast, the water remains aggressive at depth (CO_2 rich), favouring the development of endokarst. In such conditions, the uranium is particularly soluble as uranyl carbonate.

The Bakouma Uranium Deposit

The most significant uranium concentration known in karst is the deposit of Bakouma, Central African Republic. The uranium is located in phosphatic lenses deposited in a continental environment.

In the Bakouma area one can see three main geological units:

(1) an old crystalline basement with two mica gneisses or biotite gneisses, two mica granites or biotite granites. This lower part of the basement was highly metamorphosed. Less metamorphic units (two mica schists) overlie it.

(2) a sedimentary unit of Precambrian age which lies unconformably on this old basement. At the base, this unit displays dominantly detrital facies. Bougboulou Formation is followed by the Kembé and Nakondo Sandstones. These detrital formations are a few hundred metres thick. Following them, mainly pelitic units (M'Bania and Bondo) with a thickness varying from 200 m up to 500 m grade progressively to the carbonate Bakouma Unit which is 400 to 500 m thick.

This whole Precambrian sedimentary sequence underwent weak metamorphism during an early phase of the Panafrican Orogeny. Rb/Sr isotope ratios give an age between 683 and 708 Ma, which is consistent with this interpretation. A pelitic unit (Dialinga), which lies on the Bakouma and could be younger than the orogenic phase, is not present in the Bakouma area. Some volcanic dykes and sills and some lava flows with doleritic texture are also present.

(3) the third main unit begins with marine transgression and moderate erosion. As a result, the Dialinga Formation is eroded south of Bakouma and karstification occurred on the exposed limestones of the Bakouma Unit. The karstification is particularly well marked along structures which trend N 70 °E (Panafrican directions). Canyons were formed. During the Jurassic, an epeirogenic movement caused deposition of the Mouka Ouadda Sandstones. These beds are not well investigated in the Bakouma area though they were found by drilling in some depressions of the surface of the Bakouma Unit. They are well developed to the north where they contain the important diamond deposit of N'Zako. Post-Cretaceous tectonic movements are related to an E-W fault system and have induced changes in the hydrological system. Erosion has increased. Karstification is particularly

active at intersections of the E-W and N 70 °E features. On the karstified carbonates of Bakouma lies a unit containing lignite (M'Patou) which grades laterally into azoic phosphatic siltstones: it is the phosphate unit of M'Patou, which fills the karst features. As a lateral equivalent of the M'Patou lignite, the siltstone would be of Eocene age. All of these deposits are overlain by laterites.

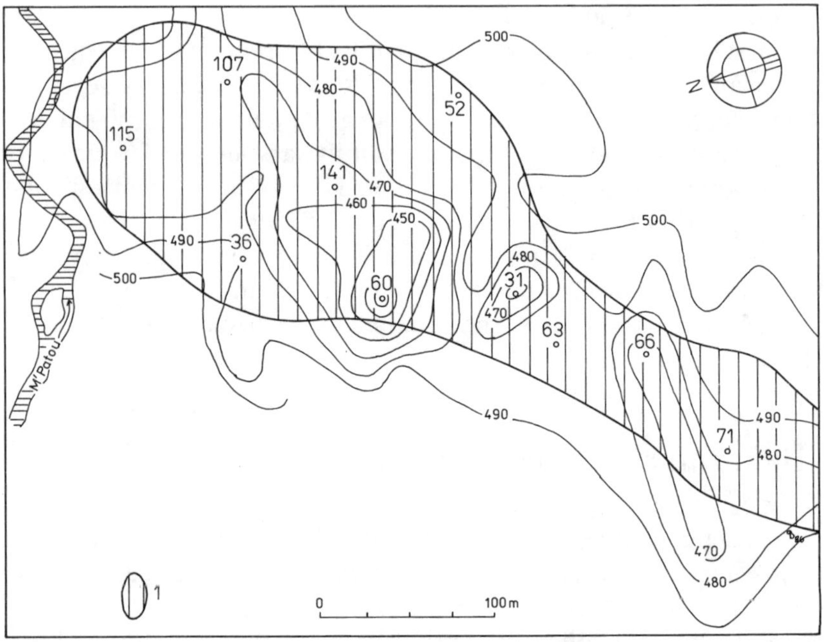

3–53. Bakouma – relation of the paleogeography of the unconformity surface and the disposition of the ore lenses (after MIAUTON, 1980). 1. ore lenses.

The development of the karst, its filling, and the importance of the uranium concentration in the karst fillings are the most interesting facts at Bakouma. The karst began to develop at the end of the Precambrian. Localization of sinkholes is related to the main tectonic features (Fig. 3–54). The post-Cretaceous tectonic event (E-W faults) gives more importance to this relationship. The main karst features are located near areas where E-W and N 70 °E faults intersect.

During the Eocene, the Bakouma basin closed and there was no more significant drainage. Lakes formed in the karst depressions. These lakes were characterized by the predominance of reducing conditions because organic matter was abundant. Black silts were deposited and the black cores of apatite crystals are rich in organic matter. Basic conditions prevail in the lakes. Various elements were provided by leaching under oxidising conditions in emerged areas (Si, Al, Cu, P, U; Fig. 3–55). Phosphorus comes obviously from the Precambrian carbonate rocks. The Palmyre Limestone (Bakouma Unit) shows P_2O_5 values between 1 and 4 % and locally as high as 10 % (MIAUTON 1980). The source of uranium could be the crystalline basement which crops out around the basin.

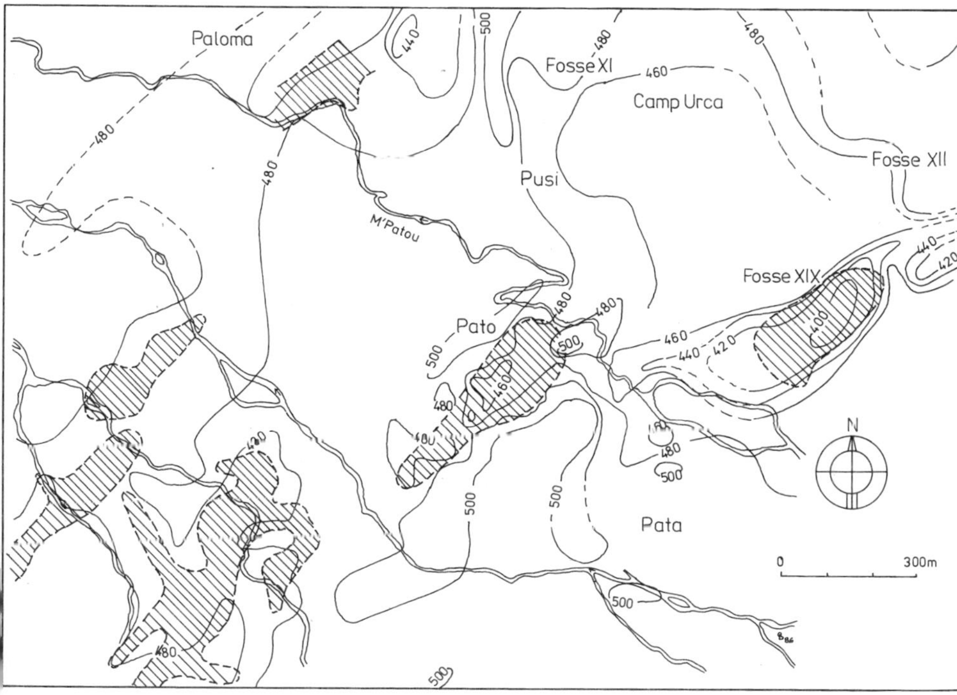

3–54. Bakouma – relation between karst features with ore lenses and the N 70°E trending tectonic features (after MIAUTON, 1980).

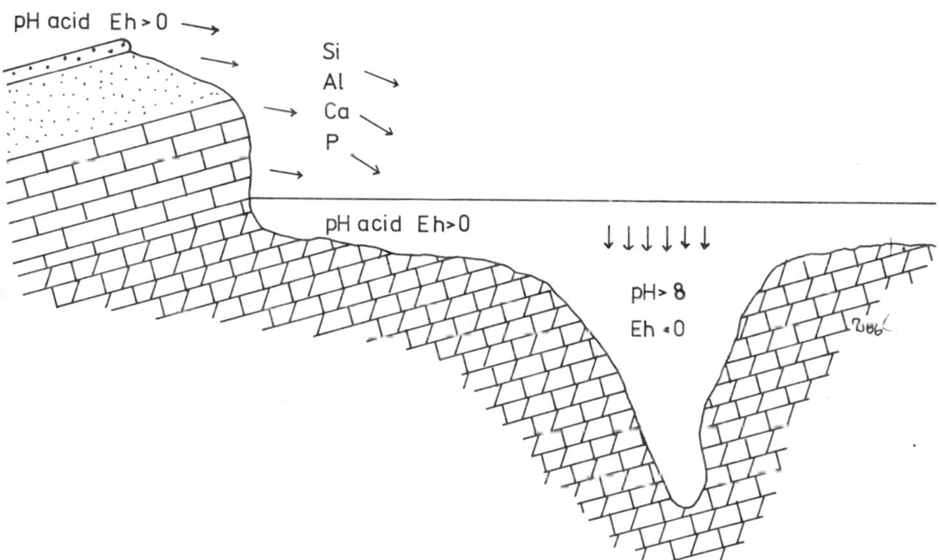

–55. Schematic cross-section through a Bakouma pond. Chemical conditions during deposition (after MIAUTON, 1980).

Apatitisation phenomena occurred in the basic and reducing environment of the lakes.

Phosphorus underwent diagenetic transformation from tricalcic phosphate, $Ca_3(PO_4)_2$, to hydroxyapatite $3Ca_3(PO_4)_2.Ca(OH)_2$, and finally to carbonate-apatite, $3Ca_3(PO_4)_2.CaCO_3$. This developed slowly and the result was an apatitic siltstone with amorphous apatite (of the order of X nm to XX nm). The apatitic formation found in karst depressions (Fig. 3–55) consists of alternating phosphatic layers and clay layers. Argillaceous layers contain numerous authigenic quartz crystals or quartz crystals with secondary overgrowths. Gypsum has been reported. During rainy periods the pH was lower. Deposition of apatite ceased and silica was precipitated as flint or as microsandstone.

Apatite normally shows low uranium content. It ranges from 0.001 to 0.01 % in apatite of magmatic origin and from 0.005 to 0.02 % in apatite of sedimentary origin. In Bakouma, the uranium content of apatite is up to 0.5 %. GONY (1969) has shown that in Bakouma 17 % to 46 % of the uranium present within the apatite is tetravalent U^{4+}. Hexavalent U^{6+} is more abundant in the upper levels as well as in the deepest level of the phosphatic lenses.

In the apatite crystal there is an isomorphic replacement of $3Ca^{2+}$ by one U^{4+} ion and two Na^{+} ions. Furthermore, substitution of Ca by Sr provides more or less deformation of the crystal lattice and favours the introduction of uranium into the apatite. The hexavalent uranium is adsorbed on the surface of the apatite crystals. This hexavalent uranium could be syngenctic, its presence being related to some changes in the environmental pH conditions. However, it is more likely to be post-syngenetic, and may have been brought by phreatic water arriving from weathering profiles and percolating through the phosphatic lenses.

After deposition of the phosphatic siltstones, the environmental conditions changed. The influence of carbonates decreased and sedimentation became more detrital. Early changes included the presence of autunite and chalcolithe, after which the conditions became more acid and iron and aluminous phosphates, such as vivianite, crandellite, and wavellite were deposited.

The Bakouma uranium deposits which are located in phosphatic siltstone within karst features thus result from a long and complicated geological evolution. The existence of a particular trap (the paleokarst) developed in P_2O_5-rich limestones favoured the formation of the uranium deposit but the apatite formative process and the uranium concentration process are not related to this karstification. Instead, they are believed to be related to a later phase of sedimentation in ponds with basic and reducing conditions that occupied these earlier karst depressions.

Uranium Deposits of the U.S.A.

Some uranium-bearing structures in the U.S.A. can be described as karst features of the general type known as breccia and pebble pipes. BRYNER (1961) described many different pipes, most of them related to hydrothermal or magmatic activity. Nevertheless, he mentioned the collapse structures in carbonate rocks described in Missouri by BRETZ (1950) and emphasized the role of circulating groundwater in their formation. Breccia pipes with uranium ore were described in Arizona by BARRINGTON and KERR (1963) and by GORNITZ and KERR (1970). These authors think that the formation of such features could be related to hydrothermal circulation but many geological features suggest that they are due to karstification processes. Most features of this type are located in the Cameron District of Arizona, near the Little Colorado River and at Orphan Mine near Grand Canyon.

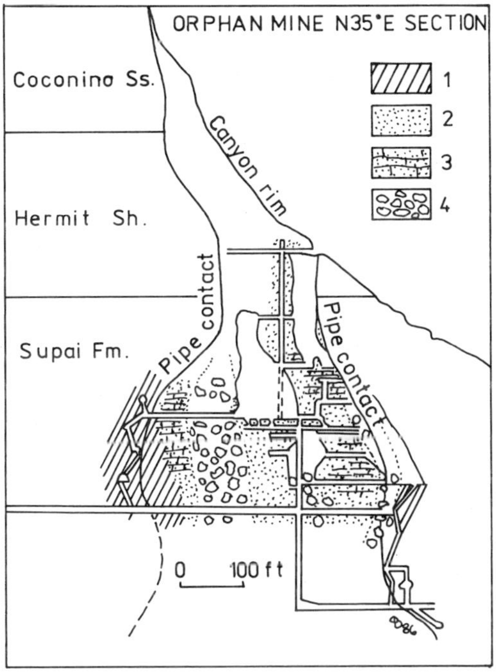

3–56. Geological cross-section of the Orphan Mine, U.S.A. (modified after GORNITZ and KERR, 1971). 1. bleached Supai Formation; 2. sandstone pipe fill; 3. calcareous sandstone pipe fill (massive); 4. breccia zone.

At Cameron, there are some silicified plugs and collapse structures (Fig. 3–56). In the collapse breccias U, Cu, S, P, Si, Fe, Mn mineralization is present. These collapse features are gently dipping in the Triassic sandstones but show steep funnel-shaped forms in the underlying Permian limestones. The silica plugs in the Moenkopi Formation (Triassic) show a sharp contact with the host -rock and have a pipelike appearance. All plugs show radioactive anomalies. Maximum activity is observed at the contact with the host-rock which is not disturbed but only bleached. The plugs consist of pebbles and cobbles in an intensively silicified matrix. The associated clay minerals are kaolinite and illite. The collapse features are circular or elliptical in form. Their diameters vary from 30 to 1 500 m. The collapse features occur close to tectonic features.

At Orphan Mine, the breccia pipe is a vertical circular structure with a diameter varying from 50 to 150 m. The pipe fill consists of downward displaced sandstone blocks, angular siltstone, shale and limestone blocks of the Permian neighbouring rocks. Its depth is about 500 m, extending down to the Redwall Limestone. The pipe is limited by a set of concentric circular features. The uranium mineralization is located in this „annular ring" within the cement of sandstones and breccias. The brecciated elements are partly bleached, partly unbleached. Intergrowths of uranium minerals with sulphides and hematite can be observed. A vertical zoning exists. Uraninite is more frequent in the upper part and increases above the level with galena. Copper mineralization has developed below the uranium. The $\delta^{34}S$ isotopic values indicate an enrichment in ^{32}S. The wide range of $\delta^{34}S$ variations and the negative values shows a strong differentiation that cannot be explained by inorganic processes alone. $^{207}Pb/^{235}U$ ratios give a minimum age of deposition of 140 Ma (Late Jurassic).

Some authors support the hypothesis of a hydrothermal origin. There is no evidence of such a process. The breccia pipes are not connected to volcanic vents nor to obvious hydrothermal systems. A karst system in the underlying Redwall Limestone could have induced the collapse features in the overlying sediments and the leaching of metals, particularly uranium, out of Triassic sandstones where ore deposits are known. A reducing environment in the lower part of the pipe would have enhanced deposition of the uraninite and some secondary sulphides, perhaps under the influence of bacterial activity.

Conclusions

Karst related uranium occurrences are numerous but with the exception of Bakouma district no economically important deposit is known yet. The reasons are the lack of uranium in karstified carbonate rock and the fact that chemical conditions within karst do not favour bulk deposition of uranium. Only in some cases such as Le Vigan (France) and Cameron District and Orphan Mine (U.S.A.), where uranium was already present in the vicinity, can significant concentrations be observed. However karstification induced more of a dilution of the uranium concentrations there than a concentration (ORGEVAL and CAPUS, 1978). Bakouma still remains an exception. Here, the karst features were developed long before the uranium mineralization occurred. The karst was only a passive feature favouring and enhancing the chemical conditions that induced the deposition of uranium.

PART IV

HYDROGEOLOGY AND ENGINEERING HAZARDS IN PALEOKARST AREAS

PALEOKARST AS AN IMPORTANT HYDROGEOLOGICAL FACTOR

Josef Zötl

Introduction

There is a close relationship between hydrogeology and paleokarst. Karst phenomena that have developed in the past have a decisive influence on the nature and extent of water permeability in carbonate rocks.

The most influential factors in the formation of paleokarst are the geological-tectonic development of an area and its climate. Together, these determine the course of morphogenesis. This in turn is responsible, among other things, for the longevity of local or regional base levels of erosion, which are a key to understanding paleokarst hydrogeological phenomena.

Today it is difficult to appreciate that in scientific karst research (starting with that on the classical karst of Istria at the beginning of this century) there was for decades a bitter fight about the *pro* and *contra* of the influence of the base level of erosion upon the development of karst hydrography (cf. GRUND, 1903; KATZER, 1909; MARTEL, 1921; BOCK, 1913; CVIJIĆ, 1918; LEHMANN, 1932; and others; see also ZÖTL, 1974). The incorrect opinion of LEHMANN (1932) that no relationship exists between base level of erosion and karst water circulation came to be accepted and for decades hindered any fruitful discussion and scientific hydrogeological karst research in Central Europe.

It was only the results of such new research approaches as extensive systematic surveys, development and performance of comprehensive tracer studies, and also hydrological isotope measurements, that led the way out of these various blind alleys. The trend to new methods in karst hydrogeology in Central Europe is to be seen, for example, in the International Symposium on Underground Water Tracing (SUWT) that has been held every four to six years since 1966. Within this context, a group of specialists from Austria, the Federal Republic of Germany, Greece, Switzerland and Yugoslavia have been continuously occupied with the further development of groundwater tracing techniques and have linked their trial use with a comprehensive cooperative study of typical karst areas in various countries.

P. Bosák, D. C. Ford, J. Głazek and I. Horáček (Editors):
Paleokarst. A Systematic and Regional Review. – Elsevier and Academia. Amsterdam and Praha. 1989.

In addition, there is a Karst Commission (current President is Dr. H. Paloc) of the IAH with members from Austria, Brazil, Czechoslovakia, Federal Republic of Germany, Greece, Hungary, Ireland, Israel, Italy, Poland, Republic of South Africa, Switzerland, the U.S.A., U.S.S.R., and Yugoslavia.

Both of these groups have published a number of works but in general there has been no distinction between the hydrogeology of paleokarst and of recent karstification.

It is thus interesting to study the development of old underground solution forms and to analyse their current hydrogeological function by describing individual examples. The examples are all taken from carbonate rocks; the hydrogeology of salt and gypsum is not considered.

Examples of Hydrogeological Phenomena of Paleokarst in the Northeastern Limestone Alps of Austria

The Hölloch in the east of the upper Muota Valley (central Switzerland) has been known for a long time and with an aggregate length of 150 km is the second longest known limestone cave system in the world (BÖGLI, 1970). But the more extensive karst regions of the Northern Limestone Alps are in the eastern massifs in Austria and have been studied better (Fig. 4–1).

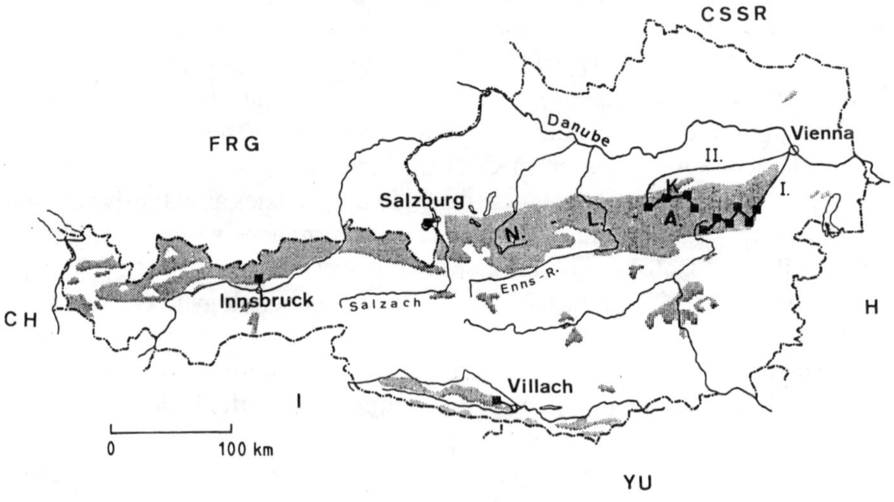

4–1. Shated areas show karst in Austria. Black squares are the catchment of large springs used for water supply to the adjacent city. Vienna I = First Viennese water supply (pipes, tunnels, aqueducts), 115 km long, constructed 1869–1873; II = Second Viennese water supply, 200 km long, constructed 1900–1912; N.-L.-A. Northeastern Limestone Alps; K. Kläffer Quelle, i. e. the largest perennial karst spring of II (see Fig. 4–3).

Within these massifs, tectonic structures originate from pre-Cenomanian, pre-Gosavian and Early Tertiary orogenic phases, characterized by extensive nappe-type folding with overthrusts. The limestone sequences (dominantly Triassic Wetterstein Limestone and Dachstein Limestone) gently dipping to the north are cut by steep faults. The low dip of thick limestone strata favoured the development of denudation plains at different altitudes during quiescent periods of the uplift of the Alps in Middle and Late Tertiary times.

The eastern karst massifs, together with the Central Alps, once formed one connected unit. This is chiefly proven by occurrences of the so-called *Augensteine* found on all the larger plateau surfaces. They are mainly well rounded quartz pebbles and quartz sands, the remnants of deposits from streams flowing from

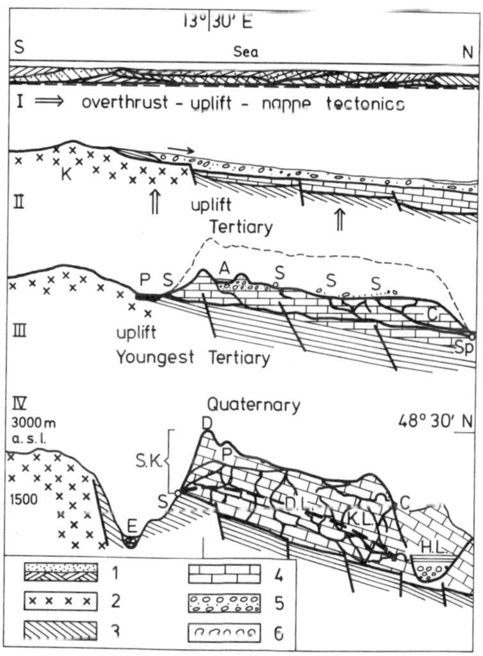

4–2. Schematic drawing of genesis and hydrogeology of paleokarst in the Northeastern Limestone Alps in Austria. Different scales of altitude and strata series. Phase I. Pre-Tertiary overthrust and nappe tectonics. Phase II. Uplift, thick Tertiary sediment cover, surface drainage towards north from the Central Alps over the later Limestone Alps. The deep W–E trending longitudinal valleys do not yet exist. Phase III. General scheme after further uplift in the Upper Tertiary. There are only remnants of sediment cover, mainly quartzose gravel and sand (*Augensteine*). Beginning of karstification of the limestone. P = polje in the eroded schist; S = sinkhole (ponor); A = Augensteine; C = cave; Sp = spring. Phase IV. Schematic draft of the recent S-N cross-section of Central Alps – Dachstein Massif.

the main Alpine Central Ridge into the sea of the Alpine Piedmont. Longitudinal valleys that now divide the Central Alps on the one side from the Northern Limestone Alps on the other did not exist at that time (Fig. 4–2). The denudation began with the progressive uplifting of the mountains. When erosion finally reached the limestone surface and the base level in the north was lower, karstification of the exposed plateau areas occurred, causing the pattern of surface rivers to disappear. The hydrogeological effect of quiet periods during the tectonic uplift can be shown in many examples.

The grouping of caves at certain elevations was discussed in detail by SCHAUBERGER (1956) for the Tennengebirge, the Dachstein Massif and the

Totes Gebirge; for the first time their development and relationships to the downward displacement of the base level of erosion were shown. This is thus paleokarst; see for contrast the tectonic development hypotheses for the large caves advanced by ARNBERGER (1955) and TRIMMEL (1955).

In the **Dachstein** area, the main entrances and systems of the giant caves, such as the Mammoth Cave and the Giant Ice Caves, are at 1 300–1 500 m a.s.l. This is some 900 m above the surface of Hallstatt Lake. Figure 4–2 shows a simplified version of the topographic development of the Northeastern Limestone Alps including the Dachstein Massif and the different elevations of the cave floors.

In the same elevation zone there are large cave horizons in other massifs (the Tantal Cave in the Hagengebirge, 1 300–1 700 m a.s.l., and the Frauenmauer-Langstein Cave in the Hochschwab Massif, 1 400–1 600 m a.s.l.). Last but not least there are extensive remnants of old valley systems that today are exposed as hanging valleys 700 m above the trunk valley floor. They afford a second example of the effect of tectonic quiet periods.

The development of such enormous cave systems must have demanded larger amounts of water than the plateau surfaces of the present-day karst massif provide. Thus, the Alps in the Late Tertiary were drained uniformly to the north, and the longitudinal Salzach and Enns Valleys were not present. It can be inferred also that the climate at that time was in a humid phase.

The loss of the catchment area in the Central Alps began with the development of the longitudinal valleys in the Late Pannonian and indicates that pronounced uplift occurred during the Upper Miocene. Glaciation in cold periods of the Pleistocene with severe erosion by thick valley glaciers did the rest of what was required to divide the Limestone Alps into separate massifs and to deepen the longitudinal valleys between the Central and Limestone Alps.

The underground drainage of individual massifs at first followed the general system of landscape development. However, in places it was very much modified by special petrographic and tectonic conditions. It is generally true that the deeper karstification along joint patterns, bedding planes and faults creates a variety of open waterways that lead to the formation of waterless plateau surfaces and to locally concentrated, large springs in the valleys.

Depending on geological and meteorological conditions and the given storage capacity, the large springs may vary greatly in discharge (Fig.4–3). With respect to storage, a very important question is whether a given Alpine karst massif displays so-called *shallow karst* or *deep karst*. We use the hydrogeological term *shallow karst* when the karst complex is underlain by non-karst rocks that perch all springs above the local base level of erosion; the *deep karst* extends below the local base level. The thickness of the karstified rock sequence has nothing to do with whether it is *shallow* or *deep*.

In the Dachstein Massif both types can occur. In the southwest the massif has the typical character of a shallow karst: the base level of karst corrosion (i.e. the altitude of underlying, nonkarstifiable schists at 1 700 m a.s.l.) does not correspond with the local base level of erosion in the Enns Valley (about 700 m a.s.l., Fig. 4–2 IV). In contrast, in the northern part of the massif, the Dachstein Limestone extends deep below the recent base level of erosion (i.e. Hallstatt Lake, 508 m a.s.l.). Here the Dachstein area is a typical example of deep karst.

4–3. Catchment house and overflow of the *Kläffer Quelle,* i.e. main spring of the 2nd Viennese High Spring Water Pipeline (Photo by M. Fürböck).

Figure 4–2 shows that in accordance with the hypothetical position of the karst water level, large karst springs emerge at the northern edge of the massif. In this relatively simple structural massif the precipitation passes more or less vertically through paleokarst caves of the vadose zone into the phreatic zone, from where it flows in paleokarst conduits to the large springs.

An exact calculation of the water budget of the Northern Limestone Alps has not yet been possible.

Of great importance is the geological structure on the **Hochschwab Massif** (summit 2 278 m a.s.l.). There are unusually strong spring outlets on the north side. At the beginning of this century (1900–1912) these were exploited for the so-called *Zweite Wiener Hochquellenleitung*/Second Viennese High-Spring-Water Pipeline (Figs. 4–1, 4–3).

The **Schneebergalpen** comprise the easternmost plateau massifs of the Northern Limestone Alps. The base of the carbonate rocks is above the local base level of erosion in the underlying Werfenian schists, so that we are again confronted with a shallow karst. Waters from the larger springs were collected from 1869–1873 for the *Erste Wiener Hochquellenleitung* and delivered to the city as its main water supply by a conduit some 100 km in length (Fig. 4–1).

In the Northern Limestone Alps, the infiltration of rainwater into the karst depends on the intensity, duration and season of precipitation. In bare karst above the tree line, the proportion collected can be as much as 90 %. Small areas of grass on clay soil in larger dolines divert water into the next sinkhole, usually only a few metres away. In wooded areas there is also no surface runoff but evapotranspiration is more important. An unknown factor is the amount of evaporation that occurs during the snowmelt season; this varies with the weather.

The horizontal caves which occur at higher altitudes are either completely lacking in a modern hydrogeological function or of subordinate importance for water storage (ice caves). SAAR (1954; p. 7) calculated an average ice volume of approximately 13 500 m^3 for the Dachstein Ice Caves. Air circulation determines the extent and course of melting and growth of the cave ice bodies during the seasonal temperature cycle. The caves are in the vadose zone of the karst massif.

Periodic and perennial overflow springs discharge water stored in the deep karst into present-day valleys (Fig. 4–2). The recent discharge is actually slowly changing the cave systems but without doubt these are to be classified as a form of paleokarst.

This relationship between water-poor plateaus and springs displaying large (variable) output is to be found in all of the carbonate massifs in the humid zone. Variations in discharge range from less than 1 $m^3 \cdot s^{-1}$ (in some cases, zero) to greater than 10 $m^3 \cdot s^{-1}$.

For the **Totes Gebirge** (590 km^2; northeast of the Dachstein Massif) tritium, deuterium and oxygen-18 content were determined for water samples taken in September 1968, March 1969, June 1969, September 1969, July 1970 and February 1971 from the most important springs at the foot of the massif (DINCER et al., 1972). The results showed that the autumn and mid-winter low water in the springs was comprised exclusively of rainwater from the previous summer. In the late winter and spring, seepage of meltwater from lower-lying areas began to appear. In early summer, the true, large karst springs carried meltwater from the high plateau exclusively, but the proportion of rainwater then increased rapidly as summer advances. This pattern is typical for the karst massifs of the eastern part of the Northern Limestone Alps.

The underground flow directions indicated by tracer experiments here were confirmed by the isotope measurements. In addition, the isotopic composition showed that in some cases where tracers showed connections between lake sinkholes and springs, the proportion of water derived from seepages within small glacial lakes and ponds was insignificant in comparison to the total output. In such cases the outlet paths suggest only short-period storage, i.e. period of residence above the outlet. The tritium and oxygen-18 content of spring waters from the Totes Gebirges shows that all of the water supplying the run-off is active, i.e. modern. This is characteristic of the type of underground karst drainage that is usually found in shallow karsts. The drainages use paleokarst waterways modified to a small extent by modern karstification.

The Deep Karst of the Vaucluse Massif

Isotope studies in the Plateau de Vaucluse have provided insight into the water regime of a sample deep karst (MARGRITA et al., 1970; PALOC, 1970).

Fontaine de Vaucluse, surely the best-known karst spring in the world, is some 25 km northeast of Avignon, where the Rhone Valley emerges from the Alps. Its spillway is a typical paleokarst form (Fig. 4–4). It drains an area of some 1 200 km^2 that extends up to 1 900 m a.s.l. For many years the discharge has

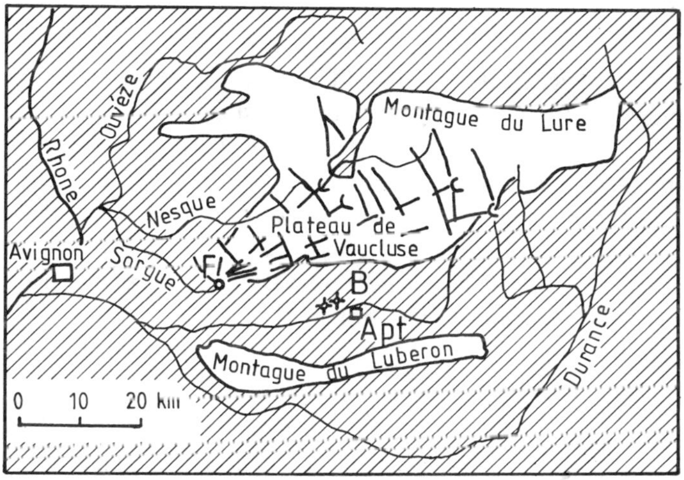

4–4. In the area of Apt, the Cretaceous limestones form a syncline and are covered by Tertiary sediments, but are known from bore-holes (B). Thick lines = faults; broken lines = results of tracing experiments; F = Fontaine de Vaucluse. Non-hatched = Cretaceous limestone of the Massif de Vaucluse (after MARGARITA et al., 1970; PALOC, 1970).

averaged 29 $m^3 \cdot s^{-1}$ but it varies between 4 and 150 $m^3 \cdot s^{-1}$. The lowest output registered since observations began in 1853 was 4.05 $m^3 \cdot s^{-1}$ recorded in October 1967 (PALOC, 1970, p. 131).

The catchment area of the spring is mainly made up of Lower Cretaceous limestones (Barremian, Bédoulian) with a thickness in the east of some 200 m, increasing to 600 m on the southern edge of the Vaucluse Massif. The limestones form a syncline under the Tertiary Apt Basin and reappear in the hilly chain of the Montagne du Luberon. A bore-hole was drilled into the water confined in the Tertiary basin and its level fluctuation was observed to follow that of the flood flow of the Vaucluse. Recent tracer studies with points of injection in the Vaucluse Massif showed traces of colour in these bore-holes before the tracer appeared in the Vaucluse spring.

The Vaucluse is a flood overflow. At low water, the water level in the flood channel (inclined at about 45°) is some 20 m below the cave portal of the paleokarst.

The unusual low-water situation in October 1967 combined with the fact that the water in the deep karst was free of tritium and poor in carbon-14 made it possible to explain the mixing ratio of old and recent waters at the spring. The discharge of Vaucluse decreased constantly from April to October 1967, when it reached its lowest known value (see above; and also in MARGRITA et al., 1970, p. 338 and Fig. 4–4). At the same time the tritium concentration decreased from 210 TU to 107 TU. It is assumed that the water flowing during the period from June to October 1967, was a mixture of waters from the spring rains (with an average T concentration of 200 TU) and so-called reserve waters (with low T content). The T content of these reserve waters could have been 107 TU at a maximum; in this case, the spring would have discharged reserve waters exclusively at the end of October 1967. Not all of the reserve waters would have had to have a T concentration of 107 TU, but only their upper layers. As later T and ^{14}C measurements in the Chene boring showed, there is actually an age stratification of the deep karst water there (MARGRITA et al., 1970, p. 346). In the minimum case, the T concentration of the reserve waters could have been zero. In this instance, though, the Vaucluse output at the end of October 1967 appears to have been composed of 50 % waters from the spring rains (with an average of 200 TU) and 50 % of old reserve waters. These two limiting cases provide a frame for possible calculations.

For the productivity of the Vaucluse at the time of the permanent decrease from 21 June to 18 October 1967, the above authors proposed the formula (p. 340):

$$Q(t) = 1.62 \exp(-0.06t) + 7.4 \exp(0.005t) \tag{1}$$

where t – time in day and $Q(t)$ – is the discharge rate in $m^3 \cdot s^{-1}$. Integration of this formula for the period of 119 days produces a water supply of 59.5 million m^3 (MARGRITA et al., 1970, p. 340).

On the other hand, the authors represented the T concentrations over this period by the equation

$$c(t) = 200 \exp(-0.004t) \tag{2}$$

where $c(t)$ – TU and t = days

Starting from the hypothetical limiting values of the T content of the reserve water (100 TU or zero TU), the contribution of the reserve water (C_0) to the total flow volume can be derived.

(1) Assume $C_0 = 0$ TU

If $\alpha(t)$ is the contribution of the reserve at time t, then

$$c(t) = 0 \cdot \alpha(t) + 200\,[1 - (t)]\,. \tag{3}$$

With $c(t) = 200 \exp(-0.004t)$ we obtain

$$\alpha(t) = 1 - \exp(-0.004t). \tag{4}$$

The contribution of the spring to the water supply during the time (119 days) when the reserves were drawn upon is thus

$$Q_1 = \int_0^{119} \alpha(t)\, q(t)\, \mathrm{d}t \cdot 86\,400 \tag{5}$$

which gives Q_1 = 10.3 million m^3, or 17 % of the total amount.

(2) Assume $C_0 = 100$ TU

$\beta(t)$ is now the contribution of the reserve waters time t and can be written

$$c(t) = 100\beta(t) + 200\,[1 - \beta(t)], \tag{6}$$

and it follows that

$$\beta(t) = 2[1 - \exp(-0.004t)]. \tag{7}$$

The result is

$$Q_2 = \int_0^{119} \beta(t)\, q(t) \cdot 86\,400 \tag{8}$$

i.e. Q_2 = 20.6 million m^3, or a contribution from the reserves to the total output of 34 % (MARGRITA et al., 1970, p. 341).

The narrowing of the possible contribution of the reserve waters to the discharge of a spring down to a range of 17–34 % is certainly not satisfactory but nonetheless is progress that could not have been made without measurements of the T content.

Owing to the rapid reaction to pressure changes, it is to be assumed that the greatest part of these reserve waters are to be found in the large voids of a deep paleokarst.

Paleokarst of Coastal Areas, Islands and Peninsulas

General Conditions

With a density of 1.00 for freshwater and of 1.02 to 1.03 for seawater, the sea level theoretically must be not only the base level of erosion but also the base level of karst corrosion for freshwater flowing from a mainland. Owing to the vertical variations of the boundary surfaces (waves and tides) this is not a sharp boundary, however, but a zone of greater or lesser amplitude. If this zone remains stable long enough, then karstification in it creates large-diameter waterways with mainly horizontal courses.

It is known that climatic changes over long periods of time have changed the sea level. The cold periods of the Pleistocene formed large bodies of ice and mountain glaciers in Eurasia and North America. It is calculated that this process caused the sea level in the Mediterranean to decrease by some 100 m during the Würm and by as much as 200 m in the Riss Ice Age. Of course confidence in these calculations decreases with the age of the phenomenon; isostatic uplift connected with erosion and other local shifts of mass make these estimates even more difficult to apply in general terms. The fact nonetheless remains that karst phases occurred during these climatically determined changes in sea level and created karst zones that today are under water. These are designated as paleokarst.

Islands and Immediate Coastal Areas

An example from my own experience is seen in stalactites taken from a subsidence cave in the east of the **Kephallinia Island.** The cave is some 100 m inland and today is filled with brackish water, but is connected to the sea. In 1961, divers retrieved stalactites from depths of 3 m and 26 m. Carbon-14 determinations by K. O. MÜNNICH (Second Physics Institute, University of Heidelberg, FRG) showed an age of about 16 000 to 20 000 years (MAURIN and ZÖTL, 1967, p. 436). Since stalactites cannot be formed in water-filled spaces, the sea level must have been lower at the time of their development. There are no signs of recent tectonic uplift and the marine terraces of the Tyrrhenian Sea are at an elevation typical for the entire area.

The presence of impermeable Tertiary sediments in the west of the island produced a W-E inclination of the underground karst drainage during the low-water periods of the Pleistocene. When the sea again rose the well-known phenomenon of the *Sea Mills of Argostoli* developed (MAURIN and ZÖTL, 1967).

4–5. Inflow of seawater into the *katavothra* (sinkholes) of Argostolion, Kephallinia (Photo by J. Zötl).

Figures 4–5 and 4–7 show the inflow of seawater into the sinkholes (katavothra) at the coast. Submarine springs are common along the eastern coastal karst of Kephallinia (Figs. 4–6 and 4–7).

ALFIREVIĆ (1966) has counted a large number of submarine springs along the **Dalmatian coast** also. They are also common along the **French, Greek, Italian** and **Turkish coasts**. All are believed to follow submarine paleokarst paths.

A special problem in karst hydrogeology is that with the spring outlets being at or below sea level, they almost always carry brackish water and the admixture of seawater and freshwater can scarcely be avoided for some distance inland as well in the coastal area. The successful sampling of freshwater from a drowned karst conduit that resulted from cooperation between specialists from the BRGM and the Société des Eaux de Marseille thus deserves to be mentioned (BRGM, 1974).

The submarine discharge of freshwater under a cliff of the **Massif des Calanques** has long been known. Preliminary studies by divers located the main gallery and secondary passages at Port Miou (east of Marseille). The length explored was more than 1 km at depth of −20 to −50 m below sea level. These explorations made it possible to cut an access gallery from the surface to an air-filled

4–6. Submarine spring near the eastern shoreline of the Kephallinia Island, Greece (Photo by J. Zötl).

chamber in the paleokarst system. From that location it was possible to observe and measure the flow pattern of lighter freshwater passing out at the top of the submerged gallery, and of denser water penetrating the aquifer underneath the freshwater.

After some measurements and calculations, a *two-stage dam* was constructed that allowed freshwater in flood to pass through and — as long as there is the slightest flow of freshwater — to trap the seawater at a level lower than the inverted dam.

The utilization of *freshwater lenses in low islands and peninsulas is more difficult*. Here, the current climatic and meteorological conditions play a decisive role.

The simplified geological map of **Ireland** shows that the coastal area in NW, N and E is mainly made up of granite, Precambrian crystalline rocks, Paleozoi

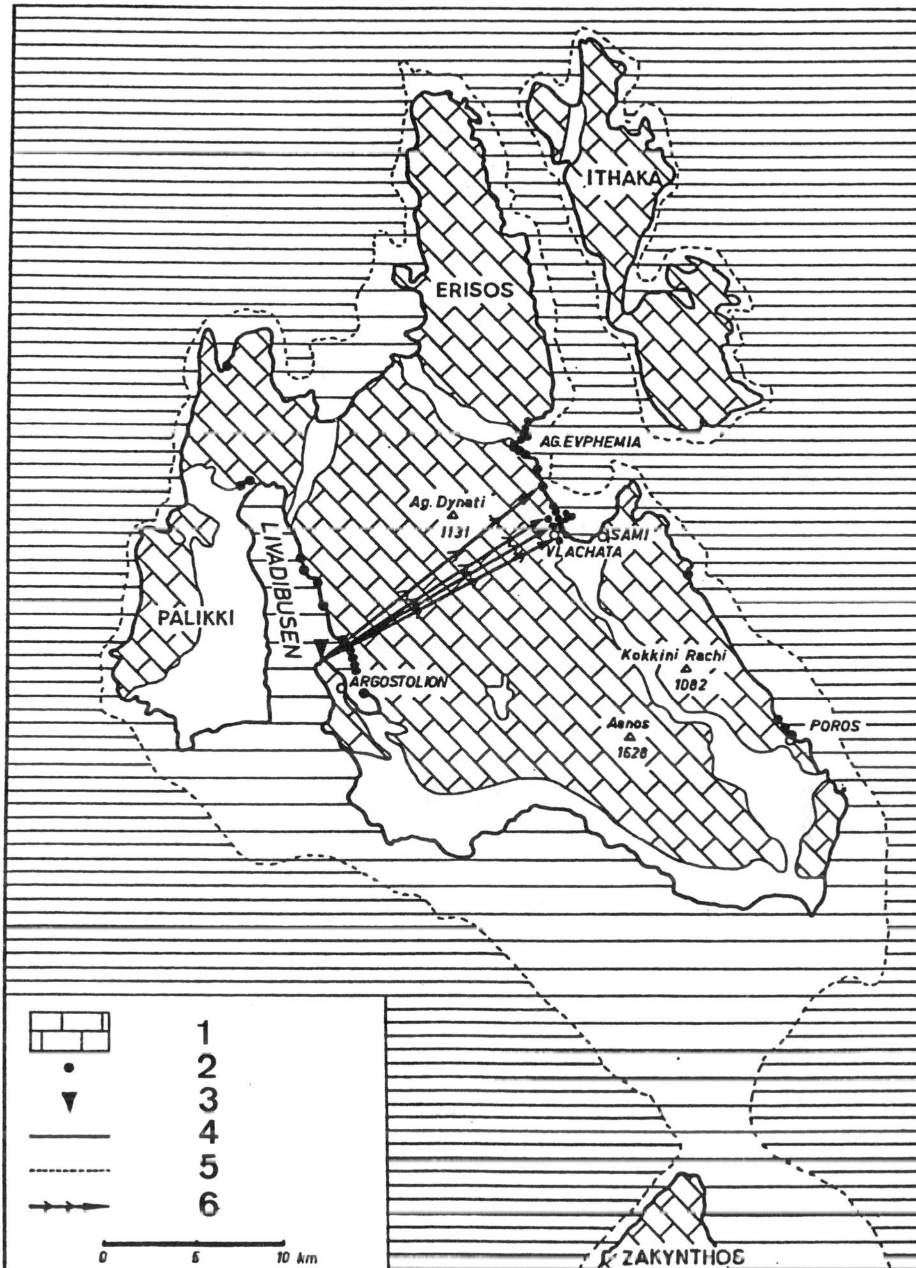

4–7. Kephallinia and Ithaka, Ionian Islands. 1. karstified Cretaceous limestone; 2. spring discharging brackish water; 3. seawater katavothra; 4. present shoreline; 5. present 100 m isobath; 6. direction of underground flow of sea and brackish water.

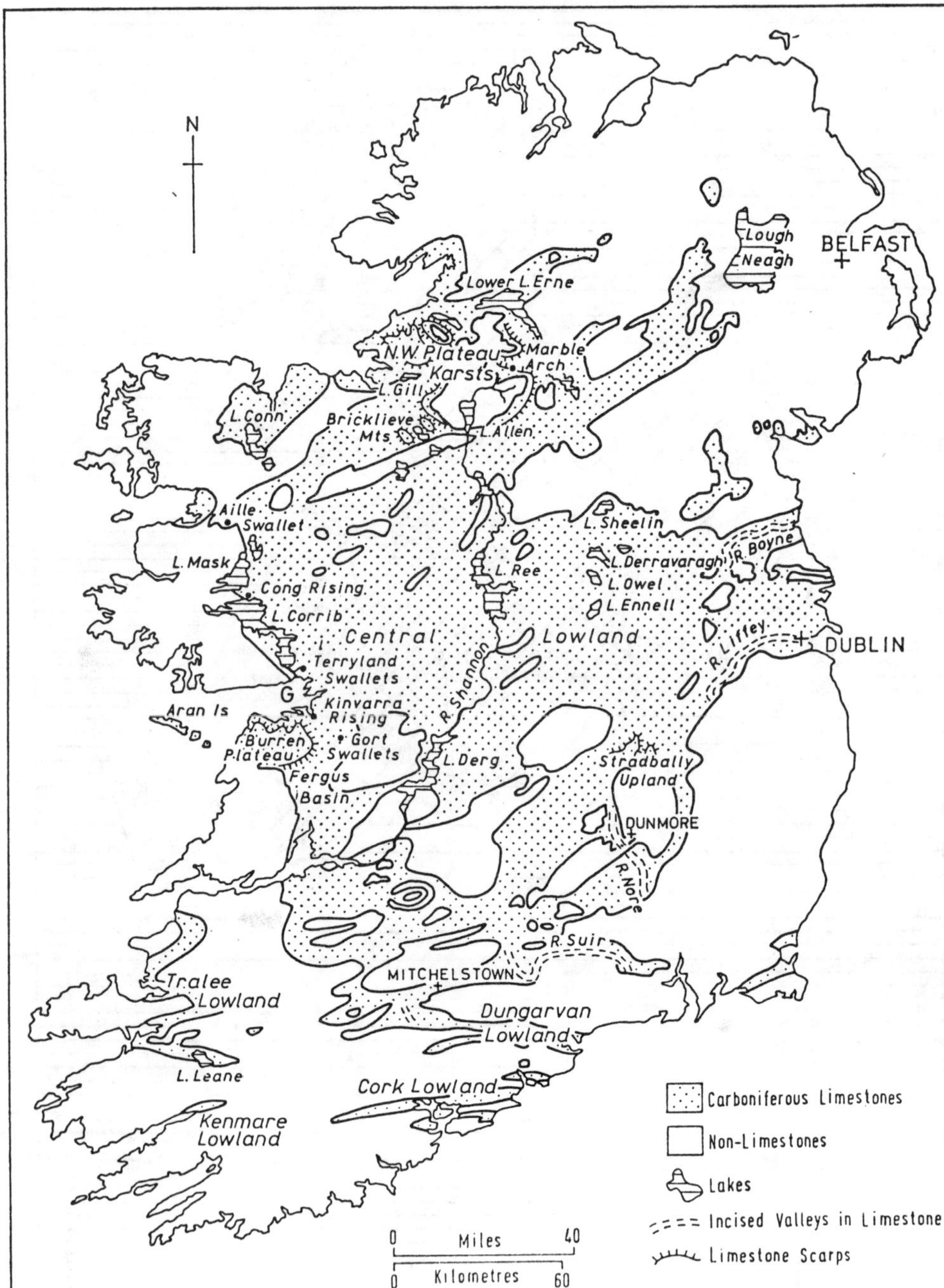

4–8. Ireland and its regions of karstified Carboniferous limestone (after P. W. WILLIAMS, 1970). G = Galway.

rocks and Tertiary volcanics. These surround the Central Lowlands and give the geologic structure a bowl shape. The Lowlands are made up mainly of Carboniferous and Permian carbonate rocks that in the southwest of Galway and in the east near Dublin also form the coast (Fig. 4–8). The carbonate rocks display old forms on their surface, and lobes of the North European ice, tectonics and sea level changes have contributed to a special kind of paleokarst. The fact that only 2 km out of 212 km of the road from Dublin to Galway do not run through Carboniferous layers emphasizes the importance of the karstification, even though the limestones show small intercallations of schists, sandstones and quartzites. Generally speaking, the Lowlands can be described as *deepland karst* with a shallow karst water table, for which the sea is the base level of erosion in the west and east, and the River Shannon in the central part.

That this is an old, deep karstification, and that this paleokarst largely determines the modern hydrogeological situation, is shown by the fact that the sea tides operate far into the Lowlands (for example in Lough (Lake) Caherglossdun, 6 km southeast of Kinvarra), with a time lag of four hours.

The most interesting phenomenon of the Lowlands, however, are without doubt the so-called *turloughs* (from the Irish tuar loch, i.e. dry lake). WILLIAMS (1970, p. 114) defined them as grass-covered surface depressions that are small at one time, extending at another time over a number of acres. They are filled with water during wet weather periods by the rise of underground waters.

As the turloughs are often lined with morainic material their direct bedrock karst channels are very seldom recognizable. Since glacial erosion may have contributed to their development one may agree with their classification as glaciokarstic landforms (WILLIAMS, 1970, p. 114).

It is not correct to describe the turloughs as poljes and to equate them to these common phenomena of Yugoslavia and Greece. The turloughs in the Irish Lowlands do not have the typical springs and ponors (katavothra) of poljes nor the seasonal change from lakes to dry basins. They also rarely have a flat floor, which is typical for normal polje. The climate and meteorological circumstances eliminate water-supply problems. Where surface water cannot be formed, wells sunk into the limestone underground never dry up.

The situation is not so favourable where there are thin freshwater lenses in flat-lying limestones, e.g. in southern **Apulia.** Here, on the heel of the Italian boot, there is a landscape of low hills and limestone plains that extends for 250 km from the Ofanto River to Cape St. Maria di Leuca. The peninsula is 35–50 km wide. It is composed mainly of chalks (and, locally Eocene-Oligocene limestone, e.g. between Otranto and Leuca) which dip under younger Tertiary and Quaternary deposits along a fault system running parallel to the Appenines toward the southwest. The Cretaceous limestones (and some less extensive dolostones) are only slightly folded but very shattered by tectonics. Sandy-clayey deposits of younger Tertiary and Quaternary age cover parts of the limestone tableland. Because these later sediments are thin, the underlying limestones are in contact with sea beyond a given depth even in places where the immediate coastline is formed by young, non-karstic layers.

Using data from numerous bore-holes COTECCHIA (1963, p. 7) emphasizes that the Cretaceous and lithologically similar Lower Tertiary limestones show a rather uniform, interlocking network of open joints as a result of deep-reaching tectonic shattering coupled with bedding planes. The groundwater movement takes place in this network. But while the tectonic joints are rather deep, karstification is mainly limited to the upper 100 to 200 m of the limestone. This is generally a paleokarst of Plio-Pleistocene age. COTECCHIA (1963, p. 8) assumes that water flow in these well-jointed and karstified limestones is mainly laminar, with movement being largely along the bedding planes, resulting in a low hydraulic gradient (0.3 to 0.5 %) and an effective Ghyben-Herzberg equilibrium. It was assumed for the exposed areas of Cretaceous limestones that 50–60 % of the precipitation (mainly from October to March) served to replenish underground water supplies.

The entire thick freshwater body, for which COTECCHIA assumes a yearly supplement of some 1 600 to 1 700 m^3 (p. 9) rests upon saltwater. It has been demonstrated that under the peninsula of Salentina, the Ionian Sea (Gulf of Tarento) communicates with the Adriatic (p. 10).

The most important results of COTECCHIA's studies were determinations of the length of the delays when infiltrating rainwater causes shifts of the freshwater/seawater boundary. Delays are brief near the coast, extending to weeks or months inland. Also of interest were related changes in the hydraulic gradient before re-establishment of static equilibrium, the interpretation of the changes in level of the karst water body and of changes in discharge of the coastal springs, and, finally, the study of the local change of the Ghyben-Herzberg equilibrium. Under the given local circumstances (with density 1 for the fresh groundwater $\varrho_s = 1.029\,9$ kp $\cdot$ 1^{-1} for the Adriatic and 1.030 7 kp $\cdot$ $1^{-1}$for the Ionian Sea) this theoretically would be around h = 32–33t (where t = height of the piezometric surface above sea level). Actually, quite different values were found to occur because: (1) there was the phenomenon of molecular salt diffusion from seawater into fresh groundwater, (2) of the mixing of salt and freshwater through vertical variations in the boundary surface, caused both by the tides (especially in the vicinity of the coast) and the seasonal variations in the piezometric surface (of the karst water level), resulting in vertical flows, and further (3) of the flow of seawater along the boundary surface with the groundwater.

All of these factors create a diffusion zone; its breadth depends on rock permeability, the thickness of the aquifer, infiltration time and quantity, sea-level variations and distance from the coast.

Numerous measurements made in Apulia showed that (with the exception of the immediate seaboard where the groundwater body is thin and the infiltration of the tides greater) the saline layering in the groundwater body is characterized by a constant, only slight, salt concentration (0.10 to 0.15 g Cl/l water) from sea level to a depth of 20 to 22 *t*. Then there is a gradual increase in salinity, with a Cl concentration of $\sim 5\ g \cdot l^{-1}$ at a depth of about 32 *t*. Only at a depth of 40 *t* is the salt concentration as great as that of seawater.

Poljes and Their Hydrogeology, an Impressive Paleokarst Phenomenon

MAULL (1958) gives a survey of the various opinions on the development of poljes. These range from the argument that dolines merge to create large karst basins, to their development from blind valleys, to the development of basins with underground drainage only in mountain synclines.

There is also disagreement up to present day with respect to which geomorphological forms can be designated as poljes. There are, for example, the uvalas which developed from the merged dolines, or the aforementioned *turloughs* in Ireland which, in my opinion, are incorrectly classified as poljes.

The synclines with underground drainage in the Swiss Jura are a transitional form. In common with the true poljes, they have no surface drainage but there is no exact boundary between the polje bottom and the flanks.

The Dinaric and Greek poljes are the typical form. The Dinaric poljes are divided into those of the Slovenian karst on the northern edge of the Dinaric mountains and the Bosnian-Herzegovinian polje-karst. The polje regions of Greece are mainly in the Peloponnesus.

Poljes on the northern edge of the Dinarides belong to the catchment are a of the Ljubljanica River (Fig. 4–9). This area has been thoroughly studied (GOSPODARIĆ and ZÖTL, 1976). The poljes form a sequence of steps: Loško Polje, approximately 700 m a.s.l., Cerkniško Polje, approximately 550 m a.s.l., Planinsko Polje, approximately 445 m a.s.l. Each polje was and usually is the base level of corrosion of the next higher one. The connections are generally via known cave systems. The springs of the Ljubljanica River (approximately 290 m a.s.l.) are the absolute local base level of corrosion and erosion. The connection and cave systems are paleokarst, and this is true of the entire polje landscape. There is no doubt that the modern, direct underground connection between Cerkniško Polje and the springs of the Ljubljanica River is younger.

The heart of the Dinaric karst is the Dalmatian-Bosnian-Herzegovinian area (loc. 2 on Fig. 4–9, Fig. 4–10). The phenomena found here stimulated the first discussion on basic factors of karst hydrology (GRUND, 1903, 1910; KATZER, 1909; etc.). In the entire area, the most intensively karstified areas consist of Cretaceous limestone. Young tectonic movements, especially young uplift

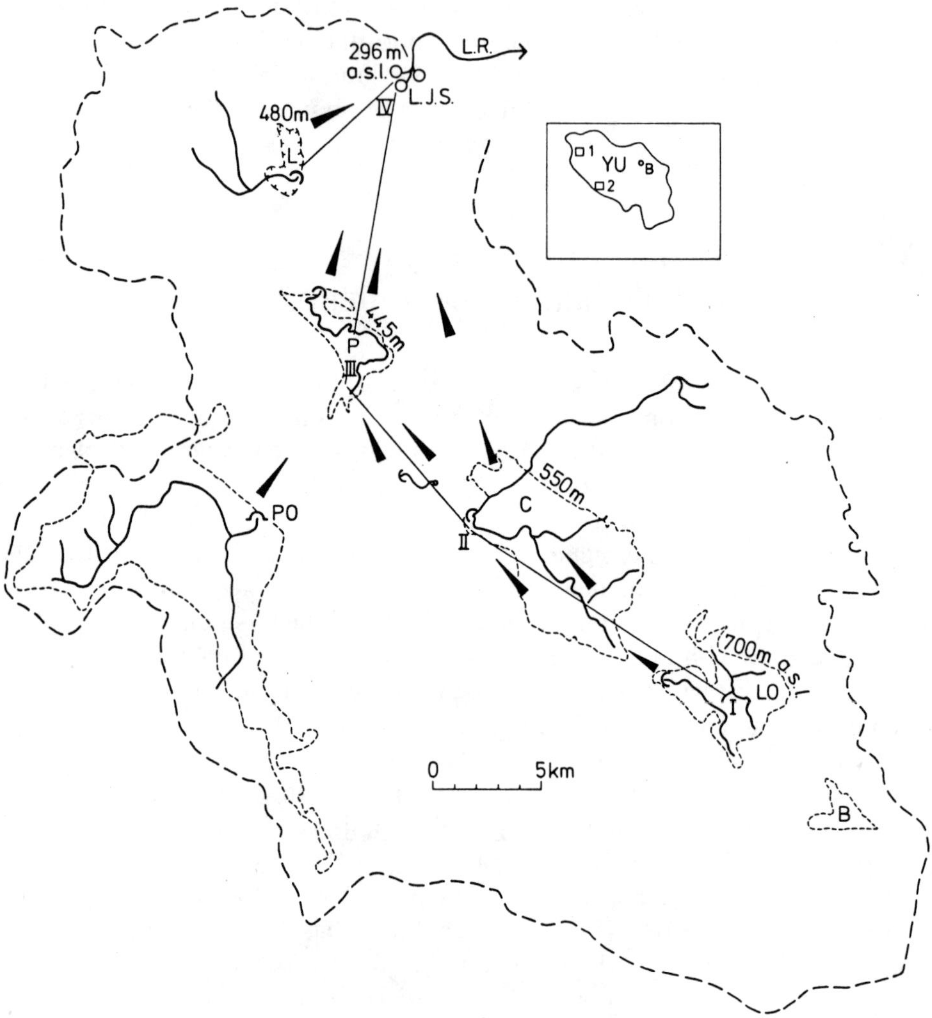

4–9. Poljes on the northern edge of the Dinarides and their underground drainage to the Ljublanica River (landmark 1 in the orientation square, B = Beograd). B. Babno polje; LO. polje of Loz; C. Cerkniško polje; P. Planinsko polje; L.J.S. springs of the Ljublanica; L.R. Ljublanica River; PO. Postojna. Arrows = flow direction in paleokarst channels and caves.

phases, and the intercalation of impermeable Tertiary sediments in the limestones have led to the formation of a landscape with the unique karst phenomena of poljes with ponors, giant springs, estavelles and a strong seasonal hydrologic rhythm.

Study of the area permits us to conclude that pre-Cretaceous tectonic effects are scarcely visible in the present karst. After an epoch in the Middle Eocene that was not so important, there were extensive uplifts in the Late Eocene and Oligocene, accompanied by folding, faulting and overthrusting. From this time onwards it is generally possible to follow the development of the underground

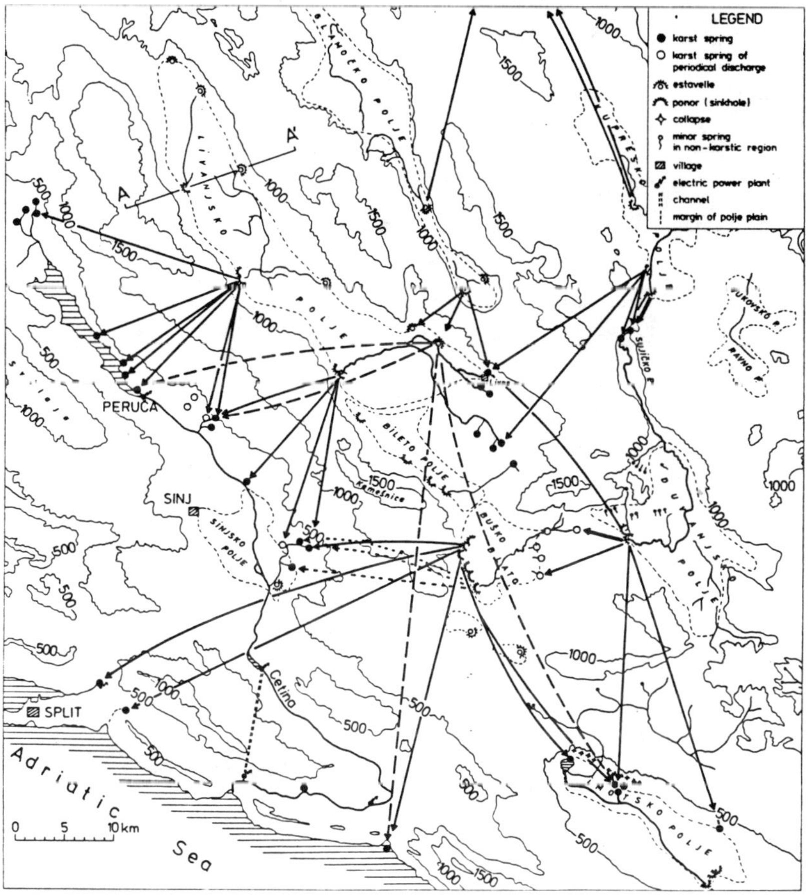

4–10. Karst drainage in the paleokarst of the Dalmatian-Bosnian area (landmark 2 on Fig. 4–9).

karstification and drainage, as well as the karst morphology, more or less clearly. It is generally assumed that in the Neogene this area lay only slightly above sea level, and that there was a high karst water table with drainage into the basin topography mainly by the numerous joints; even then there was periodic flooding and emptying.

HERAK (1972, p. 30, 31) emphasizes that the Pleistocene began not only with climatic but also with considerable tectonic changes. Tectonics and climate led to the formation of the poljes. They include an accumulation surface with a basin floor that can cover a surface from a few to more than 100 km^2.

4–11. Part of Livanjsko polje. Right side old mill and estavelle with spring water in winter, and vice versa in summer (flood water sinks into the ponor or sinkhole) (Photo by J. Zötl).

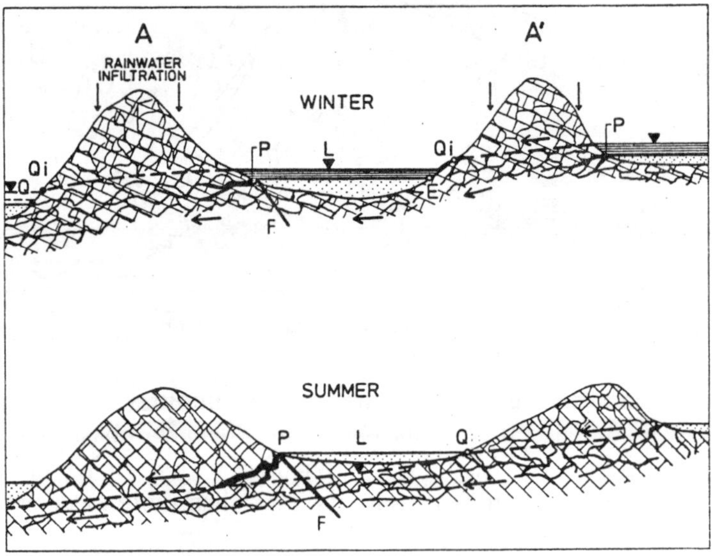

4–12. Fluctuation of the seasonal karst water level and its consequence for the polje landscape. Drawing much simplified. Q. spring; Qi. spring with intermittent outflow; P. ponor; E. estavelle; L Winter = karst water level in the flooded polje; L Summer = karst water level below the polje. Arrows = karst water flow. A–A' section in Fig. 4–10.

During the winter rains the level of the karst water body rises and floods the polje floors, to form extensive lakes (Fig. 4–11). After the rainy season, the karst water level drops and the polje floor becomes available for summer use as grazing and agricultural land (Fig. 4–12). The runoff follows shallow channels in the polje floor that lead to sinkholes (ponors) at the edges. The karst waterways along the northeast border of some poljes in the Dinaric karst can, during the winter rainy season, act as springs. During the dry summer season when runoff of the stored water in the poljes takes place, they may perform the opposite function as sinkholes (ponors). Karst phenomena with this double function are called *estavelles* (Fig. 4–11).

The use of tracers in the ponors has shown that the hydrological cycle is indeed due to a seasonal rise of the karst water level. Each lower-lying polje usually still represents the local corrosion (and erosion) level for the higher-lying polje of the staircase-like sequence that follows from the uplift phases (Fig. 4–10).

Paleokarst and Underground Water in Arid Regions

Examples of paleokarst are found in desert areas of the Old and New Worlds.

A *paleo-desert karst* in the eastern part of the Arabian Platform has been thoroughly studied. From a hydrological point of view, it is generally confined to the catchment area of the As Summan Plateau and the uplift zones east and southeast thereof and the large oasis areas of Al Hasa and Al Qatif near the coast of the Arabian Gulf. Here I will describe the As Sulb Plateau, the central part of the extensive As Summan Plateau (29°19′N to 26°41′N and 47°30′E).

The As Sulb Plateau is a real desert. It is not a flat plain but rather hilly, with ridges of Umm er Radhuma Limestone (Paleocene) partly overlain by Miocene clastics. The thick complex of limestone is highly karstified, but the karst phenomena usually lie below the surface which merely displays the openings of shafts and caves (*dahl*), which very frequently are almost imperceptible to the eye (Fig. 4–13). Numerous shafts are open to depths of more than fifty metres (FELBER et al., 1978, p. 167). Horizontal caves extend at various levels from these shafts. A remarkable feature of some of the caves, e.g. the Dahl Abu Marwan Cave southwest of Ma'aqala, is that the roof of their upper systems is no more than 20 cm thick.

It should be mentioned that the shaft openings (jama, aven) are predominantly found in flat, shallow troughs or the remnants of shallow valleys. Horizontal cave systems at the level of recent valley floors are rare.

Exploration of shafts and avens in the northern and eastern part of the As Sulb Plateau in search of calcareous tufa was unsuccessful but in Dahl Abu Marwan and nearby caves southwest of Ma'aqala, very well developed stalactites and stalagmites were found. Stalactites cemented into duricrust in the roof of Dahl Abu Marwan were most helpful in the interpretation of the morphogeny. Carbon-14 measurements and thin sections were used.

4–13. Openings of cave systems in the As Sulb Plateau (Photo by J. Zötl).

The large cave systems in the Umm er Radhuma Limestone must have developed during the Middle and Late Tertiary. Their formation would have required a humid climate. Embedding of stalactites inside the duricrust of the modern cave roofs indicates that there were climatic variations after the time when morphological changes in the land surface (relief inversion) were complete. The stalactites are beyond the limits of Carbon-14 age determination.

It is hydrologically significant that even now in the karstified As Sulb Plateau considerable amounts of water run off into the underground and enrich the aquifer when this desert experiences episodic and excessive precipitation at irregular intervals.[1]

[1] This article was written in 1985. At the present time, studies are carried out by the King Fahd University, Dhahran, Saudi Arabia, the Austrian Academy of Sciences, Vienna, and the University of Karlsruhe, F.R.G.

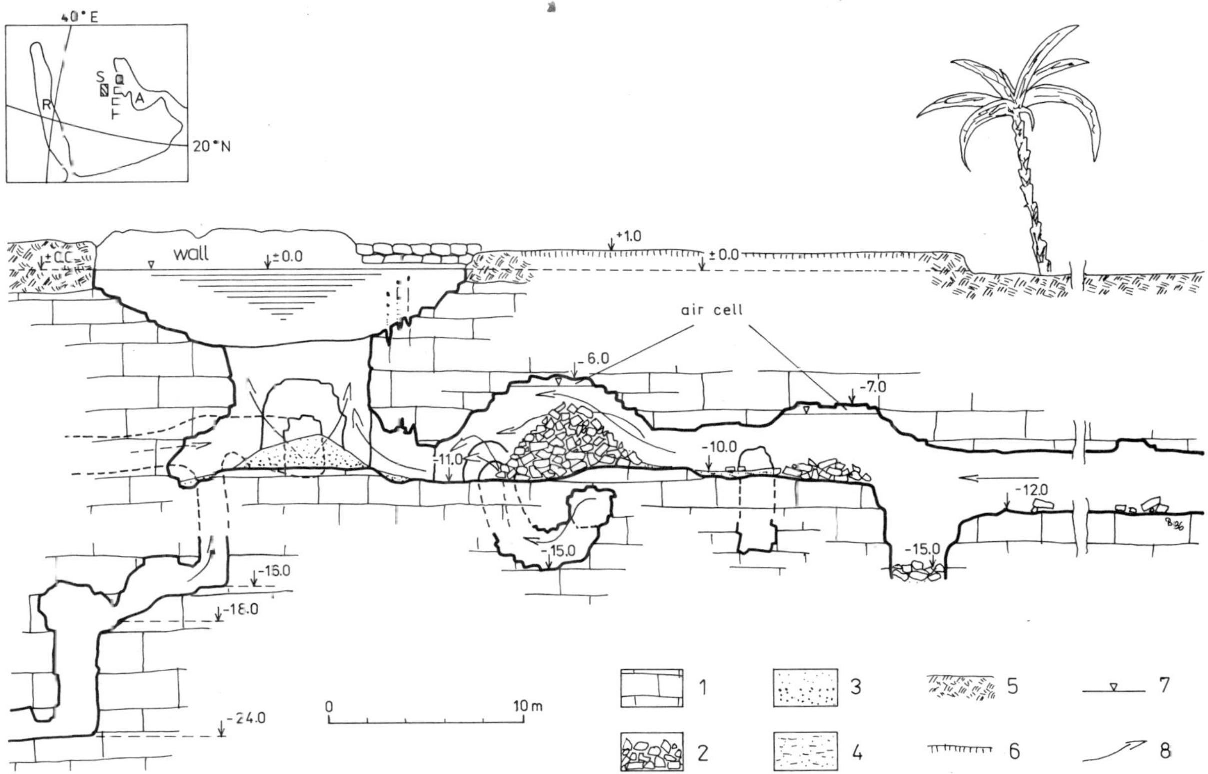

4–14. Cave system of the Ayn Khudud spring of Al Hasa oasis (orientation square S = As Sulb Plateau; H = Al Hasa oasis; Q = Qatif oasis; A = Arabian Gulf; R = Red Sea) (after a survey by the German diving group KITAG; reported by BRANDEMER and KISTENMACHER in AL-SAYARI and ZÖTL, 1978). 1. limestone and marl of the Dam Formation; 2. blocks from collapses; 3. sand and pebbles; 4. fine sand and silt; 5. land surface; 6. road; 7. water level; 8. flow direction.

The Umm er Radhuma Limestone dips toward the east. From a certain depth onwards, the pressure in the aquifer is so high that the water rises along faults and joints through denser covering layers (pressure-relief zone) and creates the basis for fertile oases with emerging springs (Al Hasa, Al Qatif). Using bore-holes, large amounts of water can be tapped artifically as artesian wells (King Faisal Project, Haradh).

The hydrogeological situation is best known at the Al Hasa oasis. UHLIG (1965) calculated the production of the 162 artesian wells as 12.4 $m^3 \cdot s^{-1}$. Here, the Umm er Radhuma Formation is at a depth of about 500 m. Chemical study of the waters shows them to be mixed waters from the Umm er Radhuma and from overlying carbonate rock formations. The Umm er Radhuma waters, however, predominate.

The upper channels of the rising waters are also paleokarst features, as is shown in the waterways for the Ayn Khudud spring measured by KITAG and shown in Fig. 4–14. The caves surveyed are developed in limestones of the Dam Formation and show a stair-like ascent following vertical joint systems and horizontal joints. The waters currently discharging are *old waters* without any admixture of recent precipitation, as is shown by the measurement of deuterium and oxygen-18.

The same situation is found for waters from the wells in the Al Qatif oasis and at the bore-holes of the King Faisal Project in Haradh.

Deep Karst Aquifers Below Tertiary Sedimentary Basins

Tectonic subsidence can cause carbonate mountains that in former geological times were high and karstified, to sink to great depths. The subsidence is seldom a downwarp; usualy it occurrs by means of fault steps of varying heights: fault throws of 1 000 m have been shown by boreholes (Fig. 4–15).

A precise demonstration of paleokarst in deep lying blocks is only possible with bore-holes. An example is in the molasse basin between the Alps and the Bohemian Massif. A large number of oil wells were sunk through the oil- and gas-bearing Tertiary basin into underlying Upper Jurassic limestones or even into the crystalline basement. Occasional large mud losses could be explained only by jointing in the limestones without significant karstification. That karstification was indeed present was shown in bore-holes where total mud loss and irrecoverable bit loss in large cavities occurred (at depths of some 2 100 to 2 200 m, among others). Associated faulting could be seen in younger, overlying Cretaceous sandstone. These large cavities can only be paleokarst. It is hard to estimate the extent to which such old karst cavities are being altered by modern (very slow) water flow. With the strong mineralization and the absence of free CO_2 it is hard to imagine there is further dissolution to any significant extent.

As far as water movement in deep karst is concerned, we must take into consideration that in the Tertiary basin impermeable clay layers 100 m and more

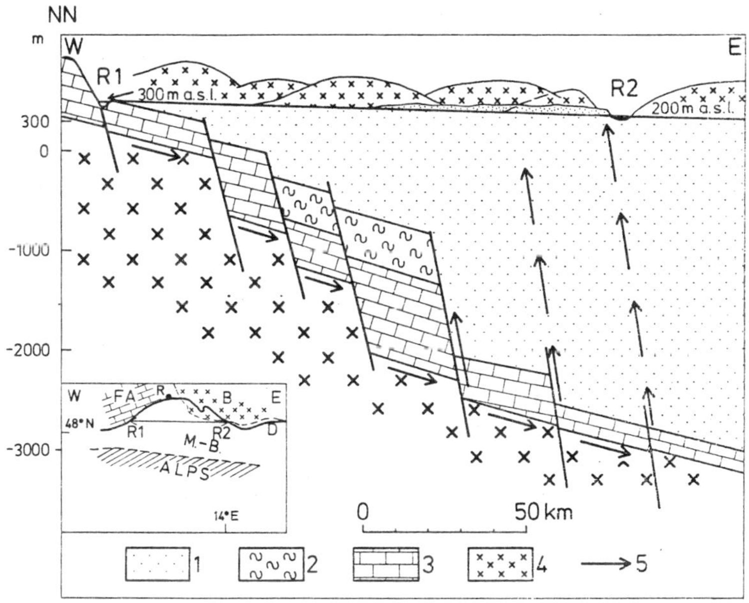

4–15. Deep karst aquifers in paleokarst of deep blocks of Jurassic limestone in the underground of the Tertiary Molasse Basin between the Alps and the Bohemian Massif (orientation square FA = Franconian Alb; R = Regensburg; D = Danube River; B = Bohemian Massif; M.-B. = Molasse Basin). R 1. infiltration area; R 2. base level of erosion for the ascending karst water; 1. Tertiary sediments of the Molasse Basin; 2. Cretaceous sandstone; 3. Jurassic limestone; 4. crystalline basement; 5. flow direction of karst water.

in thickness insulate the karst aquifer from the top. In the cross-section in Figure 4–15, we see that for 50 km from hydrostatic level R_1 there is a highly confined karst aquifer without pressure relief because the faults do not cross the Tertiary basin filling.

Hydrostatic pressure increases toward the east, where faulting also involved the lower (older) Tertiary basin filling. The buoyancy ultimately also penetrates the less permeable, higher sequences and reaches the base level of erosion (Fig. 4–15, R_2) which in the schematic cross-section given lies 100 m lower than R_1. Here, then, we are in the relief zone of the system.

Under the conditions sketched here, the covering of the karst aquifer by thick, visco-plastic aquitard strata in the filling excludes any deleterious effect of the exploitation and use of the deep groundwaters at the Earth's surface. Experience shows, however, that long-term large removals lead to a decrease in hydrostatic pressure. For this reason, re-injection is suggested in cases of large-scale removal (as, for example, the use of the geothermal energy of the deep waters).

Summary and Conclusions

Measurements made in deep, covered paleokarst have shown that even there water movement is currently taking place. To make this possible, there must also be a connection to levels where the ascending deep groundwater from the paleokarst is discharged. This means a connection to the lowest base level of erosion at the surface of the Earth. The question of the influence of the base level of erosion on karstification in general (as was emphasized in the Introduction) thus becomes an important hydrological factor for paleokarst in particular.

In conclusion, therefore, we shall again use the simplest examples to show the relationships of underground karstification to the local base level of erosion (corrosion). The most favorable conditions for this development are found in karst massifs of bedded limestones with canyons acting as surface drainage channels for the underground water.

A very good example is given in the description of the genetic relationship between caves and landforms in the Mammoth Cave National Park, Kentucky, U.S.A., by MIOTKE and PALMER (1972). The genesis of the cave system within the National Park is directly related to Tertiary and Pleistocene erosion and accumulation phases in the Ohio Valley. Pronounced cave levels were formed when the incising Green River served for a long time as the static erosion base for the formation of the caves. The main cave levels thus correspond with the river terraces in the nearby Green River Valley. The burial of the Teays River Valley to the north by the moraines of Nebraskan Ice Age (or Danube Ice Age?) added large catchment areas of the buried Teays River to the newly forming Ohio Valley. Cave levels corresponding to river terraces have been formed during the long Yarmouthian Interglacial at heights of about 500–550 feet (150–165 m).

The geological structure and variation in the rock influence only the position and form of the cave passages but not, however, the elevation of the main cave levels. Chemical studies showed that the waters reached the phreatic zone of the cave system unsaturated. Only near the erosion base (corrosion base), where the water has the longest period of time at its disposal, does the solvent activity cause a pronounced expansion of the cave passages by lateral corrosion and erosion of material by collapses.

A transition from a recent, direct connection of the local base level of erosion to a *hanging corrosion level* occurs in the Cetina Valley of Yugoslavia. Headward erosion had exceeded rates of underground corrosion. The graded section of the canyon floor thus is further upriver, while the karst water outlets even below the erosion step are still located at an old, higher level of discharge. The corrosion and erosion caused by the *hanging outlets,* however, has relocated them downward and adjusted them to the base level of erosion of the river bed. This is causing an emptying of the currently active cave system and its transformation into paleokarst; however, in the modern phase it still serves as an overflow for high water (see Fig. 84 in ZÖTL, 1974, p. 189).

Thus we come to the essential hydrological factors for the definition of paleokarst hydrogeology.

Underground karst waterways are to be designated as paleokarst when they developed under other geological and/or climatic conditions. Among the geological conditions, shifts and changes in the relationship to the base level of erosion are especially important. The hydrological effects are varied (position of the paleokarst waterways both in the present vadose and phreatic zones, different hydrostatic influences, etc.). Paleokarst resulting from climatic changes involves forms that either could not have developed to their dimensions under current climatic conditions (caves in arid areas) or that can be clearly seen to be remnants of old forms (monsoon region, tropical karst) e.g. via flow marks in places through which water presently does not constantly flow. If recent underground waters follow or cross old *fossil* karst cavities, the former can alter the latter, although the paleokarst is nonetheless to be seen as being original. Recent and modern hydrological influences can be virtually insignificant, or they can lead to a total change in the form of the paleokarst, destroying it.

For the practical exploitation of karst water it is to be borne in mind above all that in many cases modern water flows through paleokarst passages. In the Northern Limestone Alps of Central Europe, the rapid flow of precipitation through large underground cavities results in very slight total mineralization (100 200 ppm). The poor filter capacity is a question of the protection of the recharge area. If protective measures are adequate, the water is of good drinking quality.

In deep paleokarst the water often has an extremely long residence time and correspondingly high mineralization (1 000–3 000 ppm). There is a minor use of this water for balneological purposes. Most recently (and increasingly) the high water temperature is being exploited as geothermal power for remote heating and in agriculture (greenhouses, drying plants, etc. in Hungary, France, Austria and other countries).

HYDROGEOLOGICAL PROBLEMS OF OPENCAST AND UNDERGROUND MINING OF MINERAL DEPOSITS ENCOUNTERED DURING THEIR EXPLORATION, DEVELOPMENT AND EXPLOITATION STAGES

Exploitation of mineral deposits in karst areas is often very complicated. Even in terrains with only minor carbonate intercalations and lenses in the vicinity of the deposit, mining may be affected by karst and paleokarst features. The yield of karst fissure aquifers in carbonate rocks is very high in places and the hydrogeological conditions complicated and delicate. Mining operations disturb the natural karst hydrogeology and existing equilibria. Heavy water inrushes, continuous and even higher water inflow, transport of loose, water-saturated sediments into mines, inflow of surface waters and precipitation, water inrushes from shallower karst conduits, and the water saturation of mined substrate or instability of slopes in opencast mines represent the main problems for the mining companies. Moreover, extensive dewatering of mines with the loss of natural water supplies and the utilization of mine waters, increases the costs of mining. This affects the economics of the mining industry and its ability to compete in international trade.

Two case studies in the field of mining and prospecting hydrogeology are presented here, with data obtained during long-duration hydrogeological investigations of mineral deposits that were maintained during their exploration, development, and exploitation. In addition there is a discussion of the consequences of mining on the natural environment. Opencast and underground bauxite and coal mining in Hungary and the underground exploitation of Zn-Pb ores in Poland serve as good illustrations.

Pavel Bosák

P. Bosák, D. C. Ford, J. Głazek and I. Horáček (Editors):
Paleokarst. A Systematic and Regional Review. – Elsevier and Academia. Amsterdam and Praha. 1989.

HYDROGEOLOGICAL PROBLEMS OF THE CRACOW-SILESIA Zn-Pb ORE DEPOSITS

Zbigniew Wilk

Location and Geological Setting

The Cracow-Silesia Zn-Pb ore district occupies more than 1 000 km^2, the Zn deposits being among the largest known in Europe. Within this region are smaller units, the deposit areas themselves, which differ slightly from each other in geological structure and hydrogeological conditions. They have been exploited so far in the neighbourhood of Bytom, Tarnowskie Góry, Siewierz, Sławków, Chrzanów and Olkusz (Fig. 4–16). At present only four mines are in operation, including one in the region of Chrzanów. In the remaining regions the ore resources have already been exhausted, but mining is still expanding in the region of Olkusz.

The ore deposits are in Triassic carbonate rocks at the northern boundary of the Upper Silesia Coal Basin, within the Cracow-Silesia monocline. Most of the known ore resources occur at a depth not greater than 300 m. The Triassic deposits form a part of the Mesozoic cover of a platform type that lies discordantly on different structural units of the Paleozoic basement which were consolidated during the Variscan Orogeny.

The deposits lie almost flat with a slight inclination towards the North. The tectonics of the Triassic rocks are the result of various phases of the Alpine Orogeny. The dominant elements are faults, most of which are characterized by small amplitudes of throw (up to a dozen or so metres). Only a few of these dislocations, usually composed of a group of step faults, attain a total throw of some tens of metres or more (WILK et al., 1977). These displacements cut the Triassic cover into numerous graben and horsts.

As a result of erosion most of the younger deposits originally covering the Triassic have been removed. This is true also of the Triassic to some degree; its

P. Bosák, D. C. Ford, J. Głazek and I. Horáček (Editors):
Paleokarst. A Systematic and Regional Review. – Elsevier and Academia. Amsterdam and Praha. 1989.

present boundary (Fig. 4–16) is also the result of erosion. The surface of the Triassic sediments is cut by a network of preglacial or glacial buried valleys, which are some tens of metres deep and filled with fluvioglacial sands.

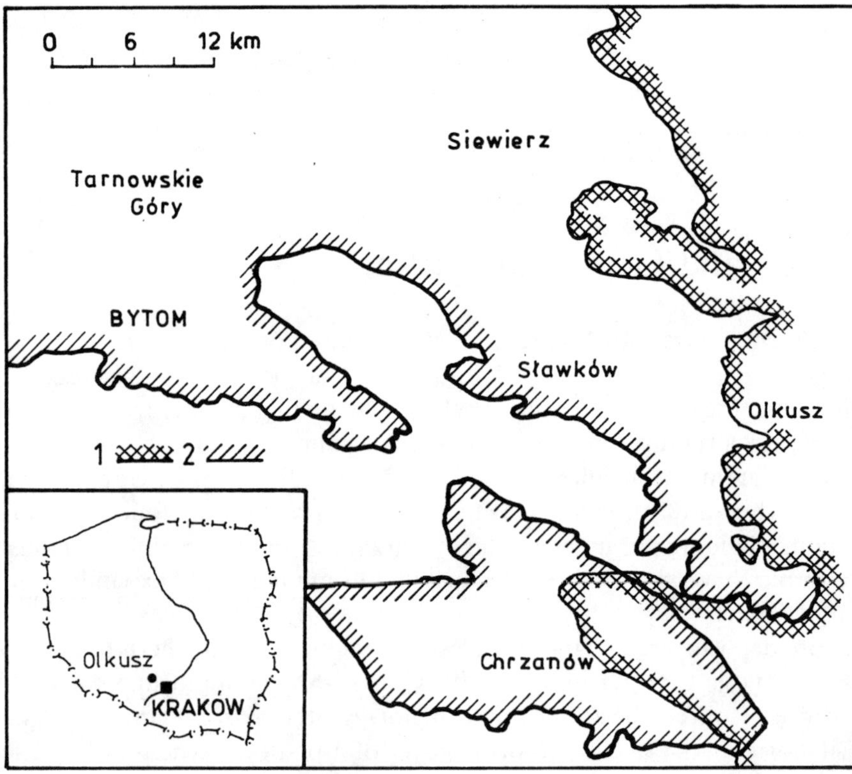

4–16. Map showing the location of main centres of Zn-Pb ore mining in the Cracow-Silesia district. 1. extent of the Jurassic; 2. extent of the Triassic.

Groundwater Problems of Zn-Pb Ore Mining in the Past

Zinc and lead ore mining in the Cracow-Silesia district has a long and eventful history, considerably older than the oldest records which come from the 12th and 13th centuries.

Within the 800 year-old history of mining in this region, there occurred periods of development, stagnation or total decline. The main reason for these changes up to the end of the 18th century was problems connected with excessive and sometimes unexpected quantities of groundwater encountered in the mine-workings. These difficulties occurred because the Triassic carbonate host-rocks are an excellent aquifer of the karst-fissure type.

To dewater mine workings situated below the groundwater table, at first simple water lifting devices, water-raising wheels and others driven by manpower or by animals, were used. The period from the middle of the 16th century up to the end of the 18th century may be called the *drainage-adit period,* as the development of ore mining in that period was associated with the construction of long drain tunnels dewatering by gravity those portions of the deposits which were situated above the local base of erosion. For example, in the 16th century, five drain adits with a total length of 16 km were driven in the Olkusz region. For those days it was an outstanding technical and organizational achievement.

As the depth of the mine workings increased with time, water inflow also increased. Control of the inflow became more and more difficult and expensive. Flooding frequently made mining unprofitable and so brought it to a halt. A vital breakthrough took place in the 19th century when steam engines were introduced to drive the pumps. It is worth mentioning that the first steam engine to be used in mining on the continent of Europe was in the ore mine Fryderyk in Tarnowskie Góry in the year 1788 (MOLENDA, 1972).

The introduction of high efficiency, multi-stage centrifugal electric pumps, special methods of sinking shafts like freezing, cementation, grouting, etc., advance drainage of deposits by means of submersible pumps, etc. enabled deep ore deposits to be developed, even those characterized by very difficult hydrogeological conditions. In spite of all these improvements, hydrogeological problems continue to be of great importance in ore mining up to the present day.

General Characteristics of the Recent Hydrogeological Problems in the Zn-Pb Ore Mining

The problems can be divided into two groups:

(1) problems connected with the influence of groundwaters on the mining operations, their methods of execution, their progress, safety and the economic results to the mining company,

(2) problems connected with the effect of mining operations on the environment and with all the aspects of water conditions in the surroundings of a mine.

These groups of problems are encountered at various stages in the prospecting, exploration, developments and exploitation of deposits. Depending on local conditions, one or other group of problems acquires greater importance, and different aspects may predominate within each of them. Recently the second group of problems has become more significant. At the stages of prospecting and exploration, hydrogeological problems consist of establishing the scope of the hydrogeological investigation, the application of appropriate methods and their realization so as to obtain — keeping the cost as small as possible — the information necessary to enable appropriate designing, development and exploitation of the deposit.

The principal methods of hydrogeological investigation in Zn-Pb ore mining are surface geological survey, hydrogeological observations in the course of drilling prospecting holes, special hydrogeological exploratory holes in which pumping tests are carried out, long-term stationary observations in especially constructed piezometric wells, hydrochemical, isotopic and geophysical investigations.

At these stages the following should be determined in particular:

(1) the horizontal extent and structure of the aquifers, their thickness, physical properties, hydrogeological parameters, catchment and drainage areas, the character of movement of the groundwater and its hydraulic connection with surface waters;

(2) the origin and quality of water and the rate of water inflow into possible excavations (including identiffication of the static and dynamic water resources) during the main development stages of the future mine;

(3) possible changes in the natural groundwater that might occur in the course of exploitation of the deposit, and the effect of these changes on the interrelationship of surface and groundwater.

While the deposits are being made accessible and developed, the basic problems are those that result from cutting across fault zones, caverns, etc. encountered en route of the main horizontal headings. Before the shaft is sunk a special exploration hole is drilled along its axis or just beside it. The hole is used to carry out specific hydrogeological investigations on the basis of which, for example, the probable volume of the inflow can be assessed. If the anticipated inflow is rather large (over 0.5 m^3/min), special methods of sinking shafts such as freezing, artificial decrease of the groundwater table, cementation, etc. are used. The possibility of applying any of these methods must be confirmed by investigating the direction and flow rate of the water, its chemical composition and temperature.

The greatest trouble with water is usually met when the first main horizontal headings are being driven. First of all water galleries of high capacity close to the shaft must be excavated in advance and efficient pumps and pipe-lines installed. The pump stations should be protected with water proof safety dams so that they can be cut off from the rest of the mine in case of emergency and the pumps can continue operation in case a momentary water inflow is greater than they can handle. The main headings are driven as double headings, one underneath which acts as a water gallery and an upper one in which the miners can continue their work. At the face of the gallery, exploration drillings are carried out to discover early enough the presence of caverns, faults and the groundwater reservoir.

The main headings in the Cracow-Silesia Zn-Pb ore district are driven below the level of the deposits. For this reason the mining itself is carried out in rocks already drained and within the range of a cone of depression created earlier. Water inflow to the mine in this situation is usually diminished.

The hydrogeological problems at this stage arise because the mining is being carried out at the relatively shallow depth of 100 to 200 m. Under these conditions in a karstified rock massif the effects of exploitation are frequently considerable. There is also the danger that the thin, aquiclude or isolating layers of Keuper clays which occur in the overburden may burst open so that water from the Quaternary formations or nearby rivers may flood into the mine.

Another problem is the need to check the extent of the cone of depression. In wells within the range of its influence, water usually disappear and natural spring vanish, since it is necessary to ascertain the area within which the mine must repair damage done to the local inhabitants or industrial companies as a result of mining operations.

This is related to the very important matter of utilizing mine waters. Since these Zn-Pb ore mines are located within an area of water shortage, and the water pumped out the Triassic formations is of high quality, it has been used for a very long time to supply the population and local industry with water (MICHAEL, 1912). At present the problem of protecting this water against contamination as well as ensuring its maximum utilization has become extremely important.

In the period following the Second World War, prospecting and exploration work was carried out in the eastern part of the Cracow-Silesia district where the Zn-Pb ore deposits occur in the Olkusz area. As a result, new deposits were discovered in the vicinity of known ones and two new mines built, the Olkusz and the Pomorzany Mines. Taking this area as an example we shall try to characterize the wide range of hydrogeological problems occurring at particular stages in the development of the new deposits.

Hydrogeological Problems of the Zn-Pb Ore Deposits in the Olkusz Region

Hydrogeological Characteristics of the Zn-Pb Ore Deposits

The ore deposit of the Olkusz area is located in the eastern outcrop of the Cracow-Silesia monocline, dipping NE and composed of Triassic, Jurassic and Cretaceous strata. In the area considered here, the Cretaceous, Jurassic and part of the Keuper cover has been eroded. The top of the Triassic carbonates has a distinct relief, mostly buried under fluvioglacial sands and gravels (Fig. 4–17).

The Zn-Pb mineralization occurs mainly within Lower Muschelkalk strata and is connected chiefly with Ore–bearing Dolomite. It is located somewhere in the middle of a lime-dolomite formation about 110 m thick. Though the par-

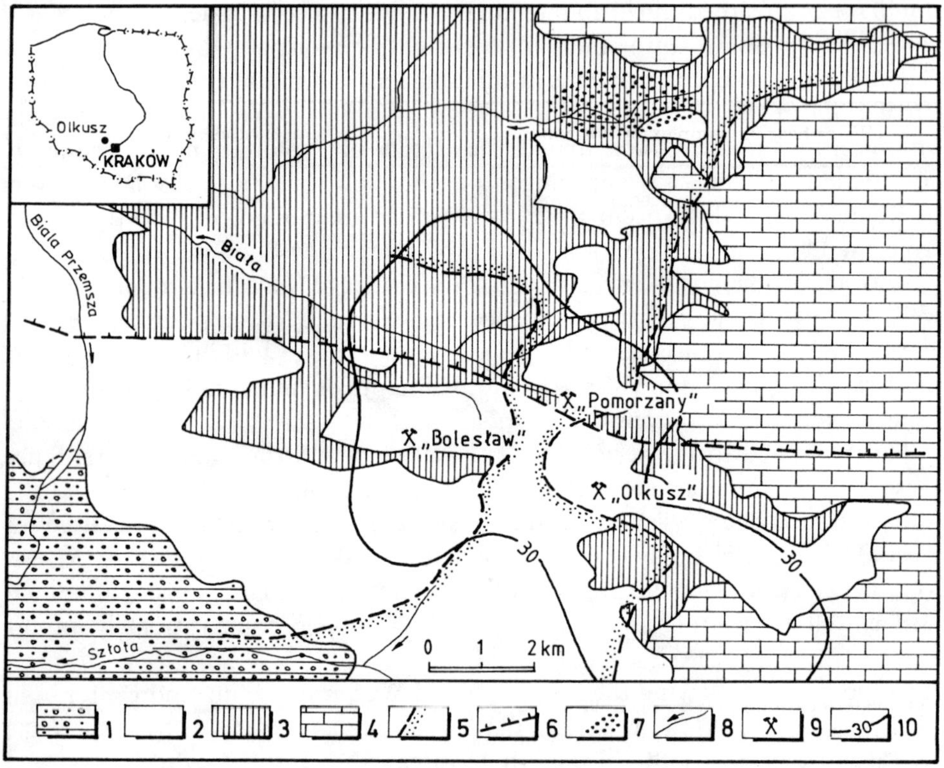

4–17. Hydrogeologic sketch of the Olkusz Zn-Pb ore mining region. 1. low permeable or impervious Upper Carboniferous, Permian and Lower Triassic (Bunter Sandstone) formations; 2. Triassic water-bearing horizon; 3. impermeable Upper Triassic (Keuper); 4. Jurassic water-bearing horizon; 5. schematic contours of preglacial valley filled with sand of significant thickness; 6. Pomorzany fault zone (schematic) separating the northern part of the Olkusz region from the southern one; 7. fluvioglacial sand polluted with lignosulphonate compounds; 8. surface streams; 9. ore mine; 10. isoline indicating the artificial lowering of the groundwater table or head by 30 m.

ticular units of the formation display different hydrogeological features the formation as a whole is a very permeable and abundant aquifer of the cavernous-fissure type. Because of the local lack of watertight clayey Keuper sediments, the Triassic aquifer had under natural conditions — before the deposit was developed — a partly confined and partly unconfined character (WILK and ZIMNY, 1976). Table 4–1 shows the vertical succession of aquifers and aquicludes.

On the basis of 240 test pumpings it has been found that the filtration coefficient (k) of the Triassic water-bearing strata exhibits a wide range of values, and its frequency distribution has a lognormal character. The most frequent value is $6.5 \cdot 10^{-5}\,m \cdot s^{-1}$ (MOTYKA and WILK, 1976). The coefficients of storage (S) also differ much from each other and their representative values for

Table 4–1

Succession of water-bearing horizons and impervious strata in the Olkusz Region (after MOTYKA and WILK, 1976)

Water-bearing complex or horizon	Impervious series
Quaternary water-bearing complex: sands and fluvioglacial gravels. Recent fluvial deposits	
	Residual clays, slope deposits, washed boulder clays, stagnant water series
Upper Jurassic water-bearing complex: massive and layered limestones	
	Marly calcareous deposits of Lower Oxfordian and Callovian, of Upper Triassic (Keuper) and, possibly, of Upper Muschelkalk
Triassic water-bearing complex – Muschelkalk water-bearing horizon: fissured, fractured and karstified *Diplopora* and Ore-bearing Dolomite and limestones	
	Marly intercalations in limestones of the Gogolin Formation
Triassic water-bearing complex – Roethian water-bearing horizon: fractured and cavernous limestones and dolostones	
	Lower Roethian and Middle Bunter Sandstone clays
Triassic water-bearing complex: Lower Bunter Sandstone horizon: sands and sandstones	
	Permian conglomerates, tuffs and clays
Paleozoic water-bearing complex: fractured Lower Carboniferous and Devonian limestones and dolostones	

unconfined conditions are $S = 9 \cdot 10^{-3}$ (WILK and MOTYKA, 1980). The coefficient of fracture porosity (n_F) varies from 0.02 % to 0.7 % whereas the areal frequency of fracture is from 0.2 to 14.7 m^{-1} (WILK, MOTYKA and JÓZEFKO, 1984). The great majority of fractures have widths less than 0.1 mm. Cross-sectional areas of observed karst passages and other openings display a wide range of values, from 0.002 to 3 m^2, although larger caves also exist (WILK, MOTYKA and JÓZEFKO, 1984).

The microporosity of the carbonate rocks, determined by using small samples, was 0.002 to 0.34, and the permeability was 0.000 4 to 638.9 mD (WILK et al., 1985).

The Triassic karst-fissured rocks represent a special type of groundwater collector. This special character is mainly due to the structure of rocks, within which four basic hydraulic systems can be distinguished, namely pores, fissures, open karst forms and secondary-filled karst solution openings (MOTYKA and WILK, 1984). These systems form the internal structure of the karst-fissured rocks and differ from each other mainly with respect to the degree of hydraulic resistance. Karst channel systems characterized by the least hydraulic resistance are the preferred routes of groundwater circulation. Groundwater circulation

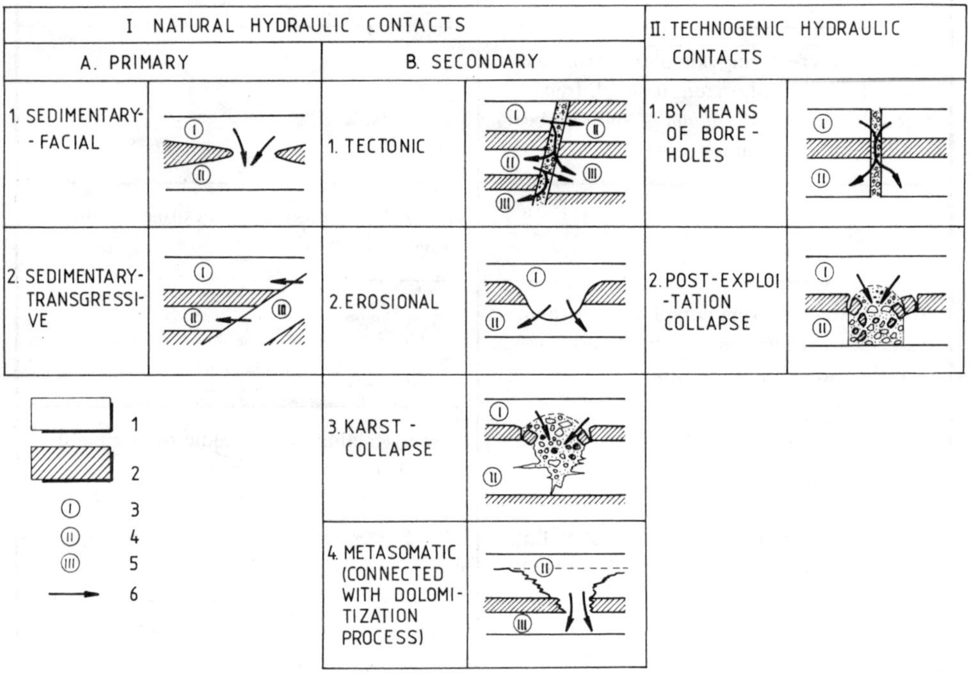

4–18. Different types of hydraulic connections between particular water-bearing horizons in the Olkusz region. 1. water-bearing horizons (aquifers); 2. impermeable series (aquicludes); 3. Quaternary water-bearing horizon; 4. Triassic water-bearing horizon; 5. Devonian water-bearing horizon; 6. direction of groundwater flow.

within karst-fissured rocks depends also on their macroheterogeneity, caused by the spatial arrangement of strata and by local as well as regional hydrogeologic conditions. The main element of the external hydraulic structure of the Triassic aquifer is the numerous disjunctive dislocations dividing the rock formations into separate hydrogeological units. Of great importance, too, is the recharge and/or the discharge of the Triassic aquifer by other water-bearing horizons via zones of hydraulic contact and by surface streams (WILK and MOTYKA, 1977). These contacts and other factors indispensable to their existence can be systematized in the way shown in Figure 4–18.

The Olkusz Zn-Pb area can be divided into two parts as far as its hydrogeology is concerned. The substantial difference between them is that in the eastern part there is a Jurassic aquifer complex. A distinctive tectonic feature is a disjunctive dislocation, the Pomorzany fault (WILK et al., 1977) and a tectonic horst separating the ore deposit into two sectors, north and south. In the southern sector the Bolesław Mine has been in operation for quite a long time and in 1957 a new mine, Olkusz, was opened. The fault and a graben separate the southern part (drained through mining) from the northern part in which natural hydrogeological conditions prevailed until 1972. They remained unchanged until the construction of a new mine, Pomorzany, in the north.

The main surface stream in this area is the Biała Przemsza River. It cuts the northern and western parts of the Olkusz region and part of it flows over the impermeable Keuper clays and part directly over outcrops of the Triassic aquifer.

Hydrogeological Problems During Exploration

Since the shallow seated deposits known for a long time in the Olkusz region had already been mined out, new prospecting work was started there in 1952. It eventually led to the discovery of deeper deposits on both sides of the Pomorzany fault. The deposits were subsequently carefully surveyed and documented and the hydrogeology was also investigated. These investigations presented difficulties because of the fissured character of the main Triassic water-bearing horizon as mentioned before, its hydraulic inhomogeneity and anisotropy, intensive faulting and the numerous hydraulic contacts between particular water-bearing horizons.

Surveying at that stage included hydrogeological mapping of an area of approximately 500 km^2 at the scale 1 : 25 000. In the course of it an inventory of about 600 domestic water wells, some industrial and municipal water supply plants and approximately 70 springs was made. In about 15 % of the wells, systematic long-term stationary observations were carried out. Hydrochemical and thermal investigations were made and the flow in surface streams measured.

Besides the hydrogeological survey, the main sources of information were the boreholes, about 1 700 in number. On account of their character and the range of information obtained from them they can be divided into: (1) nonflowing, prospecting boreholes, (2) flowing (artesian) prospecting boreholes, (3) hydrogeologic-prospecting boreholes, (4) hydrogeologic-exploratory boreholes and (5) special, large-dimension boreholes. The holes mentioned under (4) and (5) (in which over 150 pumping tests were carried out) were accompanied by observation wells. Special holes of wide dimensions were made where the mine shafts were to be located. In about 15 % of the holes piezometers for stationary observations were installed.

The information obtained enabled a preliminary prognosis of water inflow to the mine, established the extent of necessary drainage and its effect on the other water-bearing horizons, the quality of the waters and any other hydrogeological phenomena which might appear during the development and mining of the ore deposit.

Hydrogeological Problems During Development

On the basis of the geological and hydrogeological investigations plans were developed for the contruction of the Olkusz Mine, built in thc years 1959–1967, and the Pomorzany Mine, built in 1972–1974. Here we deal only with the latter.

From the prognosis, the static water resources in the Pomorzany ore deposit (i.e. the amount of water which must be pumped out before a cone of depression would form around the entire mine) were 300 million m^3. The maximum inflow to particular headings was anticipated to be about 40 to 80 m^3 per min., with sudden heavy inflows of water.

With this in mind, it was planned to excavate the main underground workings (i.e. shafts and drifts) under good conditions in dry rocks. For this reason, they were located in Permian sediments of low permeability on the uplifted southern side of the Pomorzany fault, which was the southern boundary of the deposit (Fig. 4–19 B). The first shafts to be sunk were the Chrobry and Dąbrówka. The former was sunk in the vicinity of the Olkusz Mine already in operation, which ensured that the new shaft would be sunk in those Triassic rocks already drained by the Olkusz Mine. The Dąbrówka shaft was sunk applying the freezing method. The deeper sections of both shafts were in the Permian sediments. They were connected by a main haulage drift, running parallel to the plane of the main fault on its southern side in the Permian rocks (Fig. 4–19 A). Such a location for the main drift allowed the development of an operational front broad enough to deal with the main fault and the flooded Triassic rocks behind it. At the same time, and parallel to the main haulage drift, a main water level drift was driven. The Permian sediments were, on the whole, dry with only small occasional inflow. All the time, however, there was danger of sudden flooding from the Triassic horizon, the base of which was close to the roof of the main drift (Fig. 4–19 B).

From the main drift access cross-cuts were driven every 500 m and perpendicular to the plane of the Pomorzany fault, along with parallel drain headings

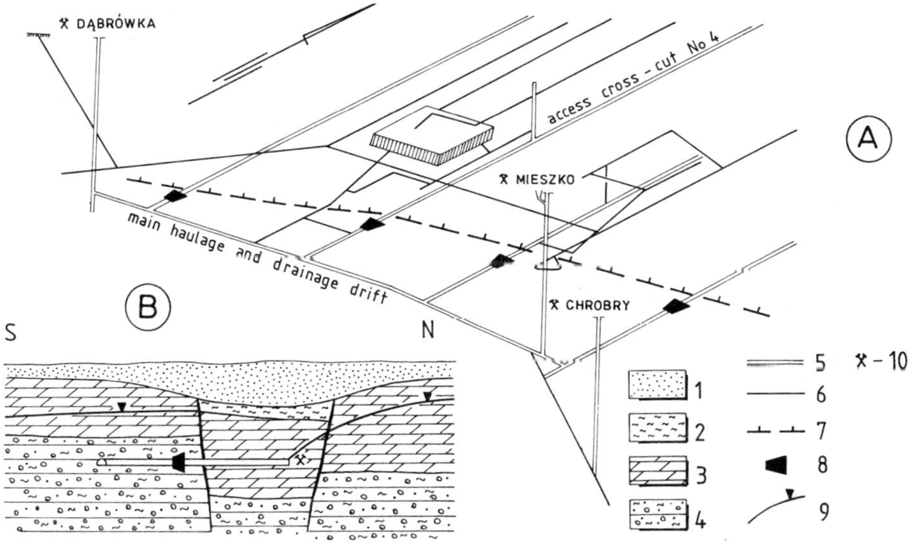

4–19. Spatial scheme of the Pomorzany ore mine (A) and schematic cross-section along the access cross-out No. 4 (B). 1. Quaternary sands; 2. Upper Triassic (Keuper) clays; 3. Middle Triassic (Muschelkalk) carbonate water-bearing horizon; 4. Permian, very low permeable conglomerates and impervious clays; 5. main access, haulage and drainage mine workings driven mostly in water-saturated rocks; 6. other mine workings driven mostly within the depression cone; 7. Pomorzany fault zone; 8. safety dams; 9. groundwater table; 10. shaft (in Fig. 4–19A) or cross-cut face (in Fig. 4–19B).

underneath them. The ore deposit behind the fault is above these cross-cuts.

The purpose of these access cross-cuts was the quick dewatering of the deposit behind the fault and the creation of several mutually supporting drainage centres. The critical moment in the development of the ore deposit came when crossing the fault and tapping the Triassic aquifer behind it. While the dislocation was being crossed it was observed to be a longitudinal, uniformly inclined, homotetic dip-slip normal fault of the brittle type (WILK et al., 1977). The fault zone was about 70 m in width and its structure was very complicated. The main fault system was accompanied by an antithetic parallel fault heading and throwing-down to the south. Together these created a local graben about 200 m wide. After the main fault had been crossed, several other marginal joints, fissures and shifting of strata were observed. The Lower Muschelkalk limestones on the downthrow side were extensively karstified, the diameter of some cavities reaching a few metres.

The main fault fracture was dry and negligible groundwater discharge occurred while the cross-cut through the disturbed brittle zone was being driven. Only after the cross-cut faces reached the karstified limestones did enormous discharges and inrushes appear.

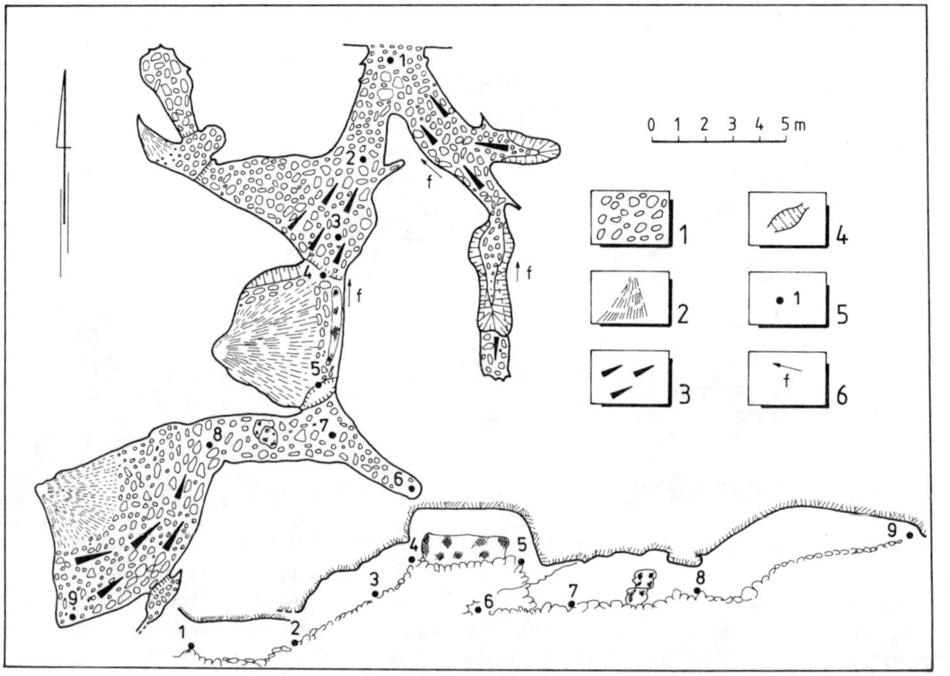

4–20. Map and cross-section of a cave encountered in the Pomorzany Mine (after WILK, MOTYKA and NIEWDANA, 1982; cave mapped by A. KOZIK). 1. blocks of carbonate rocks; 2. heaped fans of carbonate material; 3. slope direction of the cave floor; 4. threshold; 5. orientation points marking the cross-section line; 6. proved traces of groundwater movement; arrow = direction of water flow.

Since such a situation had been anticipated from the hydrogeological investigations different methods had been designed to keep the water under control. In all five access cross-cuts and in their accompanying drainage galleries, waterproof safety dams made of steel and concrete were built before the cross-cut reached the zone of the fault (Fig. 4–19 A). A network of water galleries of a capacity of approximately 30 000 m^3 was built near the shafts, and pumps were installed which could pump 420 m^3 per min. at maximum.

When all five access cross-cuts had crossed the fault, water discharges began to increase rapidly. They were accompanied by numerous sudden inrushes coming from water-filled karst features. One such cavern is shown in Figure 4–20. The yields of the successive individual flows were as follows: 26, 22, 60, 58, 30 and 50 m^3 per min. (Figs. 4–23 and 4–24). This represented from 20 % to

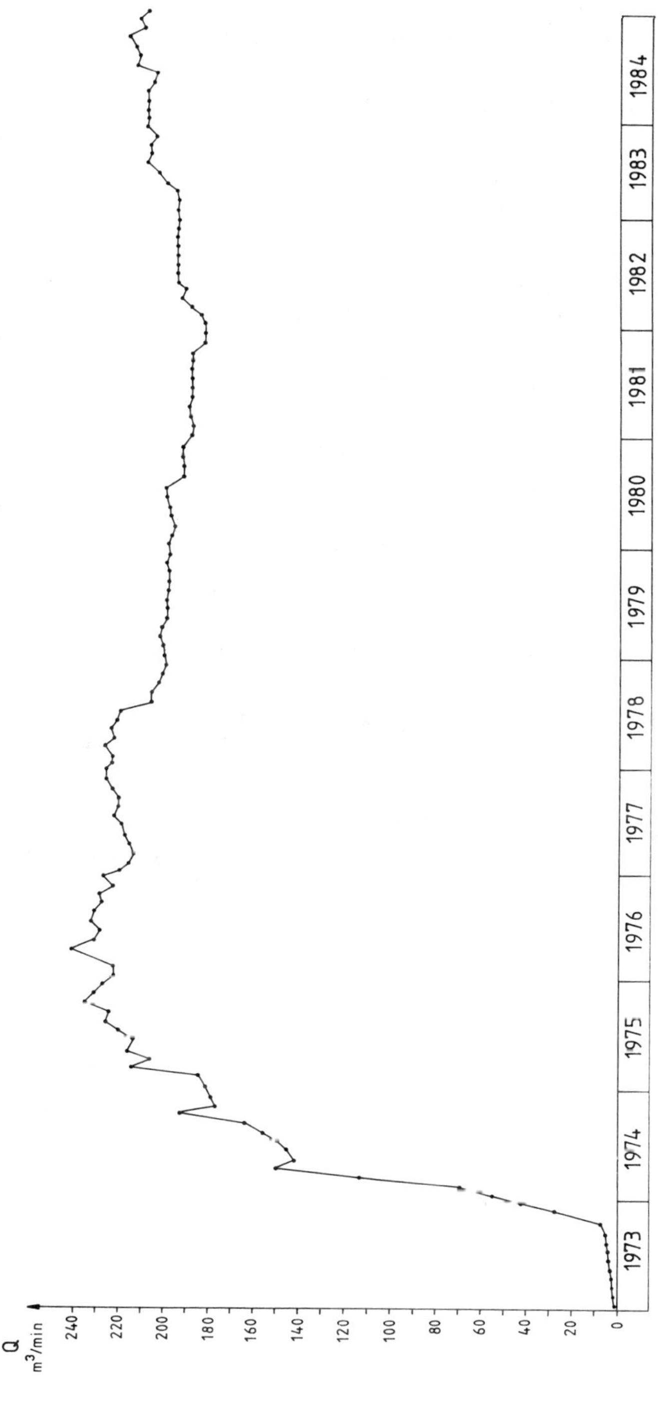

4–21. Total water inflow to the Pomorzany Mine.

over 80 % of the total water flow into the mine. The overall inflow to the mine increased correspondingly. The heaviest local outflow took place on March 6th, 1975 and was equal to about 80 m^3 per min. At that time the greatest aggregate inflow to the mine was recorded, amounting to 258 m^3 per min. (GODYŃ, KLUCZEWSKI and NIEWDANA, 1976 and Fig. 4–21).

Shortly after the Triassic aquifer had been reached by the access cross-cuts, certain regularities in the appearance of the local inflows were ascertained, which made it easier to control them. The heaviest flows, of the order of tens of cubic metres per minute, were always encountered at the boundary of the Ore-bearing Dolomite and the Gogolin Formation. Taking advantage of this fact, the sequence and advance of particular drifts and headings was arranged in such a way as to dewater the deposit in the shortest time possible, without exceeding the efficiency of the pumping devices and while still retaining the necessary reserves.

This aim was achieved. On no occasion did it prove necessary to shut the water-tight safety dams in the course of development of the Olkusz Mine (WILK, MOTYKA and NIEWDANA, 1971).

To avoid any surprises, safety measures were taken according to regulations which had been drawn up to allow for the highest category of water hazard. They included short advance boreholes and long exploration boreholes drilled perpendicular, askew and parallel to the axis of the heading being driven, the installation of an acoustic alarm system, and others.

The water flowing into the Pomorzany Mine has been used for practical purposes since its first appearance. At the development stage the problem to be solved was the fact that sudden water outflows from the caverns were accompanied by a temporary increase in the Zn content in the water. Zn occurred as brunckite (ZnS) in the form of a colloidal suspension, originally filling the caverns, pockets and other subterranean karst features.

Hydrogeological Problems During Mining

As a result of these preliminary mining operations, an extended regional cone of depression was formed and water movement on its peripheries assumed a quasi-stationary character. Since the main drain drifts had been driven below the deposit itself (Fig. 4–19) the mining was by the chamber system in an already drained rock mass and did not meet with any difficulties of a hydrogeological nature. Only at those places where the zone of cave-in over the workings included overlying karst forms did there occur local inrushes of suspended water.

At this stage two hydrogeological problems acquired special importance: (1) the effects of mining drainage on the underground and surface waters and on the surface of the area, and (2) the quality of mine waters and their maximum utilization.

Under the natural hydrogeological regime, the groundwater flow in the whole area was westwards, mainly in the direction of the principal stream, i.e. the Biała Przemsza River. The average gradient of the piezometric surface was 0.005 (Fig. 4–22 A). The situation changed when the mines began to drain the Triassic water-bearing horizon. The artificial drainage base some 150 m below the natural one created a vast depression cone.

Because of this the hydraulic gradient increased in the eastern and souteastern parts of the Olkusz region, while in the northern and western parts there was

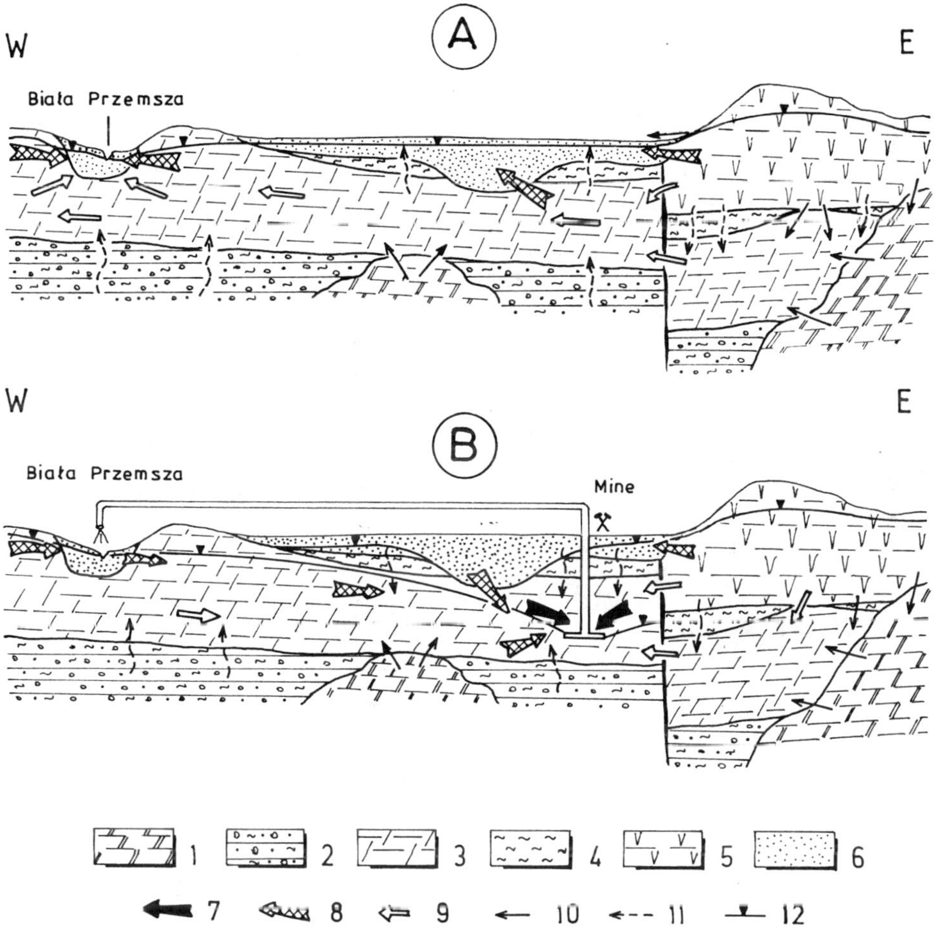

4–22. Schematic groundwater circulation diagram of the Olkusz region. A. natural conditions; B. conditions influenced by mining operations. 1. Devonian carbonate water-bearing horizon; 2. Permian low permeable conglomerates and impervious clays; 3. Middle Triassic (Muschelkalk) carbonate water-bearing horizon; 4. Upper Triassic (Keuper) impervious clays; 5. Upper Jurassic carbonate water-bearing horizon; 6. Quaternary sands; 7. intensive water circulation and rapid water exchange; 8. very easy water circulation; 9. easy water circulation; 10. difficult water circulation; 11. very difficult water circulation and slow water exchange; 12. groundwater table.

reversal of water movement from the Biała Przemsza River towards the mine drainage system (Fig. 4–22 B).

Because of the decrease of piezometric pressure in the Triassic rocks in the eastern part of the region, the difference between the elevation of the hydrostatic head in the Jurassic and in the Triassic increased. In response, the amount of water passing from the Jurassic into the Triassic strata along the contact zones of the horizons grew in direct proportion.

4–23. Groundwater outflow from a cave in the Gogolin Formation (Lower Muschelkalk) (Photo by J. Glanowski).

One can assume that such an increased difference between piezometric heads may initiate suffosion, and create direct connections between the water-bearing complexes in some fault zones where such connections did not exist before because they were filled with unconsolidated, impermeable material. The same problem applies to connections created between the Jurassic and Triassic aquifers of buried karst zones by means of poorly cased boreholes.

It appears also that the buried preglacial valley of the Biała Przemsza River, which under natural conditions had been a zone of groundwater discharge and that received Triassic artesian waters, changed into a zone of recharge and of constant supply of inflow water to the mine.

From general data one can hypothesis that this artificial decrease of the hydraulic head in the Triassic water-bearing complexes also enable water to penetrate upwards from the Devonian strata below.

4–24. Groundwater inrush from a karst passage encountered in the roof of a mine drive. Discharge rate about 50 m^3 per minute (Photo by J. Glanowski).

After the disturbance of the natural water relationships in the Olkusz region the natural springs disappeared, the flow of the Biała Przemsza River decreased (BOCHEŃSKA and PRUSSAK, 1974) and the local wells, including those of the water supply plants for the population and local industry, lost their water entirely. The mining industry had to finance the construction of waterworks and to use the water pumped from the mines. Waters from the Olkusz Mine meet the standard requirements for drinking water though the quality varies from one section of the mine to another (ADAMCZYK and WILK, 1976).

Since the waterworks had to rely upon supplies of guaranteed quality and quantity from the mine, one of the problems at this stage was to work out a likely prognosis for the behaviour of the hydrologic parameters in the future. For this purpose analogue modelling was employed (HAŁADUS et al., 1978; WILK et al., 1984). This task involved resolving various methodological problems such as the effect of turbulent flow on the accuracy of the forecasting (MOTYKA and WILK, 1982), the hydraulic characteristics of karst solution openings occurring in the Triassic rocks (WILK, MOTYKA and JÓZEFKO, 1984) and their effect on the hydrodynamic field (WILK et al., 1985) and others. The forecasts have proved remarkably accurate.

One obstacle to full use of the mine waters is a specific hydrogeological problem of mining in this region, namely the contamination of mine waters by lignosulphonate compounds, which have been disposed of into the Quaternary sands for some dozens of years by the local paper industry. The disposal point is about 6 km north of Pomorzany Mine (Fig. 4–17) in an area where there are erosional windows within the cover of the Keuper clays and hence some direct hydraulic contact between water-bearing horizons of the Triassic and of the Quaternary. As a result of draining Pomorzany Mine and so lowering the original piezometric head, the lignosulphonate compounds have been carried with water from the Quaternary deposits into the Triassic rocks and subsequently through fissures and karst passages to the mine workings. Such a situation had been anticipated at the exploration stage (WILK and ZIMNY, 1976) but about 120 000 m^3 per day of good potable water have been contaminated with toxic chemical compounds. According to the estimates (ADAMCZYK and CHAWIŃSKI, 1982) it will take some dozens of years before the lignosulphonates stored in the sands are washed away and all water flowing into the mine is fit for consumption. In this situation the problem of water contamination must be solved above all others, and various methods are being considered. For thc time being mining operations are planned in such a way that the polluted water has access to only a part of the mine. There these waters are filtered and the noxious waste matter disposed of in the river.

Another specific problem has been the appearance of sinkholes outside the mining area. A number of sinkholes first appeared about 2.5 km south-east of the nearest workings of Olkusz Mine, eight years affter the first excessive water

4–25. Sinkhole resulting from suffosion initiated by rapid artificial lowering of the groundwater table (Photo by J. Leśniak)

inflows and sudden water inrushes connected with its development. These phenomena were active for a few months, during which time several buildings were damaged. Four years after the Pomorzany fault was crossed (see above) and the first intensified inflows to the headings had occurred an increase in the rate of sinkhole formation was observed about 6 km to the north of Pomorzany Mine. In the period 1976–1981, several terrain depressions and sinkholes of different sizes (Fig. 4–25) formed here. One of them was responsible for the failure of the paper factory tailing pond. About 30 000 m^3 of sludge and Quaternary sands flowed down it into the underground karst system. The rim of the newly formed sinkhole reached almost to the bank of the Biała Przemsza River so that there was a danger that this, too, would spill into the karst system and the mine workings.

Sinkhole formation is still taking place. The origin and mechanism of formation has not been fully explained. Sinkholes are probably associated with the rapid drop of the piezometric head caused by dewatering of the ore mines. As a result of this and of the induced changes in the direction of groundwater flow suffosion was initiated which eventually led to an emptying of the secondary fillings in karst features buried under the fluvioglacial sands. This happens especially where the impermeable Keuper clays are absent due to erosion, but there has also often been some cave-in of Keuper clays that occur above karst features, as well.

Up to the present there have been no effective methods of determining when and where a catastrophic sinkhole will develop. It is often difficult to establish whether their origin is natural or man-made (WILK, MOTYKA and NIEWDANA, 1973).

HYDROGEOLOGICAL PROBLEMS OF HUNGARIAN BAUXITE AND COAL DEPOSTIS

Tivadár Böcker and Béla Vizy

Introduction

The Transdanubian Mountain Range (hereafter TMR) can be considered the industrial heart of Hungary. The most important bauxite and coal mines have been operating there for a long time.

More than 60 % of total coal production and 70 % of bauxite production comes from below the karst water table, so Hungarian mining may face major water hazards. Water inrushes into mines are of great importance in these industries. More than 80 % of the water pumped from the mines is karst water (BÖCKER, 1969).

The distribution of pumped water in the Hungarian industry in 1983 is given in Table 4–2. The distribution of mining waters according to source aquifer type in 1983 is given in Table 4–3.

It is quite apparent from these tables that the bauxite and coal deposits have a strong connection with paleokarst. Consequently, protection against flooding is one of the main tasks of the mining operations.

The karst regions of Hungary may be divided into two groups (Fig. 4–26): (1) karst rocks at the surface, and (2) covered karst formations. Covered karst

Table 4–2

Pumped water in Hungarian mining operations, 1983

Mining section	Mm^3/year	m^3/min	%
Coal mining Bauxite mining Ore mining of building materials Oil and gas mining	167.7 191.3 6.9 2.1 10.5	319 364 13 4 20	44 50 2 1 3
Total	378.5	720	100

P. Bosák, D. C. Ford, J. Głazek and I. Horáček (Editors):
Paleokarst. A Systematic and Regional Review. – Elsevier and Academia. Amsterdam and Praha. 1989.

Table 4–3

Pumped water in Hungarian mining, 1983, by

Aquifer	Mm^3/year	m^3/min	%
Waters from granular aquifers	66.5	126	18
Waters from fissured rocks	3.6	7	1
Karst waters	308.4	587	81
Total	378.5	720	100

formations occupy a substantially greater area but the different karst regions are recognised according to the surface karst expressions. An explanation for this is given in Figure 4–27.

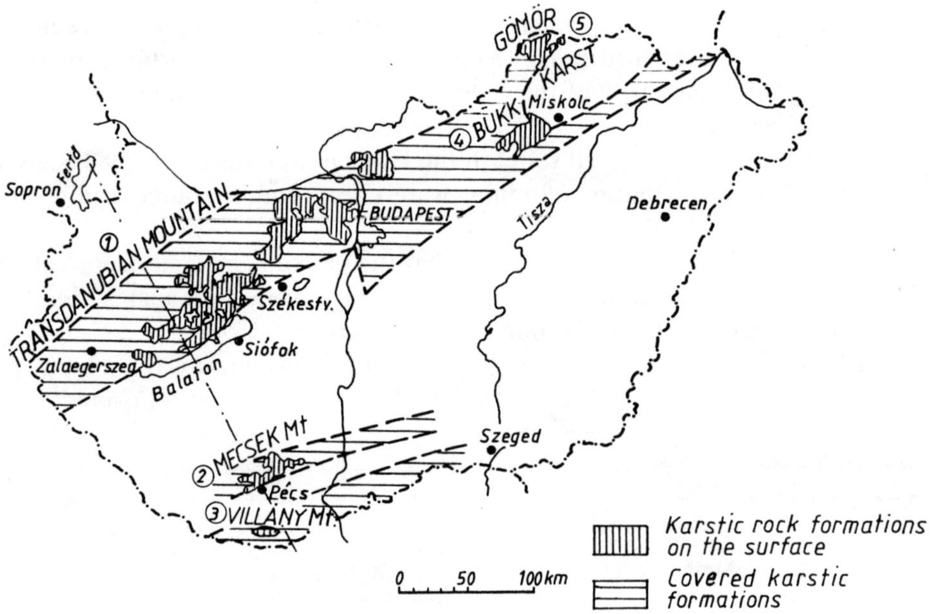

4–26. Extent of surface and covered karst in Hungary.

The carbonate rock outcrops which appear in the mountains sink gradually to greater depths along structural lines. Every karst formation is composed essentially of a surface karst recharge area and reservoir and an underground reservoir communicating with it. According to information obtained by geological and geophysical investigations these units are defined by systems of faults.

The different regional units are numbered 1 to 5 in Figure 4–26: (1) Transdanubian Mountain Range. The range is highly faulted, with frequent horizontal shear planes and longitudinal upthrusts; (2) the Mecsek Range is highly folded and faulted. It is separated tectonically from the TMR by the intervening Caledonide-Variscan depression (Fig. 4–27); (3) the

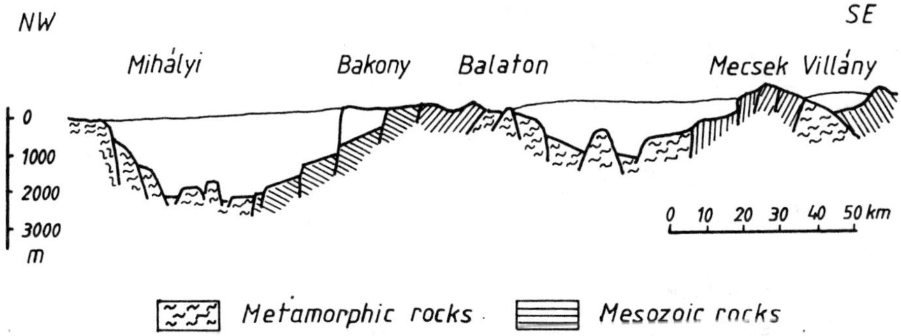

4–27. Diagrammatical profiles of exposed and covered karst (after VADÁSZ, 1952).

Villány Mountains form an unit independent of the Mecsek Range. It differs from it stratigraphically and structurally, with thrust planes rising to the north; (4) the Bükk Mountains have a character radically different from those in the TMR. The beds are highly folded and the anticlines and synclines are displaced to the south; (5) the Gömör (Gemer) karst (most of which is in Czechoslovakia) is strongly faulted. It is a rigid limestone range with extensive cave systems along the structural lines.

Table 4–4

Aerial distribution of pumped waters in mines, 1983

Areas	Mm^3/year	m^3/min	%
Transdanubian Mountain Range	302.8	576	98
Eastern Mountain Range	0.5	1	-
Other karst areas	5.1	10	2
Total	308.4	587	100

The karst area of the TMR is the most important in terms of karst water balance and storage (Table 4–4). More than 60 % of the rocks of the TMR are karstified carbonates. These rocks comprise ten distinct groups separated by paleokarst surfaces or other geological formations (Fig. 4–28). The significance of karst horizons differs from one group to another due to water balances and the amount of stored karst water. The groups are (Fig. 4–28): I. Lower Carboniferous limestones; II. Lower Triassic cellular dolostone; III. Middle Triassic limestones with ammonites and Upper Triassic limestones (Balatonfüred

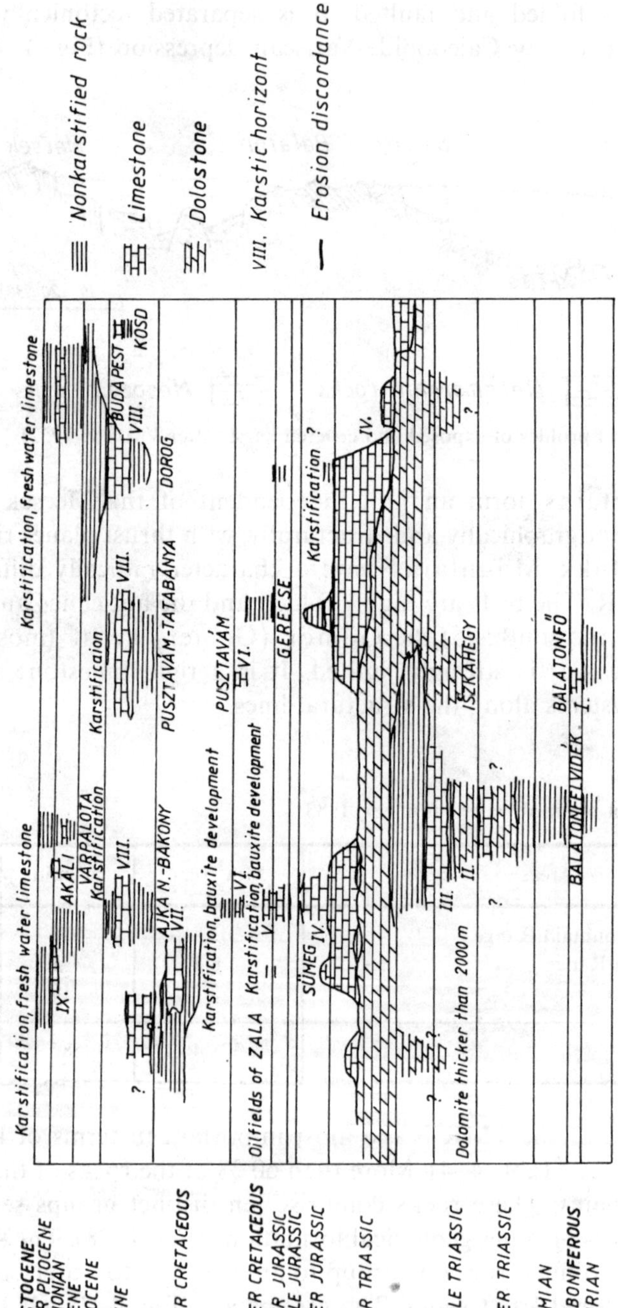

4–28. Comprehensive profile of the Transdanubian Mountain Range indicating the karstified rocks and periods (constructed by P. MÜLLER).

Formation); IV. Upper Triassic Hauptdolomit together with Upper Triassic and Lower Jurassic Dachstein Limestones. This karst horizon contains the main karst water body, a single, continuous and communicating hydraulic system (confined and unconfined) that extends to over 15 000 km^2 of the TMR; V. Upper Jurassic/Lower Cretaceous limestones; VI. Lower Cretaceous *Requienia* and *Orbitolites* limestones; VIII. Upper Cretaceous Hippurites limestones; VIII. Eocene limestones; IX. Miocene limestones, and X. Upper Pliocene travertines.

The main karst water system (horizon No. IV) has produced the greatest water hazards in the bauxite and coal mines in the TMR. Therefore, the physical and hydrogeological properties of paleokarst areas in Hungary, the relationships between karst water systems and mining, the methods of protection against water inrushes into the mines and the environmental impacts of mine dewatering will all be illustrated by examples from this karst area.

Results of Hydrogeological Exploration

Exploration of the karst carbonate rocks was undertaken for mining and water development. The following parameters were determined: (1) physical rock parameters (fissuration, porosity, permeability); (2) hydrological parameters (recharge, discharge, water level, water balance and their changes in space and time), and (3) physical and chemical karst water quality.

Physical Parameters of the Rocks

Fissuration karst porosity is determined by

$$n = f(d, \Delta) \tag{1}$$

where n = porosity, d = width of fissures, Δ = distance between fissures.

It follows from Equation (1) that the porosity created by water flow channels can be estimated reasonably well by knowledge of the statistical distribution of fissures. The distribution of channels having a width $d \geqslant 0.4$ m (caves) is given in Table 4–5 for the TMR. The data show that the difference in cave distribution between the limestone and dolostone areas is significant.

Where the widths of fissures are less than 0.4 m, an empirical equation was obtained:

$$\log d = 1.087 \log \Delta - 3.37 \tag{2}$$

Table 4–5

Distribution of caves in the TMR

Mountains and type of karst rocks	Karst surface Exposed, km^2	Number of caves	Caves per km^2
Balatonfelvidék (dolostone)	326	57	0.175
Bakony (dolostone)	340	128	0.376
Vértes (dolostone)	150	7	0.047
Gerecse (dolostone)	83	60	0.723
Pilis (limestone)	47	79	1.681
Buda (limestone)	56	93	1.661
Total for Transdanubian M. R.	1 002	424	0.423
of which dolostone area =	899	252	0.280
and limestone area =	103	172	1.670

Table 4-6

Physical parameters of rocks in the TMR

Range of fissure sizes, m	Range porosity, rp	Percentage of rp in total porosity	Summarized porosity, %
$d > 0.1$	0.01	0.4	0.4
$0.1 < d > 0.01$	0.20	8.0	8.4
$0.01 < d > 0.001$	0.50	21.2	29.6
$d < 0.001$	1.66	70.4	100.0
Total n =	2.37 %	100.0 %	

d = fissure width

The validity of this equation could be extended to the total area of the TMR. It expresses a typical relationship between fracture width and the distance separating fractures:

$$\log d = A + B \log \Delta \tag{3}$$

$$\log \Delta = A' + B' \log d \tag{4}$$

A logarithmic distribution of fissures follows from Equation (2), (3) and (4).

The porosity of the karst rocks of the TMR as determined by the fissure statistics can also be seen in Table 4–6. These studies of fissuration patterns showed that the characteristics of water flow at the regional scale in carbonate rocks were determined by fractures with widths of less than one millimetre.

Hydrogeological Parameters

To estimate the recharge from precipitation, a national register of karst springs was made in 1950. The spring yields obtained in the measurements made in early Fifties provide the basic data about the natural conditions of the karst water systems.

The amount of infiltration can be estimated by two measures, (i) moderate precipitation and (ii) the (relatively small) extent of karst rocks at the surface. According to these measures the average recharge of the TMR is less than an average yield for a karst spring (e.g. in Yugoslavia, 8–12 $m^3 \cdot s^{-1}$).

The TMR has been characterised hydraulically as being dominated by very narrow fractures; therefore, the role of precipitation sinking into the rock via sinkholes and cave channels is not significant.

The time required for a complete exchange of the stored water in the subsurface catchment area of a big yield TMR spring is approximately 30 years according to the results of the water age experiments.

Measures of infiltration rate have given different results, depending on the method used. For a forty year mean period, estimates range from 480 to 720 m^3 per minute. The mean value of this range seems to be close to reality.

The standard deviation of this average infiltration is 60 %, which indicates that the mean infiltration rate during a given year may be between 240 and 960 m^3 per minute, depending upon the amount of precipitation that year and its areal and time distribution.

Since karst water of good quality is as valuable as the bauxite and coal resources and because the environmental impacts of dewatering have been quite considerable, yearly water balance tables were prepared in 1978. These used a finite element water balance model with individual elements of 4 km^2 covering the total area of the TMR (15 000 km^2). Calculations of the yearly water balance help to determine the influence of mine dewatering on the karst water systems, to guide in the utilization of karst water resources, and to provide a basis for forecasting the impact of karst water consumption in the future. The TMR water balance tables used input from 480 waterworks, 93 mines, 155 meteorological stations and 270 observation wells.

The annual changes of water balance were found to be closely associated with the yearly changes in mineral production. Change in water resources may be caused by both natural (meteorological) and human causes. The components of the water balances (i.e. infiltration, discharge, change in storage, ets.) were estimated independently using models for closed water units bounded by subsurface watersheds. Horizontal flow is presumed not to occur between the units in this case. The condition to be accounted is

$$\text{R (recharge)} \neq \text{D (discharge)} \qquad (5)$$

The error in estimation of the hydrological parameters (i.e. infiltration) can be 30–35 %; therefore, the probability of R = D is rare. The reliability of estimates of water balance constructed in this way can be shown by:

$$E = \frac{(R - D)}{D}; \qquad E \leqslant AE \tag{6}$$

where the error (E) of the water balance is less than the acceptable error (AE) which is 10 % of the total discharge.

The model can be used to obtain water balances for open groundwater units (where there is horizontal flow between the units), taking into account condition number (2) above.

The aeral distribution of the components of water balance and their standard deviation were estimated by methods used in geostatistics. The surface available for infiltration, the morphology and the vegetation were determined for each finite element. The algorithm for infiltration was based on data from a 15-year field experiment. The infiltration during a period Δt can be estimated as follows:

$$I = P - PL \tag{7}$$

$$PL = SR + EP + \Delta w \tag{8}$$

where I = infiltration; PL = precipitation limit; SR = surface runoff; EP = evapotranspiration; Δw = change of soil humidity.

Using Equations (7) and (8), an empirical relation was determined between the yearly precipitation and infiltration for each meteorological station:

$$I = a + b \cdot \ln P \tag{9}$$

The 50-year average infiltration was estimated using Equation (9), which becomes the basis of comparision between annual water balances.

Physical and Chemical Quality of Karst Water

Water temperature and chemical composition was measured during all hydrogeological exploration drilling. According to the temperature findings, 40 % of the total stored karst water has a temperature less than 15 °C. The amount of hot water is 37 %. Geothermal investigations have given some results about geothermal gradients (*gg*). In the karst rocks *gg* are 16 °C $\cdot$ km^{-1} or greater. The *gg* of clayey rocks is 40–67 °C $\cdot$ km^{-1} in those areas where karst rocks occur at greater depth.

The geothermal conditions can be changed by water circulating in the karst rocks. For example, in areas where the carbonate rocks crop out at the surface, infiltrating waters with the regional mean annual temperature serve as a cooling factor to make the rocks comparatively cold. On the other hand, hot waters rising along tectonic lines from great depth heat the rocks significantly.

The chemical composition of karst waters can be characterised by the total dissolved solids. These ranged from 400 to 700 ppm, 70–80 % of which was $Ca(HCO_3)_2$ or $Mg(HCO_3)_2$. The content of Na and K is 2 ppm and Cl varies 1–5 ppm. The chemical composition of the karst waters varies with residence time underground (water age), with the chemical composition of aquifer strata and the water temperature. Generally, the of Na, K, Cl and SO_4 contents have been rising and the amounts of Ca, Mg and HCO_2 have decreased.

Protection of Coal and Bauxite Mines Against Underground Waters in the TMR

A *main karst water system* which comprises a single, contiguous and communicating hydraulic system is stored in the carbonate rocks of the hills. Coal mining first encountered problems of karst waters breaking into the mines early in this century and bauxite mining encountered them in the 1950s (Table 4–7).

Table 4–7

Total pumped water in bauxite and coal mines since 1910, in m^3 per minute

Year	Coal mining	Bauxite mining	Others	Total
1910	9.4	—	—	9.4
1920	12.1	—	—	12.1
1930	26.2	—	—	26.2
1940	47.6	—	—	47.6
1950	68.2	2.6	—	70.8
1960	165.4	34.2	3.9	203.5
1970	243.7	278.4	5.8	527.9
1980	184.5	397.5	4.0	586.0
1983	207.8	363.9	4.3	576.0

Protection against karst waters has been and continues to be a grave worry particularly in bauxite mining, because the bauxite rests directly upon eroded karst surfaces in the carbonate aquifer, with no impermeable barrier formations in between.

The main karst water system is contained predominantly in fissures and cavities of the Upper Triassic Dachstein Limestones and the Hauptdolomit Formation. In its undisturbed state (before the impact of mining), the rate of infiltration of precipitation was in equilibrium with the rate of discharge from karst springs and marshes in the foothills. Subsequently, pumping at increasing rates in the mines has upset the water budget of the system more and more and has depressed the original karst water table. The depression occurring within the

Table 4–8

The utilization rate of pumped mine waters in the TMR

Mine	Rate of pumping, m^3/min	Water utilized, m^3/min	%
Nyirád bauxite mines	270.5	42.5	16
Halimba bauxite mine	8.1	–	–
Kincsesbánya bauxite mines	85.3	60.3	71
Bauxite mining – total	363.9	102.8	28
Ajka coal mines	19.3	4.7	24
Dudar coal mines	1.7	0.3	18
Várpalota coal mines	19.4	2.8	14
Balinka coal mines	11.6	2.6	22
Tatabánya coal mines	137.9	36.6	27
Dorog coal mine	17.9	8.1	45
Coal mining — total	207.8	55.1	27
Urkut manganese mine	4.3	1.6	37
Total for the Transdanubian Mountain Range	576.0	159.5	28

area of influence of large-scale pumping affects springs fed by, and wells sunk into, the karst water system. Some of the water of drinking quality pumped by the mines is utilized for the water supply of nearby communities and industries (Table 4–8).

The methods used to protect mines against water inrushes depend on the hydrogeological situation. They can be summarised as follows: (1) passive protection where there is no preventive action; (2) preventive protection: (i) cut off methods: (a) rock sealing, local or regional; (b) protective barriers; (ii) active protection: (a) dewatering before starting the mines (on a local or regional scale); (b) break-pressure action. In practice appropriate combinations of these methods are applied.

Bauxite Mining at Nyirád

Most of the bauxite at Nyirád is found under Eocene and/or Miocene cover, in lenticular deposits filling depressions controlled by structural features and karst development in the Upper Triassic Hauptdolomit. The dolostone is strongly fractured, karstified and weathered on the surface and along faults. The bauxite lenses are of varying size, those containing 50 000 to 100 000 tons of ore being the most important. The Mesozoic basement (dolostone) underlying the bauxite crops out at the surface in the south and descends to depths of several hundreds of metres to the north. The bauxite explored so far lies at depths from the surface to −200 m. Bauxite quality is excellent. Over the NW part of the area, Upper Cretaceous marls and limestones occur. Here, bauxite can be found at two horizons, between the Upper Triassic dolostones and Cretaceous rocks and between the Cretaceous limestone and the Eocene rocks.

The Triassic carbonate rocks have a high water storage capacity and permeability, as revealed by effective (gravity-drained) porosities of 3–4 % and filtration coefficients of the order of 0.000 1 $m \cdot s^{-1}$ on average. There is no significant groundwater system in the upper layers of the karst (Cretaceous).

Prior to pumping, the karst water table fluctuated between 175 and 179 m a.s.l., depending on the precipitation (or, to be more precise, on the rate of infiltration). Ever since bauxite mining penetrated below the karst water table (i.e. since 1956), systematic pumping has been a major feature in the area. At first, attempts were made to seal off the inrushes of karst water that occurred. These attempts, however, were doomed to defeat because the dolostone is so strongly fissured and karstified near its top surface and along faults. This is why the decision was taken to install pumping systems which permitted the depression of the karst water table prior to the development of the mines and the extraction of their ore.

All attempts at tapping the karst water reservoir by mining methods (by driving drainage shafts and adits in the dolostone) were frustrated one by one by the tremendous inrushes that they provoked. For example, one such flood deeply silted the workings with masses of grus and other debris of altered dolostone that it swept in. This is why, since 1964, dewatering by pumping via drilled wells has gradually gained the upper hand. These wells are drilled to a depth of about 200 m, at which depth they have a drilled diameter of 2 000 mm and a casing diameter of 1 400 mm. They are pumped, depending on their yield, by one to four EMU-brand submersile pumps of 7.5 m^3 per minute capacity each.

A quarter of a century of pumping has removed some 26 Gm^3 of water from the karst reservoir, creating a comparatively flat depression that is of large horizontal extent. At its centre, where the mining is located, the water table has been lowered by 113 m on average, to 63 m level a.s.l. Thus a lowering of 1 m of the water level requires that $20 \cdot 10^6$ m^3 water be pumped (BÖCKER and VIZY, 1982).

This large amount of pumping has created a large-scale (2 500 km^2) and relatively flat drawdown cone. Over the area of its impact, natural springs and karstic bogs have dried up and water levels have dropped in wells penetrating the karst system. Piezometric head reduction also endangers valuable thermal baths at the edge of the aquifer.

Preventive regional depression of the karst water table by drilled wells is the only dewatering method being employed at Nyirád at present. The objection to its exclusive use for the dewatering of further, deeper levels is the predictable extension of the cone of depression and the resulting seriously enhanced environmental impact.

This expanding damage to the environment has set a limit to the regional water table lowering permitted for the sake of dewatering lower deposits of bauxite. That is the reason for combining this older method with the newer practice of local dewatering directly under given sets of workings.

The new method is essentially a preventive depression of the karst water table, but which is limited both in space and in time. With this method, fast pumping is necessary. It can be applied most readily by driving a gallery or system of galleries into the rock underlying the bauxite lens or cluster of lenses and drilling tapping boreholes from them. Needless to say, in the light of what has been described above, this method is applicable in the Nyirád region only in certain favourable lithological and structural conditions.

The physical-hydrogeological conditions do not prohibit the sealing of the rocks enclosing the bauxite, or at least, a significant reduction of their permeability. Rock sealing as an exclusive method of protection in bauxite mining with a water hazard is not economically viable, however, primarily because of the extensive drilling and the masses of sealing materials that are required.

In view of all this, rock sealing should be applied only as a supportive measure in combination with other means of protection. A case in point may be local dewatering combined with rock sealing, two varieties of which may be envisaged: (i) the access shaft from which the galleries to tap the karst water under a bauxite body are to be driven can be sunk with a protective seal (a mantle or casing surrounding the shaft); (ii) in cases where the galleries driven under the depressed regional water table encounter an inflow that can be handled only by exceeding the permissible pumping-rate limit or by slowing down the extraction of bauxite protected by the local depression, sealing off of zones of highest permeability (of faults principally) may permit the required depression to be attained while staying within the permissible range of pumping-rates.

Coal Mining in Tatabánya

In terms of geology and hydrogeology, the Tatabánya Coal Mines can be divided into three different basins: Tatabánya, Nagyegyháza-Csordakut and Mány. In all three, the water risk is created by the Triassic basement. Due to changes in the location of aquifers, to the varied conditions of aquiclude strata and to differences in the water-entraining capacity of the basement, different protective methods have been developed. They will be described separately for each basin or basin section.

Tatabánya Basin is further divided into three different hydrogeologic sections:

(1) in the west section, the basement is of Dachstein Limestones. The protective stratum between the Upper Triassic aquifer and the coal beds is developed only to a minimal extent and, in some places, not at all. The mining of some coal seams where the water hazard steadily rises has required the application of active protection (dewatering). Some access tunnels and drilled wells have achieved a 30 m drawdown in the basement rocks and made it possible to follow the coal resources where there are thin protective layers present.

(2) in the middle section of the Basin, successive chalky layers generally occur within the Aptian to Cenomanian succession so that the thickness of protective strata exceeds 100–150 m in some places. Thanks to the favourable protective conditions, (and despite high water pressures, up to 2.5 MPa in some places) considerable floods did not disturb the mining activity and a combination of passive + preventive protection was used.

(3) in the eastern horst, the basement is Norian Hauptdolomit. The protective stratal conditions are unfavourable. The safe exploitation of the coal has required a planned lowering of the water level, which means the introduction of active protection a groundwater.

In the **Nagyegyháza-Csordakut Basin**, mining operations are endangered by water from the Carnian dolostones that form the basement, by redeposited dolostone and freshwater limestones, plus the *Alveolina* limestone which has its own independent groundwater system. Due to environmental protection reasons, a general reduction of the active water level cannot be permitted in the area to enable general exploitation of the coal and bauxite. Therefore, a combined (so-called *instantaneous*) water protection method was adopted. The principle of this method is to design both the water and rock removal processes into one system. Water is drained during mining of economic deposits and removed via preconstructed drainage adits and shafts. Consequently, water inrushes resulting from subsequent rock displacement and stress reorganization during mining can also be collected in this preconstructed system. The spacing of drainage elements (adits or shafts) is determined by the water-delivery capacity and fracturation of the water-containing rocks.

In the **Mány Basin** a combined protection method will be introduced to control the water in the main cavernous aquifer and in redeposited dolostone located directly over this reservoir.

During the opening of new mines when the access facilities are being headed and the main system of excavations is being established, the cross-cutting of expected fissured water reservoirs is often carried out. Ensuring that the required drift driving rate is maintained necessitates the use of a drift driving technology that is mechanized in a complex way fashion. This can only be achieved in a water-free working environment. This in turn has demanded the introduction of processes to establish an artificial protecting layer in the immediate vicinity of the adit to be driven. To form this artificial protection, rock sealing or grouting methods are applied.

Bauxite Mining at Kincsesbánya

Bauxite occurs as regular beds which extend to 1–2 km^2 on the surface of the Upper Triassic dolostone. They were collected in cavities created by structural displacements and karst erosion. The cap rocks include Eocene carbonates, Oligo-Miocene clays and sands Pannonian sand and gravel. The bauxite was deposited directly onto the cavernous dolostone without any protective stratum. Therefore, due to the inflow of increasing quantities of water during the early stages of mining, the ore became saturated, its mining and transport became more difficult and it was not possible to mechanize the work. Deliberate reduction of the karst water level prior to modern mining operations was started in 1958. The present dewatering system was put into practice at the same time as the establishment Rákhegy II Mine. A drainage shaft independent of the mine was constructed into the deepest part of the bauxite and an adit system branching from this shaft was used to drain the cavernous aquifer. Water is pumped to the surface via the shaft by high efficiency RITZ plunger pumps of 15 m^3 per minute delivery. As a result of the pumping, the water level started to decrease, but because of the lower permeability (kM: $0.45 \cdot 10^{-2}\ m^2 \cdot s^{-1}$) of the rocks, a cone of depression with steep sides and relatively small volume has been formed. The quantity of water pumped so far from the main cavernous water system is 800 Mm^3. As a result, the water level has been lowered 120–250 m in the area of the mines; to achieve a water level reduction of 1 m it was necessary to pump 4–5 Mm^3 of water. The cone of depression at Kincsesbánya covers an area of 400 km^2 today.

4–29. Lowering of the karst water level in the area of the Transdanubian Mountain Range. 1. lowering in metres; 2. boundary of Paleozoic rocks; 3. Mesozoic rocks on the surface; 4. Mesozoic rocks covered; 5. dry karst springs; 6. mines; 7. average yield of pumped water in 1983 (in m^3 per minute).

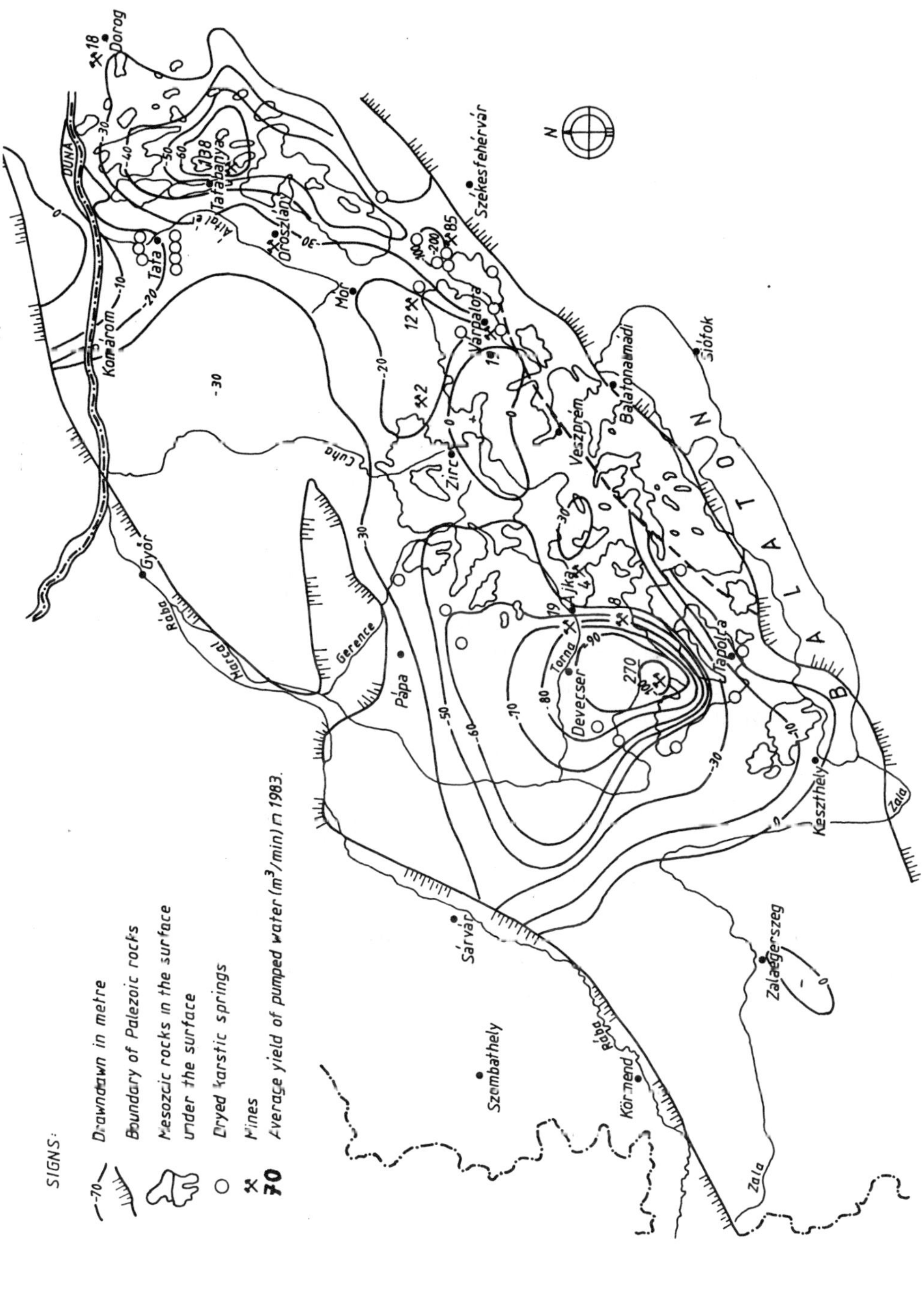
SIGNS:
Drawdown in metre
Boundary of Palezoic rocks
Mesozoic rocks in the surface under the surface
Dryed karstic springs
Mines
Average yield of pumped water (m³/min) in 1983.
DUNA
Dorog
Tatabánya
Oroszlány
Tata
Komárom
Mór
Várpalota
Székesfehérvár
Siófok
Balatonalmádi
Veszprém
Zirc
Győr
Ajka
Pápa
Devecser
Tapolca
Keszthely
Sárvár
Zalaegerszeg
Szombathely
Körmend
BALATON
Rába
Zala
Gerence
Marcal
Cuha
N

Environmental Impact

The principal impact of mining upon the environment results from the active methods of protection against subsurface waters. The purpose of active dewatering is to lower the water table and reduce water pressure. As a result

— karst-fed marshes dry out,
— springs in the beds of watercourses will disappear,
— other springs will also dry up or suffer substantial reduction of discharge,
— the output of waterworks dependent on springs or wells fed by the affected aquifers is reduced or cut off,
— the recharge of stratiform aquifers from the karst water system gradually declines. After a while, flow is reversed, so that the stratiform aquifers begin to feed into the karst aquifer, and
— as the result of all this, water reserves stored in the karst are substantially reduced.

In our systematic monitoring of these offects of pumping on the environment, the focus of interest over the last five or six years has shifted to the relationship between pumping rates in the mines and the yield of the spring which feeds Héviz Lake, and which has a temperature of 39 °C and noted medical properties. In 1969 or 1970, the discharge of this spring started to decline, a worrying development because Héviz Lake is the largest medicinal pool in Central Europe and so is an object of outstanding environmental, medical and touristic value.

Pumping at the Nyirád mines has to be adjusted so as not to cause any intolerable harm from a medical point of view. To forecast the tolerable limits it has been obseved that the reduction of the lake-spring discharge is from one to two percent of the pumping rate at Nyirád at any given time.

The pumped karst water is of drinking quality. Waterworks have been constructed from which the industrial and household water demands of Székesfehérvár, Tatabánya, Dorog, Várpalota, Ajka, Pápa, West Balaton, Kaposvár and Sümeg are served. As much as 28 % of the total quantity of water produced is marketed.

PALEOKARST IN CIVIL ENGINEERING

Adolfo Eraso

Introduction to the Problem

During large civil engineering construction projects in karst regions it is quite possible that unexpected problems will appear. The economic consequences in some known cases have been very important. The problems may be divided into three major groups: (1) foundations, (2) tunnels and (3) dams.

In principle, an increased number of exploration boreholes in the preliminary planning stages seems to be the best solution, but this is not always so. We can find cases, especially in dam building, where all the karst problems appear together. In such cases the engineering works modify the environmental conditions so that karstification processes are rejuvenated, transform themselves and seek new equilibrium conditions that are difficult to forecast in our present stage of knowledge. The karst responds to anthropogenic action like a dissipative system (in the thermodynamic sense of the term), acting as an open system that transfers energy and mass out of the system.

Modification of the local hydraulic gradients usually starts the karstification process, producing the expensive problems. The solution of these problems will be the subject of this investigation.

Dam Building in Karst Regions

When exploiting a river the engineer selects the appropriate sites for dam building (according to the hydraulic programme to be developed) using three fundamental premises: (1) the width of the canyon must be a minimum, so minimising the volume occupied by the dam as well as the construction costs; (2) the rocks at the banks must be strong enough to support the loads transmitted by the body of the dam and those from the reservoir when full; and (3) the rocks at the banks must be impermeable enough to permit the water to fill the reservoir.

P. Bosák, D. C. Ford, J. Głazek and I. Horáček (Editors):
Paleokarst. A Systematic and Regional Review. – Elsevier and Academia. Amsterdam and Praha. 1989.

Karst canyons offer excelent topographic conditions for dam building since they fulfill the first two conditions. The third condition is generally not fulfilled, which results in the dam not reaching its designed storage capacity. This phenomenon is due to leaks that allow water to escape into pre-existing karst conduits.

Very often serious water leaks appear immediately. In other cases they develop later and increase steadily due to the washing away of fillings of pre-existing karst channels, i.e. from previously infilled paleokarst (see, for example, the dam at Przeczyce, Poland, p. 000).

In both cases, the modification of the natural hydraulic gradients due to the increase of water level created by the reservoir is responsible for the leaks. The socioeconomic consequences are very important, so we have recourse to grouting operations to create impermeable curtains to stop leakage and to attain the designed functional characteristics (for example, the water reservoir for the Čierný Váh Dam, Czechoslovakia, on karstified carbonates had to be sealed by a bitumen cover and plastic foils, Eds.).

Grouting Techniques

These consist of emplacing an injected screen, by drilling wells from a tunnel excavated into the rock banks (abutments) and under the dam.

Careful planning is necessary because, given the uncertain nature of any grouting works, there always exists the risk of causing excessive pressures which may place the slurries far away from the screen. The consequence is work that is as expensive as it is unsuccessful. So it is necessary to proceed systematically as follows (Fig. 4–30):

— geological research is carried out to identify the presence of karst features and the area to be occupied by the grouting screen and the dam site is defined;

— the preferred directions of karst drainage are deduced: (1) through speleological explorations of the region, mapping the existing caverns and (2) by deducing these directions from structural geologic mapping and analysis. The preferred directions may be oriented in planes perpendicular to the strain axis (σ_3) where σ_3 is the minimum component of stress suffered by the massif;

— design the grouting screen to be orthogonal to the most probable drainage direction;

— excavate the access galleries for drilling the grouting holes, beginning beneath the dam and extending towards the interior of its banks or abutments;

in each gallery, vertical wells are drilled with modular separation (± 8 m) to the prescribed depth in the screen. If the dam is very high, it is possible to drive several gallery floors at different levels, from 50 to 70 m apart;

— every well is grouted from bottom to top in short lengths or steps (± 5 m) after being isolated between two packers. The operation is finished when back pressure appears;

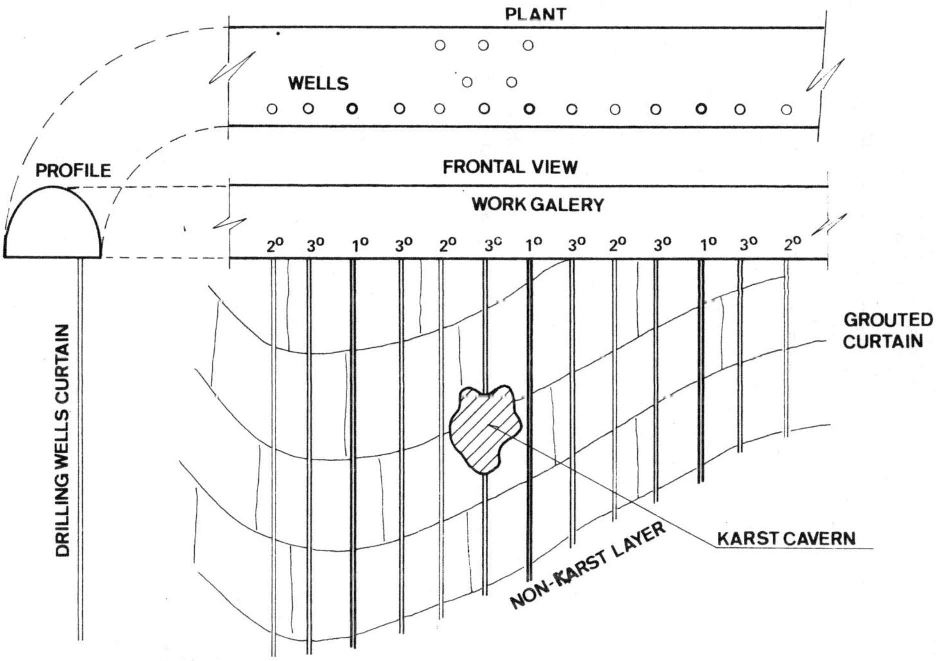

4–30. Impermeable curtain.

— before grouting each section it is necessary to carry out a LUGEON test by pumping water. Its results determines the viscosity of the first slurry;

— several slurries are prepared (Fig. 4–31), and grouting always proceeds from the more fluid ones to the more viscous ones, which rationally limits the injection of each of them if back pressure does not occur;

— once the first array of wells is grouted, we drill a new intermediate series of wells to reduce to one half the interval between any two contiguous wells;

— these operations are repeated many times. The screen is formed when the final interval between the grouting is reduced to 1 or 2 m. By that time, also, we expect to have located precisely any special sites (or singular points) where leaks are most serious due to the existence of karst conduits;

— in these special sites or points, it may be necessary to double or to treble the treatment rows in order to reduce the intervals between wells. It is necessary to fill the holes with gravel before treatment and to grout with special slurries of very short setting time;

— if the karst conduits are filled with clay (as in the case in some paleokarsts)

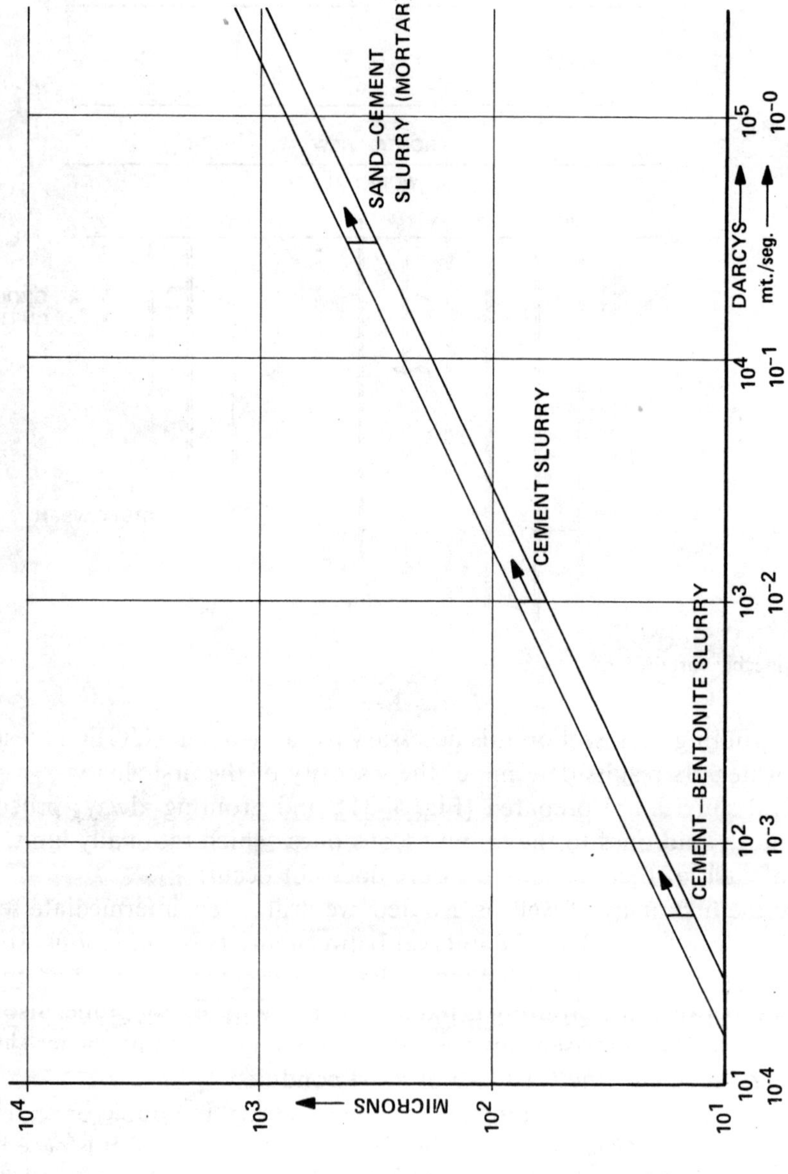

4–31. Penetrability limit of grouting slurry. Relation between rock permeability and grain-size of slurry materials.

it is necessary to wash this out before regrouting with the slurry series. Washing is carried out by drilling several parallel wells, vigorously pumping water into some of them and removing it through the others.

The grouting operation is finished when each step at each well gives back pressure.

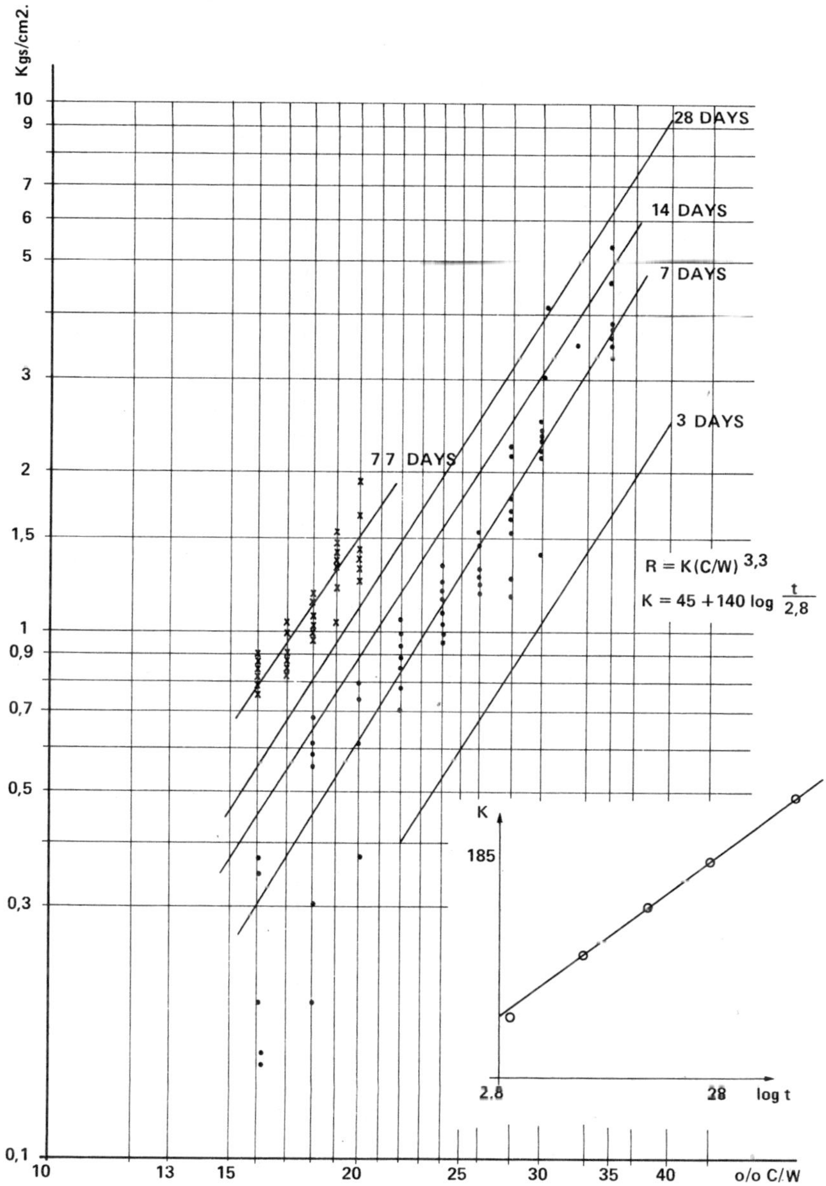

4–32. Relation between shear strength, sample age and dosages for Spanish bentonite-cement slurries.

Table 4–9

Slurry Formulation Series

Number	Dosage				Relation dry material vs water	Total weight of dry material	Quantities					Total litres of slurry	Suspension					
													of clay 1 : 1			of bentonite 1 : 5		
	Cement %	Clay %	Bentonite %	Sand %			Cement kp	Clay suspen-sion l.	Bentonite suspension l.	Water l.	Sand kp		Clay kp	Water l.	Volume l.	Bentonite kp	Water l.	Volume l.
1B	44	52	4	—	1 : 1.5	400	176	288	87	312	—	746	208	208	288	16	80	87
2B	44	52	4	—	1 : 1	800	352	576	173	224	—	1 090	416	416	576	32	160	173
3B	30	26	2	42	1 : 0.8	1 000	300	360	108	440	420	1 170	260	260	360	20	100	108
4B	30	26	2	42	1 : 0.5	1 000	300	360	108	140	420	870	260	260	360	20	100	108
Mor-tar	32	—	4	64	1 : 0.35	1 250	400	—	—	438	800	899	—	—	—	50	—	—

Slurry mixture

The choice of grouting slurries and particularly their physico-chemical and rheological characteristics (sometimes the object of patents) is the most delicate matter in a grouting operation. The types and amounts used must be programmed to attain the desired shear strength and, also, the desired shear rate values. The slurries must be strong enough to support the hydraulic gradients produced by the water in the reservoir.

Generally, slurry mixtures are composed of bentonite and cement, which give a higher penetrability (Fig. 4–31). It is necessary to test the mixture (Fig. 4–32), because the characteristics of cement differ from one producer to another and sometimes from batch to batch.

In Table 4–9, a formulation is given that solves much of the problem. This table describes four different types of slurries in which viscosity increases from the first (1B) to the last (4B) and also describes a sand-cement mortar.

The use of the slurry formulation is shown in Figures 4–33. The first slurry formulation in the series is selected according to the quantitative values of permeability obtained in U. L. (LUGEON units).

Karst Fillings and Paleokarst

In the geodynamic evolution of karst as natural hydraulic gradients diminish, water flow velocity is reduced in the karst network. That situation produces sediment deposition which becomes finer and which can lead to complete infilling.

In some known cases of paleokarst, this particular circumstance (associated sometimes with the presence of different mineral deposits) creates special problems in dam building, pre-eminently when these fillings are composed of clay minerals.

Spanish examples of such karst fillings (established by *X*-ray diffraction) are principally composed of illite, with kaolinite and/or chlorite in some cases, and always small proportions of montmorillonite..

The geomechanical characteristics resulting from this mineralogic composition include a great variation of shear stress as a function of the moisture contents or fluidity index, I_L (Fig. 4–34). Specifically, the shear stress (R) has values of 0.6 to 1.3 kp · cm^{-2} for moisture contents near the plastic limit (W_P), though the shear stress is reduced 70 to 100 times when the moisture content is near the liquid limit (W_L). When the reservoir is filled, it is evident that water in contact with these clay fillings will cause their saturation because they can be extruded out of the conduits due to the new hydraulic gradient generated by the dam. That will cause deferred leaks making the problem even worse.

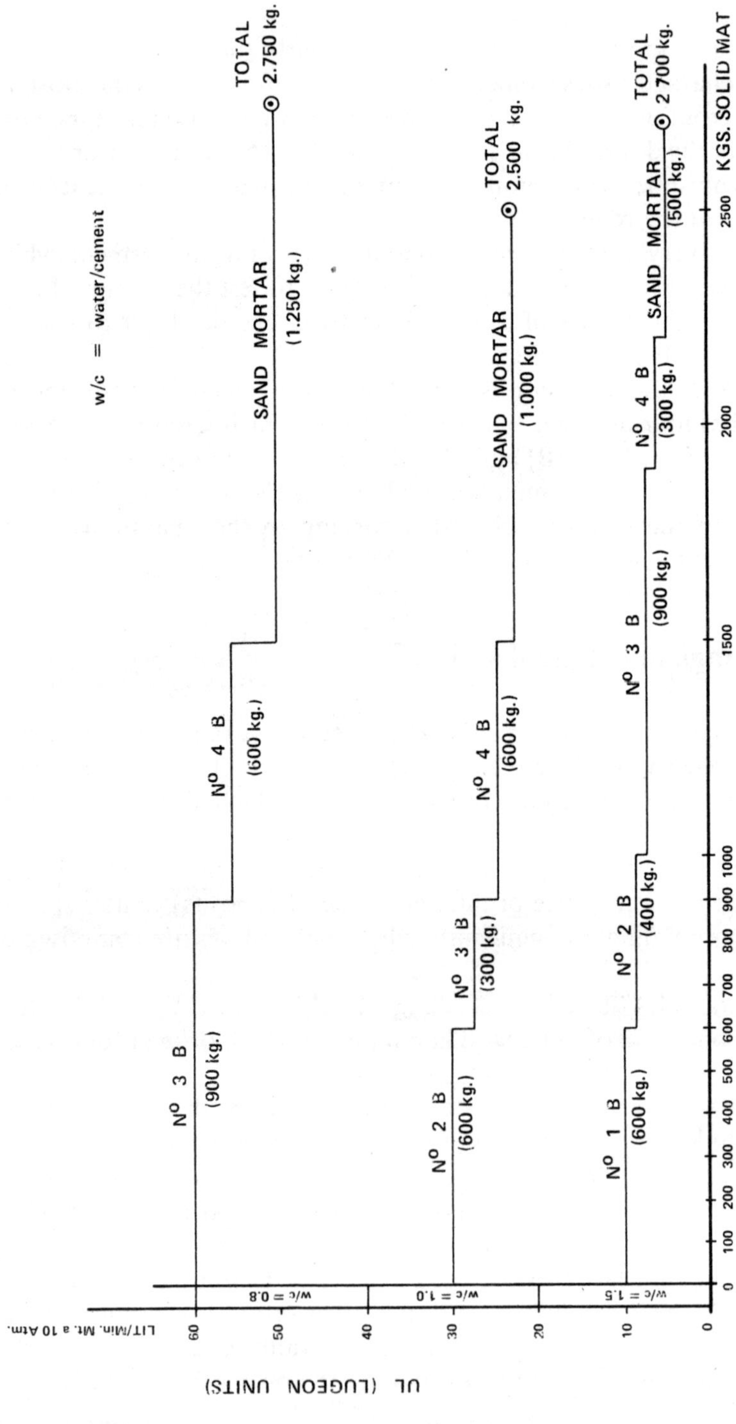

4–33. Slurry series versus permeability. Grouting job.

That risk can be avoided as follows: (1) during installation of the grouting screen, vigorously wash the filling from the karst fissures and galleries, with water circulation as noted above and, (2) when there is direct access to caverns from the reservoir, remove the filling and substitute a concrete plug.

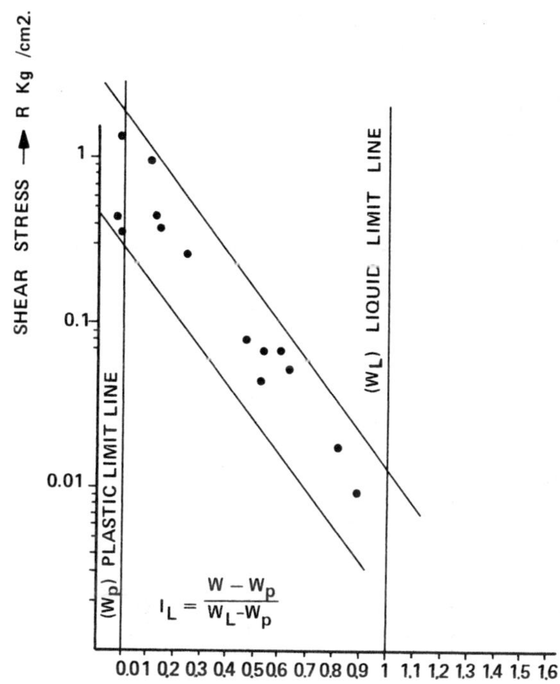

4–34. Relation between shear stress and fluidity index of several clays infilling caves.

Results

Studied cases in different karst areas of Spain and Latin America show that

— the grouting screen only intercepts a few special sites or points that are karst conduits, where consumption of grout is high. The aggregate consumption of special sites and points amounts to between 1/3 and 2/3 of the total accepted grout in the screen

Type of karst	No. of points	Grout %
mountain or Alpine	5	28
Mediterranean	7	67
alternating beds	2	37
near-surface or epigenetic	9	32

— the screen, except for the special sites and points, gives low consumption values, generally about 10 % of the average values of the screen for deep karst and about 50 % in the case of nearsurface karst;

Table 4–10

Unit costs and grouted quantity in any karst type to correct the water loss in dams built in karst areas

	$kp \cdot ml^{-1}$	$kp \cdot m^{-2}$	$ml \cdot m^{-2}$	$\$ \cdot m^{-2}$	Karst type
Min.	66	41	0.63	40	Mountain or Alpine
Max.	1 076	3 485	3.10	490	
$\bar{X}$	366	420	1.15	105	
5 special sites				28 %	
Min.	90	30	0.30	25	Mediterranean
Max.	3 250	2 925	0.90	300	
$\bar{X}$	582	343	0.59	81	
7 special sites				67 %	
Min.	22	11	0.25	12	Near-surface or epigenetic
Max.	32	66	2.10	50	
$\bar{X}$	85	24	0.28	23	
9 special sites				32 %	
Min.	3.3	1.65	0.50	5–6	Alternating beds
Max.	130	172	1.32	70–80	
$\bar{X}$	60	44	0.73	42	
2 special sites				37 %	

— in special sites the consumption may be about 400 to 900 % of the average values for the screen in deep karst and about 250 % for near-surface karst;

— unit costs of the finished screen vary with the type of karst in the range, 23 to 105 US dollars per square metre (Table 4–10);

— for more detailed data see AGROMAN (1979a, b), ERASO et al. (1984a, b) HERRERO, FERNANDEZ and PETKOVIC (1976), KRONSA (1974), THEROND (1973) and ERASO (1980a, b, c; 1981).

INTERACTION BETWEEN ENGINEERING AND ENVIRONMENT IN THE PRESENCE OF PALEOKARST: SOME CASE HISTORIES

Jerzy Głazek

The most frequent phenomenon is the unequal settling or subsidence of karstified ground underneath structures. This is often enhanced by suffosion and subrosion in the underlying paleokarst as well, when it is reactivated naturally or as a consequence of engineering works. In paleokarst there is often a shallow groundwater system comprising a network of partially filled solution channels and with groundwater resources sufficient to sustain an extensive economy. However, the increasing water demand of cities and industry accelerates circulation in the karst channels which may trigger the erosion of their deposits. Pumping lowers the piezometric level, draining great areas and causing the accelerated subsidence of regolith and overburden to form dolines. There have been considerable losses in construction works, roads, railways, etc. in many cases (JENKO, 1959; JENNINGS, 1971; GVOZDETSKY, 1972; HERAK and STRINGFIELD, 1972; ZÖTL, 1974; CALEMBERT et al., 1975; GŁAZEK and SZYNKIEWICZ, 1979; BECK, 1984).

Similar but more violent effects may be produced by the dewatering of underground mines or opencast workings in or adjoining paleokarst. The most famous examples are those of the South African gold mines in the **Far West Rand**. Karstified dolostones there are covered by an Upper Cenozoic weathering residuum up to 300 m thick. The dolostones overlie the gold-bearing quartzites. Dewatering of certain gold mines during a 10 year period created more than 200 sinkholes, some as large as 125 m wide or 100 m deep. 34 lives were lost during sudden subsidences, and approximately $ 35 millions have been spent on safety measures and compensations for damage and loss of water supplies (BEZUIDENHOUT and ENSLIN, 1969; QUINLAN, 1971; LEGGET, 1973; CALEMBERT et al., 1975).

There is anxiety concerning the brown coal mining operations that are being developed in the subrosion subsidence zone of **Bełchatów**, between the Warta and Pilica Rivers, central Poland (cf. GŁAZEK, p. 96–97; Fig. 4–35). The coal

P. Bosák, D. C. Ford, J. Głazek and I. Horáček (Editors):
Paleokarst. A Systematic and Regional Review. – Elsevier and Academia. Amsterdam and Praha. 1989.

seams occur among karstified Upper Jurassic limestones (GŁAZEK and SZYNKIEWICZ, 1979). A large paleokarst system is developed there. The dangers caused by the existence of the buried karst were disregarded initially (GŁAZEK and SZYNKIEWICZ, 1979). Karst channels and cavities are filled with clayey and

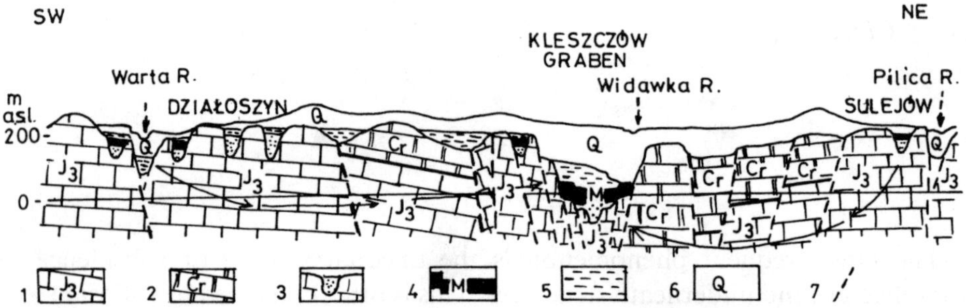

4–35. Schematic cross-section of the Łódź Synclinorium (after GŁAZEK and SZYNKIEWICZ, 1979). 1. Upper Jurassic limestones; 2. Upper Cretaceous marls and limestones, 3. Tertiary karst forms filled with clayey sands; 4. Early Miocene brown coal; 5. Middle Miocene clays; 6. fluvial and glacial Pliocene to Quaternary deposits; 7. faults.

sandy Tertiary and Quaternary deposits. Many such channels have been encountered in boreholes, extending to depths well beneath the planned level of exploitation. The surface of the limestones abound in Tertiary dolines exceeding 100 m in depth and with poljes that are some square kilometres in area, all filled with clay and sand deposits (GRADZIŃSKI, 1977; SZYNKIEWICZ, 1977; BOSÁK et al., 1979). Dolines can be observed in quarries on the Warta and Pilica River banks as well as beneath coal seams. Thus far, dewatering has not given the anticipated results. The aggregate quantity of pumped water is now several times greater than planned for in the original project. Concomitant with dewatering, there has been removal of sediments filling paleokarst channels, causing ground subsidence and collapses in infilled dolines. The Warta River is in a state of emergency due to drainage into the reactivated paleokarst system. Dams are being planned in the danger areas along this river. Saline warm waters were encountered in some dewatering wells. Moreover, mining operations induced earthquakes reaching 4.6 M (GIBOWICZ, GŁAZEK and WYSOKIŃSKI, 1981).

Similar effects are found in underground mining where karst systems are developed in the overburden or in the rock being mined. One example is the **Zonguldak Coal Mine** (Turkey). The Catalagzi Shaft, piercing karstified Cretaceous limestones, was catastrophically flooded. It required two and a half years of dewatering at a cost of about $ 1.6 million before the mine was secured again. Pumping rates were up to 24 m^3 per minute, and as much as 640 tonnes of sand and clay per day were pumped along with the water on occasion. The solids had been deposited in karst channels and dolines; their removal caused subsidence that extended to the surface (MATTES, 1975).

Satisfactory dam construction in karst country remains to some extent a matter of luck. There are many examples of collapsed dams or of total water loss from reservoirs, as well as of long lasting and very expensive technical measures being applied to stop subsurface water loss from the storage basins. Every successful modern construction has been preceded by extensive studies and then ensured by extensive safety measures. For example, at **Peruća** near Split (Yugoslavia) a deep triple grout curtain was emplaced at a cost of 6 billion dinars. The total costs of the dam and hydroelectric power station were 25 billion dinars (JENKO, 1959; ZÖTL, 1974). In the case of storage reservoirs, the gravitational energy (head) of the accumulated water has caused reactivation of groundwater circulation in paleokarst filled with unconsolidated deposits in many instances.

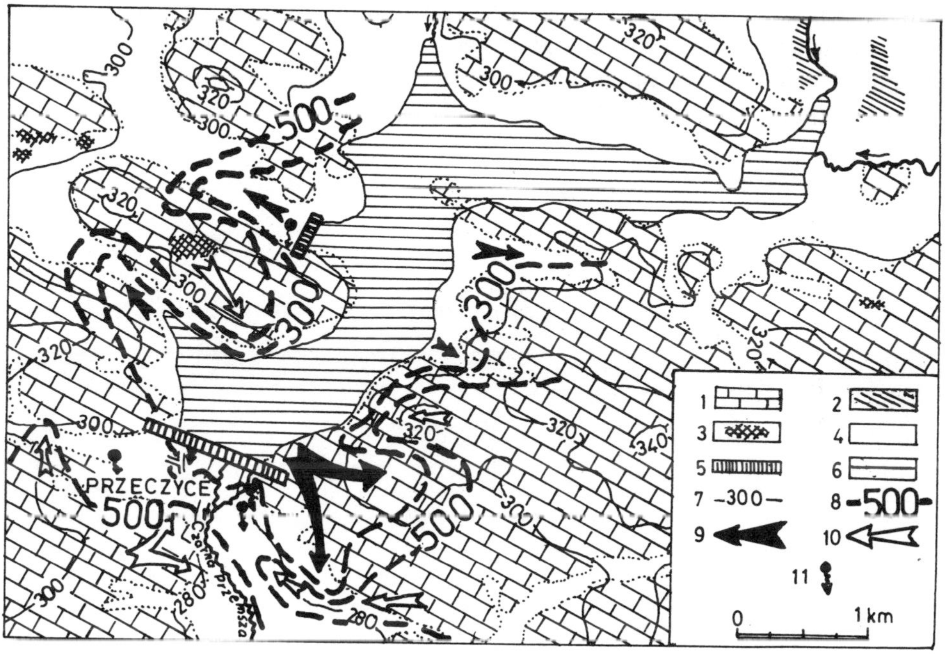

4–36. Sketch-map of the Przeczyce Reservoir (after GŁAZEK and SZYNKIEWICZ, 1979). 1. karstified carbonate-sulphate-marly deposit, Rhaetian and Muschelkalk; 2. clayey-sandy deposits, Keuper and Lower Jurassic; 3. clayey-sandy infill of paleokarst dolines; 4. sands and clays infilling valley bottom, Quaternary; 5. dams; 6. reservoir; 7. contour lines in metres; 8. isolines of groundwater mineralization in mg · l^{-1}; 9 direction of water lost from the reservoir; 11. main discharge points of water appeared after reservoir infilling.

The dam at **Przeczyce** near Zawiercie (Poland, cf. GŁAZEK, p. 77–105) was constructed at the mouth of the Czarna Przemsza River on an area composed of Triassic carbonate, sulphate and marly rocks. In the original project (in the late Fifties) karst investigations were omitted because almost all the karst forms visible at the surface were completely filled with sandy-clayey deposits. Since the dam was constructed (in the early Sixties) there has been much research and application of safety measures to prevent the stored water from escaping around and below the dam. PULINA (1974) has demonstrated that the contemporary karst denudation in the region is approximately 25 mm per 1 000 years and that the river water is nearly in equilibrium with that in the solid carbonates. Investigations of total dissolved solids in well, piezometric and spring waters showed that there was a flow of lesser mineralized water (up to 270 mg $\cdot$ l^{-1}) from the reservoir (RYŻAK, 1977). The water from the reservoir passes through Pleistocene sands on the drowned valley floor and through karst channels in carbonate rocks beneath them (Fig. 4–36). Simultaneously, there was an inflow of highly mineralized karst water into the reservoir and mixing of the highly mineralized karst water with river water directly beneath the dam. The costs of post-construction research and safety measures adopted are now greater than the entire original cost of the dam. This experience has been repeated many times in other countries.

PART V

PALEOKARST AS A SCIENTIFIC SUBJECT

SPECIAL CHARACTERISTICS OF PALEOKARST STUDIES

Ivan Horáček and Pavel Bosák

The universal pragmatic definition applicable to any area of scientific interest undoubtedly holds here: paleokarst studies are what paleokarst students do. However, a further question arises: because paleokarstology as a clearly defined branch of scientific endeavour does not exist, who are the paleokarst students?

Paleokarst research is the multidisciplinary study of the many and varied phenomena associated with karstification in past geological periods. The study encompasses the karst geomorphologists' interest in local karst genesis and the origin of recent karst landscape, the paleontologists' work on the structure of fossil communities found in karst forms, the geophysicists' attempt to elucidate irregularities in the physical properties of a karst region, as well as the problems facing engineering geologists when planning dams or mines in karst terrains and the contribution of economic geologists concerned with prospecting and exploring for mineral deposits in karst areas. Similarly, many problems of paleokarst interest occur among those investigated by hydrologists, hydrogeologists, geologists studying local tectonics and stratigraphy, as well as the petrologists' interest in the geochemistry of carbonate denudation and diagenetic processes, or the biologists' attempt to interpret an origin, phylogeny and ecology of biotic communities in caves. All these and many other branches contribute data which might be of considerable importance for a particular paleokarst study.

Richer and yet wider would be a survey of the published research topics that have arisen from data collected from paleokarst phenomena. They fill a broad span from e.g. the paleogeography of animal evolution to the rules of hydrothermal mineralization, from the dynamics of clay mineral cycles to the assessment of engineering hazards, or from relationships between tectonic activity and denudation to the techniques of quantitative geochronology.

Already these few examples reveal one of the most important and distinctive characteristics of paleokarst studies: it is truly multidisciplinary. This is true of the data as well as of the methods applied to their investigation.

P. Bosák, D. C. Ford, J. Głazek and I. Horáček (Editors):
Paleokarst. A Systematic and Regional Review. – Elsevier and Academia. Amsterdam and Praha. 1989.

Only a few of those who contribute to our knowledge of paleokarst are *professionals,* i.e. specialists in paleokarst research. But the smaller their number the more acute is the need for syntheses of all the collected data in order to understand the phenomenon of paleokarst. This, of course, necessitates understanding both the true scope of particular approaches that are relevant in the study of paleokarst phenomena and the significance of their particular data within the framework of paleokarst topics. One special characteristic is that karst is limited to relatively soluble rocks. The dissolution takes place in various conditions ranging from surficial to deep subterranean environments. Karst forms can be polycyclic and polygenetic. After they have been created they can (1) remain without any cover; (2) be partly covered; (3) be completely covered, and (4) be buried. Each of these events can fossilize karst forms or, on the other hand, can develop them further in a polygenetic fashion.

After their creation, bare karst forms can become relict structures in the landscape. Their further development may be characterized by periods of relatively limited karstification and periods of intensive evolution. In the former, chemical weathering is supplemented or supplanted by mechanical effects of weathering in dry or cold climates (destruction of karst forms). Similarly, buried karst can be enlarged and remodeled by later, deep-seated processes such as hypogene intrastratal karstification, or can serve as routes for the hydrothermal circulation of ore-bearing solutions (i.e. Pb-Zn or others) or for hydrocarbon migration and accumulation.

From these relatively simple examples arises the problem of deciding which stage of karst development represents fossil karst or paleokarst. Probably, all fossilization results from a significant change of geotectonic, geomorphologic or climatic conditions, when further evolution of the karst can continue only in different circumstances and environments. Another aspect worth mentioning is that what we designate as paleokarst represents a fossilized state of karst development just from our modern, time-limited perspective i.e. the thermodynamic system has reached some quasistationary state with local minimum dissipation. In the future, each disturbance of that stationary state will cause some rejuvenation of the forms. This operated also in the geological past; forms fossilized in a certain epoch could be rejuvenated and remodeled to such an extent that fossilization events often cannot be distinguished.

The *fossilization* represents perhaps the most striking characteristic of paleokarst and has resulted in the many different approaches taken to paleokarst topics. Fossilization preserves the phenomena existing at a certain time for later detailed studies. Fossilization includes the shape of karst forms (both subterranean and surficial), traces of animal life, fossil remains in sedimentary infills, petrological properties of karst infill, its structures and textures, changes in clay mineral composition or in the character of heavy mineral assemblages, degree of weathering, soil types, geochemistry and diagenesis of deposits, etc. It records the processes prevailing at time of fossilization e.g., type of sedimentation and/or erosion, destructive and creative karst processes, climatic and hydrologic conditions and/or the patterns of biotic environment as suggested by paleontological analysis of plant and animal fossils, etc. From these perspectives, paleokarst phenomena may be considered true *conservers* of the geologic and environmental past; but frequently they are „degraded" or „unreadable" because of some overlooked reactivation.

Paleokarst phenomena are mostly of a terrestrial nature. In contrast to the other records of terrestrial history, which tend to be indirect only (being assumed from gaps in the marine sedimentary sequences), they provide extraordinary opportunities to study phases of terrestrial development directly and in detail. The polycyclic and polygenetic nature of a great number of paleokarst forms supports the importance of paleokarst for such investigations.

Important marine records are also obtained from paleokarst, including those which show marine and terrestrial phenomena at the same place and in direct superposition. Such records are of substantial importance — they are keys to locked doors of lithostratigraphic and lithofacies correlation of different marine and continental formations and continuous geochronology of the geological past.

We can conclude without hesitation that for chronostratigraphy the paleokarst records provide „missing links" as important as the remains of the Australopithecines (obtained, by the way, from paleokarst deposits) were for the interpretation of human origin.

The study of paleokarst features both in time and in space at a global scale will enable specialists to map local anomalies in karstification, erosional and depositional history. Paleokarst records promise to become important tools for detailed studies of *tectonic history,* at both local and global scales, and a source of data for geotectonic reconstructions.

To exploit a paleokarst record in such a way necessitates, of course, the exact dating of the formative and, particularly, the fossilizing events and processes. The study of paleokarst is therefore guite closely connected to the various approaches of dating the geological past. A number of radiometric methods, thermoluminiscence dating, fission track studies and paleomagnetic investigations are applicable to the study of paleokarst. Some of them have been developed specially e.g. uranium-series dating applied to speleothems. Particularly important is biostratigraphic dating based on paleontological analyses of the fossil remains preserved in karst deposits.

Dating efforts are frequently undertaken by specialists who are not familiar with other aspects of paleokarst topics. There is a certain danger in this. It is not rare that e.g. a paleobiologist studying paleokarst deposits does not understand the details of karst sedimentation — the polycyclicity of much cave deposition and the consequences arising from that fact. Caves are not only infilled but also exhumed and perhaps rejuvenated. Erosion of any fossil infill may then be either complete or only partial, while a new sedimentation phase can follow immediately or after a long interval. As a consequence, one can encounter deposits of very different ages in the same place and in direct superposition. Moreover, as a result of similar processes (plus others such as ejection of former infills following the collapse of a plug) sediments can be from an original site and redeposited where they may cover much younger fills. Consequently, not only radical breaks in deposition but also apparently reversed superposition in sequences may appear in paleokarst deposits. Though such cases can be distinguished fairly readily in many instances, in others they cannot. This is especially true when congelifraction, burrowing animals, etc. cause such mixing. Serious complications arise where the differences between series of sediments are obscured or not distinct. It seems that this is also a relatively common effect as highly variable depositional rates are a common feature of many karst deposits. Proper relationsphips may then be distinguished only if detailed investigations are carried out by means of all possible methods, i.e. petrologic, mineralogic, geochemical, sedimentological, paleontological, numerical or other dating, etc. The more methods that are applied, the greater is the chance that the interpretation achieved will be appropriate and accurate.

Engineering geologists and economic geologists are familiar with another specific characteristic of paleokarst, i.e. its recent inactivity and the hidden nature of most forms. The influence of paleokarst on the foundations of large industrial complexes or dams or on the exploitation of mineral deposits both at the surface and in the subsurface is well documented in Parts 2, 3 and 4 of this book. However, in spite of these characteristics being now well known, many designers commonly overlook them, which leads to great economic losses.

The considerable records of paleokarst phenomena can become of real utility if their many facets are studied by teams of specialists representing all the branches which can contribute to the study of paleokarst. Paleokarst study is *multidisciplinary* in the full meaning of the word.

TECTONIC CONDITIONS FOR KARST ORIGIN AND PRESERVATION

Jerzy Głazek

Traditionally, the relationships between karstic and tectonic processes have been (and, in some cases, still are) treated on the basis of the now discarded geosynclinal theory. However, during the last decade some authors have adopted modern plate tectonic concepts in this matter e.g. GUIDA (1978). In the writer's opinion, very few studies of paleokarst take sufficient account of the wider tectonic settings. Not suprisingly therefore, these differences in treatment often lead to confusion. Some of this confusion is readily removed by comparisons of models and terminology, e.g. as presented by BOILLOT (1983, p. 120–127); other parts of it need further investigation. Matching the tectonic circumstances of paleokarst with facies is seldom easy and frequently decisions have had to be made between a subdivision of the conditions on the basis of the facies or of the tectonic structure. Moreover, tectonic interpretation of paleokarst features is a branch of earth science that is in its infancy; it is at present extremely difficult to correlate many ancient karst phenomena with particular tectonic stages. This is especially the case for many subjacent karst phenomena, e.g. those in deep (bathyphreatic) artesian confinement, or hydrothermal karst phenomena etc.

Karst phenomena are largely restricted to domains of carbonate and evaporite sedimentation on passive continental margins and some oceanic rises (microcontinents) or sea-mounts elevated above contemporaneous CCD, as well as in some epicontinental basins (Fig. 5–1). Active continental margins and island arcs are the sites of dominantly siliciclastic sedimentation. For further discussion of the relationships between carbonate and evaporite sedimentation and tectonic conditions, some modern text-books and outstanding papers are: BERNOULLI and JENKYNS (1974); WILSON (1974, 1975); READING (1978, 1982); SCHLAGER (1981).

The most extensive karst areas have been located principally on passive continental margins and adjacent platforms where there was lengthy deposition of carbonate-evaporite sequences, e.g. in pre-Pennsylvanian time on

P. Bosák, D. C. Ford, J. Głazek and I. Horáček (Editors):
Paleokarst. A Systematic and Regional Review. – Elsevier and Academia. Amsterdam and Praha. 1989.

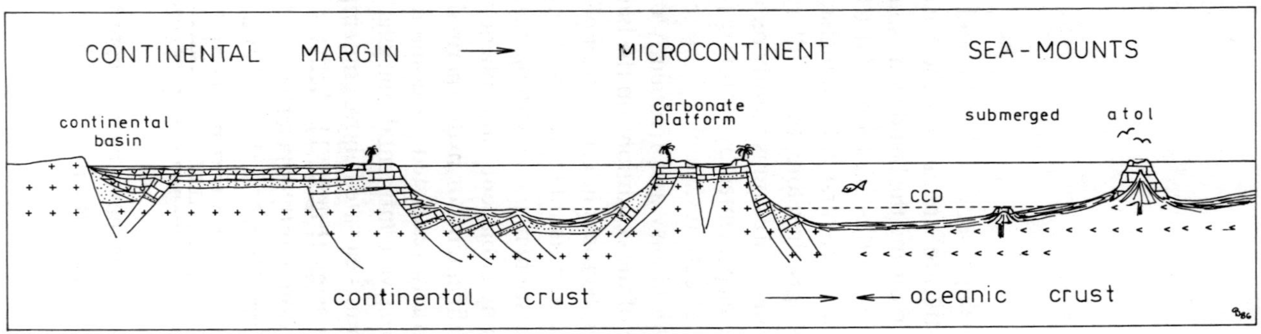

5–1. The position of carbonate depocentres in continental margin, carbonate platform and on sea-mounts.

the North American Platform or during pre-Paleogene time on the Central European Platform facing the Tethys Ocean. Less extensive, but also very important, karst areas occur within orogenic belts where carbonate platforms developed on continental blocks (microcontinents) e.g. in the Mesozoic Tethyan domain, a former narrow ocean. Within some eroded collisional orogens, metamorphosed sequences of volcanic and clastic rocks contain isolated lenses of marbles. They can be interpreted as the carbonate caps of sea-mounts involved in the collisions. Such marble lenses are often very strongly karstified as a result of the huge inflows of allochthonous waters from surrounding areas of silicate metamorphics (Fig. 5–2).

Synsedimentary and postsedimentary tectonic movements deform bodies of karstified rocks. Synsedimentary block faulting and tilting cause many changes of thickness and facies pattern inside the domains of carbonate-evaporite sedimentation. Synsedimentary faults are commonly rimmed by calcirudites on the downthrown side. These strata are favourable for karst development as consequence of their high primary (intergranular) porosity. Simultaneously, on the upthrown side, carbonate buildups frequently develop. They are even more favourable for karstification because of the primary cavernous organic framework. The occurrence of such lithologies along synsedimentary faults will have favoured karstic activity since their deposition. Later tectonic evolution may cause many changes in fluid circulation in them, leading to accumulation of metallic minerals, oil and gas.

Postsedimentary tectonic deformations result in additional surfaces or zones of rock failure and loosening. As a consequence, soluble rocks reveal complicated patterns of porosity in which the primary (sedimentary) porosity and diagenetic voids are considerably supplemented by secondary porosity along joints, fracture zones, fault planes and bedding planes. A good review of such conditions is given by T. D. FORD (in FORD and CULLINGFORD, 1976).

Aggressive fluids (water or brines able to dissolve solid rocks) transferred along these flow paths develop supplementary karst porosity by void enlargement in an open hydrochemical system. When the system becomes closed (as a result of subsidence and/or burial) the dissolution is quickly stopped and the voids are filled with solutions saturated with respect to the host-rock. Competition between different water paths leads to the development of karst conduits along the flow paths of lowest resistivity (D. C. FORD, 1971). Thus, developed karst systems are characterized by well organized conduit flow and the diminution of the importance of diffuse flow in the karst water circulation. In paleokarst systems some periods of differing direction and chemical composition of brines can often be recognized. They are separated by periods when the system was in a steady state. All of the changes are the results of tectonic instability of the karst areas or in their surroundings.

The direction and rate of water or brine flow in karst rocks depends on the

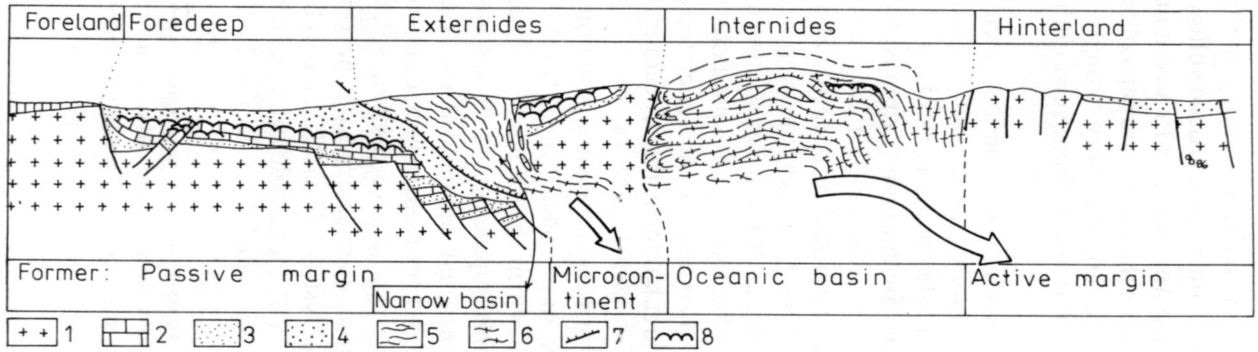

5–2. Sketch showing the relation between former domains of sedimentation (see Fig. 5–1) and the distribution of karst rocks in collisional orogen. 1. continental crust; 2. karst rocks; 3. pre-orogenic clastics of former continental margins; 4. orogenic clastics (molasse); 5. folded clastics of former extended continental crust (flysch); 6. metamorphic rocks of former oceanic basin and accretionary prism, note lenses of marbles – former sea mount deposits; 7. main overthrusts; 8. traces of paleokarst surfaces; arrows mark the direction of subduction.

pressure fields within the primitive void spaces. Fluid exchange starts as a syngenetic process responsible for early diagenesis and is discontinuously developed during the further geologic evolution. Thus, any tectonic movement changing the fluid pressure in the karst system must result in some change of its behaviour or function. To a first approximation the pressure field of karst waters is represented by the piezometric surface and measured as the topographic head. However, differences in brine density caused by the amount of dissolved matter or by temperature variations may considerably exaggerate the topographic head responsible for water circulation. This is shown by the many cases of submarine freshwater springs, and in the example of water being heated in deep, confined flow paths beneath the Hungarian Plain that is discussed by LIEBE and LORBERER (1978b) who called this phenomenon *heat lift*. Moreover, water circulation is affected by tectonic pressure and during glaciation by icesheet pressure which may expel water from deep, nearly steady aquifers. All these cases strongly affect paleokarst systems and may offer simple explanations of distant and deep water circulation in paleokarst conduits. Many such deep water circulation systems have been recognized recently (cf. D. C. FORD, p. 313–336; MÜLLER, p. 155–163; ZÖTL, p. 483–509) and are the result of tectonic deformation of paleokarst aquifers and/or glacial crustal flexing and loading (D. C. FORD, 1983a). The groundwater residence times in such deep circulation systems may be very long; some dating attempts have indicated durations of 20–50 ka or more.

The flow paths in karst are developed preferentially along fault zones, joints, bedding planes and the more porous lithosomes, in the direction of the lowest pressure in the rock body. That means along a direction perpendicular to the σ_3 axis (axis of the lowest stress). Such a rule is observed in dam construction in terrains (ERASO, 1981). It is also confirmed by many tracer tests in karst areas, where flow paths are observed to be oriented along antiform axes and perpendicular to them. This is a consequence of the two different stress patterns that existed during the folding and subsequent uplift. Thus detailed tectonic studies are very useful in determining possible flow directions in paleokarst aquifers.

For a long time karst investigations were concerned almost entirely with shallow or superficial phenomena. However, during the last few decades many examples of karst water circulation have been detected beneath not only local base level but even 2 km beneath sea level, e.g. ZÖTL (p. 483–509). Such deep circulation appears to be the result of the tectonic depression of karst aquifers with already established conduit flow. Such conduits, which developed in the geological past, frequently retain their lesser resistance to flow even after deep subsidence. Paleokarst conduits generally retain or are able to recover their function after deformation.

Deep water circulation in confined karst aquifers like those recognized beneath the Hungarian Plain (cf. MÜLLER, p. 155–163) or in mountain chains such as Erma Reka, Bulgaria (cf. STANEV and TRASHLIEV, p. 217–230) sometimes lead to the emergence of hot brines at springs. Examples of such paleokarst water circulation may be revealed in geological records by reports of mineral precipitates and the special corrosional forms that are recognized as being hydrothermal in their origin (RUDNICKI, 1979; MŮLLER, p. 155–163). The most important mineral deposits of this type are MVT zinc-lead deposits (DŻUŁYŃSKI and SASS-GUSTKIEWICZ, p. 377–397). Their age and the tectonic conditions at their formation are most difficult to determine because they have no simple genetic association with datable surface deposits. However, they occur within continental margins or carbonate platforms, in those portions that were once sunken but adjacent to elevated and eroded parts of the same platforms. Probably their development was caused by deep water circulation from the elevated blocks into the subsided ones (cf. GŁAZEK, p. 77–105).

Other results of subjacent karstification, especially of subrosion, include: rock draping or other disturbances, brecciation, tilting of blocks, occurrence of breccia pipes, etc. All of these phenomena may be treated as features of karst tectonics (BALWIERZ and DŻUŁYŃSKI, 1976; cf. DŻUŁYŃSKI and SASS-GUSTKIEWICZ, p. 377–397). Such phenomena cannot be attributed to particular orogenies but are the delayed results of uplift that led to the deep circulation of karst water. Their ages can be constrained by the ages of deposits filling any surface depressions that have developed in them. Some of these features reveal several distinct periods of subsidence e.g. Bełchatów brown coal deposit in Poland (GŁAZEK, p. 96–97). The stratigraphic records contained in such paleokarst features generally lag behind the proper age of their karstification because we are dating using deposits that fill already formed depressions or cavities, for which the earlier erosion of recessive deposits such as fine-grained siliciclastics and of overlying karst rocks is necessary. However, brief and rather insignificant emergences of carbonate platforms that result from sea level fluctuation (e.g. glacial-eustatic) tend to reveal no such retarding.

In conclusion, different scales of karstification periods should be recognized because they have differing relationships to the tectonic processes and cycles that control them.

(1) The most important and longest periods of emergence and karstification were induced by orogenies; e.g. post-Sauk and post-Kaskaskia in the North America (PALMER and PALMER, p. 341–355; D. C. FORD, p. 317–326), post-Variscan in Europe (PANOŠ,1962, 1964a, 1965; GŁAZEK, 1973, p. 85–87; BOSÁK, HORÁČEK and PANOŠ, p. 113–118). Within these long periods distinct phases and stages of karstification can be recognized. These periods are caused by lengthy post-collisional uplift of orogens and their fringes. The periods are marked by unconformities over broad areas and need not be confined to individual modern continents. The best examples of them are the post-Kaskadia and post-Variscan karstifications, which are generally synchronous within the regional units related to the present Atlantic sector of the Earth's surface.

These long periods of karstification revealed diachronicity and many lesser phases in detailed studies. They are longest in duration and most complex (as well as less recognizable due to strong erosion) at former mountain crests and become gradually shorter but better datable on the former mountain slopes and their broad fringes along adjacent continents, e.g. the Mediterranean Alpine belt and Central European and North African Platforms.

These periods resulted from the principal changes of plate motion patterns and they divided structural complexes (complexes plissécs) corresponding to orogenic cycles (GŁAZEK, 1973).

(2) A second category of shorter periods of karstification are represented as unconformities within the limited area of one past shallow marine platform and its continental fringes, or of one continent created by the collision of two plates. They do not display so many phases of karstification. They can be attributed to changes in plate motion in a segment of a spreading ocean, e.g. the Early Cretaceous period of karstification in Central Europe (cf. GŁAZEK, p. 75–105; BOSÁK, 1981). Such minor changes in part of one segment of the Earth may be induced by orogeny (principal change) in other segments. As an example, the short period of karstification at the Triassic/Jurassic boundary in Europe coincides with the greatest change of plate motion in the Pacific segment (the Indosinian Orogeny). These shorter periods of karstification may be correlated with traditional *orogenic phases* (STILLE, 1924) or a few neighbouring *phases* e.g. „Bretonic phases".

(3) The shortest karst episodes recognized on many ancient and recent carbonate platforms are disconformable or paraconformable surfaces resulting from glacial-eustatic fluctuations of sea level or from local tectonic events. Such episodes are widely known and well dated in the Pleistocene at the global scale (DALY, 1934; HARMON, 1980). Some Late Paleozoic instances of emergence probably had similar origins (cf. RAMSBOTTOM, 1979). Others occurring in limited areas can be attributed to local crustal swelling with temporary uplift as consequences of rising magma, fault activity, etc.

Further studies of such short karst episodes will contribute considerably to the discussions of event stratigraphy, both at local and global scales (e.g. JELETZKY, 1977; REYMENT and MÖRNER, 1977; RAMSBOTTOM, 1979).

Finally, paleokarst features function as traps for the stratigraphic records of long periods of continental weathering, i.e. times that are lost in the marine records. Thus detailed studies of paleokarst may be able to contribute considerably to the important debate about global or local mountain building (cf. STILLE, 1924 and GILLULY, 1949). I believe that such paleokarst studies should substantially resolve this debate in favour of GILLULY's idea, which is applicable to modern plate tectonics. In it, larger or smaller changes in the lateral movements of continents, the growth of mid-oceanic ridges, and heating or cooling of lithospheric segments can explain all „orogenic phases" that are detected, not at the global scale, but in greater or lesser, but limited, areas.

PROBLEMS OF THE ORIGIN AND FOSSILIZATION OF KARST FORMS

Pavel Bosák

Introduction

The origin and the fossilization of karst are the results of complex sets of processes with important geotectonic, geomorphologic, paleoclimatic and paleoecologic aspects. Some of the processes are observable at the present time but certain others have no modern equivalents. Karst origin and evolution has been intensively studied and the current state of the art is presented in numerous research papers, monographs and textbooks. However, no clear but short and undisputed definition of karst has been arrived at up to now, cf. e.g. ROGLIĆ (1972), JENNINGS (1971, 1985), etc. The fossilization processes, although studied at many sites, have not yet been comprehensively summarized. This is probably a consequence of the complex and diversified nature of the subject. Data concerning karst fossilization are scattered in numerous geological, paleontological or geomorphological studies where the main interest, in many cases, is in other than the karst phenomena (cf. HORÁČEK and KORDOS, p. 599–612).

Karst features result from rock/water interaction where comparatively soluble rock is dissolved in natural waters. The creation of karst „needs moving water, rain or meltwater and local relief to give power for work underground as well as on the surface are required, though it is now thought that occasionally both the water and its energy may be provided in other ways“ (JENNINGS, 1985, p. 2). Some hypogene subjacent or intrastratal paleokarst was created by expelled connate fossil waters circulating due to pressure during compaction and lithification or during tectonic movements (cf. PALMER and PALMER, p. 337–363). Similarly, hydrothermal paleokarst created by connate, petroleum or meteoric waters deeply circulating within karst massifs is very extensive. The role of juvenile thermal water is probably very limited.

Karst Rocks

Karstifiable rocks, i.e. those more soluble than others, are represented predominantly by the carbonates, although karst develops also in the evaporites

P. Bosák, D. C. Ford, J. Głazek and I. Horáček (Editors):
Paleokarst. A Systematic and Regional Review. – Elsevier and Academia. Amsterdam and Praha. 1989.

magnesite, fluorite, etc. In particular conditions, such as in hot and wet tropical climates, most other rocks are also „soluble". BÖGLI (1980) described this behaviour as a readiness to decompose rather than simply being dissolved. Aluminosilicate minerals are readily disintegrated and partly destroyed but, in contrast to the carbonates, they leave a large quantity of insoluble residues. The rock is thus affected mostly at the surface and no subterranean forms occur because incipient conduits are blocked. Only one type of silicate rocks — quartzites — can be numbered amongst the karst rocks. True superficial and subterranean karst forms occur in quartzites in Venezuela (e.g. SZCZERBAN and URBANI, 1974), in South Africa (MARTINI, 1979) or U.S.S.R. (TSYKIN, p. 268). The process of silica solution is normally very slow and so needs a long period of hot and wet climate to create karst topography. However, in moderately alkaline and highly acidic solutions silica is easily soluble (SIEVER, 1962). Such conditions are not known on the contemporary Earth's surface except at volcanic hot springs. It is possible that during the Proterozoic, environments with favourable conditions for such karstification were more common. Important karst rocks, in our experience, are those dissolved by waters where the dissolved matter is transported out of the karst terrain in solution.

Paleokarst features are preserved mostly in carbonate rocks (limestone, dolostone) and in quartzite, i.e. in rocks with „strength adequate for the support of cave roofs and walls" (JENNINGS, 1985, p. 2). Evaporites, i.e. gypsum, anhydrite and rock salt, are quite plastic and are transformed, deformed and recrystallized during their burial history; this means that only subordinate paleokarst is preserved in them. This feature is clearly reflected in the regional reviews in this book; the presentations deal mostly with features in carbonate rocks and only exceptionally with forms preserved in evaporites. Paleokarst in carbonatites is a special case.

Carbonate Rocks

For carbonate sedimentologists, karstification is a diagenetic process (cf. ESTEBAN and KLAPPA, 1983; LOHMANN, 1985). For some other specialists, karst is the product of the epigenesis of emerged carbonate rocks under the effect of subaerial exposure (PANOŠ, 1983). The karst develops after the emergence of the rocks from the sea. This process is accompanied by the creation of a freshwater lenses and freshwater diagenetic zones in near-surface environments. The later uplift of the carbonate terrains leads to the lowering of the base of erosion, enlargement of the vadose zone and stabilization of the karst hydrologic systems.

Diagenesis

A majority of carbonate rocks originated in the sea, where their diagenesis also began. Early diagenetic environments are typically characterized by a complex sequence of processes (cf. BATHURST, 1976; CHILINGAR, BISSELL and WOLF, 1979; LONGMAN, 1980, etc.), some of which lead to the creation of small-scale secondary porosity (ELF-AQUITAINE, REECKMANN and FRIEDMAN, 1982) or of larger, even karstic, vugs.

The boundary between solution vugs and pores on the one hand and karst cavities on the other may be defined either from the speleological point of view where a cave is a space large enough to be accessible to a man (JENNINGS, 1971, 1985; BÖGLI, 1980); or according to the physical properties of water flow through porous media: "a cave is a solutional conduit or other void that possesses dimensions large enough for turbulent flow of water to occur", i.e. 5–16 mm in diameter as a minimum (FORD and EWERS, 1978, p. 1784).

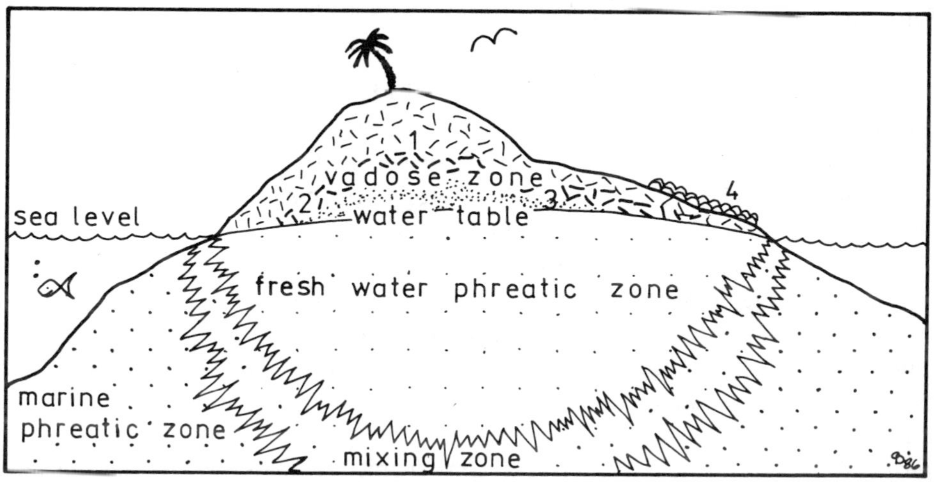

5–3. Idealized cross-section of nearsurface diagenetic zones in ideal sand island; not to scale (modified after LONGMAN, 1980). 1. intense solution near soil zone; 2. minor solution; 3. zone of precipitation (formation of calcite cements); 4. formation of caliche crusts.

Early diagenetic, near-surface environments are characterized by the development of several distinct diagenetic zones: (i) marine phreatic; (ii) marine vadose; (iii) mixing phreatic; (iv) freshwater phreatic, and (v) freshwater vadose zones (LONGMAN, 1980; Fig. 5–3). Solution in the freshwater vadose zone and at the freshwater vadose-phreatic interface is the most important for karst formation in the early stages following the emergence of carbonate rocks. Karstification in these zones occurs in emerged islands, atolls and reefs, carbonate platforms and terraces. Caves commonly form near the top of the freshwater phreatic zone and may extend over a vertical range of hundreds of metres, depending on the position of the water table and the rate of saturation with $CaCO_3$. The marine mixing phreatic zone appears to be also a place of pronounced solution (PANOŠ, 1976; RUDNICKI, 1980).

Secondary porosity originated by solution in the early stages of diagenesis is very important in hydrocarbon reservoir formation if the rock becomes buried by younger strata in the late stages of diagenesis. Late diagenesis is characterized by recrystallization, neomorphism and other processes and can continue even where the upper parts of a basin fill are exposed to subaerial conditions. For that reason the boundary between diagenesis and epigenesis (or hypergenesis) is not sharp and processes of both zones can mix. Solutional processes occur also in the stage of late diagenesis, associated with circulating connate, fossil waters or related effects. For terminology and a review of diagenetic processes see e.g. CHILINGAR, BISSELL and WOLF (1979).

Islands and Coastal Areas

Carbonate sediments in islands or coastal platforms accumulate up to high tide level and sometimes above it during times of stable sea level. Vadose diagenesis with solution is almost entirely prohibited at this stage. Sea level lowerings or uplift of the island, etc. create an extensive vadose zone in rocks deposited in any marine environment (LONGMAN, 1980).

The freshwater phreatic zone develops as a lens with a form controlled by differences in the densities of freshwater and seawater, according to the Ghyben-Herzberg theory. About 40 m of freshwater below sea level is supported by a phreatic head of 1 m above it. This implies that the freshwater can extend to depths of hundreds of metres beneath coastal regions of moderate relief. This extent also depends on factors of rainfall, local topography, distribution of permeable zones in the rocks, etc.; in rainy tropical climates the freshwater lens may be very extensive and in arid climates it may be completely missing (LONGMAN, 1980).

By these means, fluctuations of sea level influence the development of karst forms in islands and coastal areas. Lowering of the sea instigates karstification. Sea level rise cause its fossilization and even destruction, cf. effects of sea level fluctuations in the High Carbonate Islands of the Pacific Ocean (BOURROUILH-LE JAN, p. 443–471). Features of karst in reefs and emerged platforms are summarized in JENNINGS (1985). Fossil equivalents are reported e.g. from the Proterozoic of Canada (D. C. FORD, p. 315–317), the Permian of West Texas, U.S.A. (CRAIG, 1985), the Carboniferous of Great Britain (T. D. FORD, p. 51–70), from the Devonian of Moravia and Triassic of Slovakia (BOSÁK, HORÁČEK and PANOŠ, p. 111–112) and Poland (GŁAZEK, p. 85–87). Long-lasting lowering of sea level results in extensive caves and cave systems. Short emergence leads to pitted or potholed surfaces with small depressions or karren (Fig. 2–11) infilled by sediments of the next cycle of transgression.

Global sea level changes caused either by eustatic or by tectonic events influence the growth and drowning of reefs and carbonate platforms. SCHLAGER (1981) summarized examples of global sea level changes and their effects in the past. Drowned platforms typically exhibit a sequence of neritic deposits rapidly passing upwards into deep-sea deposits. Some platforms show the effects of meteoric (freshwater) diagenesis, suggesting exposure of the platform prior to drowning; e.g. a shaft in Upper Devonian reef and back-reef limestones filled with Lower Carboniferous sediments (greywackes) in the Čebín area, Czechoslovakia (cf. site 8 on Fig. 2–41, Fig. 2–42; BOSÁK, 1983).

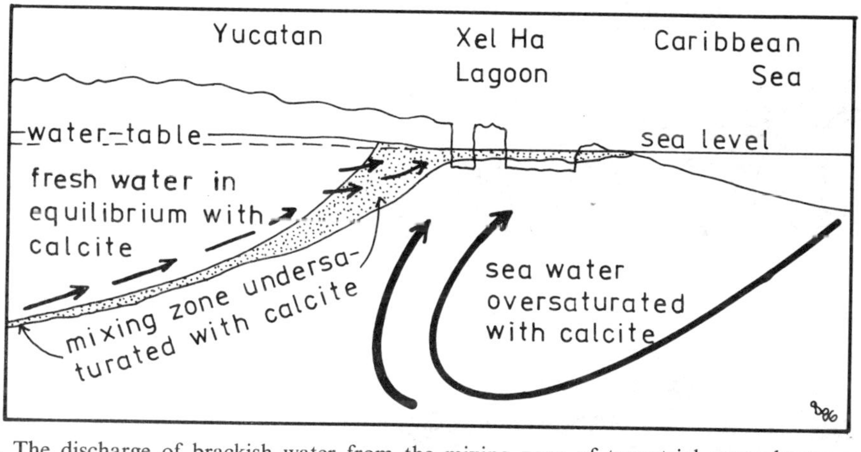

5–4. The discharge of brackish water from the mixing zone of terrestrial groundwater and a Ghyben-Herzberg zone of oceanic water into a Yucatan coastal lagoon (after BACK, HANSHAW and VAN DRIEL, 1984).

Karstification in coastal areas is related to the permeability of the rocks and the hydrostatic head controlling the shape of the freshwater lens and the character of any mixing zone. Recent studies on the Cuban, Yucatan and Apulian (Italy) coasts indicate the importance of pronounced karstification in the modern freshwater/seawater mixing zone (Fig. 5–4). RUDNICKI (1980) showed that similar situations occurred during the Pliocene and Pleistocene, creating what are now fossil cave systems on the coast of Apulia.

Inland Areas

In inland areas the location and form of the vadose and phreatic zones depends on topography, precipitation, subsurface migration of fluids, distribution of permeable rocks and other geologic and geomorphologic conditions. The concept of karst hydrologic zones and hydrogeologic systems is well studied and summarized e.g. in ZÖTL (1974); SMITH, ATKINSON and DREW (1976) or BÖGLI (1980) among others.

Karst in inland areas is influenced by geological structure, climatic conditions and geomorphologic position. Karst extends from periglacial areas (cf. BROOK and FORD, 1976, 1978b) to the tropics, from coastal lowlands to the high mountains (cf. ZÖTL, p. 483–509). Only limited karst occurs in arid areas (deserts); however, deep circulation of fossil waters there can cause intrastratal karstification that is not marked by features at the surface.

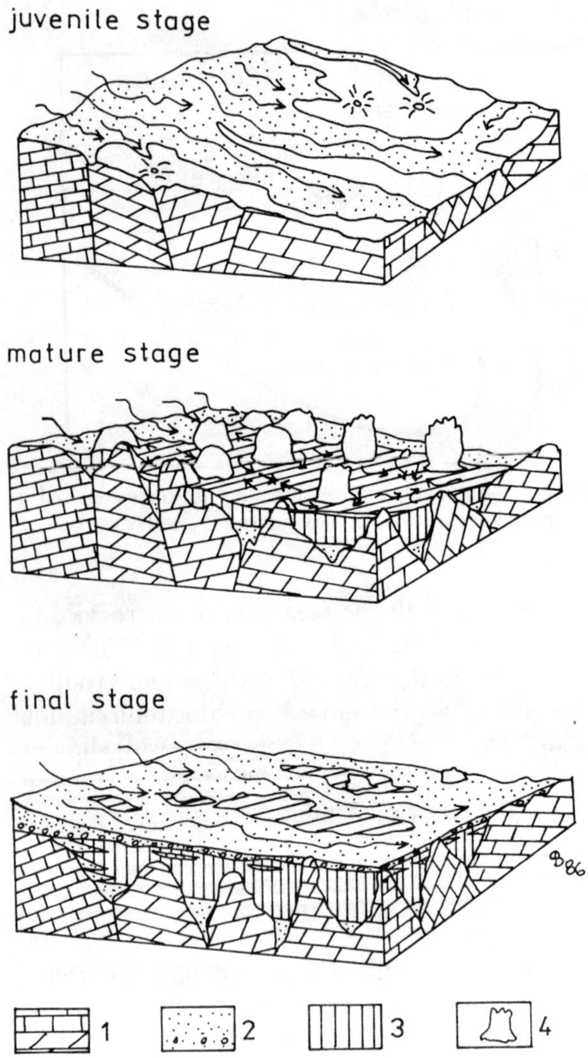

5–5. Morphological model of the evolution of the Iharkut area, Hungary, at the time of the bauxite accumulation (modified after MINDSZENTY, KNAUER and SZANTNER, 1984). 1. Triassic carbonates, dolostones prevail; 2. Middle Senonian fluviatile deposits; 3. Middle Senonian bauxite; 4. karst alluvial plain.

Theories of karst origin and the systematic or regional description of karst landforms are the subjects of many monographs, e.g. JENNINGS (1971, 1985), SWEETING (1972), T. D. FORD and CULLINGFORD (1976), HERAK and STRINGFIELD (1972), journal articles, symposia and congress proceedings (e.g. Proc. Int. Congr. Speleol.). It is beyond the scope of this book to summarize them.

Paleokarst forms, their origin and development has been intensively studied, especially by economic geologists. The shape and extent of karst forms influences the geometry, reserves and economic utilization of many related mineral deposits (see Part 3 of this book, p. 365–480). KOMLÓSSY (1985) distinguished three distinct environments of deposition for continental karst bauxites: (1) high karst plateaus; (2) erosional terraces, and (3) continental basins. Karst plateaus are characterized by deeply karstified bedrock, rapid drainage and relatively deep-seated water tables. Weakly karstified terraces tend to have karst zones up to 30 m deep, associated with a fluctuating karst water table. The morphology of continental basins is less expressive, karst forms typically reaching 6 m in depth and lakes, swamps etc. being common. MINDSZENTY, KNAUER and SZANTNER (1984) interpreted the karst evolution at the Iharkut bauxite deposit in Hungary in three stages; juvenile, mature and final (Fig. 5–5). The karst developed intensively in the mature stage and in the final stage became buried and fossilized. In the Transdanubian Mountains of Hungary, HAAS (1984) demonstrated the role of marine regression periods in karst origin (Fig. 5–6).

Karst origin and evolution in the geological past is affected by several other factors such as: (1) the duration of the continental regime, and (2) the extent and duration of marine transgressions. The longer the interval between regression and transgression, and the more humid the climate (paleotropics, paleosubtropics), the more extensive are the karst landforms that are created. Even brief emergences in wet climates can produce distinct karstification, while very long exposure to arid climates may cause only indistinct karst landforms. This situation is illustrated at many sites described in the regional parts of this book (cf. Canada, Czechoslovakia, and others).

Fossilization of Karst Forms

Karst becomes fossil or inactive when it loses its hydrological function. The general causes for this are changes of local or regional geotectonic conditions or of global sea level. Fossilization can be a result of uplift or of subsidence, of marine transgression or of mass continental (clastic facies) deposition. Continental drift may change the latitudinal position of karst areas in different geologic stages resulting in climatic changes, i.e. to arid or to humid, which may contribute to fossilization or rejuvenation.

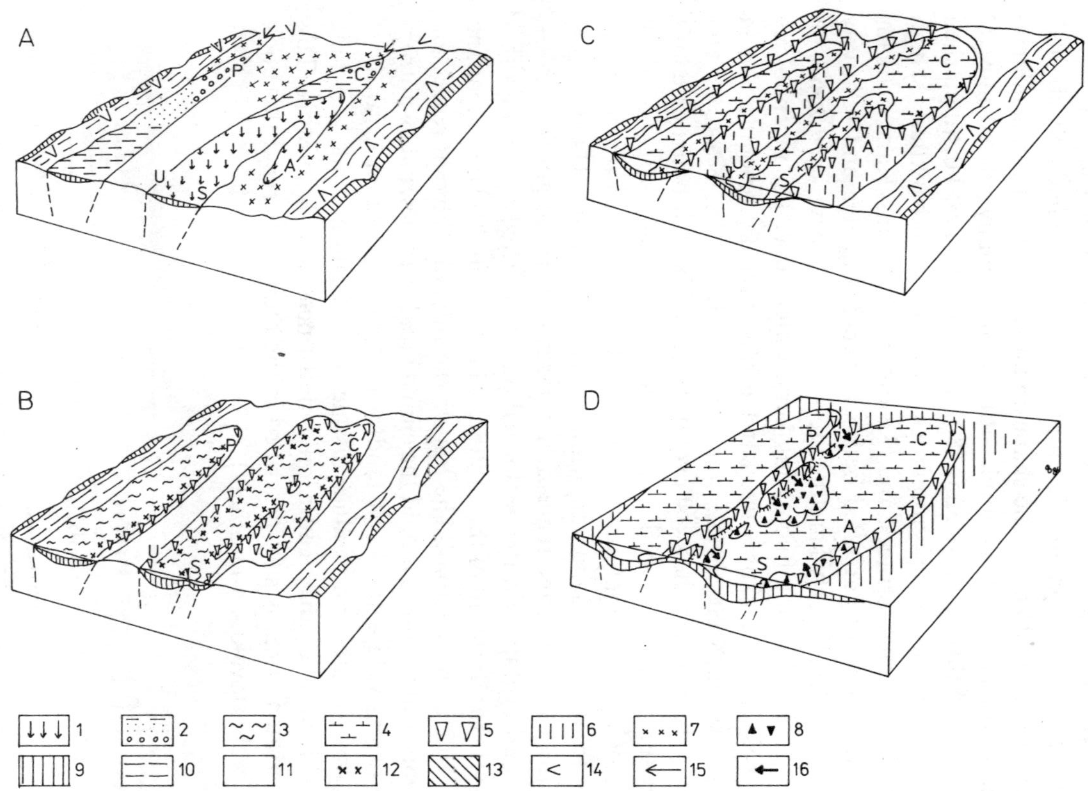

5–6. Reconstruction of the paleogeographic conditions of the karst origin and fossilization of an example of the Senonian bauxite deposits in Transdanubia, Hungary (after HAAS, 1983). A. beginning of the Senonian cycle; B. the conditions after transgression; C. situation after inundation of the karst plateaus; D. regressive phase in the upper part of the cycle. 1. freshwater coal swamp; 2. terrestrial accumulations; 3. marine marl; 4. marine clacareous clay; 5. rudistid platform deposits; 6. back-platform deposits; 7. bioclastic foreland slope deposits; 8. bioclastic slope deposits; 9. sediments deposited in the Senonian cycle; 10. Paleozoic crystalline rocks; 11. Mesozoic karst plateau; 12. karst bauxites; 13. lateritic blanket; 14. direction of laterite influx; 15. direction of transport of land-derived detritus into basin; 16. submarine sediment transport direction. P = Pápa; C = Csehbánya; A = Ajka; S = Sümeg; U = Ukk.

Marine Transgressions

Transgressions are one of the most common and typical fossilizing agents of karst. Many examples are described in the regional section of this book. Marine transgressions may produce two contrasted effects, or a mixture of both: (1) perfect karst preservation by burial and (2) karst destruction, partial or complete. Thanks to overlying marine sequences, some beautiful ancient karst forms have been preserved.

One of the most spectacular examples is described by POTY (1980) at Visé, Belgium. During ?Upper Frasnian and Fammenian times, uplift of Lower Frasnian limestones created mogote karst and caves. In the Middle and Late Tournaisian these were flooded by the sea. Lower parts were covered by marine deposits which also entered into some caves. During the ensuing Visean and Tournaisian transgression, the surviving karst relief was progressively destroyed by marine abrasion and at the end of the Visean, siliceous limestones completely covered the drowned karst topography (cf. Fig. 2–4).

Incipient stages of karst fossilization, similar to the Middle Tournaisian transgression at Visé, are exemplified by the drowned tower karst of Ha Long

5–7. Drowned mogote karst, Ha-Long Bay, Vietnam (Photo by J. Přibyl).

Bay, Vietnam, which is invading depressions among the towers and drowning cave systems at their feet. Surf action has resulted in systems of wave-cut notches at sea level (Fig. 5–7). This state does not represent the full fossilization of the tower karst because freshwater can penetrate from the exposed towers so that groundwater circulates and karstification continues in these conditions.

Coastal karst areas originated during Pleistocene sea level fluctuations at many sites and are now drowned e.g. in Florida, U.S.A. (PALMER and PALMER, p. 355–356), in Bermuda (HARMON, SCHWARCZ and FORD, 1978) or the Bahamas (GASCOYNE et al., 1979), along the Adriatic coast of Yugoslavia (cf.

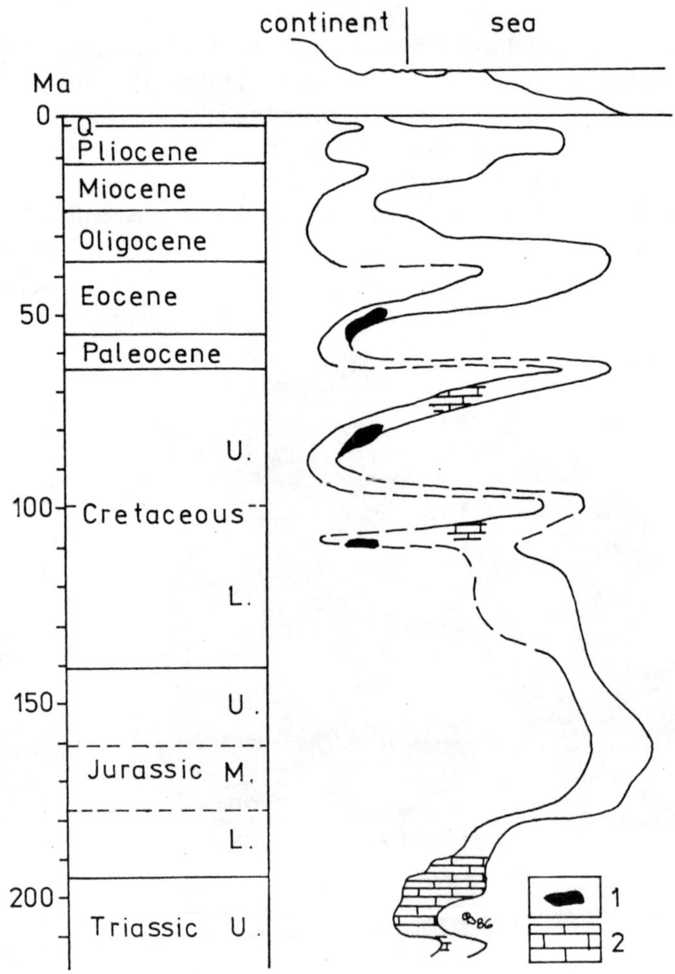

5–8. Paleogeographic position of the Transdanubian Central Range's bauxite deposits (simplified after HAAS, 1984). 1. primary bauxite accumulations; 2. carbonate rocks with features favourable for serving as bauxite footwall.

HERAK, 1972; GAVRILOVIĆ, p. 212); other examples are presented by JENNINGS (1985, p. 228–230) from Belize, Australia and Japan. At all these sites, caves with speleothems are flooded by the sea. Some speleothem age determinations indicate the Pleistocene age of speleothem formation (cf. DILL, 1977; HARMON, SCHWARCZ and FORD, 1978, or GASCOYNE et al., 1979, etc.). Fossil equivalents of flooded coastal caves are rather rare. Some examples are noted from Kotouč Hill near Štramberk (Lower Cretaceous, Czechoslovakia) by PANOŠ (1964b) and HOUŠA (1976, 1978). Other examples can be found described in the regional section of the book, e.g. Poland (GŁAZEK, p. 77–105).

Repeated transgressions and regressions in the Transdanubian Mountains of Hungary created several succesive phases of karstification and karst fossilization (cf. Fig. 5–8). Paleogeographic reconstruction of the area during the Santonian-Campanian cycle, for example, has yielded interesting data on changes in the evolution of karst depressions that served as bauxite traps.

The destructive affects of marine transgressions are illustrated by LONGMAN and BROWNLEE (1980) in an example of recent karst in the coastal zone of Palawan Island, Philippines. Wave and surf action is manifested here in the form of wave-cut notches, in the destruction and undercutting of shore cliffs up to 30 m high and in the creation of a flat abrasion platform with a few residual pinnacles more highly eroded on the seaward side. Truncated relicts of solution fissures and sinkholes are sometimes seen on the platform. The scarcity of positive paleokarst forms in the fossil record is explained by such destructive effects of transgressions.

Continental Deposits

Mass continental deposits play a similar role in fossilization. Rapid and extensive deposition of terrigeneous clastics such as red beds, eolian sands, alluvial-fluviatile complexes, glacial deposits, etc. interrupts the hydrological function of the karst. Even extensive karst landforms become mantled or buried. However, the initial stages of terrigenous deposition may be associated with the evolution of a distinctive suite of karst forms such as sand pipes (geological organs), etc. Under favourable geomorphological conditions, i.e. active underground drainage and a sufficiently wet climate, karstification can result in the creation of substantial relief in these circumstances e.g. beneath Pleistocene and Holocene volcanic ashes at Waitomo, New Zealand. Continental deposition in some cases precedes marine transgression, which then finishes the fossilization of the karst.

Polycyclic karst forms buried by continental clastics occur in North Siberia along the Anabar and Udzha Rivers, where Jurassic, Lower Cretaceous, Oligocene and Neogene karst basins, sinkholes, etc. are filled with sands, clays, gravels, etc. which bear numerous placer deposits of rare earth minerals (PROKOPCHUK et al., 1985). The landforms of each karstification phase were filled with somewhat differing deposits refflecting changes in source area and climate, i.e. the rate of weathering.

Extensive deposition of continental sands, clays, kaolins, gravels, etc. of the Rudice Formation type is noted at numerous sites (cf. BOSÁK, p. 431–442). The most conspicuous examples are described from the Cracow-Wieluń Upland, Poland (GRADZIŃSKI, 1977), Ardenne-Condroz, Belgium (CALEMBERT, 1945, 1954) or the Moravian Karst, Czechoslovakia (BOSÁK, 1980b). In these examples, the Formation covers large areas and was deposited in alluvial-fluviatile-lacustrine environments in the Tertiary or (for Moravia) in the Lower Cretaceous. Marine transgression caused complete fossilization of the karst only in Moravia (cf. BOSÁK, HORÁČEK and PANOŠ, p. 115–117).

One of the most famous examples of karst fossilization was presented by GÈZE (1938), describing polycyclic karst forms in southern Quercy, France.

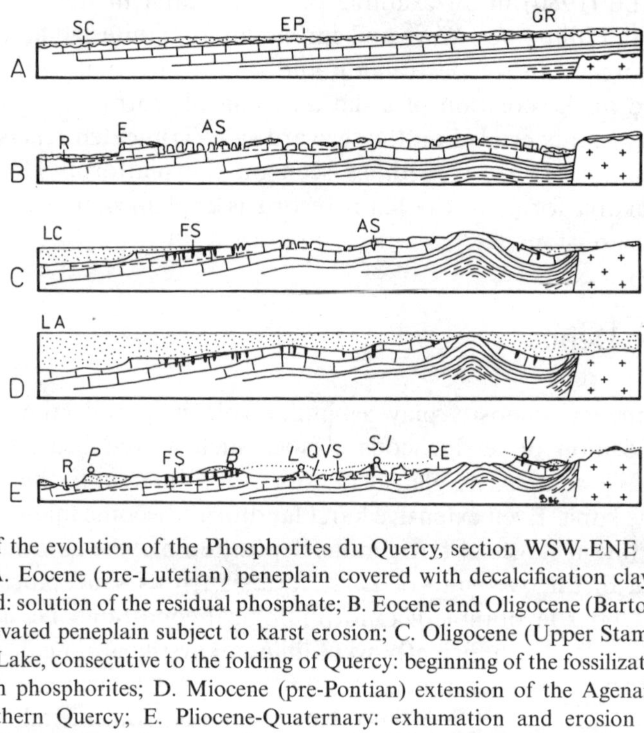

5–9. Scheme of the evolution of the Phosphorites du Quercy, section WSW-ENE (simplified after GÈZE, 1938). A. Eocene (pre-Lutetian) peneplain covered with decalcification clays and alluvium being lateritized: solution of the residual phosphate; B. Eocene and Oligocene (Bartonian-Stampian) warped and elevated peneplain subject to karst erosion; C. Oligocene (Upper Stampian) extension of the Cieurac Lake, consecutive to the folding of Quercy: beginning of the fossilization of sinkholes and shafts with phosphorites; D. Miocene (pre-Pontian) extension of the Agenais Lake entirely fossilizing southern Quercy; E. Pliocene-Quaternary: exhumation and erosion (Pliocene), and deepening and karst erosion (Quaternary); reappearance and partial rejuvenation of sinkholes and shafts with phosphorites. Villages: *P* = Puy-la-Rouque; *B* = Bach; *L* = Limognes; *SJ* = St.-Jean de Laurs; *V* = Villeneuve d'Aveyron; Other abbreviations: SC = siderolitic clay; EP = Eocene peneplain; GR = great east-west currents; R = resurgence; E = escarpment; AS = active shafts; LC = Cieurac Lake; FS = fossil shafts with Eocene phosphates; LA = Agenais Lake; QVS = Quaternary valley and sinkhole; PE = Pliocene erosion surface.

Repeated lacustrine accumulations during the Oligocene (Stampian) and the Miocene (pre-Pontian) fossilized a previous karst, beginning with deposition of phosphorites in sinkholes (Fig. 5–9).

Uplift

Uplift of karst terrains introduces two contrasting effects: (1) the fossilization of karst features, especially cave systems, due to the loss of their hydrologic function, and (2) the creation of a more extensive vadose zone with a higher gradient, creating conditions for accelerated vertical karstification. Numerous examples of both are reported from many mountain ranges, e.g. the Polish High Tatra, the Northern Limestone Alps of Austria, the Western Tien-Shan of the U.S.S.R., and the Front Ranges of the Rockies, Canada. Horizontal or sub-horizontal caves occur here at high altitudes. They are relics of ancient cave systems developed when the mountains were not so extensively divided by deep valleys into hydrologically isolated massifs. The catchment areas of such paleokarst systems were often substantially larger, as is proven by occurrences of "exotic" clastic fill from distant sources.

BAUER and ZÖTL (1972) and ZÖTL (p. 484–489) describe large cave systems concentrated at elevations of about 1 500 m a.s.l. in the Northern Limestone Alps. They are indications of a main (Pliocene) phase of subterranean flow representing the last stage of the hydrological influence of waters from the Central Alps. In the Dachstein Caves this influence is documented by sands from crystalline rocks and by scallops indicating the paleoflow direction. Valley entrenchment finished in the Late Pliocene, when the Limestone Alps became hydrologically isolated from the Central Alps. The old cave rivers no longer functioned and new, local karst drainage systems developed.

There is similar evolution of a karst plateau at Karzhantau Ridge in southern Kazakhstan. Horizontal cave passages of presumed Late Miocene to Pliocene age occur at about 2 500 m a.s.l. Rapid uplift in Pliocene and Early Quaternary times caused the incision of deep valleys (up to 600 m) and the change of the subterranean drainage system from horizontal to vertical. Middle to Late Quaternary mountain uplift (up to 1 000 m) supported the development of vertical karst drainage and the origin of deep and narrow shaft caves which partly utilized the fossil horizontal passages to drain rain and meltwater. The horizontal cave systems were dissected by numerous faults into short relics scattered at different altitudes. Correlative surface sands and gravels indicate transport from the southeast, i.e. from other mountain ridges about 30 km distant which have recently been separated by deep valleys (BOSÁK et al., 1985).

Climate Changes

Climate influences karst evolution in major ways, although its role has been over-estimated in some concepts of climatic geomorphology (cf. BÜDEL, 1977). Fossilization occurs when the amount of rainfall rapidly decreases, i.e. during arid phases or during glaciations and frost effects. The influence of glaciations is summarized by D. C. FORD (p. 328–336). For paleokarst to originate destructive and transforming processes are important. Like other fossilizing agencies such as uplift, the role of the climate is dual; fossilizing and creative, or rejuvenating.

Mineralization

Mineralization of caves is a not uncommon mode of karst fossilization. Large Mississippi Valley type zinc-lead deposits may completely fill host karst cavities or brecciated zones created by collapse into solutional voids in the rock. Although there is a long-lasting debate on the nature and interrelationship between karstification and hydrothermal mineralization (see ZUFFARDI, 1976; DŻUŁYŃSKI and SASS-GUSTKIEWICZ, p. 377–397), without any doubt the mineral filling of a karst makes it fossil. This is one of the karst processes that lacks observable recent counterparts. The process is deep-seated, hypogenetic, subjacent or intrastratal, in many cases occurring at a depth of 2–3 km or more below the surface (cf. T. D. FORD, 1984).

Metamorphism

Metamorphism is the last stage in a karst fossilization sequence. Changing P-T conditions influence not only the karst fill but also the host-rock and its karst surfaces and erosion forms. Data on karst metamorphism are rare. Metamorphic bauxites from Naxos Island, Cycladen, Greece, are believed to be originally karst bauxites of Jurassic age (FEENSTRA, 1985). The metamorphism ranges from lower greenschist to upper amphibolite facies. Geochemical data indicate that metamorphism of the bauxites was essentially isochemical, with little mass transport between the metabauxite and surrounding marbles.

Evaporites

Evaporites are easily soluble rocks. Gypsum, anhydrite and rock salt are the main examples. Transformations occurring in these rocks after burial make them unfavourable for the preservation of paleokarst. They are plastic and relatively soft which permits rapid destruction of any karst forms.

The gypsum/anhydrite cycle results from the stability relationships in calcium sulphate-water system and the normal sequence of deposition-burial-uplift-erosion. Depending on the salinity of the enclosing brine and the geothermal gradient, the process of gypsum replacement by anhydrite occurs at depths ranging from 300 to 3 000 m below the surface. Uplift and erosion may then bring the anhydrite back to the surface. During this phase the composition of the groundwater tends to be less saline than during deposition and the depth of the anhydrite/gypsum transition will be lessened (30 to 600 m; BLATT, MIDDLETON and MURRAY, 1980). When anhydrite transforms into gypsum the rock volume increases up to 1.557 times (BIESE, 1931). The opposite process leads to a volume decrease. Both processes support the destruction of karst forms in general.

Jointing and the high solubility of gypsum favour the rapid formation of caves but these qualities plus the softness of gypsum lead to their quick destruction (BÖGLI, 1980) so that few are preserved in the paleokarst record.

Rock salt is one of the most soluble rocks. Its lack of caves is caused by its high plasticity under pressure. BÖGLI (1980) illustrates this feature in the 2 500 year old Iron Age salt mine of Hallstatt, Austria. The old adits have disappeared by annealing and the only witnesses left are the well-preserved remains of mining tools now within the salt.

Subrosion

Subrosion is the process of the subterranean solution of easily soluble or disintegrated rocks, in later phases possibly accompanied by mechanical removal. It is a kind of interstratal or intrastratal karstification. Interstratal solution of rock salt is the most common case. Illustrative examples are presented in this volume, e.g. from Canada (D. C. FORD, p. 319–328; cf. Fig. 2–138). The Upper Devonian Elk Point Group salt, 10–400 m thick covers an area of approximately 500 000 km^2. Overlying carbonates and Mesozoic shales show numerous sunken areas and brecciation as a consequence of subterranean solution of salts. Some examples from Poland are summarized by GŁAZEK (p. 77–105; cf. Fig. 2–25); a particularly conspicuous effect of subrosion subsidence occurs in the Bełchatów Brown Coal Basin (cf. Fig. 2–30). TSYKIN (p. 277) describes the solution of about 560 m of salt and anhydrite in the Irkutsk Foredeep. The subrosion created large depressions that were later filled with Early Jurassic coal-bearing strata.

Very spectacular subrosion basins are documented by MEIBURG (1980) in the Hessian Depression, Federal Republic of Germany. Subrosion of salt diapirs of Zechstein, Late Bunter, Middle Muschelkalk and Middle Keuper age followed volcanic events in the Miocene. Groundwater circulation along opened tectonic lines led to deep dissolution of the salt. Subsidence formed circular basins in the limestone and marls of the Middle Bunter in which Miocene and Oligocene sands and other deposits are preserved (Fig. 5–10). Gravitational subsidence caused intensive brecciation of Triassic rocks more than 1 000 m thick. Subsidence basins have diameters of approximately 1 km. The thickness of preserved Tertiary sediments is 40 to 200 m.

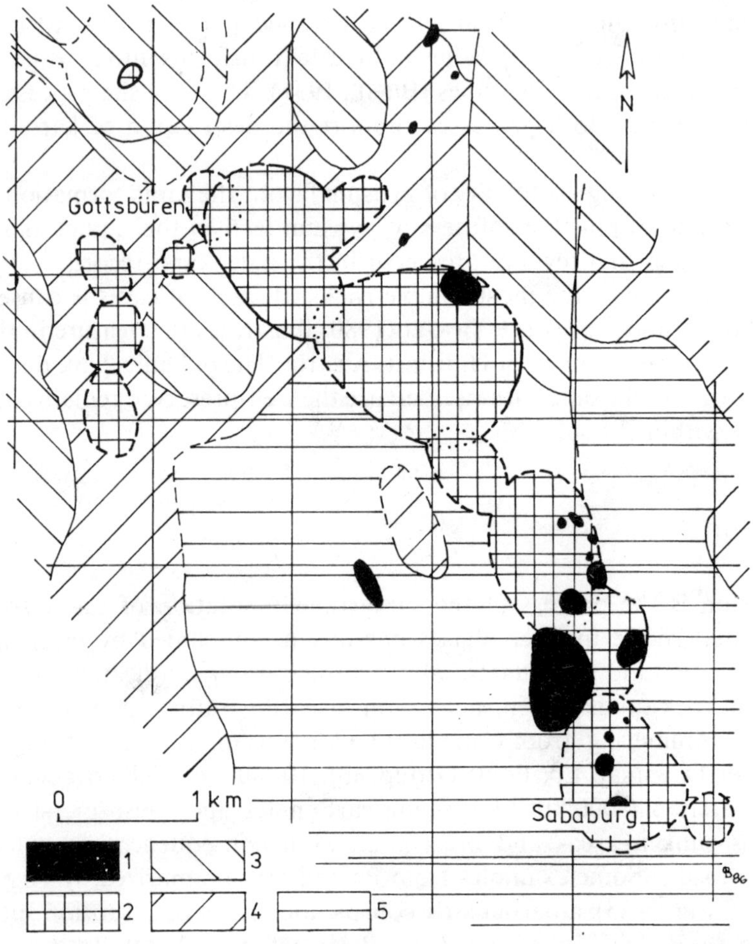

5–10. Subrosion structures in Gottsbüren-Sababurg area, Federal Republic of Germany; geological map without Quaternary cover (modified after MEIBURG, 1980). 1. basalt, tuff; 2. Miocene-Upper Oligocene; 3–4. Solling Formation (Middle Buntsandstein, Triassic), 3. Karlshafen and Trendelburg Members, 4. Wilhelmshausen Member; 5. Hardegsen Formation

Other good examples of subrosion and pipe-formation are presented in the Symposium Proceedings of the International Association of Engineering Geologists (e.g. IAEG Symposium, 1973).

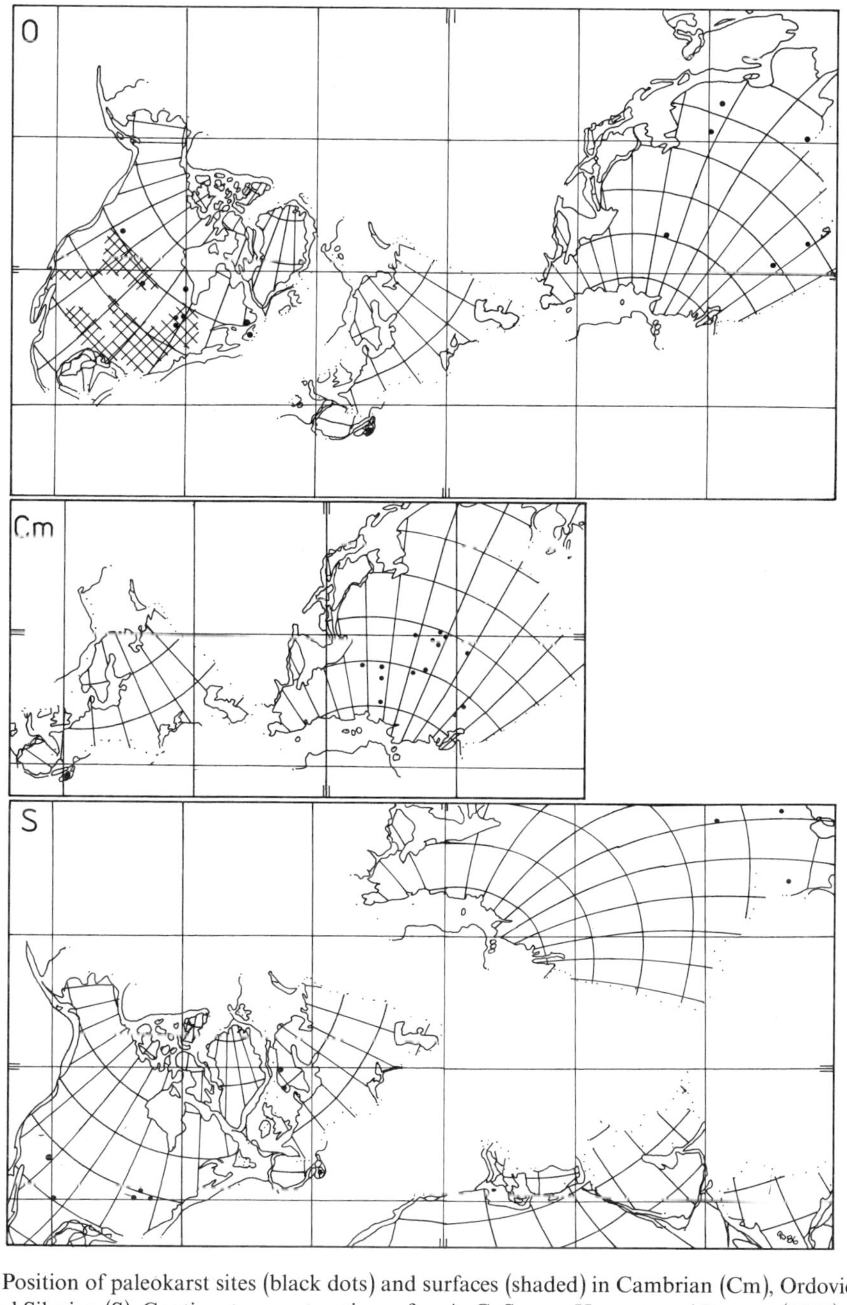

5–11. Position of paleokarst sites (black dots) and surfaces (shaded) in Cambrian (Cm), Ordovician (O) and Silurian (S). Continent reconstructions after A. G. SMITH, HURLEY and BRIDEN (1982). Cm = reconstruction for Middle Cambrian (520 Ma); O = reconstruction for Ashgillian (440 Ma); S = reconstruction for Ludlow times (400 Ma). For details see the text

The Distribution of Paleokarst in Time and Space

Reconstruction of the occurrence of paleokarst in the geological past is beset with numerous problems resulting from: (1) scarcity of data; (2) incomplete nature of much evidence; (3) incomplete knowledge of the paleogeographic and

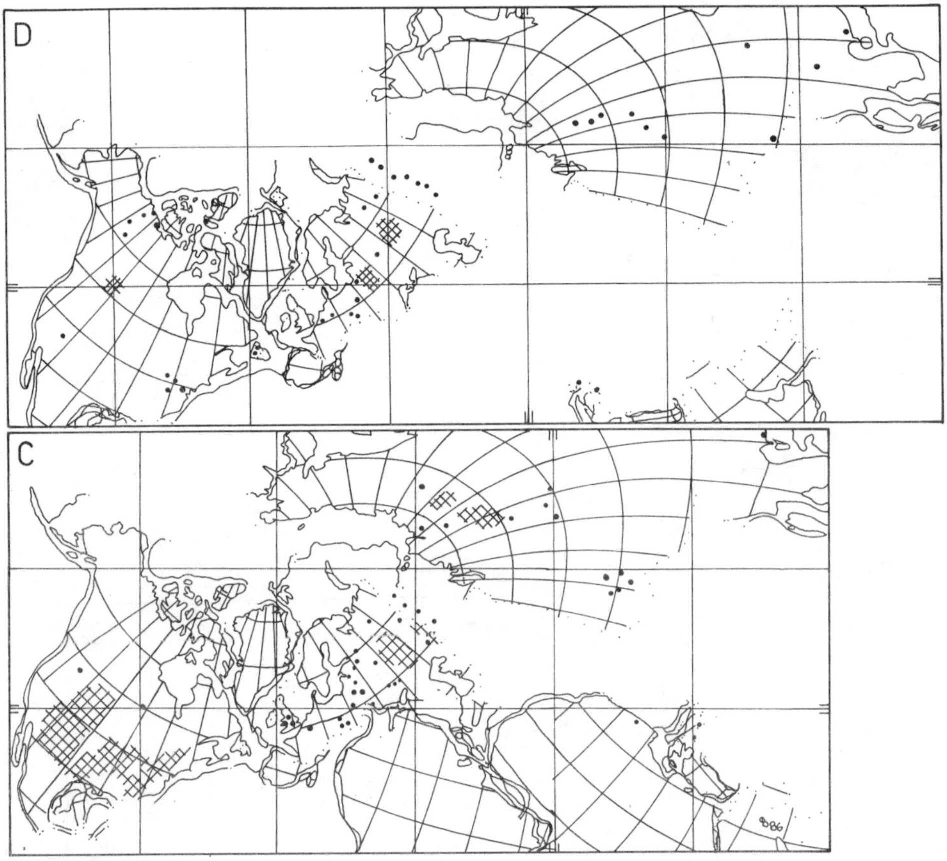

5–12. Position of paleokarst sites and surfaces in Devonian (D) and Carboniferous (C). Continental reconstructions after A. G. SMITH, HURLEY and BRIDEN (1982). D = reconstruction for Frasnian (360 Ma); C = reconstruction for Namurian (320 Ma).

continental palinspastic reconstructions, and (4) incomplete knowledge of past climates. Only a few papers deal with this problem; e.g. BOSÁK (1981c) tried to show that the distribution of European Lower Cretaceous paleokarst resulted from continental, microcontinental and cordilleran tectonics as well as from paleogeographic factors. BÁRDOSSY (1982) reconstructed the distribution of karst bauxites (Devonian to Cretaceous) in the framework of the SMITH, BRIDEN and DREWRY (1973) paleotectonic picture.

The numerous data on paleokarst that are reported in this book provide us with a good opportunity to attempt new reconstructions. Figures 5–11 to 5–15 represent the Cambrian to Paleogene continental reconstructions of SMITH,

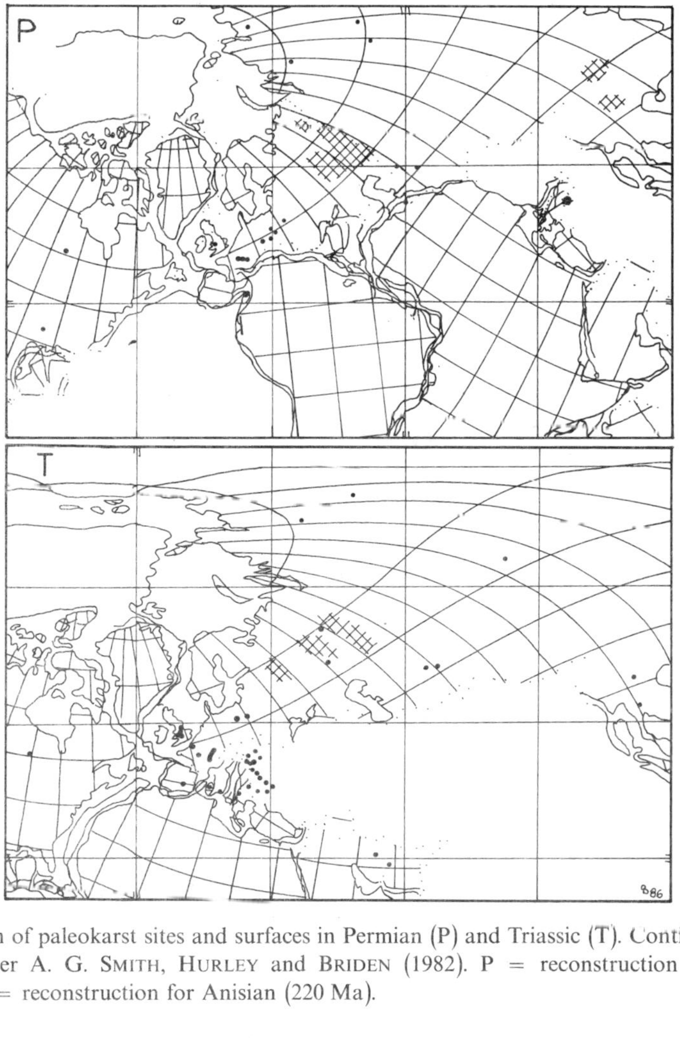

5–13. Position of paleokarst sites and surfaces in Permian (P) and Triassic (T). Continental reconstructions after A. G. SMITH, HURLEY and BRIDEN (1982). P = reconstruction for Tatarian (240 Ma); T = reconstruction for Anisian (220 Ma).

HURLEY and BRIDEN (1982) with the locations of known paleokarst sites (black dots and lines) or more extensive karst surfaces (shaded). The paleokarst sites in individual geologic periods evaluated for this reconstruction are summarized in Table 5–1.

The oldest preserved and known paleokarst features are described by MARTINI (1981) from the Transvaal, South African Republic. Karst with caves and cave deposits is 2 200 Ma old, i.e. of Early Proterozoic age. Somewhat younger are corrosional paleokarst surfaces in carbonates of Belcher Island (1 700 Ma) and of Ontario and Quebec, Canada (approximately 1 400 Ma; D. C. FORD, p. 315–316). Upper Proterozoic karsts are known from the Madoc area (Ontario) and Baffin Island, Canada (D. C. FORD, p. 316–317), Jingsha, Kaiyang and Xushui

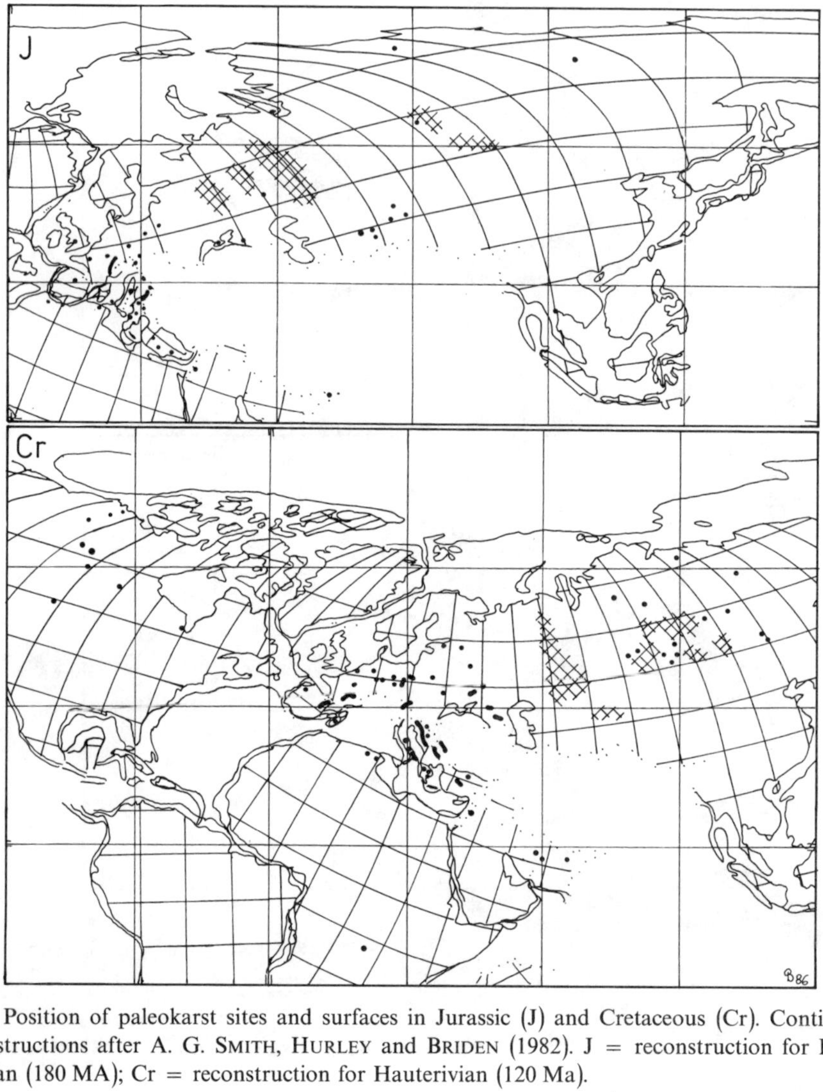

5–14. Position of paleokarst sites and surfaces in Jurassic (J) and Cretaceous (Cr). Continental reconstructions after A. G. SMITH, HURLEY and BRIDEN (1982). J = reconstruction for Pliensbachian (180 MA); Cr = reconstruction for Hauterivian (120 Ma).

Counties, China (ZHANG SHOUYUE, p. 305) and the U.S.S.R. (Riphean — in the Timan, Sayan and Ural Mountains, Vendian — in the Tunguska Syneclize, TSYKIN, p. 268–269). An Upper Proterozoic paleokarst is also reported in the Callana area, Willowran Ranges, Australia (ROWLANDS et al., 1980).

The youngest paleokarst features are connected with Holocene glaciation

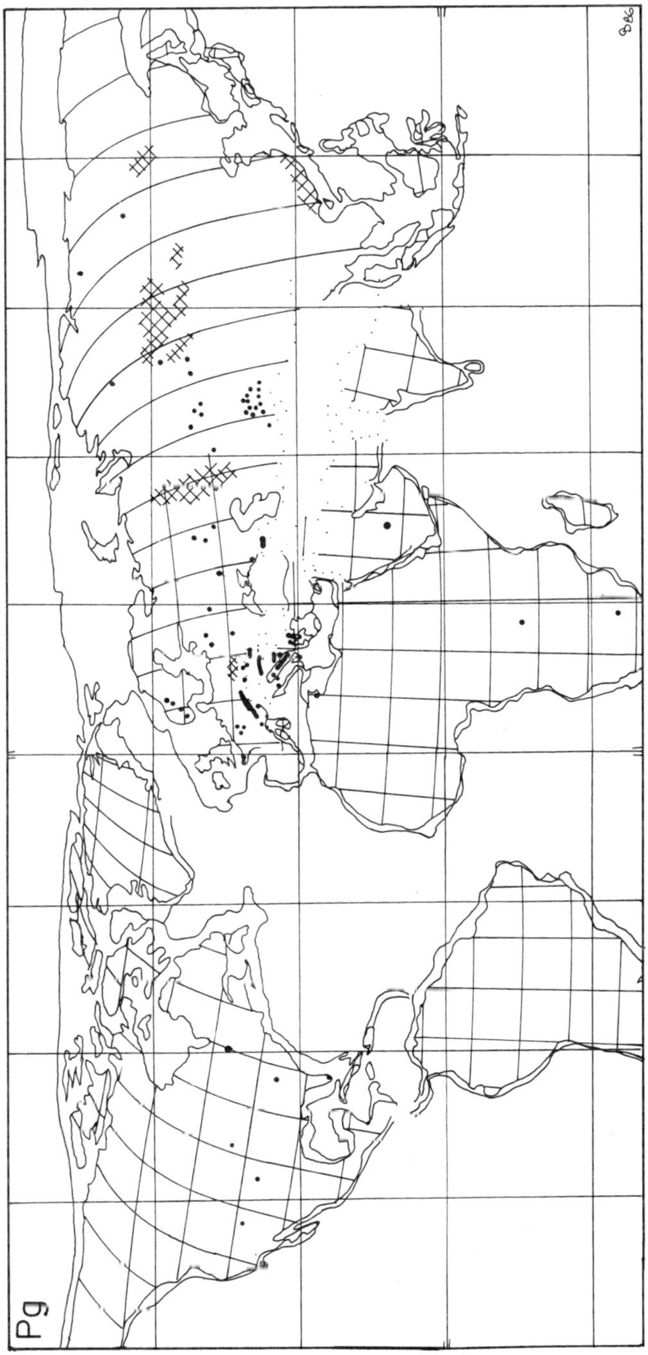

5–15 Position of paleokarst sites and surfaces in Paleogene. Continental reconstruction for Upper Eocene (40 Ma) after A. G. SMITH, HURLEY and BRIDEN (1982).

effects (cf. Canada, D. C. FORD, p. 328–336) or Pleistocene-Holocene sea level changes (cf. southern France, HORÁČEK and KORDOS, p. 609; or the High Carbonate Islands of the Pacific Ocean, BOURROUILH-LE JAN, p. 443–471).

Table 5–1

Number of paleokarst sites noted in Figures 5–11 to 5–15

Paleokarst age	Number of sites
Precambrian	12
Cambrian	9
Ordovician	20
Silurian	10
Devonian	40
Carboniferous	65
Permian	25
Triassic	70
Jurassic	70
Cretaceous	170
Paleocene	70
Eocene	100
Oligocene	110

The reconstructions of occurrences of paleokarst forms and surfaces are not completely correct. They are based on only one palinspastic map for each geologic period, although these lasted for tens of millions of years. In these long time spans there were complex sequences of marine transgressions, regressions, uplifts and subsidences, substantially changing the paleogeographic picture at each stratigraphic stage or substage. This is the first major simplification that we knowingly introduce into the reconstruction. The imprecise ages for paleokarst sites do not allow more correct and detailed interpretation. Other problems arise from the palinspastic reconstructions themselves. They contain evaluations of presently known geological and geophysical data and theories, and are general schemes rather than detailed reconstructions. The precise positions of individual microblocks and their paleolatitudes are not yet correctly established.

The reconstructions, however, show some interesting results: (1) the bulk of preserved paleokarst sites were generated between paleolatitudes 30°S and 30°N; (2) occurrences in the zone 30°N to 60°N are relatively abundant, and (3) some paleokarst sites developed up to 70°N, e.g. in the Triassic, Jurassic, Cretaceous and Paleogene. This indicates either different climatic zonation in some periods (cf. FRAKES, 1979) and/or karstification even in high latitudes as in the recent subarctic karst (cf. BROOK and FORD, 1976, 1978b).

BIOSTRATIGRAPHIC INVESTIGATIONS IN PALEOKARST

Ivan Horáček and László Kordos

Introduction

Sediments preserved in karst caves and surface depressions have long been recognised as an important source of fossils. Vertebrate bones and molluscan shells have been systematically excavated from paleokarst fills since the first half of the last century, particularly in France and Germany. As a consequence, almost all groups of fossils, both animal and plant, marine and terrestrial, macro and micro, have been found in paleokarst fills. The fossils became very important source of knowledge on paleokarst history when dating sedimentation events. Biostratigraphy and paleoecologic analysis represent now very important, widely accessible and particularly fruitful methods of paleokarst research. Of course, there are several special rules to be followed.

Interdisciplinary cooperation is necessary here. Paleontologists interested in paleobiology of ancient life forms often neglect to formulate of results in a form meaningful for a paleokarst student. The latter, being unfamiliar with details of paleontologic topic may tend to underestimate the information which fossils can actually provide him. To prevent this, paleokarst students should be informed which are the special characteristics of biostratigraphic investigation and what is to be expected and required from it.

The calcareous karst sediments provide especially favourable conditions for the fossilization of vertebrate and molluscan remains, while non-calcareous fossils, e.g. insects and most plant remains including pollen, tend to oxidize and vanish readily in this environment. Recently, advanced techniques have enabled the extraction of some pollen from calcareous deposits, but vertebrates and molluscs remain the most important fossils in paleokarst biostratigraphy. This is not only due to tradition. They appear in paleokarst fills more frequently than any other group. Moreover, the collection and further paleontologic treatment of bones and shells can be carried out without any special laboratory equipment and just during field work. Therefore, the following text deals almost exclusively with vertebrates and molluscs.

P. Bosák, D. C. Ford, J. Głazek and I. Horáček (Editors):
Paleokarst. A Systematic and Regional Review. – Elsevier and Academia. Amsterdam and Praha. 1989.

Scope of the Methods

Any application of biostratigraphic methods can be of real value for a paleokarst study only if a detailed lithologic examination of the site has clearly delimited the context of the organic deposits. The karstologic significance of a biostratigraphic datum does not arise from that datum itself but from its setting within the abiotic karst phenomena. Moreover, due to the enormously complicated deposition dynamics often occurring in karst, we can encounter sites where deposits of quite different age occur in the same place, e.g. thin Holocene deposits overlying the fossil fill, secondary deposition into cavities eroded in a lithified accumulation of substantially earlier age or cases similar to those shown in Figure 5–16.

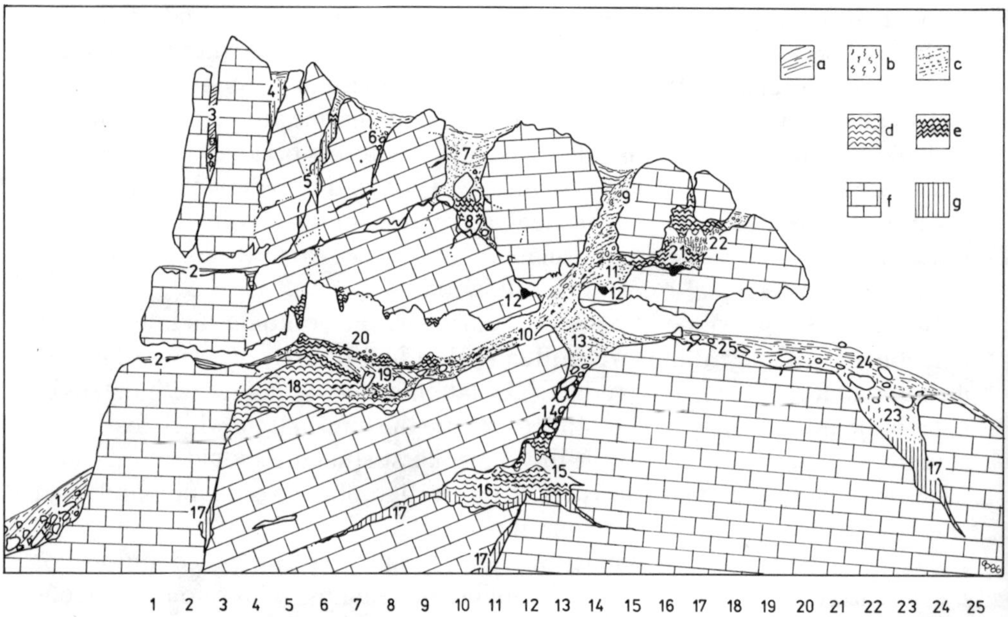

5–16. A sketch of common types of karst infills and their fossil content. Note the appearance of remains of ancient fills of intra-cave facies (12) preserved in wall niche which may lie in the direct contact with much younger deposits (11) or those preserved in different but neighbouring cavities (21 vs. 11). Collapse of sedimentary plugs and redeposition may also occur in caves (10) which may also cause serious confusion unless detailed lithological studies are done (cf. e.g. situation in the sites of 19 and 10). a. Holocene soils and related deposits; b. loess base of Holocene deposition; c. sequence of the Pleistocene and earlier surface sediments; d. former infill of the intra-cave facies, frequently fluviatile; e. flowstone sequence; f. carbonate rock; g. ancient residua of strongly weathered surface or subsurface sediments, mostly non-calcareous.

Without especially careful lithologic study of a site and without respecting this when taking the samples, a very confused picture can arise from a biostratigraphic study even though this be done very exactly. Unfortunately, this has been neglected at a great number of sites and is a cause of many erroneous interpretations. Worth mentioning, of course, is the fact that there are some localities where a mixture of different age assemblages has resulted from natural processes, e.g. repeated deposition events, redeposition, gelifluction, etc., which cannot be sorted out by even the most careful collecting methods. Fortunately, such cases are rather exceptional.

Because it is related to an infill and not to the cavern itself, the biostratigraphic datum does not determine the age of a cavity or other karst form, but the *possible maximum age of the fossilization stage* connected with the deposition of that particular layer. If the basal layer of a sedimentary sequence is under study, we can speak about the *possible maximum age* of the last fossilization phase, and similarly of its *minimum age* in the case of the top layer. If there is no evidence to the contrary, the beginning of the fossilization phase may be also expressed as the *possible minimum age of the creative karstification phase,* i.e. possible minimum age of the cave.

Paleobiologic investigations also provide data concerning the paleoenvironments in the cave and in its wider surroundings, which is often very important for the interpretation of the cave history.

A biostratigraphic study includes the following steps: (1) sampling design and extraction of fossils; (2) examination of fossils; (3) biostratigraphic interpretation, and (4) paleoecologic interpretation. While the first step can be performed well by a sufficiently experienced collector, even a layman, the remaining steps should be undertaken by specialists in particular groups of fossils who are familiar with routine techniques of biostratigraphic and paleoecologic analysis and are able to judge all special features of the studied material. Some basic rules for the individual steps are summarized below.

Sampling Design and Extraction of Fossils

Sampling a deposit for biostratigraphic purposes must always be undertaken with full regard for the results of preceding lithological investigations of the site. The volume of a sample taken for the extraction of fossils may vary according to the kind of fossils and their concentration in the deposit. Generally speaking, the amount of sample should be enough to ensure that the number of extracted fossils is sufficiently large to draw reliable biostratigraphic and/or paleoecologic conclusions, i.e. 10 500 kg per layer, as a rule.

The larger the number of fossils obtained, the more accurate the conclusions can be. From this point of view, the smaller fossils, e.g. mice, voles, shrews, molluscs, etc. which often appear in great numbers of individuals and species, are much more important for a biostratigraphic study than the seemingly more attractive remains of larger animals to which a layman's interest is more readily drawn. A common technique for the extraction of vertebrate and mollusc microfossils includes wet-sieving of previously well dried samples on a fine mesh (mesh diameter approx. 0.2–0.5 mm) and the mechanical or electrostatic picking of the fossils from a dried concentrate; for more details see, for example, LOŽEK (1964). Extraction from the cemented cave deposits with careful and dilute etching in acids is also possible.

Paleontological Examination of Fossils

The fossils obtained should be examined and assessed in the standard manner used for a given group and compared with those in other localities and described in the literature. Data required include (i) list of species appearing in the sample; (ii) percentage representation of individual species based on minimum numbers of individuals, i.e. the maximum number of fragments undoubtedly belonging

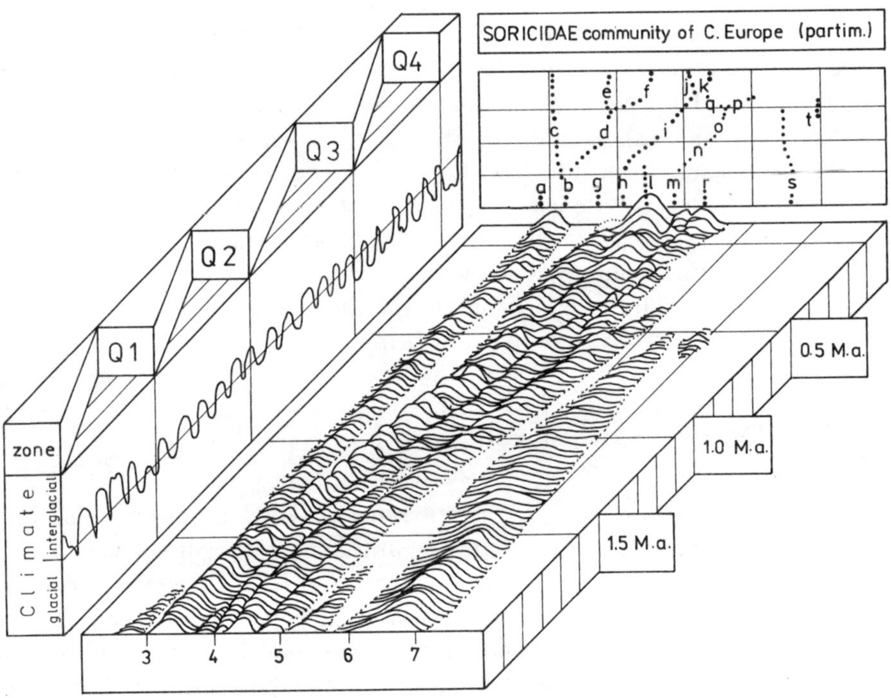

5–17. Late Pliocene (MN 17) and Quaternary changes in the structure of the mid-European community of shrews. Note shifts in both variation of metrical characters and relative representation of individual phylogenetic lines (data after HORÁČEK, 1985).

to different individuals of the same species, and (iii) data on morphologic and morphometric variation of individual species in the sample.

All three lists should be compared with similar samples from other localities. The paleontologist should then decide which of each pair of samples exhibits more advanced features and, consequently, can be understood as being younger than the other. The basis of such an assessment is knowledge of the phylogenetic differentiation in particular species groups, the course of the change in representation of individual forms in actual communities, etc. Of course, different groups possess often quite a different mode, tempo or timing of such events (cf. Fig. 5–17). Comparing them, our picture of the past becomes finer than a sole group could provide. A general expression for that type of information is the concept of a biostratigraphic system. This briefly summarizes previous experience and adapts it for routine use in biostratigraphic practice. For a paleontologist, dating means to characterize the studied faunal assemblage in terms of a biostratigraphic system.

Application of a Biostratigraphic System

Any stratigraphic system subdivides certain time periods into a sequence of shorter sections which can be distinguished clearly each from the other by differences in the state of the phenomena upon which the system is based. It is constructed either hierarchically, consisting of different rank units or, more simply, as a series of successive biozones. Each unit of a biostratigraphical system may be defined either as a *range zone,* i.e. by means of the first and the last appearance dates of suitable index forms, or as an *assemblage zone* if based on specific characteristics of community structure. A widely applicable system should consist of units which are well defined both in their range zones and their assemblage zones and can be identified also where there is only an isolated and very fragmentary community record.

A biostratigraphic system is compiled and based on comparison of numerous fossil assemblages by means of a few basic stratigraphic principles:

(1) principle of *superposition:* the assemblage in lower layers are older than those appearing in upper layers of an undisturbed sedimentary sequence. Localities where such comparision for several or many biozones is possible are of enormous stratigraphic importance.

(2) principle of *index fossils:* different assemblages come from the same stratigraphic horizon if they consist of the same forms. Especially important in this connection are widely distributed forms that occur in only one stratigraphic horizon, i.e. index fossils.

(3) principle of *facies:* different conditions (region, altitude, etc.) can at one and the same time create different assemblages, while almost the same assemb-

lage may come from different time periods i. e. the controlling conditions are the same.

Individual units of a stratigraphic system are distinguished by specific names. Their meaning is established by denoting typical localities and layers as their stratotypes. Because of the facies effects in terrestrial fossil records, a majority of biostratigraphic systems so far proposed are to be looked upon as expressing quite local specificities rather than phenomena of general significance. Their units mostly do not directly correspond to each other. The same holds, of course, with systems based on different fossil groups. Consequently, there are a great number of names used in stratigraphic practice. Some of them are abandoned during the course of time, while others may be much extended in either their application and meaning. Sometimes, names can even be used without any respect to their original definition, which can cause serious confusion, e.g. names of the Alpine glaciation system being applied in vertebrate paleontology (cf. KURTÉN, 1968).

As a consequence, one very important task of stratigraphic study is to look for the correlation points between different stratigraphic systems and records which fix the relative sequences of biostratigraphic systems to the numerical chronological scale of the geological past. Among the former, those cases which enable us to correlate terrestrial stages with those of marine stratigraphy are particularly important. Marine record, being worldwide, provides the basis for global correlation and definition of the first order units — systems, periods, eras.

The caves in **Děvínská Nová Ves** (Neudorf a. M.) near Bratislava, Czechoslovakia, serve as a good example. Their fillings are a well known source of numerous vertebrate fossils of Middle Miocene age (cf. ZAPFE, 1950, 1960; FEJFAR, 1974, etc). The faunal assemblage corresponds well to a state characterizing the stage *Orleanium* or the biozone MN 6 (sensu MEIN, 1976). At the same locality, the top of the sedimentary fill is the basal layer of a very thick marine sand accumulation which bears marine fossils. The sands correspond to the Badenian transgression of the Paratethys. Consequently, based on this locality, the mammalian biozone MN 6 can be correlated with the Badenian marine stage and, hence, dated back to approx. 16 Ma (CÍCHA et al., 1972).

By means of similar records a biostratigraphic system, originally defined in relative stages only, may become a useful tool to determine the numerical age of a deposit. Correlation of Quaternary mammal biozones with a sequence of individual glacial cycles, based primarily on faunal records in loess series (cf. HORÁČEK, 1979; RABEDER, 1981), enables us to express results of paleontologic analysis not only in terms of a biostratigraphic system, but also in terms of individual glacial cycles and with an accuracy of approx. 50–100 ka which may exceed the accuracy of conventional methods of numerical dating.

For more details and numerous references concerning the European paleokarst biostratigraphy see CÍCHA et al. (1972); FAHLBUSCH (1976, 1981); FEJFAR and HEINRICH (1983); HORÁČEK (1979); KRETZOI (1956, 1969); LOŽEK (1964); MEIN (1976); SIGÉ et al. (1979), etc.

Paleoecologic Analysis

There are further techniques which can be used for paleoenvironmental reconstruction, including a variety of numerical procedures. Interpretation of the ecologic role or requirements of fossil forms is based on the principle of *actualism,* i.e. on the assumption that their requirements correspond to those found in recent forms of similar morphology.

Paleoecologic analysis should show which part of the assemblage consists of *allochthonous* items (i.e. which were transported to the place secondarily, as by predators) and which part is formed by *autochthonous* elements, i.e. inhabiting the place at the time of their deposition. The latter are particularly important from the point of view of paleokarst study. They indicate the character of a locality at the beginning of the fossilization phase. For instance, a mass appearance of cave-dwelling bats (*Miniopterus, Rhinolopus*) proves it was a spacious cave, while a large representation of bird remains, epilithic snails or bat species roosting in rocky fissures (*Vespertilio, Tadarida, Pipistrellus, Nyctalus*) indicates an open exterior with rocky walls as well as an opening to receive the subsurface deposit (for details see Fig. 5–16). The appearance of e.g. subtropical forms (*Megadermatidae,* etc.) indicates warm conditions, while the presence of forms dependent on extensive open terrain with rocky debris (e.g. *Microtus nivalis,* numerous molluscs) also suggest the past climatic conditions of the studied locality. All these data may then become of considerable value in the interpretation of the paleokarst history. Hence, students of paleokarst should call upon paleontologists not only for biostratigraphic data but also for the paleoecologic interpretations that can be made.

Terrestrial Stages and Biozones Established With the Aid of Vertebrate Remains Found in Paleokarst: Brief Case Surveys

The orthostratigraphy of Tertiary and Quaternary terrestrial sediments is based on vertebrate bone remains, especially those of mammals. Bones and teeth often accumulate in sediments in a fairly good condition and are, therefore, suitable for use to establish representative chronological and faunal horizons.

One of the most famous paleokarst deposits containing vertebrate finds are the **Phosphorites du Quercy** (GÈZE, 1938), a sediment of Eocene-Oligocene age (38–26 Ma). Deposits up to 100 m thick and with phosphorites fill wider or narrower fissures and karst cavities in the Jurassic Causses Limestones (cf. Fig. 5–9) at the southeastern foot of the Massif Central. Paleokarst cavities are filled with red limonitic clays. Their phosphorus content is derived in part from bone decomposition. Most fissures contain rich vertebrate remains, including a number of bats. This indicates that they were spacious caves at the time of fossilization; for details see SIGÉ et al. (1979); CROCHET et al. (1981).

Sediments deposited in karst cavities in Jurassic limestones also play a great role in the study of the lengthy development of the South German Molasse area, F. R. G. (DEHM, 1961a, b). Among the number of localities in that region, one particularly worth mentioning is the fossil shaft, **Möhren 13,** filled with Oligocene deposits. In the **Burgmagerbein Quarry** near the border of the Swabian and Franconian Jura, karst features developed in Upper Jurassic limestones. There were coastal caves here during the Oligocene, while in the Lower Miocene the whole region was inundated first by Lower Marine Molasse, later by Miocene lacustrine deposits. Therefore, in this locality in just one quarry, we may find together cavities developed during the Oligocene containing Middle and Upper Oligocene terrestrial deposits with vertebrate remains, and also Early Miocene Lower Marine Molasse as well as Middle Miocene deposits with a high phosphate content. Extraordinarily rich Middle Miocene (MN 6) vertebrate assemblages have been obtained also from fossil fissures at **Steinberg** and **Goldberg** in the area of Nördlingen Ries (RACHL, 1983).

The **Vieux Colonge,** France, is a typical Lower Miocene karst filling. The caves and vertical fissures in Jurassic limestones are filled with clays (facies sidérolithique) containing also stalactites belonging to the MN 4b zone. The clay contains also an extremely important vertebrate fauna (MEIN, 1958).

A complex series of Middle Miocene karst deposits is known from **La Grive-Saint-Alban,** Isere, France. Zones MN 7 (La Grive M) and MN 8 (La Grive L3) have been defined here.

The Late Miocene and Middle Pleistocene are represented in cavity fillings in Hungary[1]. The deposits have been carefully studied. The Lower Pannonian (MN 10) fill of **Csákvár Cave** (KRETZOI, 1954) is the oldest of them. Triassic dolostones were exposed at a coast during the Early Pannonian. Vertical fissures perpendicular to the shore were widened by wave action. Later, after marine regression, these cavities became filled with fine-grained sand and clays. The deposits contain rich vertebrate faunas indicating deposition under subtropical

[1] Hungarian interpretation of the Quaternary base, i.e. 2.4 Ma differs from internationally accepted Pliocene/Quaternary boundary at 1.8 Ma. Csarnotanum represents then Middle Pliocene, Villányium Upper Pliocene and Biharium Lower Quaternary in the sense of 1.8 Ma boundary (Editors)

climate conditions. The area remained continental and a majority of the Lower Pannonian sediments were eroded during the Middle Pleistocene. The fissure walls were widened into more cavernous shapes by normal cave genetic processes. This period is characterized by the occurrence of Middle Pleistocene vertebrates. The last glacial change is recorded by Upper Pleistocene fillings.

A standard Hungarian fossil cave is the locality, **Polgárdi** (MN 13; Fig. 2–60), developed in Paleozoic limestone. Classical fossil finds were made by KORMOS (1911), in an old cave chamber. Recently, a similar fissure-like cavity with fillings has been exposed in the Polgárdi limestone quarry.

The most complete and complex paleokarst system of Hungary is found in **Osztramos Hill** in Rudabánya Range (Figs. 2–59 and 2–61). More than 20 localities yielding vertebrate remains have been uncovered by Triassic limestone quarrying. The age of the remains covers a period from the Early Pliocene to the Middle Pleistocene. Two cave horizons can be distinguished here, the upper being of Pliocene age, and the lower being developed during the Pliocene/Pleistocene transition. The cave was almost completely filled. It was then intersected by a system of Lower and Middle Pleistocene vertical fissures. Though there is even a biostratotype, Estramontium, there are localities with vertebrate remains indicating different ages in the same place (JÁNOSSY and KORDOS, 1977). Osztramos Hill lies at the southern margin of the Slovak Karst and, hence, together with sites in Czechoslovakia is very important for the reconstruction of karst history in the whole region (cf. BOSÁK, HORÁČEK and PANOŠ, p. 000–000).

The **Villány Mountains** are a classical type-region of European terrestrial biostratigraphy and chronostratigraphy. At Csarnota there is a dark brown clay filling a shaft. The Csarnotanum stratotype is located here. The stratotype of Villányium is represented by a fissure cave with speleothems in the Templomhegy Quarry. The parastratotype of Bihárium occurs in the same quarry. Vertebrate fauna accumulated in the red clayey fills of vertical solutional shafts.

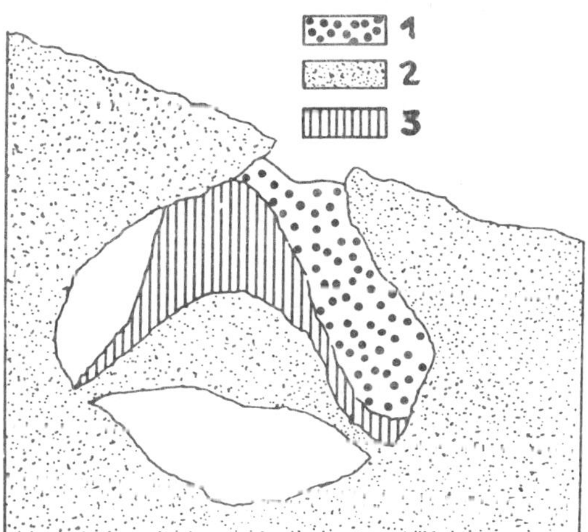

5–18. Cross-section of the cave at Swartkrans (after WOLPOFF, 1970). 1. pink breccia; 2. dolostone; 3. travertine.

Paleokarst sites are also very suitable for studies of hominid evolution. The *Paranthropus crassidens* (Broom, 1949) holotype was discovered in a fossil cave cave in dolomitic limestone at **Swartkrans,** South Africa. The cavity is divided into two stages, where an older brown breccia (Member 2) and a pink, younger breccia (Member 1) can be distinguished (Fig. 5–18). According to geomorphological investigations, the cave was created before 2.57 Ma (Partridge, 1973), while the fill is of early Middle Pleistocene age. The locality furnishes a good example of the misunderstandings often occurring with regards to the differentiation and genetic-chronological interpretation of cave fill when thorough investigations are lacking (Wolpoff, 1970).

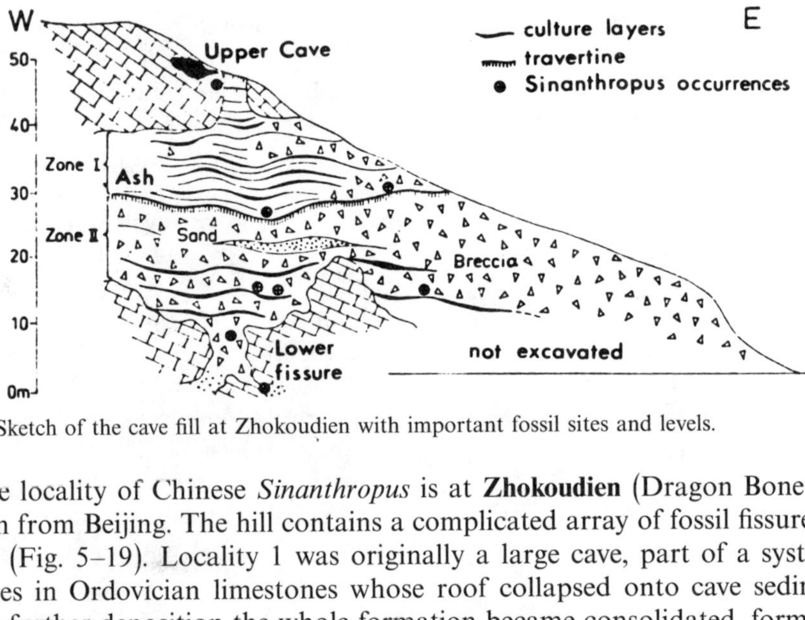

5–19. Sketch of the cave fill at Zhokoudien with important fossil sites and levels.

The locality of Chinese *Sinanthropus* is at **Zhokoudien** (Dragon Bone Hill), 42 km from Beijing. The hill contains a complicated array of fossil fissures and caves (Fig. 5–19). Locality 1 was originally a large cave, part of a system of fissures in Ordovician limestones whose roof collapsed onto cave sediments. After further deposition the whole formation became consolidated, forming in parts a stalagmitic breccia. Beijing Man lived in this cave system for many generations, leaving a vast amount of bones, stone artefacts, evidence of the use of fire (burnt bones, ash layers) and contemporaneous animal bones. Six fairly complete hominid skulls and more than 150 teeth and limb fragments have been excavated at this site. The remains represent 40 or more individuals of both sexes, young and old.

The so-called "Upper Cave" of Zhokoudien contains the bones of Upper Pleistocene man, *Neoanthropus,* and no Middle Pleistocene finds. The infilling of the cave system evidently took place in three or four stages. The *Sinanthropus* material mostly occurs in the middle period, layers 4–11. Layers 1–3 (which post-date the roof collapse) and the Upper Travertine appear to belong to a later cycle of deposition, whilst layers 12–13 may be distinctly earlier (Oakley, Campbell and Molleson, 1975).

Some Examples of the Biostratigraphy of Marine and Terrestrial Deposits in Paleokarst

Under favourable conditions, terrestrial and marine deposits and fossils can occur over each other in shoreline caves as a consequence of sea level changes. These localities are of increasing importance in the correlation of terrestrial and marine stratigraphy (see above).

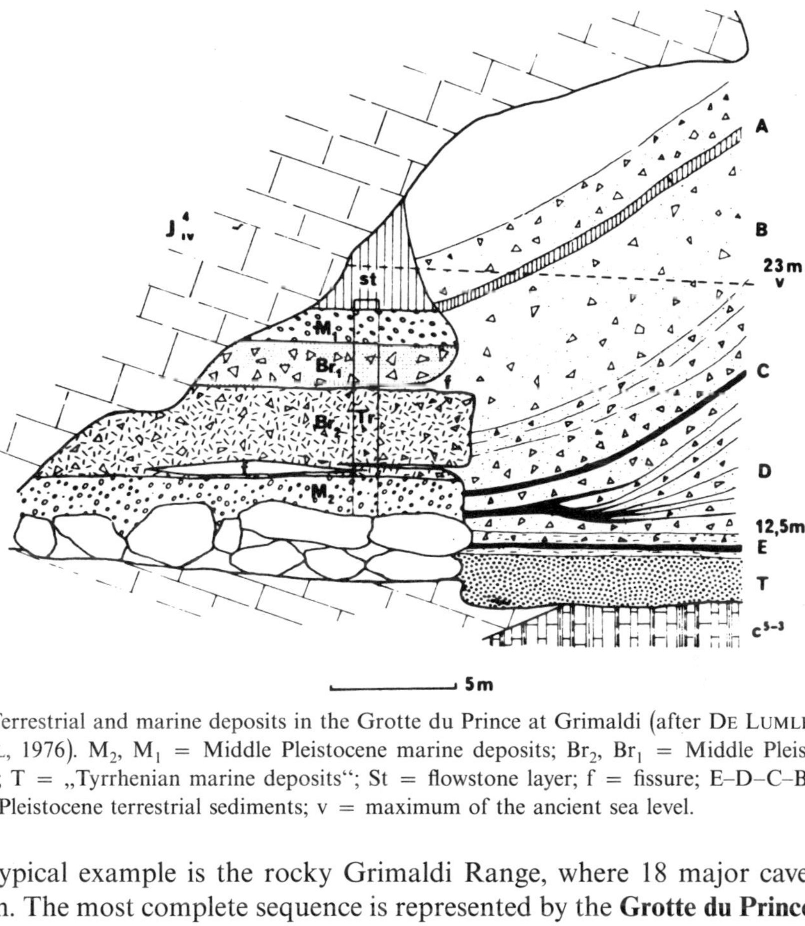

5–20. Terrestrial and marine deposits in the Grotte du Prince at Grimaldi (after DE LUMLEY and BARRAL, 1976). M_2, M_1 = Middle Pleistocene marine deposits; Br_2, Br_1 = Middle Pleistocene breccia; T = „Tyrrhenian marine deposits"; St = flowstone layer; f = fissure; E–D–C–B–A = Upper Pleistocene terrestrial sediments; v = maximum of the ancient sea level.

A typical example is the rocky Grimaldi Range, where 18 major caves are known. The most complete sequence is represented by the **Grotte du Prince** (DE LUMLEY and BARRAL, 1976). This cave developed at the contact between Jurassic limestones and Cenomanian marls. Its present floor lies 8.3 m a.s.l. The first sediment deposited in the cave was a marine abrasion sand with the molluscs *Strombus bubonis* and *Conus testudinarius* from the Tyrrhenian transgression. Three marine layers were deposited in the upper part of the cave during Middle Pleistocene interglacials (M_2, M_1 and M_3 levels) with two terrestrial levels (Br_1 and Br_2). The last member, M_1 marine conglomerate, is covered by a thick speleothem layer. The older sediments were eroded near the cave entrance and replaced by more than 10 m of Upper Pleistocene terrestrial, multistage deposits (layers A-E, Fig. 5–20).

The sedimentary sequence in the Grotte du Prince represents Middle-Upper Pleistocene events, while the nearby **Grotte du Vallonet,** Roquebrune-Cap Martin, contains preserved Lower Pleistocene marine-terrestrial contacts. The basal fill of the cave is composed of a breccia and speleothem layer formed in a period which is considered to correspond to the beginning of the Calabrian transgression. The base is overlain by marine "Calabrian" sediments with rich forams and molluscs as well as fish remains. The final terrestrial sequence contains mammals (*Allophaiomys, Ungaromys,* etc.). The Lower Pleistocene fill has been protected from later erosion by a speleothem cover. The Lower Pleistocene in the Grotte du Vallonet is at 108 m a.s.l. as compared with the 10–20 m a.s.l. of the Late Pleistocene fill in the Grotte du Prince.

Paleontological Interpretation of Transgressive Sediments in Paleokarst

It is generally true that in submerged karst, shoreline and backshore areas preserve caves and other karst forms originally developed under continental conditions and later inundated by the sea so that the cavities became filled with marine sediments. These fills contain special kinds of marine fauna. According to current biological investigations (HART, MANNING and ILIFFE, 1985), a peculiar biotope develops in submerged caves besides the usual well-known assemblages of animals indicating facies and age. Evidence from submerged caves in Bermuda, the Bahamas and Canary Islands indicate significant ties between the marine cave fauna and the deep-sea fauna; the cave faunas may contain very old elements and such caves served as refuges over very long time-spans.

These recent biologic investigations advise us to be cautious during the biostratigraphic interpretation of marine organisms found in cave deposits. Some examples demonstrating the application of marine biostratigraphic methods in the paleokarst studies are summarized below.

Peculiar paleogeographical and paleotectonic changes took place during the evolution of the karst in the **Visé** region, Belgium (see QUINIF, p. 38–40). In the Fammenian, older rocks were uplifted in a domal structure (Fig. 2–4) which was strongly karstified, with extensive mogote topography. Six significant marine transgressions followed each other in the Dinantian. The age of marine deposits was determined by the methods of classical Paleozoic paleontology using rich coral, ostracod and conodont assemblages. Marine sediments covered completely and fossilized the karst forms (POTY, 1980, 1982).

MAUTHE (1975) described interesting regressive-transgressive cyclic sedimentation in Middle to Upper Jurassic carbonates in **Soria,** northern Spain. Thin continental clastics fill corrosion surfaces in a stratified marine complex (Fig. 5–21). Fossils have been found here not in terrestrial sediments, but in the cyclic

5–21. Syngenetic debris and debris-filled paleokarst cavity in the Soria region (Spain; after MAUTHE, 1975). a = marine pelletal limestone; b = syngenetic debris in erosional contact with the lower limestone a.; c = epigenetic debris filling the paleokarst unconformity; d = cross-bedded sandstone.

marine limestones (ammonites), so that the dating of karstification episodes is based on the age of surrounding transgressive limestones. Similar features have been described in the Carboniferous of **Derbyshire** by WALKDEN (1974), OAKMAN (1984), see also T. D. FORD (p. 51–70). Such karstification in peritidal conditions and during short sea level changes is widespread in the geological record.

Interesting relations between terrestrial vertebrates, sporomorphs and marine microfossils are found at **Tytherington,** near Bristol (Great Britain). In a quarry there are several fissures and karst cavities with Triassic upland reptiles and Rhaetian marine phytoplankton as well as land-derived palynomorphs. According to MARSHALL and WHITESIDE (1980), the dissolution of the cavities resulted from the marine oscillations during the Rhaetian transgression. Therefore, seawater phytoplankton was deposited at the bottom of fissures. At times of fissure emergence above sea level, terrestrial animals living in the neighbourhood accumulated together with sporomorphs of upland origin. The mixture of land and marine fossils originated by this means. For details see also T. D. FORD (p. 51–70, Figs. 2–14 and 2–15).

Conclusions

This review, although short and incomplete, demonstrates that the appearance of paleokarst features can be extremely diversified. Therefore, each individual locality requires an individual interpretation of any paleontological data furnished by it. At the same time, biostratigraphic data interpreted within geological processes give a solid base for the determination of the time and phase of the development, filling and fossilization of paleokarst forms.

PART VI

CONCLUSIONS

Paleokarst is widespread in the geological history of the Earth. The distribution of fossilized karst features depends on the tectonic conditions and/or the differences in water dynamics, i.e. of the physical potential of the karst system. Karst is fossilized when it loses its hydrologic function. The fossilization of the karst is influenced by a complex sequence of processes which results from the change of (1) tectonic regime; (2) global sea level and/or (3) climate. Uplift, subsidence, marine transgression, large-scale continental deposition an climatic change (to an arid climate) are the main fossilization agents. Their effects are broadly discussed and summarized in numerous papers.

The oldest paleokarst features are described from the Early Proterozoic of South Africa, while the youngest paleokarsts are of Holocene age and are widespread in the peri-arctic zones with permafrost. Paleokarst is preserved mainly in carbonate rocks, carbonatites or quartzite, while paleokarst in evaporitic rocks are rare due to the physical properties of those rocks.

The paleogeographic distribution of paleokarst can be studied on the basis of global paleotectonic and palinspastic reconstructions. There is much evidence of fossilized inland paleokarst features occurring in various paleogeomorphologic circumstances. Paleokarst that developed on ancient karst plateaus with rapid and relatively well developed drainage patterns is abundant. On the contrary, coastal paleokarsts and paleokarst of ancient islands are preserved only occasionally because of the destructive effects of wave action, surf and marine transgressions. Numerous abandoned, i.e. fossilized, caves occurring in high altitudes in recent mountains represent relics of ancient subterranean hydrologic karst systems developed prior the mountain uplift.

The most intensive developments of karst in the geologic past were concentrated from paleoequators up to paleolatitude 30° N or S. However, paleokarst features in the zone 30–60° are still relatively abundant while more northerly paleokarst occurs only rarely. These reconstructions, because they are

P. Bosák, D. C. Ford, J. Głazek and I. Horáček (Editors):
Paleokarst. A Systematic and Regional Review. – Elsevier and Academia. Amsterdam and Praha. 1989.

based on global paleotectonic speculations rather than on proper reconstructions, are imperfect and have numerous systematic and other mistakes.

Paleokarst acts as a trap for former scenery and environments. If properly studied, paleokarst is a very important source of data for: (1) paleogeographic reconstructions; (2) the study of ancient climates; (3) the study of the ancient history of both abiotic and biotic environments including the phylogeny and paleoecology of various groups of animal fossils; (4) regional and stratigraphic geology; (5) archaeology and the study of the human past, etc. In platform areas, paleokarst studies contribute to our knowledge of the situations during breaks in marine deposition. It provides, therefore, a kind of *missing link* in the study of regional history. Since it has preserved subaerial processes, paleokarst may also serve as a source of very important data for reconstructions of the evolution of orogenic belts and collisional systems.

The complex nature of paleokarst as well as of its fossilization and rejuvenation events causes the need for special care during research effort carried out on it. Reversed sequences of deposits caused by internal resedimentation and other depositional irregularities or changes in sedimentation dynamics in the karst systems may even lead to serious confusion if studies were done only in a cursory style.

The dating of karstification, i.e. the locating of the formation of paleokarst within the stratigraphic column, is not simple. Generally, it is only possible to date the infill of karst by means of numerical geochronology, paleontology, lithologic and facies comparision, etc. The dating of the paleokarstification process itself is very speculative, with the exception of syngenetic karst or karst forms which originated during short-lasting emergences and depositional breaks in the cyclic sequences of shallow marine carbonate or carbonate-terrigenous deposits. In these cases, the paleokarst is dated using the surrounding rocks, from their fossil content, and not by the paleokarst fills which are commonly non-fossiliferous. Otherwise, in the majority of cases, we date the act and processes of fossilization, i.e. the age of the *preserved fills.*

The economic potential of paleokarst is very significant. The majority of economic mineral deposits in carbonate host-rocks are related to paleokarst forms and processes. Recently formed mineral deposits are rather exceptional; only some bauxite and phosphate deposits (Pacific Ocean) or polymetallic ores (Bulgaria). Paleokarst served as a trap for metals and other mineral particles carried by water either as a solid in suspension and/or in solution near the Earth's surface. The traps can be divided into: (1) mechanical, where mineral particles, minerals, etc. are deposited due to the falling energy of flowing water in superficial (dolines, poljes, karren) or subterranean (caves) karst forms or due to filtration through the karst sediments, and (2) chemical, resulting from chemical reactions with $CaCO_3$ which represents the source of active Ca^{2+} ions which can easily replace the heavy metals in solution. The origin of such deposits depends strongly on local Eh and pH conditions which in particular cases has caused the precipitation of rich ores of some metals.

The other economic aspects of paleokarst result from the exploitation of the mineral deposits and water resources and from the utilization of karst areas. For civil engineering, paleokarst represents a relatively unfavourable environment. Although sometimes the most favourable morphologic situation occurs in such karst areas, the presence of paleokarst, i.e. fossilized, inactive and latent karst features makes it very dangerous for e.g. dam construction, the foundations of extensive industrial buildings, etc. Unexpected water losses, large ground subsidences or the sudden collapse of subterranean karst hollows are not uncommon in a majority of large karst areas; they can cause not only physical problems during construction works but aso the rapid rise in the cost of construction.

The exploitation of karst-related mineral deposits is made hazardous by inundations or violent floods of water into the mines or deep open pits. Some other surficial processes and effects are related to the sudden or very fast dewatering of near-surface zones of the karst. Large ground subsidences caused by piping or suffosion of fine clastics from rejuvenated, activated paleokarst channels or systems caused numerous problems in mining districts, which in some cases were highly urbanized. The huge water inrushes are connected also with the transport of sedimentary fill of the paleokarst systems which results in problems of mine dewatering and the arising problems need a hydrological and hydrogeological solution.

The danger from water is present not only in the case of the mining of mineralized formations (i.e. the case of nearly all karst-related mineral deposits and any metasomatic mineralization in karst rocks) where karst waterways and paleokarst systems are opened directly by mining operations, but also when karst occurs in the overburden or in the underlying sequences of the mineralized and mined formation. Any interference with the natural hydrogeologic equilibrium in paleokarst leads to the disturbance of that equilibrium and fossilized and inactive systems become rejuvenated. On the other hand this can cause many other, related processes and unexpected situations.

Deep karst aquifers represent hypogenetic, subjacent or intrastratal karst features with complicated and long-lasting evolution. They can also result from the submergence of shallower karst networks developed in the geologic past and then covered and buried by thick overburdens or which developed during eustatic lowerings of sea-level. They commonly contain hot, thermal waters which can be used for urban heating, for spas, for agricultural production, etc. The water supply of large areas depends on karst waters which in many cases comes from deep paleokarst aquifers.

Paleokarst represents therefore not only a research problem and a matter of scientific interest, but also a topic of great practical importance and of great economic consequence. It is clear that theoretical and practical studies are closely interrelated. Most of the practical complications arise from poor knowledge and understanding of karst and paleokarst processes and related phenomena or even from a general under-estimation or ommission of the role of the special characteristics of karst and paleokarst. The consequences can be seen in many karst-related mineral deposits and ore districts during their prospecting, development and exploitation stages or during the utilization of karst terrains. Theoretical and practical understanding of all processes and phenomena related to karst and paleokarst is necessary for any up-to-date development of mines and open pits or engineering activity in karst. Only complex and multidisciplinary investigation can provide an adequate quantity and quality of data so that construction costs and other financial demands resulting from the "unexpected" consequences of mining or engineering operations on the karst environment can be reduced.

We may conclude that the multidisciplinary study of paleokarst, as shown in the present book, serves as an excellent example of something which is often discussed but only rarely achieved in an adequate and appropriate form in daily practice. This is the necessity for close interconnection and mutual information exchange between purely scientific and/or theoretical studies of these environmental phenomena and the ways they can be economically exploited. We have demonstrated that if the results we have obtained concerning the dynamics of karst processes are not respected, the exploitation of mineral deposits and/or the development of works in karst will be extremely complicated. Similar complications can arise if attention is paid only to subaerial processes, and the specific characteristics of extensive paleokarst deposits and/or deep karst water circulation systems are ignored. Consequently, it can be stressed that the results of theoretical approaches as well as practical experience from studied subject may produce a well balanced data base which may serve as a basis for fruitful and reasonable development of the human environment.

Pavel Bosák

PART VII

REFERENCES

ABATE, B., CATALANO, R., D'ARGENIO, B., DI STEFANO, P., LO CICERO, G., MONTANARI, L., PECORANO, C. and RENDA, P., 1982. Evoluzione delle zone di cerniera tra piattaforme carbonatiche e bacini nel Mesozoico e nel Paleogene della Sicilia occidentale. In: R. CATALANO and B. D'ARGENIO (Editors), *Guida alla geologia della Sicilia occidentale.* Soc. Geol. Ital., Guide Geol. Region, Palermo, pp. 53–81.

ADAMCZYK, A. F. and CHAWIŃSKI, S., 1982. The pollution of mining waters in the Pomorzany Mine by lignosulphate compounds (in Polish). *Przegl. Geol.,* 30(7): 348–353.

ADAMCZYK, A. F. and WILK, Z., 1976. The chemistry differentiation of mining waters in the Olkusz region (in Polish). *Spraw. Posiedz. Kom. Nauk. PAN Oddział w Krakowie,* 1976, 2: 408–410.

ADAMEK, J., DVOŘÁK, J. and KALVODA, J., 1980. The contribution to geological structure and petroleum geological evaluation of the Nikolčice-Kurdějov elevation (in Czech). *Zemný Plyn Nafta,* 25(4): 441–474.

ADAMS, A. E., 1980. Calcrete profiles in the Eyam Limestone (Carboniferous) of Derbyshire: petrology and regional significance. *Sedimentology,* 27: 651–660.

AFANASYEV, A. P., 1977. *Phanerozoic weathering crusts of the Baltic Shield* (in Russian). Nauka, Leningrad, 244 pp.

AGASSIZ, A., 1903. Reports on the Scientific results of the expedition to the tropical Pacific by the U. S. Fish Commission steamer „Albatros“ from August 1899 to March 1900. IV — The coral reefs of the tropical Pacific. *Mem. Mus. Comp. Zool.,* 28: 410 pp.

AGROMAN, 1979a. Asesoría de Geología Aplicada. *„PRESA DE ALCORLO“ (Guadalajara). Estudio del karst de la cerrada.* Unpubl. MS, Minist. Obras. Públ., 52 pp.

AGROMAN, 1979b. Asesoría de Geología Aplicada. *„PRESA DE TOUS“ (Valencia). Análisis estructural del karst de la cerrada. Estudio de las fugas, previsión del comportamiento y recomendaciones de tratamiento.* Unpubl. MS, Minist. Obras Públ., 114 pp.

ASKIN, V. et al., 1981. *Map of oil and gas fields of Yugoslavia 1 : 500 000 and the explanation.* Belgrade, 72 pp.

ALABASTER, C., 1982. The minerals of Mendip. *Somerset Mines Res. Grp.,* 1(4): 52 pp.

ALEXANDRE, J., 1976. Les surfaces de transgression exhumées et les surfaces d'applanissement. In: *Gémorphologie de la Belgique.* Lab. Géogr. Phys., Univ. de Liège, pp. 76–92.

ALEXANDROV, S., 1927: Radiographien tujamujunischer Erze. *Z. Kristallogr.,* Wiesbaden, 65: 141.

ALFIREVIĆ, S., 1966. Hydrogeological investigations of submarine springs in the Adriatic. *Congr. Assoc. Int. Hydrogeol. Belgrade 1963,* Beograd, pp. 255–265.

ALFÖLDI, L., DEÁK, J., LIEBE, P. and LORBERER, Á., 1979. Relationships between the cold and warm karstic waters in the Central Range with special regard to mine drainage requirements (in Hungarian). *VITUKI Közlem.,* 23: 300–317, 23B: 60–61.

ALFÖLDI, L. and LORBERER, Á., 1976. Untersuchung von dreidimensionellen Stromungen der Karst-Thermalwässes aufgrund von Brunnendaten (in Hungarian). *Hidrol. Köz.,* 10:433–443.

AL-SAYARI, S. S. and ZÖTL, J. G. (Editors), 1978. *Quaternary Period in Saudi Arabia,* Vol. 1. Springer, Wien, 325 pp.

AMODIO-MORELLI, M., BONARDI, G., COLONNA, V., DIETRICH, D., et al., 1976. L'arco calabro-peloritano nell'orogene appenninoco-maghrebide. *Mem. Soc. Geol. Ital.,* 17: 1–60.

AMOURI, M., 1977. *Étude des indices zincifères sous inconformité en invironnement carbonaté du Bled Zelfane (Tunisie Centrale).* Thèse Doct., 3e cycle, I. N. P. L., Nancy, 137 pp.

AMOURI, DÉVIGNE, J. P. and FUCHS, Y., 1978. Paléobactéries et minéralisations plombo-zincifères du Bled Zelfane (Tunisie Cent.). *Mineralium Depos.,* 13: 183–189.

AMSTUTZ, G. C. and BERNARD, A. J. (Editors), 1972. *Ores in Sediments.* Inter. Union Geol. Sci., Ser. A., 3, Springer, Berlin, 350 pp.

AMSTUTZ, G. C., EL GORESY, A., FRENZEL, G., KLUTH, C., MOH, G., WAUSCHKUHN, A. and ZIMMERMANN, R. A. (Editors), 1982. *Ore Genesis. The State of The Art.* Spec. Publ., 2, Soc. Geol. Applied to Mineral Deposits, Springer, Berlin, 804 pp.

ANDERSON, G. M., 1983. Some geochemical aspects of sulfide precipitation in carbonate rocks. - In: G. KISVARSANYI et al. (Editors), *Proc. Vol. Int. Conf. on Mississippi-Valley-type Pb – Zn Deposits, Rolla, Misouri,* pp. 61–76.

ANDERSON, G. M. and MACQUEEN, R. W., 1982. Ore Deposit Models. G. Mississippi Valley-Type lead-zinc deposits. *Geoscience Canada,* 9(22): 108–117.

ANDERSON, R. E., 1971. Notes on the bauxite deposits of Jamaica. *J. Geol. Soc. Jamaica, Bauxite Alumina Symp. 1971, Kingston,* pp. 9–16.

ANDRÉ, CH. K., 1821. Beschreibung des geognostischen Vorkommens und bergmännischen Abbaues der Eisenerze, welche auf den Hüttenwerken zu Blansko in Mähren verwendet werden. *Hesperus,* 30 (3): 17–21.

ANDRUSOV, D., BORZA, K., MARTINI, E. and POSPÍŠIL, A., 1958. Sur l'origine et le temps de formation des „terra rossa" de la Slovaquie Centrale et Meridionale. *Geol. sbor. Slov. akad. vied,* 9(1): 27–39.

ANELLI, F., 1959. Nomenclatura Italiana dei Fenomeni carsici. *Le Grotte d'Italia, Ser. 3[a], II, 1957–1958, Trieste,* 32 pp.

ANONYME, 1976. *Karst in China.* Inst. Hydrogeol. Engng. Geol., Chinese Acad. Geol., Sci., Shangai people's Publ. House, 75 pp.

ANTSYFEROV, A. S., 1979. About formation of Jurassic depressions of Prisayan'e (in Russian). *Dokl. Akad. Nauk SSSR,* 244(2): 421–424.

ARKHANGELSKY, A. D., 1937. Types of bauxite and their genesis in the USSR. *Trans. Conf. on Iron, Manganese and Aluminium Ore Genesis,* Izd. Akad. Nauk SSSR, Moskva.

ARMSTRONG, R. L., 1969. Control of sea level relative to the continents. *Nature,* 221: 1042–1042.

ARNBERGER, E., 1955. Höhlen und Niveaus. *Höhle, Wien,* 6 (1): 1–8.

ARPER, W. B., 1964. Effect of phosphates on limestone deposition. *Geol. Soc. Amer., Ann. Meet., Abstr., Miami,* pp. 5–6.

ASANO, D., 1942. Coral Reefs on the South Sea Islands. *Tôhoku Imp. Univ. Geol. Paleont. Inst., Rep.,* 39: 1–19.

ASSERETO, R. and BRUSCA, C., 1976. Observation a Lagny Ph. — Rouvier H., Les gisements Pb – Zn en roches carbonatées sous unconformité: gisements paléokarstiques ou gisements dans les paleokarsts? *Mém. Soc. Géol. France,* 7: 57–69.

ASSERETO, R., BRUSCA, C., GAETANI, M. and JADOUL, F., 1977. Le mineralizzazioni a Pb – Zn nel Triassico delle Dolomiti. Quadro geologico e interpretazione genetica. *Ind. Miner., Roma,* 28: 367–402.

ASSERETO, R., JADOUL, F. and OMENETTO, P., 1977. Stratigrafia e metallogenesi del settore occidentale del distretto a Pb – Zn, fluorite e barite di Gorni (Alpi Bergamasche). *Riv. Ital. Paleont. Stratigr.,* 83(3): 395–532.

BABAK, V. I. and SMIRNOVA, S. B., 1978. The new data on the age of hypergenic phosphorites of the Eastern Sayan (in Russian). *Kora vyvetrivania, Moskva,* 16: 95–105.

BACK, W., HANSHAW, B. B. and VAN DRIEL, J. N., 1984. Role of ground water in shaping the eastern coastline of Yucatan peninsula. Mexico. In: R. G. LA FLEUR (Editor), *Groundwater as a Geomorphic Agent,* Allen and Unwin, Boston, pp. 281–293.

BAIN, H. F., VAN HISE, C. R. and ADAMS, G. J., 1901. Preliminary report on the lead and zinc deposits of the Ozorac region. *US. Geol. Surv., 22 A. R.*, II: 23–229.

BAKALOWICZ, M., BALVOUX, B. and MANGIN, A., 1974. Apport du traçage isotopique naturel a la connaissance du fonctionnement d'un systeme karstique — teneurs en oxygene-18 de trois systemes des Pyrenees, France.

BALCON, J., 1981. Quelques idées sur les minéralisations plombozincifère dans les formations carbonatées en Belgique. *Bull. Soc. Belg. Géol.*, 90(1): 9–64.

BÁLDI, T. and NAGYMAROSI, A., 1976. Silicification of the Hárshegy Sandstone and its hydrothermal origin (in Hungarian). *Földt. Közl.*, 106(3): 227–275.

BAŁUK, W., 1970. The Lower Tortonian at Niskowa near Nowy Sącz, Polish Carpathians (in Polish). *Acta Geol. Pol.*, 20(1): 101–157.

BALWIERZ, J. and DŻUŁYŃSKI, S., 1976. Experiments on rock deformations produced by underground karst processes. *Rocz. Pol. Tow. Geol.*, 46(4): 419–434.

BARANIECKA, M. D. and SARNACKA, Z., 1971. The stratigraphy of the Quaternary and palaeogeography of the drainage basin of the Widawka (in Polish). *Biul. Inst. Geol.*, 254: 157–270.

BĂRBULESCU, A., 1973. Données nouvelles sur le Sénonien de Dobrogea (in Romanian). *Stud. Cerc. Geol. Geofiz. Geogr., ser. Geol.*, 18(1): 229–234.

BĂRBULESCU, A., 1974. *Stratigraphie du Jurassique dans la partie ouest de la Dobrogea Centrale.* (in Romanian). Edit. Academiei R. S. R., Bucureşti, 173 pp.

BARDET, M. G., 1974. Gélogie du diamant. T. II. *Mém. Bur. Rech. Géol. Min.*, 83: 229 pp.

BÁRDOSSY, GY., 1982. *Karst Bauxites.* Elsevier, Amsterdam, 441 pp.

BÁRDOSSY, GY., BONI, M., DALL'AGLIO, M., D'ARGENIO, B. and PANTÓ, GY., 1977. Bauxites of Peninsular Italy. Composition, Origin and Geotectonic Significance. *Monogr. Ser. Miner. Deposits,* 15, Borntraeger, Berlin, 61 pp.

BARKOV, A. S., 1932. The karst of Samarskaya Luka (in Russian). *Zemlevedenie,* 34 (1–2): 33 pp.

BARRINGTON, J. and KERR, P. F., 1963. Collapse features and silica plugs near Cameron, Arizona. *Geol. Soc. Amer. Bull.*, 75: 1237–1258.

BASS, U. B., RYABCHUN, V. K., SLAVUTSKY, M. V. and SHALYT, E. S., 1971. Bauxites of the platform part of the Ukrainian SSR (in Russian). SSR. In: *Platformennye boksity SSSR.* Nauka, Moskva, pp. 93–128.

BATEMAN, A. M., 1950. *Economic Mineral Deposits.* J. Wiley and Son, New York and Chapman and Hall, London, 916 pp.

BATHURST, R. G. C., 1976. *Carbonate Sediments and Their Diagenesis.* 2nd Ed. Elsevier, Amsterdam, 658 pp.

BATTIAU-QUENEY, Y., 1980. *Contribution a l'étude geomorphologique de Massif Gallois.* Thèse de Bretagne Occidentale. Univ. de Lille.

BATTIAU-QUENEY, Y., 1984a. In: M. M. SWEETING (Editor), *Recent Advances in Karst Research.*

BATTIAU-QUENEY, Y., 1984b. The pre-glacial evolution of Wales. *Earth Surface Processes and Landforms,* 9: 229–252.

BAUD, A. and MASSON, H., 1975. Preuves d'une tectonique liasique de distension dans le domaine briançonnais: failles conjuguées et paléokarst à Saint-Triphon (Préalpes Médianes, Suisse). *Eclogae geol. Helv.*, 68(1): 131–145.

BAUER, F. and ZÖTL, J., 1972. Karst of Austria. In. M. HERAK and V. T. STRINGFIELD (Editors), *Karst. Important Karst Regions of the Northern Hemisphere,* Elsevier, Amsterdam, pp. 225–265.

BAUGHAN, D. J. and WALSH, P. T., 1980, Paleokarst phenomena in the Carboniferous Limestone of Anglesey. *Trans. Brit. Cave Res. Assoc.*, 7: 13–30.

BAŻYŃSKI, J., 1960. Engineering geological characteristic of karst in the vicinity of Częstochowa (in Polish). *Przel. Geol.*, 8(8): 430–432.

BECK, B. F. (Editor)., 1984. *Sinkholes: their geology, engineering and environmental impact.* A. A. Balkema, Rotterdam, 429 pp.

BEDNAREK, J., GÓRECKA, E. and ZAPAŚNIK, T., 1985. Tectonically controlled development of ore mineralization in Jurassic sequence of the Silesian-Cracow Monocline (in Polish). *Rocz. Pol. Tow. Geol.*, 1983, 53 (1–4): 43–62.

BEŁKA, Z., 1987. The development and decline of the Dinantian carbonate platform: an example from the Moravia-Silesia Basin. In: J. MILLER A. E. ADAMS, and V. P. WRIGHT (Editors), *European Dinantian Environments.* J. Wiley and Sons, Chichester, pp. 177–188.

BERGOMI, G., MANFREDINI, M. and MARTELLI, G., 1975. *Note illustrative della Carta Geologica d'Italia, F. 163, Benevento.* Serv. Geol. Ital, Roma, 176 pp.

BERNOULLI, D., and JENKYNS, H. C., 1974. Alpine, Mediterranean, and Central Atlantic Mesozoic facies in relation to the early evolution of the Tethys. In: R. H. DOTT and R. H. SHAVER (Editors), *Modern and Ancient Geosynclinal Sedimentation.* Spec. Publ. Soc. Econ. Paleontologists Mineralogists, 19: 129–160.

BEZUIDENHOUT, C. A. and ENSLIN, J. F., 1969. Surface subsidence and sinkholes in the dolomitic area of the Far West Rand, Transvaal,Rep. South Africa. *Int. Symp. on Landsubsidenece, Tokyo,* pp. 482–495.

BEZVODOVÁ, B. and ZEMAN, A., 1983. Paleofeliefs of the Southern Moravia and their collector properties (in Czech). *Sbor. geol. věd, ř. G., 38: 95–149.*

BIDLÓ, G. and MAUCHA, L., 1964. The investigation of karst sediments in the vicinity of Jósvafö (in Hungarian). *E'MKE Tudom. Közl.*, 10(1): 71–82.

BIESE, W., 1931. Über Höhlenbildung. I. Teil, Entstehung der Gipshöhlen am südl. Harzrand und am Kyffhäuser. *Abh. Kön. Preuss. Geol. Landesanst., neue F.,* 137.

BJØRLYKKE, A. and OLAUSSEN, S., 1981. Silurian sediments, volcanics and mineral deposits in the Sagelvvatn area, Troms, North Norway. *Nor. Geol. Unders,* 365: 38 pp.

BLACK, T. J., 1983. Tectonics, structure, and karst in northern lower Michigan. *Michigan Basin Geol. Soc. 1983 Field. Conf.*, 11–35.

BŁASZAK, M., 1970. The character of natural materials for foundry masses in karst forms near Częstochowa (in Polish). *Biul. Inst. Geol.,* 240: 157–243.

BŁASZAK, M., 1973. Quartz sands of the pre-Quaternary age (in Polish). In: *Atlas litologiczno-surowcowy Polski. Surowce skalne, I. Surowce okruchowe,* Wyd. Geol., Warszawa, 16 pp.

BLASZAK, M., 1976. *The cataloque of deposits of moulding and glass sands of Poland* (in Polish). Wyd. Geol., Warszawa, 49 pp.

BŁASZAK, M., DANIEC, J., KOZŁOWSKI, S. and SZELĄGOWSKA-SKRZYPCZAK, E., 1976. Examinations of moulding sands in weathered Middle Jurassic sediments and Tertiary sediments of the Zębiec area near Iłża (in Polish). *Biul. Inst. Geol.,* 292: 5–60.

BLATT, H., MIDDLETON, G. and MURRAY, R., 1980. *Origin of Sedimentary Rocks.* 2nd Ed. Prentice-Hall, Englewood Cliffs, 782 pp.

BLEAHU, M., 1966. Formations périglaciaires et karst dans les Monts de Bihor. *Rev. Roum. Géol. Géophys. Géorg., Sér. Géogr.,* 10 (1): 55–64.

BLEAHU, M., 1972. Karst of Roumania. In: M. HERAK and V. T. STRINGFIELD (Editors), *Karst. Important Karst Regions of the Northern Hemisphere.* Elsevier, Amsterdam, pp. 341–353.

BLEAHU, M., LUPU, M., PATRULIUS, D., BORDEA, S., ŞTEFAN, A. and PANIN, S., 1981. The structure of the Apuseni Mountains. *Carpatho-Balkan Geol. Assoc. XII. Congr. Guide to excursion B 3. Inst Geol. Geophys., Guidebook Ser., 23, Bucureşti,* 103 pp.

BLESS, J. J., BOUCKAERT, J. and PAPROTH, E., 1980. Environmental aspects of some Pre-Permian deposits in NW Europe. *Meded. Rijks. Geol. Dienst,* 32(1): 3–13.

BOCHEŃSKA, T. and PRUSSAK, W., 1974. Natural water danger in the region of "Pomorzany" (in Polish). *Przeg. Geol.,* 22(11): 527–532.

BOCK, H., 1913. Der Karst und seine Gewässer. *Mitt. Höhlenkunde, Wien,* 6(3): 1–23.

BÖCKER, T., 1969. Karstic water research in Hungary. *Bull. Int. Assoc. Scientific Hydrogeol.,* 14(4–12), 1969: 7–20.

BÖCKER, T. and VIZY, B., 1982. Protection against underground waters of bauxite mining at Nyirád, Hungary, and its impact on the environment. *First Int. Mine Water Congr., C, Budapest.*

BOGACZ, K., DŻUŁYŃSKI, S. and HARAŃCZYK, C., 1970. Ore-filled hydrothermal karst features in the Triassic of the Cracow-Silesian Region. *Acta Geol. Pol.,* 20(2): 247–267.

BOGACZ, K., DŻUŁYŃSKI, S., HARAŃCZYK, C. and SOBCZYŃSKI, P., 1973. Cave filled with clastic dolomite and galena mineralization in disaggregated dolomites. *Rocz. Pol. Tow. Geol.,* 43: 59–72.

BÖGLI, A., 1980. *Karst Hydrogeology and Physical Speleology.* Springer, Berlin, 284 pp.

BOHNE, N., 1926. Die Insel Nauru als Typus eines korallogenen Phosphatlagers. *Pettermann's Mitt.,* 72: 52–59.

BOILLOT, G., 1983. *Géologie des marges continentales.* 2 ed. Masson, Paris, 139 pp.

BONARDI, G. and GIUNTA, G., 1982. L'estremità nord orientale della Sicilia nel quadro dell' evoluzione dell' Arco Calabro-Peloritano. In: R. CATALANO and B. D'ARGENIO (Editors), *Guida alla Geologia della Sicilia Occidentale.* Soc. Geol. Ital., Guide Geol. Region., Palermo, pp. 85–92.

BONI, M., 1974. Le argille rosse continentali del passaggio Paleocene-Miocene nella piattaforma carbonatica campano-lucan. *Boll. Soc. Geol. Ital.,* 93: 1059–1094.

BONI, M., 1979. *Zur Palaeogeographie, Mineralogie und Lagerstättenkunde der Paläokarsterze in Süd West Sardinien (Inglesiente-Sulcis).* Diss. Univ. Heidelberg, 260 pp.

BONI, M., 1982. Paleokarstic ores in SW Sardinia; some conclusive remarks. *Resoc. Assoz. Mineral. Sarda,* 87(1): 91–115.

BONI, M., 1986. The Permo-Triassic vein- and paleokarst ores in South-West Sardinia: contribution of fluid inclusions studies to their paleoenvironmental and genetic problems. *Mineralium Depos. (Berlin),* 21: 53–62.

BONI, M. and AMSTUTZ, G. C., 1982. The Permo-Triassic Paleokarst Ores of Southwestern Sardinia (Iglesiente-Sulcis). An Attempt at a Reconstruction of Paleokarst Conditions. In G. C. AMSTUTZ et al. (Editors), *Ore Genesis. The State of the Art,* Springer, Berlin, pp. 73–82.

BONI, M., STANZIONE, D. and ZENONE, F., 1979. The Senonian flint-clays of the internal carbonate platform (Southern Apennines). *Sedimentary Geol.,* 24: 17–24.

BONTE, A., 1970. Mise en place et evolution des bauxites sur mur calcaire. *Magy. Áll. Földt. Intéz. Évk,* 54: 29–51.

BORODAEVSKAYA, M. B. and ROZHKOV, J. S., 1978. Gold deposits. In: *Rudnye mestorozhdenia SSSR* (Ore deposits of the USSR) (in Russian), 3. Nedra, Moskva, pp. 5–76.

BORZA, K., 1977. Cyclic deposition of Dachstein Limestones of the Muráň Plateau (in Slovakian). *Geol. práce, Spr.,* 67: 23–52.

BORZA, K., 1980. Lirhologic-microfacies characteristics of Upper Jurassic and Lower Cretaceous sediments of the Belianská Group (Strážov Upland) (in Slovakian). *Geol. práce, Spr.,* 74: 33–56.

BORZA, K., KÖHLER, E., BEGAN, A. and SAMUEL, O., 1980. The occurrence of the Belianská Group west of Bošáca (in Slovakian). *Geol. práce, Spr.,* 74: 57–63.

BOSÁK, P., 1978a. The genesis and age of the Rudice Formation in the Moravian Karst, Czechoslovakia (in Polish). *Kras i speleol.,* 2(11): 27–37.

BOSÁK, P., 1978b. The Rudice Plateau in the Moravian Karst-part III. Petrography and diagenesis of carbonates and silicites of the Jurassic relict near Olomučany (in Czech). *Čas. Morav. muz., vědy přír.,* 58: 7–28.

BOSÁK, P., 1980a. Karst phenomena in the vicinity of Čebín in the Tišnov area and their age (in Czech). *Čs. kras, 1978,* 30: 125–128.

BOSÁK, P., 1980b. Lower Cretaceous fossil karst of the Rudice Plateau in the Moravian Karst (in Czech). *Čs. kras, 1979,* 31: 57–67.

BOSÁK, P., 1980c. Mineral deposits in carbonate areas related to karst phenomena (in Polish). *Kras i speleol.*, 3(12): 50–56.

BOSÁK, P., 1981a. The Lower Cretaceous Paleokarst in the Moravian Karst (Czechoslovakia). *Proc. 8th. Int. Congr. Speleol., Part I, Huntsville, Alabama,* p. 164–166.

BOSÁK, P., 1981b. The terminology and the investigation of old karst phenomena (in Czech). *Čes. kras (Beroun)*, 6: 58–66.

BOSÁK, P. 1981c. The development of the Lower Cretaceous karst. A comparision with the plate tectonics. *Proc. 8th Int. Speleol. Congr., Part I, Huntsville, Alabama,* pp. 170–173.

BOSÁK, P., 1983. Polycyclic development of karst phenomena in the Tišnov area (Western Moravia, Czechoslovakia). *Proc. Europ. Conf. Speleol., 2, Sofia,* pp. 253–256.

BOSÁK, P. 1986. Hydrothermal and superficial paleokarst of the Oshskie gorki (Kirghizia, USSR). *Proc. 9th. Int. Congr. Speleol.,* 1, *Barcelona,* pp. 198.

BOSÁK, P. in press. Economic Mineral Deposits Associated with Karst: A Review of the Problem. *Proc. Symp. New Trends in Speleol., II.* Czech Speleol. Soc., Praha.

BOSÁK, P., ČADEK, J., HORÁČEK, I., LOŽEK, V., TŮMA, S. and ULRYCH, J., 1984. Die Karsterscheinungen des Turold-Berges bei Mikulov (in Czech). *Studie ČSAV,* 5: 108 pp.

BOSÁK, P., ČADEK, J., HORÁČEK, I., TŮMA, S. and ULRYCH, J., 1983. The development of the Na Turoldu Cave (Southern Moravia, Czechoslovakia). *Proc. Region. Europ. Conf. Speleol., 2, Sofia,* pp. 247–252.

BOSÁK, P., GŁAZEK, J., GRADZIŃSKI, R. and WÓJCIK, Z., 1979. Genesis and age of sediments of the Rudice type in fossil-karst depressions. *Čas. mineral. geol.,* 24(2): 147–154.

BOSÁK, P., GŁAZEK, J., HORÁČEK, I. and SZYNKIEWICZ, A., 1982. New locality of Early Pleistocene vertebrates — Żabia Cave at Podlesice, Central Poland. *Acta Geol. Pol.,* 32 (3–4): 217–226.

BOSÁK, P. and HORÁČEK, I., 1981. Investigation of Old Karst Phenomena of the Bohemian Massif in Czechoslovakia: A Preliminary Regional Evaluation. *Proc. 8th. Int. Congr. Speleol., Part I, Huntsville, Alabama,* pp. 167–169.

BOSÁK, P., HORÁČEK, I., SÝKORA, J. and TŮMA, S., 1985. Karst phenomena of the Karzhantau Ridge, Western Tchien-Shan, Southern Kazakhstan (in Czech). *Čs. kras,* 36: 85–96.

BOULTER, M. C., 1971. A palynological study of two of the Neogene plant beds in Derbyshire. *Bull. Brit. Mus. Natur. Hist., Geol.,* 19(7): 361–400.

BOULTER, M. C., FORD, T. D., IJTABA, M. and WALSH, P. T., 1971. The Brassington Formation — a newly recognized Tertiary Formation in the Southern Pennines. *Nature,* 231: 134–136.

BOURQUE, M. S., 1982. Stratigraphy and Sedimentology of Carbonate Metasediments within the Grenville Soupergroup in the Havelock — Madoc — Bancroft area. *Ontario Geol. Surv., Misc. Pap.,* 106: 89–91.

BOURROUILH-LE JAN F. G., 1977. Géomorphologie de quelques atolls dits „soulevés“ du Pacifique W et SW, origine et évolution des formes récifales actuelles. *Mém. Bur. Rech. géol. Min.,* 89: 419–439.

BOURROUILH-LE JAN, F. G, 1980. Phosphates, sols bauxitique et karsts dolomitiques di Centre et SW Pacifique. Comparaisons sédimentologiques et géochimiques. *Colloq. Int., Doc. Bur. Rech. géol. Min.,* 24: 113–128.

BOURROUILH-LE JAN, F. G., 1982. Les étapes de la dolomitisation. Géochimie et sédimentologie d'un atoll soulévé. Livre Jubilaire G. Lucas, *Mém. geol. Univ. Dijon,* pp. 3–20.

BOURROUILH-LE JAN, F. G., 1983. Modern phosphate sedimentation in the enclosed lagoon of Clipperton Island, ENE Pacific. *P. I. C. G. Project 156-Phosphorites, Sixième Réun. Int.*

BOURROUILH-LE JAN F. G., 1985. An attempt of datation by isotopic stratigraphy of Holocene and Pleistocene notches and deposits of some so-called uplifted atolls (High Carbonate Islands). *Proc. 5th Int. Coral Reef Congr., Vol. 3, Tahiti:* 125–130.

BOURROUILH-LE JAN, F. G., CARSIN, J. M., NIAUSSAT, P. M. and THOMMERET, Y., 1985. Sedimentation phosphatée actuelle dans le lagon confiné de l'ile de Clipperton, Océan Pacifique, sédimentologie et géochimie. *Sci. Géol. Mém.*

BOURROUILH-LE JAN, F. G. and HOTTINGER, L., 1988. Occurrence of rhodolites in the tropical Pacific – a consequence of Mid-Miocene paleo-oceanographic change. *Sedimentary Geol.*, 60: 355–367.

BRAITHWAITE, G. J. R., 1968. Diagenesis of phosphatic carbonate rocks on Remire, Amirantes, Indian Ocean. *J. sed. Petrology*, 38(4): 1194–1212.

BRETZ, J. H., 1940. Solution cavities in the Joliet Limestone of Northeastern Illinois. *J. Geol.*, 48(4): 337–384.

BRETZ, J. H., 1950. Origin of filled sink-structures and circle deposits of Missouri. *Geol. Soc. Amer. Bull.*, 61: 789–834.

BRIDGES, P. H., 1982. The origin of cyclothems in the late Dinantian platform carbonates at Crich, Derbyshire. *Proceedings (Yorkshire Geol. Soc.)*, 44: 159–180.

BRIGO, L. and DI COLBERTALDO, D., 1972. Un nuovo orizzonte metallifero nel Paleozoico delle Alpi orientali. *Proc. 2th. Int. Symp. Mineral. Depos. Alps, Ljubljana*, pp. 109–124.

BRIGO, L. and OMENETTO, P., 1978. The lead and zinc ores of the Raibl (Cave del Predil, Northern Italy) zone: new metallogenic data. *Verh. Geol. Bundesanst, Wien*, 3: 241–247.

BRODTKORB, M. K., 1982. Vanadium Oxides in the Urcal Deposit, Argentina. — In: G. G. AMSTUTZ et al. (Editors), *Ore Genesis. The State of the Art.* Springer, Berlin, pp. 221–229.

BROOK, G. A., 1976. *Geomorphology of the North Karst, South Nahanni River region, N. W. T., Canada.* PhD. Diss., McMaster Univ.

BROOK, G. A. and FORD, D. C., 1976. The Nahanni North Karst: A question mark on the validity of the morphoclimatic concept of karst development. *Proc. 6th. In. Congr. Speleol., 2, Olomouc*, pp. 43–57.

BROOK, G. A. and FORD, D. C., 1978a. The nature of labyrinth karst and its implication for climaspecific models of tower karst. *Nature*, 280: 383–385.

BROOK, G. A. and FORD, D. C., 1978b. The origin of labyrinth and tower karst and the climatic conditions necessary for their development. *Nature*, 275(5680): 493–496.

BRUSCA, C., GAETANI, M., JADOUL, F. and VIEL, G., 1981. Paleogeografia Ladino-Carnica e metalligenesi del Sudalpino. *Mem. Soc. Geol. Ital.*, 22: 65–82.

BRYNER, L., 1961. Breccia and pebble columns associated with epigenetic ore deposits. *Econ. Geol.*, 56: 488–508.

BUCHART, B., 1978. Oxygen isotope paleotemperatures from the Tertiary period in the North Sea area. *Nature*, 257(5676): 121–123.

BÜDEL, J., 1977. *Klima Geomorphologie.* Borntraeger, Berlin, 304 pp.

BUDYKO, M. I., 1982: *The Earth's climate: past and future.* Academic Press, London, 307 pp.

BUŁA, D. and JURA, D., 1983. Lithostratigraphy of the Carpathian Foredeep sediments, region of Cieszyn, Silesia (in Polish). *Zesz. Nauk. Akad. Górniczo-Hutnicza, Geol.*, 9(1): 5–27.

BULLA, B., 1956. The evolutionary rhytms in highlands of Hungary in the light of new geomorphological ideas (in Hungarian). *MTA Társ. Tört. Tudom. Oszt. Közl*, 7.

BURCHART, J., 1972. Fission-track age determinations of accessory apatite from the Tatra Mts., Poland. *Earth. Planet. Sci. Lett.*, 15(4): 418–422.

BUREAU de Recherches Géologiques et Minières and Société des eaux de Marseille, 1974. *Recherche d'eau douce .. en mer.* Marseille, 4 pp.

BURKHARDT, R., 1962. Traces of fossil karstification near Kadov in Southwestern Moravia (in Czech). *Kras v Československu*, 1–2: 30.

BURKHARDT, R., 1974. The Rudice Plateau in the Moravian Karst-part I. The contribution to the theory of fossil karst and geological evolution (in Czech). *Čas. Morav. muz., vědy přír.*, 59: 37–58.

BUSHINSKY, G. I., 1975. *The geology of bauxites* (in Russian). Nedra, Moskva, 366 pp.

BUTCHER, N. J. D. and HEDGES, J. D., 1987. Exploration and extraction of structurally and lithostratigraphically controlled fluorite deposits in the Castleton-Bradwell area of the South Pennine orefield, England. *Trans. Inst. Min. Metallurg., Sect. B*, 96: 149–155.

BYSTRICKÝ, J., MAZÚR, E. and JAKÁL, J., 1972. Karst of Czechoslovakia. In: M. HERAK and V. T. STRINGFIELD (Editors), *Karst. Important Karst Regions of the Northern Hemisphere.* Elsevier, Amsterdam, pp. 297–325.

CALEMBERT, L., 1941. Un gisement de terres plastiques constitué par l'altération du Namurien, à Loyers. *Ann. Soc. Géol. Belg.,* 65(2): B 76–84.

CALEMBERT, L., 1945. *Les gisements de terres plastiques et réfractaires d'Andenne et du Condroz.* H. Vaillant-Carmanne, Liège, 204 pp.

CALEMBERT, L., 1947. Les argiles plastiques et réfractaires de la Haute-Belgique. *Centenaire de l'A. I. Lg. Congr., Sect. Géol,* pp. 283–302.

CALEMBERT, L., 1948. Observations sur led depôts tertiaires lacustres de La Triche (Andenne) et des régions voisiens. *Ann. Soc. Géol. Belg.,* 72: B 115–152.

CALEMBERT, L., 1954. Les formations tertiaires de la Haute Belgique. In: *Prodôme d'une description géologique de la Belgique.* Liège, pp. 510–532.

CALEMBERT, L. and VAN LECKWIJCK, W., 1943. Etude d'une brêche viséenne spéciale découverte à Marche-en-pré (Province de Namur). Comparaison avec les autres brêches du calcaire carbonifère de Belgique. *Ann. Soc. Géol. Belg.,* 67: 445–470.

CALEMBERT, L. et al., 1975. Problémes de géologie de l'ingénieur en regions karstiques. *Bull. Int. Assoc. Engng. Geol.,* 12: 93–132.

CAMPBELL, N. P., 1977. Possible exhumed fossil caverns in the Madison Group (Mississippian) of the northern Rocky Mountains: a discussion. *Nat. Speleol. Soc. Bull.,* 39: 43–54.

CARANNANTE, G., COCOZZA, T., D'ARGENIO, B. and SALVADORI, I., 1974. Caratteri deposizionali e diagenetici della „dolomia rigata" del Cambrico inferiore della Sardegna. *Rc. Soc. Mineral. Ital.,* 30(2): 1159–1173.

CARANNANTE, G., D'ARGENIO, B., FERRERI, V., HUERTAS, F., LINARES, J., PALUMBO, P., PONGITORE, F., POZZUOLI, A., SIMONE, I. and STANZIONE, D., 1978. La cavità paleocarsiche cretaciche dell' Appennino campano: stratigrafia, mineralogia, geochimica. *Paleokarst and Neokarst in Italy, Napoli,* 11 pp.

CARANNANTE, G., FERRERI, V. and SIMONE, L., 1974. Le cavità paleocarsiche cretaciche di Dragoni (Campania). *Boll. Soc. Nat. Napoli,* 83: 151–161.

CASTANY, G., 1967. *Traité pratique des eaux souterraines.* 2ème Éd. Dunod, Paris, 657 pp.

CASTELLARIN, A., 1972. Evoluzione paleotectonica sinsedimentaria del limite tra „Piattaforma Veneta" e „Bacino Lombardo" a nord di Riva del Garda. *G. Geol.,* Ser. 2, 38: 11–212.

CASTELLARIN, A., 1982. Tettonica sinsedimentaria ed associazioni di facies carbonatiche (con principali riferimenti al Giurossico Sudalpino). In: A. CASTELLARIN and G. B. VAI (Editors), *Guida alla Geologia nel Sudalpino Centro-Orientale.* Soc. Geol. Ital., Guide Region., Bologna, pp. 23–55.

CASTLEDEN, R., 1982. Classic landforms of the Sussex Coast. *Geogr. Assoc. Landform Guide,* 2: 39 pp.

CATALANO, R. and D'ARGENIO, B., 1982. Schema Geologico della Sicilia. In: R. CATALANO and B. D'ARGENIO (Editors), *Guida alla Geologia della Sicilia Occidentale.* Soc. Geol. Ital., Guide Region., Palermo, pp. 9–41.

CATALANO, R., D'ARGENIO, B. and LO CICERO, G., 1974. I ciclotemi triassici di Capo Rama (Monti di Palermo). Studi sulle Piattaforme Carbonatiche siciliane. *Geol. Romana,* 13. 125–145.

CAVELIER, C., CHATEAUNEUF, O., POMEROL, C., RABUSSIER, D., RENARD, M. and VERGNAUD-GRAZZINI, C., 1981. The geological events at the Eocene/Oligocene boundary. *Palaeogeogr. Palaeoclimatol. Palaeoecol.,* 36(3–4): 223–248.

CECCARINI, C., OGGIANO, S. and SALVADORI, I., 1984. Nurra (Sardinia, Italy) bauxite deposit. *Proc. Bauxite Symp. Los Angeles,* pp. 525–538.

CHEKHA, V. P., 1982. Ancient karst of North Siberian bauxite province (in Russian). *Geomorfol,* 2: 48–55.

CHERCHI, A. and TREMOLIERES, P., 1984. Nouvelles données sur l'evolution structurale on Mesozoique et on Cenozoique de la Sardaigne et leurs implications géodynamiques dans le cadre méditerranéen. *C. R. Acad. Sci., Paris,* 298: 889–894.

CHERKASOV, G. N., 1978. The geology, composition and genesis of bauxites and gibbsite-allophane rocks of Western Yakutia and Pribaikal'e (in Russian). *Litol. Polezn. Iskop.,* 6: 120–127.

CHERNOV, N. I., 1973. Upper Jurassic paleokarst of the platform area of Western Uzbekistan (in russian). In: *Karstovye kollektory nefti i gaza.* Perm Univ. Press, Perm, pp. 147–148.

CHIKISHEV, A. G., 1964. The karst of Central Ural and its importance in the national economy (in Russian). In: *Karst i ego narodokhozyaistvennoe znachenie* (Karst and its national economic importance). Nauka, Moskva, pp. 5–33.

CHIKISHEV, A. G., 1978. *Karst of the Russian lowland* (in Russian). Nauka, Moskva, 191 pp.

CHILINGAR, G. V., BISSEL, H. J. and WOLF, K. H., 1979. Diagenesis of carbonate sediments and epigenesis (or catagenesis). In: G. LARSEN and G. V. CHILINGAR (Editors), *Diagenesis in Sediments and Sedimentary Rocks.* Developments in sedimentology, 25A. Elvier, Amsterdam, pp. 247–422.

CHIRVINSKY, P. N., 1925. Tyuyamunite from the Tyuya-Muyun radium mine in Fergana. *Mineral. mag.,* 20(108): 286–295.

CHLUPÁČ, I., 1967. Early Paleozoic of the Bohemian Massif. *Int. Geol. Congr., 23rd. Sess., Guide to Excursion 11 AC:* 43 pp.

CHLUPÁČ, I. and ZIKMUNDOVÁ, J., 1976. The Devonian and Lower Carboniferous in the Nepasice bore in East Bohemia. *Věst. Ústř. úst. geol.,* 51(5): 269–278.

CHRISTIANSEN, E. A., 1967. Collapse structures near Saskatoon, Saskatchewan, Canada. *Canad. J. Earth Sci.,* 4: 757–767.

CHRISTIANSEN, E. A., 1971. Geology of the Crater Lake collapse structure in southeastern Saskatchewan. *Canad. J. Earth Sci.,* 8(12): 1505–1513.

CHRISTIANSEN, S., 1964. On lake Tegano, Rennel Island, and some remarks on the problem of Rennell's origin. *Saertryk Geogr. Tidsskr.,* 63: 99–111.

CÍCHA, I., FAHLBUSCH, V. and FEJFAR, O., 1972. Die biostratigraphische Korrelation einiger jungtertiärer Wirbeltierfaunen Mitteleuropas. *N. Jb. Geol. Paläont. Abh.,* 140(2): 129–145.

CIGNA, A. (Editor), 1978. *Manuale di speleologia.* Longanesi, Milano, 567 pp.

CIRY, R., 1962. Le role du froid dans la spéléogenese. *Spelunca Mém.,* 2(4): 29–34.

CIUK, E., 1974. Lithostratigraphic schemes of the Paleogene in Poland, except for the Carpathians and the Carpathian Foredeep (in Polish). *Biul. Inst. Geol.,* 281: 7–48.

CIUK, E., 1975. A development of lithology and sedimentation of Tertiary sediments from the Tanowo vicinity, north-west of Szczecin (in Polish). *Biul. Inst. Geol.,* 284: 133–169.

CIUK, E. and PIWOCKI, M., 1980. Geology of Tertiary deposits in the Kleszców Graben and its vicinity (in Polish). *Przewodnik 52 Zjazdu Polskiego Towarzystwa Geologicznego, Bełchatów 1980,* Inst. Geol. Warszawa, pp. 56–70.

CLARK, L. D., 1965. Bauxite deposits of the Springvale District, Georgia. *US Geol. Surv. Bull.,* 1199–F: 24 pp.

CLARKE, O. M., 1984. Bauxite deposits of the United States. *Proc. Bauxite Symp. Los Angeles,* pp. 245–161.

CLOUD, P. E. Jr., 1966. Bauxite deposits of the Anniston, Fort Payne and Ashville Areas, Northeast Alabama. *US Geol. Surv. Bull.,* 1190–O: 35 pp.

CLOUD, P. E. Jr., 1967. Geology and bauxite deposits of the Rock Run and Goshen Valley areas, Northeast Alabama. *US Geol. Surv. Bull.,* 1199–N: 74 pp.

COCCO, G. and PECORINI, G., 1959. Osservazioni sulle bauxiti della Nurra (Sardegna Nord Occidentale). *Atti Accad. Naz. Lincei,* 5: 175–314

COCOZZA, T. and GANDIN, S., 1976. Età e significato ambientale delle facies detritico-carbonatiche dell'altopiano di Campumari (Sardegna sud-occidentale). *Boll. Soc. Geol. Ital.,* 95(6): 1521–1540.

COCOZZA, T. and JACOBACCI, A., 1975. Geological Outline of Sardinia. In: C. SQUIRREL (Editor), *Geology of Italy*. Earth Sci. Soc. Lybian Arab Republ., pp. 49–81.

COCOZZA, T., JACOBACCI, A., NARDI, R. and SALVADORI, I., 1974. Schema stratigrafico-strutturale del massiccio sardo-corso e minerogenesi della Sardegna. *Mem. Soc. Geol. Ital.*, 13: 85–186.

COLACICCHI, R., 1967. Geologia della Marsica Orientale. *Geol. Romana*, 6: 189–316.

COLACICCHI, R. and PIALLI, G., 1967. Dati a conferna di una lacuna dovuta ad emersione nel Giuresse del Monte Cucco (Appennino Umbro). *Boll. Soc. Geol. Ital.*, 86: 179–192.

COLLINS, J. A. and SMITH, L., 1975. Zinc deposits related to diagenesis and intrakarstic sedimentation in the Lower Ordovician St. George Formation, western Newfoundland. *Bull. Canad. Petrol. Geol.*, 23(3): 393–427.

COMES, P. J., 1969. Recherches sur la genèse des bauxites dans le Nord-Est de l'Espagne, le Languedoc et l'Ariège (France). *Mém. Cent. Ét. Rech. Geol. Hydrogeol., Montpellier*, 342 pp.

COMER, J. B., 1974. Genesis of Jamaican bauxite. *Econ. Geol.*, 69: 1251–1264.

CORRÀ, G., 1975. Osservazioni sui fenomeni paleocarsici terziari nei calcari eocenici delle coline di Verona. *Atti Tav. Rot. Int. Carsologia, S. Trent. Sci. Nat.*, 54.

COTECCHIA, V., 1963. Geohydrological aspects of the Cretaceous limestone aquifer in Apulia, and their bearing on the practical avoidance of sea water contamination in extraction from wells and springs. *Quad. Geofis. Appl., Milano*, 24: 1–24.

COTECCHIA, V. and DELL'ANNA, L., 1959. Contributo alla conoscenza delle bauxiti e terre rosse del Salento. *Mem. e Note (Ist. Geol. Appl., Napoli)*, 7: 1–20.

CRAIG, D. H., 1985. Paleokarst in the San Andres dolomite, Yates Field, West Texas. *Sòc. Econ. Paleontologists Mineralogists A. Midyear Meet., Abstr., Vol. II:* 21.

CROCHET, G.-Y. et al., 1981. Les nouvelles faunes de vertebres anteieure à la „Grande coupure" de couvertes dans les phosphorites du Quercy. *Bull. Mus. Nat. Hist. Natur, 4ème Sér., Paris*, 3(3): 245–266.

CROS, P. and LAGNY, Ph., 1969. Paléokarst dans le Trias moyen et supérieur des Dolomites et des Alpes Carniques occidentales. Importance stratigraphique et paléogéographique. *Sci. de la Terre*, 16(2): 139–195.

CSÁSZÁR, G., 1984. *Stratigraphy of the Transdanubian Central Range formations of Middle Cretaceous age and their connections with the bauxite development* (in Hungarian). PhD. Thesis, Budapest.

CVIJIĆ, J., 1918. Hydrographie souterraine et evolution morphologique du Karst. *Rec. Trav. Inst. Geogr. Alp.*, 6(4): 376–420.

CZARNIK, J., 1972. Paleogeography of the environs of Turek in the Upper Tertiary and Pleistocene (in Polish). *Stud. Geol. Pol.*, 40: 160 pp.

CZUDEK, T. and DEMEK, J., 1970. Pleistocene cryopedimentation in Czechoslovakia. *Acta Geogr. Łodż.*, 24: 101–108.

DALY, R. A., 1934. *The Changing World of the Ice Age*. Yale Univ. Press, New Haven, 271 pp.

DAMIEAN, G., 1956. La sédimentation depuis la transgressions dévonienne jusqu'au Viséen dans la région de Huccorgne (flanc nord du synclinal de Namur). *Ann. Soc. Géol. Belg.*, 79: B 365–382.

DANA, J. D., 1849. *Geology of the United States Exploring Expedition 1938–1842*. Philadelphia.

DARÁNYI, F., 1972. Karst of Hungary. In: M. HERAK and V. T. STRINGFIELD (Editors), *Karst. Important Karst Regions of the Northern Hemisphere*. Elsevier, Amsterdam, pp. 267–295.

D'ARGENIO, B., 1967. Geologia nel Taburno-Camposauro (Campania). *Atti Accad. Sci. Fis. Mat., Napoli*, 6: 35–220.

D'ARGENIO, B., 1969. Central and Southern Italy Cretaceous bauxites: stratigraphy and paleogeography. *Magy. Áll. Földt. Intéz. Évk.*, 54 (3): 221–233.

D'ARGENIO, B., CHANNELL, J. E. T. and HORVÁTH, F., 1980. Adria, the African Promontory in Mesozoic Mediterranean Paleogeography. *Earth Sci. Rev.*, 15: 313–292.

D'ARGENIO, B., PESCATORE, T. and SCANDONE, P., 1973. Schema geologico dell'Appennino Campano-Lucano. atti Conv.: *Moderne vedute sulla geologia dell'Appennino, quad. 183*, Acc. Naz. Lincei, pp. 49–72.

DARWIN, Ch., 1842. *The structure and distribution of coral reefs*. Smith, Elder et Co., London, 214 pp.

DAVIES, W. E., 1951. Mechanics of breakdown. *Nat. Speleol. Soc. Bull.*, 13: 36–42.

DAVIES, W. E. and LE GRAND H. E., 1972. Karst of the United States. In: M. HERAK and V. T. STRINGFIELD (Editors), *Karst. Important Karst Region of the Northern Hemisphere*. Elsevier, Amsterdam, pp. 473–481.

DAVIS, W. M., 1928. *The coral reef problem*. Spec. Publ. Amer. Geogr. Soc., 9: 596 pp.

DEDINSZKY, J., 1972. Nagylengyel–an interesting reservoir. *Proc. Int. Petrol. Conf. Zadar, 1972, March 9–12.*, pp. 314–321.

DEHM, R., 1961a. Spaltenfüllungen als Lagerstätten fossiler Landwirbeltiere. *Mitt. Bayer. Staatsslg. Pal. hist. Geol.*, 1: 57–72.

DEHM, R., 1961b. Über neue tertiäre Spaltenfüllungen des süddeutschen Jura und Muschelkalk-Gebietes. *Mitt. Bayer. Staatsslg. Pal. hist. Geol.*, 1: 27–56.

DEJONGHE, L., 1985. Belgium. In:. *Mineral Deposits of Europe*. Inst. Min. Metallurg. and Mineral Soc.

DEJONGHE, L. and JANS, D., 1983. Les gisements zincifères de l'Est de la Belgium. *Chron. Mines Rech. Min.*, 470: 3–24.

DE LAPPARENT, J., 1930. Les bauxites de la France méridionale. *Mém. Carte Géol. France, Paris*, 187 pp.

DELEON, G., GOJKOVIĆ, S. and VUKASOVIĆ, M., 1961. Measurement of the absolute geological age of known granite bodies of Yuagoslavia (in Slovenian). *III. Kongres Geologa Jugoslaviji. Titograd*, 1: 417–434.

DELEPINE, G., 1911. Recherches sur le calcaire carbonifère de la Belgium. *Mem. Trav. Fac. Cathol Lille, fasc.* 7: 400 pp.

DELMER, A., 1977. Le Bassin du Hainaut el le sondage de Saint-Ghislain. *Prof. Paper Serv. Géol. Belg.*, 6(143): 12 pp.

DELMER, A., LECLERQ, V., MARLIÈRE, R. and ROBASZYNSKI, F., 1982. Le géothermie du Hainaut et le sondage de Ghlin (Mons-Belgique). *Ann. Soc. Géol. Nord, TCI:* 189–206.

DELMER, A., and VAN WICHELEN, P., 1980. Répertoire des puits naturels connus en terrain houiller du Hainaut. *Prof. Paper, Serv. Géol. Belg.*, 172.

DE LUMLEY, H. and BARRAL, L. (Editors), 1976. Sites paléolithiqies de la région de Nice et Grottes de Grimaldi. *UISPP IXe Congr., Livret-Guide de l'Excursion B 1:* 157 pp.

DE MARTONNE, E., 1910. *Traité de Géographie Physique, Vol. II.* Paris, 670 pp.

DEMEK, J. (Editor), 1965. *The geomorphology of Czech countries* (in Czech). NČSAV, Praha, 336 pp.

DE MILLE, G., SHOULDICE, J. R. and NELSON, H. W., 1964. Collapse structures related to evaporites of the Prairie Formation, Saskatchewan. *Bull. Geol. Soc. Amer.*, 75: 307–316.

DEMINA, V. N. and KUZMINA, O. V., 1975. Bauxites of Central Timan, their mineralogy and some questions of genesis (in Russian). In: *Problemy genezisa boksitov* (Problems of bauxite genesis). Nauka, Moskva, pp. 126–135.

DENSON, N. M. and WAAGÉ, K. M., 1966. Some bauxite and clay deposits in Northeastern Alabama. *US Geol. Surv. Bull.*, 1199–P: 27 pp.

DESROCHERS, A. and JAMES, N. P., 1985. Early Paleozoic surface and subsurface paleokarst; Middle Ordovician carbonates, Mingan Island, Quebec. *Soc. Econ. Paleontologist Mineralogists, A. Midyear Meet., Abstr.*, vol. 2: 23.

DE VOTO, R. H. (Editor), 1985. Sedimentology, dolomitization, karstification, and mineralization of the Leadville Limestone (Mississippian), central Colorado. *Soc. Econ. Paleontologists Mineralogists, A. Midyear Meet., Guidebook for field trip 6:* 180 pp.

DE WEISSE, G., 1964. Bauxite lateritique et bauxite karstique. *Symp. ICSOBA Zagreb*, Vol. 1: 7–29.

DE WEISSE, G., 1970. Bauxite sur un atoll du Pacifique. *Mineralium Depos.*, 5: 181–183.

DILL, R. F., 1977. The Blue Holes — geologically significant submerged sink holes and caves off British Honduras and Andros, Bahama Islands. *Proc. 3rd Int. Coral Reef Symp. Miami, Florida, 2, Geology:* 238–242.

DINCER, T., PAYNE, B. R., YEN, C. K. and ZÖTL, J. G., 1972. Das Tote Gebirge als Entwässerungstypus der Karstmassive der nordöstlichen Kalkhochalpen (Ergebnisse von Isotopenmessungen). *Steier. Beitr. Hydrogeol.*, 24: 71–109.

DJURIČKOVIĆ, A., 1982. The metallogeny of Brasina-Zajača-Dobri potok ore field (in Slovenian). *Vesnik, Ser. A. (Zav. Geol. Geofiz. Istraž.)*, 40: 17–53.

DOLFUSS, G., 1904. Observation à la note de F. Laur. *Bull. Soc. Géol. France, 4e Sér.*, IV: 171.

DREW, D. P. and INGLE SMITH, D., 1972. An unconformity cave. *Proc. Univ. Bristol Speleol. Soc.*, 13(1): 89–103.

DUBYANSKY, A. A., 1937. Fossil karst of Upper Cretaceous deposits (in Russian). *Bjull. Mosk. Obshch. Ispyt. Prir., Otd. Geol.*, 15(4): 301–303.

DUDEK, A., 1980. The crystalline basement block of the Outer Carpathians in Moravia: Bruno-Vistulicum. *Rozpr. Čes. akad. věd., Ř. matem. přír. věd.*, 90(8): 85 pp.

DUNHAM, R. J., 1962. Classification of carbonate rocks according to depositional texture. In W. E. HAM (Editor), *Classification of carbonate rocks. A symposium.* Amer. Assoc. Petrol. Geol., Mem., 1: 108–121.

DUNLAP, J. C., BERGQUIST, G. R., CRAIG, L. C. and OVERSTREET, E. F., 1965. Bauxite deposits of Tennessee. *US Geol. Surv. Bull.*, 1199–L: 37 pp.

DUPUY, J. J., 1984. *Étude du gîte plombifère de Oued Mekta (Maroc Oriental); Essai d'explication de la répartition des métaux Pb – Zn (Cu, Ba) dans le district de Touissit Bou Beker.* Thèse 3e cycle, Orléans, 264 pp.

DUTHOU, J. L., EMBERGER, A. and LASSERRE, M., 1976. Résultats graphiques et interprétations des mesures isotopiques du plomb des galènes et des minéraux oxydés du Maroc. *C. R. Somm. Séanc. Soc. Géol. France,* 7: 221–226.

DVOŘÁK, J., 1953. To the genesis of Rudice layers (in Czech). *Čs. kras,* 6: 218–219.

DVOŘÁK, J., 1978. The geology of the Paleozoic underlying Carpathians to southeast of the Drahany Upland (in Czech). *Zemný Plyn Nafta,* 23(2): 185–203.

DVOŘÁK, J. and FRIÁKOVÁ, O., 1981. Paleogeography of Fammenian and Tournaisian in the northern part of the Moravian Karst, based on conodont faunas (in Czech). *Čas. mineral. geol.,* 26(3): 301–306.

DŻUŁYŃSKI, S., 1976. Hydrothermal karst and Zn – Pb sulfide ores. *Rocz. Pol. Tow. Geol.,* 46: 217–230.

DŻUŁYŃSKI, S., HENKIEL, A., KLIMEK, K. and POKORNY, J., 1966. The development of valleys in the southern part of the Cracow upland (in Polish). *Rocz. Pol. Tow. Geol.,* 36(4): 329–343.

DŻUŁYŃSKI, S., KRYSOWSKA-IWASZKIEWICZ, M., OSZAST, J. and STARKEL, L., 1968. On Early Quaternary gravels in the Sandomierz Depression (in Polish). *Stud. Geomorphol. Carpatho-Balcanica,* 2: 63–76.

DŻUŁYŃSKI, S. and KUBICZ, A., 1971. Recrystallized and disaggregated limestones in the Triassic of Silesia. *Rocz. Pol. Tow. Geol.,* 41: 521–530.

DŻUŁYŃSKI, S. and SASS-GUSTKIEWICZ, M., 1977. Comments on the genesis of the Eastern Alpine Zn – Pb deposits. *Mineralium Depos.,* 12: 219–233.

DŻUŁYŃSKI, S. and SASS-GUSTKIEWICZ, M., 1982. The role of hydrothermal karst processes in the emplacement of sulfide ores. *Kras i Speleol.,* 4(13): 21–32.

DŻUŁYŃSKI, S. and SASS-GUSTKIEWICZ, M., 1985. Hydrothermal karst phenomena as a factor in the formation of Mississippi Valley-type deposits. In: K. H. WOLF (Editor), *Handbook of Strata-Bound and Stratiform Ore Deposits, 13.* Elsevier, Amsterdam, pp. 391–439.

EK, C., 1976. Les phénomènes karstiques. *Lab. Géogr. Phys., Univ. Liège:* 138–156.

EK, C. and POTY, E., 1982. Esquisse d'une chronologie des phénomènes karstiques en Belgium. *Rev. Belg. Géogr.,* 106(1): 73–85.

ELF-AQUITAINE, REECKMANN, A. and FRIEDMAN, G. M., 1982. *Exploration for Carbonate Petroleum Reservoirs.* Willey-Intersci. Publ., New York, 220 pp.

ELIÁŠ, M., 1981. Facies and paleogeography of the Jurassic of the Bohemian Massif. *Sbor. geol. věd, ř. G.,* 35: 75–144.

ELIÁŠ, M., 1983. Notes to the origin of accumulation of the Štramberk Limestone (in Czech). *Věst. ústř. úst. geol.,* 58(4): 235–239.

ELIÁŠ, M. and ELIÁŠOVÁ, H., 1983. Facies and paleogeography of the Jurassic in the western part of the Outer Flysch Carpathians in Czechoslovakia. *Sbor. geol. věd., ř. G.,* 39: 105–170.

ELLIS, A. F., 1936. *Ocean Island and Nauru; their story.* Angus and Robertson, Sydney, 319 pp.

ELSCHNER, C., 1913. *Corrallogene Phosphat-Inseln Austral-Oceaniens und ihre Produkte.* Max Schmidt, Lübeck, 188 pp.

ENTSOV, G. I., 1982. "Lower Carboniferous" bauxite of the western slope of Central Ural (in Russian). *Geol. Rud. Mestorozhd.,* 26(6): 94–101.

ENTSOV, I. I. and SHAITOV, U. I., 1973. Oil- and gas-bearing carbonate collectors of Upper Devonian and Tournaisian of the Verkhnepechora region of the Timan-Pechora Province (in Russian). In: *Karstovye kollektory nefti i gaza* (Karst collectors of oil and gas). Perm Univ. Press., Perm, pp. 118–120.

ERASO, A., 1975. Le role des facteurs physico-chimique dans le processus de la karstification (Conférence inaugurale). *Ann. Spéléol.,* 30(4): 567–580.

ERASO, A., 1980a. Problemas que se presentan en la construcción de presas en regiones kársticas. Ideas para su resolución. *Rev. OBRAS,* 139: 42–49.

ERASO, A., 1980b. Problematique de la construction des barrages dans des pays karstiques. *Proc. Symp-Int. Utiliz. Aree Carsiche,* Trieste, pp. 113–123.

ERASO, A., 1980c. Problemes of dam building in karstic countries. First ideas for price estimation at a project stage. *Abstr. Europ. Region. Speleol. Conf., Sofia 1980.*

ERASO, A., 1981. New contribution to the problem of dam building in karstic regions. *Proc. 8th Int. Congr. Speleol,* 1: 348–350.

ERASO, A., NAVARRO, J. V., PARRA, F., LOPEZ ACEVEDO, V., JERCIER, P., VAL, J., ECHEGARAY, M., COCERO, A., DOMINGUEZ, S. and ENRILE, A., 1984. Study of the karst at the Alcorlo Dam (Spain). *Kras i Speleol.,* 5(14): 45–53.

ERASO, A., PARRA, F., SAINT-AUBIN, J., NAVARRO, J. V., CANO, P., VAL, J., BERMEJO, F., VALDES, C. and RIVELLES, J., 1984. Problemes poses par le karst dans le barrage de Tous (Espagne). *Kras i Speleol.,* 5(14): 53–62.

ERDOSH, G. and SANDVIK, P. O., 1976. The apatite deposit of the Cargill carbonatite complex, Canada. *25th Int. Geol. Congr., Sydney,* 1: 160–161.

ESTEBAN, M. and KLAPPA, C. I., 1983. Subaerial exposure environments. In: P. A. SCHOLLE, D. G. BEBOUT and C. H. MOORE (Editors), *Carbonate Depositional Environments.* Amer. Assoc. Petrol. Geologists, pp. 1–54.

FAHLBUSCH, V., 1976. Report on the International Symposium on mammalian stratigraphy of the European Tertiary. *Newsl. Stratigr.,* 5(2–3): 160–167.

FAHLBUSCH, V., 1981. Miozän und Pliozän was ist was? *Mitt. Bayer. Staatsslg. Paleont. hist. Geol.,* 21: 121–127.

FAZZUOLI, M. and SGUAZZONI, G., 1981. Presenza di facies di tipo „Rosso Ammonitico" e di forme paleocarsiche al tetto dei marmi in località Pianellacci (M. te Pisanino-Alpi Apuane). *Bol. Soc. Geol. Ital.,* 100: 555–566.

FEENSTRA, A., 1985. Mineralogy and geochemistry of the metamorphic bauxites from Naxos, Greece. *ISCOBA Int. Symp. on Bauxite Mining and Prospection, Tapolca, Hungary, Abstr.:* 22.

FEJFAR, O., 1974. Die Eomyiden und Cricetiden (Rodentia, Mammalia) des Miozäns der Tschechoslowakei. *Paleontographica, Abt. A,* 146: 180 pp.

FEJFAR, O., 1976. Plio-Pleistocene mammal sequences. *IGCP 73/1/24, Rep. 3:* 351–366.

FEJFAR, O. and HEINRICH, W. - D., 1983. Arvicoliden Sukzession und Biostratigraphie der Oberpliozäns und Quartärs in Europa. *Schr. - Reihe Geol. Wiss.*, 19/20: 61–109.

FEJFAR, O. and HORÁČEK, I., 1983. Zum Entwicklung der Kleinsäugerfaunen im Villányium and Altbiharium auf dem Gebiet der ČSSR. *Schr. – Reihe Geol. Wiss.*, 19/20: 111–208.

FELBER, H., HÖTZL, H., MOSER, H., RAUERT, W. and ZÖTL, J. G., 1978. Karstification and geomorphology of As Sulg Plateau. In: S. S. AL-SAYARI and J. G. ZÖTL (Editors), *Quaternary Period in Saudi Arabia.* Springer, Wien, pp. 166–172.

FELDBARG, N. E. and RYBAKOVA, N. O., 1974. New data on the time of formation of fossil placers of the Yenisei Range (in Russian). *Vest. Mosk. Univ., Geol.*, 6: 81–83.

FÉNELON, P., 1968. Vocabulaire français des phénomènes karstiques. *Mém. et Doc., Centre Rech. Docum. Cartogr. Géogr., N. S.* 4: 13–68.

FERSMAN, A. E., 1924. The Tyuya-Muyun radium mine (in Russian). *Priroda,* 1–6: 58–87.

FERSMAN, A. E., 1927. To the morphology and geochemistry of Tyuya-Muyun (in Russian). *Trudy Izuch. radia, Leningrad,* 3: 93 pp.

FERSMAN, A. E., 1930. Geochemische Migration der Elemente und deren wissenschaftliche und wirtschaftliche Bedeutung erläutert an vier Mineralvorkommen: Chibina-Tundren-Smaragdgruben, Uran-Grube Tuja-Mujun-Wüste Karakumy, Teil. II. *Abh. prakt. geol. Bergwirtsch.-Lehre,* 19: 1–52.

FILIPOVIĆ, B., 1975. Hydrogeological characteristics of "Kosturi" bauxite deposit (in Slovenian). *Zbor. Rad. Rud.-Geol. Fak. (Beograd),* 18: 139–174.

FINK, M. H. (Editor), 1973. *Mehrsprachiges Lexikon der Karst- und Höhlenkunde (Speläologie) (Entwurf).* UIS Kommission für Dokumentation, 45 pp.

FINK, M. H., 1976. Zum Stand der phänomenologischen und typologischen Karstforschung. *Mitt. Öster. Geogr. Ges.,* 118 (II): 211–236.

FLICOTEAUX, R., NAHON, D. and PAQUET, H. 1977. Genèse des phosphates à partir des sédiments argillo-phosphatés du Tertiaire de Lam-Lam (Sénégal). Suite minéralogique. Permanence et changements des structures. *Sci. Géol. Bull. (Strasbourg),* 30(3): 153–174.

FÖLDVÁRI, A., 1933. The pre-Eocene karst of the Transdanubian Mts. *Földt. Közl.,* 63: 49–56.

FOMINSKY, V. I., 1974. Phosphorite-bearing Vendian of the Mrassk anticlinorium, Altay-Sayan region (in Russian). In: *Genezis i zakonomernosti razmeshchenia fosforitov* (The genesis and regularities in phosphorite distribution). Nauka, Moskva, pp. 44–77.

FORD, D. C., 1971. Geologic structure and a new explanation of limestone cavern genesis. *Trans. Cave Res. Grp. Great Britain,* 13 (2): 81–94.

FORD, D. C., 1983a. Effects of glaciations upon karst aquifers in Canada. *J. Hydrol.,* 61: 149–158.

FORD, D. C., 1983b. Alpine karst systems at Crowsnest Pass, Alberta-British Columbia, Canada. *J. Hydrol.,* 61: 187–192.

FORD, D. C. (Editor), 1983c. Castleguard Cave and Karst, Columbia Icefields Area, Rocky Mountains of Canada: A Symposium. *Arctic Alpine Res.,* 15(4): 425–554.

FORD, D. C., 1983d. Karstic interpretation of the Winnipeg Aquifer. *J. Hydrol.,* 61: 177–180.

FORD, D. C., 1984. Karst groundwater activity and landform genesis in modern permafrost regions of Canada. In: R. G. LA FLEUR (Editor), *Groundwater as a Geomorphic Agent.* Allen and Unwin, London, pp. 340–350.

FORD, D. C., 1986. Comment upon "Genesis of Paleokarst and Strata-Bound Zinc-Lead Sulfide Deposits in a Proterozoic Dolostone, Northern Baffin Island, Canada" by R. A. Olson. *Econ. Geol.*

FORD, D. C., ANDREWS, J. C., DAY, T. E., HARRIS, S. A., MAC PHERSON, J. B., OCCHIETTI, S., RANNIE, W. F. and SLAYMAKER, H. O., 1984. Canada: How many glaciations? *Canad. Geographer,* 38(3): 205–225.

FORD, D. C. and EWERS, R. O., 1978. The development of cave systems in dimensions of length and depth. *Can. J. Earth Sci.,* 15(11): 1783–1798.

FORD, D. C., SCHWARCZ, H. P., DRAKE, J. J., GASCOYNE, M., HARMON, R. S. and LATHAM, A. G., 1981. Estimates of the age of the existing relief within the southern Rocky Mountains of Canada. *Arctic Alpine Res.*, 13: 1–10.

FORD, T. D., 1963. The dolomite tors of Derbyshire. *E.Midland Geographer,* 3: 148–153.

FORD, T. D., 1964. Fossil karst in Derbyshire. *Proc. Brit. Speleol.* Assoc., (2): 49–62.

FORD, T. D., 1967. A quartz-rock-filled sink-hole on the Carboniferous Limestone near Castleton, Derbyshire. *Mercian Geol.,* 2: 57–62.

FORD, T. D., 1969a. Dolomite tors and sand-filled sink-holes in the Carboniferous Limestone of Derbyshire, England. In: T. L. PÉWÉ (Editor), *The Periglacial Environment.* McGill Univ. Press, Montreal.

FORD, T. D., 1969b. The Blue John fluorspar deposits of Treak Cliff, Derbyshire, in relation to the boulder bed. *Proceedings (Yorkshire Geol. Soc.),* 37: 153–158.

FORD, T. D., 1972a. Evidence of early stages in the evolution of the Derbyshire karst. *Trans. Cave Res. Grp. Great Britain,* 14: 73–77.

FORD, T. D., 1972b. Field meeting in the Peak District. *Proc. Geol. Assoc. London,* 83: 231–236.

FORD, T. D., 1976. The Geology of Caves. In: T. D. FORD and C. H. D. CULLINGFORD (Editors), *The Science of Speleology.* Academic Press, London, pp. 11–60.

FORD, T. D., 1977. *Limestones and caves of the Peak.* Geo-Books, Norwich, 469 pp.

FORD, T. D., 1984. Paleokarsts in Britain. *Cave Sci. (Trans. Brit. Cave Res. Assoc.),* 11(4): 246–264.

FORD, T. D. and CULLINGFORD, C. H. D. (Editors), 1976. *The Science of Speleology.* Academic Press, London, 593 pp.

FORD, T. D. and KING, R. J., 1965. Epigenetic bedded galena-baryte deposits in the Golconda Mine, Brassington, Derbyshire. *Econ. Geol.,* 60: 1686–1702.

FORD, T. D. and KING, R. J., 1966. The Golconda Mine caverns, Brassington, Derbyshire. *Trans. Cave Res. Grp. Great Britain,* 7: 91–114.

FOURNEAU, R., 1972. La Basse-Marlagne calcaire. *Bull. Soc. Géogr. Liège,* (8): 23–67.

FRAKES, L. A., 1979. *Climates throughout geological time.* Elsevier, Amsterdam, 310 pp.

FRASER, J. K. and HENOCH, W. S., 1959. Notes on the glaciation of King William Island and Adelaide Peninsula, N. W. T. *Dep. Mines Tech. Surv., Geogr. Branch Pap.,* 22: 54 pp.

FRASER, N. C. and WALKDEN, G. M., 1983. The ecology of a late Triassic reptile assemblage from Gloucestershire, England. *Palaeogeogr., Palaeoclimatol., Palaeoecol.,* 42: 341–365.

FUSCHS, Y., 1973. Sur les relations entre émersion et concentration métallifère. Quelques exemples tunisiens. *Livre Jubilaire M. Solignac, Ann. Mines Géol., Tunis,* 26: 479–509.

FYFE, W. S., PRICE, N. J. and THOMPSON, A. B., 1978. *Fluids in the Earth's crust.* Developments in Geochemistry, 1, Elsevier, Amsterdam, 383 pp.

GALEWSKI, K. and GŁAZEK, J., 1973. An unusual occurrence of the *Dytiscidae (Coleoptera)* in the siliceous flowstone of the Upper Miocene cave at Przeworno, Lower Silesia, Poland. *Acta Geol. Pol,* 23(3): 445–461.

GALEWSKI, K. and GŁAZEK, J., 1978. Upper Miocene *Dytiscidae (Coleoptera)* from Przeworno (Lower Silesia) and the problem of Dytiscidae evolution. *Bull. Acad. Pol. Sci., Sér. Sci. Biol.,* 25(12): 781–789.

GAMS, I., 1973. *Slovenian karst terminology* (in Slovenian). Kraška terminologija Jugoslovanskih narodov, Knjiga 1, Lubljana, pp 1–35.

GARETSKY, R. G., 1972. *Tectonics of young platforms of Eurasia* (in Russian). Trudy Geol. Inst. Akad. Nauk SSSR, 226: 298 pp.

GARYAINOV, A. A., 1980. *Exogeneous structures and their prospection significance* (in Russian). Saratov Univ. Press, Saratov, 208 pp.

GARYAINOV, V. A., VASILYEVA, M. A. and AGANIN, U. D., 1978. Paleokarst of the Akkermanovsk iron deposit (in Russian). In: *Voprosy geologii Yuzhnogo Urala i Povolzhyu* (Question of geology of Southern Ural and Povolzh'e), 16. Saratov Univ. Press, Saratov, pp. 18–40.

GASCOYNE, M., BENJAMIN, G. J., SCHWARCZ, H. P. and FORD, D. C., 1979. Sea-level lowering during the Illinoian Glaciation: Evidence from a Bahama "Blue Hole". *Science,* 205: 806–808.

GASTUCHE, M. C., FRIPIAT, J. J. and SOKOLSKI, S., 1963. Fixation du phosphore par les hydroxydes de fer et d'aluminium amorphes et cristallisées. *Pédologie, Gand,* 13: 150–180.

GATTER, I., 1984. Fluid inclusion study of vein-filling and cave concretions (in Hungarian). *Karszt és Barlang,* I: 9–18.

GAVEN, C. and BOURROUILH-LE JAN, F. G., 1980. Géochronologie (230Th-234U-238U) et sédimentologie des faciès récifaux pléistocènes à Maré, Archipel des Loyauté, SW Pacifique. *26ème Congr. Géol. Int., Paris, Rés.* 3: 843.

GAVEN, C. and BORROUILH-LE JAN, F. G., 1981. Géochronologie, sédimentologie et néotectonique des faciès récifaux pléistocènes à Maré, Archipel des Loyauté. *Oceanis,* 7(5): 327–487.

GAVRILOVIĆ, D., 1969. Kegelkarst-elemente im Relief des Gebirges Beljanica (Jugoslawien). In: *Problems of karst denudation.* Stud. Geogr., Brno, 5: 159–166.

GAVRILOVIĆ, D., 1976. The karst of Serbia. *Mem. Serbian Geogr. Soc.,* 13: 3–28.

GERASIMOV, I. P. and SIDORENKO, A. V. (Editors), 1974. *Planation surfaces and weathering crusts in the USSR* (in Rusian). Nedra, Moskva, 443 pp.

GÈZE, B., 1938. Contribution à la connaissance des phosphorites du Quercy. *Bull. Soc. Géol. France,* 5, 8: 123–146.

GÈZE, B., 1973. Lexique des termes français de Spéléologie physique et de karstologie. *Ann. Spéléol,* 28(1): 1–20.

GÈZE, B., 1981. *Eleménts de rapport pour la France, suivant les directives de Sofia 1980.* Unpubl. Rep. Comm. Spéléochronol. Paleokarst UIS, 4 pp.

GIBOWICZ, S. J., GŁAZEK, J. and WYSOKIŃSKI, L., 1981. Seismic phenomena in the region of Bełchatów brown coal mine (in Polish). *Przegl. Geol.,* 29(5): 246–250.

GIERWIELANIEC, J. and SZTUK, T., 1977. Fossil karst of the neighbourhood of Ludowa near Duszniki Zdrój (Lower Silesia) (in Polish). *Prace Geol. Min., Wroclaw,* 6: 115–122.

GILEWSKA, S., 1964. Fossil karst in Poland. *Erdkunde,* 18(2): 124–135.

GILEWSKA, S., 1965. The evolution of karst phenomena in the Southern part of the Silesian Upland. In: O. ŠTELCL (Editor), *Problems of Speleological Research,* 1. Academia, Praha, pp. 37–45

GILEWSKA, S., 1971. The paleogeographic conditions of karst evolution in Poland (with Europe as a backround). *Stud. Geomorphol. Carpatho-balcanica,* 5: 5–24.

GILLULY, J., 1949. Distribution of mountain building in geological times. *Bull. Geol. Soc. Amer.,* 60(4): 561–590.

GINSBURG, I. I., 1966. Karst und Erzbildung. *S. Angew. Geol.,* 12(2): 67–71.

GIROLIMETTO, F., 1982. L'origine des dèpôts cénozoîques „Om" et „On" à l'Ouest de la Meuse de Dinant. *Bull. Soc. Géogr. Liège,* 18: 49–57.

GJEMS, O., 1963. Kaolin as a weathering product of Eocambrian sandstone (Sparagmite) in the Rondane Mountains, East Norway. *Nor. geol. Tidsskr.,* 43: 537–538.

GJESSING, J., 1967. Norway's Paleic Surface. *Nor. geogr. Tidsskr.,* 21: 69–132.

GŁAZEK, J., 1965. Recent oncolites in streams of North Vietnam and of the Polish Tatra Mts. *Rocz. Pol. Tow. Geol.,* 35(2): 221–242.

GŁAZEK, J., 1966. On the karst phenomena in North Vietnam. *Bull. Acad. Sci. Pol., Sér. Sci. Géol. Géogr.,* 14(1): 45–51.

GŁAZEK, J., 1973. The importance of karst phenomena in paleogeographic and paleotectonic reconstructions (in Polish). *Przegl. Geol.,* 21(10): 517–523.

GŁAZEK, J., 1975. Fossil karst in the paleogeographic development of Poland. *Proc. 6th Int. Congr. Speleol., Olomouc, 1973,* 1: 423–426.

GŁAZEK, J., 1984. First isotope datings of speleothems from Tatra caves and their bearing on Pleistocene stratigraphy of the Tatra Mts (in Polish). *Przegl. Geol.,* 32(1): 39–43.

Głazek, J., Bednarek, J., Szynkiewicz, A. and Wierzbowski, A., 1979. Origin of the Szachownica cave–the greatest cave system in the Cracow-Wieluń Upland (in Polish). *Kras i Speleol.*, 2(11): 38–50.

Głazek, J., Dąbrowski, T. and Gradziński, R., 1972. Karst of Poland. In: M. Herak and V. T. Stringfield (Editors), *Karst. Important Karst Regions of the Northern Hemisphere*. Elsevier, Amsterdam, pp. 327–340.

Głazek, J., Galewski, K., Oberc, J. and Sulimski, A., 1975. Miocene fossil karst at Przeworno with vertebrate and beetle faunas (Lower Silesia, Poland). *Proc. 6th Int. Congr. Speleol., Olomouc 1973*, 1: 427–434.

Głazek, J., Galewski, K. and Wysoczański-Minkowicz, T., 1977. New data on fossil karst at Przeworno (in Polish). *Kras i Speleol.*, 1(10): 81–88.

Głazek, J. and Harmon, R. S., 1981. Radiometric Dating of Polish Cave Speleothems: Current Results. *Proc. 8th Int. Congr. Speleol, Part 2, Huntsville, Alabama*: 424–427.

Głazek, J. and Harmon, R. S. in press., Stratigraphic and geomorphic implications of $^{230}Th/^{234}U$ dates on speleothems from Poland. *Acta Geol. Pol.*, 39.

Głazek, J. and Juskowiak, O., 1964. Sur la stratigraphie et la métamorphose des roches de la zone de Bao-ha — Yen-bay (Vietnam). *Bull. Acad. Pol. Sci., Sér. Sci. Géol. Géogr.*, 12(3): 195–204.

Głazek, J., Karwowski, Ł., Racki, G. and Wrzołek, T., 1982. The Early Devonian continental/marine succession at Chęciny in the Holy Cross Mts., and its paleogeographic and tectonic significance. *Acta Geol. Pol.*, 1981, 31(3–4): 233–250.

Głazek, J., Lindner, L., Kowalski, K., Młynarski, M., Stworzewicz, E., Tucholka, P. and Wysoczański-Minkowicz, T., 1977. Cave deposits at Kozi Grzbiet (Holy Cross Mts, Central Poland) with vertebrate and snail faunas of the Mindelian I/Mindelian II interglacial and their stratigraphic correlations. *Proc. 7th Int. Congr. Speleol., Sheffield:* 211–214.

Głazek, J., Lindner, L. and Wysoczański-Minkowicz, T., 1976. Interglacial Mindel I/Mindel II in fossil-bearing karst at Kozi Grzbiet in the Holy Cross Mts. *Acta Geol. Pol.*, 26(3): 377–393.

Głazek, J., Marcinowski, R. and Wierzbowski, A., 1971. Lower Cenomanian trace fossils and transgressive deposits in the Cracow Upland. *Acta Geol. Pol.*, 21(3): 433–448.

Głazek, J., Oberc, J. and Sulimski, A., 1971. Miocene vertebrate faunas from Przeworno (Lower Silesia) and their geological setting. *Acta Geol. Pol.*, 21(3): 473–516.

Głazek, J. and Radwański, A., 1970. The Tortonian cave near Chmielnik (in Polish). *Speleologia*, 5(1–2): 23–32.

Głazek, J., Rudnicki, J. and Szynkiewicz, A., 1977. Proglacial caves — a special genetic type of cave in glaciated areas. *Proc. 7th Int. Congr. Speleol., Sheffield*, pp. 215–217.

Głazek, J., Sulimski, A., Szynkiewicz, A. and Wyszoczański-Minkowicz, T., 1976. Middle Pleistocene karst deposits with *Ursus spelaeus* at Draby near Działoszyn, Central Poland. *Acta Geol. Pol.*, 26(3): 451–466.

Głazek, J., Sulimski, A. and Wyszoczański-Minkowicz, T., 1975. On the stratigraphic position of Węże 1 locality (Middle Poland). *Proc. 6th Int. Congr. Speleol., Olomouc 1973, 1:* 435–442.

Głazek, J. and Szynkiewicz, A., 1979. Fossil karst as a determining factor of environmental changes during engineering. In: *Symposium I. A. E. G. Changes of the geological environment under the influence of man's activity, Poland, Kraków*, I: 39–49.

Głazek, J. and Szynkiewicz, A., 1980. The karst between the Pilica and the Warta Rivers and its practical meaning (in Polish). *Przewodnik 52 Zjazdu Pol. Tow. Geol., Belchatów*. Inst. Geol., Warszawa, pp. 84–99.

Głazek, J. and Szynkiewicz, A., 1987. Stratigraphy of the Late Tertiary and Early Quaternary karst deposits in Poland and their paleogeographic implications (in Polish). *Proc. Conf. Problemy mlodszego neogenu i eoplejstocenu w Polsce, Wrocław 1985*. Ossolineum, Wrocław, pp. 113–130.

Głazek, J., Trammer, J. and Zawidzka, K., 1973. The Alpine microfacies with *Globospira densa* (Pantić) in the Muschelkalk of Poland and some related paleogeographical and geotectonic problems. *Acta Geol. Pol.*, 23(3): 463–482.

GLAZKOVSKY, A. A., GORBUNOV, G. I. and SYSOEV, F. A., 1978. Nickel deposits (in Russian). In: *Rudnye mestorozhdeniya v SSSR* (Ore deposits of the USSR), 2. Nedra, Moskva, pp. 5–76.

GODYŃ, J., KLUCZEWSKI, L. and NIEWDANA, J., 1976. The protection of mine in circumstances of great water inflow on exampíle of Pomorzany Mine (in Polish). *Wiad. Gór., Katowice,* 27(5): 134–138.

GOGINA, N. I. and LEONOV, B. I., 1984. Paleokarst in lower course of the Lena and its significance in prospection of raw materials (in Russian). *Geol. i Geofiz.,* 3: 41–47.

GONY, J. N., 1969. *Étude cristallochimique du phosphate uranifère de Bakouma.* Thèse Dr. Ing. I. N. P. L., Nancy.

GÓRECKA, E., 1972. Ore mineralization in Palaeozoic rocks of the Silesian-Cracow area (in Polish). *Acta Geol. Pol.,* 22(2): 275–339.

GORNITZ, V. and KERR, P. F., 1970. Uranium mineralization and alteration, Orphan Mine, Grand Canyon, Arizona. *Econ. Geol.,* 65: 751–768.

GORZYŃSKI, Z., 1963. Methods and preliminary results of search works for clayey raw materials in the Upper Silesia area (in Polish). *Przegl. Geol.,* 11(11): 483–486.

GOSPODARIĆ, R., 1981. *Report of the Socialist Republic of Slovenia.* Unpubl. Rep. Comm. Spéléochronol. Paleokarst UIS, 8 pp.

GOSPODARIĆ, R. and ZÖTL, J. G. (Editors), 1976. Markierung unterirdischer Wässer (Untersuchungen in Slowenien 1972–1975). *Steier. Beitr. Hydrogeol., neue F.,* 28: 7–258.

GRADZIŃSKI, R., 1962. Origin and development of subterranean karst in the southern part of the Cracow Upland (in Polish). *Rocz. Pol. Tow. Geol.,* 32(4): 429–492.

GRADZIŃSKI, R., 1977. Sedimentation of "moulding sands" on karstified limestone surface in the middle part of Kraków-Wieluń Upland (in Polish). *Kras i Speleol.,* 1(10): 59–70.

GRADZIŃSKI, R. and WÓJCIK, Z., 1966. Fossil karst in Poland (in Polish). *Prace Muz. Ziemi, Prace Geol.,* 9: 151–222.

GRAULICH, J. M., 1962. La phase sudète de l'Orogène varisque dans le synclinorium de Namur à l'Est du Samson. *Bull. Soc. Géol. Paleont. Hydrol.,* 71(2): 181–199.

GRECULA, P. and ROTH, Z., 1978. The kinematic model of West Carpathians in complex section (in Czech). *Sbor. geol. věd, Geol.,* 32: 49–73.

GREENLY, E., 1901. Sandstone pipes in the Carboniferous Limestone at Dwlban Point. *Geol. Mag., Decade V, 7*: 20–24.

GROESSENS, E., CONIL, R. and HENNEBERT, M., 1979. Le Dinantien du sondage Saint-Ghislain. *Mém. Explan. Cartes Géol. Min. Belg.,* 22: 137 pp.

GROSS, P. and KÖHLER, E. (Editors), 1980. *The geology of the Liptov Basin* (in Slovakian). Geol. úst. Dionýza Štúra, Bratislava, 242 pp.

GRUBIĆ, A., 1957. One interesting example of fossil relief (in Slovenian). *Vesnik (Zav. Geol. Geofiz. Istraž.), Beograd,* 13: 111–114.

GRUBIĆ, A., 1975. *Geology of Yugoslav bauxites.* Serbian Acad. Sci. Arts Monogr., 44: 181 pp.

GRUND, A., 1903. Die Karsthydrographie. *Geogr. Abh. hsg. A. Penck,* 7(3): 103–300.

GRUND, A., 1910. Beiträge zur Morphologie des Dinarischen Gebirges. *Geogr. Abh. hsg. A. Penck,* 9(3): 175 pp.

GUDDEN, H., 1984. Zur Entstehung der nordbayerischen Kreide Eisenerzlagerstätten. *Geol. Jb., D,* 66: 3–49.

GUIDA, 1978. *Guida alle escursioni 1–3. aprile 1978. Processi paleokarsici e neokarsici e loro importanza economica nell'Italia meridionale.* Ist. Geol. Geofic. Univ. di Napoli, Gruppo Speleol. C. A. I. di Napoli, Napoli, 210 pp.

GUIDE, 1978. *Guide of Excursions, Symposium on Central European Permian, 2, Holy Cross Mts.* Geol. Inst., Warszawa, 113 pp.

GUTERCH, A., 1977. Structure and physical properties of the Earth's crust in Poland in the light of new data of DSS. *Publ. Inst. Geophys. Pol. Acad. Sci.,* A-4(115): 347–357.

GVOZDETSKY, N. A., 1972. *Problems of karst investigation and practice* (in Russian). Mysl, Moskva, 392 pp.

GVOZDETSKY, N. A., 1981. *Karst* (in Russian). Mysl, Moskva, 214 pp.

GYÖRGY, A., 1923. Halimba bauxite deposit near Veszprém (in Hungarian). *Bányászati Kohászati Lap.*, 66: 57–63, 73–77.

HAAS, J., 1983. Senonian cycle in the Transdanubian Central Range. *Acta Geol. Hung.*, 26(1–2): 21–40.

HAAS, J., 1984. Paleogeographic and geochronologic circumstances of bauxite generation in Hungary. *Acta Geol. Hung.*, 27(1–2): 23–39.

HABIĆ, P., 1978. The distribution of karst soils in the Dinaric Karst (in Slovenian). *Geogr. Vest., Ljubljana*, 50: 17–31.

HAGNI, R. D., 1976. Tri-State ore deposits: the character of their host rocks and their genesis. In: K. H. WOLF (Editor), *Handbook of Strata-Bound and Stratiform Deposits, 6*. Elsevier, Amsterdam, pp. 457–494.

HAŁADUS, A., MOTYKA, J., SZCZEPAŃSKI, A. and WILK, Z., 1978. Forecasting of ground-water inflow into mines in fissure-karst horizons using electric analog simulation method. *Rocz. Pol. Tow. Geol.*, 48(3–4): 559–587.

HALBICHOVÁ, I. and JANČAŘÍK, A., 1982–83. Conseguenze del cambiamento della morfologia e del microclima in alcuni riempimenti minerari delle grotte di Koněprusy. *Not. Sezionale C. A. I., Sezione di Napoli, N. S.* 37(1): 51–55.

HALLET, B., 1976. Deposits formed by subglacial precipitation of $CaCO_3$. *Bull. Geol. Soc. Amer.*, 87: 1003–1015.

HALSTEAD, L. B. and NICOLL, P. G., 1971. Fossilized caves of Mendip. *Stud. Speleol.*, 2(3–4): 93–102.

HANKEN, N. M., 1974. *En undersokelse av 5b sedimentene på Ullerntangen, Ringerike. An investigation of the Upper Ashgill sediments at Ullerntangen, Ringerike*. PhD. Thesis, Univ. of Olso, 131 pp.

HANZLÍKOVÁ, E. and BOSÁK, P., 1978. Microfossils and microfacies of the Jurassic relict near Olomučany (Blansko district). *Věst. Ústř. úst. geol.*, 52: 73–79.

HARMON, R. S., 1980. Paleoclimatic information from isotopic studies of speleothems: A review. In: W. C. MAHANEY (Editor), *Quaternary Paleoclimate*. Geo-Books, Norwich, pp. 299–318.

HARMON, R. S., FORD, D. C. and SCHWARCŹ, H. P., 1977. Interglacial chronology of the Rocky and Mackenzie mountains based upon $^{224}Th/^{234}U$ dating of calcite speleothems. *Canad. J. Earth Sci.*, 14(11): 1730–1738.

HARMON, R. S., SCHWARCZ, H. P. and FORD, D. C., 1978. Late Pleistocene sea level history of Bermuda. *Quaternary Res.*, 9: 205–218.

HARRIS, L. D., 1971. A lower Paleozoic paleoaquifer–the Kingsport Formation and Mascot Dolomite of Tennessee and southwest Virginia. *Econ. Geol.*, 66: 735–743.

HARRIS, T. M., 1957. A Liasso-Rhaetic flora in South Wales. *Proc. Roy. Soc., Ser. B.*, 147: 289–308.

HART, C. W., MANNING, R. B. and ILIFFE, T. M., 1985. The fauna of Atlantic marine caves: Evidence of dispersal by sea floor spreading while maintaining ties to deep waters. *Proc. Biol. Soc. Washington*, 98(1): 288–292.

HEAP, D., 1970. *Report of the Speleological Expedition to Arctic Norway, 1970*. William Hulme's Grammar School Rep., Manchester, 21 pp.

HENNINGSMOEN, G., 1960. Cambro-Silurian deposits of the Oslo region. *Norg. Geol. Unders.*, 208: 130–150.

HERAK, M., 1972. Karst of Yugoslavia. In: M. HERAK and V. T. STRINGFIELD (Editors), *Karst. Important Karst Region of the Northern Hemisphere*. Elsevier, Amsterdam, pp. 25–83.

HERAK, M. and STRINGFIELD, V. T. (Editors), 1972. *Karst. Important Karst Regions of the Northern Hemisphere*. Elsevier, Amsterdam, 551 pp.

HERBOSCH, A., DEWITTE, S. M. and PREAT, A., 1970. Recherches sur les indices de minéralisations uranifère dans la région de Visé. *Prof. Pap. Serv. Géol. Belg.*, 162 : 90 pp.

HERCMAN, H., PAZDUR, M. F. and WYSOCZAŃSKI-MINKOWICZ, T. 1987. Reconstruction of climatic changes in the Tatra Mts, based on dating of deposits from selected caves. *Stud. Geomorphol. Carpaho-Balcanica,* 21: 59–75.

HERMS, U., 1982. *Untersuchungen zur Schwermetalllösligkeit in kontaminierten Böden und kompostierten Siedlungsabfällen in Abhängigkeit von Bodenreaktion, Redoxbedingungen und Stoffbestand.* Diss. Univ. Kiel, 269 pp.

HERRERO, E., FERNANDEZ, F. and PETKOVIC, T., 1976. Impermeabilización del macizo kárstico de CANELLES bajo una elevada presión de embalse. *3rd Int. Congr. Rock Mechan., Denver.*

HEVESI, A., 1978. The origin of Bükk and a sketch of the surface evolution (in Hungarian). *Földrajzi Ért.,* 27(2): 169–203.

HEVESI, A., 1980. Contributions à l'image paléogeographique quaternaire de la Montagne Bükk. *Földt. Közl.,* 110: 540–550.

HILL, C. A., 1976. *Cave Minerals.* Natl. Speleol. Soc., Huntsville, 137 pp.

HILL, V. G. and OSTOJIĆ, S., 1982. The bauxite deposits of Jamaica: A distinctive karstic type. *J. Geol. Soc. Jamaica, Spec. Issue, Proc. Bauxite Symp., V, Kingston,* pp: 9–18.

HILL, W. T., MORRIS, R. G. and HAGEGEORGE, C. G., 1971. Ore controls and related sedimentary features at the Flat Gap mine, Treadway, Tennessee. *Econ. Geol.,* 66: 748–756.

HITCHON, B., 1969. Fluid flow in the Western Canada Sedimentary Basin. *Wat. Resour. Res.,* 5(1): 186–195 and 5(2): 460–469.

HOFFET, J.-H., 1939. Sur la génese des gisements de fer de la rive droite du Fleuve Rouge (Tonkin). *CR Acad. Sci. Paris,* 208(2): 112–114.

HOMOLA, V., 1950. The rests of fossil tropical soils on limestones of the western part of the Barrandean and their geologic age and importance for investigation of karst processes (in Czech). *Čs. kras,* 3: 97–107.

HORÁČEK, I., 1979. Comments on the lithostratigraphic context of the Early Pleistocene mammal biozones of Central Europe. *IGCP 73/1/24, Rep. 6:* 99–118.

HORÁČEK, I., 1980a. *Finds of Young Cenozoic faunas in the Bohemian Karst and their importance for the knowledge of the evolution of this area* (in Czech). Unpubl. Rep., Inst. Geol. Geotechn. Czechoslovak Acad. Sci., Praha, 31 pp.

HORÁČEK, I., 1980b. The new paleontological locality on the Golden Horse Hill near Koněprusy (in Czech). *Čs. kras,* 1979, 31: 105–106.

HORÁČEK, I., 1980c. Včeláre 3 — new locality of Early Pleistocene fauna in the slovak Karst (in Czech). *Slov. kras,* 18: 183–192.

HORÁČEK, I., 1982. The investigation of fossil vertebrates in protected area of the Bohemian Karst (in Czech). *Památky Příroda,* 2: 106–111.

HORÁČEK, I., 1985. Survey of the fossil vertebrate localities Včeláre 1–7. *Čas. mineral. geol.,* 30(4): 353–366.

HORÁČEK, I. and LOŽEK, V., 1984. The find of Early Pleistocene fauna in karst cavities near Brekov (in Czech). *Čs. kras,* 35: 100–102.

HORVÁTH, F. and ROYDEN, L., 1981. Mechanism for the formation of the intra-Carpathian basins. *Earth Evolution Sci.,* 1(3–4): 307–316.

HORVÁTH, I. and ODÓR, L. 1984. Alkaline ultrabasic rocks and associated silicocarbonatites in the NE part of the Transdanubian Mts. (Hungary). *Mineralia Slov.,* 16(1): 115–119.

HOSE, H. R., 1963. Jamaica type bauxites developed on limestones. *Econ. Geol.,* 58: 62–69.

HOUŠA, V., 1976. The Lower Cretaceous formations associated with bodies of Tithonian limestones near Štramberk (in Czech). *Acta Mus. Silesiae, Ser. A.,* 25: 63–85, 119–131.

HOUŠA, V., 1978. Calpionels of the Štrambek Tithonian and Lower Cretaceous (in Czech). *Zemný Plyn Nafta,* 23: 525–531.

HSÜ, K. J., 1978. When the Black Sea was drained. *Sci. Amer.,* 238(5): 53–63.

HSÜ, K. J. and GIOVANOLI, F., 1979. Messinian event in the Black Sea. *Palaeogeogr. Palaeoclimatol. Palaeoecol.,* 29(1–2): 75–93.

HUANG WANBO, 1982. Geologic age of fossil fauna in some typical caves in Liuzhou District, Guangxi (in Chinese). In: *Selec. Pap. 2nd all-China Symp. on Karst Sponsored by the Geol. Soc. China.* Sci. Press, Beijing, pp. 322–327.

HUNT, T. S., 1855. Mineralogical notes. Dolomite of Matea. *Amer. J. Sci. Arts, 2nd Ser.,* 19: 428–429.

HUTCHINSON, G. E., 1950. The biogeochemistry of vertebrate excretion. *Bull. Amer. Mus. Natur. Hist.,* 96: 554 pp.

HYPR, D., 1982. Cave levels in the northern and central part of the Moravian Karst (in Czech). *Sbor. Okres. muz. v Blansku,* 1980, 12: 65–79.

IAEG Symposium, 1973. *Proceedings. Symposium on sink-holes and subsidence; engineering-geological problems related to soluble rocks, Hannover 1973.* Dtsch. Gesel. Erd- und Grundbau, Essen.

IANNONE, A. and PIERI, P., 1978. *Osservazioni sul carsismo fossile in Puglia.* Unpubl. Rep. Symp. Paleokarst and Neokarst in Italy, Napoli, 1978, 11 pp.

IKKONIKOV, A. B., 1984. Notes on Geology of bauxite deposits in China. *Proc. Bauxite Symp. Los Angeles,* pp. 539–555.

INESON, P. R. and FORD, T. D., 1982. The South Pennine Orefields: its genetic theories and eastward extension. *Mercian Geol.,* 8: 285–303.

INESON, P. R. and WALTERS, S. G., 1983. Dinantian extrusive activity in the South Pennines. *Mercian Geol.,* 9(2): 88–98.

IVANOV, R., 1981. The deep-seated general Rhodope nappe and interference tectonics of the Rhodope crystalline basement. *Geologica Balc., Sofia,* 11(3): 47–64.

IVANOV, R. and STANEV, I., 1982. Paleohydrogeologic relations between the genesis of polymetallic ore deposits and volcanism in the Rhodope Mountains (in Bulgarian). *Spis. Balg. Geol. Druzh., Sofia,* 43(1): 69–73.

JAKÁL, J., 1975. *Karst of the Silica Plateau* (in Slovakian). Osveta, Martin, 152 pp.

JAKUCS, L., 1977a. The history of evolution of karst types in Hungary (in Hungarian). *Karszt és Barlang,* 1–2: 1–16.

JAKUCS, L., 1977b. Genetic types of the Hungarian karst. *Karszt és Barlang, Spec. Issue:* 3–18.

JAKUCS, L., 1977c. *Morphogenetics of karst regions. Variants of karst evolution.* Akadémiai Kiadó, Budapest, 284 pp.

JÁMBOR, Á., 1957. Geology of the region southeast of Szilvásvárad. *Földt. Intéz. Évi Jel.,* 1957-röl: 89–93.

JAMES, N. P. and CHOQUETTE, P. W. (Editors), 1988. *Paleokarst.* Springer, New York, 420pp.

JÁNOSSY, D. and KORDOS, L., 1977. The faunistical and karst-morphological review of paleontological localities for vertebrate at Osztramos (Northern Hungary). *Fragm. Mineral. Palaeontol., Budapest,* 8: 39–67.

JARMOŁOWICZ SZULC, K., 1984. Geochronological study of a part of the northern cover of the Karkonosze granite by fission track methods (in Polish). *Arch. Mineral, Warszawa,* 39(2): 139–183.

JELETZKY, J. A., 1977. Causes of Cretaceous oscillations of sea level in Western and Arctic Canada and some general geotectonic Implications. In: *Mid-Cretaceous Events, Hokkaido Symp., 1976.* Palaeontol. Soc. Japan, Spec. Pap., 21: 233–246.

JENKINS, H. G., 1970. The Jurassic of Western Sicily. In: *Geology and History of Sicily.* Petrol. Explor. Soc. Libya, pp. 245–254.

JENKO, R., 1959. *The hydrogeology and water economy of karst.* Državna Založba Slovenije, Zagreb.

JENNINGS, J. N., 1967. The surface and underground geomorphology. In. J. R. DUNKLEY and T. M L. WIGLEY (Editors), *Caves of the Nullarbor.* Sydney, pp. 13–31.

JENNINGS, J. N., 1968. Syngenetic Karst in Australia. In: *Contributions to the Study of Karst.* Univ. Canberra Press, Canberra, pp. 41–110.

JENNINGS, J. N., 1971. *Karst.* The M. I. T. Press, Cambridge, 252 pp.

JENNINGS, J. N., 1985. *Karst Geomorphology.* Basil Blackwell, Oxford, 293 pp.

JURCSAK, T. and POPA, E., 1978. Dinosauriens Eocrétacés dans les bauxites de Cornet, Bihor, Roumania (in Romanian). *Nymphaea, Oradea,* 6: 61–64.

JUVE, G., 1967. Zinc and Lead deposits in the Haafjell syncline, Ofoten, Northern Norway. *Norg. Geol. Unders.,* 54: 199–202.

KARANOV, Ch., 1965. Die Bauxite aus dem Gebiet von Trăn (in Bulgarian). *Spis. Balg. Geol. Druzh.,* 26(1): 21–40.

KARASZEWSKI, W., 1966. Traces of earlier marine Tertiary in the Iłżanka River Basin (in Polish). *Kwart. Geol.,* 10(1): 153–156.

KARAVANOV, K. P., 1973. Karst and karst waters of the Priamur'e and Priokhot'e (in Russian). In: *Prirodnye vody Dalnego Vostoka* (Natural waters of the Far East). Vladivostotskoe knizhnoe izdatelstvo, Vladivostok, pp. 284–302.

KARNKOWSKI, P., 1985. Miocene delta in the Carpathian Foredeep. *Proc. 13 Congr. Carpatho-Balcan Geol. Assoc., Kraków,* 2: 310–313.

KAROLYI, M. S. and FORD, D. C., 1983. The Goose Arm Karst, Newfoundland, Canada. *J. Hydrol.,* 61: 181–185.

KÁRPÁT, J., 1982. The Well-cave at Cserszegtomaj. *Karszt és Barlang,* 1: 35–40.

KARST RESEARCH GROUP, 1979. *Research of China karst* (in Chinese). Sci. Press, Beijing, 336 pp.

KATZER, F., 1909. *Karst und Karsthydrographie.* Zur Kunde der Balkanhalbinsel, 8 : 94 pp.

KAZAKEVICH, U. P., 1972. *Conditions of the origin and preservation of complex buried gold placers* (in Russian). Nedra, Moskva, 216 pp.

KELLER, W. D., WESTCOTT, J. F. and BLEDSOE, A. O., 1954. The origin of Missouri fire clays. *Proc. 2nd Nat. Conf. Clays and Clay Minerals,* pp. 7–46.

KEREKES, J., 1948. Die Periglaziälen bildungen Ungarns. *Földt. Intéz. Évk.:* 57 pp.

KETTNER, R., 1960. The morphologic evolution of the Moravian Karst (in Czech). *Čs. kras, 1959,* 12: 47–84.

KIRKALDY, J. F., 1950. Solution of chalk in the Mimms Valley, Hertfordshire. *Proc. Geol. Assoc. London,* 61: 219–224.

KIRPAL, G. R., 1976. Bauxite deposits in Kazakhstan (in Russian). Nedra, Moskva, 205 pp.

KIRPAL, G. R. and TENIAKOV, V. A., 1978. Bauxite deposits. In: *Ore deposits of the USSR, I.* Nedra, Moscow, pp. 262–336.

KIŠPATIĆ, M., 1912. Bauxite des Kroatischen Karstes und ihre Entstehung. *N. J. Mineral., Bull.,* 34: 513–522.

KLAU, W. and MOSTLER, H., 1983. Alpine Middle and Upper Triassic Pb-Zn deposits. *Int. Conf. Mississippi Valley-type Pb-Zn deposits, Rolla,* pp. 113–128.

KLEMM, D. D. and SCHNEIDER, H.-J. (Editors), 1977. *Time- and Strata-Bound Ore Deposits.* Springer, Berlin, 444 pp.

KOBLUK, D. R., PEMBERTON, S. G., KAROLYI, M. S. and RISK, M. J., 1977. The Silurian-Devonian disconformity in southeastern Ontario. *Bull. Canad. Petrol. Geol.,* 25(6): 1157–1186.

KOCHNEV, E. A., TITOVA, A. P., GENTSHKE, O. L. and CHASTNIKOVA, L. S., 1975. The genesis and evaluation of perspectives of bauxite-bearing Paleozoic of the Western Uzbekistan. In: *Problemy genezisa boksitov* (Problems of bauxite genesis). Nauka, Moskva, pp. 225–234.

KODYM, O., 1923. The southernmost relicts of Cretaceous near Prague (in Czech). *Rozpr. II. Čes. akad. věd, Umění, Tř. mat.-přírodověd.,* 32(6): 16 pp.

KODYM, O., 1960. To the genesis of the Hranice Karst (in Czech). *Čas. mineral. geol.,* 5(3): 262–264.

KOLOKOLTSEV, V. G., 1975. Petrographic characteristics of bauxite origin of Kedva-Vapovsk deposits in Timan (in Russian). In: *Problemy genezisa boksitov* (Problems of bauxite genesis). Nauka, Moskva, pp. 135–151.

KOMLÓSSY, Gy., 1976. Mineralogie, géochimie et genetique des bauxites du Vietnam du Nord. *Acta Geol. Acad. Sci. Hung.,* 20: 199–244.

KOMLÓSSY, G., 1985. Some remarks on the bauxite genesis and paleogeography. *ICSOBA Int. Symp. Bauxite Mining and Prospection, Tapolca, Hungary, Abstr.:* 37.

KOPECKÝ, L., 1978. Neoidic taphrogenic evolution and young alkaline volcanism of the Bohemian Massif. *Sbor. geol. věd, ř. Geol.*, 31: 91–107.

KORDOS, L., 1976. Theoretical and practical questions of speleochronology. *Karszt és Barlang*, 1–2: 15–20.

KORDOS, L., 1982. The Prehominid Locality of Rudabánya (NE Hungary) and its Neighbourhood: A Palaeogeographic Reconstruction. *Földt. Intéz. Évi Jel.*, 1980-tól: 395–406.

KORDOS, L., in press. Lower Turolian *Anomalospalax* n. g. from Hungary and its phylogenetic position. *Fragm. Mineral. Palaeontol., Budapest.*

KORMOS, T., 1911. Der pliozäne Knochenfunde bei Polgárdi. *Földt. Közl.*, 41: 48–64.

KORMOS T., 1943. Bauxitablagerung in Höhlen. *Földt. Közl.*, 73: 296–305.

KORPÁS, L., 1981. Oligocene-Lower Miocene formations of the Transdanubian Central Mountains in Hungary. *Földt. Intéz. Évk.*, 64: 140 pp.

KORZHUEV, S. S., 1972. Ancient karst and karstification cycles of the Siberian Platforms (in Russian). In: *Karst v karbonatnykh porodakh* (Karst in carbonate rocks). Moscow Univ. Press, Moskva, pp. 141–151.

KOVÁCS, J. and MÜLLER, P., 1984. Origin of the hydrothermal karstic phenomena in the buda-Hills (Hungary). *Kras i speleol.*, 5(14): 77–84.

KOWALSKI, K., 1962. Les Micro-Mammifères du Pliocène et du Pleistocène Inférieur de la Pologne. *Colloq. Int. CNRS, Paris*, 104: 409–416.

KOWALSKI, K., 1964. Paleoecology of mammals from the Pliocene and Early Pleistocene of Poland (in Polish). *Acta Theriologica, Białowieża*, 8(4): 73–88.

KOWALSKI, K., 1974. Remains of *Gerbillinae* (Rodentia, Mammalia) from the Pliocene of Poland. *Bull. Acad. Sci. Pol., Sér. Sci. Biol.*, 22(9): 591–595.

KRAJEWSKI, K., 1984. *Sedimentology of the Albian phosphate-bearing condensed sequence in the Polish Tatra Mountains, Western Carpathians.* Ph. D. Thesis, Inst. Geol. Sci., Pol. Acad. Sci, Warsaw, 80 pp.

KRASOŃ, J. and WÓJCIK, Z., 1965. The deep karst of the Bolesławiec syncline in the Sudetes Mountains (in Polish). *Acta Geol. Pol.*, 15(2): 179–212.

KRASTEV, T., 1974. Morphology of the buried karst in the Ludogorie (in Bulgarian). *Probl. Geogr., Sofia*, 1: 43–54.

KRASTEV, T., 1975. The role of tectonic movements in the evolution of buried karst of Ruse and Razgrad (in Bulgarian). *Probl. Geogr., Sofia*, 1(3): 29–34.

KRENNER, J., 1867. The origin of "terra rossa" (in Hungarian). *Természettudom. Közl.*, 8: 224–228.

KRETZOI, M., 1951. The *Hipparion* fauna from Csákvár. *Földt. Közl.*, 81: 402–417.

KRETZOI, M., 1954. Rapport final des fouilles paleontologiques dans la grotte de Csákvár. *Magy. Áll. Földt. Intéz. Évi Jelent.*, 1952: 37–68.

KRETZOI, M., 1956. Die altpleistozänen Wirbeltierfaunen des villanyer Gebirges. *Geologica Hung., Ser. Palaeontol.*, 27: 264 pp.

KRETZOI, M., 1969. Sketch of the Late Cenożoic (Pliocene and Quaternary) terrestrial stratigraphy of Hungary. *Külonlen Földt. Kozl.*, 3. 179–204

KRETZOI, M., 1971. In: Á. JÁMBOR, L. KORPÁS, M. KRETZOI and I. PÁLFALVY, Stratigraphische Probleme des Transdanubischen Oligozäns. *Magy. Áll. Földt. Intéz. Évi Jelent.*, 1969-röl: 141–154.

KRETZOI, M. and KROLOPP, E., 1972. Oberpliozäne und quartäre Stratigraphie des Alföld (Grosse Ungarische Tiefebene) auf Grund paläontologischer Angabe. *Földrajzi Ért.*, 21(2–3): 133–158.

KRETZOI, M. and PÉCSI, M., 1982. Pliocene and Quaternary chronostratigraphy and continental surface development of the Pannonian Basin. *Quaternary Studies in Hungary. INQUA Hung. Nat. Comm., Budapest*, pp. 11–42.

KRIVTSON, A. I., 1968–69. *Mesozoic and Cenozoic bauxites of the USSR, their genesis and industrial importance* (in Russian). Nedra, Moskva, 1: 367 pp. (1968) and 2: 327 pp. (1969).

Krolopp, E., Schweitzer, R., Scheuer, Gy., Dénes, Gy., Kordos, L., Skoflek, I. and Jánossy, D., 1976. Quaternary formations of the Castle Hill in Buda (in Hungarian). *Földt. Közl.*, 106(3): 193–228.

Kronsa, 1974. *Trabajos de impermeabilización en la ladera izquierda del salto de CANELLES.* Unpubl. Rep., Engher. Barcelona, Vols. I to XII.

Kropachev, A. M., 1972. To the genetic classification of deposits in karst cavities and depressions (in Russian). In: *Karst Srednei Azii i gornykh stran* (Karst of Central Asia and mountainous regions). Tashkent, pp. 171–173.

Kruber, A. A., 1915. *Karst region of Mountainous Crimea* (in Russian). Mosk. tovarishchestva I. N. Kushnerev i K° izd., Moskva, 103 pp.

Krulc, Z. and Vidović, N., 1960. Application of resistance method in prospection of buried bauxite deposits of Istria (in Slovenian). *Vesnik, Ser. C (Zav. Geol. Geofiz. Istraž.), Beograd,* 1: 81–97.

Krysowska-Iwaskiewicz, M., 1974. Mineralogic-petrographic investigation of Cenozoic terrestrial sediments in the Cracow Upland (in Polish). *Prace mineral., PAN Oddz. Kraków*, 35: 69 pp.

Krystek, I., 1966. Zum Entstehung und zum Alter der ruditzer Schichten. In: O. Štelcl (Editor), *Problems of the Speleological Research, 2.* Academia, Praha, pp. 79–82.

Krystek, I., 1972. The preliminary report on the find of Lower to Middle Cretaceous limestones near Kuřim (in Czech). *Geol. práce, Spr.*, 58: 251–256.

Książkiewicz, M., Oberc, J. and Pożaryski, W., 1978. *Geology of Poland, Vol. IV Tectonics.* Inst. Geol., Warszawa: 718 pp.

Kubiak, H., 1982. Die miozäne Wirbeltierfunde von Przeworno (Dolny Śląsk, VR Polen). *Z. Geol. Wiss.*, 10(7): 997–1007.

Kucheruk, A. D., 1976. *Karst of Povolzh'e* (in Russian). Naukova Dumka, Kiev, 197 pp.

Kunský, J., 1957. Thermomineral karst caves of Zbrashov, northern Moravia. *Sbor. Čs. zeměpis. spol.*, 62: 306–351.

Kurtén, B., 1968. *Pleistocene Mammals of Europe.* Aldine, Chicago, 317 pp.

Kuszneruk, J., 1984. The structure of Mesozoic bedrock and the problem of dewatering of the Bełchatów mine (in Polish). *Techn. Poszuk. Geol.*, 23(2): 32–34.

Kutek, J. and Głazek, J., 1972. The Holy Cross area, Central Poland, in the Alpine cycle. *Acta Geol. Pol.*, 22(4): 603–653.

Kutyrev, E. I. and Lyakhnitsky, U. S., 1982. Role of karst in the origin of lead, zinc, antimony, mercury and fluorite deposits (in Russian). *Litol. Polezn. Iskop.*, 2: 54–69.

Kyle, J. R., 1983. Economic aspect of subaerial carbonates. In: P. A. Scholle, D. G. Bebout and C. H. Moore (Editors), *Carbonate Depositional Environments.* Amer. Assoc. Petrol. Geologists, Tulsa, pp. 73–92.

Ladd, H. S. and Hoffmeister, J. E., 1945. Geology of Lau, Fiji. *Bull. Bernice P. Bishop Mus.*, 181: 399 pp.

Lagny, Ph., 1971. Les minéralisations plombo-zincifère de la region d'Auronzo (province de Belluno, Italie): remplissage paléokarstique d'age Anisien superieur. *C. R. Acad. Sci., sér. D,* 273: 1539–1542.

Lagny, Ph., 1974. Emersion successives, karstification et sédimentation continentale au Trias moyen dans le région de Sappada (province de Belluno, Italie). *Sci. de la Terre,* 19(3): 193–233.

Lagny, Ph., 1975. Le gisement plombo-zincifère de Salafossa (Alpes italiennes orientales): Remplissage d'un paléokarst triasique. *Mineralium depos.*, 10: 345–361.

Lagny, Ph. and Rouvier, H., 1976. Les gisements Pb – Zn en roches carbonatées sous inconformité: Gisements paléokarstiques ou gisements dans les paléokarst? *Mém. Soc. Géol. France,* 7: 57–69.

Lajoinie, J. P. and Laville, P., 1979. Les formations bauxitiques de la Provence et du Languedoc. Dimensions et distribution des gisements. *Mém. Bur. Rech. Géol. Min.*, 100: 146 pp.

Laurant, A., 1976. La formation du réseau hydrographique de la Belgique. In: *Géomorphologie de la Belgique.* Lab. Géogr. Phys., Univ. Liège, pp. 29–49.

LAURITZEN, S. E., 1983. *Arctic and Alpine Karst Symposium, August 1–15, 1983. Program and field guide.* Dept. Chemistry, Univ. Oslo, 89 pp.

LAURITZEN, S. E., 1984. Speleothem Dating in Norway: an interglacial chronology. *Nor. Geogr. Tidsskr.,* 38: 198 pp.

LAURITZEN, S. E. and GASCOYNE, M., 1980. The first radiometric dating of Norwegian stalagmites-evidence of pre-Weichselian karst caves. *Nor. Geogr. Tidsskr.,* 34: 77–82.

LAZNICKA, P., 1985. *Empirical Metallogeny. Depositional Environments, Lithologic Associations and Metallic Ores. Vol. 1: Phanerozoic Environments, Associations and Deposits.* Developments in Econ. Geol., 19 A, B, Elsevier, Amsterdam: 1758 pp.

LEBEDEV, L. M., 1974. Recent ore-forming hydrothermal systems (in Russian). Nedra, Moskva, 260 pp.

LEÉL-ÖSSY, S., 1953. Cserszegtomaj well-cave (in Hungarian). *Hidrol. Közl.,* 33: 309–313.

LEGGET, R. F., 1973. *Cities and Geology.* Mc Graw Hill, New York, 396 pp.

LEHMANN, O., 1932. *Die Hydrographie des Karstes.* Enzykl. Erdkde., Leipzig, 6: 212 pp.

LEIPTSIG, A. V. and LEVINA, A. P., 1975. Mesozoic and cenozoic rhytms and characteristics of bauxite deposition of Nizhnoe Priangar'e (in Russian). In: *Problemy genezisa boksitov* (Problems of bauxite genesis). Nauka, Moskva, pp. 186–192.

LEMMENS, M., LORRAIN, R. and HAREN, J., 1982. Isotopic composition of ice and subglacially precipitated calcite in an alpine area. *Z. Gletchekde. Glazialgeol.,* 18(2): 151–159.

LETOUZY, J., BIJU-DUVAL, B., DORKEL, A., GONNARD, R., KRISCHEV, K., MONTADERT, L. and SUNGURLU, O., 1977. The Black Sea: A marginal basin. Geophysical and geological data. *Int. Symp. Struct. Hist. Mediterranean Basin.* Ed. Technip, Paris, pp. 363–367.

LEWARNE, G. C. and PALLOT, J. M., 1957. Mesozoic plants from fissures in the Carboniferous Limestone of South Wales. *Ann. Mag. Nat. Hist., Ser. 12,* 10: 72–79.

LIEBE, P. and LORBERER, Á., 1978a. A study of the flow and temperature conditions in the karst reservoirs of thermal water (in Hungarian). *Int. Symp. Karst Hydrol., Budapest, Proc.,* 1: 79–110.

LIEBE, P. and LORBERER, Á., 1978b. The interrelation between flow and temperature conditions in karstified rocks conatining thermal water. *VITUKI Közlem., Beszámoló 1975,* 3: 162–175.

LINDGREN, W., 1918. Volume changes in metamorphism. *J. Geol.,* 26: 542–544.

LINDNER, L., 1984. An outline of Pleistocene chronostratigraphy in Poland. *Acta Geol. Pol.,* 34(1–2): 27–49.

LIPIARSKI, I., 1971. On fossil remains of Arthropods and plants, and organic matter from cavities in the triassic rocks of the Cracow area. *Bull. Acad. Sci. Pol., Sér. Sci. de la Terre,* 19(2): 79–83.

LIS, J. and WÓJCIK, Z., 1960. Triassic bone breccia and karst forms in Stare Gliny Quarry near Olkusz (Cracow region). *Kwart. Geol.,* 4(1): 55–74.

LLOPIS LLADO, N., 1953. Karst holofossile et merofossile. *Actes 1er Congr. Int. Spéléol., T. II, Paris,* pp. 44–50.

LOHMANN, K. C., 1985. The geochemical record of karst formations. *Soc. Econ. Paleontologists Mineralogists, A. Midyear Meet., Abstr.* Vol. II: 55.

LONGMAN, M. V., 1980. Carbonate Diagenetic Texture from Nearsurface Diagenetic Environments. *Amer. Assoc. Petrol. Geol. Bull.,* 64(4): 461–487.

LONGMAN, M. V. and BROWNLEE, D. N., 1980. Characteristics of karst topography, Palawan, Philippines. *Z. geomorphol., N. F.,* 24(3): 299–317.

LOŽEK, V., 1963. To the question of the origin and the age of vertical corrosion cavities in the Bohemian Karst (in Czech). *Čs. kras,* 15: 125–127.

LOŽEK, V., 1964. Quartärmollusken der Tschechoslowakei. *Rozpr. Ústř. úst. geol.,* 31: 374 pp.

LOŽEK, V. and ZÁRUBA, Q., 1965. Pleistocene slope breccias in karst region of Slovakia (in Czech). *Čs. kras,* 16 (1964): 67–76.

LUCIUS, M., 1952. Étude sur le gisement de fer du Grand Duché de Luxembourg. *19 Congr. Int. Geol., Alger, Symp. sur les gisements de fer du monde,* Vol. II: 349–387.

ŁUCZKOWSKA, E. and DYJOR, S., 1971. Tertiary microfauna of the Poznań Clays in Lower Silesia (in Polish). *Rocz. Pol. Tow. Geol.,* 41(2): 337–358.

LUKNIŠ, M., 1962. Review of geomorphology. In: O. FUSÁN et al., *Vysvětlivky k prehl'adnej geologickej mape ČSSR* (Explanatory booklet to the geological map of Czechoslovakia) 1 : 200 000, M-34-XXXIII Rimavská Sobota. Geofond in NSAV, Bratislava, pp. 81–88.

LUNSGERSGAUZEN, G. F., 1966. Influvium – a special genetic type of continental deposits (in Russian). *Dok. Akad. Nauk SSSR,* 171(3): 690–693.

LYEW-AYEE, P. A., 1984. Bauxites of the Caribbean. *Proc. Bauxite Symp. Los Angeles,* pp. 262–296.

LYSENKO, V., 1982. Polyphase origin of caves in the Bohemian Karst (in Czech). In: V. KRÁL (Editor), *Geomorfologická konference, Praha 1980.* Univ. Karlova, Praha, pp. 185–190.

Mac NEIL, F. S., 1954. The shape of atolls: an inheritance from subaerial erosion forms. *Amer. J. Sci.,* 252: 402–427.

MADEYSKA, T., 1981. Le milieu naturel de l'Homme du Paléolithique moyen et supérieur en Pologne à la lumière des recherches géologiques (in Polish). *Stud. Geol. Pol.,* 69: 125 pp.

MADEYSKA, T., 1982. Late Pleistocene cave deposits in Poland. *Kras i Speleol.,* 4(13): 43–66.

MADEYSKA-NIKLEWSKA, T., 1969. Upper Pleistocene deposits in the caves of the Cracow Upland (in Polish). *Acta Geol. Pol.,* 19(2): 341–392.

MAHJOUB, M. N., 1973. *Étude des minéralisations vanadifères dans les poches karstiques du Jebel Goraa (Jebba, Tunisie).* Thèse 3e cycle, Orsay, 77 pp.

MAKSIMOVIĆ, B., 1956. *Geologic and tectonic relations of coal-bearing Senjsko-resavski Mine and surrounding areas* (in Slovenian). Geol. Inst. „Jovan Žujović", Beograd, 104 pp.

MAKSIMOVICH, G. A., 1963. Principles of the study of karst (in Russian). Permskoe knizh. izd., Perm, 444 pp.

MAKSIMOVICH, G. A., 1969. Principles of the study of karst (in Russian). Geogr. Obshch. USSR, 2, 560 pp.

MAKSIMOVICH, G. A. and ARMISHEV, V. M., 1962. Paleokarst collectors of oil and gas (in Russian). *Gidrogeol. ikarstovedenie,* 1: 7–52.

MAKSIMOVICH, G. A. and BYKOV, V. N., 1978. *Karst of carbonate oil- and gas-bearing formations (in Russian).* Perm Univ. Press, Perm, 96 pp.

MALAKHOV, A. A., 1947. Karst of Chusovsk region (in Russian). *Tezisy dokl. karstovoy konf.* Perm Univ. Press, Perm, pp. 105–107.

MALICH, M. S., 1975. *Tectonic evolution of the mantle of the Siberian Platform* (in Russian). Nedra, Moskva, 215 pp.

MALKOVSKÝ, M., 1979. Tectogenesis of the platform cover of the Bohemian Massif (in Czech). *Knih. Ústř. úst. geol.,* 53: 176 pp.

MAMATKULOV, M. M., 1979. *Karst of the Southern and Western Tchien-Shan* (in Russian). Fan, Tashkent, 164 pp.

MANEV, D. and IGNATOVSKI, P., 1981. La structure du secteur d'Erma reka du champ métallifère de Madan (in Bulgarian). *Rudoobraz. Procesi Miner. Nachodišta, Sofia,* 14–15: 76–88.

MANGIN, A., 1976. Les systemes karstiques et leur methodologie d'investigation. *Ann. Sci. Univ. Besançon, Deuxième Coloq. Hydrol. en pays calcaire, Géol., Fasc. 25, 3ème Sér.:* 263–273.

MARCINOWSKI, R., 1970. The Cretaceous transgressive deposits east of Częstochowa (Polish Jura Chain). *Acta Geol. Pol.,* 20(3): 413–449.

MARCINOWSKI, R., 1974. The transgressive Cretaceous (Upper Albian through Turonian) deposits of the Polish Jura chain. *Acta Geol. Pol.,* 24(1): 117–217.

MARGRITA, R., EVIN, J., FLANDRIN, J. and PALOC, H., 1970. Contribution des mesures isotopiques à l'etude de la Fontaine de Vaucluse. *Isotope Hydrology 1970.* IAEA, Vienna: pp. 333–348.

MARLIÊRE, R., 1932. Une roche peu connue provenant d'un puits naturel de la région de Bernissart. *Ann. Soc. Géol. Belg.,* 55: B 184–189.

MARSHALL, J. E. A. and WHITESIDE, D. I., 1980. Marine influence in the Triassic uplands. *Nature,* 287(5783): 627–628.

MARTEL, E. A., 1921. *Nouveau Traité des Eaux Souterraines.* Gaston Doin, Paris, 824 pp.

MARTIN, J., 1965. Quelques types de dépressions karstiques du Moyen Atlas Central. *Rev. Géogr. Maroc.*, 7: 95–106.

MARTINI, J. E. J., 1979. Karst in Black Reef Quartzite near Kaapsehoog, Eastern Transvaal. *Ann. Geol. Opname, Pretoria,* 13: 115–128.

MARTINI, J. E. J., 1981. Early Proterozoic Paleokarst of the Transvaal, South Africa. *Proc. 8th Int. Congr. Speleol. Part 1, Huntsville, Alabama:* 4–5.

MARUASHVILI, L. I., 1969. Morphological analysis of karst caves (in Russian). In: *Ocherki po fizicheskoy geografii Gruzii* (Assays on physical geography of Georgia). Mecniereba, Tbilisi, pp. 3–74.

MARUSZCZAK, H., 1966. Phénomènes karstiques dans les roches du Crétacé supérieur entre la Vistule et le Bug (type de karst crayeux) (in Polish). *Przegl. Geogr.,* 38(3): 339–370.

MASLOV, V. P., 1945. To the question of phases of deposition and karstification of buried massifs of Ishimbay (in Russian). *Izv. Akad. Nauk SSSR, Ser. Geol.:* 88–100.

MASLYN, R. M., 1976. *Late Mississippian paleokarst in the Aspen, Colorado, area.* M. S. Thesis, Colorado School of Mines, 96 pp.

MATTES, G., 1975. How engineers beat shaft flood at Zonguldak Coal Mine. *World Coal,* 1(3): 28–32.

MAULL, O., 1958. *Handbuch der Geomorphologie.* Deuticke, Wien, 600 pp.

MAURIN, V. and ZÖTL, J. G., 1967. Saltwater encroachment in the low altitude karst water horizons of the island of Kephallinia. *Proc. Dubr. Symp. 1965, Publ. 73, Int. Assoc. Sci. Hydrol., 1, Paris,* pp. 436–455.

MAUTHE, F., 1975. Paläokarst im Jura der Iberischen Ketten (Prov. Soria, Nordspanien). *N. Jb. Geol. Paläontol., Abh.,* 150(3): 354–372.

MAZÚR, E., 1965. Major features of West Carpathians as a result of young tectonic movements. In: E. MAZÚR (Editor), *Problems of West Carpathian Geomorphology.* Bratislava, pp. 9–53.

MCKEE, E. D. and GUTCHICK, R. C., 1969. History of the Redwall Limestone of northern Arizona. *Geol. Soc. Amer. Mem.,* 114: 726 pp.

MCCORMICK, J. E., EVANS, L. L., PALMER, R. A. and RASNICK, F. D., 1971. Environments of the zinc deposits of the Mascot-Jefferson City district, Tennessee. *Econ. Geol.,* 66: 757–763.

MCKNIGHT, E. T. and FISCHER, R. P., 1970. Geology and ore deposits of the Picher Field, Oklahoma and Kansas. *US Geol. Surv. Profess. Pap.,* 588: 169 pp.

MEIBURG, P., 1980. Subrosions-Stockwerke in Nordhessischen Bergland. *Aufschluss,* 31: 265–287.

MEIN, P., 1958. Les mammifères de la faune sidérolithique de Vieux-Collonges. *Nouv. Arch. Mus. Hist. Natur., Lyon,* 5: 122 pp.

MEIN, P., 1976. Biozonation du Néogene Mediterranéen à partir des Mammifères. *Proc. 6th Cong. R. C. M. N. S. Bratislava 1975,* 2: 18.

MELLO, J. and SNOPKOVÁ, P., 1973. Upper Cretaceous filling in the cavities of Triassic limestones in the Gombasek Quarry. *Geol. práce, Spr.,* 61: 239–253.

MIAUTON, J. D., 1980. *Bakouma, genèse et gîtologie d'un gisement néoformé continental phosphatà-uranifère.* Thèse 3e cycle, INPL, Nancy, 160 pp.

MICHAEL, R., 1912. Die geologische Position der Wasserwerke im Oberschlesischen Industriebezirk. *Jb. Preuss. Geol. Landesanst.,* 33(1): 77–107.

MICHOT, P., 1980. Belgique. In: *Géologie des pays européens; France, Belgique, Luxembourg.* Dunod, Paris, pp. 485–576.

MIJATOVIĆ, B., 1983. Relation of hydrodynamic regime and budget to deep circulation in the Lelić Karst (in Slovenian). *Zbor. Rad. Odb. Speleol. SANU, Beograd,* 1: 1–24.

MILANOVIĆ, P. T., 1981. *Karst Hydrogeology.* Water Resources Publications, Littleton, Colorado, 434 pp. (A translation of „*Hidrogeologija karsta i metode istraživanja*". Trebinje, 1979).

MILOJEVIĆ, N., FILIPOVIĆ, B. and DIMITRIJEVIĆ, N., 1968. The influence of siphonal circulation to the karstification process in the Vrdnić Karst (in Slovenian). *Cvijićev Zbor. SANU, Beograd,* pp. 161–168.

MILOVANOVIĆ, B., 1965. Epeirogenic and orogenic dynamics of Dinarides and problem of paleokarstification and geological evolution of holokarst (in Slovenian). *Vesnik, Ser. B (Zav. Geol. Geofiz. Istraz.), Beograd,* 4/5: 5–44.

MINDSZENTY, A., KNAUER, J. and SZANTNER, F., 1984. Sedimentological characteristics of Iharkút bauxite and conditions of its deposition (in Hungarian). *Földt. Közl.,* 114(1): 19–48.

MIOŠIĆ, N., 1982. Relation between thermomineral and cold waters in Miocene karst environment of Gradačac. *Naš krš, Sarajevo,* 12/13: 23–48.

MIOTKE, F.-D. and PALMER, A. N., 1972. Genetic Relationship between Caves and Landforms in the Mammoth Cave National Park Area. *Veröff. Geogr. Instit. Tech. Univ. Hannover,* 69 pp.

MIŠÍK, M., 1978. Continental, brackish and hypersaline facies in the Mesozoic of Central Carpathians and a question of emerged areas (in Slovakian). In: J. VOZÁR et al. (Editors), *Paleogeografický vývoj Západných Karpát* (Paleogeographic evolution of West Carpathians). Geol. úst. Dionýza Štúra, Bratislava, pp. 35–48.

MIŠÍK, M., 1980. Miocene sinter crusts (speleothems) and calcrete deposits from neptunian dykes, Malé Karpaty Mts. *Geol. zbor. (Geologia Carpathica),* 31(4): 495–512.

MIŠÍK, M. and SÝKORA, M., 1980. Jura der Silica-Einheit, rekonstruiert aus Geröllen und oberkretazische Süsswasserkalke des Gemerikums. *Geol. zbor. (Geologia Carpathica),* 31(2): 239–261.

MIŠÍK, M. and SÝKORA, M., 1981. Pieniny exotic elevation reconstructed from pebbles of carbonate rocks in Cretaceous conglomerates of the klippen belt and Maniny unit (in Slovakian). *Západné Karpaty, Ser. Geol.,* 7: 7–111.

MITSUY Mining and Smelting Company, 1970. *Feasibility Study, Rennell Island bauxite project.* Tokyo.

MOJSKI, J. E., 1985. *Geology of Poland, Vol. I. Stratigraphy, Pt. 3b. Cenozoic, Quaternary.* Inst. Geol., Warszawa, 224 pp.

MOLENDA, D., 1972. Lead-ore mines in the region of Silesian-Cracow deposits in XVI to XVIII centuries (in Polish). In: *Z dziejów postępu technicznego w eksploatacji kruszców,* Inst. Hist. Kult. Mater. PAN, Warszawa.

MONROE, W. H., 1970. A Glossary of Karst Terminology. *US Geol. Surv. Water -Supply Pap.,* 1899–K: 26 pp.

MOORBATH, S. and VOKES, F. M., 1963. Lead isotope abundance studies of galena occurrences in Norway. *Nor. Geol. Tidsskr.,* 43: 283–344.

MOREHOUSE, D. F., 1968. Cave development via the sulfuric reaction. *Nat. Speleol. Soc. Bull.,* 30(1): 1–10.

MOTYKA, J. and WILK, Z., 1976. Vertical differentiation in the water permeability of carbonate Triassic rocks in the light of a statistical analysis of the results of pumping tests. *Kwart. Geol.,* 20(2): 381–397.

MOTYKA, J. and WILK, Z., 1982. The neglection of turbulent features and its influence to the results of water inflows to mines (in Polish). *XI Sesja Nauk., 45 Zjazd Nauk. Wychowanków Akad. Gór. Hutn., Kraków.*

MOTYKA, J. and WILK, Z., 1984. Hydraulic structure of karst fissured Triassic rocks in the vicinity of Olkusz (southern Poland. near Cracow). *Kras i Speleol.,* 5(14): 11–23.

M'RABET, A., 1981. Differentiation of environment of dolomite formation, Lower Cretaceous of Tunisia. *Sedimentology,* 28: 331–352.

MROCZKOWSKI, J., 1982. On the role of climatic factor in Zechstein marine transgression. *Bull. Acad. Pol. Sci., Sér. Sci. de la Terre,* 1981, 29(4): 303–310.

MÜLLER, P., 1971. Karstkorrosionwirkung des Kohlendioxid metamorphen Ursprungs (in Hungarian). *Karszt és Barlang,* 1971(2): 53–56.

MÜLLER, P., 1974. On the origin of thermal caves and spherical niches (in Hungarian). *Karszt és Barlang,* 1974(1): 7–10.

MÜLLER, P. and SÁRVÁRY, I., 1977. Some aspects of development in Hungarian Speleology Theories during the last 10 years. *Karszt és Barlang, Spec. Issue:* 53–60.

MUREŞAN, M., 1970. Sur la présence des dépôts Mésozoïques continentaux pré-vracono-cénomaniens (Albiens?) dans l'unité épimetamorphique du Massif Poiana Ruscă (in Romanian). *C. R. Séanc. Inst. Geol.,* 56(4): 59–75.

NAGY, B. and PELIKÁN, P., 1976. Metacinnabar and cinnabar occurring at the Rókahegy in the Csillaghegy area (in Hungarian). *Magy. Áll. Földt. Intéz. Évi Jelent.,* 1973–röl: 51–55.

NARKIEWICZ, M., 1979. Stratigraphy and facies development of the Upper Devonian in the Olkusz-Zawiercie area, southern Poland (in Polish). *Acta Geol. Pol.,* 1978, 28(4): 415–470.

NARKIEWICZ, M., 1985. Studies of the Devonian carbonate shelf of southern Poland (in Polish). *Przegl. Geol.,* 33(5): 253–258.

NICOLIS, C. I. and PRIGONINE, I., 1977. *Self-organization in non-equilibrium systems.* Wiley, New York.

NOEL, M., SHAW, R. P. and FORD, T. D., 1984. A palaeomagnetic reversal in early Quaternary sediments in Masson Hill, Matlock, Derbyshire. *Mercian Geol.,* 9(4): 235–242.

OAKLEY, K. P., CAMPBELL, B. G. and MOLLESON, T. I., 1975. *Catalogue of Fossil Hominids. Pt. III: Americas, Asia, Australia.* British Mus. Natur. Hist., London, pp. 52–77.

OAKMAN, C. D., 1984. *The sedimentology and palaeoenvironmental analysis of the Dinantian limestones of the Wirksworth-Grangemill area, Derbyshire.* PhD. Thesis, Univ. Aberdeen.

OBELLIANNE, J. M., 1963. Le gisement de phosphate tricalcique de Makatea (Polynésie française, Pacifique Sud). *Sci. de la Terre,* 9(1): 5–60.

ODIN, G. S. (Editor), 1982.*Numerical dating in stratigraphy.* J. Wiley and Sons, Chichester, 1040 pp.

ODRZYWOLSKA-BIEŃKOWA, E., POŻARYSKA, K. and MARTINI, E., 1978. Middle Oligocene microfossils from the Polish Lowlands; their stratigraphical and paleogeographical significance. *Acta. Palaeont. Pol.,* 23(3): 249–291.

OFTEDAHL, C., 1980. *Geology of Norway.* Nor. Geol. Unders., 356, Bull., 54: 114 pp.

OHLE, E. L., 1951. The influence of permeability on ore distribution in limestone and dolomite. *Econ. Geol.,* 46: 667–706.

OHLE, E. L., 1980. Some considerations in determining the origin of ore deposits of the Mississippi Valey type-part III. *Econ. Geol.,* 75: 161–172.

OLSON, R. A., 1984. Genesis of Paleokarst and Strata-Bound Zinc-Lead Sulfide deposits in a Proterozoic Dolostone, Northern Baffin Island, Canada. *Econ. Geol.,* 79: 1056–1103.

OMENETTO, P. and BRIGO, L., 1974. Metallogenesi nel quadro dell'orogene ercinico delle Alpi (con particolare riguardo al versante italiano). *Mem. Soc. Geol. Ital.,* 13(4): 1–24.

OMENETTO, P. and VAILATI, G., 1977. Ricerche geo-minerarie nel settore centrale del distretto a Pb, Zn, fluorite e barite di Gorno (Lombardia). *Ind. Miner., Roma,* 28: 25–44.

OPALINSKI, P. R., 1977. *The stratigraphy and sedimentology of the Mjøsa Limestone (Middle Ordovician), Norway.* PhD. Thesis, Univ. Liverpool.

OPALINSKI, P. R. and HARLAND, R. L., 1981. The Middle Ordovician of the Oslo Region, Norway 29. Stratigraphy of the Mjøsa Limestone in the Toten and Nes-Hamar areas. *Nor. Geol. Tidsskr.,* 61: 58–78.

ORGEVAL, J. J. and CAPUS, G., 1978. Existence d'un horizon phosphaté uranifère à la base des formations carbonatées cambriennes de la région du Vigan (Gard-France); son remaniement karstique. *C. R. Somm. Séanc. Soc. Géol. France,* 3: 115–117.

ORŁOWSKI, S., 1975. Lower Cambrian trilobites from Upper Silesia (Goczałkowice borehole). *Acta Geol. Pol.,* 25(3): 377–383.

OSIKA, R., 1961. Travaux de l'Expédition Géologique Polonaise au Vietnam. *Przegl. Geol.,* 9(1): 47–53.

OSIKA, R. (Editor), 1980. *La Géologie et gites mineraux en Pologne.* Biul. Inst. Geol., 251: 307 pp.

OSIKA, R., GŁAZEK, J. and JUSKOWIAK, O. (Editors), 1967. Travaux de l'Expédition Géologique Polonaise au Vietnam. *Biul. Inst. Geol.,* 177: 376 pp.

Osmólski, T., 1976. Karst and the origin of sulfur deposits in Poland (in Polish). *Kwart. Geol.*, 20(3): 559–573.

Ossowski, G., 1882. Etat des recherches dans les cavernes en Pologne. *Mater. Hist. Primitive Naturelle Homme, Toulouse*, 1: 1–24.

Ovcharenko, A. V., 1978. Perspectives of oil and gas potential of carbonate deposits of the Siberian Platform (in Russian). *Sov. Geol.*, 11: 23–34.

Overstreet, E. F., 1964. Geology of the southeastern bauxite deposits. *US Geol. Surv. Bull.*, 1199–A: 19 pp.

Owen, T. R. and Jones, D. G., 1966. The nature of the Millstone Grit-Carboniferous Limestone junction of a part of the North Crop of the South Wales coalfield. *Proc. Geol. Assoc. London*, 72: 239–250.

Ozoray, G., 1961. The mineral filling of the thermal spring caves of Budapest. *Atti Symp. Speleol., Mem.*, 2: 152–170.

Ozoray, G., 1977. The Athabasca carbonate and evaporite buried karst. *Int. Assoc. Hydrogeol., Mem.* 12: 85–98.

Padalino, G., Pretti, S., Tamburrini, D., Tocco, S., Uras, I., Violo, M. and Zuffardi, P., 1973. Ore deposition in karst formations with examples from Sardinia. In: G. C. Amstutz and A. J. Bernard (Editors), *Ores in sediments*. Springer, Berlin, pp. 209–220.

Padalino, G., Tocco, S. and Violo, M., 1980. Different genetic environments may produce similar occurrences of stratiform barite: Examples from Sardinia (Italy). In: J. D. Ridge (Editor), *Proc. Fifth Quadren. IAGOD Symp., Snowbird, Utah, U.S.A.* Schweitzerbarts'che Verlagsbuchhandlung (Nägele u. Obermiller), Stuttgart, pp. 459–446.

Palmer, A. N., 1984. *Jewel Cave, a gift from the past*. Wind Cave Natur. Hist. Assoc., Hot Springs, 41 pp.

Palmer, T. J. and Fürsich, F. T., 1974. The ecology of a Middle Jurassic hardground and crevice fauna. *Palaeontology*, 17: 507–523.

Palmer, T. J., McKerrow, W. S. and Cowie, J. W., 1980. Sedimentological evidence for a stratigraphic break in the Durness Group. *Nature, 287:* 720–722.

Paloc, H., 1970. La Fontaine de Vaucluse et son bassin d'alimentation. *Ann. Soc. Hort. Hist. Natur. Herault, Montpellier*, 110.

Panek, S. and Szuwarzyński, M., 1976. Pre-Tortonian valley with Tertiary fill near Chrzanów (Cracow Upland, Poland) (in Polish). *Rocz. Pol. Tow. Geol.*, 46(4): 503–523.

Panin, S., Bleahu, M., Tomescu, C., Mirauta, E. and Stefanescu, M., 1982. Structure of the Vaşcău Plateau. *C. R. Séanc., Inst. Géol. Geophys.*, 67(5): 109–124.

Panoš, V., 1960. Die thermale Karst des Gebirges von Buda und dessen eigentümliche Ablagerungen (in Hungarian). *Hidrol. Közl.*, 40(5): 391–395.

Panoš, V., 1962. Fossil destructive forms of the eastern part of the Czech Highland (in Czech). *Geogr. čas.*, 14(3): 181–204.

Panoš, V., 1962–63. To the question of the origin and age of planation surfaces in the Moravian Karst (in Czech). *Čs. kras*, 14: 29–41.

Panoš, V., 1963. The Sloup valley-polje and its draining caves, Moravian Karst (in Czech). *Kras v Československu*, 1–2: 1–10.

Panoš, V., 1964a. Der Urkarst in Ostflügel der Bohmischen Masse. *Z. Geomorphol., N. F.*, 8(2): 105–162.

Panoš, V., 1964b. Report on fossil karst forms research in reef limestones of the Štramberk-Highland (West Carpathians). *Zpr. Geogr. úst. ČSAV, Opava*, 5: 1

Panoš, V., 1965. Genetic features of a specific type of the karst in the Central European climate morphogenic area. In: O. Štelcl (Editor), *Problems of Speleological Research, 1.* NČSAV, Praha, pp. 11–23.

PANOŠ, V., 1970. Die Rolle der Erosionsniveaus bei der Entwicklung der oberflächlichen und unterirdischen Entwässerung des Mährischen Karstes. *Acta Univ. Palackianae Olomouc., Geol.-Geogr., Praha,* 10: 77–127.

PANOŠ, V., 1976. Some notes on the coastal karst development. In: I. GAMS (Editor), *Karst Processes and Relevant Landforms.* Ljubljana Univ., Ljubljana, pp. 181–187.

PANOŠ, V., 1978. To the karst typology I to III (in Czech). *Acta Univ. Palackianae Olomouc., Geol.-Geogr., Praha,* 17: 83–132.

PANOŠ, V., 1983. Karstification – a part of carbonate rocks' epigenesis (in Czech). *Acta Univ. Palackianae, Geogr.-Geol., Praha,* 21: 31–50.

PANTIĆ, N., 1958. Fossil flora from white bauxites in western Crna Gora (in Slovenian). *Geol. An. Balkan. Poluostr., Beograd,* 25: 77–84.

PAPIU, V. C., IOSOF, V., MÎNZATU, S., GIUŞCA, R. and JACOTA, G., 1975. La constitution chimico-minéralogique des bauxites de la région de Schodol-Cîmpeni, district of Bihor (in Romanian). *C. R. Séanc., Inst. Géol. Géophys.,* 61(2): 37–69.

PAPIU, V. C. and MINZATU, S., 1969. Hypothèse sur la genèse des bauxites du Massif de Pădurea Craiului, Monts Apuseni (in Romanian). *Annu. Com. Etat. Géol., Bucuresti,* 37: 37–70.

PAPIU, V. C., MÎNZATU, S., IOSOF, V., UDRESCU, C. and GIUSCA, R., 1971. Constitution chimico-minéralogique de la formation bauxitifère du Bassin de Haţeg (in Romanian). *C. R. Séanc., Inst. Géol. Géophys.,* 57(2): 77–123.

PAPIU, V. C., STEFANUT, V., ISOSOF, V., MÎNZATU, S., POPESCU, F., NEACSU, V. and UDRESCU, C., 1984. Les bauxites de Poiana Ruscă. *C. R. Séanc., Inst. Géol. Géophys.,* 68(2): 103–139.

PARASCHIV, D., 1966. Sur l'évolution paleomorphologique de la Plaine Roumanie. *Rev. Roum. Géol. Géophys. Géogr., Sér. Géogr.,* 10(1): 47–54.

PARASCHIV, D., 1979. *The Moesian Platform and its hydrocarbon deposits* (in Romanian). Ed. Academiei R.S.R., Bucuresti, 195 pp.

PARASCHIV, D. and BUCUR, I., 1975. Oil-bearing zone west of Bucureşti (in Romanian).*Mine Petrol Gaze, Bucuresti,* 26(10): 471–475.

PARK, W. C. and AMSTUTZ, G. C., 1968. Primary „cut-and-fill" channels and gravitational diagenetic features. *Mineralium depos.,* 3: 66–80.

PARTRIDGE, T. C., 1973. Geomorphological dating of cave openings at Makapansgat, Sterkfontein, Swartkrans and Taung. *Nature,* 246: 75–79.

PASSENDORFER, E., 1930. Étude stratigraphique et paléontologique du Crétacé de la Serie Hauttatrique dans les Tatras. *Prace Państw. Inst. Geol.,* 2(4): 511–676.

PASZKOWSKI, M. and WIECZOREK, J., 1982. Fossil karst with Mesozoic bone breccia in Czatkowice (Cracow Upland, Poland). *Kras i Speleol.,* 4(13): 32–40.

PATRULIUS, D., MARINESCU, F. and BALTRES, A., 1983. Dinosauriens ornitopodes dans les bauxites néocomiennes de l'unité de Bihor (Monts Apuseni). *Annu. Inst. Géol. Géophys., Bucuresti,* 59: 109–117.

PAVIĆ, A., 1956. Stratigraphic position of bauxite between Nikšićko Polje and Rovač Plateau of the Morač canyon, Montenegro (in Slovenian). *Vesnik (Zav. Geol. Geofiz. Istraž.), Beograd,* 12: 291–320.

PECORINI, G., 1956. Stratigrafia e distribuzione della Bauxiti della Nurra (Sardegna Nord-occi.). *Symp. Assoc. Mineral. Sarda, Sez. 1, Cagliari,* pp. 2–15.

PECORINI, 1965. Stratigrafia e distribuzione delle bauxiti della Nurra (Sardegna Nord-Occidentale). *Symp. Geomineral. Sardo, Assoc. Mineral. Sarda, Sez. 1,* B–6: 1–15.

PÉCSI,K M., 1963. Pediment surfaces in the Hungarian Mountains. *Földrajzi Köz.,* 9(3): 195–212.

PEIVE, A. V., KHAIN, V. E., MOURATOV, M. V. and DELANY, F (Editors), 1981. *Tectonics of Europe and adjacent areas. Cratons, Baikalides, Caledonides. Explanatory note to the International Tectonic Map of Europe and adjacent areas, Scale 1 : 2 500 000.* Nauka, Moscow, 415 pp.

PEIVE, A. V., KHAIN, V. E., MOURATOV, M. V. and DELANY, F. (Editors), 1982. *Tectonic map of Europe and adjacent areas. Variscides, epi-Paleozoic platforms, Alpides. Explanatory note to the International Tectonic Map of Europe and adjacent areas, Scale 1 : 2 500 000*. Nauka, Moscow, 627 pp.

PEI WEN-CHUNG, 1965. Excavation of Liucheng *Gigantopithecus* cave and exploration of other caves in Kwangsi. *Inst. Vertebr. Palaeontol. Palaeoanthropol. Acad. Sinica, Mem.*, 7: 54 pp.

PELIETEK, E. I., 1971. Bauxite deposits of the Yenisei Range and the Siberian Platform (Angara-Yenisei province) (in Russian). In: *Platformennye boksity SSSR* (Platform bauxites of the USSR). Nauka, Moskva, pp. 221–262.

PELÍŠEK, J., 1975. Fossil weathering products and pedosediments in crystalline limestones of the Moravicum, northwestern Moravia (in Czech). *Čas. mineral. geol.*, 20(2): 179–184.

PELÍŠEK, J., 1976. Geochemistry of sediments and pedosediments in sinkholes of the Moravian Karst (in Czech). *Čs. kras*, 28: 49–57.

PETRBOK, J., 1950. Chronological paradoxon of terrace deposits in the Bohemian Karst (in Czech). *Čs. kras*, 3(6–7): 176–177.

PENEAUD, P., 1978. *La paragenèse phosphatée de la grotte de la Caune de l'Arago (Pyrénées orientales)*. Thèse 3e cycle, Univ. Paris VII.

PERNA, G., 1975. Giacimenti minerari carsici. *Proc. 6th Int. Congr. Speleol.*, Pt. I: 523–543.

PERYT, T. M., 1984. Sedimentation and early diagenesis of Zechstein limestones in western Poland (in Polish). *Prace Inst. Geol.*, 109: 80 pp.

PETROVIĆ, J., 1969. The occurrence of deep karstification in southern part of old massif of Kopaonik (in Slovenian). *Zbor. Prir. Nauk. Matice Srpske, Novi Sad*, 36: 97–108.

PINCZÉS, Z., 1970. Problems of Relief Planation. *Stud. Hung. Geogr.*, 8: 65–151.

PIRLET, H., 1970. L'influence d'un karst sous-jacent sur la sédimentation calcaires et l'intérêt de l'étude des paléokarsts. *Ann. Soc. Géol. Belg.*, 93: 247–254.

PISERA, A., 1985. Paleoecology and lithogenesis of the Middle Miocene (Badenian) algal-vermetid reefs from Roztocze Hills, south-eastern Poland. *Acta Geol. Pol.*, 35(1–2): 89–155.

PIWOCKI, M., 1975. The Tertiary of the Rawicz vicinity and its coal-bearing properties (in Polish). *Biul. Inst. Geol.*, 284: 73–133.

PLUMMER, L. N., 1977. Defining reactions and mass transfer in part of the Florida aquifer. *Wat. Resour. Res., Washington*, 13: 801–812.

PLUMMER, L. N. and BUSENBERG, E., 1982. The solubilities of calcite, aragonite and vaterite in CO_2–H_2O solutions between 0° and 90 °C and an evaluation of the aqueous model for the system $CaCO_3$–CO_2–H_2O. *Geochim. Cosmochim. Acta*, 46: 1011–1040.

POBORSKI, J., 1957. Ausbildung der Gipshutes und Entwicklung der Karstphänomene aus dem Salzstock von Inowrocław (in Polish). *Arch. Gór. Hutn.*, 2(4): 225–248.

POBORSKI, J., 1975. On halogenic karst features in Upper Permian in Łeba elevation (in Polish). *Przegl. Geol.*, 23(7): 325–328.

POLJAK, J., 1952. The occurrence of old karst forms in deposits near Debeljak in northern Velebit (in Slovenian). *Geol. Vjesn., Zagreb*, 2/4: 99–110.

POLŠAK, A., 1976. Paleotemperature conditions in Jurassic and Cretaceous of Dinarides and Alps based on oxygen isotopic method (in Slovenian). *8 Congr. Geologa Jugoslavije*, 2. Ljubljana.

POLŠAK, A. and PEZDIČ, J., 1978. Paleotemperature conditions in Carboniferous and Permian of Dinarides and Alps based on oxygen isotopic method and their role in paleogeography (in Slovenian). *Geol. Vjesn., Zagreb*, 30(1): 167–187.

POP, G. and MÎRZA, I., 1977. Les paléokarst éocrétacé des Monts Pădurea Craiului et sa signification dans le mécanisme de genèse des bauxites. *Bull. Sci. Géol. Strasbourg*, 30(1): 51–58.

POPOV, A., 1968. Les types morphologiques et la répartition des gisements de plomb-zinc en Algérie. *A. S. G. A. Congr., Tunis 1966, Ann. Mines Géol. Tunis*, 23: 103–203.

POŠEPNÝ, F., 1894. Über die Entstehung der Blei- und Zinc-lagerstätten in auflöslichen Gesteinen. *Berg- und Hüttenmänn. Mh.*, 42: 77–130.

Poty, E., 1980. Evolution and drowning of paleokarst in Frasnian carbonates at Visé, Belgium. *Meded. Rijks. Geol. Dienst.*, 32(7): 53–55.

Poty, E., 1982. Paleokarst et breches d'effondrement dans le Frasnien moyen des environs de Visé. Leur influence dans le paléogéographie Dinantienne. *Ann. Soc. Géol. Belg.*, 105: 315–337.

Pożaryska, K. and Odrzywolska-Bieńkowa, E., 1977. On the Upper Eocene in Poland (in Polish). *Kwatr. Geol.*, 21(2): 59–72.

Pożarski, W. and Brochwicz-Lewiński, W., 1978. On the Polish Trough. *Geol. en Mijnb.*, 57(4): 545–557.

Pozzuoli, A., Stanzione, D., D'Argenio, B., Fenoll, P., Huertas, F. and Linares, J., 1977. Studio mineralogico e geochimico preliminare dei riempimenti delle cavità carsiche cretaciche di Vitulano (Appennino meridionale, Italia). *Geol. Apl. Idrogeol.*, 12(2): 201–212.

Pretti, S., Salvadori, I. and Violo, M., 1978. Anew approach to the investigation of karst and ore bodies in Cambrian carbonatic rocks of Sardinia. *Assoc. Mineral. Sarda,* 84(1): 67–85.

Prodrôme, 1954, *Prodrôme d'une description géologique de la Belgique publié sous la direction de P. Fourmarier.* H. Vaillant-Carmanne, Liège.

Prokopchuk, B. I., Levin, M. P. and Metelkina, M. P., 1982. Lithological- structural and geochemical control of ancient karst of the Siberian Platform (in Russian). *Geol. i Geofiz.*, 9: 54–59.

Prokopchuk, B. I., Levin, V. I., Metelkina, M. P. and Shofman, I. L., 1985. *Ancient karst and its placer mineralogeny* (in Russian). Nauka, Moskva, 175 pp.

Prosová, M., 1952. The Štramberk Karst (in Czech). *Přírodověd. sbor. ostrav. kraje, Opava,* 13: 417–446.

Pulina, M., 1975. Chemical denudation on the carbonate karst areas. *Prace Geogr., Geogr, Inst. Pol. Acad. Sci,* 105: 160 pp.

Purdy, E. G., 1974. Reef configuration: cause and effect. In: L. F. Laporte (Editor), *Reefs in Time and Space.* Spec. Publ. Soc. Econ. Paleontologists Mineralogist, 18: 9–76.

Quinet, G. E., 1969. *Bernissart ... il y a 125 000 000 d'années!* I. R. Sc. N. B., Bruxelles.

Quinif, Y., 1983. Éléments d'une approche énergétique du karst. Application a quelques examples réels de karsts. *Karstologia,* 1(1): 47–54.

Quinlan, J. F., 1971. Collapse sinkholes and subsidence depressions caused by groundwater pumpage in an urban area in the Transvaal karst, S. Africa: origin and detection of incipient development. *Cave and Karst, Castro Valley,* 13(6): 48.

Quinlan, J. F., 1972. Karst-related mineral deposits and possible criteria for the recognition of paleokarsts: A review of preservable characteristics of Holocene and older karst terranes. *Proc. 24th Int. Geol. Congr. Montreal, Proc.* 6: 156–168.

Quinlan, J. F., 1978. *Types of karst, with emphasis on cover beds in their classification and development.* Ph. D. Thesis, Univ. Texas, Austin. 323 pp.

Rabeder, G., 1981. *Arvicoliden* (Mammalia, Rodentia) aus der Pliozän und Altpleistozän von Österreich. *Beitr. Paläont. Österreich,* 8: 373 pp.

Rachl, R., 1983. *Die Chiroptera(Mammalia) aus den mittelmiozänen Kalk des Nördlichen Ries (Süddeutschland).* Inaugural Diss., Univ. München, 284 pp.

Radwański, A., 1969. Lower Tortonian transgression onto the southern slopes of the Holy Cross Mts (in Polish). *Acta Geol. Pol.,* 19(1): 164 pp.

Radwański, A., 1973. Lower Tortonian transgression onto the southeastern and eastern slopes of the Holy Cross Mts (in Polish). *Acta. Geol. Pol.,* 23(3): 375–434.

Rajlich, P., Legierski, J. and Šmejkal, V., 1983. Stable isotope study of base metal deposits from the Eastern High Atlas, Morocco. *Mineralium Depos.,* 18: 161–171.

Ramsbottom, W. H. C., 1973. Transgressions and regressions in the Dinantian: a new synthesis of British stratigraphy. *Proceedings (Yorkshire Geol. Soc.),* 39: 567–607.

Ramsbottom, W. H. C., 1979. Rates of transgression and regression in the Carboniferous of NW Europe. *J. Geol. Soc. London,* 136: 147–153.

READING, H. G. (Editor), 1978. *Sedimentary Environments and Facies*. Blackwell, Oxford, 557 pp.

READING, H. G., 1982. Sedimentary basin and global tectonics. *Proc. Geol. Assoc. London*, 93(4): 321–350.

REICHENBACH, K., 1834. *Geologische Mittheilungen aus Mähren. Geognostische Darstellung der Umgegenden von Blansko*. J. C. Heubner, Wien, 218 pp.

RENDER, F. W., 1970. Geohydrology of the metropolitan Winnipeg area as related to groundwater supply and construction. *Canad. Geotechn. J.*, 7(3): 243–274.

REN MEI'E, 1982. A preliminary study on the rules of karst development in China (in Chinese). *Selec. Pap. 2nd all-China Symp. Karst Sponsored by Geol. Soc. China*. Sci. Press, Beijing, pp. 1–4.

REUSCH, H., 1903. Norske kaolinforekomster. *Naturen (norw.)*, 27: 129–132.

REUSS, A. E., 1854. Beiträge zur geognostischen Kentniss Mährens. *Jb. K.-Kön. Geol. Reichsanst., Wien*, pp. 659–765.

REYMET, R. A. and MÖRNER, N. A., 1977. Cretaceous transgressions and regressions exemplified by the South Atlantic. In: *Mid-Cretaceous Events, Hokkaido Symp. 1976*. Palaeontol. Soc. Japan, Spec. Pap., 21: 247–261.

RHODES, D., LANTOS, E. A., LANTOS, J. A., WEBB, R. J. and OWENS, D. C., 1984. Pine Point orebodies and their relationship to structure, dolomitization and karstification of the Middle Devonian barrier complex. *Econ. Geol.*, 70: 991–1055.

RICHTER, D., 1966. On the New Red Sandstone Neptunian Dykes of the Tor Bay Area, Devonshire. *Proc. Geol. Assoc. London*, 77: 173–186.

RIDGE, J. D. (Editor), 1980. *Proceedings of Fifth Quadrenial IAGOD Symposium, Snowbird, Utah, U.S.A.* Schweitzerbarts'che Verlagsbuchhandlung/Nägele u. Obermiller, Stuttgart.

RIVIÊRE, M., RAUTUREAU, M., BESSON, G., STEINBERG, M. and AMOURI, M., 1985. Complémentarité des rayons X et de la microscopie électronique pour la détermination des diverses phases d'une argile zincifère. *Clay Miner., Oxford*, 20.

ROALDSET, E., PETTERSEN, E., LONGVA, O. and MANGERUD, J., 1982. Remnants of preglacial weathering in western Norway. *Nor. Geol. Tidsskr.*, 62: 169–178.

ROBASZYNSKI, F., 1979. Comparison between the Middle Cretaceous of Belgium and some French Regions. In: *Aspekte der Kreide Europas*. Int. Union Geol. Sci., Ser. A, Stuttgart, 6: 543–561.

ROBASZYNSKI, F. and DUPUIS, C., 1983. *Belgique-Guides géologiques régionaux*. Masson, Paris, 204 pp.

ROBINSON, P. L., 1957. The Mesozoic fissures of the Bristol Channel area and their vertebrate faunas. *J. Linnean Soc.*, 43: 260–282.

ROBINSON, P. L., 1971. A problem of faunal replacement on Permo-Triassic continents. *Palaeontology*, 14: 131–153.

RODIONOV, N. V., 1963. *Karst of the European part of the USSR, Ural and Caucasus* (in Russian). Gosgeoltekhizdat, Moskva, 175 pp.

ROE, K. K., BURNETT, W. C. and LEE, A. I. N., 1983. Uranium disequilibrium dating of phosphate deposits from Lau group, Fiji. *Nature*, 302(5909): 603–606.

ROEDDER, E., 1971. Fluid inclusion evidence on the environment of formation of mineral deposits of the southern Appalachian Valley. *Econ. Geol.*, 66: 777–791.

ROGLIĆ, J., 1972. Historical review of morphologic concepts. In: M. HERAK and V. T. STRINGFIELD (Editors), *Karst. Important Karst Regions of the Northern Hemisphere*. Elsevier, Amsterdam, pp. 1–18.

RÖHLICH, P. and CHLUPÁČ, I., 1951. The relicts of marine Cenomanian near Svatý Jan pod Skalou (in Czech). *Čas. Nár. muz., Odd. Přírodověd.*, 1949–50, 118–119.

RÖMER, F., 1883. Die Knochenhöhlen von Ojców in Polen. *Palaeontographica*, 29: 41 pp.

ROTH, Z., 1980. West Carpathians-the Tertiary structure of central Europe (in Czech). *Knih. ústř. úst. geol.*, 55: 128 pp.

ROUSSET, C., 1967. Sur la nature et la genèse des minerais de fer du Haut Var; leurs relations avec la paleogeographie locale. Rapport et differences avec les formations bauxitiques et les „Terra Rossa". *Bull. Soc. Géol. France,* 7(9): 640–5-651.

ROUTHIER, P., 1963. *Traité des gîtes métallifères. 2 vol.* Masson, Paris, 1282 pp.

ROUVIER, H., 1971. Minéralisation plombo-zincifère et phénomène karstique. Exemple tunisien: le gisement du Djebel Hallouf. *Mineralium Depos.,* 6(3): 196–208.

ROWLANDS, N. J., BLIGHT, P. G., JARVIS, D. M. and VON DER BORCH, C. C., 1980. Sabkha and playa environments in late Proterozoic grabens, Willouran Ranges, South Australia. *J. Geol. Soc. Austral.,* 27(1): 55–68.

RÓŻYCKI, S. Z., 1979. From Mochty to a synthesis of the Polish Pleistocene (in Polish). *Rocz. Pol. Tow. Geol.,* 1978, 48(3–4): 445–478.

RUBINOWSKI, Z., 1971. The non-ferrous metal ores of the Świętokrzyskie Mountains and their metallogenic position (in Polish). *Biul. Inst. Geol.,* 247: 166 pp.

RUDNICKI, J., 1967. Origin and age of the Western Tatra caverns (in Polish). *Acta Geol. Pol.,* 17(4): 521–591.

RUDNICKI, J., 1979. Role of convection in shaping subterranean karst forms. *Kras i Speleol.,* 2(11); 1978: 92–101.

RUDNICKI, J., 1980. Subsurface karst processes in coastal area, based on the example of Apulia, Southern Italy (in Polish). *Stud. Geol. Pol.,* 65: 60 pp.

RUPPERT, H., 1984. Physiko-chemische Betrachtungen zur Entstehung der Kreide Eisenerz Lagerstätten in Nord-Ost Bayern. *Geol. Jb.,* D, 66: 51–76.

RUTTNER, A. W., 1970. Die Bauxit-Vorkommen der Oberkreide in den Ost-Alpen und deren paläogeographische Bedeutung. *Magy. Áll. Földt. Intéz. Évk.,* 54: 131–134.

RYŻAK, R., 1977. Method for measuring the electrical resistivity of waters in the investigation of filtration phenomena in the region of dams and river reservoirs. *Gosp. Wod., Warszawa,* 37(1): 15–20.

RZHECHITSKY, P. B., 1966. Karst of polymetallic deposits of Priangun'e, Eastern Transbaikal (in Russian). *Trudy Irkutsk. Politekhn. Inst., Geol.,* pp. 155–168.

SAAR, R., 1954. Beiträge zur Meteorologie der dynamichen Wetterhöhlen. *Mitt. Höhlenkomm., Wien,* 1953 (1): 4–52.

SADOWSKA, H., 1977. Vegetation and stratigraphy of Upper Miocene coal seams of the southwestern Poland (in Polish). *Acta palaeobot., Kraków,* 18(1): 87–122.

SAKAĆ, K., 1966. On paleorelief and pseudopaleorelief of karst underlying bauxites (in Slovenian). *Geol. Vesnik, Zagreb,* 19: 123–130.

SAMSON, P., 1969. La stratigraphie de Quaternaire supérieur en Roumanie. *8e Congr. INQUA, Paris,* 2: 629–639.

SAMSONOWICZ, J., 1934. Sur les phénomènes karstiques et la brèche osseuse de Wężе près de Dzialoszyn sur la Warta (in Polish). *Zabytki Przyr. Nieożyw., Warszawa,* 3: 151–162.

SANDFORD, B. V., 1964. Subsurface stratigraphy of Silurian rocks in southwestern Ontario. *Geol. Surv. Canada, Pap.* 64–2: 14–19.

SANDO, W. J., 1974. Ancient solution phenomena in the Madison Limestone (Mississippian) of north-central Wyoming. *J. Res. US Geol. Surv.,* 2: 133–141.

SĂNDULESCU, M., 1975. Etude géologique de la partie centrale et septentrionale du synclinal de Hăghimaş – Carpates Orientales (in Romanian). *Annu. Inst. Géol. Géophys.,* 45: 6–200.

SANDVIK, P. O. and ERDOSH, G., 1977. Geology of the Cargill Phosphate Deposit in Northern Ontario. *Bull. Canad. Inst. Min.,* pp. 90–98.

SARTONI, S. and COLALONGO, M., 1964. Sul Cretacico dei dintorni di Caiazzo (Caserta). *Mem. Soc. Geol. Ital.,* 4(2).

SASS-GUSTKIEWICZ, M., 1974. Collapse breccia in the ore-bearing dolomite of the Olkusz mine (Cracow-Silesian region). *Rocz. Pol. Tow. Geol.,* 44: 217–226.

SASS-GUSTKIEWICZ, M., 1975a. Stratified sulfide ores in karst cavities in the Olkusz mine (Cracow-Silesian region). *Roc. Pol. Tow. Geol.*, 45: 63–68.

SASS-GUSTKIEWICZ, M., 1975b. Zinc and lead mineralization in collapse breccias of the Olkusz mine (Cracow-Silesian region). *Rocz. Pol. Tow. Geol.*, 45: 303–326.

SASS-GUSTKIEWICZ, M., 1983. Zinc-lead ore structures from Upper Silesian region in the light of solution transfer. In: G. KISVARSANYI et al. (Editors), *Proc. Vol. Int. Conf. Mississippi Valley-type Pb – Zn depos., Rolla, Missouri*, pp. 20–26.

SASS-GUSTKIEWICZ, M., 1985. Upper Silesian Zn-Pb ore deposits in the light of solution transfer (in Polish). *Zesz. Nauk. Akad. Gór. Hutn., Geol., Kraków*, 31: 1–119.

SASS-GUSTKIEWICZ, M., DŻUŁYŃSKI, S. and RIDGE, J. D., 1982. The emplacement of zinc-lead sulphide ores in the Upper Silesian district: a contribution to the understanding of Mississippi Valley-type deposits. *Econ. Geol.*, 77: 392–412.

SAWICKI, L., 1908. Skizze des slowakischen Karstes und der geographische Zyklus im Karst überhaupt (in Polish). *Kosmos, Lwów*, 6–7: 395–444.

SAWICKI, L., 1909. Ein Beitrag zum geographischen Zyklus im Karst. *Geogr. Z., Wien*, 15: 185–204, 259–281.

SCHAFARZIK, F., 1928. Rückblicke auf die Entwicklungsgeschichte der budapester Thermen. *Hidrol. Közl.*, 1921 (1): 9–14, 57–61.

SCHAUBERGER, O., 1956. Über die vertikale Verteilung der alpinen Karsthöhlen. *Mitt. Höhlenkomm., Wien*, 1955(1): 21–28.

SCHERF, E., 1928. Hydrothermales Gesteinmetamorphose im Buda-Piliser Gebirge. *Hidrol. Közl.*, 1922(2): 19–88.

SCHEUER, Gy. and SCHWEITZER, F., 1974. The development of the Hungarian travertine-sequence in the Quaternary. *Földrajzi Közlem.*, 21(2): 133–141.

SCHEUER, Gy. and SCHWEITZER, F., 1980. Entwicklunsgeschichte der Budaer Thermalquellen von oberen Pannon bis in unsere Tage. *Hidrol. Közl.*, 60: 492–501.

SCHEUER, Gy. and SCHWEITZER, F., 1981. Rekonstruktion der paläokarsthydrologischen Verhältnisse des Gerecse-Gebirge von oberen Pannon bis zu unseren Tagen. *Hidrol. Közl.*, 61: 33–38.

SCHLAGER, W., 1981. The paradox of drowned reefs and carbonate platforms. *Geol. Soc. Amer. Bull.*, Pt. I, 92: 197–211.

SCHNEIDER, H., J. (Editor), 1983. *Mineral Deposits of the Alps and the Alpine Epoch in Europe*. Proc. IV ISMIDA 1981. Springer, Berlin, 402 pp.

SCHOTT, W. and STOPPEL, D., 1976. *International map of natural gas fields in Europe – 1 : 2 500 000. Explanatory Note.* Hannover.

SHULZ, O., 1982. Karst or thermal mineralization interpreted in the light of sedimentary ore fabrics. In: G. C. AMSTUTZ et al. (Editors), *Ore Genesis. The State of the Art.* Springer, Berlin, pp. 108–117.

SELLI, R., 1974. Appunti sulla geologia del Mar Tirreno. In: Paleogeografia del Terziario Sardo nell'Ambito del Mediterraneo Occidentale. *Ren. Semin. Fac. Sci. Univ. Cagliari, Suppl.*, 43: 327–351.

SELLWOOD, B. W., 1978. Hardground Community. In: W. S. MCKERROW, (Editor), *The Ecology of Fossils.* Duckworth, London, pp. 226–228.

SEYFERT, C. K. and SIRKIN, L. A., 1979. *Earth history and plate tectonics.* Harper and Row, New York, 600 pp.

SHACKLETON, N. J., 1976. Oxygen-isotope and paleomagnetic stratigraphy of Pacific core V 28–239 Late Pliocene to latest Pleistocene. *Geol. Soc. Amer., Mem.* 145: 449–464.

SHOFMAN, I. L., 1973. Mesozoic continental deposits of the Patoma Highland (in Russian). *Dokl. Akad. Nauk SSSR, Geol.*, 213(6): 1389–1391.

SIBSON, R. H., MOORE, J. MCM. and RANKIN, A. H., 1975. Seismic pumping–a hydrothermal fluid transport meachanism. *Quart. J. Geol. Soc. London*, 131: 653–659.

SIEDLECKI, S. and WIESER, T., 1948. A porphyr in the Czernka valley. *Rocz. Pol. Tow. Geol.*, 17: 121–135.

SIEVER, R., 1962. Silica solubility, 0–200 °C, and the diagenesis of siliceous sediments. *J. Geol.*, 70: 127–150.

SIGÉ, B. et al., 1979. Catalogue des mammifères du Quercy. In: F. WESTPHAL (Editor), *Fossilium Catalogus*. Vol. I. Animalia, Pt. 126. Jung, Den Haag, 100 pp.

SIGOV, A. P., 1947. Geomorphology of Ural and general epochs of karstification (in Russian). In: *Tezisi dokladov karstovoy konf.*, Perm, pp. 83–84.

SIGOV, A. P., 1969. *Metallogeny of Mesozoic and Cenozoic of Ural*. Nedra, Moskva, 296 pp.

ŠILAR, J., 1968. Tectonic uplifts and their influence to the karst evolution near Poniky southeast of Banská Bystrica (in Czech). *Čs. kras*, 1967, 19: 69–80.

SIMPSON, I. M. and BROADHURST, F. M., 1969. A boulder bed at Treak Cliff, north Derbyshire. *Proceedings (Yorkshite Geol. Soc.)*, 37: 141–152.

SKÁCEL, J., 1954. Limestones of the Inner Phyllite Unit on the Bílý potok and karst developed in them (in Czech). *Čs. kras*, 7: 172–178.

ŠKERLJ, Ž. and ŠKERLJ, J., 1973. *Phosphates of Istria* (in Slovenian). *Geologija, Ljubljana*, 16: 335–341.

SKJESETH, S., 1963. Contributions to the geology of the Mjøsa District and the classical sparagmite area in southern Norway. *Norg. Geol. Unders.*, 220: 125 pp.

SKOMPSKI, S., 1985. *Sedimentation and microfacies of Upper Viséan limestones in NE border zone of Lublin Coal Basin (in Polish)*. Ph. D. Thesis, Inst. Geol., Univ. Warsaw, 148 pp.

SKŘIVÁNEK, F., 1965. Die Karstentwicklung der Plešivec-Plateaus im Südslowakischen Karst (in Czech). *Čs. kras*, 17: 42–58.

SKŘIVÁNEK, F., 1968. Die Karstentwicklung der Plešivec-Plateaus im Südslowakischen Karst. *Actes IVe Congr. Int. Spéléol. Ljubljana*.

ŚLĄCZKA, A., 1976. New data on the structure of the basement of the Carpathians south of Wadowice (in Polish). *Rocz. Pol. Tow. Geol.*, 46(3): 337–350.

SLEZÁK, L., 1954. To the genesis in literature unknown karst features on the Květnice Hill near Tišnov (in Czech). *Čs. kras*, 7: 63–64.

SLOSS, L. L., 1963. Sequences in the cratonic interior of North America. *Geol. Soc. Amer. Bull.*, 74: 93–114.

SLUKIN, A. D., 1971. Weathering crust and bauxite types of the Chadobets Uplift (in Russian). In: *Platformennye boksity SSSR* (Platform bauxites of the USSR). Nauka, Moskva, pp. 276–296.

SLUKIN, A. D. and ZHIVTSOV, D. A., 1975. Gibbsite-halloysite-allophane mineralization in the basin of the Maya River (in Russian). In: *Problemy genezisa boksitov* (Problems of bauxite genesis). Nauka, Moskva, pp. 183–201.

SMIRNOV, V. I., 1975. *Geology of mineral deposits*. Mir, Moskva, 520 pp.

SMITH, A. G., BRIDEN, J. C. and DREWRY, G. E., 1973. *Phanerozoic world maps*. Spec. Pap. Paleontol., 12: 42 pp.

SMITH, A. G., HURLEY, A. M. and BRIDEN, J. C., 1982. *Paläokontinentale Weltkarten des Phanerozoikums*. Enke, Stuttgart: 102 pp.

SMITH, D. B., 1970. The palaeogeography of the English Zechstein. In: J. L. RAU and L. F. DELLWIG (Editors), *3rd Symp. on Salt, 1*. Northern Ohio Geol. Soc, Cleveland, pp. 20–33.

SMITH, D. I., ATKINSON, T. C. and DREW, D. P., 1976. The hydrology of limestone terrains. In: T. D. FORD and C. H. D. CULLINGFORD (Editors), *The Science of Speleology*. Academic Press, London, pp. 179–212.

SMITH, R. M. and MARTELL, A. C., 1976. *Critical stability constants, 4. Inorganic complexes*. Plenum Press, New York.

SMOLYANINOVA, N. N., 1970. Some data on mineralogy and genesis of Tyuya-Muyun deposit (in Russian). In: *Ocherki po geologii i geokhimii rudnykh mestorozhdeniy* (Assays on geology and geochemistry of ore deposits). Nauka, Moskva, pp. 58–90.

SOBÁNYI, L., 1896. The evolution history of the surroundings of Kanyapta basin (in Hungarian). *Földrajzi Közl.*, 24.

SOBCZYŃSKI, P. and SZUWARZYŃSKI, M., 1975. Sphalerite dripstone from Trzebionka mine (Cracow-Silesian region). *Rocz. Pol. Tow. Geol.*,45: 171–175.

SOKOLOV, G. A. and GRIGOREV, V. M., 1977. Deposits of iron. In: V. I. SMIRNOV (Editor), *Ore deposits of the USSR, Vol. I.* Pitman, London, pp. 7–113.

SOMERVILLE, I. D., 1979a. Minor sedimentary cyclicity in late Asbian (upper D_1) limestones in the Llangollem District of North Wales. *Proceedings (Yorkshire Geol. Soc.)*, 42: 317–342.

SOMERVILLE, I. D., 1979b. A sedimentary cyclicity in early Asbian (lower D_1) limestones in the Llangollen District of North Wales. *Proceedings (Yorkshire Geol. Soc.)*, 42: 397–404.

SOUCHEZ, R., 1963. Le relief de Couvin-Nismes en temps que paléokarst de climat chaud et humide. *Bull. Soc. Belge. Ét. Géogr.*, 32: 269–280.

SOYER, J., 1979. Les sables tertiaires de l'Entre-Sambre-et-Meuse condruzin. *Ann. Soc. Géol. Belg.*, 101: 93–100.

SPALLETTA, C., VAI, G. B. and VENTURINI, G., 1981. Controllo ambientale e stratigrafico delle mineralizzazioni in calcari devono-dinantiani delle Alpi Carniche. *Mem. Soc. Geol. Ital.*, 22: 101–110.

SPECZIK, S., 1985. Metallogeny of the pre-Zechstein basement of the fore-Sudetic monocline, SW Poland (in Polish). *Geol. Sudetica*, 20(1): 37–111.

SPRINGER, J. S., 1983. Ontario Precambrian dolomite as refractory raw material. *Ontario Geol. Surv. Misc. Pap.*, 116: 303–312.

STAHL, A., 1933. Das Relief des oberschlesischen Steinkohlengebirges im Lichte der Paläogeographie. *Jb. Preuss. Geol. Landesanst.*, 1932, 53: 95–111.

STANEV, I., 1970. Ancient and recent karst in Upper Jurassic-Lower Cretaceous aquifer in North Bulgaria (in Bulgarian). *Spis. Balg. Geol. Druzh., Sofia*, 31(2): 241–248.

STANEV, I. and PANAYOTOV, V., 1976. Tithonian hydrothermal mineralizations in Middle-Upper Triassic aquifer in Bulgaria and their dependance on paleohydrogeological conditions (in Bulgarian). *Rudoobraz. Procesi Miner. Nachodishta*, Sofia, 5: 3–19.

STANOIU, I., 1967. Nouvelles données stratigraphiques sur le Jurassique de Valea Tătarca-Valea Lucava, partie septentrionale du synclinal de Rarău, Carpates Orientales (in Romanian). *C. R. Séanc. Inst. Géol. Géophys.*, 53(1): 465–470.

STEINBERG, M., 1970. Contribution de la sédimentation et de la géochimie à l'étude des formations continentales azoïques, faciès sidérolithique du Détroit Poitevin. *Com. Trav. Hist. Sci., Mem. Sect. Sci., Paris*, 174 pp.

STEPANOV, I. S., 1967. Origin of diamond placers of western slopes of Ural (in Russian). *Sov. Geol.*, 2: 75–82.

STILLE, H., 1924. *Grundfragen der vergleichenden Tektonik.* Borntraeger, Berlin, 443 pp.

STÖRR, M., KUŽVART, M. and NEUŽIL, J., 1978. Genesis of the weathering crust of the Bohemian Massif. *Schr.-Reihe Geol. Wiss.*, 11: 265–281.

STRAND, T., 1960. The pre-Devonian rocks and structures in the region of Caledonian deformation. *Norg. Geol. Unders.*, 208: 170–284.

STUPISHIN, A. V., 1967. *Lowland karst and regularities of its evolution on example of Central Povolzh'e* (in Russian). Kazan' Univ. Press, Kazan', 291 pp.

STUTZER, O. and WETZEL, W., 1932. *Phosphat- Nitrat. 2 Ed.* Borntraeger, Berlin, 390 pp.

SUGDEN, D. A., 1978. Glacial erosion by the Laurentide ice sheet. *J. Glaciol.*, 20(83): 367–392.

SUK, M. (Editor), 1984. *Geological History of the Territory of the Czech Socialist Republic.* Ústř. úst. geol., Praha, 400 pp.

SULIMSKI, A., 1960. *Hystrix primigenia* (Wagner) in the Pliocene from Węże. *Acta Palaeontol. Pol.*, 5(3): 319–336.

SULIMSKI, A., 1964. Pliocene Lagomorpha and Rodentia from Węże 1 (Poland). *Acta Palaeontol. Pol.*, 9(2): 149–261.

SULTANOV, Z. S., 1972. Subterranean karst forms of Southern Ferghana and their origin (in Russian). *Peshchery,* 12–13: 88–95.

SWEETING, M. M., 1972. *Karst landforms.* Macmillan, London, 362 pp.

SWYSEN, L., 1971. La lapiès du „Fondry des Chiens" à Nismes (Belgique). *Ann. Soc. Géol. Belg.,* 94: 165–171.

SZABÓ, P. Z., 1956. Les relations historico-climatique des formes karstiques de Hongrie. *Földrajzi Közlem.,* 4(2): 183–188.

SZABÓ, P. Z.,1957. Karst as a climatic-morphological problem (in Hungarian). *Dunántuli Tudom. Gyüjtemény, Ser. Geogr.,* 15(8): 1–24.

SZABÓ, P. Z., 1960. Karstic landscape forms in Hungary in the light of climatic history. *Ét. Sci. Géogr. Hung.:* 39–55.

SZABÓ, P. Z., 1964. *Neue Daten und beobachtungen zur Kenntniss der Paleokarsterscheinungen in Ungarn.* Erdkunde, 18: 135–142.

SZABÓ, Z., GRASSELLY, Gy. and CZEH-NÉMETH, J., 1981. Some conceptial questions regarding the origin of manganese in the Urkut deposit, Hungary. *Chem. Geol.,* 34: 19–29.

SZÁDECZKY-KARDOSS, E., 1955. *Geochemistry* (in Hungarian). Akademiai Kiadó, Budapest, 680 pp.

SZCZERBAN, E. and URBANI, F., 1974. Carsos de Venezuela, 4: Formas carsicas en areniscas precambricas de Territorio Federal Amazonas y Estato Bolivar. *Bol. Soc. Venezolana Espeleol.,* 5(1): 27–54.

SZENTHE, I., 1969. *Evolution of karst and karst morphology at Nagy-Kevély Hill, Pills Mts.* (in Hungarian). Unpubl. Rep., Budapest, 75 pp.

SZULCZEWSKI, M., 1971. Upper Devonian conodonts, stratigraphy and facies development in the Holy Cross Mts. *Acta Geol. Pol.,* 21(1): 1–129.

SZULCZEWSKI, M., 1973. Fammenian-Tournaisian neptunian dykes and their conodont fauna from Dalnia in the Holy Cross Mts. *Acta Geol. Pol.,* 23(1): 15–58.

SZULCZEWSKI, M., 1978. The nature of unconformities in the Upper Devonian-Lower Carboniferous condensed sequence in the Holy Cross Mts. *Acta Geol. Pol.,* 28(3): 283–298.

SZYDEŁ, Z. and SZYDEŁ, R., 1981. Lias deposits in the Lublin Coal Basin (in Polish). *Przegl. Geol.,* 29(11): 586–571.

SZYNKIEWICZ, A., 1977. Infilled dolines in northern part of the Polish Jura Region. *Proc. 7th Int. Congr. Speleol., Sheffield,* pp. 391–392.

TARLO, L. B., 1959. A new Middle Triassic reptile fauna from fissures in the Middle Devonian limestones in Poland. *Proc. Geol. Soc. London,* 1538: 63–64.

TATSUMOTO, M. and GOLDBERG, E. D., 1959. Some aspects of the marine geochemistry of uranium. *Geochim. Cosmochim. Acta,* 17: 201–208.

TAYLOR, G. R. and HUGHES, G. W., 1975. Biogenesis of the Rennell bauxite. *Econ. Geol.,* 79: 542–546.

TEILHARD DE CHARDIN, P., YOUNG, C. C., PEI, W. C. and CHANG, H. C., 1935. On the Cenozoic formation of Kwangsi and Kwangtung. *Bull. Geol. Soc. China,* 14: 179–205.

TELEGDI RÓTH, K., 1923. Transdanubian bauxite deposits (in Hungarian). *Földt. Szemle,* 1: 95–103.

TELEGDI RÓTH, K., 1927. Die Bauxitlager des Transdanubischen Mittelgebirges in Ungarn. *Földt. Szemle,* 1: 33–45.

TENU, A., DAVIDESCU, F. D. and SLAVESCU, A., 1983. Quelques conclusions d'hydrogéologie isotopique de la Dobrogea du sud. *Annu. Inst. Géol. Géophys.,* 63: 257–264.

TERCINIER, G., 1971. Contribution à la connaissance des phénomènes de bauxitisation et d'allitisation: les sols des karsts d'atolls surélevés du SW Pacifique. *Cah. O. R. S. T. O. M., Sér. Pédol.,* 9(3): 307–334.

TERCINIER. G., 1972. La crandallite, phosphate naturel répandu dans les sols et souvent abondant dans les produits de remplissage des karsts. *C. R. Acad. Sci. Paris , Sér. D.,* 274: 1445–1448.

THEROND, R., 1973. *Recherche sur l'etancheite des barrages en pays karstiques.* Thesis Doct., EY-ROLLE-S, Paris, 144 pp.

THIBIÉROS, J., 1974. *Hamman Zriba et Hamman Djedidi (région de Zaghouan, Tunisie). Etude géologique et minière de deux gisements stratiformes dans le cadre de la province fluorée. L'association des concentrations fluorées aux surfaces d'émersion.* Thèse 3e cycle, Univ. P. et M. Curie, Paris, 376 pp.

THOMAS, T. M., 1952. Notes on the structure of some minor outlying occurrences of littoral Trias in the Vale of Glamorgan. *Geol. Mag.*, 89: 153–162.

THOMAS, T. M., 1963. Solution subsidence in southeast Carmarthenshire and southwest Breconshire. *Inst. Brit. Geogrs. Trans.*, 33: 45–60.

THOMAS, T. M., 1971. Gash–breccia of South Pembrokeshire: fossil karstic phenomena? *Inst. Brit. Geogrs. Trans*, 54: 89–100.

THOMAS, T. M., 1973. Solution subsidence mechanism and end-products in southeast Breconshire. *Inst. Brit. Geogrs. Trans.*, 60: 69–86.

THOMAS, T. M., 1974. The South Wales interstratal karst. *Trans. Brit. Cave Res. Assoc.*, 1: 131–152.

THOREZ, J. and BOURGUIGNON, P., 1984. Matériaux argileux des paléokarsts in „Le karst belge". *Kölner Geogr. Arb.*, 45: 47–57.

THOREZ, J., BULLOCK, P., CATT, J. A. and WEIR, A. H., 1971. The petrography and origin of deposits filling solution pipes in the chalk near South Mimms, Hertfordshire. *Geol. Mag.*, 108: 413–423.

THRAILKILL, J., 1976. Chemical and hydrologic factors in the excavation of limestone caves. *Bull. Geol. Soc. Amer.*, 79: 19–45.

TIN MO, SUTTLE, A. D. and SACKETT, W. D., 1973. Uranium concentration in marine sediments. *Geochim. Cosmochim. Acta*, 37: 35–51.

TODOROVA, T. and STEFANOV, D., 1973. Mineral composition of bauxites, clays and terra rossa from the Trăn area (in Bulgarian). *Izv. Geol. Inst., Ser. Stratigr. Litol., Sofia*, 22: 195–212.

TODOROVA, T. and STEFANOV, D., 1975a. Mineralogy and potential bauxite content of the continental formations developed along the boundary of Tithonian and Barremian-Aptian in the Belogradchik district (in Bulgarian). *Geochim. Mineral. Petrologiya, Sofia*, 1: 102–112.

TODOROVA, T. and STEFANOV, D., 1975b. Mineralogy of continental deposits at the contact of Upper Barremian-Lower Aptian limestones and Turonian conglomerates in the Trăn area (in Bulgarian). *Geochim. Mineral. Petrologiya, Sofia*, 2: 102–110.

TOUAHRI, B., 1983. *Le gîte de plomb-zinc d'El Abed, Wilaya de Tlemeen, Algérie. Une mineralisation „sous inconformité" en environnement dolomitique épicontinental. Contribution à une recherche des guides de prospection.* Thèse 3e cycle, Orsay, 194 pp.

TOUBAL, A., 1984. *Contribution à l'étude des minéralisations antimonifères du Nord-Est algériens. Une province métallogénique hétérochrone.* Thèse 3e cycle, Univ. P. et M. Curie, Paris, 141 pp.

TOUHAMI, A., 1979. *Contribution à l'étude géologique et métallogénique de la province fluorée tunisienne. Rôle de l'altération superficielle dans la formation des concentrations fluorées.* Thèse 3e cycle, Nancy, 321 pp.

TRASHLIEV, S., 1979. Lower Sarmatian paleokarst in gypsum rocks near village Koshava, Vidin district (in Bulgarian). *Rudoobraz. Procesi Miner. Nachodishta, Sofia*, 11: 24–33.

TRETTIN, H. P., 1969. Lower Paleozoic sediments of northwestern Baffin Island, District of Franklin. *Geol. Surv. Canada, Bull*, 157: 70 pp.

TRIMMEL, H., 1955. Höhlen und Niveaus. *Die Höhle, Wien*, 6(1): 5–8.

TRIMMEL, H. (Editor), 1965. *Speläologisches Fachwörterbuch.* Akten 3 Int. Kongr. Speläol., Wien, Bd. C: 109 pp.

TSYKIN, R. A., 1980. On conditions of the genesis of covered karst (in Russian). *Geol. i Geofiz.*, 8: 52–58.

TSYKIN, R. A., 1985. *Deposits and raw materials of fossil karst* (in Russian). Nauka, Sibir. otd., Novosibirsk, 166 pp.

TSYKIN, R. A., and KOSTENENKO, L. P., 1984. Mesozoic-Cenozoic mineralization of the Porozhinsk manganese deposit (in Russian). In: *Margantsevoe rudoobrazovanie na territorii SSSR* (Manganese metallogeny on the territory of the USSR). Nauka, Moskva, pp. 234–242.

TSYKIN, R. A. and TSYKINA, Zh. L., 1978. *Karst of eastern part of Altay-Sayan fold belt (in Russian).* Nauka, Novosibirsk, 104 pp.

TUĆAN, R., 1912. Terra rossa, deren Natur und Entstehung. *Neu. Jb. Mineral. Beil,* 34: 401–430.
TUCKER, M. E., 1977. The marginal Triassic deposits of South Wales: continental facies and palaeography. *Geol. J.,* 12(2): 169–188.
TURNOVEC, I., 1980. Karst form under Cenomanian sediments near Zadní Kopanina (in Czech). *Čs. kras,* 1980, 31: 105.
TURNOVEC, I. and PECHOVÁ, J., 1980. Clay deposit near Zadní Kopanina (in Czech). *Geol. Průzk.,* 22(3): 90–91.
TYCZYŃSKA, M., 1958. The pre-Tortonian surface of karst planation in the vicinity of Cracow. *Bull. Acad. Pol. Sci., Sér. Sci. Géol. Géogr.,* 399–401.
TYRÁČEK, J., 1962. Fossil cone karst near Hranice na Moravě (in Czech). *Čas. mineral. geol.,* 2: 176–185.
UHLIG, D., 1965. Das Be-und Entwässerungsprojekt der Al-Hassa-Oase in Saudi Arabien. *Wasserwirtschaft, Stuttgart,* 55: 53–57.
UHLIG, V., 1881. *Die Jurabildungen in der Umgebung von Brünn.* A. Hölder, Wien, 72 pp.
VADÁSZ. Die geologische Entwicklung und das Alter der ungarischen Bauxitvorkommen. *Magy. Áll. Földt. Intéz. Évk.,* 38(2): 173–230.
VADÁSZ, E., 1951. *Bauxite science* (in Hungarian). Akadémiai Kiadó, Budapest, 127 pp.
VAI, G. B., 1979. Una palinspastica permiana della catena paleocarnica. *Rc. Soc. Geol. Ital,* 1978: 25–27.
VAIL, P. R., MITCHUM, R. M. and THOMPSON, S., 1977. Seismic stratigraphy and global changes of sea level, Part 4: Global cycles of relative changes of sea level. In: C. E. PAYTON (Editor), *Seismic Stratigraphy – Applications to Hydrocarbon Exploration.* Amer. Assoc. Petrol. Geologists, Mem. 26: 83–97.
VALKOVIČOVÁ, J., 1979. *Structural analysis of the northern part of the Moravian Karst* (in Czech). RNDr. Thesis, Univ. Karlova, Praha, 60 pp.
VAN BREEMEN, O., AFTALION, M., BOWES, D. R., DUDEK, A., MÍSAŘ, Z., POVONDRA, P. and VRÁNA, S., 1982. Geochronological studies of the Bohemian Massif, Czechoslovakia, and their significance in the evolution of Central Europe. *Trans. Roy. Soc. Edinburgh, Earth Sci.,* 73: 89–108.
VAN DEN BROECK, E., MARTEL, E.-A. and RAHIR, E., 1910. *Les cavernes et rivières souterraines de la Belgique.* Lamertin, Bruxelles, T. 1: 786 pp., T. 2: 787–1592.
VAN EVERDINGEN, R. O., 1981. Morphology, hydrology and hydrochemistry of karst in permafrost near Great Bear Lake, Northwest Territories. *Nat. Hydrol. Res. Inst. Canad, Pap.,* 11.
VARSANOF'YEFA, V. A., 1915. Karst phenomena in the northern part of the Ufimsk plateau (in Russian). *Zemlevedenie,* 22(4): 1–45.
VENDEL, M. and KISHÁZI, P., 1964. Relationships of karst waters and hot springs in the Transdanubian Range (in Hungarian). I–II. *MTA Müszaki Tudom. Osztál. Közlem.,* 32: 393–417 and 33: 205–234.
VENERANDI-PIRRI, I., 1978. La paragenesi a Zn-Cu-Pb-Sb-Hg-Ni-As-fluorite e barite nel Devonico della Catena Carnica. *Rc. Soc. Ital. Mineral. Petrologia,* 33(2): 821–844.
VITÁLIS, Gy. and HEGYI-PATKÓ, J., 1974. Hydrothermal and metasomatic phenomena in the Triassic limestone areas bordering on the Andesite Mountain of the Danube. *Acta Mineral. Petrogr., Szeged,* 21(2): 219–227.
VITÁLIS, Gy. and HEGYI-PATKÓ, J., 1977. Contribution to the problem of metasomatic dolomitization. *Acta Geol. Acad. Sci. Hung.,* 21(1–3): 91–98.
VOLOGODSKY, G. P., 1975. *Karst of Irkutsk Amphitheatre* (in Russian). Nauka, Moskva, 124 pp.
VUČIĆ, A. and PAVLOVIĆ, B., 1958. The age of travertine beds near Nikšićka Banja based on radium content (in Slovenian). *Vesnik Društva Mat. Fiz., Beograd,* 10.
VYSOTSKY, I. V. (Editor), 1976. *The guide on gas and oil deposits of foreign countries* (in Russian). Nedra, Moskva, T. 1 : 600 pp. and T. 2: 584 pp.
WALKDEN, G. M., 1974. Palaeokarstic surfaces in Upper Viséan (Carboniferous) limestones of the Derbyshire Block, England. *J. Sed. Petrology,* 44: 1234–1247.

WALKDEN, G. M., 1977. Volcanic and erosive events on an Upper Viséan carbonate platform, North Derbyshire. *Proceedings (Yorkshire Geol. Soc.)*, 41: 347–366.

WALKDEN, G. M., 1982. *Field guide to the Lower Carboniferous rocks of the Southeast Margin of the Derbyshire block.* Univ. Aberdeen Geol. Mineral. Dept. Publ., 20 pp.

WALKDEN, G. M. and DAVIES, J., 1983, Polyphase erosion of subaerial omission surfaces in the Late Dinantian of Anglesey, North Wales. *Sedimentology,* 30: 861–878.

WALKER, G. P. L., 1972. In: I. A. COWPERTHWAITE, F. J. FITCH, J. A. MILLER et al., Sedimentation, petrogenesis and radio-isotopic age of Cretaceous Fuller's Earth of Southern England. *Clay Miner.*, 9: 309–327.

WALKER, R. T., 1928. *Deposition of ore in preexisting limestone caves.* AIME Techn. Publ., 154: 43 pp.

WALSH, P. T., BOULTER, M. C., IJTABA, M. and URBANI, D. M., 1972. The preservation of the Neogene Brassington Formation of the southern Pennines and its bearing on the evolution of Upland Britain. *J. Geol. Soc. London,* 128: 519–559.

WALSH, P. T. and BROWN, E. T., 1971. Solution subsidence outliers containing probable Tertiary sediments in northeast Wales. *Geol. J.*, 7: 299–320.

WALSH, P. T., COLLINS, P., IJTABA, M., NEWTON, J. P., SCOTT, N. H. and TURNER, P. R., 1980. Palaeocurrent directions and their bearing on the origin of the Brassington Formation (Miocene-Pliocene) of the southern Pennines, Derbyshire, England. *Mercian Geol.*, 8: 47–62.

WALTHAM, A. C., 1970. Cave development in the limestone of the Ingleborough District. *Geogr. J.*, 136: 574–585.

WALTHAM, A. C., 1971. Shale units in the Great Scar Limestone of the Southern Askrigg Block. *Proceedings (Yorkshire Geol. Soc.)*, 38: 285–292.

WALTHAM, A. C., 1974. *Limestones and Caves of Northwest England.* David and Charles, Newton Abbot, 477 pp.

WALTERS, S. G. and INESON, P. R., 1981. A review of the distribution and correlation of igneous rocks in Derbyshire, England. *Mercian Geol.*, 8(2): 81–132.

WANG CHUNG-HSING, 1965. Genetic conditions of bauxite deposits, central Kweichow. *Int. Geol. Rev.*, 11: 677–682.

WANKEL, H., 1882. *Bilder aus der Mährischen Schweitz und ihrer Vergangenheit.* Holzhauser, Wien, 422 pp.

WARDLAW, N. C. and REINSON, G. E., 1971. Carbonate and evaporite deposition and diagenesis, Middle Devonian Winnipegosis and Prairie Evaporite formations of south-central Saskatchewan. *Amer. Assoc. Petrol. Geol. Bull.*, 55(10): 1759–1786.

WARIN, O. N., 1967. Deposits of phosphate rocks in Oceania. *Semin. Sources of Mineral raw materials for fertilizer industry in Asia and the Far East, 4–11. Dec. 1967, Bangkok.* Econ. Comm. Asia Far East Comm. Ind. Natur. Res.

WARREN, W. C., BRIDGE, J. and OVERSTREET, E. F., 1965. Bauxite deposits of Virginia. *US Geol. Surv. Bull.*, 1199–K: 17 pp.

WARREN, W. C. and CLARK, L. D., 1965. Bauxite deposits of the Eufaula District, Alabama. *US Geol. Surv. Bull.*, 1199–E: 31 pp.

WEGNER, R. N., 1913. Tertiaer und umgelagerte Kreide bei Oppeln (Oberschlesien). *Palaeontographica,* 60: 175–274.

WENDT, J., 1971. Genese und Fauna submariner sedimentärer Spaltenfüllungen in mediterranean Jura. *Palaeontographica,* 136: 121–192.

WHITE, E. L. and WHITE, W. B., 1968. Processes of cave breakdown. *Nat. Speleol. Soc. Bull.*, 31: 83–96.

WHITE, W. S., DENSON, N. M., DUNLAP, J. C. and OVERSTREET, E. F., 1966. Bauxite deposits of Northwest Georgia. *US Geol. Surv. Bul.*, 1199–M: 42 pp.

WHITESIDE, D. I. and ROBINSON, D., 1983. A glauconitic clay-mineral from a speleological deposit of late Triassic age. *Palaeogeogr. Palaeoclimatol. Palaeoecol.*, 41: 81–85.

WILDER, G. P., 1934. The flora of Makatea. *Bull. Bernice P. Bishop Mus.*, 120: 49 pp.

WILK, Z., ADAMCZYK, A. F., BIERNACKI, J. and MOTYKA, J., 1977. Geological and hydrogeological characteristics of the Pomorzany fault zone. *Rocz. Pol. Tow. Geol.*, 47(3): 459–481.

WILK, Z., HAŁADUS, A., MOTYKA, J. and SZCZEPAŃSKI, A., 1984. New prognoses of inflow to mines draining Triassic of northeastern boundary of Upper Silesian Coal Basin (in Polish). *Geologia, Kraków*, 9(3): 5–24.

WILK, Z., KAWALEC, T., MOTYKA, J. and ZUBER, K., 1985. The influence of karst forms to the shape of hydrodynamic field in surrounding rocks (in Polish). *Kras i Speleol.*, 6(15).

WILK, Z. and MOTYKA, J., 1977. Contacts between water-bearing horizons (Olkusz ore mining region, near Cracow). *Rocz. Pol. Tow. Geol.*, 47(1): 115–143.

WILK, Z. and MOTYKA, J., 1980. Ground-water storativity of karst fissured Triassic rocks in the eastern section of the Cracow-Silesian monocline. *Rocz. Pol. Tow. Geol.*, 50(3–4): 447–484.

WILK, Z., MOTYKA, J., BORCZAK, S. and MAKOWSKI, Z., 1985. Microhydraulic properties of the Muschelkalk and Rhaetian rocks of the southern section of the Cracow-Silesian monocline (Poland). *Ann. Soc. Geol. Pol.*, 55(3–4).

WILK, Z., MOTYKA, J. and JÓZEFKO, I., 1984. Investigations of some hydraulic properties of karst solution openings and fractures. *Ann. Soc. Geol. Pol.*, 54(1–2): 15–43.

WILK, Z., MOTYKA, J. and NIEWDANA, J., 1971. The character and classification of water appearances in mine active in fissure-karst aquifer (in Polish). *Geologia, Kraków*, 14: 43–60.

WILK, Z., MOTYKA, J. and NIEWDANA, J., 1973. Geological and hydrogeological causes of origin of depressions in mining region of one of Pb-Zn mines (in Polish). *Biul. Inst. Geol.*, 277: 359–378.

WILK, Z., MOTYKA, J. and NIEWDANA, J., 1982. Fissure-karst waters in Olkusz Triassic (in Polish). In: A. RÓŻKOWSKI and J. SLÓSARZ (Editors), *Przewodnik 54 Zjazdu Pol. Tow. Geol.* Wydaw. Geol., Warszawa, pp. 214–239.

WILK, Z. and ZIMNY, W., 1976. Hydrogeological problems connected with the development of mining in the Olkusz region (in Polish). *Geologia, Kraków*, 17: 117–134.

WILLIAMS, P. W., 1970. Limestone Morphology in Ireland. In: N. STEPHENS and R. E. GLASSCOCK (Editors), *Irish Geographical Studies in Honour of Estin Evans*. Belfast, pp. 102–120.

WILSON, J. L., 1974. Characteristics of carbonate-platform margins. *Amer. Assoc. Petrol. Geol. Bull.*, 58(4): 810–824.

WILSON, J. L., 1975. *Carbonate Facies in Geologic History*. Springer, Berlin, 471 pp.

WILSON, J. L., 1985. Petroleum reservoirs and karst. *Soc. Econ. Mineralogists Paleontologists, A. Midyear Meet., Golden, Abstr.*, Vol. II: 97–98.

WÓJCIK, Z., 1966. On the origin and age of clastic deposits in the Tatra caves (in Polish). *Prace Muz. Ziemi*, 9: 3–130.

WOLPOFF, M. H., 1970. The evidence for multiple hominid taxa at Swartkrans. *Amer. Anthropol.* 72(3): 576–607.

WOO, M.-K. and MARSH, P. E., 1977. Effect of vegetation on limestone solution in a small High Arctic basin. *Canad. J. Earth Sci.*, 14(4): 571–581.

WORLEY, N. E., 1976. Lithostratigraphic control of mineralization in the Blende Vein, Magpie Mine, Sheldon, near Bakewell, Derbyshire. *Proceedings (Yorkshire Geol. Soc.)*, 41: 95–106.

WRIGHT, S. C., 1983. The Sangomore Formation of North-West Scotland lower Ordovician hypersaline peritidal carbonates (Abstract). *4th Int. Assoc. Sedimentologists, Split*, pp. 170–171.

WRIGHT, V. P., 1982. The recognition and interpretation of palaeokarsts: two examples from the Lower Carboniferous of South Wales. *J. Sed. Petrology*, 52: 83–94.

WRIGHT, V. P., 1984a. The significance of needle-fibre calcite in a Lower Carboniferous paleosol. *Geol. J.*, 19: 23–32.

WRIGHT, V. P., 1984b. The polyphase karstification of the Carboniferous Limestone in South Wales. In: M. M. SWEETING and K. PATTERSON (Editors), *New Trends in Karst Geomorphology*. Geo Books, Norwich.

WRIGHT, V. P., RAVEN, M. R. and BURCHETTE, T. P., 1981. *Field Guide to the Lower Carboniferous Rocks near Abergavenny.* Univ. Coll. Wales, Cardiff, Geol. Dept. Publ., 76 pp.

WYSOCZAŃSKI-MINKOWICZ, T., 1969. An attempt at relative age determination of fossil bones by fluorine-chlorine-apatite method (in Polish). *Stud. Geol. Pol.,* 28: 76 pp.

YAKUSHOVA, A. F., 1949. Karst in Paleozoic carbonate rocks of the Russian lowland (in Russian). *Uchen. Zap. Mosk. Univ.,* 136(3): 55–109.

YAMANARI, H., 1935. Aluminium phosphate deposit in Kita-Daitô-Zima. *Tôhoku Imp. Univ. Res. Contr. Inst. Palaeontol.,* 15: 65 pp.

YORATH, C. J., BALKWILL, H. R. and KLASSEN, R. W., 1968. Geology of the eastern part of the Northern Interior and Arctic Coastal Plains. *Geol. Surv. Canada, Pap.,* 68–27: 44 pp.

YORKE, C., 1954–61. The Pocket Deposits of Derbyshire. Private Publ., Birkenhead.

ZANIN, U. N., 1969. *Geology of phosphate-bearing weathering crusts and connected phosphate deposits* (in Russian). Nauka, Moskva, 158 pp.

ZAPFE, H., 1950. Die Fauna der miozänen Spaltenfühllung von Neudorf and der March (ČSR). Carnivora, Chiroptera. *Sitz.-Ber. K. Akad. Wiss., math.-naturwiss. Kl., Wien,* 159: 51–64, 109–141.

ZAPFE, H., 1960. Die Primatenfunde aus der miozänen Spaltenfüllung von Neudorf an der March (Děvínská Nová Ves, Tschechoslowakei). *Schweiz. Paläont. Abh.,* Basel, 78: 1–293.

ZAPP, A. D., 1965. Bauxite deposits of the Andersonville District. *US Geol. Surv. Bull.,* 1199–G: 37 pp.

ZAPP, A. D., and CLARK, L. D., 1965. Bauxite in areas adjacent to and between the Springvale and Andersonville Districts, Georgia. *US Geol. Surv. Bull.,* 1199–H: 10 pp.

ZELENKA, P., 1980. Upper Cretaceous deposits near Slivenec south of Prague (in Czech). *Čes. kras, Beroun,* 5: 7–18.

ZELENKA, P., 1984. Cretaceous sediments in karst depression on Dívčí Hrady (in Czech). *Čes. kras, Beroun,* 10: 51–55.

ŻELICHOWSKI, A., JUSKOWIAKOWA, M. and MIŁACZEWSKI, L., 1974. Covers of the Lower Carboniferous weathered loams in the Central area of the Lublin region, SE Poland. *Kwart. Geol.,* 18(3): 506–514.

ZEMAN, A., 1980. Pre-Miocene relief and weathering products of the Carpathian Foredeep and Moravian Carpathians in prospection of oil and gas deposits (in Czech). *Věst. Ústř. úst. geol.,* 55(6): 357–366.

ZHABIN, A. G., SAMSONOVA, N. S. and ASTAKHOV, G. N., 1984. Facies of ore deposition depending on the contrast composition of enclosing rocks (in Russian). *Geol. Rud. Mestorozhd.,* 26(3): 73–86.

ZHANG SHOUYUE, 1980. Carbonate rock formations and karst zonation in China (in Chinese). *Sci. Pap. Geol. Int. Exchange, Beijing,* 5: 91–97.

ZHANG SHOUYUE, 1984. Karst and subsidence in China. In: B. F. BECK (Editor), *Sinkholes, their geology, engineering and environmental impact.* A. A. Balkema, Rotterdam, pp. 97–104.

ZHANG SHOUYUE, 1986. On Paleoenvironment of Karst Development and Plate Tectonics. *Com. 9° Congr. Int. Espeleol., Barcelona,* Vol. 1: 223–226.

ZHANG SHOUYUE and YUAN DAPXIAN, 1982. Progress in karst research (in Chinese). *Geol. Rev., Beijing,* 28(3): 281–285.

ZHANG SHOUYUE and ZHAO SHUSEN, in press. $^{230}Th/^{234}U$ ages of speleothems in eastern Mainland of China and paleoenvironment.

ZHANG ZHIGAN, 1980. Karst types in China. *Geo-Journal, Wiesbaden,* 4(6): 541–570.

ZHELNIN, S. G., 1977. Ancient placers in the Far East of the USSR (in Russian). In: *Drevnie i pogrebennye rossypy SSSR* (Ancient and buried placers of the USSR). Naukova Dumka, Kiev, Pt. 2: 13–26.

ZHERO, O. G., SMIRNOV, L. V. and SURKOV, V. S., 1977. Geological structure and oil and gas content of Paleozoic deposits in the Nurolsk Bassin (in Russian). In: *Problemy geologii i neftegazonosnosti*

Zapadno Sibirskoy plity (Problems of geology and oil and gas deposits of Western Siberian Platform). Trudy Sibirs. Nauch. Issled. Inst. Geol. Geofiz. Miner. Syrya, Novosibirsk, 225: 22–38.

ZHUKOV, V. V., 1974. Belinsk bauxite deposit (in Russian). In: *Geneticheskaya klassifikaciya i tipy boksitovykh mestorozhdeniy v SSSR* (Genetic classification and types of bauxite deposits in the USSR). Nauka, Moskva, pp. 265–272.

ZHIYI, Z., 1979. Chinese karst compared. *Geotimes,* 24: 20–22.

ZÖTL, J. G., 1974. *Karsthydrogeologie*. Springer, Wien, 291 pp.

ZUFFARDI, P., 1976. Karst and economic mineral deposits. In K. H. WOLF (Editor), *Handbook of Strata-Bound and Stratiform Ore Deposits,* 5. Elsevier, Amsterdam, pp. 175–212.

ZUKALOVÁ, V., 1981. Stromatoporoidea, Foraminifera and Algae from the Givetian and Frasnian of the Krásná-1 borehole. *Sbor. geol. věd., ř. Paleontol.,* 24: 63–94.

ZUKALOVÁ, V. and CHLUPÁČ, I., 1982. Stratigraphic classification of unmetamorphosed Devonian of the Moravo-Silesian region (in Czech). *Čas. mineral. geol.,* 27(3): 226–241.

PART VIII
INDEXES

AUTHOR INDEX

Indexes by Pavel Bosák

GEOGRAPHICAL INDEX

SUBJECT INDEX

Abbreviations of deposits:

Ag – silver
Ba – baryte
bx – bauxite
C – diamonds
co – coal
Cu – copper
F – fluorite
Fe – iron
Hg – mercury
ka – kaolin
Mn – manganese
Ni – nickel
Pb – lead
Ph – phosphate, phosphorite
pm – polymetalls
ps – precious stones
Sb – antimony
U – uranium
V – vanadium
Zn – zinc